Second Edition

FINITE ELEMENT ANALYSIS

Thermomechanics of Solids

Second Edition

FINITE ELEMENT ANALYSIS

Thermomechanics of Solids

David W. Nicholson

CRC Press
Taylor & Francis Group
Boca Raton London New York

CRC Press is an imprint of the
Taylor & Francis Group, an **informa** business

CRC Press
Taylor & Francis Group
6000 Broken Sound Parkway NW, Suite 300
Boca Raton, FL 33487-2742

First issued in paperback 2019

ISBN-13: 978-1-4200-5095-0 (hbk)
ISBN-13: 978-0-367-38743-3 (pbk)

Library of Congress Cataloging-in-Publication Data

Nicholson, D. W.
Finite element analysis : thermomechanics of solids / by David W. Nicholson.
-- 2nd ed.
p. cm.
Includes bibliographical references and index.
ISBN-13: 978-1-4200-5095-0
ISBN-10: 1-4200-5095-8
1. Thermal stresses--Mathematical models. 2. Finite element method. I. Title.

TA418.58.N53 2008
620.1'121--dc22 2007042632

Visit the Taylor & Francis Web site at
http://www.taylorandfrancis.com

and the CRC Press Web site at
http://www.crcpress.com

Dedication

To Linda and Mike,
with the deepest love

Contents

Preface to the Second Edition

The first edition of *Finite Element Analysis: Thermomechanics of Solids* was intended to give a unified but very concise presentation of the finite element method applied to thermomechanics of solids, together with supporting mathematics and continuum mechanics. Of necessity the presentation was selective, attempting to include selected important results while remaining concise and focused on the unity of the method. Coverage was provided on variational and incremental variational principles ensuing from mechanical and thermal field equations and from internal constraints, together with their realization in element formulations. Emphasis was placed on the role of tensors, to which end Kronecker product notation was used throughout. The scope embraced mechanical, thermal, and coupled thermomechanical problems; compressible, incompressible, and near-incompressible materials; static and dynamic problems as well as rotational effects; linear problems as well as problems with material, geometric, and boundary condition nonlinearity; and important numerical methods.

The second edition is nearly 170% as long as the first. It is intended to achieve greater integration and balance between introductory and advanced levels, with many more fully solved examples and with advanced materials more concentrated in the latter portion of the monograph. Of course there are a number of worthy topics that have not been possible to include, for example, meshless methods and reduced integration. New coverage includes selected developments in numerical methods, detailing accelerated computations in eigenstructure extraction, time integration, and stiffness matrix triangularization. There is a much more extensive coverage of the arc length method for nonlinear problems. The treatment of rotating bodies and of buckling has been significantly expanded and enhanced. As in the first edition, the intent still has been on presenting and explicating topics in a way that shows the highly unified structure of the finite element method.

The preface to the first edition listed monographs that have excellent coverage of many topics not addressed or done justice to. Since then the author has become familiar with two additional monographs that the reader may find beneficial: M.A. Crisfield (1991) and J.N. Reddy (2004).

The finite element method has matured to a point that it can accurately and reliably be used, by a careful analyst, for an amazingly wide range of applications. Nonlinear problems with nonlinearities due to geometry, material properties, or boundary conditions such as contact are in many cases accessible to analysis. Problems with instabilities, such as buckling, can be treated. Dynamic problems can be solved, with the caution that eigenvalues and eigenvectors beyond the lowest few natural frequencies should be viewed with skepticism. Enormous progress has been made of the vexatious problem of sliding contact. Several areas in which additional progress would be welcome include more accurate treatment of singularities such as occurring at corners, and integrated rather than staggered treatments of mixed field problems. The impact of FEA will accelerate as integration with computer-aided design and solid modeling systems progresses and as algorithmic and computational resources permit "attacking" ever more complex and larger-scale applications while insisting on high resolution.

Preface to the First Edition

Thousands of engineers use finite element codes such as ANSYS for thermomechanical and nonlinear applications. Most academic departments offering advanced degrees in mechanical engineering, civil engineering and aerospace engineering offer a *first course* in the finite element method, and by now almost undergraduates of such programs have some exposure to the finite element method. A number of departments offer a second course. It is hoped that the current monograph will appeal to instructors of such courses. Of course it hopefully will also be helpful to engineers engaged in self-study on nonlinear and thermomechanical finite element analysis.

The principles of the finite element method are presented for application to the mechanical, thermal and thermomechanical response, both static and dynamic, of linear and nonlinear solids. It is intended to provide an integrated treatment of

- basic principles, material models and contact models (for example linear elasticity, hyperelasticity and thermohyperelasticity)
- computational, numerical and software design aspects (such as finite element data structures)
- modeling principles and strategies (including mesh design)

The text is designed *for a second course, as a reference work and for self study.* Familiarity is assumed with the finite element method at the level of a first graduate or advanced undergraduate course.

A first course in the finite element method, for which many excellent books are available, barely succeeds in covering static linear elasticity and linear heat transfer. There is virtually no exposure to nonlinear methods, which are considered topics for a second course. Nor is there much emphasis on *coupled* thermomechanical problems. However, it is believed that many engineers would benefit from a monograph culminating in nonlinear problems and the associated continuum thermomechanics. Such as text may be used in a formal class or for self study. Many important applications have significant nonlinearity, making nonlinear finite element modeling necessary. As a few examples we mention polymer processing, metalforming, rubber components such as tires and seals, biomechanics and crashworthiness. Many applications combine thermal and mechanical response, such as rubber seals in hot engines. Engineers coping with such applications have access to powerful finite element codes and computers. But they often lack and urgently need an in-depth but compact exposition of the finite element method, providing a foundation for addressing problems. It is hoped the current monograph fills this need.

Of necessity a selection to topics has been made and topics are given coverage proportional to the author's sense of their importance to the understanding of the reader. Topics have been selected with the intent of giving a unified and complete but still compact and tractable presentations. Several other excellent texts and monographs have appeared over the years, from which the author has benefited. Four previous texts to which the author is indebted are:

1. Zienkiewicz and Taylor, *The Finite Element Method* Vols 1,2, McGraw-Hill, 1989.
2. Kleiber, M., *Incremental Finite Element Modeling in Nonlinear Solid Mechanics*, Ellis Horwood, Ltd, 1989.
3. Bonet, J. and Wood, R.D., *Nonlinear Continuum Mechanics for Finite Element Analysis*, Cambridge University Press, 1997.
4. Belytschko, T., Lui, W.K, and Moran, B., *Nonlinear Finite Elements for Continua and Structures*, J. Wiley and Sons, 2000.

The current monograph has the following characteristics:

1. emphasis on use of Kronecker product notation instead of tensor, tensor-indicial, Voigt, or traditional finite element matrix–vector notation;
2. emphasis on integrated and coupled thermal and mechanical effects;
3. inclusion of elasticity, hyperelasticity, plasticity and viscoelasticity *with thermal effects*;
4. inclusion of nonlinear boundary conditions, including contact, in an integrated *incremental variational* formulation.

Regarding (1) Kronecker product algebra (KPA) has been widely used in control theory for many years (Graham, (1982)). It is very compact and satisfies very simple rules: for example the inverse of a Kronecker product of two nonsingular matrices is the Kronecker product of the inverses. Recently a number of extensions of KPA have been introduced and shown to permit compact expressions for otherwise elaborate quantities in continuum and computational mechanics. Examples include:

1. compact expressions for the tangent modulus tensors in hyperelasticity (invariant based and stretch-based; compressible, incompressible and near-incompressible), thermohyperelasticity and finite strain plasticity;
2. a general, compact expression for the tangent stiffness matrix in nonlinear FEA, including nonlinear boundary conditions such as contact.

KPA with recent extensions can completely replace other notations in most cases of interest here. In the experience of the author, students experience little difficulty in gaining a command of it.

The first three chapters concern mathematical foundations, and Kronecker product notation for tensors is introduced. The next four chapters cover relevant linear and nonlinear continuum thermomechanics, to enable a unified account of the finite element method. Chapters 8 through 15 represent a compact presentation of the finite element method in linear elastic, thermal and thermomechanical media, including solution methods. The final five chapters address nonlinear problems, based on a unified set of incremental variational principles. Material nonlinearity is treated, as is geometric nonlinearity and nonlinearity due to boundary conditions. Several numerical issues in nonlinear analysis are discussed, such as iterative triangularization of stiffness matrices.

Author

David W. Nicholson, PhD, is a professor in the Department of Mechanical, Materials, and Aerospace Engineering at the University of Central Florida (UCF), Orlando, Florida, where he has served as chair (1990–1995) and interim chair (2000–2002). He also holds an appointment in the UCF Department of Mathematics. His teaching has included the finite element method at introductory and advanced levels, continuum mechanics, and advanced dynamics. He earned his SB degree at Massachusetts Institute of Technology in 1966 and his PhD at Yale University in 1971. His 36 years of experience since his doctorate have included research in industry (Goodyear, seven years) and government (the Naval Surface Weapons Center, six years), and faculty positions at Stevens Institute of Technology (1984–1990) and UCF. He has authored over 155 articles on the finite element method, continuum mechanics, fracture mechanics, and dynamics, has authored two monographs, and has served as coeditor of four proceedings volumes. He has served as a technical editor for *Applied Mechanics Reviews*, as an associate editor of *Tire Science and Technology*, and as a member of scientific advisory committees for a number of international conferences. He has been an instructor in over 25 short courses and workshops. In 2002, he served as executive chair of the XXIst Southeastern Conference on Theoretical and Applied Mechanics, and as coeditor of the proceedings volume *Developments in Theoretical and Applied Mechanics*, volume XXI.

Acknowledgments

Finite Element Analysis: Thermomechanics of Solids, second edition, has been made possible through the love and patience of my wife Linda and son Michael, to whom it is dedicated. To my profound love I cheerfully add my immense gratitude. Most deserving of acknowledgment is Mr. Ashok Balasubramaniam, who developed many of the examples appearing in the second edition. His careful attention to the first edition greatly helped to make it achieve my objectives. Much of my understanding of the advanced topics addressed arose through the highly successful research of my then doctoral student, Dr. Baojiu Lin. He continued his invaluable support by providing comments and corrections on the manuscript. A number of new topics in the second edition are the result of our continuing collaboration. Thanks are due to hundreds of graduate students in my courses in continuum mechanics and finite elements over the years. I tested and refined the materials through the courses and benefited immensely from their strenuous efforts to gain command of the course materials. Finally, I would like to express my deep appreciation to Taylor & Francis/CRC Press for giving me the opportunity to prepare this second edition, which I regard as the culminating achievement of my academic career.

Acknowledgments

1 Introduction to the Finite Element Method

1.1 INTRODUCTION

This monograph is intended to present a concise and unified *two-part* treatment of finite element analysis (FEA) in thermomechanics. The first part encompasses topics typically found in a first course in FEA. Included are elementary mathematical foundations, an introduction to linear variational principles, the stiffness and mass matrices in linear mechanical and thermal elements, assemblage, eigenstructure determination, and numerical procedures. The second part continues into topics which are appropriate for a second course in FEA. It addresses nonlinearity due to material behavior, geometry and boundary conditions, as well as associated advanced mathematical and numerical topics, incremental variational principles including thermal effects and incompressibility constraints, and accommodation of hyperelasticity, plasticity, viscoplasticity, and damage mechanics.

In thermomechanical analysis of members and structures, FEA is an essential resource for computing *displacement and temperature* fields from known *applied loads and heat fluxes*. FEA has emerged in recent decades as critical to mechanical and structural designers. Its use is often mandated by standards such as the ASME Pressure Vessel Code, by insurance requirements, and even by law. Its pervasiveness has been promoted by rapid progress in related computer hardware and software, especially computer-aided design (CAD) systems. A large number of comprehensive "user friendly" finite element codes are available commercially.

In FEA practice, a design file developed using a CAD system is often "imported" into finite element codes, from which point little or no additional effort often suffices to develop the finite element model consisting of a mesh together with material, constraint, boundary condition, and initial value data. The model is communicated to an analysis module to perform sophisticated thermomechanical analysis and simulation. CAD integrated with an analysis tool such as FEA is an example of computer-aided engineering (CAE). CAE possesses the potential of identifying design problems and improvements much more efficiently, rapidly, and "cost-effectively" than purely by "trial and error."

A major FEA application is the determination of stresses and temperatures in a component or member in locations where failure is thought most likely. If the stresses or temperatures exceed allowable or safe values, the product can be redesigned and then reanalyzed. Analysis can also be diagnostic, supporting interpretation of product failure data. Analysis can be used to assess performance, for example, by determining whether the design stiffness coefficient for a rubber spring is attained. FEA can serve to minimize weight and cost without loss of structural integrity or reliability.

1.2 OVERVIEW OF THE FINITE ELEMENT METHOD

Consider a thermoelastic body with force and heat applied to its exterior boundary. The finite element method serves to determine the displacement vector $\mathbf{u}(\mathbf{X}, t)$ and the temperature $\mathrm{T}(\mathbf{X}, t)$ as functions of the undeformed position $\mathbf{X}$ and time t. The process of creating a finite element model to support design of a mechanical system may be viewed as having (at least) eight steps:

1. The body is first discretized, i.e., it is modeled as a mesh of finite elements connected at nodes.
2. Within each element interpolation models are introduced to provide approximate expressions for the unknowns, typically $\mathbf{u}(\mathbf{X}, t)$ and $\mathrm{T}(\mathbf{X}, t)$, in terms of their nodal values, which now become the unknowns in the finite element model.
3. The strain–displacement relation and its thermal analog are applied to the approximations for $\mathbf{u}$ and T to furnish approximations for the (Lagrangian) strain and the thermal gradient.
4. The stress–strain relation and its thermal analog (Fourier's Law) are applied to obtain approximations to stress S and heat flux $\mathbf{q}$ in terms of the nodal values of $\mathbf{u}$ and T.
5. Equilibrium principles in variational form are applied using the various approximations within each element, leading to *element equilibrium equations*.
6. The element equilibrium equations are assembled to provide a *global equilibrium equation*.
7. Prescribed kinematic and temperature conditions on the boundaries (*constraints*) are applied to the global equilibrium equations, thereby reducing the number of degrees of freedom and eliminating "rigid body" modes.
8. The resulting global equilibrium equations are then solved using computer algorithms.

The output is postprocessed. Initially the output should be compared to data or benchmarks or otherwise validated, to establish that the model correctly represents the underlying mechanical system. If not satisfied, the analyst may revise the finite element model and repeat the computations. When the model is validated, postprocessing, with heavy reliance on graphics, serves to interpret the results, for example, determining whether the underlying design is satisfactory. If problems with the design are identified, the analyst may then choose to revise the design. The revised design is then modeled, and the process of validation and interpretation is repeated.

1.3 MESH DEVELOPMENT

Finite element simulation has classically been viewed as having three stages: *preprocessing*, *analysis*, and *postprocessing*. The input file developed at the preprocessing stage consists of several elements:

1. Control information (type of analysis, etc.)
2. Material properties (e.g., elastic modulus)
3. Mesh (element types, nodal coordinates, connectivities)
4. Applied force and heat flux data
5. Supports and constraints (e.g., prescribed displacements)
6. Initial conditions (dynamic problems)

In problems without severe stress concentrations, much of the mesh data can be developed conveniently using automatic mesh generation. With the input file developed, the analysis processor is activated. "Raw" output files are generated. The postprocessor module typically contains (interfaces to) graphical utilities, facilitating display of output in the form chosen by the analyst, for example, contours of the Von Mises stress. Two problems arise at this stage: *validation* and *interpretation*. The analyst may use benchmark solutions, special cases, or experimental data to validate the analysis. With validation, the analyst gains confidence in, say, the mesh. He/she still may face problems of interpretation, particularly if the output is voluminous. Fortunately, current graphical display systems make interpretation easier and more reliable, such as by displaying high stress regions in vivid colors. Postprocessors often allow the analyst to "zoom in" on regions of high interest, for example, where rubber is highly confined. More recent methods based on virtual reality technology enable the analyst to "fly through" and otherwise become immersed in the model.

The goal of mesh design is to select the number and location of finite element nodes and element types so that the associated analyses are sufficiently accurate. Several methods include automatic mesh generation with adaptive capabilities which serve to produce and iteratively refine the mesh, based on a user-selected error tolerance. Even so, satisfactory meshes are not necessarily obtained, so that model editing by the analyst may be necessary. Several practical rules are given below:

1. Nodes should be located where concentrated loads and heat fluxes are applied.
2. Nodes should be located where displacements and temperatures are constrained or prescribed in a concentrated manner, for example, where "pins" prevent movement.
3. Nodes should be located where concentrated springs and masses and their thermal analogues are present.
4. Nodes should be located along lines and surface patches over which pressures, shear stresses, compliant foundations, distributed heat fluxes, and surface convection are applied.

5. Nodes should be located at boundary points where the applied tractions and heat fluxes experience discontinuities.
6. Nodes should be located along lines of symmetry.
7. Nodes should be located along interfaces between different materials or components.
8. Element aspect ratios (ratio of largest to smallest element dimensions) should be no greater than, say, five.
9. Symmetric configurations should have symmetric meshes.
10. The density of elements should be greater in domains with high gradients.
11. Interior angles in elements should not be excessively acute or obtuse, for example, less than 45° or greater than 135°.
12. Element density variations should be gradual rather than abrupt.
13. Meshes should be uniform in subdomains with low gradients.
14. Element orientations should be staggered to prevent "bias."

In modeling a configuration, a good practice is initially to develop the mesh locally in domains expected to have high gradients, and thereafter to develop the mesh in the intervening low gradient domains, thereby "reconciling" the high gradient domains. There are two classes of errors in FEA:

Modeling error ensues from inaccuracies in such input data as the material properties, boundary conditions, and initial values. In addition, there often are compromises in the mesh, for example, modeling sharp corners as rounded.

Numerical error is primarily due to truncation and roundoff. As a practical matter, error in a finite element simulation is often assessed by comparing solutions from two meshes, the second of which is a refinement of the first.

The sensitivity of finite element computations to error is to some extent controllable. If the condition number of the stiffness matrix (the ratio of the maximum to the minimum eigenvalue) is modest, sensitivity is reduced. Typically, the condition number increases rapidly as the number of nodes in a system grows. In addition, highly irregular meshes tend to produce high condition numbers. Models mixing soft components, for example, made of rubber, with stiff components, such as steel plates, are also likely to have high condition numbers. Where possible, the model should be designed to reduce the condition number.

2 Mathematical Foundations: Vectors and Matrices

2.1 INTRODUCTION

This chapter gives a review of mathematical relations which will prove to be useful in the subsequent chapters. A more complete development is given in Chandrasekharaiah and Debnath (1994).

2.1.1 RANGE AND SUMMATION CONVENTION

Unless otherwise noted, repeated Latin indices will imply summation over the range 1 to 3. For example,

$$a_i b_i = \sum_{i=1}^{3} a_i b_i = a_1 b_1 + a_2 b_2 + a_3 b_3 \tag{2.1}$$

$$a_{ij} b_{jk} = a_{i1} b_{1k} + a_{i2} b_{2k} + a_{i3} b_{3k} \tag{2.2}$$

The repeated index is "summed out" and therefore "dummy". The quantity $a_{ij}b_{jk}$ in Equation 2.2 has two free indices i and k (and later will be shown to be the ikth entry of a second-order tensor). Note that Greek indices do *not* imply summation. Thus $a_\alpha b_\alpha = a_1 b_1$ if $\alpha = 1$.

2.1.2 SUBSTITUTION OPERATOR

The quantity δ_{ij}, later to be called the Kronecker tensor, has the property that

$$\delta_{ij} = \begin{cases} 1, & i = j \\ 0, & i \neq j \end{cases} \tag{2.3}$$

For example, $\delta_{ij} v_j = 1 \times v_i$, thereby illustrating the substitution property.

2.2 VECTORS

2.2.1 Notation: Scalar and Vector Products

Throughout this and the following chapters, orthogonal coordinate systems will be used. Figure 2.1 shows such a system, with base vectors $\mathbf{e}_1$, $\mathbf{e}_2$, and $\mathbf{e}_3$. The scalar product of vector analysis satisfies

$$\mathbf{e}_i \cdot \mathbf{e}_j = \delta_{ij} \tag{2.4}$$

The *vector product* satisfies

$$\mathbf{e}_i \times \mathbf{e}_j = \begin{cases} \mathbf{e}_k, & i \neq j \text{ and } ijk \text{ in right-handed order} \\ -\mathbf{e}_k, & i \neq j \text{ and } ijk \text{ not in right-handed order} \\ \mathbf{0}, & i = j \end{cases} \tag{2.5}$$

The vector cross product enables introducing the alternating operator ε_{ijk}, also known as the *ijk*th entry of the permutation tensor:

$$\begin{aligned} \varepsilon_{ijk} &= [\mathbf{e}_i \times \mathbf{e}_j] \cdot \mathbf{e}_k \\ &= \begin{cases} 1, & ijk \text{ distinct and in right-handed order} \\ -1, & ijk \text{ distinct but not in right-handed order} \\ 0, & ijk \text{ not distinct} \end{cases} \end{aligned} \tag{2.6}$$

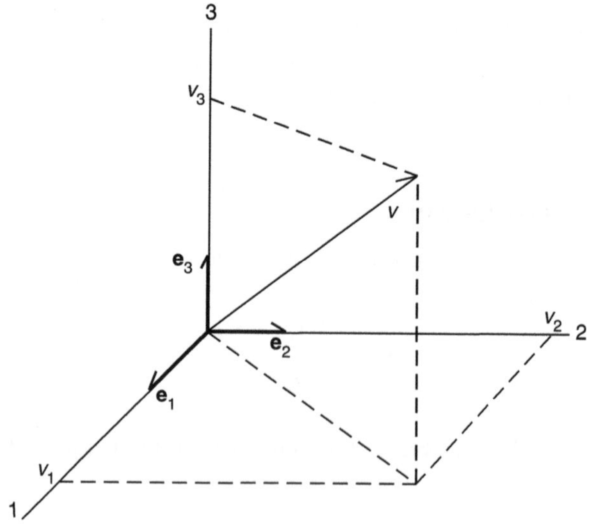

FIGURE 2.1 Rectangular coordinate system.

EXAMPLE 2.1

For three 3 × 1 vectors **a**, **b**, **c**, we prove that $\mathbf{a} \cdot (\mathbf{b} \times \mathbf{c}) = \mathbf{b} \cdot (\mathbf{c} \times \mathbf{a})$.

Solution

Writing the triple product in indicial notation gives

$$\mathbf{a} \cdot \mathbf{b} \times \mathbf{c} \rightarrow a_i \in_{ijk} b_j c_k$$

But $\in_{ijk} = \in_{jki}$ (cyclic order), and hence

$$\begin{aligned} a_i \in_{ijk} b_j c_k &= a_i \in_{jki} b_j c_k \\ &= b_j \in_{jki} c_k a_i \\ &= \mathbf{b} \cdot (\mathbf{c} \times \mathbf{a}) \end{aligned}$$

Similarly $\mathbf{a} \times \mathbf{b} \cdot \mathbf{c} = \mathbf{a} \cdot \mathbf{b} \times \mathbf{c}$

EXAMPLE 2.2

Verify that

$$\in_{kij} \in_{klm} = \delta_{il}\delta_{jm} - \delta_{im}\delta_{jl}$$

Solution

A moment's reflection shows that ij and ℓm must both differ from k and each other in any nonvanishing instances of $\in_{kij}\in_{klm}$. But $\delta_{il}\delta_{jm} - \delta_{im}\delta_{jl}$ likewise vanishes under these conditions.

Furthermore, if $i = \ell$ and $j = m$ but $i \neq j$, $\in_{kij}\in_{klm} = 1$, and $\delta_{il}\delta_{jm} - \delta_{im}\delta_{jl} = 1 - 0 = 1$. If $i = m$ and $j = \ell$ but $i \neq j$, then $\in_{kij} = -\in_{klm}$ and $\in_{kij}\in_{klm} = -1$. But in this case $\delta_{il}\delta_{jm} - \delta_{im}\delta_{jl} = 0 - 1 = -1$. The relation to be verified is thereby satisfied in conditions exhausing all cases.

Now consider two vectors **v** and **w**. It will prove convenient to use two different types of notation. In **tensor-indicial notation**, denoted by (*T), **v** and **w** are represented as

*T) $$\mathbf{v} = v_i \mathbf{e}_i, \quad \mathbf{w} = w_i \mathbf{e}_i \tag{2.7}$$

Occasionally base vectors are not displayed, so that **v** is denoted by v_i. By displaying base vectors tensor-indicial notation is explicit and minimizes confusion and ambiguity. However, it is also cumbersome.

In the current text, the "default" is **matrix–vector** (*M) notation, illustrated by

*M) $$\mathbf{v} = \begin{pmatrix} v_1 \\ v_2 \\ v_3 \end{pmatrix}, \quad \mathbf{w} = \begin{pmatrix} w_1 \\ w_2 \\ w_3 \end{pmatrix} \tag{2.8}$$

It is very compact, but also risks confusion by not displaying the underlying base vectors. In *M notation the transposes $\mathbf{v}^T$ and $\mathbf{w}^T$ are also introduced; they are displayed as "row vectors"

*M) $$\mathbf{v}^T = \{v_1 \quad v_2 \quad v_3\}, \quad \mathbf{w}^T = \{w_1 \quad w_2 \quad w_3\} \tag{2.9}$$

The scalar product of **v** and **w** is written as

*T) $$\begin{aligned}\mathbf{v}\cdot\mathbf{w} &= (v_i\mathbf{e}_i)\cdot(w_j\mathbf{e}_j)\\ &= v_iw_j\mathbf{e}_i\cdot\mathbf{e}_j\\ &= v_iw_j\delta_{ij}\\ &= v_iw_i\end{aligned} \tag{2.10}$$

The magnitude of **v** is defined by

*T) $$|\mathbf{v}| = \sqrt{\mathbf{v}\cdot\mathbf{v}} \tag{2.11}$$

The scalar product of **v** and **w** satisfies

*T) $$\mathbf{v}\cdot\mathbf{w} = |\mathbf{v}||\mathbf{w}|\cos\theta_{vw} \tag{2.12}$$

in which θ_{vw} is the angle between the vectors **v** and **w**. The scalar or dot product is written in matrix–vector notation as

*M) $$\mathbf{v}\cdot\mathbf{w} \rightarrow \mathbf{v}^T\mathbf{w} \tag{2.13}$$

The vector or cross product is written as

*T) $$\begin{aligned}\mathbf{v}\times\mathbf{w} &= v_iw_j\mathbf{e}_i\times\mathbf{e}_j\\ &= \varepsilon_{ijk}v_iw_j\mathbf{e}_k\end{aligned} \tag{2.14}$$

Additional results on vector notation will be presented in the next section, which introduces matrix notation. Finally, the vector product satisfies

*T) $$|\mathbf{v}\times\mathbf{w}| = |\mathbf{v}||\mathbf{w}|\sin\theta_{vw} \tag{2.15}$$

and **n** is the unit normal vector perpendicular to the plane containing **v** and **w**. The area of the triangle defined by the vectors **v** and **w** is given by $\frac{1}{2}|\mathbf{v}\times\mathbf{w}|$.

2.2.2 Gradient, Divergence, and Curl in Rectilinear Coordinates

The derivative $\mathrm{d}\varphi/\mathrm{d}\mathbf{x}$ of a scalar $\phi(\mathbf{x})$ with respect to a vector $\mathbf{x}$ is defined implicitly by

*M) $$\mathrm{d}\varphi = \frac{\mathrm{d}\varphi}{\mathrm{d}\mathbf{x}}\mathrm{d}\mathbf{x} \tag{2.16}$$

and it is a row vector whose ith entry is $\mathrm{d}\phi/\mathrm{d}x_i$. In three-dimensional rectangular coordinates the gradient and divergence operators are defined by

*M)
$$\nabla(\ ) = \begin{pmatrix} \frac{\partial(\ )}{\partial x} \\ \frac{\partial(\ )}{\partial y} \\ \frac{\partial(\ )}{\partial z} \end{pmatrix} \tag{2.17}$$

and clearly

*M)
$$\left(\frac{\mathrm{d}}{\mathrm{d}\mathbf{x}}\right)^T (\ ) = \nabla(\ ) \tag{2.18}$$

The gradient of a scalar function ϕ satisfies the following integral relation:

$$\int \nabla\phi\, \mathrm{d}V = \int \mathbf{n}\phi\, \mathrm{d}S \tag{2.19}$$

The expression $\nabla\mathbf{v}^T$ will be seen to be a *tensor* (see Chapter 3). Clearly

$$\nabla\mathbf{v}^T = [\nabla v_1\ \nabla v_2\ \nabla v_3] \tag{2.20}$$

from which Equation 2.19 may be invoked to obtain the integral relation

$$\int \nabla\mathbf{v}^T\, \mathrm{d}V = \int \mathbf{n}\mathbf{v}^T\, \mathrm{d}S \tag{2.21}$$

Next, a most important relation is the *divergence theorem*. Let V denote the volume of a closed domain, with surface S. Let $\mathbf{n}$ denote the exterior surface normal to S and let $\mathbf{v}$ denote a vector-valued function of $\mathbf{x}$, the position of a given point within the body. The divergence of $\mathbf{v}$ satisfies the integral relation

*M)
$$\oint \nabla^T\mathbf{v}\, \mathrm{d}V = \oint \mathbf{n}^T\mathbf{v}\, \mathrm{d}S \tag{2.22}$$

The curl of the vector $\mathbf{v}$, $\nabla \times \mathbf{v}$, is expressed by

$$(\nabla \times \mathbf{v})_i = \varepsilon_{ijk} \frac{\partial}{\partial x_j} v_k \tag{2.23}$$

which is the conventional cross product except that the first vector is replaced by the divergence operator. The curl satisfies an integral theorem, analogous to the divergence theorem (Schey, 1973), namely:

$$\int \nabla \times \mathbf{v} \, \mathrm{d}V = \int \mathbf{n} \times \mathbf{v} \, \mathrm{d}S \tag{2.24}$$

Finally, the reader may verify with some effort that, for a vector $\mathbf{v}(\mathbf{X})$, and a path $\mathbf{X}(S)$ in which S is the arc length along the path

$$\int \mathbf{v} \cdot \mathrm{d}\mathbf{X}(S) = \int \mathbf{n} \cdot \nabla \times \mathbf{v} \, \mathrm{d}S \tag{2.25}$$

The integral between fixed endpoints is single valued if it is path-independent, in which case $\mathbf{n} \cdot \nabla \times \mathbf{v}$ must vanish. But $\mathbf{n}$ is arbitrary since the path is arbitrary, giving the condition for $\mathbf{v}$ to have a path-independent integral as

$$\nabla \times \mathbf{v} = \mathbf{0} \tag{2.26}$$

EXAMPLE 2.3

Verify that the relation $\nabla^2 \mathbf{v} = \nabla(\nabla \cdot \mathbf{v}) - \nabla \times \nabla \times \mathbf{v}$ is satisfied in rectangular coordinates.

Here $\nabla^2 \mathbf{v} = (\nabla \cdot \nabla)\mathbf{v}$ is called the Laplacian of $\mathbf{v}$, and $\nabla(\nabla \cdot \mathbf{v})$ is recognized as the gradient of the divergence of $\mathbf{v}$.

SOLUTION

First note that $\nabla(\nabla \cdot \mathbf{v}) = \frac{\partial}{\partial x_i}\frac{\partial v_j}{\partial x_j}$. Next

$$\begin{aligned} \nabla \times \nabla \times \mathbf{v} &= \epsilon_{ijk} \frac{\partial}{\partial x_j} \epsilon_{klm} \frac{\partial v_m}{\partial x_l} \\ &= \epsilon_{ijk}\epsilon_{klm} \frac{\partial}{\partial x_j}\frac{\partial v_m}{\partial x_l} \\ &= \epsilon_{kij}\epsilon_{klm} \frac{\partial}{\partial x_j}\frac{\partial v_m}{\partial x_l} \end{aligned}$$

But Example 2.2 presents the relation $\epsilon_{kij}\epsilon_{klm} = \delta_{il}\delta_{jm} - \delta_{im}\delta_{jl}$.

Accordingly,

$$\begin{aligned} \epsilon_{kij}\epsilon_{klm} \frac{\partial}{\partial x_j}\frac{\partial v_m}{\partial x_l} &= \left(\delta_{il}\delta_{jm} - \delta_{im}\delta_{jl}\right) \frac{\partial}{\partial x_j}\frac{\partial v_m}{\partial x_l} \\ &= \frac{\partial}{\partial x_i}\frac{\partial v_j}{\partial x_j} - \frac{\partial}{\partial x_j}\frac{\partial v_i}{\partial x_j} \end{aligned}$$

Verification is seen by recognizing that $\nabla(\nabla \cdot \mathbf{v}) = \frac{\partial}{\partial x_i}\frac{\partial v_j}{\partial x_j}$ and $\nabla^2 \mathbf{v} = \frac{\partial}{\partial x_j}\frac{\partial v_i}{\partial x_j}$.

EXAMPLE 2.4

Verify the divergence theorem using the square plate in Figure 2.2, and using

$$\mathbf{v} = \begin{bmatrix} x - y \\ x + y \end{bmatrix}$$

Solution

$$\operatorname{div}(\mathbf{v}) = \nabla \cdot \mathbf{v} = \nabla^T \mathbf{v} = \left(\frac{\partial}{\partial x}\ \frac{\partial}{\partial y}\right)\begin{pmatrix} x - y \\ x + y \end{pmatrix} = 2$$

and

$$\int_V \nabla \cdot \mathbf{v}\, \mathrm{d}V = \int_V 2\, \mathrm{d}x\, \mathrm{d}y = 2$$

For faces (1) and (3), $\mathbf{n}_1 = \mathbf{e}_1$, $\mathbf{n}_3 = -\mathbf{e}_1$.

$$\begin{aligned} \int_{S_1} (\mathbf{v} \cdot \mathbf{n}_1)\, \mathrm{d}S + \int_{S_3} (\mathbf{v} \cdot \mathbf{n}_3)\, \mathrm{d}S &= \int_{x=1,\ 0<y<1} [(x - y)\mathbf{e}_1 + (x + y)\mathbf{e}_2] \cdot \mathbf{e}_1\, \mathrm{d}y \\ &\quad + \int_{x=0,\ 0<y<1} [(x - y)\mathbf{e}_1 + (x + y)\mathbf{e}_2] \cdot (-\mathbf{e}_1)\, \mathrm{d}y \\ &= \int_0^1 (1 - y)\, \mathrm{d}y + \int_0^1 y\, \mathrm{d}y = 1 \end{aligned}$$

For faces (2) and (4), $\mathbf{n}_2 = \mathbf{e}_2$, $\mathbf{n}_4 = -\mathbf{e}_2$.

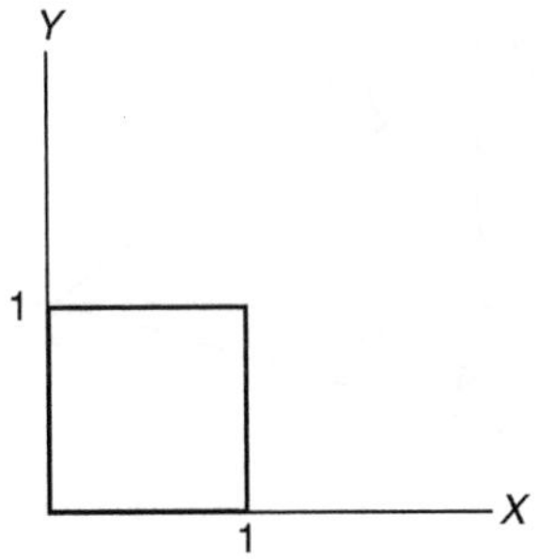

FIGURE 2.2 Test figure for the divergence theorem.

$$\int_{S_2} (\mathbf{v}\cdot\mathbf{n}_2)\,dS + \int_{S_4} (\mathbf{v}\cdot\mathbf{n}_4)\,dS = \int_{0<x<1,\ y=1} [(x-y)\mathbf{e}_1 + (x+y)\mathbf{e}_2]\cdot\mathbf{e}_2\,dx$$

$$+ \int_{0<x<1,\ y=0} [(x-y)\mathbf{e}_1 + (x+y)\mathbf{e}_2]\cdot(-\mathbf{e}_2)\,dx$$

$$= \int_0^1 (x+1)\,dx - \int_0^1 x\,dx = 1$$

Consequently,

$$\int_S (\mathbf{v}\cdot\mathbf{n})\,dS = \int_{S_1} (\mathbf{v}\cdot\mathbf{n}_1)\,dS + \int_{S_2} (\mathbf{v}\cdot\mathbf{n}_2)\,dS + \int_{S_3} (\mathbf{v}\cdot\mathbf{n}_3)\,dS + \int_{S_4} (\mathbf{v}\cdot\mathbf{n}_4)\,dS = 2$$

and the divergence theorem has been verified in the case in question.

EXAMPLE 2.5

In the tetrahedron shown in Figure 2.3, A_1, A_2, and A_3 denote the areas of the faces whose normal vectors point in the $-\mathbf{e}_1$, $-\mathbf{e}_2$, and $-\mathbf{e}_3$ directions, and let A and $\mathbf{n}$ denote the area and normal vector of the inclined face. Prove that

$$\mathbf{n} = \frac{A_1}{A}\mathbf{e}_1 + \frac{A_2}{A}\mathbf{e}_2 + \frac{A_3}{A}\mathbf{e}_3$$

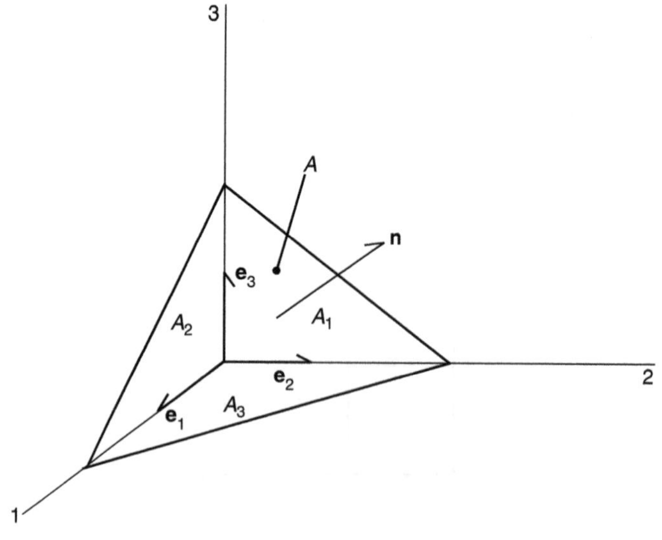

FIGURE 2.3 Geometry of a tetrahedron.

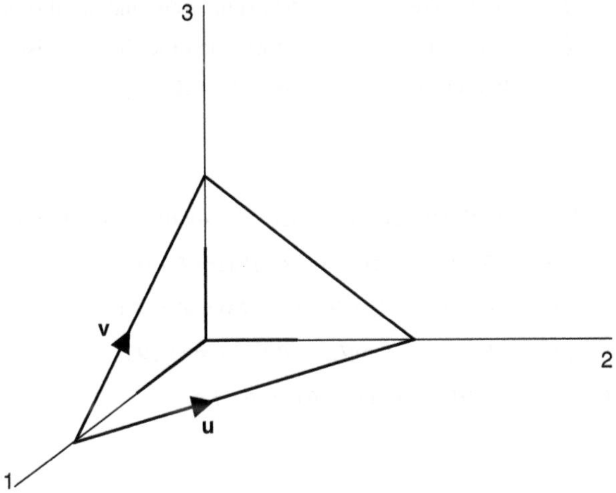

FIGURE 2.4 Illustration of the base vectors.

Solution

From the definition of the normal vector and referring to the Figure 2.4, we have

$$\mathbf{n} = \frac{\mathbf{u} \times \mathbf{v}}{|\mathbf{u} \times \mathbf{v}|}$$

Let a, b, and c represent the length of the edges of the tetrahedron along 1-, 2-, and 3-axes, respectively. Now

$$\mathbf{u} = -a\mathbf{e}_1 + b\mathbf{e}_2 \quad \text{and} \quad \mathbf{v} = -a\mathbf{e}_1 + c\mathbf{e}_3$$

$$\therefore \ \mathbf{u} \times \mathbf{v} = bc\mathbf{e}_1 + ca\mathbf{e}_2 + ab\mathbf{e}_3$$

But $bc = 2A_1$; $ca = 2A_2$; $ab = 2A_3$. Also $|\mathbf{u} \times \mathbf{v}| = 2A$. Hence

$$\mathbf{n} = \frac{A_1}{A}\mathbf{e}_1 + \frac{A_2}{A}\mathbf{e}_2 + \frac{A_3}{A}\mathbf{e}_3$$

EXAMPLE 2.6

Prove that if $\boldsymbol{\sigma}$ is a symmetric matrix with entries σ_{ij}, that

$$\varepsilon_{ijk}\sigma_{jk} = 0, \quad i = 1, 2, 3$$

Solution

It is given that $\boldsymbol{\sigma}$ is a symmetric tensor. Therefore $\sigma_{jk} = \sigma_{kj}$. We know that,

$$\varepsilon_{ijk} = \begin{cases} 1 & \text{if } ijk \text{ take values in the cyclic order and are distinct} \\ -1 & \text{if } ijk \text{ take values in the acyclic order but are distinct} \\ 0 & \text{two or all of } i, j, k \text{ take same values} \end{cases}$$

Hence,

$$\begin{aligned} \varepsilon_{ijk}\sigma_{jk} &= \varepsilon_{111}\sigma_{11} + \varepsilon_{112}\sigma_{12} + \varepsilon_{113}\sigma_{13} + \varepsilon_{121}\sigma_{21} + \varepsilon_{122}\sigma_{22} + \varepsilon_{123}\sigma_{23} + \varepsilon_{131}\sigma_{31} \\ &\quad + \varepsilon_{132}\sigma_{32} + \varepsilon_{133}\sigma_{33} + \varepsilon_{211}\sigma_{11} + \varepsilon_{212}\sigma_{12} + \varepsilon_{213}\sigma_{13} + \varepsilon_{221}\sigma_{21} + \varepsilon_{222}\sigma_{22} \\ &\quad + \varepsilon_{223}\sigma_{23} + \varepsilon_{231}\sigma_{31} + \varepsilon_{232}\sigma_{32} + \varepsilon_{233}\sigma_{33} + \varepsilon_{311}\sigma_{11} + \varepsilon_{312}\sigma_{12} + \varepsilon_{313}\sigma_{13} \\ &\quad + \varepsilon_{321}\sigma_{21} + \varepsilon_{322}\sigma_{22} + \varepsilon_{323}\sigma_{23} + \varepsilon_{331}\sigma_{31} + \varepsilon_{332}\sigma_{32} + \varepsilon_{333}\sigma_{33} \\ &= (\sigma_{23} - \sigma_{32}) - (\sigma_{13} - \sigma_{31}) + (\sigma_{12} - \sigma_{21}) \\ &= 0 \end{aligned}$$

EXAMPLE 2.7

If $\mathbf{v}$ and $\mathbf{w}$ are 3×1 vectors, prove that $\mathbf{v} \times \mathbf{w}$ may be written as

$$\mathbf{v} \times \mathbf{w} = \mathbf{V}\mathbf{w}$$

in which $\mathbf{V}$ is an antisymmetric tensor and $\mathbf{v}$ is called the axial vector of $\mathbf{V}$. Derive the expression for $\mathbf{V}$.

Solution

Now

$$\mathbf{v} \times \mathbf{w} = \varepsilon_{ijk} v_j w_k = V_{ik} w_k$$

where

$$V_{ik} = \varepsilon_{ijk} v_j, \quad [\mathbf{V}]_{ik} = V_{ik}$$

Now $V_{11} = \varepsilon_{1j1} v_j = 0$. Similarly $V_{22} = V_{33} = 0$. Also

$$\begin{aligned} V_{12} &= \varepsilon_{132} v_3 = -v_3 \\ V_{21} &= \varepsilon_{231} v_3 = v_3 \\ V_{23} &= \varepsilon_{213} v_1 = -v_1 \\ V_{32} &= \varepsilon_{312} v_1 = v_1 \\ V_{31} &= \varepsilon_{321} v_2 = -v_2 \\ V_{13} &= \varepsilon_{123} v_2 = v_2 \end{aligned}$$

and so

$$\mathbf{V} = \begin{bmatrix} 0 & -v_3 & v_2 \\ v_3 & 0 & -v_1 \\ -v_2 & v_1 & 0 \end{bmatrix}$$

Also, $\mathbf{V}$ is antisymmetric, i.e., $\mathbf{V} = \mathbf{V}^T$.

2.3 MATRICES

An $n \times n$ matrix is simply an array of numbers arranged in rows and columns, in which case we call it a second-order array. For the matrix $\mathbf{A}$ the entry a_{ij} occupies the intersection of the ith row and the jth column. We may also introduce the $n \times 1$ first-order array $\mathbf{a}$ in which a_i denotes the ith entry. We likewise refer to the $1 \times n$ array $\mathbf{a}^T$ as first order. In the current context, a first-order array is not a vector unless it is associated with a coordinate system and certain transformation properties to be introduced shortly. In the following all matrices will be real unless otherwise noted. Several properties of first- and second-order arrays are now introduced:

(i) Sum of two $n \times n$ matrices $\mathbf{A}$ and $\mathbf{B}$ is a matrix $\mathbf{C}$ in which $c_{ij} = a_{ij} + b_{ij}$.

(ii) Product of a matrix $\mathbf{A}$ and a scalar q is a matrix $\mathbf{C}$ in which $c_{ij} = qa_{ij}$.

(iii) Transpose of a matrix $\mathbf{A}$, denoted $\mathbf{A}^T$, is a matrix in which $a_{ij}^T = a_{ji}$. $\mathbf{A}$ is called *symmetric* if $\mathbf{A} = \mathbf{A}^T$, and it is called *antisymmetric (or skew symmetric)* if $\mathbf{A} = -\mathbf{A}^T$.

(iv) Product of two matrices $\mathbf{A}$ and $\mathbf{B}$ is the matrix $\mathbf{C}$ for which

*T) $$c_{ij} = a_{ik} b_{kj} \tag{2.27}$$

Matrix multiplication may be easily visualized as follows. Let the first-order array $\mathbf{a}_i^T$ denote the ith row of $\mathbf{A}$, while the first-order array $\mathbf{b}_j$ denotes the jth column of $\mathbf{B}$. Then c_{ij} may be written as

*T) $$c_{ij} = \mathbf{a}_i^T \mathbf{b}_j \tag{2.28}$$

(v) Product of a matrix $\mathbf{A}$ and a first-order array $\mathbf{c}$ is the first-order array $\mathbf{d}$ in which the ith entry is $d_i = a_{ij} c_j$.

(vi) ijth entry of the identity matrix $\mathbf{I}$ is δ_{ij}. Thus it exhibits ones on the diagonal positions ($i = j$) and zeroes off-diagonal ($i \neq j$). Clearly $\mathbf{I}$ is the matrix counterpart of the substitution operator.

(vii) Determinant of $\mathbf{A}$ is given by

*T) $$\det(\mathbf{A}) = \tfrac{1}{6} \varepsilon_{ijk} \varepsilon_{pqr} a_{ip} a_{jq} a_{kr} \tag{2.29}$$

Suppose $\mathbf{a}$ and $\mathbf{b}$ are two nonzero first-order $n \times 1$ arrays. If $\det(\mathbf{A}) = 0$ the matrix $\mathbf{A}$ is singular, in which case there is no solution to equations of the form $\mathbf{Aa} = \mathbf{b}$. However, if $\mathbf{b} = \mathbf{0}$, there may be multiple solutions. If $\det(\mathbf{A}) \neq 0$, then there is a unique nontrivial solution $\mathbf{a}$.

(viii) Let $\mathbf{A}$ and $\mathbf{B}$ be $n \times n$ nonsingular matrices. The determinant has the following useful properties:

*M)
$$\begin{aligned} \det(\mathbf{AB}) &= \det(\mathbf{A})\det(\mathbf{B}) \\ \det(\mathbf{A}^T) &= \det(\mathbf{A}) \\ \det(\mathbf{I}) &= 1 \end{aligned} \tag{2.30}$$

(ix) If $\det(\mathbf{A}) \neq 0$, then $\mathbf{A}$ is nonsingular, and there exists an inverse matrix, $\mathbf{A}^{-1}$, for which

*M)
$$\mathbf{AA}^{-1} = \mathbf{A}^{-1}\mathbf{A} = \mathbf{I} \tag{2.31}$$

(x) The transpose of a matrix product satisfies

*M)
$$(\mathbf{AB})^T = \mathbf{B}^T\mathbf{A}^T \tag{2.32}$$

(xi) The inverse of a matrix product satisfies

*M)
$$(\mathbf{AB})^{-1} = \mathbf{B}^{-1}\mathbf{A}^{-1} \tag{2.33}$$

(xii) If $\mathbf{c}$ and $\mathbf{d}$ are two 3×1 vectors the vector product $\mathbf{c} \times \mathbf{d}$ generates the vector $\mathbf{c} \times \mathbf{d} = \mathbf{Cd}$ in which $\mathbf{C}$ is an antisymmetric matrix given by

*M)
$$\mathbf{C} = \begin{bmatrix} 0 & -c_3 & c_2 \\ c_3 & 0 & -c_1 \\ -c_2 & c_1 & 0 \end{bmatrix} \tag{2.34}$$

Recalling that $\mathbf{c} \times \mathbf{d} = \varepsilon_{ikj}c_k d_j$ and noting that $\varepsilon_{ikj}c_k$ denotes the (ij)th component of an antisymmetric tensor, it is immediate that $[\mathbf{C}]_{ij} = \varepsilon_{ikj}c_k$.

(xiii) If $\mathbf{c}$ and $\mathbf{d}$ are two vectors the outer product $\mathbf{cd}^T$ generates the matrix $\mathbf{C}$ given by

*M)
$$\mathbf{C} = \begin{bmatrix} c_1d_1 & c_1d_2 & c_1d_3 \\ c_2d_1 & c_2d_2 & c_2d_3 \\ c_3d_1 & c_3d_2 & c_3d_3 \end{bmatrix} \tag{2.35}$$

We will see later that $\mathbf{C}$ is a second-order tensor if $\mathbf{c}$ and $\mathbf{d}$ have the transformation properties of vectors.

(xiv) An $n \times n$ matrix $\mathbf{A}$ may be decomposed into symmetric and antisymmetric matrices using

$$\mathbf{A} = \mathbf{A}_s + \mathbf{A}_a, \quad \mathbf{A}_s = \tfrac{1}{2}\left[\mathbf{A} + \mathbf{A}^T\right], \; \mathbf{A}_a = \tfrac{1}{2}\left[\mathbf{A} - \mathbf{A}^T\right] \tag{2.36}$$

EXAMPLE 2.8

Find the transposes of the matrices

$$\mathbf{A} = \begin{bmatrix} 1 & -1/2 \\ -1/3 & 1/4 \end{bmatrix}, \quad \mathbf{B} = \begin{bmatrix} 1 & 1/3 \\ 1/2 & 1/4 \end{bmatrix}$$

(a) Verify that $\mathbf{AB} \neq \mathbf{BA}$.
(b) Verify that $(\mathbf{AB})^T = \mathbf{B}^T\mathbf{A}^T$.

Solution

Here

$$\mathbf{A}^T = \begin{bmatrix} 1 & -1/3 \\ -1/2 & 1/4 \end{bmatrix}, \quad \mathbf{B}^T = \begin{bmatrix} 1 & 1/2 \\ 1/3 & 1/4 \end{bmatrix}$$

Also

$$\mathbf{AB} = \begin{bmatrix} 1 & -1/2 \\ -1/3 & 1/4 \end{bmatrix} \begin{bmatrix} 1 & 1/3 \\ 1/2 & 1/4 \end{bmatrix} = \begin{bmatrix} 3/4 & 5/24 \\ -5/24 & -7/144 \end{bmatrix}$$

$$\mathbf{BA} = \begin{bmatrix} 1 & 1/3 \\ 1/2 & 1/4 \end{bmatrix} \begin{bmatrix} 1 & -1/2 \\ -1/3 & 1/4 \end{bmatrix} = \begin{bmatrix} 8/9 & -5/12 \\ 5/12 & -3/16 \end{bmatrix}$$

and clearly $\mathbf{AB} \neq \mathbf{BA}$. Now

$$(\mathbf{AB})^T = \begin{bmatrix} 3/4 & -5/24 \\ 5/24 & -7/144 \end{bmatrix}$$

$$\mathbf{B}^T\mathbf{A}^T = \begin{bmatrix} 1 & 1/2 \\ 1/3 & 1/4 \end{bmatrix} \begin{bmatrix} 1 & -1/3 \\ -1/2 & 1/4 \end{bmatrix} = \begin{bmatrix} 3/4 & -5/24 \\ 5/24 & -7/144 \end{bmatrix}$$

and it is seen that $(\mathbf{AB})^T = \mathbf{B}^T\mathbf{A}^T$.

EXAMPLE 2.9

For the matrices in Example 2.8 find the inverses and verify that

$$(\mathbf{AB})^{-1} = \mathbf{B}^{-1}\mathbf{A}^{-1}$$

Solution

$$\mathbf{A}^{-1} = \frac{1}{1/12}\begin{bmatrix} 1/4 & 1/2 \\ 1/3 & 1 \end{bmatrix} = \begin{bmatrix} 3 & 6 \\ 4 & 12 \end{bmatrix}$$

$$\mathbf{B}^{-1} = \frac{1}{1/12}\begin{bmatrix} 1/4 & -1/3 \\ -1/2 & 1 \end{bmatrix} = \begin{bmatrix} 3 & -4 \\ -6 & 12 \end{bmatrix}$$

$$\mathbf{AB} = \begin{bmatrix} 3/4 & 5/24 \\ -5/24 & -7/144 \end{bmatrix} \rightarrow (\mathbf{AB})^{-1} = \frac{1}{1/144}\begin{bmatrix} -7/144 & -5/24 \\ 5/24 & 3/4 \end{bmatrix}$$

$$= \begin{bmatrix} -7 & -30 \\ 30 & 108 \end{bmatrix}$$

Now

$$\mathbf{B}^{-1}\mathbf{A}^{-1} = \begin{bmatrix} 3 & -4 \\ -6 & 12 \end{bmatrix}\begin{bmatrix} 3 & 6 \\ 4 & 12 \end{bmatrix} = \begin{bmatrix} -7 & -30 \\ 30 & 108 \end{bmatrix}$$

$$\Rightarrow (\mathbf{AB})^{-1} = \mathbf{B}^{-1}\mathbf{A}^{-1}$$

EXAMPLE 2.10

Consider a matrix $\mathbf{C}$ given by

$$\mathbf{C} = \begin{bmatrix} a & b \\ c & d \end{bmatrix}$$

Verify that its inverse is given by

$$\mathbf{C}^{-1} = \frac{1}{ad - bc}\begin{bmatrix} d & -b \\ -c & a \end{bmatrix}$$

Solution

The cofactor matrix is given by

$$\text{cof } \mathbf{C} = \begin{bmatrix} d & -c \\ -b & a \end{bmatrix}$$

The adjoint matrix is given by

$$\text{adj } \mathbf{C} = \begin{bmatrix} d & -b \\ -c & a \end{bmatrix}$$

The determinant of the matrix is $|\mathbf{C}| = ad - bc$. Hence

$$\mathbf{C}^{-1} = \frac{\text{adj } \mathbf{C}}{|\mathbf{C}|} = \frac{1}{ad - bc}\begin{bmatrix} d & -b \\ -c & a \end{bmatrix}$$

as expected. Alternatively, note that

$$\mathbf{CC}^{-1} = \frac{1}{ad-bc}\begin{bmatrix} d & -b \\ -c & a \end{bmatrix}\begin{bmatrix} a & b \\ c & d \end{bmatrix} = \frac{1}{ad-bc}\begin{bmatrix} ad-bc & 0 \\ 0 & ad-bc \end{bmatrix} = \begin{bmatrix} 1 & 0 \\ 0 & 1 \end{bmatrix}$$

EXAMPLE 2.11

Using the matrix **C** from Example 2.10, and introducing the vectors (one-dimensional arrays)

$$\mathbf{a} = \begin{pmatrix} q \\ r \end{pmatrix}, \quad \mathbf{b} = \begin{pmatrix} s \\ t \end{pmatrix}$$

verify that

$$\mathbf{a}^T\mathbf{C}\mathbf{b} = \mathbf{b}^T\mathbf{C}^T\mathbf{a}$$

SOLUTION

$$\begin{aligned} \mathbf{a}^T\mathbf{C}\mathbf{b} &= (q \quad r)\begin{bmatrix} a & b \\ c & d \end{bmatrix}\begin{pmatrix} s \\ t \end{pmatrix} = (q \quad r)\begin{pmatrix} as+bt \\ cs+dt \end{pmatrix} \\ &= aqs + bqt + crs + drt \\ &= (as+bt \quad cs+dt)\begin{pmatrix} q \\ r \end{pmatrix} \\ &= (s \quad t)\begin{bmatrix} a & c \\ b & d \end{bmatrix}\begin{pmatrix} q \\ r \end{pmatrix} \\ &= \mathbf{b}^T\mathbf{C}^T\mathbf{a} \end{aligned}$$

and accordingly $\mathbf{a}^T\mathbf{C}\mathbf{b} = \mathbf{b}^T\mathbf{C}^T\mathbf{a}$.

EXAMPLE 2.12

For the geometry of Example 2.4, verify that

$$\int \mathbf{n} \times \mathbf{A}\,\mathrm{d}S = \int \nabla \times \mathbf{A}^T\,\mathrm{d}V$$

using

$$\begin{aligned} a_{11} &= x + y + x^2 + y^2 \\ a_{12} &= x + y + x^2 - y^2 \\ a_{21} &= x + y - x^2 - y^2 \\ a_{22} &= x - y - x^2 - y^2 \end{aligned}$$

Solution

Given

$$\mathbf{A} = \begin{bmatrix} a_{11} & a_{12} \\ a_{21} & a_{22} \end{bmatrix}$$

let

$$\mathbf{A} = [\boldsymbol{\beta}_1 \quad \boldsymbol{\beta}_2], \quad \text{where } \boldsymbol{\beta}_1 = \begin{pmatrix} a_{11} \\ a_{21} \end{pmatrix} \quad \text{and} \quad \boldsymbol{\beta}_2 = \begin{pmatrix} a_{12} \\ a_{22} \end{pmatrix}$$

Now

$$\nabla \times \boldsymbol{\beta}_1 = \left[\frac{\partial}{\partial x}\mathbf{e}_1 + \frac{\partial}{\partial y}\mathbf{e}_2\right] \times [a_{11}\mathbf{e}_1 + a_{21}\mathbf{e}_2] = -2(x+y)\mathbf{e}_3$$

$$\nabla \times \boldsymbol{\beta}_2 = \left[\frac{\partial}{\partial x}\mathbf{e}_1 + \frac{\partial}{\partial y}\mathbf{e}_2\right] \times [a_{12}\mathbf{e}_1 + a_{22}\mathbf{e}_2] = -2(x-y)\mathbf{e}_3$$

$$\therefore \int_V \nabla \times \mathbf{A}^T \mathrm{d}V = \int_V [\nabla \times \boldsymbol{\beta}_1 \quad \nabla \times \boldsymbol{\beta}_2]\mathrm{d}V = \int \begin{bmatrix} 0 & 0 \\ 0 & 0 \\ -2(x+y) & -2(x-y) \end{bmatrix} \mathrm{d}x\,\mathrm{d}y = \begin{bmatrix} 0 & 0 \\ 0 & 0 \\ -2 & 0 \end{bmatrix}$$

For faces (1) and (3), $\mathbf{n}_1 = \mathbf{e}_1$, $\mathbf{n}_3 = -\mathbf{e}_1$. Therefore

$$\int_{S_1} (\mathbf{n}_1 \times \mathbf{A})\,\mathrm{d}S + \int_{S_3} (\mathbf{n}_3 \times \mathbf{A})\,\mathrm{d}S = \int_{x=1,\,0<y<1} \mathbf{e}_1 \times [\boldsymbol{\beta}_1 \quad \boldsymbol{\beta}_2]\,\mathrm{d}S - \int_{x=0,\,0<y<1} \mathbf{e}_1 \times [\boldsymbol{\beta}_1 \quad \boldsymbol{\beta}_2]\,\mathrm{d}S$$

$$= \int_{x=1,\,0<y<1} \begin{bmatrix} 0 & 0 \\ 0 & 0 \\ x+y-x^2-y^2 & x-y-x^2-y^2 \end{bmatrix} \mathrm{d}y$$

$$- \int_{x=0,\,0<y<1} \begin{bmatrix} 0 & 0 \\ 0 & 0 \\ x+y-x^2-y^2 & x-y-x^2-y^2 \end{bmatrix} \mathrm{d}y$$

$$= \int_0^1 \begin{bmatrix} 0 & 0 \\ 0 & 0 \\ y-y^2 & -y-y^2 \end{bmatrix} \mathrm{d}y - \int_0^1 \begin{bmatrix} 0 & 0 \\ 0 & 0 \\ y-y^2 & -y-y^2 \end{bmatrix} \mathrm{d}y = \mathbf{0}$$

For faces (2) and (4), $\mathbf{n}_2=\mathbf{e}_2$, $\mathbf{n}_4=-\mathbf{e}_2$. Hence

$$\int_{S_2}(\mathbf{n}_2\times\mathbf{A})\,dS+\int_{S_4}(\mathbf{n}_4\times\mathbf{A})\,dS=\int_{0<x<1,y=1}\mathbf{e}_2\times[\boldsymbol{\beta}_1\;\;\boldsymbol{\beta}_2]\,dS-\int_{0<x<1,y=0}\mathbf{e}_2\times[\boldsymbol{\beta}_1\;\;\boldsymbol{\beta}_2]\,dS$$

$$=-\int_{0<x<1,y=1}\begin{bmatrix}0&0\\0&0\\x+y+x^2+y^2&x-y+x^2-y^2\end{bmatrix}dx$$

$$+\int_{0<x<1,y=0}\begin{bmatrix}0&0\\0&0\\x+y+x^2+y^2&x-y+x^2-y^2\end{bmatrix}dx$$

$$=-\int_0^1\begin{bmatrix}0&0\\0&0\\x+x^2+2&x+x^2\end{bmatrix}dx+\int_0^1\begin{bmatrix}0&0\\0&0\\x+x^2&x+x^2\end{bmatrix}dx=\begin{bmatrix}0&0\\0&0\\-2&0\end{bmatrix}$$

Therefore

$$\int_S(\mathbf{n}\times\mathbf{A})\,dS=\int_{S_1}(\mathbf{n}_1\times\mathbf{A})\,dS+\int_{S_2}(\mathbf{n}_2\times\mathbf{A})\,dS+\int_{S_3}(\mathbf{n}_3\times\mathbf{A})\,dS+\int_{S_4}(\mathbf{n}_4\times\mathbf{A})\,dS$$

$$=\begin{bmatrix}0&0\\0&0\\-2&0\end{bmatrix}$$

Hence $\int_S(\mathbf{n}\times\mathbf{A})\,dS=\int_V\nabla\times\mathbf{A}^T\,dV$ is verified.

2.4 EIGENVALUES AND EIGENVECTORS

Again, $\mathbf{A}$ is an $n\times n$ matrix. We now introduce the eigenvalue equation

$$(\mathbf{A}-\lambda_j\mathbf{I})\mathbf{x}_j=\mathbf{0} \tag{2.37}$$

The solution for $\mathbf{x}_j$ is trivial unless $\mathbf{A}-\lambda_j\mathbf{I}$ is singular, in which event det $(\mathbf{A}-\lambda_j\mathbf{I})=0$. There are n possibly complex roots. If the magnitude of the eigenvectors is set to unity, they may likewise be determined. As an example consider

$$\mathbf{A}=\begin{bmatrix}2&1\\1&2\end{bmatrix} \tag{2.38}$$

The equation det $(\mathbf{A}-\lambda_j\mathbf{I})=0$ is expanded as $(2-\lambda_j)^2-1$, with roots $\lambda_{1,2}=1, 3$, and

$$\mathbf{A}-\lambda_1\mathbf{I}=\begin{bmatrix}1&1\\1&1\end{bmatrix},\quad \mathbf{A}-\lambda_2\mathbf{I}=\begin{bmatrix}-1&1\\1&-1\end{bmatrix} \tag{2.39}$$

Note that in each case the rows are multiples of each other, so that only one row can be considered independent. We next determine the eigenvectors. It is easily seen that magnitudes of the eigenvectors are arbitrary. For example, if $\mathbf{x}_1$ is an eigenvector, so is $10\mathbf{x}_1$. Accordingly, the magnitudes are arbitrarily set to unity. For $\mathbf{x}_1 = \{x_{11} \quad x_{12}\}^T$,

$$x_{11} + x_{12} = 0, \quad x_{11}^2 + x_{12}^2 = 1 \tag{2.40}$$

from which we conclude that $\mathbf{x}_1 = \{1 \quad -1\}^T/\sqrt{2}$. A parallel argument furnishes $\mathbf{x}_2 = \{1 \quad 1\}^T/\sqrt{2}$.

If $\mathbf{A}$ is symmetric, the eigenvalues and eigenvectors are real and the eigenvectors are orthogonal to each other: $\mathbf{x}_i^T\mathbf{x}_j = \delta_{ij}$. The eigenvalue equations can be "stacked up" as follows:

$$\mathrm{A}[x_1 \vdots x_2 \vdots \ldots x_n] = [x_1 \vdots x_2 \vdots \ldots x_n]\begin{bmatrix} \lambda_1 & 0 & . & . & . \\ 0 & \lambda_2 & . & . & . \\ . & . & . & . & . \\ . & . & . & \lambda_{n-1} & 0 \\ . & . & . & 0 & \lambda_n \end{bmatrix} \tag{2.41}$$

With obvious identifications,

$$\mathbf{AX} = \mathbf{X\Lambda} \tag{2.42}$$

and $\mathbf{X}$ may be called the modal matrix. Let y_{ij} is the ijth entry of $\mathbf{Y} = \mathbf{X}^T\mathbf{X}$. Now

$$y_{ij} = \mathbf{x}_i^T\mathbf{x}_j = \delta_{ij} \tag{2.43}$$

so that $\mathbf{Y} = \mathbf{I}$. We conclude that $\mathbf{X}$ is an orthogonal tensor: $\mathbf{X}^T = \mathbf{X}^{-1}$. Further

$$\mathbf{X}^T\mathbf{AX} = \mathbf{\Lambda}, \quad \mathbf{A} = \mathbf{X\Lambda X}^T \tag{2.44}$$

and $\mathbf{X}$ may be interpreted as representing a rotation from the reference axes to the principal axes.

2.5 COORDINATE TRANSFORMATIONS

Suppose that the vectors $\mathbf{v}$ and $\mathbf{w}$ are depicted in a second coordinate system whose base vectors are denoted by $\mathbf{e}_j'$. Now $\mathbf{e}_j'$ may be represented as a linear sum of the base vectors $\mathbf{e}_i$:

*T)

$$\mathbf{e}_j' = q_{ji}\mathbf{e}_i \tag{2.45}$$

But then $\mathbf{e}_i \cdot \mathbf{e}_j' = q_{ij} = \cos(\theta_{ij'})$. It follows that $\delta_{ij} = \mathbf{e}_i' \cdot \mathbf{e}_j' = (q_{ik} \ \mathbf{e}_k) \cdot (q_{jl} \ \mathbf{e}_l) = q_{ik}q_{jl}\delta_{kl}$, so that

*T)
$$q_{ik}q_{jk} = q_{ik}q^T_{kj} \\ = \delta_{ij}$$

In *M) notation this is written as

*M)
$$\mathbf{QQ}^T = \mathbf{I} \tag{2.46}$$

in which case the matrix **Q** is called *orthogonal*. An analogous argument serves to prove that $\mathbf{Q}^T\mathbf{Q} = \mathbf{I}$. From Equation 2.30, $1 = \det(\mathbf{QQ}^T) = \det(\mathbf{Q})\det(\mathbf{Q}^T) = \det^2(\mathbf{Q})$. Right-handed rotations satisfy $\det(\mathbf{Q}) = 1$, in which case **Q** is called *proper orthogonal*.

EXAMPLE 2.13

Consider the matrix

$$\mathbf{Q} = \begin{bmatrix} \cos\theta & \sin\theta \\ -\sin\theta & \cos\theta \end{bmatrix}$$

Verify that

(a) $\mathbf{QQ}^T = \mathbf{Q}^T\mathbf{Q}$
(b) $\mathbf{Q}^T = \mathbf{Q}^{-1}$
(c) For any 2×1 vector **a**

$$|\mathbf{Qa}| = |\mathbf{a}|$$

(The relation in (c) is general, and the vector **Qa** represents a rotation of **a**.)

Solution

First consider

$$\mathbf{QQ}^T = \begin{bmatrix} \cos\theta & \sin\theta \\ -\sin\theta & \cos\theta \end{bmatrix}\begin{bmatrix} \cos\theta & -\sin\theta \\ \sin\theta & \cos\theta \end{bmatrix}$$
$$= \begin{bmatrix} \cos^2\theta + \sin^2\theta & \cos\theta(-\sin\theta) + \sin\theta\cos\theta \\ -\sin\theta\cos\theta + \cos\theta\sin\theta & \cos^2\theta + \sin^2\theta \end{bmatrix} = \begin{bmatrix} 1 & 0 \\ 0 & 1 \end{bmatrix}$$

$$\mathbf{Q}^T\mathbf{Q} = \begin{bmatrix} \cos\theta & -\sin\theta \\ \sin\theta & \cos\theta \end{bmatrix}\begin{bmatrix} \cos\theta & \sin\theta \\ -\sin\theta & \cos\theta \end{bmatrix}$$
$$= \begin{bmatrix} \cos^2\theta + \sin^2\theta & \cos\theta\sin\theta - \sin\theta\cos\theta \\ \sin\theta\cos\theta + \cos\theta(-\sin\theta) & \cos^2\theta + \sin^2\theta \end{bmatrix} = \begin{bmatrix} 1 & 0 \\ 0 & 1 \end{bmatrix}$$

and $\mathbf{QQ}^T = \mathbf{Q}^T\mathbf{Q} = \mathbf{I}$, as expected.

Next, from the foregoing relations

$$\mathbf{Q}^{-1} = \frac{1}{\cos\theta\cos\theta + \sin\theta\sin\theta}\begin{bmatrix}\cos\theta & -\sin\theta \\ \sin\theta & \cos\theta\end{bmatrix} = \mathbf{Q}^T$$

Let

$$\mathbf{a} = \begin{pmatrix} a_1 \\ a_2 \end{pmatrix}$$

Now

$$\mathbf{Qa} = \begin{bmatrix}\cos\theta & \sin\theta \\ -\sin\theta & \cos\theta\end{bmatrix}\begin{pmatrix} a_1 \\ a_2 \end{pmatrix} = \begin{pmatrix} a_1\cos\theta + a_2\sin\theta \\ -a_1\sin\theta + a_2\cos\theta \end{pmatrix}$$

from which

$$|\mathbf{Qa}| = \sqrt{(a_1\cos\theta + a_2\sin\theta)^2 + (-a_1\sin\theta + a_2\cos\theta)^2} = \sqrt{a_1^2 + a_2^2} = |\mathbf{a}|$$

showing that $|\mathbf{Qa}| = |\mathbf{a}|$.

2.5.1 Transformations of Vectors

The vector $\mathbf{v}'$ is the same as the vector $\mathbf{v}$ except that $\mathbf{v}'$ is referred to $\mathbf{e}_j'$ while $\mathbf{v}$ is referred to $\mathbf{e}_i$. Now

*T)

$$\begin{aligned}\mathbf{v}' &= v_j'\mathbf{e}_j' \\ &= v_j' q_{ji}\mathbf{e}_i \\ &= \mathrm{v}_i\mathbf{e}_i\end{aligned} \tag{2.47}$$

It follows that $v_i = v_j' q_{ji}$, and hence

*M)

$$\mathbf{v} = \mathbf{Q}^T\mathbf{v}', \quad \mathbf{v}' = \mathbf{Q}\mathbf{v} \tag{2.48}$$

in which q_{ij} is the jith entry of $\mathbf{Q}^T$.

We now state an alternate definition of a vector as a first-order tensor. Let $\mathbf{v}$ be an $n \times 1$ array of numbers referred to a coordinate system with base vectors $\mathbf{e}_i$. It is a vector if and only if, upon a rotation of the coordinate system to base vectors $\mathbf{e}_j'$, $\mathbf{v}'$ transforms according to Equation 2.48.

Since $\left(\frac{\mathrm{d}\phi}{\mathrm{d}\mathbf{x}}\right)'\mathrm{d}\mathbf{x}'$ is likewise equal to $\mathrm{d}\phi$,

*M)

$$\left(\frac{\mathrm{d}\phi}{\mathrm{d}\mathbf{x}}\right)' = \left(\frac{\mathrm{d}\phi}{\mathrm{d}\mathbf{x}}\right)\mathbf{Q}^T \tag{2.49}$$

for which reason $\mathrm{d}\phi/\mathrm{d}\mathbf{x}$ is called a covariant vector in a more general presentation, while $\mathbf{v}$ is properly called a contravariant vector.

Finally, to display the base vectors to which the tensor $\mathbf{A}$ is referred (i.e., in tensor-indicial notation), we introduce the *outer product*

$$\mathbf{e}_i \wedge \mathbf{e}_j \tag{2.50}$$

and it is recognized as the matrix–vector counterpart $\mathbf{e}_i\mathbf{e}_j^T$. Now

$$\mathbf{A} = a_{ij}\mathbf{e}_i \wedge \mathbf{e}_j \tag{2.51a}$$

Note the useful result that

$$\mathbf{e}_i \wedge \mathbf{e}_j \cdot \mathbf{e}_k = \mathbf{e}_i\delta_{jk} \tag{2.51b}$$

In this notation, given a vector $\mathbf{b} = b_k\mathbf{e}_k$,

$$\begin{aligned}\mathbf{Ab} &= a_{ij}\mathbf{e}_i \wedge \mathbf{e}_j \cdot b_k\mathbf{e}_k \\ &= a_{ij}b_k\mathbf{e}_i \wedge \mathbf{e}_j \cdot \mathbf{e}_k \\ &= a_{ij}b_k\mathbf{e}_i\delta_{jk} \\ &= a_{ij}b_j\mathbf{e}_i\end{aligned} \tag{2.52}$$

as expected.

2.6 ORTHOGONAL CURVILINEAR COORDINATES

The position vector of a point P referred to a three-dimensional rectilinear coordinate system is expressed in tensor-indicial notation as $\mathbf{R}_\mathrm{P} = x_i\,\mathbf{e}_i$. The position vector connecting two "sufficiently close" points P and Q is given by

$$\Delta\mathbf{R} = \mathbf{R}_\mathrm{P} - \mathbf{R}_\mathrm{Q} \approx \mathbf{dR}_\mathbf{x} \tag{2.53}$$

where

$$\mathbf{dR}_\mathrm{x} = \mathrm{d}x_i\,\mathbf{e}_i \tag{2.54}$$

with differential arc length

$$\mathrm{d}S_x = \sqrt{\mathrm{d}x_i\,\mathrm{d}x_i} \tag{2.55}$$

Suppose now that the coordinates are transformed to coordinates $y_j : x_i = x_i(y_j)$. The same position vector, now referred to the transformed system, is

$$\mathbf{d}\boldsymbol{R}_y = \sum_{1}^{3} \mathrm{d}y_\alpha\, \mathbf{g}_\alpha$$
$$\mathbf{g}_\alpha = h_\alpha \boldsymbol{\gamma}_\alpha \tag{2.56}$$
$$h_\alpha = \sqrt{\frac{\mathrm{d}x_j}{\mathrm{d}y_\alpha}\frac{\mathrm{d}x_j}{\mathrm{d}y_\alpha}}$$

$$\boldsymbol{\gamma}_\alpha = \frac{\dfrac{\mathrm{d}x_i}{\mathrm{d}y_\alpha}}{\sqrt{\dfrac{\mathrm{d}x_j}{\mathrm{d}y_\alpha}\dfrac{\mathrm{d}x_j}{\mathrm{d}y_\alpha}}}\mathbf{e}_i$$
$$= \frac{1}{h_\alpha}\frac{\mathrm{d}x_i}{\mathrm{d}y_\alpha}\mathbf{e}_i$$

in which h_α is called the scale factor. Recall that the use of Greek letters for indices implies no summation. Clearly $\boldsymbol{\gamma}_\alpha$ is a unit vector. Conversely, if the transformation is reversed

$$\mathbf{dR}_y = \mathbf{g}_i\, \mathrm{d}y_i$$
$$= \frac{\mathrm{d}y_i}{\mathrm{d}x_j}\mathbf{g}_i\, \mathrm{d}x_j \tag{2.57}$$

with the consequence that

$$\mathbf{e}_j = \frac{\mathrm{d}y_i}{\mathrm{d}x_j}\mathbf{g}_i = \sum_\alpha \frac{\mathrm{d}y_\alpha}{\mathrm{d}x_j}h_\alpha \boldsymbol{\gamma}_\alpha \tag{2.58}$$

We restrict attention to orthogonal coordinate systems y_j with the property that

$$\boldsymbol{\gamma}_\alpha^T \boldsymbol{\gamma}_\beta = \delta_{\alpha\beta} \tag{2.59}$$

The length of the vector $\mathbf{dR}_y$ is now

$$\mathrm{d}S_y = h_\alpha \sqrt{\mathrm{d}y_i\, \mathrm{d}y_i} \tag{2.60}$$

Under restriction to orthogonal coordinate systems, the initial base vectors $\mathbf{e}_i$ may be expressed in terms of $\boldsymbol{\gamma}_\alpha$ using

$$\mathbf{e}_i = \left(\boldsymbol{\gamma}_j^T \mathbf{e}_i\right)\boldsymbol{\gamma}_j$$
$$= \frac{1}{h_i}\frac{\partial x_i}{\partial y_j}\boldsymbol{\gamma}_j$$
$$= \frac{1}{h_i h_j}\frac{\partial x_i}{\partial y_j}\frac{\partial x_k}{\partial y_j}\mathbf{e}_k \tag{2.61}$$

furnishing

$$\frac{1}{h_i h_j} \frac{\partial x_i}{\partial y_j} \frac{\partial x_k}{\partial y_j} = \delta_{ik} \tag{2.62}$$

Also of interest is the volume element: the volume determined by the vector $\mathbf{dR}_y$ is given by the vector triple product

$$\begin{aligned} \mathrm{d}V_y &= (h_1 \mathrm{d}y_1 \boldsymbol{\gamma}_1) \cdot [h_2 \mathrm{d}y_2 \boldsymbol{\gamma}_2 \times h_3 \mathrm{d}y_3 \boldsymbol{\gamma}_3] \\ &= h_1 h_2 h_3 \, \mathrm{d}y_1 \, \mathrm{d}y_2 \, \mathrm{d}y_3 \end{aligned} \tag{2.63}$$

and $h_1 h_2 h_3$ is called the *Jacobian of the transformation*. For cylindrical coordinates using r, θ, and z as shown in Figure 2.5, $x_1 = r \cos \theta$, $x_2 = r \sin \theta$, and $x_3 = z$. Simple manipulation furnishes that $h_r = 1$, $h_\theta = r$, $h_z = 1$, and

$$\mathbf{e}_r = \cos\theta \, \mathbf{e}_1 + \sin\theta \, \mathbf{e}_2, \quad \mathbf{e}_\theta = -\sin\theta \, \mathbf{e}_1 + \cos\theta \, \mathbf{e}_2, \quad \mathbf{e}_z = \mathbf{e}_3 \tag{2.64}$$

which of course are orthonormal vectors. Also of later interest are the relations $\mathbf{de}_r = \mathbf{e}_\theta \, \mathrm{d}\theta$ and $\mathbf{de}_\theta = -\mathbf{e}_r \, \mathrm{d}\theta$.

Transformation of the coordinate system from rectilinear to cylindrical coordinates may be viewed as a rotation of the coordinate system through θ. Thus if the vector $\mathbf{v}$ is referred to the reference rectilinear system and $\mathbf{v}'$ is the same vector referred to a cylindrical coordinate system, in two dimensions

$$\mathbf{v}' = \mathbf{Q}(\theta)\mathbf{v}, \qquad \mathbf{Q}(\theta) = \begin{bmatrix} \cos\theta & \sin\theta & 0 \\ -\sin\theta & \cos\theta & 0 \\ 0 & 0 & 1 \end{bmatrix} \tag{2.65}$$

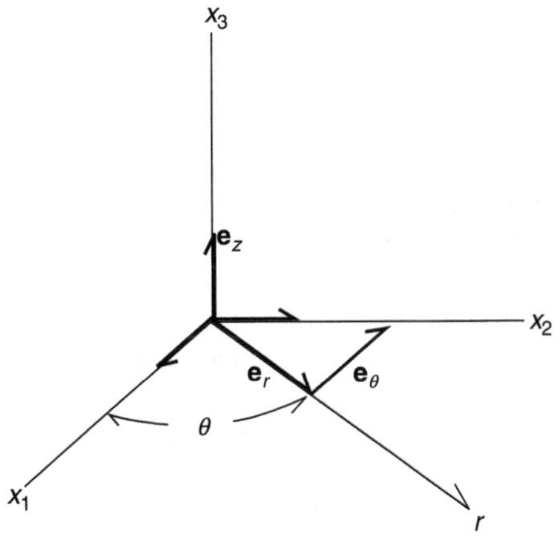

FIGURE 2.5 Cylindrical coordinate system.

If $\mathbf{v}'$ is differentiated, for example, with respect to time t, there is a contribution from the rotation of the coordinate system: as an illustration let $\mathbf{v}$ and θ be functions of time t,

$$\begin{aligned}\frac{\mathrm{d}}{\mathrm{d}t}\mathbf{v}' &= \mathbf{Q}(\theta)\frac{\mathrm{d}}{\mathrm{d}t}\mathbf{v} + \frac{\mathrm{d}\mathbf{Q}(\theta)}{\mathrm{d}t}\mathbf{v}\\ &= \frac{\partial}{\partial t}\mathbf{v}' + \frac{\mathrm{d}\mathbf{Q}(\theta)}{\mathrm{d}t}\mathbf{Q}^T(\theta)\mathbf{v}'\end{aligned} \tag{2.66}$$

where the partial derivative implies differentiation with θ instantaneously held fixed, and

$$\frac{\mathrm{d}\mathbf{Q}(\theta)}{\mathrm{d}t} = \begin{bmatrix} -\sin\theta & \cos\theta & 0\\ -\cos\theta & -\sin\theta & 0\\ 0 & 0 & 1\end{bmatrix}\frac{\mathrm{d}\theta}{\mathrm{d}t} \tag{2.67}$$

Now $\frac{\mathrm{d}\mathbf{Q}(\theta)}{\mathrm{d}t}\mathbf{Q}^T(\theta)$ is an antisymmetric matrix $\boldsymbol{\Omega}$ (to be identified later as a *tensor*) since

$$\mathbf{0} = \frac{\mathrm{d}}{\mathrm{d}t}\left(\mathbf{Q}(\theta)\mathbf{Q}^T(\theta)\right) = \frac{\mathrm{d}\mathbf{Q}(\theta)}{\mathrm{d}t}\mathbf{Q}^T(\theta) + \left[\frac{\mathrm{d}\mathbf{Q}(\theta)}{\mathrm{d}t}\mathbf{Q}^T(\theta)\right]^T \tag{2.68}$$

In fact

$$\frac{\mathrm{d}\mathbf{Q}(\theta)}{\mathrm{d}t}\mathbf{Q}^T(\theta) = \begin{pmatrix} 0 & 1 & 0\\ -1 & 0 & 0\\ 0 & 0 & 0\end{pmatrix}\frac{\mathrm{d}\theta}{\mathrm{d}t} \tag{2.69}$$

It follows that

$$\frac{\mathrm{d}}{\mathrm{d}t}\mathbf{v}' = \frac{\partial}{\partial t}\mathbf{v}' + \boldsymbol{\omega}\times\mathbf{v}' \tag{2.70}$$

in which $\boldsymbol{\omega}$ is the axial vector of $\boldsymbol{\Omega}$.

Referring to Figure 2.6, spherical coordinates r, θ, ϕ are introduced by the transformation

$$x_1 = r\cos\theta\cos\phi,\quad x_2 = r\sin\theta\cos\phi,\quad x_3 = r\sin\phi \tag{2.71}$$

The position vector is stated in spherical coordinates as

$$\begin{aligned}\mathbf{r} &= x_1\mathbf{e}_1 + x_2\mathbf{e}_2 + x_3\mathbf{e}_3\\ &= r\cos\theta\cos\phi\,\mathbf{e}_1 + r\sin\theta\cos\phi\,\mathbf{e}_2 + r\sin\phi\,\mathbf{e}_3\end{aligned} \tag{2.72}$$

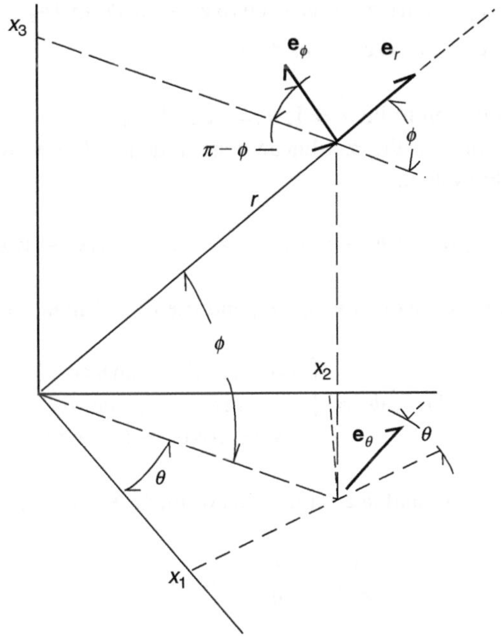

FIGURE 2.6 Spherical coordinate system.

Of course $\mathbf{e}_r$ has the same direction as the position vector: $\mathbf{r} = r\mathbf{e}_r$. It follows that

$$\mathbf{e}_r = \cos\theta\cos\phi\,\mathbf{e}_1 + \sin\theta\cos\phi\,\mathbf{e}_2 + \sin\phi\,\mathbf{e}_3 \tag{2.73}$$

Following the general procedure in the foregoing paragraphs:

$$\begin{aligned}
&\frac{\partial x_1}{\partial r} = \cos\theta\cos\phi, \quad \frac{\partial x_1}{\partial\theta} = -r\sin\theta\cos\phi, \quad \frac{\partial x_1}{\partial\phi} = -r\cos\theta\sin\phi \\
&\frac{\partial x_2}{\partial r} = \sin\theta\cos\phi, \quad \frac{\partial x_2}{\partial\theta} = r\cos\theta\cos\phi, \quad \frac{\partial x_2}{\partial\phi} = -r\sin\theta\sin\phi \\
&\frac{\partial x_3}{\partial r} = \sin\phi, \quad \frac{\partial x_3}{\partial\theta} = 0, \quad \frac{\partial x_3}{\partial\phi} = r\cos\phi
\end{aligned} \tag{2.74}$$

The differential of the position vector furnishes

$$\begin{aligned}
\mathrm{d}\mathbf{r} &= \mathrm{d}r\,\mathbf{e}_r + r\cos\phi\,\mathrm{d}\theta\,\mathbf{e}_\theta + r\,\mathrm{d}\phi\,\mathbf{e}_\phi \\
\mathbf{e}_r &= \cos\theta\cos\phi\,\mathbf{e}_1 + \sin\theta\cos\phi\,\mathbf{e}_2 + \sin\phi\,\mathbf{e}_3 \\
\mathbf{e}_\theta &= -\sin\theta\,\mathbf{e}_1 + \cos\theta\,\mathbf{e}_2 \\
\mathbf{e}_\phi &= -\sin\phi\,[\cos\theta\,\mathbf{e}_1 + \sin\theta\,\mathbf{e}_2] + \cos\phi\,\mathbf{e}_3 \\
\mathbf{e}_1 &= \cos\theta\cos\phi\,\mathbf{e}_r - \sin\theta\,\mathbf{e}_\theta - \sin\phi\cos\theta\,\mathbf{e}_\phi
\end{aligned} \tag{2.75}$$

$$\mathbf{e}_2 = \sin\theta\cos\phi\,\mathbf{e}_r + \cos\theta\,\mathbf{e}_\theta - \sin\phi\sin\theta\,\mathbf{e}_\phi$$
$$\mathbf{e}_3 = \sin\phi\,\mathbf{e}_r + \cos\phi\,\mathbf{e}_\phi$$

The scale factors are seen to be $h_r = 1$, $h_\theta = r\cos\phi$, $h_\phi = r$.

Consider a vector $\mathbf{v}$ in the rectilinear system, denoted as $\mathbf{v}'$ when referred to a spherical coordinate system:

$$\mathbf{v} = v_1\mathbf{e}_1 + v_2\mathbf{e}_2 + v_3\mathbf{e}_3, \qquad \mathbf{v}' = v_r\mathbf{e}_r + v_\theta\mathbf{e}_\theta + v_\phi\mathbf{e}_\phi \tag{2.76}$$

Eliminating $\mathbf{e}_1$, $\mathbf{e}_2$, $\mathbf{e}_3$ in favor of $\mathbf{e}_r$, $\mathbf{e}_\theta$, $\mathbf{e}_\phi$ and use of *M notation permits writing

$$\mathbf{v}' = \mathbf{Q}(\theta,\phi)\mathbf{v}, \qquad \mathbf{Q}(\theta,\phi) = \begin{bmatrix} \cos\theta\cos\phi & \sin\theta\cos\phi & \sin\phi \\ -\sin\theta & \cos\theta & 0 \\ -\sin\phi\cos\theta & -\sin\phi\sin\theta & \cos\phi \end{bmatrix} \tag{2.77}$$

Suppose now that $\mathbf{v}(t)$, θ, and ϕ are functions of time. As in cylindrical coordinates,

$$\frac{\mathrm{d}}{\mathrm{d}t}\mathbf{v}' = \frac{\partial}{\partial t}\mathbf{v}' + \boldsymbol{\omega}\times\mathbf{v}' \tag{2.78}$$

where $\boldsymbol{\omega}$ is the axial vector of $\frac{\mathrm{d}\mathbf{Q}(\theta)}{\mathrm{d}t}\mathbf{Q}^T(\theta)$. After some manipulation,

$$\frac{\mathrm{d}\mathbf{Q}(\theta)}{\mathrm{d}t} = \begin{bmatrix} -\sin\theta\cos\phi & \cos\theta\cos\phi & 0 \\ -\cos\theta & -\sin\theta & 0 \\ \sin\theta\sin\phi & -\cos\theta\sin\phi & 0 \end{bmatrix}\frac{\mathrm{d}\theta}{\mathrm{d}t} + \begin{bmatrix} -\cos\theta\sin\phi & -\sin\theta\sin\phi & \cos\phi \\ 0 & 0 & 0 \\ -\cos\theta\cos\phi & -\sin\theta\cos\phi & -\sin\phi \end{bmatrix}\frac{\mathrm{d}\phi}{\mathrm{d}t}$$

and

$$\frac{\mathrm{d}\mathbf{Q}(\theta)}{\mathrm{d}t}\mathbf{Q}^T(\theta) = \begin{bmatrix} 0 & \cos\phi & 0 \\ -\cos\phi & 0 & \sin\phi \\ 0 & -\sin\phi & 0 \end{bmatrix}\frac{\mathrm{d}\theta}{\mathrm{d}t} + \begin{bmatrix} 0 & 0 & 1 \\ 0 & 0 & 0 \\ -1 & 0 & 0 \end{bmatrix}\frac{\mathrm{d}\phi}{\mathrm{d}t} \tag{2.79}$$

2.7 GRADIENT OPERATOR IN ORTHOGONAL COORDINATES

The gradient operator in orthogonal coordinates is of great interest owing to its role in formulating the strain tensor, a topic to be encountered in Chapter 5. In rectilinear coordinates, let ψ be a scalar-valued function of $\mathbf{x}$: $\psi(\mathbf{x})$. Starting with the chain rule

*T)

$$\begin{aligned} \mathrm{d}\psi &= \frac{\partial\psi}{\partial x_i}\mathrm{d}x_i \\ &= [\nabla\psi]\cdot\mathrm{d}\mathbf{r}, \quad \mathrm{d}\mathbf{r} = \mathbf{e}_i\,\mathrm{d}x_i, \quad \nabla\psi = \mathbf{e}_i\frac{\partial\psi}{\partial x_i} \end{aligned} \tag{2.80}$$

Clearly $\mathrm{d}\psi$ is a scalar and is unaffected by a coordinate transformation. Now suppose that $\mathbf{x} = \mathbf{x}(\mathbf{y})$: $\mathrm{d}\mathbf{r}' = \mathbf{g}_i\,\mathrm{d}y_i$. Observe that

$$
\begin{aligned}
d\psi &= \frac{\partial \psi}{\partial x_i} dx_i \\
&= \sum_\alpha \frac{1}{h_\alpha} \frac{\partial \psi}{\partial y_\alpha} h_\alpha \, dy_\alpha \\
&= \left[\sum_\alpha \frac{\boldsymbol{\gamma}_\alpha}{h_\alpha} \frac{\partial \psi}{\partial y_\alpha} \right] \cdot \left[\sum_\beta h_\beta \left(dy_\beta \boldsymbol{\gamma}_\beta \right) \right]
\end{aligned} \tag{2.81}
$$

implying the identification

$$
(\nabla \psi)' = \sum_\alpha \frac{\boldsymbol{\gamma}_\alpha}{h_\alpha} \frac{\partial \psi}{\partial y_\alpha} \tag{2.82}
$$

For cylindrical coordinates in tensor-indicial notation with $\mathbf{e}_r = \boldsymbol{\gamma}_r$, $\mathbf{e}_\theta = \boldsymbol{\gamma}_\theta$, $\mathbf{e}_z = \boldsymbol{\gamma}_z$,

$$
\nabla \psi = \mathbf{e}_r \frac{\partial \psi}{\partial r} + \frac{\mathbf{e}_\theta}{r} \frac{\partial \psi}{\partial \theta} + \mathbf{e}_z \frac{\partial \psi}{\partial z} \tag{2.83}
$$

and in spherical coordinates

$$
\nabla \psi = \mathbf{e}_r \frac{\partial \psi}{\partial r} + \frac{\mathbf{e}_\theta}{r \cos \phi} \frac{\partial \psi}{\partial \theta} + \frac{\mathbf{e}_z}{r} \frac{\partial \psi}{\partial \phi} \tag{2.84}
$$

2.8 DIVERGENCE AND CURL OF VECTORS IN ORTHOGONAL COORDINATES

Under orthogonal transformations, the divergence and curl operators are invariant and satisfy the divergence and curl theorems, respectively. Unfortunately, the transformation properties of the divergence and curl operators are elaborate. The reader is referred to texts in continuum mechanics such as Chung (1988). The development is given in Appendix I. Here we simply list the results: let $\mathbf{v}$ be a vector referred to rectilinear coordinates, and let $\mathbf{v}'$ denote the same vector referred to orthogonal coordinates. The divergence and curl satisfy

$$
(\nabla \cdot \mathbf{v})' = \frac{1}{h_1 h_2 h_3} \left[\frac{\partial}{\partial y_1} \left(h_2 h_3 v_1' \right) + \frac{\partial}{\partial y_2} \left(h_3 h_1 v_2' \right) + \frac{\partial}{\partial y_3} \left(h_1 h_2 v_3' \right) \right] \tag{2.85}
$$

$$
\begin{aligned}
(\nabla \times \mathbf{v})' = \frac{1}{h_1 h_2 h_3} \Bigg[& h_1 \left\{ \frac{\partial}{\partial y_2} \left(h_3 v_3' \right) - \frac{\partial}{\partial y_3} \left(h_2 v_2' \right) \right\} \boldsymbol{\gamma}_1 \\
& - h_2 \left\{ \frac{\partial}{\partial y_1} \left(h_3 v_3' \right) - \frac{\partial}{\partial y_3} \left(h_1 v_1' \right) \right\} \boldsymbol{\gamma}_2 + h_3 \left\{ \frac{\partial}{\partial y_1} \left(h_2 v_2' \right) - \frac{\partial}{\partial y_2} \left(h_1 v_1' \right) \right\} \boldsymbol{\gamma}_3 \Bigg]
\end{aligned} \tag{2.86}
$$

In cylindrical coordinates:

$$(\nabla \cdot \mathbf{v})' = \frac{1}{r}\frac{\partial(rv_r)}{\partial r} + \frac{1}{r}\frac{\partial v_\theta}{\partial \theta} + \frac{\partial v_z}{\partial z} \tag{2.87}$$

$$(\nabla \times \mathbf{v})' = \frac{1}{r}\left[\left\{\frac{\partial v_z}{\partial \theta} - \frac{\partial(rv_\theta)}{\partial z}\right\}\mathbf{e}_r - \left\{\frac{\partial v_z}{\partial r} - \frac{\partial v_r}{\partial z}\right\} r\,\mathbf{e}_\theta + \left\{\frac{\partial(rv_\theta)}{\partial r} - \frac{\partial v_r}{\partial \theta}\right\}\mathbf{e}_z\right] \tag{2.88}$$

2.9 APPENDIX I: DIVERGENCE AND CURL OF VECTORS IN ORTHOGONAL CURVILINEAR COORDINATES

Derivatives of Base Vectors: In tensor-indicial notation, a vector $\mathbf{v}$ may be represented in rectilinear coordinates as $\mathbf{v} = v_k\mathbf{e}_k$. In orthogonal curvilinear coordinates it is written as

$$\mathbf{v}' = \sum_\alpha v'_\alpha \boldsymbol{\gamma}_\alpha = \sum_\alpha v'_\alpha \frac{\mathbf{g}_\alpha}{h_\alpha}$$

A line segment $\mathrm{d}\mathbf{r} = \mathrm{d}x_i\, \mathbf{e}_i$ transforms to $\mathrm{d}\mathbf{r}' = \mathrm{d}y_k\, \mathbf{g}_k$. Recall that

$$\begin{aligned}\mathbf{e}_k &= \frac{\partial y_l}{\partial x_k}\mathbf{g}_l = \sum_\beta h_\beta \frac{\partial y_\beta}{\partial x_k}\boldsymbol{\gamma}_\beta \\ \mathbf{g}_\alpha &= \frac{\partial x_\alpha}{\partial y_k}\mathbf{e}_k, \quad h_\alpha = \sqrt{\mathbf{g}_\alpha \cdot \mathbf{g}_\alpha}\end{aligned} \tag{2A.1}$$

From Equation 2A.1,

$$\begin{aligned}\frac{\partial \mathbf{g}_\alpha}{\partial y_j} &= \frac{\partial^2 x_\alpha}{\partial y_k \partial y_j}\mathbf{e}_k \\ &= \sum_\beta \begin{bmatrix} \alpha & & \beta \\ & j & \end{bmatrix} h_\beta \boldsymbol{\gamma}_\beta, \quad \begin{bmatrix} \alpha & & \beta \\ & j & \end{bmatrix} = \frac{\partial^2 x_\alpha}{\partial y_k \partial y_j}\frac{\partial y_\beta}{\partial x_k}\end{aligned} \tag{2A.2}$$

The bracketed quantities are known as Christoffel symbols. From Equations 2A.1 and 2A.2

$$\frac{\mathrm{d}h_\alpha}{\mathrm{d}y_j} = \boldsymbol{\gamma}_\alpha \cdot \frac{\mathrm{d}\mathbf{g}_\alpha}{\mathrm{d}y_j} = \begin{bmatrix} \alpha & & \alpha \\ & j & \end{bmatrix} h_\alpha \tag{2A.3}$$

Continuing,

$$\begin{aligned}\frac{\partial \boldsymbol{\gamma}_\alpha}{\partial y_j} &= \frac{1}{h_\alpha}\frac{\partial \mathbf{g}_\alpha}{\partial y_j} - \frac{\boldsymbol{\gamma}_\alpha}{h_\alpha}\frac{\partial h_\alpha}{\partial y_j} \\ &= \sum_\beta c_{\alpha j\beta}\boldsymbol{\gamma}_\beta, \quad c_{\alpha j\beta} = \frac{1}{h_\alpha}\left(1 - \delta_{\alpha\beta}\right)\begin{bmatrix} \alpha & & \beta \\ & j & \end{bmatrix} h_\beta\end{aligned} \tag{2A.4}$$

Divergence: The development below is based on the fact that

$$\nabla \cdot \mathbf{v} = \nabla' \cdot \mathbf{v}' = tr\left(\frac{\mathrm{d}\mathbf{v}'}{\mathrm{d}\mathbf{r}'}\right) = tr\left(\frac{\mathrm{d}\mathbf{v}}{\mathrm{d}\mathbf{r}}\right) \tag{2A.5}$$

The differential of $\mathbf{v}'$ is readily seen to be

$$\mathrm{d}\mathbf{v}' = \mathrm{d}v_j\,\boldsymbol{\gamma}_j + v_j\,\mathrm{d}\boldsymbol{\gamma}_j \tag{2A.6}$$

First note that

$$\begin{aligned}
\mathrm{d}v_j\,\boldsymbol{\gamma}_j &= \frac{\partial v_j}{\partial y_k}\boldsymbol{\gamma}_j\,\mathrm{d}y_k \\
&= \sum_\alpha \left(\frac{1}{h_\alpha}\frac{\partial v_j}{\partial y_\alpha}\boldsymbol{\gamma}_j\right)(h_\alpha\,\mathrm{d}y_\alpha) \\
&= \sum_\alpha \left(\frac{1}{h_\alpha}\frac{\partial v_j}{\partial y_\alpha}\boldsymbol{\gamma}_j \wedge \boldsymbol{\gamma}_\alpha\right)\cdot \sum_\beta (h_\beta \boldsymbol{\gamma}_\beta\,\mathrm{d}y_\beta) \\
&= \sum_\alpha \left(\frac{1}{h_\alpha}\frac{\partial v_j}{\partial y_\alpha}\boldsymbol{\gamma}_j \wedge \boldsymbol{\gamma}_\alpha\right)\cdot \mathrm{d}\mathbf{r}'
\end{aligned} \tag{2A.7}$$

Similarly

$$\begin{aligned}
v_j\,\mathrm{d}\boldsymbol{\gamma}_j &= v_j\frac{\partial \boldsymbol{\gamma}_j}{\partial y_k}\,\mathrm{d}y_k \\
&= \sum_\alpha \left(\frac{v_j}{h_\alpha}\frac{\partial \boldsymbol{\gamma}_j}{\partial y_\alpha}\right)(h_\alpha\,\mathrm{d}y_\alpha) \\
&= \sum_\alpha \left(\frac{v_j}{h_\alpha}\frac{\partial \boldsymbol{\gamma}_j}{\partial y_\alpha}\wedge \boldsymbol{\gamma}_\alpha\right)\cdot \sum_\beta (h_\beta \boldsymbol{\gamma}_\beta\,\mathrm{d}y_\beta) \\
&= \sum_\alpha \left(\frac{v_j}{h_\alpha}\frac{\partial \boldsymbol{\gamma}_j}{\partial y_\alpha}\wedge \boldsymbol{\gamma}_\alpha\right)\cdot \mathrm{d}\mathbf{r} \\
&= \sum_\alpha \left[\frac{v_j}{h_\alpha}\left(\sum_\beta c_{j\alpha\beta}\boldsymbol{\gamma}_\beta \wedge \boldsymbol{\gamma}_\alpha\right)\right]\cdot \mathrm{d}\mathbf{r}'
\end{aligned} \tag{2A.8}$$

Consequently,

$$\left[\frac{\mathrm{d}v}{\mathrm{d}r}\right]_{\beta\alpha} = \frac{1}{h_\alpha}\left[\frac{\partial v_j}{\partial y_\alpha}\delta_{j\beta} + v_j c_{j\beta\alpha}\right], \quad \nabla \cdot v = \sum_\alpha \frac{1}{h_\alpha}\left[\frac{\partial v_\alpha}{\partial y_\alpha} + v_j c_{j\alpha\alpha}\right] \tag{2A.9}$$

Curl: In rectilinear coordinates, the individual entries of the curl can be expressed as a divergence as follows. For the ith entry

$$\begin{aligned}[\nabla \times \mathbf{v}]_i &= \varepsilon_{ijk}\frac{\partial v_k}{\partial x_j} \\ &= \frac{\partial}{\partial x_j}w_j^{(i)}, \quad w_j^{(i)} = \varepsilon_{jki}v_k \\ &= \nabla \cdot \mathbf{w}^{(i)} \end{aligned} \tag{2A.10}$$

Consequently, the curl of $\mathbf{v}$ may be written as

$$\nabla \times \mathbf{v} = \begin{pmatrix} \nabla \cdot \mathbf{w}^{(1)} \\ \nabla \cdot \mathbf{w}^{(2)} \\ \nabla \cdot \mathbf{w}^{(3)} \end{pmatrix} \tag{2A.11}$$

The transformation properties of the curl can be readily induced from Equation 2A.9.

EXAMPLE 2.14

Obtain the expressions for the gradient, divergence, and curl in spherical coordinates.

Solution

In spherical coordinates

$$\begin{aligned} &\boldsymbol{\gamma}_1 = \mathbf{e}_r, \quad \boldsymbol{\gamma}_2 = \mathbf{e}_\theta, \quad \boldsymbol{\gamma}_3 = \mathbf{e}_\phi \\ &y_1 = r, \quad y_2 = \theta, \quad y_3 = \phi \\ &h_r = 1, \quad h_\theta = r\cos\phi, \quad h_\phi = r \end{aligned}$$

On substituting these relations into the expressions for the gradient and divergence operators in the text, we have

$$\begin{aligned} \nabla\psi &= \frac{\partial\psi}{\partial r}\mathbf{e}_r + \frac{1}{r\cos\phi}\frac{\partial\psi}{\partial\theta}\mathbf{e}_\theta + \frac{1}{r}\frac{\partial\psi}{\partial\phi}\mathbf{e}_\phi \\ \nabla\cdot\mathbf{v} &= \frac{1}{r^2\cos\phi}\left[\frac{\partial}{\partial r}\left(r^2\cos\phi\, v_r\right) + \frac{\partial}{\partial\theta}(rv_\theta) + \frac{\partial}{\partial\phi}\left(r\cos\phi\, v_\phi\right)\right] \\ &= \frac{1}{r^2}\frac{\partial(r^2 v_r)}{\partial r} + \frac{1}{r\cos\phi}\frac{\partial v_\theta}{\partial\theta} + \frac{1}{r\cos\phi}\frac{\partial(\cos\phi\, v_\phi)}{\partial\phi} \end{aligned}$$

$$\nabla \times \mathbf{v} = \frac{1}{r^2 \cos\phi} \begin{pmatrix} \left\{ \dfrac{\partial}{\partial \theta}(r v_\phi) - \dfrac{\partial}{\partial \phi}(r \cos\phi v_\theta) \right\} \mathbf{e}_r \\ -r \cos\phi \left\{ \dfrac{\partial}{\partial r}(r v_\phi) - \dfrac{\partial}{\partial \phi}(v_r) \right\} \mathbf{e}_\theta \\ +r \left\{ \dfrac{\partial}{\partial r}(r \cos\phi v_\theta) - \dfrac{\partial}{\partial \theta}(v_r) \right\} \mathbf{e}_\phi \end{pmatrix}$$

$$= \frac{1}{r \cos\phi} \begin{pmatrix} \left\{ \dfrac{\partial v_\phi}{\partial \theta} - \dfrac{\partial(\cos\phi v_\theta)}{\partial \phi} \right\} \mathbf{e}_r \\ -\cos\phi \left\{ \dfrac{\partial(r v_\phi)}{\partial r} - \dfrac{\partial v_r}{\partial \phi} \right\} \mathbf{e}_\theta \\ +\left\{ \dfrac{\partial(r \cos\phi v_\theta)}{\partial r} - \dfrac{\partial v_r}{\partial \theta} \right\} \mathbf{e}_\phi \end{pmatrix}$$

3 Mathematical Foundations: Tensors

3.1 TENSORS

We now consider two $n \times 1$ vectors $\mathbf{v}$ and $\mathbf{w}$ and an $n \times n$ matrix $\mathbf{A}$ such that $\mathbf{v} = \mathbf{A}\mathbf{w}$. The important assumption is made that the underlying information in this relation is preserved under rotation of the coordinate system. In particular, simple manipulation furnishes that

*M)

$$\begin{aligned} \mathbf{v}' &= \mathbf{Q}\mathbf{v} \\ &= \mathbf{Q}\mathbf{A}\mathbf{w} \\ &= \mathbf{Q}\mathbf{A}\mathbf{Q}^T\mathbf{Q}\mathbf{w} \\ &= \mathbf{Q}\mathbf{A}\mathbf{Q}^T\mathbf{w}' \end{aligned} \tag{3.1}$$

The square matrix $\mathbf{A}$ is now called a second-order tensor if and only if $\mathbf{A}' = \mathbf{Q}\mathbf{A}\mathbf{Q}^T$.

Let $\mathbf{A}$ and $\mathbf{B}$ be second-order $n \times n$ tensors. The manipulations below demonstrate that A^T, $(\mathbf{A} + \mathbf{B})$, $\mathbf{AB}$, and $\mathbf{A}^{-1}$ are likewise tensors.

$$\begin{aligned} (\mathbf{A}^T)' &= (\mathbf{Q}\mathbf{A}\mathbf{Q}^T)^T \\ &= \mathbf{Q}^{TT}\mathbf{A}^T\mathbf{Q}^T \end{aligned} \tag{3.2}$$

$$\begin{aligned} \mathbf{A}'\mathbf{B}' &= (\mathbf{Q}\mathbf{A}\mathbf{Q}^T)(\mathbf{Q}\mathbf{B}\mathbf{Q}^T) \\ &= \mathbf{Q}\mathbf{A}(\mathbf{Q}\mathbf{Q}^T)\mathbf{B}\mathbf{Q}^T \\ &= \mathbf{Q}\mathbf{A}\mathbf{B}\mathbf{Q}^T \end{aligned} \tag{3.3}$$

$$\begin{aligned} (\mathbf{A} + \mathbf{B})' &= \mathbf{A}' + \mathbf{B}' \\ &= \mathbf{Q}\mathbf{A}\mathbf{Q}^T + \mathbf{Q}\mathbf{B}\mathbf{Q}^T \\ &= \mathbf{Q}(\mathbf{A} + \mathbf{B})\mathbf{Q}^T \end{aligned} \tag{3.4}$$

$$\begin{aligned} \mathbf{A}'^{-1} &= (\mathbf{Q}\mathbf{A}\mathbf{Q}^T)^{-1} \\ &= \mathbf{Q}^{T-1}\mathbf{A}^{-1}\mathbf{Q}^{-1} \\ &= \mathbf{Q}\mathbf{A}^{-1}\mathbf{Q}^T \end{aligned} \tag{3.5}$$

Let $\mathbf{x}$ denote an $n \times 1$ vector. The outer product $\mathbf{x}\mathbf{x}^T$ is a second-order tensor since

$$\begin{aligned}(\mathbf{x}\mathbf{x}^T)' &= \mathbf{x}'\mathbf{x}'^T \\ &= (\mathbf{Q}\mathbf{x})(\mathbf{Q}\mathbf{x})^T \\ &= \mathbf{Q}(\mathbf{x}\mathbf{x}^T)\mathbf{Q}^T \end{aligned} \tag{3.6}$$

Next

$$\mathrm{d}^2\varphi = \mathrm{d}\mathbf{x}^T\mathbf{H}\,\mathrm{d}\mathbf{x}, \qquad \mathbf{H} = \left(\frac{\mathrm{d}}{\mathrm{d}\mathbf{x}}\right)^T\left(\frac{\mathrm{d}\varphi}{\mathrm{d}\mathbf{x}}\right) \tag{3.7}$$

But

$$\begin{aligned}\mathrm{d}\mathbf{x}'^T\mathbf{H}'\,\mathrm{d}\mathbf{x}' &= (\mathbf{Q}\mathrm{d}\mathbf{x})^T\mathbf{H}'\mathbf{Q}\,\mathrm{d}\mathbf{x} \\ &= \mathrm{d}\mathbf{x}^T\left(\mathbf{Q}^T\mathbf{H}'\mathbf{Q}\right)\mathrm{d}\mathbf{x}\end{aligned} \tag{3.8}$$

from which we conclude that the Hessian $\mathbf{H}$ is a second-order tensor.

Finally, let $\mathbf{u}$ be a vector-valued function of $\mathbf{x}$. Then, $\mathrm{d}\mathbf{u} = \frac{\partial\mathbf{u}}{\partial\mathbf{x}}\mathrm{d}\mathbf{x}$ ($\mathrm{d}u_i = \frac{\partial u_i}{\partial x_j}\mathrm{d}x_j$) from the chair rule of calculus, from which we conclude that

$$\mathrm{d}\mathbf{u}^T = \mathrm{d}\mathbf{x}^T\left(\frac{\partial\mathbf{u}}{\partial\mathbf{x}}\right)^T \tag{3.9}$$

But, also from the chain rule,

$$\mathrm{d}\mathbf{u}^T = \mathrm{d}\mathbf{x}^T\frac{\partial\mathbf{u}^T}{\partial\mathbf{x}^T} \tag{3.10}$$

We conclude that

$$\left(\frac{\partial\mathbf{u}}{\partial\mathbf{x}}\right)^T = \frac{\partial\mathbf{u}^T}{\partial\mathbf{x}^T} \tag{3.11}$$

Furthermore, if $\mathrm{d}\mathbf{u}'$ is a vector generated from $\mathrm{d}\mathbf{u}$ by rotation in the opposite sense from the coordinate axes (i.e., clockwise if the axes rotate counterclockwise), then $\mathrm{d}\mathbf{u}' = \mathbf{Q}\,\mathrm{d}\mathbf{u}$ and $\mathrm{d}\mathbf{x} = \mathbf{Q}^T\mathrm{d}\mathbf{x}'$. Hence $\mathbf{Q}$ is a tensor which may be viewed the counterclockwise rotation of the axes. (Note that $\mathbf{x}$ and $\mathbf{Q}\mathbf{x}$ are vectors, implying that $\mathbf{Q}$ is a tensor.) Also, since $\mathrm{d}\mathbf{u}' = \frac{\partial\mathbf{u}'}{\partial\mathbf{x}'}\mathrm{d}\mathbf{x}'$, it is apparent that

$$\frac{\partial\mathbf{u}'}{\partial\mathbf{x}'} = \mathbf{Q}\frac{\partial\mathbf{u}}{\partial\mathbf{x}}\mathbf{Q}^T \tag{3.12}$$

from which we conclude that $\frac{\partial\mathbf{u}}{\partial\mathbf{x}}$ is a tensor. We may similarly show that $\mathbf{I}$ and $\mathbf{0}$ are tensors, albeit of a special type (isotropic) owing to the property $\mathbf{I}' = \mathbf{I}$, $\mathbf{0}' = \mathbf{0}$.

3.2 DIVERGENCE OF A TENSOR

Suppose $\mathbf{A}$ is a tensor and $\mathbf{b}$ is an arbitrary spatially constant vector of compatible dimension.

The divergence of a vector has already been defined. For later purposes we will need to extend the definition of the divergence to the tensor $\mathbf{A}$. Recall the divergence theorem for the vector $\mathbf{c}(\mathbf{x})$: $\int \mathbf{c}^T\mathbf{n}\,\mathrm{d}S = \int \nabla^T\mathbf{c}\,\mathrm{d}V$. Let $\mathbf{c} = \mathbf{A}^T\mathbf{b}$ in which $\mathbf{b}$ is an arbitrary *constant* vector. Now

$$\begin{aligned}\mathbf{b}^T\int \mathbf{A}\mathbf{n}\,\mathrm{d}S &= \int \nabla^T(\mathbf{A}^T\mathbf{b})\,\mathrm{d}V \\ &= \int \left(\nabla^T\mathbf{A}^T\right)\mathrm{d}V\,\mathbf{b} \\ &= \mathbf{b}^T\int \left[\nabla^T\mathbf{A}^T\right]^T\mathrm{d}V\end{aligned} \tag{3.13}$$

Consequently, we seek to define the divergence of $\mathbf{A}$ such that

*M)
$$\int \mathbf{A}\mathbf{n}\,\mathrm{d}S - \int \left[\nabla^T\mathbf{A}^T\right]^T\mathrm{d}V = 0 \tag{3.14}$$

In tensor-indicial notation

$$\int b_i a_{ij} n_j\,\mathrm{d}S - \int b_i\left[\left[\nabla^T\mathbf{A}^T\right]^T\right]_i\mathrm{d}V = 0 \tag{3.15}$$

Application of the divergence theorem to the vector $c_j = b_i a_{ij}$ furnishes

$$b_i\int\left[\frac{\partial}{\partial x_j}a_{ij} - \left[\left[\nabla^T\mathbf{A}^T\right]^T\right]_i\right]\mathrm{d}V = 0 \tag{3.16}$$

Since $\mathbf{b}$ is arbitrary, we conclude that

$$\left[\nabla^T\mathbf{A}^T\right]_i = \frac{\partial}{\partial x_j}a_{ij} = \frac{\partial}{\partial x_j}a^T_{ji} \tag{3.17}$$

and hence, if we are to write $\nabla\cdot\mathbf{A}$ as a (column) vector (mixing tensor and matrix-vector notation) we have

$$\nabla\cdot\mathbf{A} = [\nabla^T\mathbf{A}^T]^T \tag{3.18}$$

It should be evident that $(\nabla\,\cdot)$ has a more elaborate meaning when applied to a tensor as opposed to a vector.

Now, suppose $\mathbf{A}$ is written in the form

$$\mathbf{A} = \begin{bmatrix}\boldsymbol{\alpha}_1^T \\ \boldsymbol{\alpha}_2^T \\ \boldsymbol{\alpha}_3^T\end{bmatrix} \tag{3.19}$$

in which $\boldsymbol{\alpha}_i^T$ corresponds to the ith row of $\mathbf{A}$: $[\boldsymbol{\alpha}_i^T]_j = a_{ij}$. It is easily seen that

$$\nabla^T\mathbf{A}^T = (\nabla^T\boldsymbol{\alpha}_1 \quad \nabla^T\boldsymbol{\alpha}_2 \quad \nabla^T\boldsymbol{\alpha}_3) \tag{3.20}$$

3.3 INVARIANTS

Letting $\mathbf{A}$ denote a nonsingular symmetric 3×3 tensor, the equation $\det(\mathbf{A} - \lambda\mathbf{I}) = 0$ can be expanded as

$$\lambda^3 - I_1\lambda^2 + I_2\lambda - I_3 = 0 \tag{3.21}$$

in which

$$I_1 = tr(\mathbf{A}), \quad I_2 = \tfrac{1}{2}\left[tr^2(\mathbf{A}) - tr(\mathbf{A}^2)\right], \quad I_3 = \det(\mathbf{A}) \tag{3.22}$$

Here $tr(\mathbf{A}) = \delta_{ij}a_{ij}$ denotes the *trace* of $\mathbf{A}$. Equation 3.21 implies *the Cayley–Hamilton theorem*

$$\mathbf{A}^3 - I_1\mathbf{A}^2 + I_2\mathbf{A} - I_3\mathbf{I} = \mathbf{0} \tag{3.23}$$

from which

$$\begin{aligned} I_3 &= \tfrac{1}{3}\left[tr(\mathbf{A}^3) - I_1\, tr(\mathbf{A}^2) + I_2\, tr(\mathbf{A})\right] \\ \mathbf{A}^{-1} &= I_3^{-1}\left[\mathbf{A}^2 - I_1\mathbf{A} + I_2\mathbf{I}\right] \end{aligned} \tag{3.24}$$

Now the trace of any $n \times n$ symmetric tensor $\mathbf{B}$ is invariant under orthogonal transformations (rotations): $tr(\mathbf{B}') = tr(\mathbf{B})$ since

$$\begin{aligned} a'_{pq}\delta_{pq} &= q_{pr}q_{qs}a_{rs}\delta_{pq} \\ &= a_{rs}q_{pr}q_{qs} \\ &= a_{rs}\delta_{rs} \end{aligned} \tag{3.25}$$

Likewise $tr(\mathbf{A}^2)$ and $tr(\mathbf{A}^3)$ are invariant since $\mathbf{A}$, $\mathbf{A}^2$, and $\mathbf{A}^3$ are tensors, and hence I_1, I_2, and I_3 are invariants. Derivatives of invariants are of interest and will be presented in Section 3.6.8.

3.4 POSITIVE DEFINITENESS

In the finite element method, an attractive property of some symmetric tensors is positive definiteness, defined as follows. The symmetric $n \times n$ tensor $\mathbf{A}$ is positive definite, written $\mathbf{A} > \mathbf{0}$, if the quadratic product $\Omega(\mathbf{A},\mathbf{x}) = \mathbf{x}^T\mathbf{A}\mathbf{x} > 0$ for all nonvanishing $n \times 1$ vectors $\mathbf{x}$. The importance of this property is shown in the following example. Let $\Pi = \mathbf{x}^T\mathbf{A}\mathbf{x} - \mathbf{x}^T\mathbf{f}$, in which $\mathbf{f}$ is known and $\mathbf{A} > \mathbf{0}$. After some simple manipulation,

$$
\begin{aligned}
d^2\Pi &= d\mathbf{x}^T \left[\left(\frac{d}{d\mathbf{x}} \right)^T \frac{d}{d\mathbf{x}} \Pi \right] d\mathbf{x} \\
&= d\mathbf{x}^T \mathbf{A}\, d\mathbf{x}
\end{aligned}
\tag{3.26}
$$

It follows that Π is a globally convex function which attains a (global) minimum when $\mathbf{Ax} = \mathbf{f}$ ($d\Pi = 0$). This fact can be invoked in the finite element method in classical elasticity to show that the solution of the finite element equations under static conditions represents a minimum.

The foregoing definition is equivalent to the statement that the symmetric $n \times n$ tensor $\mathbf{A}$ is positive definite if and only if its eigenvalues are all positive. For the sake of demonstration, if $\mathbf{X}$ is the matrix (tensor) which diagonalizes $\mathbf{A}$ (cf. Chapter 2),

$$
\begin{aligned}
\mathbf{x}^T \mathbf{Ax} &= \mathbf{x}^T \mathbf{X} \boldsymbol{\Lambda} \mathbf{X}^T \mathbf{x} \\
&= \mathbf{y}^T \boldsymbol{\Lambda} \mathbf{y}, \quad (\mathbf{y} = \mathbf{X}^T \mathbf{x}) \\
&= \sum_i \lambda_i y_i^2
\end{aligned}
\tag{3.27}
$$

The last expression can be positive for arbitrary $\mathbf{y}$ (arbitrary $\mathbf{x}$) only if $\lambda_i > 0$, $i = 1, 2, \ldots, n$. The matrix $\mathbf{A}$ is semidefinite if $\mathbf{x}^T \mathbf{Ax} \geq 0$, and negative definite (written $\mathbf{A} < \mathbf{0}$) if $\mathbf{x}^T \mathbf{Ax} < 0$. If $\mathbf{B}$ is a nonsingular tensor, then $\mathbf{B}^T \mathbf{B} > \mathbf{0}$, since $\Omega(\mathbf{B}^T\mathbf{B}, \mathbf{x}) = \mathbf{x}^T \mathbf{B}^T \mathbf{Bx} = \mathbf{y}^T \mathbf{y} > 0$ in which $\mathbf{y} = \mathbf{Bx}$. If $\mathbf{B}$ is singular, for example, if $\mathbf{B} = \mathbf{yy}^T$ where $\mathbf{y}$ is an $n \times 1$ vector, $\mathbf{B}^T \mathbf{B}$ is positive semidefinite since a nonzero eigenvector $\mathbf{x}$ of $\mathbf{B}$ can be found for which the quadratic product $\Omega(\mathbf{B}^T\mathbf{B}, \mathbf{x})$ vanishes.

Now suppose that $\mathbf{B}$ is a nonsingular antisymmetric tensor. Then multiplying through $\mathbf{Bx}_j = \lambda_j \mathbf{x}_j$ with $\mathbf{B}^T$ furnishes

$$
\begin{aligned}
\mathbf{B}^T \mathbf{Bx}_j &= \lambda_j \mathbf{B}^T \mathbf{x}_j \\
&= -\lambda_j \mathbf{Bx}_j \\
&= -\lambda_j^2 \mathbf{x}_j
\end{aligned}
\tag{3.28}
$$

Since $\mathbf{B}^T\mathbf{B}$ is positive definite it follows that $-\lambda_j^2 > 0$ and hence λ_j is imaginary: $\lambda_j = i\mu_j$ using $i = \sqrt{-1}$. Consequently, $\mathbf{B}^2 \mathbf{x}_j = \lambda_j^2 \mathbf{x}_j = -\mu_j^2 \mathbf{x}_j$, demonstrating that $\mathbf{B}^2$ is *negative definite.*

3.5 POLAR DECOMPOSITION THEOREM

For an $n \times n$ nonsingular matrix $\mathbf{B}$, $\mathbf{B}^T \mathbf{B} > 0$. If the *modal matrix* of $\mathbf{B}$ is denoted by $\mathbf{X}_b$, we may write

$$
\begin{aligned}
\mathbf{B}^T \mathbf{B} &= \mathbf{X}_b^T \boldsymbol{\Delta}_b \mathbf{X}_b \\
&= \mathbf{X}_b^T (\boldsymbol{\Delta}_b)^{\frac{1}{2}} \mathbf{Y} \mathbf{Y}^T (\boldsymbol{\Delta}_b)^{\frac{1}{2}} \mathbf{X}_b \\
&= \left(\mathbf{X}_b^T (\boldsymbol{\Delta}_b)^{\frac{1}{2}} \mathbf{Y} \right) \left(\mathbf{X}_b^T (\boldsymbol{\Delta}_b)^{\frac{1}{2}} \mathbf{Y} \right)^T
\end{aligned}
\tag{3.29a}
$$

in which $\mathbf{Y}$ is an (unknown) orthogonal tensor. Accordingly, we may in general write

$$\mathbf{B} = \mathbf{Y}^T(\boldsymbol{\Delta}_b)^{\frac{1}{2}}\mathbf{X}_b \tag{3.29b}$$

To "justify" Equation 3.29b we introduce the tensor-valued square root $\sqrt{\mathbf{B}^T\mathbf{B}}$ using

$$\sqrt{\mathbf{B}^T\mathbf{B}} = \mathbf{X}_b^T\boldsymbol{\Delta}_b^{\frac{1}{2}}\mathbf{X}_b, \quad \boldsymbol{\Delta}_b^{\frac{1}{2}} = \begin{bmatrix} \sqrt{\lambda_1} & 0 & . & . & . \\ 0 & \sqrt{\lambda_2} & . & . & . \\ . & . & . & . & . \\ . & . & . & . & 0 \\ . & . & . & 0 & \sqrt{\lambda_n} \end{bmatrix} \tag{3.29c}$$

in which the positive square roots are used. It is elementary to verify that $\left(\sqrt{\mathbf{B}^T\mathbf{B}}\right)^2 = \mathbf{B}$, and also that $\sqrt{\mathbf{B}^T\mathbf{B}} > 0$. Now note that

$$\begin{aligned}\left[\mathbf{B}\left(\sqrt{\mathbf{B}^T\mathbf{B}}\right)^{-\frac{1}{2}}\right]\left[\mathbf{B}\left(\sqrt{\mathbf{B}^T\mathbf{B}}\right)^{-\frac{1}{2}}\right]^T &= \left[\left(\sqrt{\mathbf{B}^T\mathbf{B}}\right)^{-\frac{1}{2}}\right][\mathbf{B}^T\mathbf{B}]\left[\left(\sqrt{\mathbf{B}^T\mathbf{B}}\right)^{-\frac{1}{2}}\right] \\ &= \mathbf{I}\end{aligned} \tag{3.29d}$$

Thus, $\mathbf{B}\left(\sqrt{\mathbf{B}^T\mathbf{B}}\right)^{-\frac{1}{2}}$ is an orthogonal tensor, say $\mathbf{Z}$, and hence we may write

$$\begin{aligned}\mathbf{B} &= \mathbf{Z}\sqrt{\mathbf{B}^T\mathbf{B}} \\ &= \mathbf{Z}\mathbf{X}_b^T\boldsymbol{\Delta}_b^{\frac{1}{2}}\mathbf{X}_b\end{aligned} \tag{3.29e}$$

Finally, noting that $(\mathbf{Z}\mathbf{X}_b^T)(\mathbf{Z}\mathbf{X}_b^T)^T = \mathbf{Z}(\mathbf{X}_b^T\mathbf{X}_b)\mathbf{Z}^T = \mathbf{Z}\mathbf{Z}^T = \mathbf{I}$, we make the identification $\mathbf{Y}^T = \mathbf{Z}\mathbf{X}_b^T$ in Equation 3.29b. Equations 3.29a through 3.29e play a major role in the interpretation of strain tensors to be introduced in subsequent chapters.

3.6 KRONECKER PRODUCTS OF TENSORS

3.6.1 *VEC* Operator and the Kronecker Product

Let $\mathbf{A}$ be an $n \times n$ (second-order) tensor. Kronecker product notation (Graham 1981) reduces $\mathbf{A}$ to a first-order $n \times 1$ tensor (vector) as follows:

$$VEC(\mathbf{A}) = \{a_{11} \quad a_{21} \quad a_{31} \quad . \quad . \quad . \quad a_{n,n-1} \quad a_{nn}\}^T \tag{3.30}$$

The inverse *VEC* operator (*IVEC*) is introduced by the obvious relation $IVEC(VEC(\mathbf{A})) = \mathbf{A}$. The *Kronecker product* of an $n \times m$ matrix $\mathbf{A}$ and an $r \times s$ matrix $\mathbf{B}$ generates an $nr \times ms$ matrix as follows:

$$\mathbf{A} \otimes \mathbf{B} = \begin{bmatrix} a_{11}\mathbf{B} & a_{12}\mathbf{B} & . & . & a_{1m}\mathbf{B} \\ a_{21}\mathbf{B} & . & . & . & . \\ . & . & . & . & . \\ . & . & . & . & . \\ a_{n1}\mathbf{B} & . & . & . & a_{nm}\mathbf{B} \end{bmatrix} \tag{3.31}$$

Now if m, n, r, and s are equal to n and if $\mathbf{A}$ and $\mathbf{B}$ are tensors, then $\mathbf{A} \otimes \mathbf{B}$ transforms as a second-order $n^2 \times n^2$ tensor in a sense to be explained subsequently. Equation 3.31 implies that the $n^2 \times 1$ Kronecker product of two $n \times 1$ vectors $\mathbf{a}$ and $\mathbf{b}$ is written as

$$\mathbf{a} \otimes \mathbf{b} = \begin{pmatrix} a_1\mathbf{b} \\ a_2\mathbf{b} \\ \vdots \\ a_n\mathbf{b} \end{pmatrix} \tag{3.32}$$

3.6.2 Fundamental Relations for Kronecker Products

Six basic relations are introduced, followed by a number of subsidiary relations. The proofs of the first five relations are based on Graham (1981).

Relation 1: Let $\mathbf{A}$ denote an $n \times m$ real matrix, with entry a_{ij} in the ith row and jth column. Let $I = (j-1)n + i$ and $J = (i-1)m + j$. Let $\mathbf{U}_{nm}$ denote the $nm \times nm$ matrix, independent of $\mathbf{A}$, satisfying

$$u_{JK} = \begin{cases} 1, & K = I, \\ 0, & K \neq I, \end{cases} \qquad u_{IK} = \begin{cases} 1, & K = J \\ 0, & K \neq J \end{cases} \tag{3.33}$$

Then

$$VEC(\mathbf{A}^T) = \mathbf{U}_{nm} VEC(\mathbf{A}) \tag{3.34}$$

Note that $u_{JK} = u_{JI} = 1$ and $u_{IK} = u_{IJ} = 1$, with all other entries vanishing. Hence if $m = n$, $u_{JI} = u_{IJ}$, so that $\mathbf{U}_{nm}$ is symmetric if $m = n$.

Relation 2: If $\mathbf{A}$ and $\mathbf{B}$ are second-order $n \times n$ tensors, then

$$tr(\mathbf{AB}) = VEC^T(\mathbf{A}^T) VEC(\mathbf{B}) \tag{3.35}$$

Relation 3: If $\mathbf{I}_n$ denotes the $n \times n$ identity matrix and if $\mathbf{B}$ denotes an $n \times n$ tensor, then

$$\mathbf{I}_n \otimes \mathbf{B}^T = (\mathbf{I}_n \otimes \mathbf{B})^T \tag{3.36}$$

Relation 4: Let $\mathbf{A}$, $\mathbf{B}$, $\mathbf{C}$, and $\mathbf{D}$, respectively, denote $m \times n$, $r \times$ s, $n \times p$, and $s \times q$ matrices, then

$$(\mathbf{A} \otimes \mathbf{B})(\mathbf{C} \otimes \mathbf{D}) = \mathbf{AC} \otimes \mathbf{BD} \tag{3.37}$$

Relation 5: If $\mathbf{A}$, $\mathbf{B}$, and $\mathbf{C}$ are $n \times m$, $m \times r$, and $r \times s$ matrices, then

$$VEC(\mathbf{ACB}) = \mathbf{B}^T \otimes \mathbf{A} VEC(\mathbf{C}) \tag{3.38}$$

Relation 6: If $\mathbf{a}$ and $\mathbf{b}$ are $n \times 1$ vectors, then

$$\mathbf{a} \otimes \mathbf{b} = VEC\left(\left[\mathbf{ab}^T\right]^T\right) \tag{3.39}$$

As proof of Relation (6), if $I = (j-1)n + i$, the Ith entry of $VEC(\mathbf{ba}^T)$ is $b_i a_j$. It is likewise the Ith entry of $\mathbf{a} \otimes \mathbf{b}$. Hence $\mathbf{a} \otimes \mathbf{b} = VEC(\mathbf{ba}^T) = VEC([\mathbf{ab}^T]^T)$.

Symmetry of $\mathbf{U}_{nn}$ was established in Relation (1): $\mathbf{U}_{nn} = \mathbf{U}_{n^2}$. Note that $VEC(\mathbf{A}) = \mathbf{U}_v VEC(\mathbf{A}^T) = \mathbf{U}^2_{n^2}\ VEC(\mathbf{A})$ if A is $n \times n$, and hence the matrix $\mathbf{U}_{nn}$ satisfies

$$\mathbf{U}^2_{n^2} = \mathbf{I}_{n^2}, \qquad \mathbf{U}_{n^2} = \mathbf{U}^T_{n^2} = \mathbf{U}^{-1}_{n^2} \tag{3.40}$$

$\mathbf{U}_{nn}$ is hereafter called the permutation tensor for $n \times n$ tensors. If $\mathbf{A}$ is symmetric $(\mathbf{U}_{n^2} - \mathbf{I}_{n^2})VEC(\mathbf{A}) = \mathbf{0}$. If $\mathbf{A}$ is antisymmetric $(\mathbf{U}_{n^2} + \mathbf{I}_{n^2})VEC(\mathbf{A}) = \mathbf{0}$.

If $\mathbf{A}$ and $\mathbf{B}$ are second-order $n \times n$ tensors, then

$$\begin{aligned} tr(\mathbf{AB}) &= VEC^T(\mathbf{B})VEC(\mathbf{A}^T) \\ &= VEC^T(\mathbf{B})\mathbf{U}_{n^2}VEC(\mathbf{A}) \\ &= [\mathbf{U}_{n^2}VEC(\mathbf{B})]^T VEC(\mathbf{A}) \\ &= VEC^T(\mathbf{B}^T)VEC(\mathbf{A}) \\ &= tr(\mathbf{BA}) \end{aligned} \tag{3.41}$$

thereby recovering a well-known relation.

If $\mathbf{I}_n$ is the $n \times n$ identity tensor and $\mathbf{i}_n = VEC(\mathbf{I}_n)$, $VEC(\mathbf{A}) = \mathbf{I}_n \otimes \mathbf{A}\mathbf{i}_n$ since $VEC(\mathbf{A}) = VEC(\mathbf{AI}_n)$. If $\mathbf{I}_{n^2}$ is the identity tensor in an n^2-dimensional (Euclidean vector) space, $\mathbf{I}_n \otimes \mathbf{I}_n = \mathbf{I}_{n^2}$ since $VEC(\mathbf{I}_n) = VEC(\mathbf{I}_n\mathbf{I}_n) = \mathbf{I}_n \otimes \mathbf{I}_n VEC(\mathbf{I}_n)$. But $\mathbf{i}_n = \mathbf{I}_n\mathbf{i}$ and hence $\mathbf{I}_n \otimes \mathbf{I}_n = \mathbf{I}_{n^2}$.

If $\mathbf{A}$, $\mathbf{B}$, and $\mathbf{C}$ denote $n \times n$ tensors,

$$\begin{aligned} VEC(\mathbf{ACB}^T) &= \mathbf{I}_n \otimes \mathbf{A}VEC(\mathbf{CB}^T) \\ &= (\mathbf{I}_n \otimes \mathbf{A})(\mathbf{B} \otimes \mathbf{I}_n)VEC(\mathbf{C}) \\ &= \mathbf{B} \otimes \mathbf{A}VEC(\mathbf{C}) \end{aligned} \tag{3.42}$$

But by a parallel argument

$$\begin{aligned} VEC(\mathbf{ACB}^T)^T &= VEC(\mathbf{BC}^T\mathbf{A}^T) \\ &= \mathbf{A} \otimes \mathbf{B}VEC(\mathbf{C}^T) \\ &= \mathbf{A} \otimes \mathbf{B}\mathbf{U}_{n^2}VEC(\mathbf{C}) \end{aligned} \tag{3.43}$$

However, the permutation tensor $\mathbf{U}_{n^2}$ arises in the n^2-dimensional space in the relation

$$VEC(\mathbf{ACB}^T)^T = \mathbf{U}_{n^2}VEC(\mathbf{ACB}^T) \tag{3.44}$$

Consequently, if $\mathbf{C}$ is arbitrary,

$$\mathbf{U}_{n^2}\mathbf{B} \otimes \mathbf{A}VEC(\mathbf{C}) = \mathbf{A} \otimes \mathbf{B}\mathbf{U}_{n^2}VEC(\mathbf{C}) \tag{3.45}$$

and, upon using the relation $\mathbf{U}_{n^2} = \mathbf{U}_{n^2}^{-1}$, we obtain an important result

$$\mathbf{B} \otimes \mathbf{A} = \mathbf{U}_{n^2}\mathbf{A} \otimes \mathbf{B}\mathbf{U}_{n^2} \tag{3.46}$$

If $\mathbf{A}$ and $\mathbf{B}$ are nonsingular $n \times n$ tensors,

$$\begin{aligned}(\mathbf{A} \otimes \mathbf{B})(\mathbf{A}^{-1} \otimes \mathbf{B}^{-1}) &= \mathbf{A}\mathbf{A}^{-1} \otimes \mathbf{B}\mathbf{B}^{-1} \\ &= \mathbf{I}_n \otimes \mathbf{I}_n \\ &= \mathbf{I}_{n^2}\end{aligned} \tag{3.47}$$

The Kronecker sum and difference appear frequently (e.g., in control theory) and are defined as follows:

$$\mathbf{A} \oplus \mathbf{B} = \mathbf{A} \otimes \mathbf{I}_n + \mathbf{I}_n \otimes \mathbf{B}, \qquad \mathbf{A} \ominus \mathbf{B} = \mathbf{A} \otimes \mathbf{I}_n - \mathbf{I}_n \otimes \mathbf{B} \tag{3.48}$$

The Kronecker sum and difference of two $n \times n$ tensors are $n^2 \times n^2$ tensors in a sense explained below.

3.6.3 Eigenstructures of Kronecker Products

Let α_j and β_k denote the eigenvalues of $\mathbf{A}$ and $\mathbf{B}$, and let $\mathbf{y}_j$ and $\mathbf{z}_k$ denote the corresponding eigenvectors. The Kronecker product, sum, and difference have the following eigenstructures:

Expression	*jk*th **eigenvalue**	*jk*th **eigenvector**	
$\mathbf{A} \otimes \mathbf{B}$	$\alpha_j\beta_k$	$\mathbf{y}_j \otimes \mathbf{z}_k$	
$\mathbf{A} \oplus \mathbf{B}$	$\alpha_j + \beta_k$	$\mathbf{y}_j \otimes \mathbf{z}_k$	(3.49)
$\mathbf{A} \ominus \mathbf{B}$	$\alpha_j - \beta_k$	$\mathbf{y}_j \otimes \mathbf{z}_k$	

As proof,

$$\begin{aligned}\alpha_j\mathbf{y}_j \otimes \beta_k\mathbf{z}_k &= \alpha_j\beta_k\mathbf{y}_j \otimes \mathbf{z}_k \\ &= \mathbf{A}\mathbf{y}_j \otimes \mathbf{B}\mathbf{z}_k \\ &= (\mathbf{A} \otimes \mathbf{B})(\mathbf{y}_j \otimes \mathbf{z}_k)\end{aligned} \tag{3.50}$$

Now the eigenvalues of $\mathbf{A} \otimes \mathbf{I}_n$ are $1 \times \alpha_j$, while the eigenvectors are $\mathbf{y}_j \otimes \mathbf{w}_k$ in which $\mathbf{w}_k$ is an arbitrary unit vector (eigenvector of $\mathbf{I}_n$). The corresponding quantities for $\mathbf{I}_n \otimes \mathbf{B}$ are $\beta_k \times 1$ and $\mathbf{v}_j \times \mathbf{z}_k$, in which $\mathbf{v}_j$ is an arbitrary eigenvector of $\mathbf{I}_n$. Upon selecting $\mathbf{w}_k = \mathbf{z}_k$ and $\mathbf{v}_j = \mathbf{y}_j$, the Kronecker sum is seen to have eigenvalues $\alpha_j + \beta_k$ with corresponding eigenvectors $\mathbf{y}_j \otimes \mathbf{z}_k$.

3.6.4 Kronecker Form of Quadratic Products

Let $\mathbf{R}$ be a second-order $n \times n$ tensor. The quadratic product $\mathbf{a}^T\mathbf{R}\mathbf{b}$ is easily derived: if $\mathbf{r} = VEC(\mathbf{R})$

$$\begin{aligned}\mathbf{a}^T\mathbf{R}\mathbf{b} &= tr\left[\mathbf{b}\mathbf{a}^T\mathbf{R}\right]\\ &= VEC^T\left(\left[\mathbf{b}\mathbf{a}^T\right]^T\right)VEC(\mathbf{R})\\ &= VEC^T\left(\mathbf{a}\mathbf{b}^T\right)VEC(\mathbf{R})\\ &= \mathbf{b}^T \otimes \mathbf{a}^T\mathbf{r}\end{aligned} \tag{3.51}$$

3.6.5 Kronecker Product Operators for Fourth-Order Tensors

Of course fourth-order tensors, and to a lesser extent third-order tensors, play a critical role in continuum thermomechanics and the finite element method. For example, they encompass the stiffnesses relating the stress tensor to the strain tensor.

Let $\mathbf{A}$ and $\mathbf{B}$ be second-order $n \times n$ tensors and let $\mathbf{C}$ be an $n \times n \times n \times n$ matrix. Suppose that $\mathbf{A} = \mathbf{C}\mathbf{B}$, which is equivalent to $a_{ij} = c_{ijkl}b_{kl}$ in which the range of i, j, k, and l is $(1,n)$. In this case, $\mathbf{C}$ is called a fourth-order tensor if there exists such that $\mathbf{A}' = \mathbf{C}'\mathbf{B}'$. In indicial notation the entries of $\mathbf{C}'$ are related to those of $\mathbf{C}$ by $c'_{pqmn} = q_{pi}q_{qj}q_{km}q_{ln}c_{ijkl}$.

The $TEN22$ operator is introduced implicitly using

$$VEC(\mathbf{A}) = TEN22(\mathbf{C})VEC(\mathbf{B}) \tag{3.52}$$

It "collapses" a fourth-order tensor relating two second-order tensors into a second-order tensor in $n^2 \times n^2$-dimensional space. Note that

$$\begin{aligned}TEN22(\mathbf{ACB})VEC(\mathbf{D}) &= VEC(\mathbf{ACBD})\\ &= \mathbf{I}_n \otimes \mathbf{A}VEC(\mathbf{CBD})\\ &= \mathbf{I}_n \otimes \mathbf{A}TEN22(\mathbf{C})VEC(\mathbf{BD})\\ &= \mathbf{I}_n \otimes \mathbf{A}TEN22(\mathbf{C})\mathbf{I}_n \otimes \mathbf{B}VEC(\mathbf{D})\end{aligned} \tag{3.53}$$

and hence $TEN22(\mathbf{ACB}) = \mathbf{I}_n \otimes \mathbf{A}\ TEN22(\mathbf{C})\mathbf{I}_n \otimes \mathbf{B}$. Upon writing $\mathbf{B} = \mathbf{C}^{-1}\mathbf{A}$, it is immediate that $VEC(\mathbf{B}) = TEN22(\mathbf{C}^{-1})VEC(\mathbf{A})$. But $TEN22(\mathbf{C})VEC(\mathbf{B}) = VEC(\mathbf{A})$ and hence $VEC(\mathbf{B}) = [TEN22(\mathbf{C})]^{-1}VEC(\mathbf{A})$. We conclude that $TEN22(\mathbf{C}^{-1}) = TEN22^{-1}(\mathbf{C})$. Also, writing $\mathbf{A}^T = \hat{\mathbf{C}}\mathbf{B}^T$, it is immediate that $\mathbf{U}_{n^2}\mathbf{a} = TEN22(\mathbf{C})\mathbf{U}_{n^2}\mathbf{b}$, and hence $TEN22(\hat{\mathbf{C}}) = \mathbf{U}_{n^2}TEN22(\mathbf{C})\mathbf{U}_{n^2}$. The inverse of the $TEN22$ operator is introduced using the obvious relation $ITEN22(TEN22(\mathbf{C})) = \mathbf{C}$.

3.6.6 Transformation Properties of *VEC*, *TEN*22, *TEN*21, and *TEN*12

Suppose that $\mathbf{A}$ and $\mathbf{B}$ are real second-order $n \times n$ tensors and $\mathbf{C}$ is a fourth-order $n \times n \times n \times n$ tensor such that $\mathbf{A} = \mathbf{C}\mathbf{B}$. All are referred to a coordinate system denoted as Y. Let the unitary matrix (orthogonal tensor) $\mathbf{Q}_n$ represent a rotation which gives rise to a coordinate system Y′ and let $\mathbf{A}'$, $\mathbf{B}'$, and $\mathbf{C}'$ denote the counterparts of $\mathbf{A}$, $\mathbf{B}$, and $\mathbf{C}$. Now, since $\mathbf{A}' = \mathbf{Q}_n\mathbf{A}\mathbf{Q}_n^T$,

$$VEC(\mathbf{A}') = \mathbf{Q} \otimes \mathbf{Q}VEC(\mathbf{A}) \tag{3.54}$$

But note that $(\mathbf{Q} \otimes \mathbf{Q})^T = \mathbf{Q}^T \otimes \mathbf{Q}^T = \mathbf{Q}^{-1} \otimes \mathbf{Q}^{-1} = (\mathbf{Q} \otimes \mathbf{Q})^{-1}$. Hence $\mathbf{Q} \otimes \mathbf{Q}$ is a unitary matrix in an n^2-dimensional vector space. However, not all rotations in n^2-dimensional space can be expressed in the form $\mathbf{Q} \otimes \mathbf{Q}$. It follows that $VEC(\mathbf{A})$ transforms as an $n^2 \times 1$ vector *under rotations of the form* $\mathbf{Q}_{n^2} = \mathbf{Q} \otimes \mathbf{Q}$.

Now write $\mathbf{A}' = \mathbf{C}'\mathbf{B}'$ and observe that

$$\mathbf{Q} \otimes \mathbf{Q}VEC(\mathbf{A}) = TEN22(\mathbf{C}')\mathbf{Q} \otimes \mathbf{Q}VEC(\mathbf{B}) \tag{3.55}$$

It follows that

$$TEN22(\mathbf{C}') = \mathbf{Q} \otimes \mathbf{Q}TEN22(\mathbf{C})(\mathbf{Q} \otimes \mathbf{Q})^T \tag{3.56}$$

and hence $TEN22(\mathbf{C})$ transforms a second-order $n^2 \times n^2$ tensor *under rotations of the form* $\mathbf{Q} \otimes \mathbf{Q}$.

Finally, let $\mathbf{C}_a$ and $\mathbf{C}_b$ denote third-order $n \times n \times n$ tensors, respectively, which satisfy the relations of the form $\mathbf{A} = \mathbf{C}_a\mathbf{b}$ and $\mathbf{b} = \mathbf{C}_b\mathbf{A}$. We introduce the operators $TEN21(\mathbf{C}_a)$ and $TEN12(\mathbf{C}_b)$ using $VEC(\mathbf{A}) = TEN21(\mathbf{C})\mathbf{b}$ and $\mathbf{b} = TEN12(\mathbf{C})VEC(\mathbf{A})$. The operators satisfy the transformation properties

$$\begin{aligned} TEN21\left(\mathbf{C}_a'\right) &= \mathbf{Q} \otimes \mathbf{Q}TEN21(\mathbf{C}_a)\mathbf{Q}^T \quad n^2 \times n \\ TEN12\left(\mathbf{C}_b'\right) &= \mathbf{Q}TEN12(\mathbf{C}_b)\mathbf{Q}^T \otimes \mathbf{Q}^T \quad n \times n^2 \end{aligned} \tag{3.57}$$

for which reason we say that $TEN21$ and $TEN12$ are tensors of order (2,1) and (1,2), respectively.

3.6.7 Kronecker Expressions for Symmetry Classes in Fourth-Order Tensors

Let $\mathbf{C}$ denote a fourth-order tensor with entries c_{ijkl}. If the entries observe

$$c_{ijkl} = c_{jikl} \tag{3.58a}$$

$$c_{ijkl} = c_{ijlk} \tag{3.58b}$$

$$c_{ijkl} = c_{klij} \tag{3.58c}$$

we say that $\mathbf{C}$ is *totally symmetric*. A fourth-order tensor $\mathbf{C}$ satisfying Equation 3.58a but not Equations 3.58b and 3.58c will be called *symmetric*.

Kronecker product conditions for symmetry are now stated. The fourth-order tensor $\mathbf{C}$ is totally symmetric if and only if

$$TEN22(\mathbf{C}) = TEN22^T(\mathbf{C}) \tag{3.59a}$$

$$\mathbf{U}_{n^2}TEN22(\mathbf{C}) = TEN22(\mathbf{C}) \tag{3.59b}$$

$$TEN22(\mathbf{C})\mathbf{U}_{n^2} = TEN22(\mathbf{C}) \tag{3.59c}$$

Equation 3.59a is equivalent to symmetry with respect to exchange of *ij* and *kl* in **C**. Total symmetry also implies that, for any second-order $n \times n$ tensor **B**, the corresponding tensor $\mathbf{A} = \mathbf{CB}$ is symmetric. Thus, if $\mathbf{a} = VEC(\mathbf{A})$ and $\mathbf{b} = VEC(\mathbf{B})$, then $\mathbf{a} = TEN22(\mathbf{C})$. Also $\mathbf{U}_{n^2}\mathbf{a} = TEN22(\mathbf{C})\mathbf{b}$. Multiplying through the later expression with $\mathbf{U}_{n^2}$ implies Equation 3.58b. Next, for any $n \times n$ tensor **A**, the tensor $\mathbf{B} = \mathbf{C}^{-1}\mathbf{A}$ is symmetric. It follows that $\mathbf{b} = TEN22(\mathbf{C}^{-1})\mathbf{a} = TEN22^{-1}(\mathbf{C})\mathbf{a}$, *and* $\mathbf{U}_{n^2}\mathbf{b} = TEN22^{-1}(\mathbf{C})\mathbf{a}$. Thus, $TEN22(\mathbf{C}^{-1}) = \mathbf{U}_{n^2}TEN22^{-1}(\mathbf{C})$. Also $TEN22(\mathbf{C}) = [\mathbf{U}_{n^2}TEN22^{-1}(\mathbf{C})]^{-1} = TEN22(\mathbf{C})\mathbf{U}_{n^2}$. The conclusion is immediate that $\mathbf{U}_{n^2}TEN22(\mathbf{C})\mathbf{U}_{n^2} = TEN22(\mathbf{C})$ if **C** is totally symmetric.

We next prove the following:

$$\mathbf{C}^{-1} \text{ is totally symmetric, if } \mathbf{C} \text{ is totally symmetric} \tag{3.60}$$

Note that $TEN22(\mathbf{C})\mathbf{U}_{n^2} = TEN22(\mathbf{C})$ implies that $\mathbf{U}_{n^2}TEN22(\mathbf{C}^{-1}) = TEN22(\mathbf{C}^{-1})$, while $\mathbf{U}_{n^2}TEN22(\mathbf{C}) = TEN22(\mathbf{C})$ implies that $TEN22(\mathbf{C}^{-1})\,\mathbf{U}_{n^2} = TEN22(\mathbf{C}^{-1})$.

Finally, we prove the following: for a nonsingular $n \times n$ tensor **G**,

$$\mathbf{GCG}^T \text{ is totally symmetric if and only if } \mathbf{C} \text{ is totally symmetric} \tag{3.61}$$

First, Equation 3.56 implies that $TEN22(\mathbf{GCG}^{-T}) = \mathbf{I} \otimes \mathbf{G}TEN22(\mathbf{C})\,\mathbf{I} \otimes \mathbf{G}^T$, so that $TEN22(\mathbf{GCG}^T)$ is certainly symmetric.

Next consider whether $\mathbf{A}'$ given by

$$\mathbf{A}' = \mathbf{GCG}^T\mathbf{B}' \tag{3.62}$$

is symmetric in which $\mathbf{B}'$ is a second-order nonsingular $n \times n$ tensor. But we may write

$$\mathbf{G}^{-1}\mathbf{A}'\mathbf{G}^{-T} = \mathbf{CG}^{-1}\mathbf{B}'\mathbf{G}^{-T} \tag{3.63}$$

Now $\mathbf{G}^{-1}\mathbf{A}'\mathbf{G}^{-T}$ is symmetric since **C** is totally symmetric, and therefore $\mathbf{A}'$ is symmetric. Next consider whether $\mathbf{B}'$ given by the following is symmetric:

$$\mathbf{B}' = \mathbf{G}^{-T}\mathbf{C}^{-1}\mathbf{G}^{-1}\mathbf{A}' \tag{3.64}$$

But we may write

$$\mathbf{G}^T\mathbf{B}'\mathbf{G} = \mathbf{C}^{-1}\mathbf{G}^{-1}\mathbf{A}'\mathbf{G} \tag{3.65}$$

Since $\mathbf{C}^{-1}$ is totally symmetric, it follows that $\mathbf{G}^T\,\mathbf{B}'\mathbf{G}$ is symmetric, and hence $\mathbf{B}'$ is symmetric. We conclude that $\mathbf{GCG}^T$ is totally symmetric. The "only if" argument follows as a consequence of

$$\mathbf{C} = \mathbf{G}^{-1}(\mathbf{GCG}^T)\mathbf{G}$$

3.6.8 Differentials of Tensor Invariants

Let **A** be a symmetric 3×3 tensor, with invariants $I_1(\mathbf{A})$, $I_2(\mathbf{A})$, and $I_3(\mathbf{A})$. For a scalar-valued function $f(\mathbf{A})$,

$$df(\mathbf{A}) = \frac{\partial f}{\partial a_{ij}} da_{ij} = tr\left(\frac{\partial f}{\partial \mathbf{A}} d\mathbf{A}\right), \quad \left(\frac{\partial f}{\partial \mathbf{A}}\right)_{ij} = \frac{\partial f}{\partial a_{ij}} \tag{3.66}$$

But with $\mathbf{a} = VEC(\mathbf{A})$, we may also write

$$\begin{aligned} df(\mathbf{A}) &= VEC^T\left(\left[\frac{\partial f}{\partial \mathbf{A}}\right]^T\right) VEC(d\mathbf{A}) \\ &= \frac{\partial f}{\partial \mathbf{a}} d\mathbf{a} \end{aligned} \tag{3.67}$$

Continuing,

$$\begin{aligned} \frac{\partial I_1}{\partial \mathbf{a}} &= \frac{\partial}{\partial \mathbf{a}}\left(\mathbf{i}^T\mathbf{a}\right) = \mathbf{i}^T \\ \frac{\partial I_2}{\partial \mathbf{a}} &= \frac{\partial}{\partial \mathbf{a}}\left[\frac{1}{2}\left(\mathbf{i}^T\mathbf{a}\right)^2 - \mathbf{a}^T\mathbf{a}\right] \\ &= I_1\mathbf{i}^T - \mathbf{a}^T \end{aligned} \tag{3.68}$$

and

$$\begin{aligned} dI_3 &= tr\left(\mathbf{A}^2 d\mathbf{A}\right) - I_1\, tr(\mathbf{A}\, d\mathbf{A}) + I_2\, d\mathbf{A} \\ &= tr\left(I_3\mathbf{A}^{-1} d\mathbf{A}\right) \end{aligned} \tag{3.69}$$

so that

$$\frac{\partial I_3}{\partial \mathbf{a}} = I_3 VEC\left(\mathbf{A}^{-1}\right) \tag{3.70}$$

3.7 EXAMPLES

EXAMPLE 3.1

Given a symmetric $n \times n$ tensor $\boldsymbol{\sigma}$, prove that

$$tr(\boldsymbol{\sigma} - tr(\boldsymbol{\sigma})I_n/n) = 0$$

SOLUTION

$$tr(\boldsymbol{\sigma} - tr(\boldsymbol{\sigma})\boldsymbol{I}_n/n) = tr\,\boldsymbol{\sigma} - tr(\boldsymbol{\sigma})\,tr(\boldsymbol{I}_n)/n. \quad \text{But } tr(\boldsymbol{I}_n) = n.$$

$$\text{Hence} \quad tr(\boldsymbol{\sigma} - tr(\boldsymbol{\sigma})\boldsymbol{I}_n/n) = tr\,\boldsymbol{\sigma} - tr(\boldsymbol{\sigma}) = 0$$

EXAMPLE 3.2

Verify using 2×2 tensors that

$$tr(\mathbf{AB}) = tr(\mathbf{BA})$$

SOLUTION

Let

$$\mathbf{A} = \begin{bmatrix} a & b \\ c & d \end{bmatrix}, \quad \mathbf{B} = \begin{bmatrix} e & f \\ g & h \end{bmatrix}$$

Now

$$\mathbf{AB} = \begin{bmatrix} ae + bg & af + bh \\ ce + dg & cf + dh \end{bmatrix} \rightarrow tr(\mathbf{AB}) = ae + bg + cf + dh$$

$$\mathbf{BA} = \begin{bmatrix} ae + cf & be + df \\ ag + ch & bg + dh \end{bmatrix} \rightarrow tr(\mathbf{BA}) = ae + bg + cf + dh$$

Hence $tr(\mathbf{AB}) = tr(\mathbf{BA})$.

EXAMPLE 3.3

Express I_3 as a function of I_1 and I_2.

SOLUTION

We know that $I_1 = tr(\mathbf{A})$. Also

$$I_2 = \tfrac{1}{2}\left[tr^2(\mathbf{A}) - tr\left(\mathbf{A}^2\right)\right] = \tfrac{1}{2}\left[I_1^2 - tr\left(\mathbf{A}^2\right)\right]$$

from which $tr(\mathbf{A}^2) = I_1^2 - 2I_2$. From the Cayley–Hamilton relation $\mathbf{A}^3 - I_1\mathbf{A}^2 + I_2\,\mathbf{A} - I_3\mathbf{I} = \mathbf{0}$, it follows that

$$\begin{aligned} I_3 &= \tfrac{1}{3}\left[tr\left(\mathbf{A}^3\right) - I_1 tr\left(\mathbf{A}^2\right) + I_2 tr(\mathbf{A})\right] \\ &= \tfrac{1}{3}\left[tr\left(\mathbf{A}^3\right) - I_1^3 + 3I_1I_2\right] \end{aligned}$$

Now using the relations $I_1 = a_1 + a_2 + a_3$, $I_2 = a_1a_2 + a_2a_3 + a_3a_1$, $I_3 = a_1a_2a_3$, we find

$$tr(\mathbf{A}^3) = I_1^3 - 3I_3 - I_1I_2$$

EXAMPLE 3.4

Using 2×2 tensors and 2×1 vectors, verify the six relations given for Kronecker products.

Solution

Let

$$\mathbf{A} = \begin{bmatrix} a_{11} & a_{12} \\ a_{21} & a_{22} \end{bmatrix}, \quad \mathbf{B} = \begin{bmatrix} b_{11} & b_{12} \\ b_{21} & b_{22} \end{bmatrix}, \quad \mathbf{C} = \begin{bmatrix} c_{11} & c_{12} \\ c_{21} & c_{22} \end{bmatrix}, \quad \mathbf{D} = \begin{bmatrix} d_{11} & d_{12} \\ d_{21} & d_{22} \end{bmatrix}$$

$$\mathbf{a} = \begin{pmatrix} a_1 \\ a_2 \end{pmatrix}, \quad \mathbf{b} = \begin{pmatrix} b_1 \\ b_2 \end{pmatrix}$$

Relation 1: $VEC(\mathbf{A}^T) = \mathbf{U}_{n^2}VEC(\mathbf{A})$

$$\begin{pmatrix} a_{11} \\ a_{12} \\ a_{21} \\ a_{22} \end{pmatrix} = \begin{bmatrix} u_{11} & u_{12} & u_{13} & u_{14} \\ u_{21} & u_{22} & u_{23} & u_{24} \\ u_{31} & u_{32} & u_{33} & u_{34} \\ u_{41} & u_{42} & u_{43} & u_{44} \end{bmatrix} \begin{pmatrix} a_{11} \\ a_{21} \\ a_{12} \\ a_{22} \end{pmatrix}$$

implying that

$$\mathbf{U}_{22} = \begin{bmatrix} 1 & 0 & 0 & 0 \\ 0 & 0 & 1 & 0 \\ 0 & 1 & 0 & 0 \\ 0 & 0 & 0 & 1 \end{bmatrix}$$

Relation 2: $tr(\mathbf{AB}) = VEC^T(\mathbf{A}^T)VEC(\mathbf{B})$

First

$$\mathbf{AB} = \begin{bmatrix} a_{11}b_{11} + a_{12}b_{21} & a_{11}b_{12} + a_{12}b_{22} \\ a_{21}b_{11} + a_{22}b_{21} & a_{21}b_{12} + a_{22}b_{22} \end{bmatrix}$$

$$tr(\mathbf{AB}) = a_{11}b_{11} + a_{12}b_{21} + a_{21}b_{12} + a_{22}b_{22} \qquad \text{(S2.1)}$$

Next

$$\mathbf{A}^T = \begin{bmatrix} a_{11} & a_{21} \\ a_{12} & a_{22} \end{bmatrix} \rightarrow VEC^T(\mathbf{A}^T) = \{a_{11} \quad a_{12} \quad a_{21} \quad a_{22}\}$$

and finally

$$VEC^T(\mathbf{A}^T)VEC(\mathbf{B}) = (a_{11} \quad a_{12} \quad a_{21} \quad a_{22})\begin{pmatrix} b_{11} \\ b_{21} \\ b_{12} \\ b_{22} \end{pmatrix}$$

$$= a_{11}b_{11} + a_{12}b_{21} + a_{21}b_{12} + a_{22}b_{22} \tag{S2.2}$$

thereby verifying Relation 2.

Relation 3: $\mathbf{I}_n \otimes \mathbf{B}^T = (\mathbf{I}_n \otimes \mathbf{B})^T$

Here

$$\mathbf{I}_2 = \begin{bmatrix} 1 & 0 \\ 0 & 1 \end{bmatrix} \quad \text{and} \quad \mathbf{B}^T = \begin{bmatrix} b_{11} & b_{21} \\ b_{12} & b_{22} \end{bmatrix}$$

from which

$$\mathbf{I}_n \otimes \mathbf{B}^T = \begin{bmatrix} b_{11} & b_{21} & 0 & 0 \\ b_{12} & b_{22} & 0 & 0 \\ 0 & 0 & b_{11} & b_{21} \\ 0 & 0 & b_{12} & b_{22} \end{bmatrix} \tag{S2.3}$$

Now

$$(\mathbf{I}_n \otimes \mathbf{B})^T = \begin{bmatrix} b_{11} & b_{21} & 0 & 0 \\ b_{12} & b_{22} & 0 & 0 \\ 0 & 0 & b_{11} & b_{21} \\ 0 & 0 & b_{12} & b_{22} \end{bmatrix}$$

$$\mathbf{I}_n \otimes \mathbf{B} = \begin{bmatrix} b_{11} & b_{12} & 0 & 0 \\ b_{21} & b_{22} & 0 & 0 \\ 0 & 0 & b_{11} & b_{12} \\ 0 & 0 & b_{21} & b_{22} \end{bmatrix} \tag{S2.4}$$

and it is immediate that

$$(\mathbf{I}_n \otimes \mathbf{B})^T = \begin{bmatrix} b_{11} & b_{21} & 0 & 0 \\ b_{12} & b_{22} & 0 & 0 \\ 0 & 0 & b_{11} & b_{21} \\ 0 & 0 & b_{12} & b_{22} \end{bmatrix}$$

$$= \mathbf{I}_n \otimes \mathbf{B}^T$$

as expected.

Relation 4: $(\mathbf{A} \otimes \mathbf{B})(\mathbf{C} \otimes \mathbf{D}) = \mathbf{AC} \otimes \mathbf{BD}$

$$\mathbf{A} \otimes \mathbf{B} = \begin{bmatrix} a_{11}\mathbf{B} & a_{12}\mathbf{B} \\ a_{21}\mathbf{B} & a_{22}\mathbf{B} \end{bmatrix} = \begin{bmatrix} a_{11}b_{11} & a_{11}b_{12} & a_{12}b_{11} & a_{12}b_{12} \\ a_{11}b_{21} & a_{11}b_{22} & a_{12}b_{21} & a_{12}b_{22} \\ a_{21}b_{11} & a_{21}b_{12} & a_{22}b_{11} & a_{22}b_{12} \\ a_{21}b_{21} & a_{21}b_{22} & a_{22}b_{21} & a_{22}b_{22} \end{bmatrix}$$

Similarly,

$$\mathbf{C} \otimes \mathbf{D} = \begin{bmatrix} c_{11}d_{11} & c_{11}d_{12} & c_{12}d_{11} & c_{12}d_{12} \\ c_{11}d_{21} & c_{11}d_{22} & c_{12}d_{21} & c_{12}d_{22} \\ c_{21}d_{11} & c_{21}d_{12} & c_{22}d_{11} & c_{22}d_{12} \\ c_{21}d_{21} & c_{21}d_{22} & c_{22}d_{21} & c_{22}d_{22} \end{bmatrix}$$

Now

$$(\mathbf{A} \otimes \mathbf{B})(\mathbf{C} \otimes \mathbf{D}) = \begin{bmatrix} x_1y_1 & x_1y_2 & x_2y_1 & x_2y_2 \\ x_1y_3 & x_1y_4 & x_2y_3 & x_2y_4 \\ x_3y_1 & x_3y_2 & x_4y_1 & x_4y_2 \\ x_3y_3 & x_3y_4 & x_4y_3 & x_4y_4 \end{bmatrix}$$

in which

$$x_1 = a_{11}c_{11} + a_{12}c_{21}, \quad x_2 = a_{11}c_{12} + a_{12}c_{22}, \quad x_3 = a_{21}c_{11} + a_{22}c_{21}, \quad x_4 = a_{21}c_{12} + a_{22}c_{22}$$
$$y_1 = b_{11}d_{11} + b_{12}d_{21}, \quad y_2 = b_{11}d_{12} + b_{12}d_{22}, \quad y_3 = b_{21}d_{11} + b_{22}d_{21}, \quad y_4 = b_{21}d_{12} + b_{22}d_{22}$$

But

$$\mathbf{BD} = \begin{bmatrix} b_{11} & b_{12} \\ b_{21} & b_{22} \end{bmatrix} \begin{bmatrix} d_{11} & d_{12} \\ d_{21} & d_{22} \end{bmatrix} = \begin{bmatrix} y_1 & y_2 \\ y_3 & y_4 \end{bmatrix}$$

and now

$$(\mathbf{A} \otimes \mathbf{B})(\mathbf{C} \otimes \mathbf{D}) = \begin{bmatrix} x_1\mathbf{BD} & x_2\mathbf{BD} \\ x_3\mathbf{BD} & x_4\mathbf{BD} \end{bmatrix}$$

Also

$$\mathbf{AC} = \begin{bmatrix} a_{11} & a_{12} \\ a_{21} & a_{22} \end{bmatrix} \begin{bmatrix} c_{11} & c_{12} \\ c_{21} & c_{22} \end{bmatrix} = \begin{bmatrix} x_1 & x_2 \\ x_3 & x_4 \end{bmatrix}$$

and hence

$$(\mathbf{A} \otimes \mathbf{B})(\mathbf{C} \otimes \mathbf{D}) = \mathbf{AC} \otimes \mathbf{BD}$$

as expected.

Relation 5: $VEC(\mathbf{ACB}) = \mathbf{B}^T \otimes \mathbf{A}VEC(\mathbf{C})$
Here

$$\mathbf{ACB} = \begin{bmatrix} a_{11} & a_{12} \\ a_{21} & a_{22} \end{bmatrix} \begin{bmatrix} c_{11} & c_{12} \\ c_{21} & c_{22} \end{bmatrix} \begin{bmatrix} b_{11} & b_{12} \\ b_{21} & b_{22} \end{bmatrix}$$

$$= \begin{bmatrix} a_{11} & a_{12} \\ a_{21} & a_{22} \end{bmatrix} \begin{bmatrix} b_{11}c_{11} + b_{21}c_{12} & b_{12}c_{11} + b_{22}c_{12} \\ b_{11}c_{21} + b_{21}c_{22} & b_{12}c_{21} + b_{22}c_{22} \end{bmatrix}$$

$$= \begin{bmatrix} a_{11}b_{11}c_{11} + a_{11}b_{21}c_{12} + a_{12}b_{11}c_{21} + a_{12}b_{21}c_{22} & a_{11}b_{12}c_{11} + a_{11}b_{22}c_{12} + a_{12}b_{12}c_{21} + a_{12}b_{22}c_{22} \\ a_{21}b_{11}c_{11} + a_{21}b_{21}c_{12} + a_{22}b_{11}c_{21} + a_{22}b_{21}c_{22} & a_{21}b_{12}c_{11} + a_{21}b_{22}c_{12} + a_{22}b_{12}c_{21} + a_{22}b_{22}c_{22} \end{bmatrix}$$

Hence

$$VEC(\mathbf{ACB}) = \begin{bmatrix} (a_{11}b_{11})c_{11} + (a_{12}b_{11})c_{21} + (a_{11}b_{21})c_{12} + (a_{12}b_{21})c_{22} \\ (a_{21}b_{11})c_{11} + (a_{22}b_{11})c_{21} + (a_{21}b_{21})c_{12} + (a_{22}b_{21})c_{22} \\ (a_{11}b_{12})c_{11} + (a_{12}b_{12})c_{21} + (a_{11}b_{22})c_{12} + (a_{12}b_{22})c_{22} \\ (a_{21}b_{12})c_{11} + (a_{22}b_{12})c_{21} + (a_{21}b_{22})c_{12} + (a_{22}b_{22})c_{22} \end{bmatrix}$$

$$= \begin{bmatrix} a_{11}b_{11} & a_{12}b_{11} & a_{11}b_{21} & a_{12}b_{21} \\ a_{21}b_{11} & a_{22}b_{11} & a_{21}b_{21} & a_{22}b_{21} \\ a_{11}b_{12} & a_{12}b_{12} & a_{11}b_{22} & a_{12}b_{22} \\ a_{21}b_{12} & a_{22}b_{12} & a_{21}b_{22} & a_{22}b_{22} \end{bmatrix} \begin{pmatrix} c_{11} \\ c_{21} \\ c_{12} \\ c_{22} \end{pmatrix}$$

$$= \begin{bmatrix} b_{11}\mathbf{A} & b_{12}\mathbf{A} \\ b_{21}\mathbf{A} & b_{22}\mathbf{A} \end{bmatrix} VEC(\mathbf{C})$$

$$= \mathbf{B}^T \otimes \mathbf{A}VEC(\mathbf{C})$$

as expected.

Relation 6: $\mathbf{a} \otimes \mathbf{b} = VEC([\mathbf{ab}^T]^T)$
First

$$\mathbf{a} \otimes \mathbf{b} = \begin{pmatrix} a_1\mathbf{b} \\ a_2\mathbf{b} \end{pmatrix} = \begin{pmatrix} a_1b_1 \\ a_1b_2 \\ a_2b_1 \\ a_2b_2 \end{pmatrix} \tag{S2.5}$$

Now

$$\mathbf{ab}^T = \begin{pmatrix} a_1 \\ a_2 \end{pmatrix} (b_1 \quad b_2) = \begin{bmatrix} a_1b_1 & a_1b_2 \\ a_2b_1 & a_2b_2 \end{bmatrix}$$

and also

$$[\mathbf{ab}^T]^T = \begin{bmatrix} a_1b_1 & a_2b_1 \\ a_1b_2 & a_2b_2 \end{bmatrix}, \quad VEC\left([\mathbf{ab}^T]^T\right) = \begin{pmatrix} a_1b_1 \\ a_1b_2 \\ a_2b_1 \\ a_2b_2 \end{pmatrix} \tag{S2.6}$$

serving to verify Relation 6.

EXAMPLE 3.5

Write out the 9×9 quantity $TEN22(\mathbf{C})$ in $c_{ijkl} = 2\mu(\delta_{ik}\delta_{jl}) + \lambda\delta_{ij}\delta_{kl}$, which appears in the Lamé form of the stress–strain relation in linear isotropic elasticity under small strain, namely $\sigma_{ij} = c_{ijkl}\varepsilon_{kl}$.

Solution

It is readily verified that

$$\mathbf{C} = 2\mu\mathbf{I}_n \otimes \mathbf{I}_n + \lambda\mathbf{i}_n\mathbf{i}_n^T, \quad \mathbf{i}_n = VEC(\mathbf{I}_n)\,(9 \times 1)$$

Here $n = 3$.
Expansion gives

$$\mathbf{C} = (2\mu)\begin{bmatrix} 1\cdot\mathbf{I}_3 & 0\cdot\mathbf{I}_3 & 0\cdot\mathbf{I}_3 \\ 0\cdot\mathbf{I}_3 & 1\cdot\mathbf{I}_3 & 0\cdot\mathbf{I}_3 \\ 0\cdot\mathbf{I}_3 & 0\cdot\mathbf{I}_3 & 1\cdot\mathbf{I}_3 \end{bmatrix} + \lambda\begin{Bmatrix} 1 \\ 0 \\ 0 \\ 0 \\ 1 \\ 0 \\ 0 \\ 0 \\ 1 \end{Bmatrix}\{1 \quad 0 \quad 0 \quad 0 \quad 1 \quad 0 \quad 0 \quad 0 \quad 1\}$$

$$= (2\mu)\begin{bmatrix} 1&0&0&0&0&0&0&0&0 \\ 0&1&0&0&0&0&0&0&0 \\ 0&0&1&0&0&0&0&0&0 \\ 0&0&0&1&0&0&0&0&0 \\ 0&0&0&0&1&0&0&0&0 \\ 0&0&0&0&0&1&0&0&0 \\ 0&0&0&0&0&0&1&0&0 \\ 0&0&0&0&0&0&0&1&0 \\ 0&0&0&0&0&0&0&0&1 \end{bmatrix} + \lambda\begin{bmatrix} 1&0&0&0&1&0&0&0&1 \\ 0&0&0&0&0&0&0&0&0 \\ 0&0&0&0&0&0&0&0&0 \\ 0&0&0&0&0&0&0&0&0 \\ 1&0&0&0&1&0&0&0&1 \\ 0&0&0&0&0&0&0&0&0 \\ 0&0&0&0&0&0&0&0&0 \\ 0&0&0&0&0&0&0&0&0 \\ 1&0&0&0&1&0&0&0&1 \end{bmatrix}$$

EXAMPLE 3.6

Prove that the rows of a 3×3 tensor A are *row vectors*.

Solution

First note that if $\mathbf{f}$ and $\mathbf{g}$ are 3×1 vectors, the scalar product $\mathbf{f}^T\mathbf{g}$ is invariant: $\mathbf{f}'^T\mathbf{g}' = (\mathbf{Qf})^T(\mathbf{Qg}) = \mathbf{f}^T\mathbf{g}$. Alternately, let $\mathbf{g}$ be a 3×1 vector but suppose initally that $\mathbf{f}$ is simply a 3×1 *array*. If the matrix-algebraic product $(\mathbf{Qf})^T(\mathbf{Qg})$ is equal to $\mathbf{f}^T\mathbf{g}$ for all orthogonal 3×3 transformation matrices $\mathbf{Q}$, we conclude that the array $\mathbf{f}$ is in fact a 3×1 *vector*.

To support this statement, suppose instead that $\mathbf{f}' = \mathbf{Qf} + \boldsymbol{\alpha}$ in which $\boldsymbol{\alpha}$ represents the deviation from the vectorial relationship. But then $\mathbf{f}'^T\mathbf{g}' = (\mathbf{Qf} + \boldsymbol{\alpha})^T(\mathbf{Qg})$, and consequently it must be the case that $\boldsymbol{\alpha}^T\mathbf{Qg} = 0$. But this must hold for arbitrary $\mathbf{Q}$, and there is a particular instance of $\mathbf{Q}$, say $\mathbf{Q}_\alpha$, which rotates $\mathbf{g}$ to be colinear with $\boldsymbol{\alpha}$. If follows that $\boldsymbol{\alpha}^T\mathbf{Q}_\alpha\mathbf{g}$ can only vanish if $\boldsymbol{\alpha} = 0$.

We now consider whether the rows are transposed vectors. By hypothesis A is a 3×3 tensor. We write

$$\mathbf{A} = \begin{bmatrix} \mathbf{a}_1^T \\ \mathbf{a}_2^T \\ \mathbf{a}_3^T \end{bmatrix}$$

and consider the equation

$$\mathbf{Ab} = \mathbf{c}$$

in which $\mathbf{b}$ and $\mathbf{c}$ are 3×1 vectors. Now simple manipulation serves to verify that

$$\mathbf{AQ}^T = \begin{bmatrix} \mathbf{a}_1^T\mathbf{Q}^T \\ \mathbf{a}_2^T\mathbf{Q}^T \\ \mathbf{a}_3^T\mathbf{Q}^T \end{bmatrix}$$

and so we may write $\mathbf{AQ}^T(\mathbf{Qb}) = \mathbf{c}$. Clearly,

$$\begin{bmatrix} \mathbf{a}_1^T\mathbf{Q}^T \\ \mathbf{a}_2^T\mathbf{Q}^T \\ \mathbf{a}_3^T\mathbf{Q}^T \end{bmatrix} \mathbf{b}' = \begin{bmatrix} \mathbf{a}_1^T \\ \mathbf{a}_2^T \\ \mathbf{a}_3^T \end{bmatrix} \mathbf{b}$$

Since the scalar product of two 3×1 vectors is invariant under orthogonal transformations we conclude that

$$\begin{bmatrix} \mathbf{a}_1^T\mathbf{Q}^T \\ \mathbf{a}_2^T\mathbf{Q}^T \\ \mathbf{a}_3^T\mathbf{Q}^T \end{bmatrix} = \begin{bmatrix} \mathbf{a}_1'^T \\ \mathbf{a}_2'^T \\ \mathbf{a}_3'^T \end{bmatrix}$$

and in consequence that the rows of a 3×3 tensor are row vectors. Of course by using $\mathbf{A}^T$ we may similarly conclude that the columns of $\mathbf{A}$ are column vectors.

EXAMPLE 3.7

Let $\mathbf{x}$ denote the 3×1 position vector in a body. Also let $\boldsymbol{\gamma}$ be an $n \times 1$ vector, while $\mathbf{A}$ $(\mathbf{x})$ is an $n \times n$ matrix dependent on position and $\mathbf{P}_0$ is a constant $n \times n$ matrix. Prove that $\int \boldsymbol{\gamma}^T\mathbf{A}^T(\mathbf{x})\mathbf{P}_0\mathbf{A}(\mathbf{x})\boldsymbol{\gamma}\,dV$ may be written in a form such that the integration does not depend on $\mathbf{P}_0$.

Solution

From the properties of Kronecker products,

$$\boldsymbol{\gamma}^T\mathbf{A}^T(\mathbf{x})\mathbf{P}_0\mathbf{A}(\mathbf{x})\boldsymbol{\gamma} = \boldsymbol{\gamma}^T \otimes \boldsymbol{\gamma}^T VEC(\mathbf{A}^T(\mathbf{x})\mathbf{P}_0\mathbf{A}(\mathbf{x}))$$
$$= \boldsymbol{\gamma}^T \otimes \delta\boldsymbol{\gamma}^T\mathbf{A}^T(\mathbf{x}) \otimes \mathbf{A}^T(\mathbf{x})VEC(\mathbf{P}_0)$$

The integral now reduces to

$$\boldsymbol{\gamma}^T \otimes \boldsymbol{\gamma}^T\left[\int \mathbf{A}^T(\mathbf{x}) \otimes \mathbf{A}^T(\mathbf{x})\,\mathrm{d}V\right]VEC(\mathbf{P}_0) = \delta\boldsymbol{\gamma}^T\mathbf{K}\boldsymbol{\gamma}$$

$$\mathbf{K} = IVEC\left(\left[\int \mathbf{A}^T(\mathbf{x}) \otimes \mathbf{A}^T(\mathbf{x})\mathrm{d}V\right]VEC(\mathbf{P}_0)\right)$$

and the integration in the matrix **K** is independent of the matrix $\mathbf{P}_0$. This relation will be of interest later when finite element analysis of buckling is considered.

4 Introduction to Variational Methods

4.1 INTRODUCTORY NOTIONS

In this section we introduce the central notion of the variation. Recall that Chapter 1 described one step in FEA as expressing equilibrium equations as integral equations using variational calculus.

Let $\mathbf{u}(\mathbf{x})$ be a vector-valued function of position vector $\mathbf{x}$, and consider a vector-valued functional $\mathbf{F}(\mathbf{u}(\mathbf{x}),\mathbf{u}'(\mathbf{x}),\mathbf{x})$, in which $\mathbf{u}'(\mathbf{x}) = \partial\mathbf{u}/\partial\mathbf{x}$. (Just like the definite integral, a functional maps functions, say of $\mathbf{x}$, into numbers.) Next, let $\mathbf{v}(\mathbf{x})$ be a function such that $\mathbf{v}(\mathbf{x}) = \mathbf{0}$ whenever $\mathbf{u}(\mathbf{x}) = \mathbf{0}$, and also $\mathbf{v}'(\mathbf{x}) = \mathbf{0}$ when $\mathbf{u}'(\mathbf{x}) = \mathbf{0}$. Otherwise, $\mathbf{v}(\mathbf{x})$ is *arbitrary*. The differential $\mathbf{dF}$ measures how much $\mathbf{F}$ changes if $\mathbf{x}$ changes.

The variation $\delta\mathbf{F}$, defined below, measures how much $\mathbf{F}$ changes if $\mathbf{u}$ and $\mathbf{u}'$ change at fixed $\mathbf{x}$. Following Ewing (1985) we introduce the vector-valued function $\boldsymbol{\Phi}(e{:}\mathbf{F})$ as follows:

$$\boldsymbol{\Phi}(e{:}\mathbf{F}) = \mathbf{F}(\mathbf{u}(\mathbf{x}) + e\mathbf{v}(\mathbf{x}),\mathbf{u}'(\mathbf{x}) + e\mathbf{v}'(\mathbf{x}),\mathbf{x}) - \mathbf{F}(\mathbf{u}(\mathbf{x}),\mathbf{u}'(\mathbf{x}),\mathbf{x}) \tag{4.1}$$

in which e is a scalar "modulating" the difference between $\mathbf{u}$ and $\mathbf{u} + e\mathbf{v}$ as well as between $\mathbf{u}'$ and $\mathbf{u}' + e\mathbf{v}'$. The variation $\delta\mathbf{F}$ is defined by

$$\delta\mathbf{F} = e\left(\frac{\mathrm{d}\boldsymbol{\Phi}}{\mathrm{d}e}\right)_{|e=0} \tag{4.2}$$

with $\mathbf{x}$ fixed. Elementary manipulation using differential calculus serves to demonstrate that

$$\delta\mathbf{F} = \frac{\partial\mathbf{F}}{\partial\mathbf{u}}e\mathbf{v} + tr\left(\frac{\partial\mathbf{F}}{\partial\mathbf{u}'}e\mathbf{v}'\right) \tag{4.3}$$

in which $tr\left(\frac{\partial\mathbf{F}}{\partial\mathbf{u}'}e\mathbf{v}'\right) = \frac{\partial\mathbf{F}}{\partial u'_{ij}}ev'_{ij}$. If $\mathbf{F} = \mathbf{u}$, then $\delta\mathbf{F} = \delta\mathbf{u} = e\mathbf{v}$. If $\mathbf{F} = \mathbf{u}'$, then $\delta\mathbf{F} = \delta\mathbf{u}' = e\mathbf{v}'$. This suggests the notational convention $e\mathbf{v} \to \delta\mathbf{u}$ and $e\mathbf{v}' \to \delta\mathbf{u}'$, leading to the expression

$$\delta\mathbf{F} = \frac{\partial\mathbf{F}}{\partial\mathbf{u}}\delta\mathbf{u} + tr\left(\frac{\partial\mathbf{F}}{\partial\mathbf{u}'}\delta\mathbf{u}'\right) \tag{4.4}$$

4.2 PROPERTIES OF THE VARIATIONAL OPERATOR δ

The variational operator exhibits five important properties:

1. $\delta(\cdot)$ commutes with linear differential operators and integrals. For example, if S denotes a prescribed contour of integration

$$\int \delta(\,)\,\mathrm{d}S = \delta\left[\int(\,)\,\mathrm{d}S\right] \tag{4.5}$$

2. $\delta(f)$ vanishes when its argument f is prescribed.
3. $\delta(\cdot)$ satisfies the same operational rules as $\mathrm{d}(\cdot)$. For example, if the scalars q and r are both subject to variation,

$$\delta(qr) = q\delta(r) + \delta(q)r \tag{4.6}$$

4. If f is a prescribed function of (scalar) x and if $u(x)$ is subject to variation, then

$$\delta(fu) = f\delta u \tag{4.7}$$

5. Other than for (2), the variation is arbitrary. For example, for two vectors $\mathbf{v}$ and $\mathbf{w}$, $\mathbf{v}^T\mathrm{d}\mathbf{w} = 0$ simply implies that $\mathbf{v}$ and $\mathbf{w}$ are orthogonal to each other. However, $\mathbf{v}^T\delta\mathbf{w} = 0$ implies that $\mathbf{v} = \mathbf{0}$, since only the zero vector can be orthogonal to an arbitrary vector.

In the current monograph, attention will be restricted to variations with respect to position. There will be no consideration of variation with respect to time. Accordingly $\delta[f(t)u(x)] = f(t)\delta u(x)$.

4.3 EXAMPLE: VARIATIONAL EQUATION FOR A CANTILEVERED ELASTIC ROD

Determine the variational principle for the system in Figure 4.1 which depicts a rod of length L, cross-sectional area A, and elastic modulus E. At $x = 0$, it is built in while at $x = L$ the tensile force P is applied. Inertia is neglected. In terms of displacement u, stress S, and (linear) strain E, the governing equations are given by

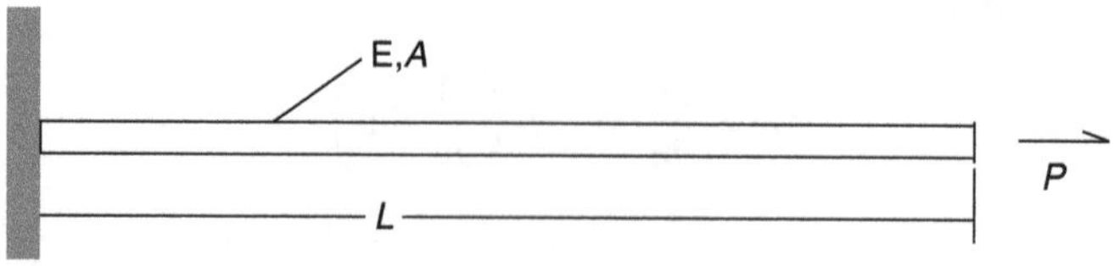

FIGURE 4.1 Rod under uniaxial tension.

$$\begin{aligned} &\text{Strain} - \text{displacement } \mathrm{E} = \frac{\mathrm{d}u}{\mathrm{d}x} \\ &\text{Stress} - \text{strain } S = \mathsf{E}\mathrm{E} \\ &\text{Equilibrium } \frac{\mathrm{d}S}{\mathrm{d}x} = 0 \end{aligned} \tag{4.8}$$

Combining the equations furnishes

$$\mathsf{E}A\frac{\mathrm{d}^2u}{\mathrm{d}x^2} = 0 \tag{4.9}$$

The steps below serve to derive a variational equation which is equivalent to the foregoing differential equation and endpoint conditions (*boundary conditions* and *constraints*).

Step 1: Multiply by the variation of the variable to be determined (u) and integrate over the domain.

$$\int_0^L \delta u \mathsf{E}A\frac{\mathrm{d}^2u}{\mathrm{d}x^2}A\,\mathrm{d}x = 0 \tag{4.10}$$

Differential equations to be satisfied at every point in the domain have now been replaced with an integral equation whose integrand includes an arbitrary function.

Step 2: Integrate by parts as needed to render the argument in the domain integral *quadratic*.

$$\int_0^L \left[\frac{\mathrm{d}}{\mathrm{d}x}\left[\delta u \mathsf{E}A\frac{\mathrm{d}u}{\mathrm{d}x}\right] - \left(\frac{\mathrm{d}\delta u}{\mathrm{d}x}\right)\mathsf{E}A\frac{\mathrm{d}u}{\mathrm{d}x}\right]\mathrm{d}x = 0 \tag{4.11}$$

To determine whether an integrand is quadratic for variational purposes, disregard the variational operator and derivatives with respect to time. If what is left is quadratic, the integrand is positive definite. We will see that in the finite element method terms with this property give rise to positive definite matrices.

Now the first term in Equation 4.11 is the integral of a derivative, so that

$$\int_0^L \left[\left(\frac{\mathrm{d}\delta u}{\mathrm{d}x}\right)\mathsf{E}A\frac{\mathrm{d}u}{\mathrm{d}x}\right]\mathrm{d}x = \delta u \mathsf{E}A\frac{\mathrm{d}u}{\mathrm{d}x}\Big|_0^L \tag{4.12}$$

Step 3: Identify the primary and secondary variables.

The primary variable is present in the endpoint terms (rhs) under the variational symbol and is u in the current example. The secondary variable conjugate to u is $\mathsf{E}A\frac{\mathrm{d}u}{\mathrm{d}x}$.

Step 4: Satisfy the constraints and boundary conditions.

At $x = 0$, u is prescribed and hence $\delta u = 0$. At $x = L$, the load $P = \mathsf{E}A\frac{du}{dx}$ is prescribed. Also, note that $\left(\frac{d\delta u}{dx}\right)\mathsf{E}A\frac{du}{dx} = \delta\left(\frac{1}{2}\mathsf{E}A\left(\frac{du}{dx}\right)^2\right)$. The right-hand term represents the reason why the left-hand term is deemed quadratic.

Step 5: Form the variational equation.

The foregoing equations and boundary conditions are consolidated into one integral equation as $\delta F = 0$ where

$$F = \int_0^L \frac{1}{2}\mathsf{E}A\left(\frac{du}{dx}\right)^2 dx - Pu(L) \tag{4.13}$$

4.4 HIGHER ORDER VARIATIONS

We now consider variations of order higher than unity. The jth variation of a vector-valued quantity $\mathbf{F}$ is defined by

$$\delta^j\mathbf{F} = e^j\left(\frac{d^j\boldsymbol{\Phi}}{de^j}\right)_{|e=0} \tag{4.14}$$

It follows that $\delta^2\mathbf{u} = \mathbf{0}$ and $\delta^2\mathbf{u}' = \mathbf{0}$. Now restricting $\mathbf{F}$ to being a scalar-valued function F and letting $\mathbf{x}$ reduce to x, we obtain

$$\delta^2 F = \{\delta\mathbf{u}^T \quad \delta\mathbf{u}'^T\}\mathbf{H}\begin{pmatrix}\delta\mathbf{u}\\ \delta\mathbf{u}'\end{pmatrix}, \quad \mathbf{H} = \begin{bmatrix}\left(\frac{\partial}{\partial\mathbf{u}}\right)^T\frac{\partial}{\partial\mathbf{u}}F & \left(\frac{\partial}{\partial\mathbf{u}}\right)^T\frac{\partial}{\partial\mathbf{u}'}F\\ \left(\frac{\partial}{\partial\mathbf{u}'}\right)^T\frac{\partial}{\partial\mathbf{u}}F & \left(\frac{\partial}{\partial\mathbf{u}'}\right)^T\frac{\partial}{\partial\mathbf{u}'}F\end{bmatrix} \tag{4.15}$$

and $\mathbf{H}$ is known as the Hessian matrix.

Now consider G given by

$$G = \int F(\mathbf{x},\mathbf{u}(\mathbf{x}),\mathbf{u}'(\mathbf{x}))\, dV + \int \mathbf{h}^T(\mathbf{x})\mathbf{u}(\mathbf{x})\, dS \tag{4.16}$$

in which V again denotes the volume of a domain and S denotes its surface area. In addition, $\mathbf{h}$ is a prescribed (known) function on S. G is called a *functional* since it generates a number for every function $\mathbf{u}(\mathbf{x})$. We first limit attention to a three-dimensional rectangular coordinate system and suppose that $\delta G = 0$, as in the *Principle of Stationary Potential Energy* in elasticity. Note that

$$\frac{\partial}{\partial\mathbf{x}}\left[\frac{\partial F}{\partial\mathbf{u}'}\delta\mathbf{u}\right] = tr\left(\frac{\partial F}{\partial\mathbf{u}'}\delta\mathbf{u}'\right) + \frac{\partial}{\partial\mathbf{x}}\left(\frac{\partial}{\partial\mathbf{u}'}F\right)\delta\mathbf{u} \tag{4.17}$$

The first and third terms in Equation 4.17 may be recognized as divergences of vectors. We now invoke the divergence theorem to obtain

$$
\begin{aligned}
0 &= \delta G \\
&= \int \left[\frac{\partial F}{\partial \mathbf{u}} \delta \mathbf{u} + tr\left(\frac{\partial F}{\partial \mathbf{u}'} \delta \mathbf{u}' \right) \right] \mathrm{d}V + \int \mathbf{h}^T(\mathbf{x}) \delta \mathbf{u}(\mathbf{x})\, \mathrm{d}S \\
&= \int \left[\left(\frac{\partial F}{\partial \mathbf{u}} - \frac{\partial}{\partial \mathbf{x}} \frac{\partial F}{\partial \mathbf{u}'} \right) \delta \mathbf{u} \right] \mathrm{d}V + \int \mathbf{n}^T \frac{\partial F}{\partial \mathbf{u}'} \delta \mathbf{u}\, \mathrm{d}S + \int \mathbf{h}^T(\mathbf{x}) \delta \mathbf{u}(\mathbf{x})\, \mathrm{d}S
\end{aligned} \tag{4.18}
$$

For suitable continuity properties of $\mathbf{u}$, arbitrariness of $\delta \mathbf{u}$ implies that $\delta G = 0$ is equivalent to the following *Euler equation, boundary conditions and constraints* (the latter two are not uniquely determined by the variational principle):

$$\frac{\partial F}{\partial \mathbf{u}} - \frac{\partial}{\partial \mathbf{x}} \frac{\partial F}{\partial \mathbf{u}'} = \mathbf{0}^T \tag{4.19}$$

$$
\begin{cases}
\mathbf{u}(\mathbf{x}) \text{ prescribed,} & x \text{ on } \mathrm{S}_1 \\
\mathbf{n}^T \dfrac{\partial F}{\partial \mathbf{u}'} + \mathbf{h}_1^T(\mathbf{x}) = \mathbf{0}^T, & x \text{ on } \mathrm{S} - \mathrm{S}_1
\end{cases}
$$

Let $\mathbf{D} > 0$ denote a constant positive definite and symmetric second order tensor and let $\boldsymbol{\pi}$ denote a vector which is a *nonlinear* function of a second vector $\mathbf{u}$ that is subject to variation. The function $F = \frac{1}{2} \boldsymbol{\pi}^T(\mathbf{u}) \mathbf{D} \boldsymbol{\pi}(\mathbf{u})$ satisfies

$$\delta F = \delta \mathbf{u}^T \left(\frac{\partial \boldsymbol{\pi}}{\partial \mathbf{u}} \right)^T \mathbf{D} \boldsymbol{\pi} \tag{4.20}$$

$$\delta^2 F = \delta \mathbf{u}^T \left(\frac{\partial \boldsymbol{\pi}}{\partial \mathbf{u}} \right)^T \mathbf{D} \frac{\partial \boldsymbol{\pi}}{\partial \mathbf{u}} \delta \mathbf{u} + \left[\delta \mathbf{u}^T \left(\frac{\partial}{\partial \mathbf{u}} \left(\frac{\partial \boldsymbol{\pi}}{\partial \mathbf{u}} \right)^T \right) \delta \mathbf{u} \right] \mathbf{D} \boldsymbol{\pi} \tag{4.21}$$

Since $\mathbf{D} > 0$, if only the first right-hand term were present, the expression would be quadratic with the implication that $\delta^2 F > 0$. However, for a general functional relation between $\boldsymbol{\pi}$ and $\mathbf{u}$ the second right-hand term is not quadratic. Accordingly, the specific vector $\mathbf{u}^*$ satisfying $\delta F = 0$ may correspond to a stationary point which does not constitute a minimum.

4.5 EXAMPLES

EXAMPLE 4.1

Directly apply variational calculus to F given by

$$F = \int_0^L \frac{1}{2} \mathrm{EA} \left(\frac{\mathrm{d}u}{\mathrm{d}x} \right)^2 \mathrm{d}x - Pu(L)$$

to verify that $\delta F = 0$ gives rise to the Euler equation

$$\mathsf{EA}\frac{d^2u}{dx^2} = 0$$

What endpoint conditions (not unique) are compatible with $\delta F = 0$?

Solution

Given that

$$F = \int_0^L \frac{1}{2}\mathsf{EA}\left(\frac{du}{dx}\right)^2 dx - Pu(L)$$

variational operations furnish

$$\begin{aligned} \delta F &= \int_0^L \delta\left[\frac{1}{2}\mathsf{EA}\left(\frac{du}{dx}\right)^2\right] dx - P\delta u(L) \\ &= \int_0^L \mathsf{EA}\delta u' u' \, dx - P\delta u(L) \end{aligned} \tag{4.22}$$

Now consider $\frac{d}{dx}(\delta u u') = \delta u' u' + \delta u u''$. After some manipulation

$$\int_0^L \mathsf{EA}\frac{d}{dx}(\delta u u')\, dx = \int_0^L \mathsf{EA}\delta u' u' \, dx + \int_0^L \mathsf{EA}\delta u u'' \, dx$$

in which

$$\int_0^L \mathsf{EA}\delta u' u' \, dx = [\mathsf{EA}\delta u u']_0^L - \int_0^L \mathsf{EA}\delta u u'' \, dx$$

On substituting Equation 4.23 into Equation 4.22, we have

$$\delta F = [\mathsf{EA}\delta u u']_0^L - \int_0^L \mathsf{EA}\delta u u'' \, dx - P\delta u(L)$$

Now $\delta F = 0$ is seen to imply that

$$\mathsf{EA}u'(L)\delta u(L) - \mathsf{EA}u'(0)\delta u(0) - \int_0^L \delta u \mathsf{EA}u'' \, dx - P\delta u(L) = 0$$

from which

$$[\mathsf{E}Au'(L) - P]\delta u(L) - \mathsf{E}Au'(0)\delta u(0) - \int_0^L \delta u \mathsf{E}Au''\,\mathrm{d}x = 0$$

The domain integral must vanish, and the endpoint expressions must vanish. Hence

$$\begin{aligned}
&\text{(i)} \quad [\mathsf{E}Au'(L) - P]\delta u(L) = 0 \\
&\text{(ii)} \quad \mathsf{E}Au'(0)\delta u(0) = 0 \\
&\text{(iii)} \quad \mathsf{E}Au'' = 0
\end{aligned}$$

Note that (iii) states the Euler equation implied by $\delta F = 0$.

Finally, since δu is arbitrary (i) and (ii) have the following implications:

$$\text{(i)} \quad \mathsf{E}Au'|_{x=L} = P \quad \text{so that} \quad \left(\frac{\mathrm{d}u}{\mathrm{d}x}\right)_{x=L} = P/\mathsf{E}A$$

$$\text{(ii)} \quad \mathsf{E}Au'|_{x=0} = 0 \quad \text{from which} \quad \left(\frac{\mathrm{d}u}{\mathrm{d}x}\right)_{x=0} = 0$$

EXAMPLE 4.2

The governing equation for an Euler–Bernoulli beam in Figure 4.2 below is

$$\mathsf{E}I\frac{\mathrm{d}^4 w}{\mathrm{d}x^4} = 0$$

in which w is the vertical displacement of the neutral (centroidal) axis. The shear force V and the bending moment M satisfy

$$M = -\mathsf{E}I\frac{\mathrm{d}^2 w}{\mathrm{d}x^2}, \quad V = -\mathsf{E}I\frac{\mathrm{d}^3 w}{\mathrm{d}x^3}$$

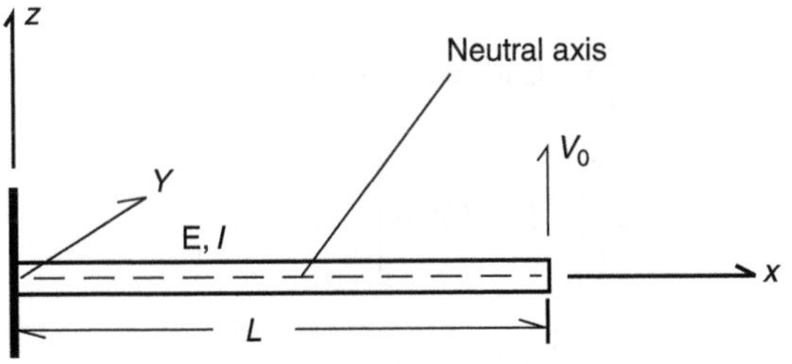

FIGURE 4.2 Cantilevered beam.

Using integration by parts twice, obtain the function F such that $\delta F = 0$ is equivalent to the foregoing differential equation together with the boundary conditions for a cantilevered beam of length L:

$$w(0) = w'(0) = 0, \quad M(L) = 0, \quad V(L) = V_0$$

Solution

On multiplying $\mathsf{E}Iw^{iv} = 0$ by δw and integrating over the domain, $0 \le x \le L$ we obtain

$$\int_0^L \delta w\left[\mathsf{E}Iw^{iv}\right] \mathrm{d}x = 0$$

Also

$$\frac{\mathrm{d}}{\mathrm{d}x}[\delta w' \mathsf{E}Iw''] = \delta w' \mathsf{E}Iw''' + \delta w'' \mathsf{E}Iw''$$

Combining the last two equations furnishes

$$\frac{\mathrm{d}}{\mathrm{d}x}[\delta w \mathsf{E}Iw'''] - \frac{\mathrm{d}}{\mathrm{d}x}[\delta w' \mathsf{E}Iw''] = -\delta w'' \mathsf{E}Iw''$$

Continuing,

$$\begin{aligned}\int_0^L \delta w'' \mathsf{E}Iw'' \mathrm{d}x &= -\int_0^L \frac{\mathrm{d}}{\mathrm{d}x}[\delta w \mathsf{E}Iw''']\,\mathrm{d}x + \int_0^L \frac{\mathrm{d}}{\mathrm{d}x}[\delta w' \mathsf{E}Iw'']\,\mathrm{d}x \\ &= -[\delta w \mathsf{E}Iw''']_0^L + [\delta w' \mathsf{E}Iw'']_0^L\end{aligned}$$

Note that $w(0) = w'(0) = 0$. This implies that $\delta w(0) = \delta w'(0) = 0$. Also evident are the conditions

$$-\mathsf{E}Iw''(L) = M(L) = 0 \quad \text{and} \quad -\mathsf{E}Iw'''(L) = V(L) = V_0$$

Some manipulation serves to establish that

$$\int_0^L \delta w'' \mathsf{E}Iw''\,\mathrm{d}x = \delta w(L) V_0$$

$$\delta\left[\int_0^L \frac{1}{2}\mathsf{E}I(w'')^2\,\mathrm{d}x\right] = V_0 \delta w(L)$$

$$\delta\left[\int_0^L \frac{1}{2}\mathsf{E}I(w'')^2\,\mathrm{d}x - V_0 w(L)\right] = 0$$

Finally, since $\delta F = 0$ we conclude that

$$F = \int_0^L \frac{1}{2} EI(w'')^2 \, dx - V_0 w(L)$$

EXAMPLE 4.3

Equation combining rod and beam behavior

What are the primary variables and the corresponding secondary variables in the following equation?

$$B\frac{\partial^4 w}{\partial x^4} - A\frac{\partial^2 w}{\partial x^2} + Cw + D\frac{\partial^2 w}{\partial t^2} = 0$$

in which $0 \leq x \leq L$.

SOLUTION

The first step is to write

$$\int_0^L \delta w \left[B\frac{\partial^4 w}{\partial x^4} - A\frac{\partial^2 w}{\partial x^2} + Cw + D\frac{\partial^2 w}{\partial t^2} \right] dV = 0$$

The third and fourth terms in the integrand, namely $\int_0^L \delta w C w \, dV$ and $\int_0^L \delta w D \frac{\partial^2 w}{\partial t^2} \, dV$ are already quadratic. The second and first terms give, respectively,

$$\int_0^L \delta w \left[-A\frac{\partial^2 w}{\partial x^2} \right] dV = -\delta w (Aw')|_0^L + \int_0^L \delta w' A w' \, dV$$

$$\int_0^L \delta w \left[B\frac{\partial^4 w}{\partial x^4} \right] dV = -\delta w(-Bw''')|_0^L - (-\delta w')(-Bw'')|_0^L + \int_0^L \delta w'' A w'' \, dV$$

Combining the last two equations reveals the primary variables to be $w, -w'$. The corresponding secondary variables are $-Aw' + Bw'''$ and Bw''.

EXAMPLE 4.4

Two-dimensional heat conduction

Obtain the variational equations corresponding to the following equation in two dimensions. What are the primary variables and the conjugate secondary variables?

The first equation is for unsteady heat conduction. The second equation has the same form as the wave equation and the biharmonic equation in classical elasticity.

$$k\nabla \cdot \nabla T = \rho c_e \frac{\partial T}{\partial t}$$

$k =$ thermal conductivity
$\rho =$ mass density
$c_e =$ specific heat at constant strain

Solution

The first step is to write

$$\int_V \delta T \left[k\nabla \cdot \nabla T - \rho c_e \frac{\partial T}{\partial t} \right] \mathrm{d}V = 0$$

Applying integration by parts to the first term furnishes

$$\int_V \delta T[k\nabla \cdot \nabla T]\,\mathrm{d}V = \int_V \nabla \cdot (\delta T[k\nabla T])\,\mathrm{d}V - \int_V \delta \nabla T \cdot [k\nabla T]\,\mathrm{d}V$$

Using the divergence theorem on the first right-hand term furnishes

$$\int_V \nabla \cdot (\delta T[k\nabla T])\,\mathrm{d}V = \int_S \delta T \mathbf{n} \cdot k\nabla T\,\mathrm{d}S$$

The primary variable is seen to be T owing to its presence under the variational operator in the surface integral. The corresponding secondary variable is $\mathbf{n} \cdot k\nabla T$, which equals the negative of the normal projection of the heat flux vector $\mathbf{q} = -k\nabla T$.

EXAMPLE 4.5

Navier equation and plate equation in elasticity

Find variational forms of the following equations. What are the primary and secondary variables?

$$(\mu + \lambda)\frac{\partial}{\partial x_i}\frac{\partial u_j}{\partial x_j} + \mu \frac{\partial^2 u_i}{\partial x_j^2} = \rho u_i \quad \text{Navier's equation}$$

$$\frac{\partial^2}{\partial x_i^2}\frac{\partial^2 w}{\partial x_j^2} = 0 \quad \text{Elastic Plate equation (static)}$$

Solution

Navier's equation

The first step is to write

$$\int \delta u_i\left((\mu+\lambda)\frac{\partial}{\partial x_i}\frac{\partial u_j}{\partial x_j}+\mu\frac{\partial^2 u_i}{\partial x_j^2}-\rho\ddot{u}_i\right)\mathrm{d}V=0$$

First note that the third term in the integrand, $-\delta u_i\rho\ddot{u}_i$, is already quadratic. Next consider the middle term in the integrand:

$$\frac{\partial}{\partial x_j}\left(\delta u_i\mu\frac{\partial u_i}{\partial x_j}\right)=\frac{\partial\delta u_i}{\partial x_j}\mu\frac{\partial u_i}{\partial x_j}+\delta u_i\mu\frac{\partial^2 u_i}{\partial x_j^2}$$

and

$$\delta u_i\mu\frac{\partial^2 u_i}{\partial x_j^2}=\frac{\partial}{\partial x_j}\left(\delta u_i\mu\frac{\partial u_i}{\partial x_j}\right)-\frac{\partial\delta u_i}{\partial x_j}\mu\frac{\partial u_i}{\partial x_j}$$

Consequently, using the divergence theorem, the integral of the middle term gives

$$\int \delta u_i\mu\frac{\partial^2 u_i}{\partial x_j^2}\mathrm{d}V=\int \delta u_i n_j\mu\frac{\partial u_i}{\partial x_j}\mathrm{d}S-\int\frac{\partial\delta u_i}{\partial x_j}\mu\frac{\partial u_i}{\partial x_j}\mathrm{d}V$$

The first term in the integrand may be rewritten as

$$\frac{\partial}{\partial x_i}\left(\delta u_i(\mu+\lambda)\frac{\partial u_j}{\partial x_j}\right)=\frac{\partial\delta u_i}{\partial x_i}(\mu+\lambda)\frac{\partial u_j}{\partial x_j}+\delta u_i\left((\mu+\lambda)\frac{\partial}{\partial x_i}\frac{\partial u_j}{\partial x_j}\right)$$

with the result that

$$\delta u_i\left((\mu+\lambda)\frac{\partial}{\partial x_i}\frac{\partial u_j}{\partial x_j}\right)=\frac{\partial}{\partial x_i}\left(\delta u_i(\mu+\lambda)\frac{\partial u_j}{\partial x_j}\right)-\frac{\partial\delta u_i}{\partial x_i}(\mu+\lambda)\frac{\partial u_j}{\partial x_j}$$

Upon integration and application of the divergence theorem,

$$\int_V \delta u_i\left((\mu+\lambda)\frac{\partial}{\partial x_i}\frac{\partial u_j}{\partial x_j}\right)\mathrm{d}V=\int_V n_i\delta u_i(\mu+\lambda)\frac{\partial u_j}{\partial x_j}\mathrm{d}S-\int_V\frac{\partial\delta u_i}{\partial x_i}(\mu+\lambda)\frac{\partial u_j}{\partial x_j}\mathrm{d}V$$

Collecting the surface terms gives

$$\int_S \delta u_i\left(n_i(\mu+\lambda)\frac{\partial u_j}{\partial x_j}+n_j\mu\frac{\partial u_i}{\partial x_j}\right)\mathrm{d}S\rightarrow\int_S \delta\mathbf{u}^T\left((\mu+\lambda)(\nabla\cdot\mathbf{u})\mathbf{n}+\mu\frac{\mathbf{du}}{\mathbf{dx}}\cdot\mathbf{n}\right)\mathrm{d}S$$

We now identify $\mathbf{u}$ as the primary variable and $\left((\mu+\lambda)(\nabla\cdot\mathbf{u})\mathbf{n}+\mu\frac{\mathbf{du}}{\mathbf{dx}}\cdot\mathbf{n}\right)$ as the conjugate secondary variable.

We now examine the domain terms. Note that

$$\int\left(\frac{\partial\delta u_i}{\partial x_j}\frac{\partial u_i}{\partial x_j}+\frac{\partial\delta u_i}{\partial x_i}(\mu+\lambda)\frac{\partial u_j}{\partial x_j}+\delta u_i\rho\ddot{u}_i\right)\mathrm{d}V$$

$$=\int\left(tr\left(\left(\frac{\mathbf{d}\delta\mathbf{u}}{\mathbf{dx}}\right)^T\mu\frac{\mathbf{du}}{\mathbf{dx}}\right)+(\mu+\lambda)(\nabla\cdot\delta\mathbf{u})(\nabla\cdot\mathbf{u})+\delta\mathbf{u}\rho\ddot{\mathbf{u}}\right)\mathrm{d}V$$

All three terms in the integrand are quadratic.

Finally the variational statement of Navier's equation is recapitulated as

$$\int\left(tr\left(\left(\frac{\mathbf{d}\delta\mathbf{u}}{\mathbf{dx}}\right)^T\mu\frac{\mathbf{du}}{\mathbf{dx}}\right)+(\mu+\lambda)(\nabla\cdot\delta\mathbf{u})(\nabla\cdot\mathbf{u})+\delta\mathbf{u}^T\rho\ddot{\mathbf{u}}\right)\mathrm{d}V$$

$$=\int_S\delta\mathbf{u}^T\left((\mu+\lambda)(\nabla\cdot\mathbf{u})\mathbf{n}+\mu\frac{\mathbf{du}}{\mathbf{dx}}\cdot\mathbf{n}\right)\mathrm{d}S$$

Solution

Plate equation

The first step, again, is

$$\int_V\delta w\frac{\partial^2}{\partial x_i^2}\frac{\partial^2 w}{\partial x_j^2}\mathrm{d}V=0$$

Integration by parts must be performed twice. The first step is

$$\frac{\partial}{\partial x_i}\left(\delta w\frac{\partial}{\partial x_i}\frac{\partial^2 w}{\partial x_j^2}\right)=\delta w\frac{\partial^2}{\partial x_i^2}\frac{\partial^2 w}{\partial x_j^2}+\frac{\partial\delta w}{\partial x_i}\frac{\partial}{\partial x_i}\frac{\partial^2 w}{\partial x_j^2}\tag{4.23}$$

and the second step is

$$\frac{\partial}{\partial x_j}\left(\frac{\partial\delta w}{\partial x_i}\frac{\partial}{\partial x_i}\frac{\partial w}{\partial x_j}\right)=\left(\frac{\partial}{\partial x_j}\frac{\partial\delta w}{\partial x_i}\right)\left(\frac{\partial}{\partial x_j}\frac{\partial w}{\partial x_i}\right)+\frac{\partial\delta w}{\partial x_i}\frac{\partial}{\partial x_i}\frac{\partial^2 w}{\partial x_j^2}\tag{4.24}$$

Subtracting Equation 4.24 from Equation 4.23 gives

$$\frac{\partial}{\partial x_i}\left(\delta w\frac{\partial}{\partial x_i}\frac{\partial^2 w}{\partial x_j^2}\right)-\frac{\partial}{\partial x_j}\left(\frac{\partial\delta w}{\partial x_i}\frac{\partial}{\partial x_i}\frac{\partial w}{\partial x_j}\right)=\delta w\frac{\partial^2}{\partial x_i^2}\frac{\partial^2 w}{\partial x_j^2}-\left(\frac{\partial}{\partial x_j}\frac{\partial\delta w}{\partial x_i}\right)\left(\frac{\partial}{\partial x_j}\frac{\partial w}{\partial x_i}\right)$$

which becomes upon integration

$$\int_V\left(\frac{\partial}{\partial x_j}\frac{\partial\delta w}{\partial x_i}\right)\left(\frac{\partial}{\partial x_j}\frac{\partial w}{\partial x_i}\right)\mathrm{d}V=\int_S\left(\delta w\left(-n_i\frac{\partial}{\partial x_i}\frac{\partial^2 w}{\partial x_j^2}\right)+\left(-\frac{\partial\delta w}{\partial x_i}\right)\left(-n_j\frac{\partial}{\partial x_j}\frac{\partial w}{\partial x_i}\right)\right)\mathrm{d}S$$

Examination of the domain terms reveals that

$$\int_V \left(\frac{\partial}{\partial x_j}\frac{\partial \delta w}{\partial x_i}\right)\left(\frac{\partial}{\partial x_j}\frac{\partial w}{\partial x_i}\right) \mathrm{d}V = \int_V tr\left(\left(\left(\frac{\mathrm{d}}{\mathbf{dx}}\right)^T \frac{\mathrm{d}\delta w}{\mathbf{dx}}\right)^T \left(\left(\frac{\mathrm{d}}{\mathbf{dx}}\right)^T \frac{\mathrm{d}w}{\mathbf{dx}}\right)\right) \mathrm{d}V$$

and the surface terms are rewritten as

$$\int_S \left(\delta w\left(-n_i \frac{\partial}{\partial x_i}\frac{\partial^2 w}{\partial x_j^2}\right) + \left(-\frac{\partial \delta w}{\partial x_i}\right)\left(-n_j \frac{\partial}{\partial x_j}\frac{\partial w}{\partial x_i}\right)\right) \mathrm{d}S$$

$$= \int_S \left(\delta w\left(-(\mathbf{n}\cdot\nabla)\nabla^2 w\right) + (-\nabla\delta w)(-(\mathbf{n}\cdot\nabla)\nabla w)\right) \mathrm{d}S$$

The primary variables are now identified as $w, -\nabla w$, and the conjugate secondary variables are respectively $(-(\mathbf{n}\cdot\nabla)\nabla^2 w)$ and $(-(\mathbf{n}\cdot\nabla)\nabla w)$.

EXAMPLE 4.6

Consider a one-dimensional system described by a sixth-order differential equation:

$$Q\frac{\mathrm{d}^6 q}{\mathrm{d}x^6} = 0, \quad Q \text{ a constant.}$$

Consider an element from x_e to x_{e+1}. Using the natural coordinate $\xi = -1$ when $x = x_e = +1$ when $x = x_{e+1}$, for an interpolation model with the minimum order that is meaningful, obtain expressions for $\boldsymbol{\varphi}(\xi)$, $\boldsymbol{\Phi}$, and $\boldsymbol{\gamma}$, serving to express $\mathbf{q}$ as

$$\mathbf{q}(\xi) = \boldsymbol{\varphi}^T(\xi)\boldsymbol{\Phi}\boldsymbol{\gamma}$$

Solution

Since the given equation is sixth order, the lowest order interpolation model consistent with six integration constants is a fifth-order polynomial, in the form (physical coordinates),

$$q(x,t) = \boldsymbol{\varphi}^T(x)\boldsymbol{\Phi}\boldsymbol{\gamma}(t)$$

in which

$$\boldsymbol{\gamma}(t) = \begin{pmatrix} q_e(t) \\ -q_e'(t) \\ q_e''(t) \\ q_{e+1}(t) \\ -q_{e+1}'(t) \\ q_{e+1}''(t) \end{pmatrix}, \quad \boldsymbol{\varphi}^T(x) = \begin{pmatrix} 1 & x & x^2 & x^3 & x^4 & x^5 \end{pmatrix}$$

We seek to identify $\boldsymbol{\Phi}$ in terms of the nodal values of $\mathbf{q}$.

Letting $q_e = q(x_e)$, $q_{e+1} = q(x_{e+1})$, $q'_e = q'(x_e)\ldots$ furnishes the following in the matrix–vector notation:

$$\begin{pmatrix} q_e(t) \\ -q'_e(t) \\ q''_e(t) \\ q_{e+1}(t) \\ -q'_{e+1}(t) \\ q''_{e+1}(t) \end{pmatrix} = \begin{bmatrix} 1 & x_e & x_e^2 & x_e^3 & x_e^4 & x_e^5 \\ 0 & -1 & -2x_e & -3x_e^2 & -4x_e^3 & -5x_e^4 \\ 0 & 0 & 2 & 6x_e & 12x_e^2 & 20x_e^3 \\ 1 & x_{e+1} & x_{e+1}^2 & x_{e+1}^3 & x_{e+1}^4 & x_{e+1}^5 \\ 0 & -1 & -2x_{e+1} & -3x_{e+1}^2 & -4x_{e+1}^3 & -5x_{e+1}^4 \\ 0 & 0 & 2 & 6x_{e+1} & 12x_{e+1}^2 & 20x_{e+1}^3 \end{bmatrix} \mathbf{\Phi}\boldsymbol{\gamma}(t)$$

Hence

$$\mathbf{\Phi} = \begin{bmatrix} 1 & x_e & x_e^2 & x_e^3 & x_e^4 & x_e^5 \\ 0 & -1 & -2x_e & -3x_e^2 & -4x_e^3 & -5x_e^4 \\ 0 & 0 & 2 & 6x_e & 12x_e^2 & 20x_e^3 \\ 1 & x_{e+1} & x_{e+1}^2 & x_{e+1}^3 & x_{e+1}^4 & x_{e+1}^5 \\ 0 & -1 & -2x_{e+1} & -3x_{e+1}^2 & -4x_{e+1}^3 & -5x_{e+1}^4 \\ 0 & 0 & 2 & 6x_{e+1} & 12x_{e+1}^2 & 20x_{e+1}^3 \end{bmatrix}^{-1}$$

Now, on converting this to the natural coordinate $\xi = -1$ when $x = x_e, = +1$ when $x = x_{e+1}$, we have

$$a = \frac{2}{l_e}$$
$$\frac{\partial}{\partial x} = a\frac{\partial}{\partial \xi}$$

from which

$$\frac{\partial^6}{\partial x^6} = a^6\frac{\partial^6}{\partial \xi^6} = \frac{64}{l_e^2}\frac{\partial^6}{\partial \xi^6}$$

Hence, the governing equation becomes

$$\frac{64}{l_e^2}\frac{\mathrm{d}^6 q}{\mathrm{d}\xi^6} = 0$$

The interpolation model now becomes

$$q(\xi(x),t) = \boldsymbol{\varphi}^T(\xi(x))\mathbf{\Phi}\boldsymbol{\gamma}(t)$$

and

$$\boldsymbol{\varphi}^T(\xi(x)) = \begin{pmatrix} 1 & \xi(x) & \xi(x)^2 & \xi(x)^3 & \xi(x)^4 & \xi(x)^5 \end{pmatrix}$$
$$\boldsymbol{\gamma}^T(t) = \begin{pmatrix} q_e(t) & -q'_e(t) & q''_e(t) & q_{e+1}(t) & -q'_{e+1}(t) & q''_{e+1}(t) \end{pmatrix}$$

Substituting $\xi(x_e) = -1$ and $\xi(x_e) = -1$ yields the desired result at

$$\Phi = \begin{bmatrix} 1 & -1 & 1 & -1 & 1 & -1 \\ 0 & -1 & 2 & -3 & 4 & -5 \\ 0 & 0 & 2 & -6 & 12 & -20 \\ 1 & 1 & 1 & 1 & 1 & 1 \\ 0 & -1 & -2 & -3 & -4 & -5 \\ 0 & 0 & 2 & 6 & 12 & 20 \end{bmatrix}^{-1}$$

5 Fundamental Notions of Linear Solid Mechanics

In this chapter we provide a condensed review of basic solid mechanics typically presented in upper-level undergraduate courses. Our emphasis is on formal relations as well as examples.

5.1 DISPLACEMENT VECTOR

Figure 5.1 depicts a body in both its reference and current configurations. The former is considered to be the *undeformed configuration*, and the latter is called the *deformed configuration*—it reflects the deformation induced by the forces applied to the undeformed configuration. Consider a material particle occupying "point" P in the undeformed position and point Q in the deformed position. In the undeformed configuration its position determined the *undeformed position vector*, given in rectilinear coordinates as

$$\mathbf{X} = X_1\mathbf{i} + X_2\mathbf{j} + X_3\mathbf{k} \tag{5.1}$$

while the same particle in the deformed configuration gives rise to the *deformed position vector*

$$\mathbf{x} = x_1\mathbf{i} + x_2\mathbf{j} + x_3\mathbf{k} \tag{5.2}$$

referred to the same base vectors **i**, **j**, **k**. It is assumed that x_i are functions of X_j and time t.

The vector difference between **x** and **X** is called the *displacement vector*. In rectilinear coordinates

$$\mathbf{u} = \mathbf{x} - \mathbf{X} \tag{5.3}$$

and in alternate notation

$$\mathbf{u} = \begin{Bmatrix} u \\ v \\ w \end{Bmatrix} = \begin{Bmatrix} x - X \\ y - Y \\ z - Z \end{Bmatrix} \tag{5.4}$$

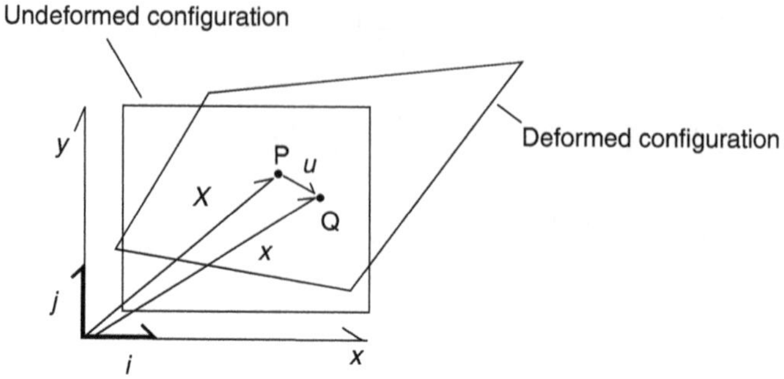

FIGURE 5.1 Illustration of the displacement vector.

5.2 LINEAR STRAIN AND ROTATION TENSORS

Suppose there is a slight change in the position of Q. From the chain rule of calculus, in the $X-Y$ plane the displacements $u(X,Y)$ and $v(X,Y)$ satisfy $\mathrm{d}u = \frac{\mathrm{d}u}{\mathrm{d}X}\mathrm{d}X + \frac{\mathrm{d}u}{\mathrm{d}Y}\mathrm{d}Y$ and $\mathrm{d}v = \frac{\mathrm{d}v}{\mathrm{d}X}\mathrm{d}X + \frac{\mathrm{d}v}{\mathrm{d}Y}\mathrm{d}Y$. Elementary calculus furnishes

$$\begin{Bmatrix} \mathrm{d}u \\ \mathrm{d}v \end{Bmatrix} = \begin{bmatrix} \frac{\mathrm{d}u}{\mathrm{d}X} & \frac{\mathrm{d}u}{\mathrm{d}Y} \\ \frac{\mathrm{d}v}{\mathrm{d}X} & \frac{\mathrm{d}v}{\mathrm{d}Y} \end{bmatrix} \begin{Bmatrix} \mathrm{d}X \\ \mathrm{d}Y \end{Bmatrix}$$
$$= \left[\begin{bmatrix} \mathrm{E}_{xx} & \mathrm{E}_{yx} \\ \mathrm{E}_{xy} & \mathrm{E}_{yy} \end{bmatrix} + \begin{bmatrix} \omega_{xx} & \omega_{yx} \\ \omega_{xy} & \omega_{yy} \end{bmatrix} \right] \begin{Bmatrix} \mathrm{d}X \\ \mathrm{d}Y \end{Bmatrix} \tag{5.5}$$

in which

$$\mathrm{E}_{xx} = \frac{1}{2}\left(\frac{\mathrm{d}u}{\mathrm{d}X} + \frac{\mathrm{d}u}{\mathrm{d}X}\right), \quad \mathrm{E}_{yx} = \mathrm{E}_{xy} = \frac{1}{2}\left(\frac{\mathrm{d}v}{\mathrm{d}X} + \frac{\mathrm{d}u}{\mathrm{d}Y}\right), \quad \mathrm{E}_{yy} = \frac{1}{2}\left(\frac{\mathrm{d}v}{\mathrm{d}Y} + \frac{\mathrm{d}v}{\mathrm{d}Y}\right)$$
$$\omega_{xx} = 0, \quad \omega_{xy} = -\omega_{yx} = \frac{1}{2}\left(\frac{\mathrm{d}v}{\mathrm{d}X} - \frac{\mathrm{d}u}{\mathrm{d}Y}\right), \quad \omega_{yy} = 0$$

Of course E_{xx}, E_{yy}, and E_{xy} denote the linear strains and will be seen to comprise a measure of local stretching. Further, ω_{xy} represents rotation.

Suppose, for example, that $\mathbf{x} = \mathbf{Q}(t)\mathbf{X} + \mathbf{b}(t)$ in which $\mathbf{Q}(t)$ is an orthogonal tensor independent of $\mathbf{X}$ and the vector $\mathbf{b}(t)$ is likewise independent of $\mathbf{X}$. $\mathbf{Q}(t)$ is said to represent rigid body rotation and $\mathbf{b}(t)$ represents rigid body translation. To first order in displacements and their derivatives, the linear strains are unaffected by the rigid body component of the deformation. However, the rotation is strongly affected by rigid body rotation.

In rectilinear coordinates, the linear strains and rotations are given in three dimensions by

$$E_{ij} = \frac{1}{2}\left(\frac{\partial u_i}{\partial X_j} + \frac{\partial u_j}{\partial X_i}\right) \rightarrow \begin{cases} E_{XX} = \dfrac{\partial u}{\partial X} & \text{normal strain} \\ E_{YY} = \dfrac{\partial v}{\partial Y} & \text{normal strain} \\ E_{ZZ} = \dfrac{\partial w}{\partial Z} & \text{normal strain} \\ E_{XY} = \dfrac{1}{2}\left(\dfrac{\partial u}{\partial Y} + \dfrac{\partial v}{\partial X}\right) & \text{shear strain} \\ E_{YZ} = \dfrac{1}{2}\left(\dfrac{\partial y}{\partial Z} + \dfrac{\partial w}{\partial Y}\right) & \text{shear strain} \\ E_{ZX} = \dfrac{1}{2}\left(\dfrac{\partial w}{\partial X} + \dfrac{\partial u}{\partial Z}\right) & \text{shear strain} \end{cases} \tag{5.6}$$

$$\omega_{ij} = \frac{1}{2}\left(\frac{\partial u_i}{\partial X_j} - \frac{\partial u_j}{\partial X_i}\right) \rightarrow \begin{cases} \omega_{XX} = \omega_{YY} = \omega_{ZZ} = 0 \\ -\omega_{YX} = \omega_{XY} = \dfrac{1}{2}\left(\dfrac{\partial u}{\partial Y} - \dfrac{\partial v}{\partial X}\right) \\ -\omega_{ZY} = \omega_{YZ} = \dfrac{1}{2}\left(\dfrac{\partial v}{\partial Z} - \dfrac{\partial w}{\partial Y}\right) \\ -\omega_{XZ} = \omega_{ZX} = \dfrac{1}{2}\left(\dfrac{\partial w}{\partial X} - \dfrac{\partial u}{\partial Z}\right) \end{cases} \tag{5.7}$$

In alternate notation we may write in general that

$$\begin{aligned} \mathrm{d}\mathbf{u} &= \frac{\mathrm{d}\mathbf{u}}{\mathrm{d}\mathbf{X}}\mathrm{d}\mathbf{X} \\ &= [\mathbf{E}_L + \boldsymbol{\omega}]\mathrm{d}\mathbf{X} \end{aligned} \tag{5.8}$$

in which

$$\mathbf{E}_L = \frac{1}{2}\left(\frac{\mathrm{d}\mathbf{u}}{\mathrm{d}\mathbf{X}} + \left(\frac{\mathrm{d}\mathbf{u}}{\mathrm{d}\mathbf{X}}\right)^T\right), \quad \boldsymbol{\omega} = \frac{1}{2}\left(\frac{\mathrm{d}\mathbf{u}}{\mathrm{d}\mathbf{X}} - \left(\frac{\mathrm{d}\mathbf{u}}{\mathrm{d}\mathbf{X}}\right)^T\right)$$

The counterpart of $\mathbf{E}_L$ and $\boldsymbol{\omega}$ in tensor-indicial notation is

$$E_{ij} = \frac{1}{2}\left(\frac{\partial u_i}{\partial X_j} + \frac{\partial u_j}{\partial X_i}\right), \quad \omega_{ij} = \frac{1}{2}\left(\frac{\partial u_i}{\partial X_j} + \frac{\partial u_j}{\partial X_i}\right) \tag{5.9}$$

Since $\mathrm{d}\mathbf{u}$ is a 3×1 vector, it follows from Chapter 3 that $\frac{\mathrm{d}\mathbf{u}}{\mathrm{d}\mathbf{X}}$ is a 3×3 tensor. Likewise $\left(\frac{\mathrm{d}\mathbf{u}}{\mathrm{d}\mathbf{X}}\right)^T$, $\mathbf{E}_L$ and $\boldsymbol{\omega}$ are 3×3 tensors.

5.3 EXAMPLES OF LINEAR STRAIN AND ROTATION TENSORS

EXAMPLE 5.1

The plate shown is initially square and 6 cm on a side (Figure 5.2a). It is deformed as shown in Figure 5.2b. Find the displacement and the strain fields.

SOLUTION

Assume that the deformed coordinates can be expressed in terms of the undeformed coordinates using the expressions

$$x = a + bX + cY + dXY \tag{5.10}$$

$$y = e + fX + gY + hXY \tag{5.11}$$

Our goal is to detect the coefficients a through h by fitting the coordinates at the four corners of the element.

Along the bottom face of the block $Y = 0$ from which $x = a + bX$ and $y = e + fX$. It follows that $y = \frac{f}{b}x - e + \frac{a}{b}$, implying that the lower face remains a line after deformation. On the right face $X = 1$ so that $x = a + b + (c + d)Y$ and $y = e + f + (g + h)Y$. Now Y may be eliminated from these expressions to yield a linear relation between the deformed coordinates x and y.

Again the side of the element remains a line in the deformed cofiguration. Similar results are are immediate for the upper face and the left-hand face.

Clearly the assumed relations (5.10, 5.11) are suitable if the sides of the square remain straight after deformation. Eight nodal relations are used to determine the eight coefficients in (5.10, 5.11).

Lower left node:

For $(X,Y) = (0,0)$, $(x,y) = (0,0)$ and hence $a = 1$, $e = 1$.

Lower right node:

For $(X,Y) = (6,0)$, $x = 7.1$ and $y = 1.1$. Hence $b = 6.1/6, f = 0.1/6$

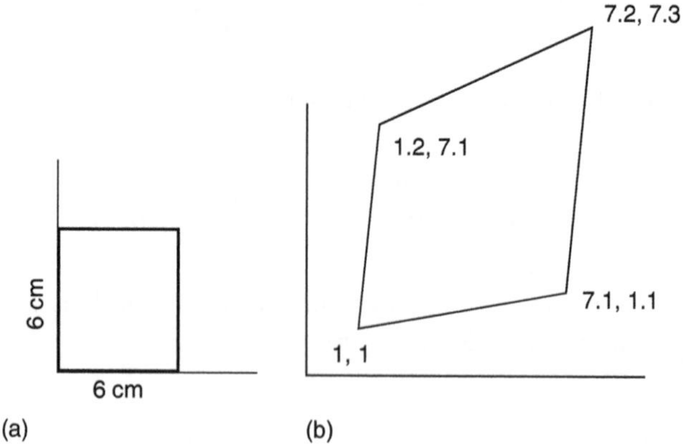

FIGURE 5.2 Example of strain and rotation.

Upper left:

For $(X,Y)=(0,6)$, $x=1.2$ and $y=7.1$, giving $c=1.2/6$ and $g=6.1/6$.

Apex:

For $(X,Y)=(6,6)$, $x=7.2$ and $y=7.3$. Accordingly,

$$
\begin{aligned}
7.2 &= 1+b*6+c*6+d*36 \quad \text{from which } d=-1.3/36 \\
7.3 &= 1+f*6+g*6+h*36 \quad \text{from which } h=+0.1/36
\end{aligned}
$$

The displacements are found as

$$
\begin{aligned}
u &= x-X = 1+(-0.1/6)X+(1.2/6)Y+(-1.3/36)XY \\
v &= y-Y = 1+(0.1/6)X+(0.1/6)Y+(0.1/36)XY
\end{aligned}
$$

Upon applying the formulae for the strains and rotations we obtain

$$
\begin{aligned}
\mathrm{E}_{xx} &= \frac{\partial u}{\partial X} = 0.1/6-1.3Y/36 \\
\mathrm{E}_{yy} &= \frac{\partial v}{\partial Y} = 0.1/6+0.1X/36 \\
\mathrm{E}_{xy} &= \frac{1}{2}\left(\frac{\partial u}{\partial Y}+\frac{\partial v}{\partial X}\right) = \frac{1}{2}(1.3/6-1.3X/6+0.1Y/36) \\
\omega_{xy} &= \frac{1}{2}\left(\frac{\partial u}{\partial Y}-\frac{\partial v}{\partial X}\right) = \frac{1}{2}(1.1/6-1.3X/36-0.1Y/36)
\end{aligned}
$$

EXAMPLE 5.2

A rectangular block in Figure 5.3 is rotated through θ degrees. Find the displacements and strains and rotations. What happens if the rotation angle is very small?

Solution

The vector **R** rotates into the vector **r**. Both have length R. Thus, coordinates of the endpoint of **R** are $X=R\cos\phi$, $Y=R\sin\phi$. The deformed coordinates, displacement, strains, and rotation are now found to be

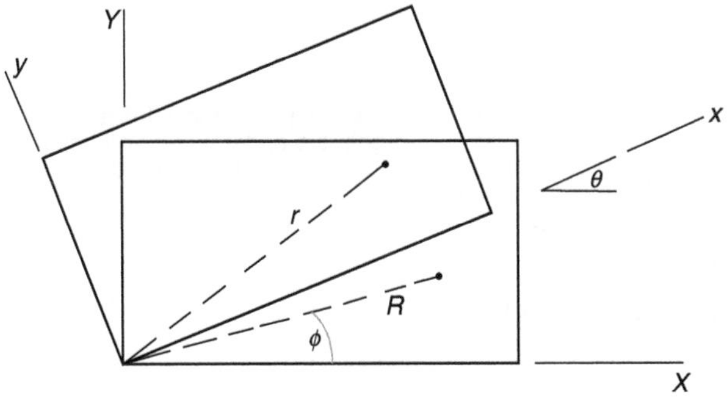

FIGURE 5.3 Strain and rotation in body experiencing rigid body motion.

$$
\begin{aligned}
x &= R\cos(\phi+\theta) & y &= R\sin(\phi+\theta)\\
&= R[\cos\phi\cos\theta - \sin\phi\sin\theta] & &= R[\sin\phi\cos\theta + \cos\phi\sin\theta]\\
&= X\cos\theta - Y\sin\theta & &= X\sin\theta + Y\cos\theta\\
u &= (\cos\theta - 1)X - Y\sin\theta & v &= X\sin\theta + (\cos\theta - 1)Y\\
E_{XX} &= \cos\theta - 1 & E_{YY} &= \cos\theta - 1\\
E_{XY} &= 0 & \omega_{XY} &= -\omega_{YX} = -\sin\theta
\end{aligned}
$$

At small values of θ, to first order in θ $E_{XX} \approx 0$, $E_{YY} \approx 0$, $\omega_{XY} \approx -\theta$. Clearly, to first order in θ, the strains vanish and the nonzero rotation is given by θ. Of course θ represents *rigid body rotation*.

EXAMPLE 5.3

Find the displacements, strains and rotations in deformed body shown in Figure 5.4.

SOLUTION

The deformed and undeformed positions of the nodes are given in Table 5.1.
We again assume that the deformed coordinates may be expressed in terms of the undeformed coordinates using

$$x = a + bX + cY + dXY, \quad y = e + fX + gY + hXY$$

It was shown in the previous example that this form is capable of mapping the straight sides in the undeformed configuration onto straight sides in the deformed configuration. Following the same procedures as in the previous example given

$$
\begin{array}{llr}
\text{Node 1:} & 0 = a + b0 + c0 + d0 & : a = 0\\
 & 0 = e + f0 + g0 + h0 & : e = 0\\
\text{Node 2:} & \cos\theta = a + b1 + c0 + d0 & : b = \cos\theta\\
 & \sin\theta = e + f1 + g0 + h0 & : f = \sin\theta\\
\text{Node 4:} & \sin\theta = a + b0 + c1 + d0 & : c = \sin\theta\\
 & \cos\theta = e + f0 + g1 + h0 & : g = \cos\theta\\
\text{Node 3:} & \cos\theta + \sin\theta = a + b1 + c1 + d1 = \cos\theta + \sin\theta + d & : d = 0\\
 & \sin\theta + \cos\theta = e + f1 + g1 + h1 = \sin\theta + \cos\theta + h & : h = 0
\end{array}
$$

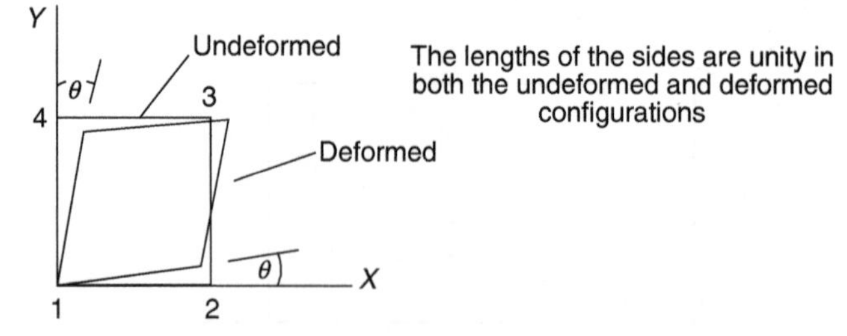

FIGURE 5.4 Illustration of shear strain.

TABLE 5.1
Deformed and Undeformed Nodal Coordinates

Node	Undeformed Position	Deformed Position
1	0, 0	0, 0
2	1, 0	$\cos\theta$, $\sin\theta$
3	1, 1	$\cos\theta+\sin\theta$, $\sin\theta+\cos\theta$
4	0, 1	$\sin\,\theta$, $\cos\,\theta$

We have now determined that

$$x = \cos\theta\, X + \sin\theta\, Y, \quad y = \sin\theta\, X + \cos\theta\, Y$$

Now assuming that θ is small enough to permit neglecting quadratic and higher-order terms,

$$\begin{aligned} u &= (\cos\theta - 1)X + \sin\theta\, Y & v &= \sin\theta\, X + (\cos\theta - 1)Y \\ &\approx \theta Y & &\approx \theta X \\ \mathrm{E}_{XX} &= \cos\theta - 1 & \mathrm{E}_{YY} &= \cos\theta - 1 \\ &\approx -\theta^2/2 & &\approx -\theta^2/2 \\ \mathrm{E}_{XY} &= \sin\theta & \omega_{XY} &= 0 \\ &\approx \theta & & \end{aligned}$$

In contrast to the case of rigid body rotation, for small angle θ and to first order in θ, the only nonzero strain is the shear strain, which is (approximately) equal to θ. The shear strain is a measure of how much the sides rotate relative to each other.

5.4 TRACTION AND STRESS

Consider a differential "brick element" emanating from an origin of the (X,Y,Z) coordinate system shown.

The differential forces on the faces of the element can be displayed as in Figures 5.5 and 5.6,

$dF_j^{(i)}$ = differential force component in the jth direction acting on face dA_i

dA_i = differential area whose normal points in the ith direction

The stresses are introduced by

$$\mathrm{S}_{ij} = \frac{\partial F_j^{(i)}}{\partial A_i}$$

and if X, Y, and Z denote the coordinate axes,

$$\begin{aligned} \mathrm{S}_{xx} &= \frac{dF_x^{(x)}}{dA_x}, & \mathrm{S}_{xy} &= \mathrm{S}_{yx} = \frac{dF_y^{(x)}}{dA_x} = \frac{dF_x^{(y)}}{dA_y} \\ \mathrm{S}_{yy} &= \frac{dF_y^{(y)}}{dA_y}, & \mathrm{S}_{yz} &= \mathrm{S}_{zy} = \frac{dF_z^{(y)}}{dA_y} = \frac{dF_y^{(z)}}{dA_z} \\ \mathrm{S}_{zz} &= \frac{dF_z^{(z)}}{dA_z}, & \mathrm{S}_{zx} &= \mathrm{S}_{xz} = \frac{dF_x^{(z)}}{dA_z} = \frac{dF_z^{(x)}}{dA_x} \end{aligned} \tag{5.12}$$

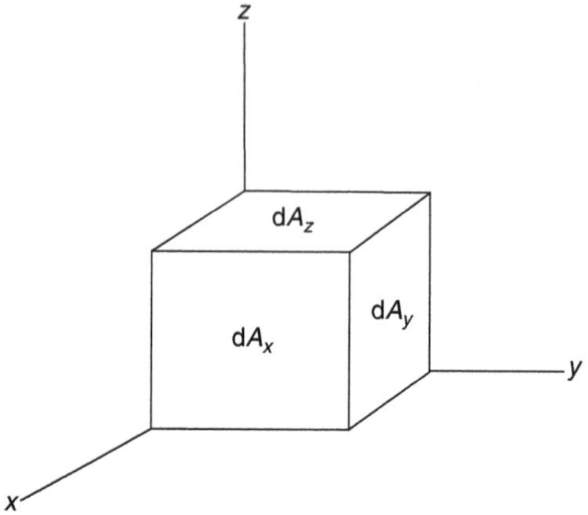

FIGURE 5.5 Differential brick element.

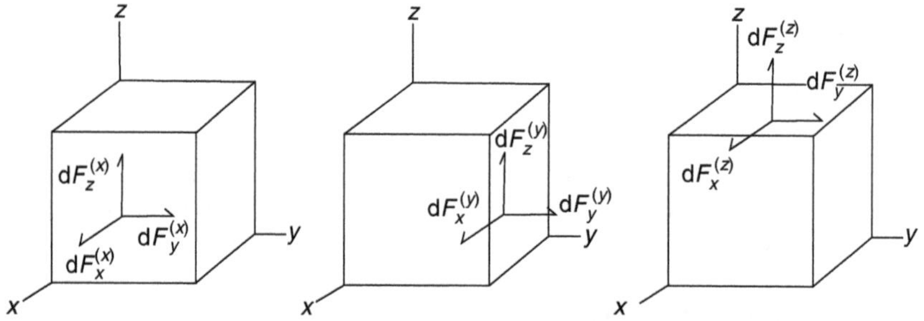

FIGURE 5.6 Forces on differential brick element.

The first index i represents the differential area on which the force acts, while the second index j represents the direction in which the force is acting. The total differential force vector acting on the ith face is given by $\mathbf{dF}^{(i)} = \mathbf{t}^{(i)}\,\mathrm{d}A_i$ in which the traction vector $\mathbf{t}^{(i)}$ on the ith face is given by

$$\mathbf{t}^{(i)} = \frac{\mathrm{d}F_1^{(i)}}{\mathrm{d}A_i}\mathbf{e}_1 + \frac{\mathrm{d}F_2^{(i)}}{\mathrm{d}A_i}\mathbf{e}_2 + \frac{\mathrm{d}F_3^{(i)}}{\mathrm{d}A_i}\mathbf{e}_3 \tag{5.13}$$

More generally, we consider a differential tetrahedron enclosing the point $\mathbf{x}$ in the deformed configuration (Figure 5.7). The area of the inclined (shaded) face is $\mathrm{d}S$, and $\mathrm{d}S_i$ is the area of the face on the back of the tetrahedron whose exterior normal vector is $-\mathbf{e}_i$. Simple vector analysis serves to derive that $n_i = \mathrm{d}S_i/\mathrm{d}S$, see Example 2.5 in Chapter 2. Next let $\mathbf{dF}$ denote the force on a surface element $\mathrm{d}S$, and let $\mathbf{dF}^{(i)}$ denote the force on area $\mathrm{d}S_i$ on the back of the tetrahedron. The traction vector acting on the inclined face is introduced by $\mathbf{t} = \mathbf{dF}/\mathrm{d}S$. As the tetrahedron shrinks to a point, the contribution of volume forces such as inertia decays faster than that of surface forces.

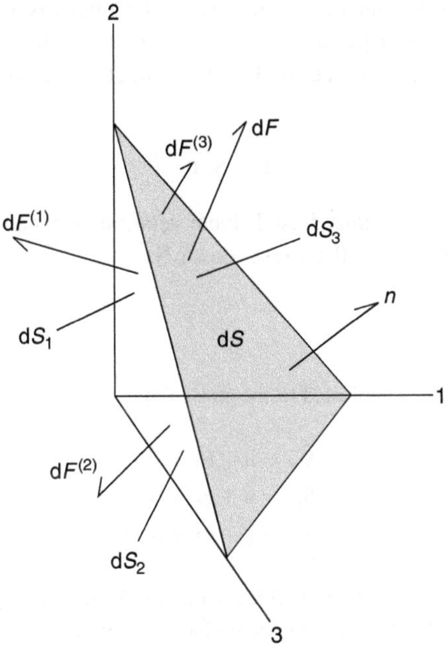

FIGURE 5.7 Forces on a differential tetrahedron.

Balance of forces on the tetrahedron now requires that

$$\mathbf{dF} = \sum_i \mathbf{dF}^{(i)} \tag{5.14}$$

The traction vector acting on the inclined face is defined by

$$\mathbf{t} = \frac{\mathbf{dF}}{\mathrm{d}S} \tag{5.15}$$

which together with the equilibrium Equation 5.14 furnishes

$$\begin{aligned} t_j &= \sum_i \frac{\mathrm{d}F_j^{(i)}}{\mathrm{d}S} \\ &= \sum_i \frac{\mathrm{d}F_j^{(i)}}{\mathrm{d}S_i} \frac{\mathrm{d}S_i}{\mathrm{d}S} \\ &= \mathrm{S}_{ij} n_i \end{aligned} \tag{5.16}$$

in which

$$\mathrm{S}_{ij} = \frac{\mathrm{d}F_j^{(i)}}{\mathrm{d}S_i} \tag{5.17}$$

It is readily seen that S_{ij} can be interpreted as the intensity of the force acting in the j direction on the facet pointing in the $-i$ direction. It is the ijth entry in the tensor $\mathbf{S}$. In matrix–vector notation the stress–traction relation Equation 5.16 is written as

$$\mathbf{t} = \mathbf{S}^T\mathbf{n} \tag{5.18}$$

Since $\mathbf{S}^T$ appears in a physically based linear relation between vectors of the same dimension, it is a tensor, as is the stress tensor $\mathbf{S}$.

EXAMPLE 5.4

The plate in Figure 5.8 is subjected to the stress field

$$S_{xx} = a + bx + cy$$
$$S_{yy} = d + ex + fy$$
$$S_{xy} = g + hx + jy$$

Find the traction vector and the total force acting on the top and right faces. Consider the moment (M_z) exerted by the tractions on these two faces about the origin.

On the top face consider the interval dx at x. The total force on this interval is $d\mathbf{F}_{\text{top}} = S_{yx}\,dx\,\mathbf{e}_x + S_{yy}\,dx\,\mathbf{e}_y$. The total force on the face is

$$\begin{aligned}
\mathbf{F}_{\text{top}} &= \int_0^L \left[S_{yx}\,dx\,\mathbf{e}_x + S_{yy}\,dx\,\mathbf{e}_y\right] \\
&= \int_0^L \left[(g + hx + jH)\,dx\,\mathbf{e}_x + (d + ex + fH)\,dx\,\mathbf{e}_y\right] \\
&= \left(g + jH + \tfrac{1}{2}hL\right)L\,\mathbf{e}_x + \left(d + \tfrac{1}{2}eL + fH\right)L\,\mathbf{e}_y
\end{aligned}$$

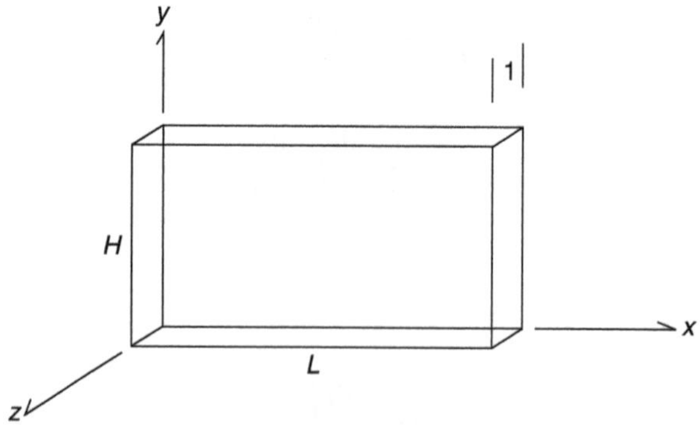

FIGURE 5.8 Forces and moments determined by tractions.

The moment due to the force on the interval is

$$\begin{aligned} \mathrm{d}\mathbf{M}_{\mathrm{top}} &= (x\mathbf{e}_x + H\mathbf{e}_y) \times \mathrm{d}\mathbf{F}_{\mathrm{top}} \\ &= (x\mathbf{e}_x + H\mathbf{e}_y) \times (g + hx + jH)\,\mathrm{d}x\,\mathbf{e}_x + (d + ex + fH)\,\mathrm{d}x\,\mathbf{e}_y \\ &= [[x(d + ex + fH)\,\mathrm{d}x] - H(g + hx + jH)]\,\mathrm{d}x\,\mathbf{e}_k \end{aligned}$$

and integration gives

$$M_{\mathrm{top}} = \tfrac{1}{2}(fH + d)L^2 + \tfrac{1}{3}eL^3 - (H(g + h)L - \tfrac{1}{2}HhL^2$$

On the right face consider the interval dy at y. The total force on the interval is

$$\mathrm{d}\mathbf{F}_{\mathrm{right}} = (\mathrm{S}_{xx}\,\mathrm{d}y\,\mathbf{e}_x + \mathrm{S}_{xy}\,\mathrm{d}x\,\mathbf{e}_y)$$

The total force on the face is now

$$\begin{aligned} \mathbf{F}_{\mathrm{right}} &= \int_0^H \left[\mathrm{S}_{xx}\,\mathrm{d}y\,\mathbf{e}_x + \mathrm{S}_{xy}\,\mathrm{d}y\,\mathbf{e}_y\right] \\ &= \int_0^H \left[(a + bL + cy)\,\mathrm{d}y\,\mathbf{e}_x + (g + hL + jy)\,\mathrm{d}y\,\mathbf{e}_y\right] \\ &= \left(a + bL + \tfrac{1}{2}cH\right)H\mathbf{e}_x + \left(g + hL + \tfrac{1}{2}jH\right)H\mathbf{e}_y \end{aligned}$$

EXAMPLE 5.5

At point (0,0,0) the tractions $\mathbf{t}_1$, $\mathbf{t}_2$, $\mathbf{t}_3$ act on planes with normal vectors $\mathbf{n}_1$, $\mathbf{n}_2$, and $\mathbf{n}_3$. Find the stress tensor $\mathbf{S}$, given that

$$\begin{aligned} \mathbf{n}_1 &= \tfrac{1}{\sqrt{3}}[\mathbf{e}_1 + \mathbf{e}_2 + \mathbf{e}_3], & \mathbf{t}_1 &= \tfrac{1}{\sqrt{3}}[2\mathbf{e}_1 - 5\mathbf{e}_2 + 6\mathbf{e}_3] \\ \mathbf{n}_2 &= \tfrac{1}{\sqrt{2}}[\mathbf{e}_1 + \mathbf{e}_2], & \mathbf{t}_2 &= \tfrac{1}{\sqrt{2}}[\mathbf{e}_1 - \mathbf{e}_2 + \mathbf{e}_3] \\ \mathbf{n}_3 &= \tfrac{1}{\sqrt{3}}[\mathbf{e}_1 - \mathbf{e}_2 - \mathbf{e}_3], & \mathbf{t}_3 &= -\tfrac{1}{\sqrt{3}}[6\mathbf{e}_1 - 1\mathbf{e}_2 + 2\mathbf{e}_3] \end{aligned}$$

Solution

On applying stress–traction relation $\mathbf{t}_1 = \mathbf{S}^T\mathbf{n}_1$ we find

$$\frac{1}{\sqrt{3}}\begin{pmatrix} 2 \\ -5 \\ 6 \end{pmatrix} = \frac{1}{\sqrt{3}}\begin{bmatrix} \mathrm{S}_{11} & \mathrm{S}_{21} & \mathrm{S}_{31} \\ \mathrm{S}_{12} & \mathrm{S}_{22} & \mathrm{S}_{32} \\ \mathrm{S}_{13} & \mathrm{S}_{23} & \mathrm{S}_{33} \end{bmatrix}\begin{pmatrix} 1 \\ 1 \\ 1 \end{pmatrix}$$

and

$$S_{11} + S_{21} + S_{31} = 2 \tag{5.19}$$

$$S_{12} + S_{22} + S_{32} = -5 \tag{5.20}$$

$$S_{13} + S_{23} + S_{33} = 6 \tag{5.21}$$

Similarly, $\mathbf{t}_2 = \mathbf{S}^T\mathbf{n}_2$ implies that

$$S_{11} + S_{21} = 1 \tag{5.22}$$

$$S_{12} + S_{22} = -1 \tag{5.23}$$

$$S_{13} + S_{23} = 1 \tag{5.24}$$

Finally, $\mathbf{t}_3 = \mathbf{S}^T\mathbf{n}_3$ yields

$$S_{11} - S_{21} - S_{31} = -6 \tag{5.25}$$

$$S_{12} - S_{22} - S_{32} = 1 \tag{5.26}$$

$$S_{13} - S_{23} - S_{33} = -2 \tag{5.27}$$

Solving the nine equations (Equations 5.19 through 5.27) by sequential elimination gives the stress tensor as

$$\mathbf{S} = \begin{bmatrix} -2 & -2 & 2 \\ 3 & 1 & -1 \\ 1 & -4 & 5 \end{bmatrix}$$

5.5 EQUILIBRIUM

We now consider equilibrium in a more general way and include inertial effects, under the restriction that the deformation is "small enough" to neglect effects of deformation on the stress or the equilibrium relation. Neglecting gravity and other "body forces," the total external force $\mathbf{F}$ on a body is assumed to be exerted on its boundaries. It is given by the integral $\mathbf{F} = \int \mathbf{t}\,dS$. However, recalling that $\mathbf{t} = \mathbf{S}^T\mathbf{n}$, the divergence theorem may be applied to obtain $\mathbf{F} = \int \nabla \cdot \mathbf{S}^T dV$. The total force must equal the rate of change of the total linear momentum $\boldsymbol{L}$ of the body: $\boldsymbol{L} = \int \rho\dot{\mathbf{x}}\,dV$ in which the mass density is denoted by ρ. Now owing to Rayleigh's transport theorem (see Chandrasekharaiah and Debnath, 1994), $\frac{d\boldsymbol{L}}{dt} = \frac{d}{dt}\int \rho\dot{\mathbf{x}}\,dV = \int \rho\ddot{\mathbf{x}}\,dV$. Accordingly, $\mathbf{F} = \frac{d\boldsymbol{L}}{dt}$ implies the integral (*global*) equation $\int \left(\nabla \cdot \mathbf{S}^T - \rho\ddot{\mathbf{x}}^T\right) dV = \mathbf{0}^T$.

In most cases of interest in the current monograph there is a fixed point in the body so that the undeformed position $\mathbf{X}$ can be taken to be independent on time. In this event $\ddot{\mathbf{u}} = \ddot{\mathbf{x}}$. (In Chapter 12, consideration will be given to bodies in which there is no fixed point, with the consequence that the analysis is referred to a translating and/or rotating coordinate system.) The integral equation must hold for the whole body or any closed sub-body within the body. We conclude that *the local balance of linear momentum equation*

$$\nabla \cdot \mathbf{S}^T = \rho \ddot{\mathbf{u}}^T \tag{5.28}$$

holds pointwise (locally) throughout the body.

It is also necessary to consider balance of angular momentum. The differential moment exerted by the traction vector on a surface patch is given by

$$\begin{aligned} \mathrm{d}\mathbf{M} &= \mathbf{X} \times \mathbf{t}\,\mathrm{d}S \qquad & \mathrm{d}M_i &= \in_{ijk} X_j t_k \,\mathrm{d}S \\ &= \mathbf{X} \times \mathbf{S}^T \mathbf{n}\,\mathrm{d}S & &= \in_{ijk} X_j S^T_{kl} n_l \,\mathrm{d}S \end{aligned} \tag{5.29}$$

Invoking the divergence theorem (Chapter 2), the total moment on the body is given by

$$\begin{aligned} M_i &= \int \in_{ijk} X_j S^T_{kl} n_l \,\mathrm{d}S \\ &= \int \frac{\partial}{\partial X_l} \left(\in_{ijk} X_j S^T_{kl} \right) \mathrm{d}V \\ &= \int \left(\in_{ijk} \frac{\partial X_j}{\partial X_l} S^T_{kl} + \in_{ijk} X_j \frac{\partial S^T_{kl}}{\partial X_l} \right) \mathrm{d}V \end{aligned} \tag{5.30}$$

Now $\frac{\partial X_j}{\partial X_l}$ is recognized as δ_{jl}, and $\in_{ijk} X_j \frac{\partial S^T_{kl}}{\partial X_l} = \in_{ijk} X_j \ddot{u}_k$ by virtue of the balance of linear momentum. We may rewrite the foregoing equation as

$$M_i = \int \left(\in_{ijk} S^T_{kj} + \rho \in_{ijk} X_j \ddot{u}_k \right) \mathrm{d}V \tag{5.31}$$

Invoking Rayleigh's Transport Theorem to be introduced in Chapter 13, the rate of change of the angular momentum is given by

$$\begin{aligned} \frac{\mathrm{d}\mathbf{H}}{\mathrm{d}t} &= \frac{\mathrm{d}}{\mathrm{d}t} \int \rho \mathbf{X} \times \dot{\mathbf{u}} \,\mathrm{d}V \\ &= \int \left[\rho \left(\frac{\mathrm{d}}{\mathrm{d}t} \mathbf{X} \right) \times \dot{\mathbf{u}} + \rho \mathbf{X} \times \ddot{\mathbf{u}} \right] \mathrm{d}V \end{aligned} \tag{5.32}$$

Note that $\left(\frac{\mathrm{d}}{\mathrm{d}t}\mathbf{X}\right) = \dot{\mathbf{u}}$ and hence $\left(\frac{\mathrm{d}}{\mathrm{d}t}\mathbf{X}\right) \times \dot{\mathbf{u}} = \mathbf{0}$. The global equation for the balance of angular momentum now becomes, in indicial notation,

$$M_i = \frac{\mathrm{d}H_i}{\mathrm{d}t} = \int \rho \ni_{ijk} X_j \ddot{u}_k \,\mathrm{d}V \tag{5.33}$$

Combining Equations 5.31 and 5.32, and invoking the fact that the result applies for arbitrary closed subvolumes we have the pointwise relation

$$\in_{ijk} S^T_{kj} = 0 \tag{5.34}$$

The tensor $\mathbf{S}^T$ may be expressed in terms of its symmetric and antisymmetric part. From Example 2.6 in Chapter 2, it was shown that $\in_{ijk} a_{jk} = 0$ if the corresponding tensor **A** is symmetric. By a similar argument it may be shown that $\in_{ijk} a_{jk} \neq 0$ if **A** is antisymmetric. Consequently, Equation 5.34 cannot be satisfied if $\mathbf{S}^T$ possesses a nonvanishing antisymmetric part. Accordingly $\mathbf{S}^T$ is symmetric: $\mathbf{S} = \mathbf{S}^T$.

EXAMPLE 5.6

Assume that the differential rectangle shown is a unit thickness (Figure 5.9). Prove from (static) moment equilibrium that $\mathbf{S}_{xy} = \mathbf{S}_{yx}$.

Solution

Take the moments about the lower left-hand corner.

1. The shear stresses on the bottom and left-hand faces do not create a moment relative to this point since the line of action goes through the origin.
2. Assuming that the forces due to the normal stresses are considered to act at the midpoint of the faces on which they act, they do not contribute to the moment since the positive forces have the same line of action as the negative forces.
3. This leaves only the shear stresses on the right-hand and top faces. Now moment balance $M = 0$ implies

$$\begin{aligned} 0 &= (S_{xy}\,dy)\,dx - (S_{yx}\,dx)\,dy \\ &= \left(S_{xy} - S_{yx}\right) dx\,dy \end{aligned}$$

with the consequence that $S_{yx} = S_{xy}$. This solution illustrates the fact that the stress tensor is symmetric. (The linear strain tensor is symmetric since it is the symmetric part of the tensor $\mathbf{du}/\mathbf{dX}$.)

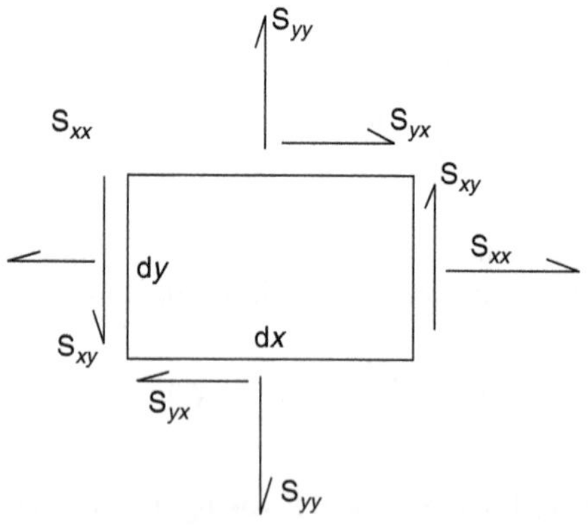

FIGURE 5.9 Net moment on a differential element.

EXAMPLE 5.7

Suppose that, in a differential plate element as shown, the stresses vary slightly across the element. Find the equations for static equilibrium of forces in the x- and y-directions, treating the figure shown as a *free body diagram* at the differential level (Figure 5.10).

SOLUTION

The total force in the x-direction is

$$\mathrm{d}F_x = \left(\cancel{S_{xx}} + \frac{\mathrm{d}S_{xx}}{\mathrm{d}x}\mathrm{d}x - \cancel{S_{xx}}\right)\mathrm{d}y + \left(\cancel{S_{yx}} + \frac{\mathrm{d}S_{yx}}{\mathrm{d}y}\mathrm{d}y - \cancel{S_{yx}}\right)$$

But the equilibrium condition $\mathrm{d}F_x = 0$ requires that $\left(\frac{\mathrm{d}S_{xx}}{\mathrm{d}x} + \frac{\mathrm{d}S_{yx}}{\mathrm{d}y}\right)\mathrm{d}x\,\mathrm{d}y = 0$, which becomes

$$\frac{\mathrm{d}S_{xx}}{\mathrm{d}x} + \frac{\mathrm{d}S_{yx}}{\mathrm{d}y} = 0$$

The total force in the y-direction is

$$\mathrm{d}F_y = \left(\cancel{S_{xy}} + \frac{\mathrm{d}S_{xy}}{\mathrm{d}x}\mathrm{d}x - \cancel{S_{xy}}\right)\mathrm{d}y + \left(\cancel{S_{yy}} + \frac{\mathrm{d}S_{yy}}{\mathrm{d}x}\mathrm{d}y - \cancel{S_{yy}}\right)\mathrm{d}x$$

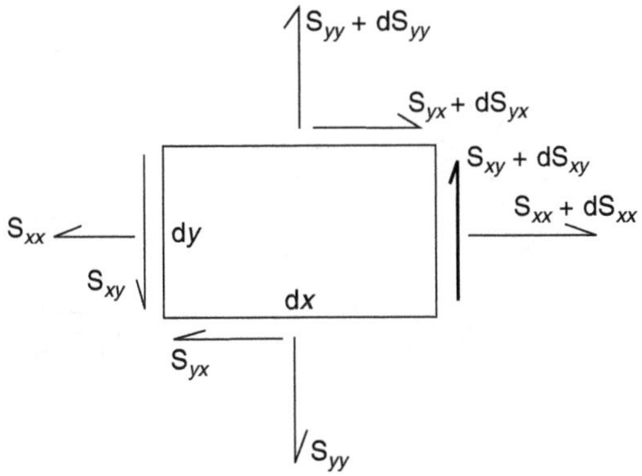

Note: $\mathrm{d}S_{xy} = \frac{\mathrm{d}S_{xy}}{\mathrm{d}x}\mathrm{d}x$, etc.

FIGURE 5.10 Balance of forces on a differential element.

and $dF_y = 0$ immediately furnishes that

$$\frac{dS_{xy}}{dx} + \frac{dS_{yy}}{dx} = 0$$

5.6 STRESS AND STRAIN TRANSFORMATIONS

The plate shown below is under a state of stress referred to the x- and y-axis.

A small cutout is shown whose lower side is inclined at an angle θ from the x-axis. We embed a rotated coordinate system $X' - Y'$ in the cutout, and seek to express the stresses in the rotated system (Figure 5.11).

We already know from the relations of Chapter 3 that the stresses represent a tensor and that the stress tensor satisfies the relation $\mathbf{S}' = \mathbf{QSQ}^T$ in which $\mathbf{Q}$ is the orthogonal matrix rotating the base vectors $\mathbf{e}_X$, $\mathbf{e}_Y$ to the $\mathbf{e}_{X'}$, $\mathbf{e}_{Y'}$. Furthermore, we already know that, in the plane,

$$\mathbf{Q} = \begin{bmatrix} \cos\theta & \sin\theta \\ -\sin\theta & \cos\theta \end{bmatrix} \tag{5.35}$$

Elementary matrix multiplications and the double-angle formulae of trigonometry suffice to verify that

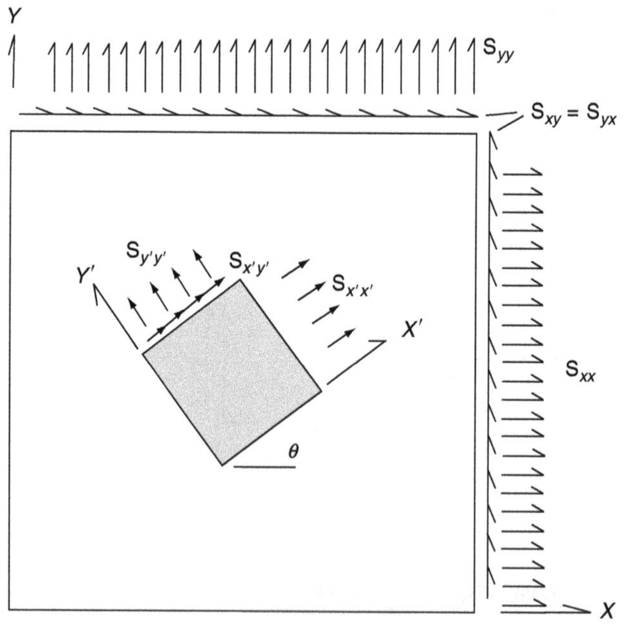

FIGURE 5.11 Stresses referred to a rotated coordinate system.

$$
\begin{aligned}
S_{X'X'} &= \frac{S_{XX} + S_{yy}}{2} + \frac{S_{XX} - S_{yy}}{2}\cos 2\theta + S_{XY}\sin 2\theta \\
S_{Y'Y'} &= \frac{S_{XX} + S_{YY}}{2} - \frac{\sigma_{XX} - S_{YY}}{2}\cos 2\theta - S_{XY}\sin 2\theta \\
S_{X'Y'} &= -\frac{S_{XX} - S_{YY}}{2}\sin 2\theta + S_{XY}\cos 2\theta
\end{aligned}
\tag{5.36}
$$

The strains induced by the stresses likewise represent a tensor and transform exactly the same way as the stresses:

$$
\begin{aligned}
E_{X'X'} &= \frac{E_{XX} + E_{YY}}{2} + \frac{E_{XX} - E_{YY}}{2}\cos 2\theta + E_{XY}\sin 2\theta \\
E_{Y'Y'} &= \frac{E_{XX} + E_{YY}}{2} - \frac{E_{XX} - E_{YY}}{2}\cos 2\theta - E_{XY}\sin 2\theta \\
E_{X'Y'} &= -\frac{E_{XX} - E_{YY}}{2}\sin 2\theta + E_{XY}\cos 2\theta
\end{aligned}
\tag{5.37}
$$

Several simple examples are now given illustrating the use of coordinate transformations.

EXAMPLE 5.8

Relative to the x–y axes the stresses are

$$S_{xx} = 0, \quad S_{yy} = 0, \quad S_{xy} = 25\ ksi$$

What are the stresses referred to axes x' and y' which are rotated by 45° from x and y? Do the same for stresses referred to axes x'' and y'' rotated by −45°.

SOLUTION

Using the stress transformations through +45°

$$
\begin{aligned}
S_{x'x'} &= \frac{S_{xx} + S_{yy}}{2} + \frac{S_{xx} - S_{yy}}{2}\cos(2*45) + S_{xy}\sin(2*45) = 25\ ksi \\
S_{y'y'} &= \frac{S_{xx} + S_{yy}}{2} - \frac{S_{xx} - S_{yy}}{2}\cos(2*45) - S_{xy}\sin(2*45) = -25\ ksi \\
S_{x'y'} &= -\frac{S_{xx} - S_{yy}}{2}\sin(2*45) + S_{xy}\cos(2*45) = 0
\end{aligned}
$$

Doing the stress transformations through −45°

$$
\begin{aligned}
S_{x'x'} &= \frac{S_{xx} + S_{yy}}{2} + \frac{S_{xx} - S_{yy}}{2}\cos(2*45) + S_{xy}\sin(2*45) = -25\ ksi \\
S_{y'y'} &= \frac{S_{xx} + S_{yy}}{2} - \frac{S_{xx} - S_{yy}}{2}\cos(2*45) - S_{xy}\sin(2*45) = +25\ ksi \\
S_{x'y'} &= -\frac{S_{xx} - S_{yy}}{2}\sin(2*45) + \sigma_{xy}\cos(2*45) = 0
\end{aligned}
$$

Since the elements at ±45° have no shear stress, their normal stresses are the principal stresses.

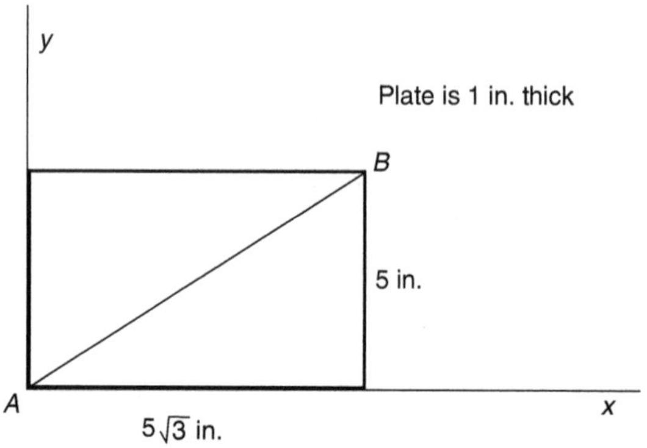

FIGURE 5.12 Length of the diagonal in a plate under strain.

EXAMPLE 5.9

The stresses in the square plate shown (Figure 5.12) are uniform and are given by

$$S_{xx} = 10\ ksi, \quad S_{yy} = -10\ ksi, \quad S_{xy} = 0$$

What is the total force acting transverse to the line AB, and what is the total moment of this force relative to point A?

SOLUTION

$$S_{x'x'} = \frac{S_{xx} + S_{yy}}{2} + \frac{S_{xx} - S_{yy}}{2} \cos 60 + S_{xy} \sin 60 = 5\ ksi$$

$$S_{y'y'} = \frac{S_{xx} + S_{yy}}{2} - \frac{S_{xx} - S_{yy}}{2} \cos 60 - S_{xy} \sin 60 = -5\ ksi$$

$$S_{x'y'} = -\frac{S_{xx} - S_{yy}}{2} \sin 60 + S_{xy} \cos 60 = -5\sqrt{3}\ ksi$$

$$\begin{aligned} F_{AB} &= S_{y'y'} * \text{ area} \\ &= -5\ ksi * 1 * 10 \\ &= -50000 \text{ lb} \end{aligned} \qquad \begin{aligned} M_{AB} &= -50000 \times 5 \\ &= -250000 \text{ in. lb} \end{aligned}$$

EXAMPLE 5.10

A plate element is under the strains $E_{xx} \neq 0$, $E_{yy} \neq 0$, $E_{xy} = 0$. What are the strains in an element rotated +90°, −90°, and 180°?

SOLUTION

$$E_{x'x'} = \frac{E_{xx} + E_{yy}}{2} + \frac{E_{xx} - E_{yy}}{2} \cos(180) + E_{xy} \sin(180) = \varepsilon_{yy}$$

$$E_{y'y'} = \frac{E_{xx} + E_{yy}}{2} - \frac{E_{xx} - E_{yy}}{2} \cos(180) - \varepsilon_{xy} \sin(180) = \varepsilon_{xx}$$

$$E_{x'y'} = -\frac{E_{xx} - E_{yy}}{2} \sin(180) + E_{xy} \cos(180) = 0$$

$\theta = -90°$:

$$E_{x'x'} = \frac{E_{xx} + E_{yy}}{2} + \frac{E_{xx} - E_{yy}}{2}\cos(-180) + \cancel{E_{xy}}\sin(-180) = E_{yy}$$

$$E_{y'y'} = \frac{E_{xx} + E_{yy}}{2} - \frac{\varepsilon_{xx} - \varepsilon_{yy}}{2}\cos(-180) - \cancel{E_{xy}}\sin(-180) = E_{xx}$$

$$E_{x'y'} = -\frac{E_{xx} - E_{yy}}{2}\cancel{\sin(-180)} + \cancel{E_{xy}}\cos(-180) = 0$$

$\theta = 180°$:

$$\varepsilon_{x'x'} = \frac{E_{xx} + E_{yy}}{2} + \frac{E_{xx} - E_{yy}}{2}\cos(360) + \cancel{E_{xy}}\sin(360) = \varepsilon_{xx}$$

$$\varepsilon_{y'y'} = \frac{E_{xx} + E_{yy}}{2} - \frac{E_{xx} - E_{yy}}{2}\cos(360) - \cancel{E_{xy}}\sin(360) = \varepsilon_{yy}$$

$$\varepsilon_{x'y'} = -\frac{E_{xx} - E_{yy}}{2}\cancel{\sin(360)} + \cancel{E_{xy}}\cos(360) = 0$$

EXAMPLE 5.11

Transformation of stresses in two dimensions (2D)

Verify the stress transformation relations using force balance in two dimensions (Figure 5.13).

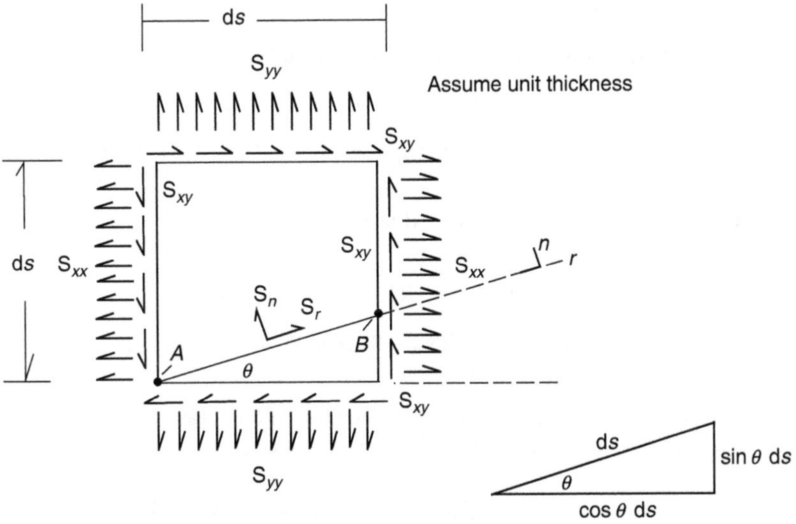

FIGURE 5.13 Transformation of stress.

Solution

In the current and succeeding example we will make use of the relations $\cos^2\theta = \frac{1}{2}(1+\cos 2\theta)$, $\sin^2\theta = \frac{1}{2}(1-\cos 2\theta)$, and $2\sin\theta\cos\theta = \sin 2\theta$. Given the stresses on a square element, we seek the normal and tangential stresses on the line AB at angle θ from the x-axis.

The unit vectors defining the normal and tangential directions of the line AB are given by

$$\mathbf{n} = -\sin\theta\,\mathbf{i} + \cos\theta\,\mathbf{j}, \quad \boldsymbol{\tau} = \cos\theta\,\mathbf{i} + \sin\theta\,\mathbf{j}$$

The vectors representing the normal, tangential, and total force on the inclined line are in terms of the normal and tangential stresses as

$$\begin{aligned}
\mathbf{dF}_n &= (-\sin\theta\,\mathrm{S}_n\,\mathrm{d}s)\mathbf{i} + \cos\theta\,\mathrm{S}_n\,\mathrm{d}s\,\mathbf{j}\\
\mathbf{dF}_t &= \cos\theta\,\mathrm{S}_t\,\mathrm{d}s\,\mathbf{i} + \sin\theta\,\mathrm{S}_t\,\mathrm{d}s\,\mathbf{j}\\
\mathbf{dF} &= (-\sin\theta\,\mathrm{S}_n\,\mathrm{d}s + \cos\theta\,\mathrm{S}_n\,\mathrm{d}s)\mathbf{i} + (\cos\theta\,\mathrm{S}_n\,\mathrm{d}s + \mathrm{S}_t\sin\theta\,\mathrm{d}s)\mathbf{j}
\end{aligned}$$

Balance of forces in the x and y directions requires that

$$\begin{aligned}
-\sin\theta\,\mathrm{S}_n\,\mathrm{d}s + \cos\theta\,\mathrm{S}_t\,\mathrm{d}s + \sin\theta\,\mathrm{S}_{xx}\,\mathrm{d}s - \cos\theta\,\mathrm{S}_{xy}\,\mathrm{d}s &= 0\\
\cos\theta\,\mathrm{S}_n\,\mathrm{d}s + \sin\theta\,\mathrm{S}_t\,\mathrm{d}s + \sin\theta\,\mathrm{S}_{xy}\,\mathrm{d}s - \cos\theta\,\mathrm{S}_{yy}\,\mathrm{d}s &= 0
\end{aligned}$$

After some manipulations involving double-angle formulae from trigonometry, we have

$$\begin{aligned}
-\sin\theta\,\mathrm{S}_n + \cos\theta\,\mathrm{S}_t &= -\mathrm{S}_{xx}\sin\theta + \mathrm{S}_{xy}\cos\theta\\
\cos\theta\,\mathrm{S}_n + \sin\theta\,\mathrm{S}_t &= -\mathrm{S}_{xy}\sin\theta + \mathrm{S}_{yy}\cos\theta\\
\sin^2\theta\,\mathrm{S}_n - \sin\theta\cos\theta\,\mathrm{S}_t &= \mathrm{S}_{xx}\sin^2\theta - \mathrm{S}_{xy}\sin\theta\cos\theta\\
\cos^2\theta\,\mathrm{S}_n + \sin\theta\cos\theta\,\mathrm{S}_t &= -\mathrm{S}_{xy}\sin\theta\cos\theta + \mathrm{S}_{yy}\cos^2\theta\\
\mathrm{S}_n &= \mathrm{S}_{yy}\cos^2\theta + \mathrm{S}_{xx}\sin^2\theta - 2\sin\theta\cos\theta\,\mathrm{S}_{xy}
\end{aligned}$$

and hence the stress normal to the line AB is given by

$$\begin{aligned}
\mathrm{S}_n &= \tfrac{1}{2}(1+\cos 2\theta)\mathrm{S}_{yy} + \tfrac{1}{2}(1-\sin^2\theta)\mathrm{S}_{xx} - \sin 2\theta\,\mathrm{S}_{xy}\\
&= \tfrac{1}{2}(\mathrm{S}_{xx}+\mathrm{S}_{yy}) - \tfrac{1}{2}(\mathrm{S}_{xx}-\mathrm{S}_{yy})\cos 2\theta - \sin 2\theta\,\mathrm{S}_{xy}
\end{aligned}$$

in agreement with the previously reported transformation formulae.

Similar operations are performed for the tangential stress.

$$\begin{aligned}
-\sin\theta\cos\theta\,\mathrm{S}_n + \cos^2\theta\,\mathrm{S}_t &= -\mathrm{S}_{xx}\sin\theta\cos\theta + \mathrm{S}_{xy}\cos^2\theta\\
\sin\theta\cos\theta\,\mathrm{S}_n + \sin^2\theta\,\mathrm{S}_t &= -\mathrm{S}_{xy}\sin^2\theta + \mathrm{S}_{yy}\sin\theta\cos\theta
\end{aligned}$$

The stress tangential to the line AB is now found to be

$$\sigma_t = -(\sigma_{xx} - \sigma_{yy})\sin\theta\cos\theta + \sigma_{xy}(\cos^2\theta - \sin^2\theta)$$

$$\sigma_t = -\frac{(\sigma_{xx} - \sigma_{yy})}{2}\sin 2\theta + \sigma_{xy}\cos 2\theta$$

likewise in agreement with the previously reported transformation formulae.

EXAMPLE 5.12

Transformations of strains in 2D.
In two dimensions show that $E_{x'x'} = \frac{\partial u'}{\partial X'}$, $E_{y'y'} = \frac{\partial v'}{\partial Y'}$, and $E_{x'y'} = \frac{1}{2}\left(\frac{\partial u'}{\partial Y'} + \frac{\partial v'}{\partial X'}\right)$.

Solution

Figure 5.14 depicts the undeformed position vector **X** and the deformed position vector **x**. It also shows **X**′ and **x**′ which are **X** and **x**′ rotated through $-\theta$. Recall that a positive rotation of the coordinate system has the same effect as a negative rotation of a vector relative to the same coordinate system.

The undeformed and deformed coordinates in the rotated vectors are given by

$$X' = \cos(\varphi - \theta)R \qquad Y' = \sin(\varphi - \theta)R$$
$$= X\cos\theta + Y\sin\theta \qquad = -X\sin\theta + Y\cos\theta$$

and

$$x' = x\cos\theta + y\sin\theta, \quad y' = -x\sin\theta + y\cos\theta$$

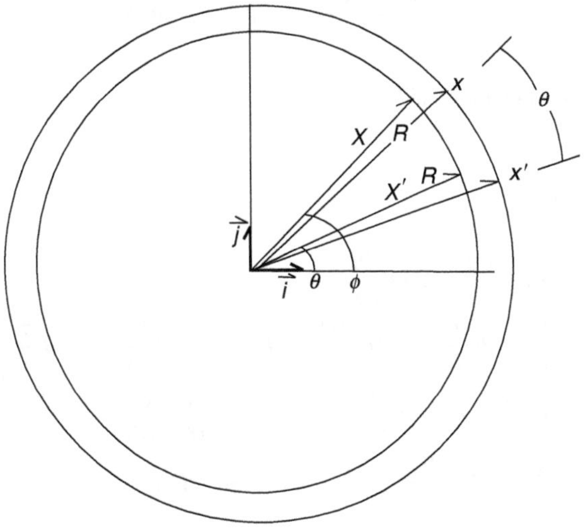

FIGURE 5.14 Undeformed position vector before and after rotation.

The relations just presented may be easily inverted to furnish

$$X = X' \cos\theta - Y' \sin\theta, \qquad x = x' \cos\theta - y' \sin\theta$$
$$Y = X' \sin\theta + Y' \cos\theta, \qquad y = x' \sin\theta + y' \cos\theta$$

The chain rule of calculus yields the useful relations

$$\begin{aligned}\frac{\partial}{\partial x'} &= \frac{\partial x}{\partial x'}\frac{\partial}{\partial x} + \frac{\partial y}{\partial x'}\frac{\partial}{\partial y} & \frac{\partial}{\partial y'} &= \frac{\partial x}{\partial y'}\frac{\partial}{\partial x} + \frac{\partial y}{\partial y'}\frac{\partial}{\partial y}\\ &= \cos\theta\frac{\partial}{\partial x} + \sin\theta\frac{\partial}{\partial y} & &= -\sin\theta\frac{\partial}{\partial x} + \cos\theta\frac{\partial}{\partial y}\end{aligned}$$

It further follows that

$$\begin{aligned}u' &= x' - X' & v' &= y' - Y'\\ &= (x - X)\cos\theta + (y - Y)\sin\theta & &= -(x - X)\sin\theta + (y - Y)\cos\theta\\ &= u\cos\theta + v\sin\theta & &= -u\sin\theta + v\cos\theta\end{aligned}$$

Next,

$$\begin{aligned}\mathrm{E}'_{XX} &= \cos\theta\frac{\partial u}{\partial X'} + \sin\theta\frac{\partial v}{\partial X'}\\ &= \left(\cos\theta\frac{\partial u}{\partial X} + \sin\theta\frac{\partial u}{\partial Y}\right)\cos\theta + \left(\cos\theta\frac{\partial v}{\partial X} + \sin\theta\frac{\partial v}{\partial Y}\right)\sin\theta\\ &= \cos^2\theta\frac{\partial u}{\partial X} + \sin^2\theta\frac{\partial v}{\partial Y} + \cos\theta\sin\theta\left(\frac{\partial u}{\partial Y} + \frac{\partial v}{\partial X}\right)\\ &= \tfrac{1}{2}(1 + \cos 2\theta)\mathrm{E}_{XX} + \tfrac{1}{2}(1 - \cos 2\theta)\mathrm{E}_{YY} + \sin 2\theta\mathrm{E}_{XY}\\ &= \frac{\mathrm{E}_{XX} + \mathrm{E}_{YY}}{2} + \frac{\mathrm{E}_{XX} - \mathrm{E}_{YY}}{2}\cos 2\theta + \mathrm{E}_{XY}\sin 2\theta\end{aligned}$$

in agreement with Equation 5.37.
Similarly,

$$\begin{aligned}\mathrm{E}_{Y'Y'} &= \frac{\partial v'}{\partial Y'}\\ &= -\sin\theta\frac{\partial u}{\partial Y'} + \cos\theta\frac{\partial v}{\partial Y'}\\ &= -\left[-\sin\theta\frac{\partial u}{\partial X} + \cos\theta\frac{\partial u}{\partial Y}\right]\sin\theta + \left[-\sin\theta\frac{\partial v}{\partial X} + \cos\theta\frac{\partial v}{\partial Y}\right]\cos\theta\\ &= \sin^2\theta\frac{\partial u}{\partial X} + \cos^2\theta\frac{\partial v}{\partial Y} - \sin\theta\cos\theta\left(\frac{\partial u}{\partial Y} + \frac{\partial v}{\partial X}\right)\\ &= \frac{\mathrm{E}_{XX} + \mathrm{E}_{YY}}{2} - \frac{\mathrm{E}_{XX} - \mathrm{E}_{YY}}{2}\cos\theta - \sin 2\theta\mathrm{E}_{XY}\end{aligned}$$

again in agreement with Equation 5.37.

Finally,

$$
\begin{aligned}
\mathrm{E}_{X'Y'} &= \frac{1}{2}\left(\frac{\partial u'}{\partial Y'}+\frac{\partial v'}{\partial X'}\right)\\
&= \frac{1}{2}\frac{\partial}{\partial Y'}(u\cos\theta+v\sin\theta)+\frac{1}{2}\frac{\partial}{\partial X'}(-u\sin\theta+v\cos\theta)\\
&= \frac{1}{2}\left(-\sin\theta\frac{\partial}{\partial X}+\cos\theta\frac{\partial}{\partial Y}\right)(u\cos\theta+v\sin\theta)+\frac{1}{2}\left(\cos\theta\frac{\partial}{\partial X}+\sin\theta\frac{\partial}{\partial Y}\right)(-u\sin\theta+v\cos\theta)\\
&= \frac{1}{2}\left(2\sin\theta\cos\theta\left(\frac{\partial v}{\partial Y}-\frac{\partial u}{\partial X}\right)+\left(\cos^2\theta-\sin^2\theta\right)\left(\frac{\partial u}{\partial Y}+\frac{\partial v}{\partial X}\right)\right)\\
&= -\frac{\mathrm{E}_{XX}-\mathrm{E}_{YY}}{2}\sin 2\theta+\mathrm{E}_{XY}\cos 2\theta
\end{aligned}
$$

as expected.

5.7 PRINCIPAL STRESSES AND STRAINS

In the plane, upon rotation of the coordinate axes through the angle θ_0 satisfying $\tan 2\theta_0 = \frac{2\varepsilon_{xy}}{\varepsilon_{xx}-\varepsilon_{yy}}$, the shear strain and the shear stress both vanish: $\varepsilon_{X'Y'}=0$, $\sigma_{X'Y'}=0$.

The corresponding normal stresses (strains) are called the *principal stresses* (strains), denoted by σ_{I}, σ_{II} (ε_{I}, $\varepsilon_{\mathrm{II}}$). Of course from Section 2.4 of Chapter 2 we know that the eigenvectors of the stress tensor form the columns of an orthogonal tensor $\mathbf{Q}_o(\sigma)$ which serves to diagonalize $\mathbf{S}$. The diagonal entries are simply the eigenvalues, which in the current context are called the principal values. In the plane the determinant equations for the principal stresses and strains are simply

$$
\mathrm{E}_{I,II} = \frac{\mathrm{E}_{xx}+\mathrm{E}_{yy}}{2} \pm \sqrt{\left(\frac{\mathrm{E}_{xx}-\mathrm{E}_{yy}}{2}\right)^2+\mathrm{E}_{xy}^2}
$$

$$
\mathrm{S}_{I,II} = \frac{\mathrm{S}_{XX}+\mathrm{S}_{YY}}{2} \pm \sqrt{\left(\frac{\mathrm{S}_{XX}-\mathrm{S}_{YY}}{2}\right)^2+\mathrm{S}_{XY}^2}
$$

The rotated axes in the current context are called the *principal axes.*

EXAMPLE 5.13

The strain ellipsoid

A three-dimensional body has a uniform state of strain in which

$$
\mathrm{E}_{xx}=0,\quad \mathrm{E}_{yy}=0.01,\quad \mathrm{E}_{zz}=0,\quad \mathrm{E}_{xy}=0,\quad \mathrm{E}_{yz}=0,\quad \mathrm{E}_{zx}=0.01
$$

Imagine a small sub-body, which is spherical with radius ρ in the undeformed configuration. When the strain is imposed, it becomes an ellipsoid (Figure 5.15). What are the lengths of the three semiaxes *a, b, c* of the ellipsoid, and what are their orientations relative to the *x–y–z* axes? What are the principal strains?

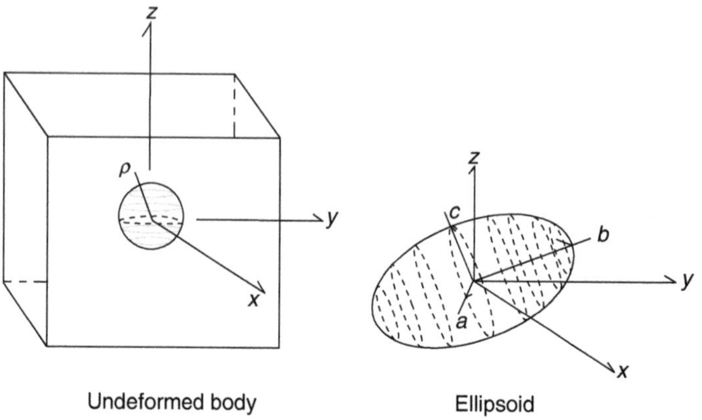

FIGURE 5.15 Deformation of a small sphere.

Solution

The strain tensor $\mathbf{E} = \begin{bmatrix} 0 & 0 & 0.01 \\ 0 & 0.1 & 0 \\ 0.01 & 0 & 0 \end{bmatrix}$ is diagonalized by a rotation of the z–x plane about the y-axis, namely

$$\mathbf{E}' = \begin{bmatrix} \cos\frac{\pi}{4} & 0 & \sin\frac{\pi}{4} \\ 0 & 1 & 0 \\ -\sin\frac{\pi}{4} & 0 & \cos\frac{\pi}{4} \end{bmatrix} \begin{bmatrix} 0 & 0 & 0.01 \\ 0 & 0.1 & 0 \\ 0.01 & 0 & 0 \end{bmatrix} \begin{bmatrix} \cos\frac{\pi}{4} & 0 & -\sin\frac{\pi}{4} \\ 0 & 1 & 0 \\ \sin\frac{\pi}{4} & 0 & \cos\frac{\pi}{4} \end{bmatrix}$$

$$= 0.01 \begin{bmatrix} \sin\frac{\pi}{4} & 0 & \cos\frac{\pi}{4} \\ 0 & 1 & 0 \\ \cos\frac{\pi}{4} & 0 & -\sin\frac{\pi}{4} \end{bmatrix} \begin{bmatrix} \cos\frac{\pi}{4} & 0 & -\sin\frac{\pi}{4} \\ 0 & 1 & 0 \\ \sin\frac{\pi}{4} & 0 & \cos\frac{\pi}{4} \end{bmatrix}$$

$$= 0.01 \begin{bmatrix} 2\sin\frac{\pi}{4}\cos\frac{\pi}{4} & 0 & \cos^2\frac{\pi}{4} - \sin^2\frac{\pi}{4} \\ 0 & 1 & 0 \\ \cos^2\frac{\pi}{4} - \sin^2\frac{\pi}{4} & 0 & -2\sin\frac{\pi}{4}\cos\frac{\pi}{4} \end{bmatrix}$$

$$= 0.01 \begin{bmatrix} 1 & 0 & 0 \\ 0 & 1 & 0 \\ 0 & 0 & -1 \end{bmatrix}$$

The lengths of the three semiaxes are: 1.01ρ, 1.0ρ, 0.99ρ. The principal strains are then relative length changes of the semiaxes. In other words,

$$\mathrm{E}_I = \frac{1.01\rho - \rho}{\rho} = 0.01, \qquad \mathrm{E}_{II} = \frac{1.01\rho - \rho}{\rho} = 0.01, \qquad \mathrm{E}_{III} = \frac{0.99\rho - \rho}{\rho} = -0.01$$

EXAMPLE 5.14

The Von Mises stress and the Tresca stress

Often stresses calculated by finite element analysis are compared to failure stresses measured in the laboratory. Typically, the laboratory tests are one-dimensional, either simple tension/compression or simple shear. However, the calculated stresses are often multiaxial. In order to enable comparison with failure stresses it is common to introduce "equivalent stresses" of which primary examples are the Von Mises stress and the Tresca stress introduced as follows using principal stresses:

$$S_{VM} = \frac{1}{\sqrt{2}}\sqrt{(S_I - S_{II})^2 + (S_{II} - S_{III})^2 + (S_{III} - S_I)^2} \qquad : \text{Von Mises}$$

$$\tau_T = \tfrac{1}{2}\max(|S_I - S_{II}|, |S_{II} - S_{III}|, |S_{III} - S_I|) \qquad : \text{Tresca}$$

In uniaxial tension or compression the S_{VM} is equal to the actual stress, thereby providing "motivation" for regarding it as the counterpart of the uniaxial tensile/compressive failure stress. On the other hand, in simple shear τ_T is equal to the stress, and accordingly we view it as the counterpart of the failure stress in simple shear.

In the case in which there is only one normal stress S_{XX} and one shear stress S_{XY}, S_{VM} and τ_T, we derive the following simple formulae:

$$S_{VM} = \sqrt{S_{XX}^2 + 3S_{XY}^2}, \quad \tau_T = \frac{1}{2}\sqrt{S_{XX}^2 + 4S_{XY}^2}$$

The principal stresses are given by

$$S_{I,II} = S_{xx} + \frac{\cancel{S_{yy}}}{2} \pm \sqrt{\left(\frac{S_{xx} + \cancel{S_{yy}}}{2}\right)^2 + S_{xy}^2} = \frac{S_{xx}}{2} \pm \sqrt{\left(\frac{S_{xx}}{2}\right)^2 + S_{xy}^2}, \quad S_{III} = 0$$

For the Von Mises stress, we find

$$\begin{aligned}
S_{VM}^2 &= \frac{1}{2}\left((S_I - S_{II})^2 + \left(S_{II} - \cancel{S_{III}}\right)^2 + \left(\cancel{S_{III}} - S_I\right)^2\right) \\
&= \frac{1}{2}\left(\left[2\sqrt{\left(\frac{S_{xx}}{2}\right)^2 + S_{xy}^2}\right]^2 + \left[\frac{S_{xx}}{2} + \sqrt{\left(\frac{S_{xx}}{2}\right)^2 + S_{xy}^2}\right]^2 + \left[\frac{S_{xx}}{2} - \sqrt{\left(\frac{S_{xx}}{2}\right)^2 + S_{xy}^2}\right]^2\right) \\
&= \frac{1}{2}\left(4\left[\left(\frac{S_{xx}}{2}\right)^2 + S_{xy}^2\right] + 2\left[\left(\frac{S_{xx}}{2}\right)^2 + \left(\left(\frac{S_{xx}}{2}\right)^2 + S_{xy}^2\right)\right]\right) \\
&= S_{xx}^2 + 3S_{xy}^2
\end{aligned}$$

For the Tresca stress,

$$\begin{aligned}
\tau_T &= \frac{1}{2}\max\left[|S_I - S_{II}|, \left|S_{II} - \cancel{S_{III}}\right|, \left|\cancel{S_{III}} - S_I\right|\right] \\
&= \frac{1}{2}\max\left[\left|2\sqrt{\left(\frac{S_{xx}}{2}\right)^2 + S_{xy}^2}\right|, \left|\frac{S_{xx}}{2} + \sqrt{\left(\frac{S_{xx}}{2}\right)^2 + S_{xy}^2}\right|, \left|\frac{S_{xx}}{2} - \sqrt{\left(\frac{S_{xx}}{2}\right)^2 + S_{xy}^2}\right|\right]
\end{aligned}$$

Now, taking the positive square root and noting that $\sqrt{\left(\frac{S_{xx}}{2}\right)^2 + S_{xy}^2} \geq \left|\frac{S_{xx}}{2}\right|$, the magnitudes satisfy

$$\left|2\sqrt{\left(\frac{S_{xx}}{2}\right)^2 + S_{xy}^2}\right| = 2\sqrt{\left(\frac{S_{xx}}{2}\right)^2 + S_{xy}^2}$$

$$\left|\frac{S_{xx}}{2} + \sqrt{\left(\frac{S_{xx}}{2}\right)^2 + S_{xy}^2}\right| = \frac{S_{xx}}{2} + \sqrt{\left(\frac{S_{xx}}{2}\right)^2 + S_{xy}^2}$$

$$\left|\frac{S_{xx}}{2} - \sqrt{\left(\frac{S_{xx}}{2}\right)^2 + S_{xy}^2}\right| = \sqrt{\left(\frac{S_{xx}}{2}\right)^2 + S_{xy}^2} - \frac{S_{xx}}{2}$$

regardless of the sign of S_{xx}. But, also,

$$2\sqrt{\left(\frac{S_{xx}}{2}\right)^2 + S_{xy}^2} - \left(\frac{S_{xx}}{2} + \sqrt{\left(\frac{S_{xx}}{2}\right)^2 + S_{xy}^2}\right) = \sqrt{\left(\frac{S_{xx}}{2}\right)^2 + S_{xy}^2} - \frac{S_{xx}}{2} \geq 0$$

$$2\sqrt{\left(\frac{S_{xx}}{2}\right)^2 + S_{xy}^2} - \left(-\frac{S_{xx}}{2} + \sqrt{\left(\frac{S_{xx}}{2}\right)^2 + S_{xy}^2}\right) = \sqrt{\left(\frac{S_{xx}}{2}\right)^2 + S_{xy}^2} + \frac{S_{xx}}{2} \geq 0$$

Consequently,

$$\max\left[\left|2\sqrt{\left(\frac{S_{xx}}{2}\right)^2 + S_{xy}^2}\right|, \left|\frac{S_{xx}}{2} + \sqrt{\left(\frac{S_{xx}}{2}\right)^2 + S_{xy}^2}\right|, \left|\frac{S_{xx}}{2} - \sqrt{\left(\frac{S_{xx}}{2}\right)^2 + \sigma_{xy}^2}\right|\right]$$
$$= 2\sqrt{\left(\frac{S_{xx}}{2}\right)^2 + S_{xy}^2}$$

from which we obtain the expected relation

$$\tau_T = \frac{1}{2}\sqrt{(S_{xx})^2 + 4S_{xy}^2}$$

5.8 STRESS–STRAIN RELATIONS

In linear elasticity, the stresses are linear functions of strain. Furthermore, if the material is *isotropic*, its properties are characterized by two coefficients: the *elastic modulus* E and the *Poisson's ratio* ν. To illustrate isotropy using Figure 5.16, in one-dimensional tension specimens cut from a single plate of a given material, the material is isotropic if the measured stress–strain curves are the same and independent of the orientation at which they are cut. Otherwise it exhibits anisotropy, but may still exhibit limited types of symmetry such as transverse isotropy or orthotropy.

The isotropic stress–strain relations of linear elasticity are given in alternate forms, one of which is the Lamé form

$$\mathbf{S} = 2\mu\mathbf{E} + \lambda\, tr(\mathbf{E})\mathbf{I} \tag{5.38}$$

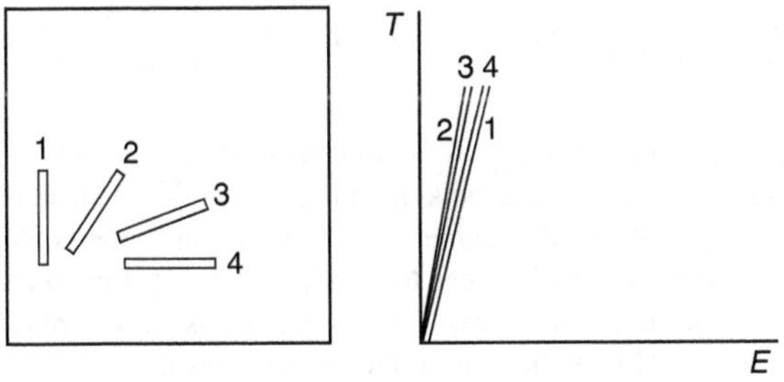

FIGURE 5.16 Illustration of isotropy.

As indicated in Example 3.6 of Chapter 3, in the Lamé relation there exists an orthogonal tensor $\mathbf{Q}_o$ which simultaneously diagonalizes $\mathbf{S}$ and $\mathbf{E}$. This is a formal statement of the property of isotropy. The Lamé form may be inverted to furnish

$$\mathbf{E} = \frac{1}{2\mu}\mathbf{S} - \frac{1}{2\mu}\frac{\lambda}{2\mu + 3\lambda} tr(\mathbf{S})\mathbf{I} \tag{5.39}$$

EXAMPLE 5.15

Invert the relation in Equation 5.36 to express the strain as a function of stress.

Solution

Applying the trace operator to Equation 5.38 results in

$$tr(\mathbf{S}) = 2\mu\, tr(\mathbf{E}) + 3\lambda\, tr(\mathbf{E})$$

from which we conclude that

$$tr(\mathbf{E}) = \frac{tr(\mathbf{S})}{2\mu + 3\lambda}$$

Now

$$\mathbf{S} = 2\mu\mathbf{E} + \lambda\frac{tr(\mathbf{S})}{2\mu + 3\lambda}\mathbf{I}$$

from which it is immediate that

$$\mathbf{E} = \frac{1}{2\mu}\left(\mathbf{S} - \frac{\lambda}{2\mu + 3\lambda} tr(\mathbf{S})\mathbf{I}\right)$$

Now returning to the main development, in the case of uniaxial tension the Lamé relation yields

$$\mathrm{E}_{XX} = \frac{1}{2\mu}\left(1 - \frac{\lambda}{2\mu + 2\lambda}\right)\mathrm{S}_{XX}, \qquad \mathrm{E}_{YY} = \mathrm{E}_{ZZ} = -\frac{1}{2\mu}\frac{\lambda}{2\mu + 2\lambda}\mathrm{E}_{XX} \tag{5.40}$$

Letting $\mathsf{E} = \mathrm{S}_{XX}/\mathrm{E}_{XX}$ and $\nu = -\frac{\mathrm{E}_{YY}}{\mathrm{E}_{XX}}$, it is immediate that $\mathsf{E} = \frac{\mu(2\mu+3\lambda)}{\mu+\lambda}$, $\nu = \frac{\lambda}{2(\mu+\lambda)}$. Another important property is the bulk modulus $\kappa = \frac{1}{3}\frac{tr(\mathrm{S})}{tr(\mathrm{E})}$, which is found to be given by $\kappa = \frac{\mathsf{E}}{3(1-2\nu)} = \frac{2\mu+3\lambda}{3}$. It becomes infinite in the incompressible case $\nu = 1.2$. Finally, the shear modulus is readily seen to satisfy $\frac{1}{\mathsf{E}} = \frac{1}{2\mu} - \frac{\nu}{\mathsf{E}}$, from which $\mu = \frac{\mathsf{E}}{2(1+\nu)}$. The inverse Lamé form of the stress–strain relations is now written in the form $\mathbf{E} = \frac{1+\nu}{\mathrm{E}}\mathbf{S} - \frac{\nu}{\mathrm{E}}tr(\mathbf{S})\mathbf{I}$, and is written out in the six relations as

$$\mathrm{E}_{xx} = \tfrac{1}{\mathsf{E}}[\mathrm{S}_{xx} - \nu(\mathrm{S}_{yy} + \mathrm{S}_{zz})]$$

$$\mathrm{E}_{yy} = \tfrac{1}{\mathsf{E}}[\mathrm{S}_{yy} - \nu(\mathrm{S}_{zz} + \mathrm{S}_{xx})]$$

$$\mathrm{E}_{zz} = \tfrac{1}{\mathsf{E}}[\mathrm{S}_{zz} - \nu(\mathrm{S}_{xx} + \mathrm{S}_{yy})]$$

$$\mathrm{E}_{xy} = \frac{1+\nu}{\mathsf{E}}\mathrm{S}_{xy}$$

$$\mathrm{E}_{yz} = \frac{1+\nu}{\mathsf{E}}\mathrm{S}_{yz}$$

$$\mathrm{E}_{zx} = \frac{1+\nu}{\mathsf{E}}\mathrm{S}_{zx}$$

EXAMPLE 5.16

The $2'' \times 2'' \times 2''$ cube shown below is confined on its sides facing the $\pm x$ faces by rigid frictionless walls. The sides facing the $\pm z$ faces are free. The top and bottom faces are subjected to a compressive force of 100 lb (Figure 5.17). Take $\mathsf{E} = 10^7$ psi and $\nu = 1/3$. Find all nonzero stresses and strains. Find all principal stresses and strains.

Solution

Since parallel sides of the block remain parallel, there is no shear strain:

$$\mathrm{E}_{xy} = 0, \quad \mathrm{E}_{yz} = 0, \quad \mathrm{E}_{zx} = 0$$

It follows from the stress–strain relations that the shear stresses vanish.

$$\mathrm{S}_{xy} = 0, \quad \mathrm{S}_{yz} = 0, \quad \mathrm{S}_{zx} = 0$$

It is immediate that the normal stresses and strains are also the principal stresses and strains.

$$\mathrm{S}_{I} = \mathrm{S}_{xx}, \quad \mathrm{S}_{II} = \mathrm{S}_{yy}, \quad \mathrm{S}_{III} = \mathrm{S}_{zz}, \quad \mathrm{E}_{I} = \mathrm{E}_{xx}, \quad \mathrm{E}_{II} = \mathrm{E}_{yy}, \quad \mathrm{E}_{III} = \mathrm{E}_{zz}$$

We now consider the normal stresses and strains. From the description of the problem

$$\mathrm{E}_{yy} = -\frac{F_y}{A_y}, \quad \mathrm{E}_{zz} = 0, \quad \mathrm{E}_{xx} = 0$$

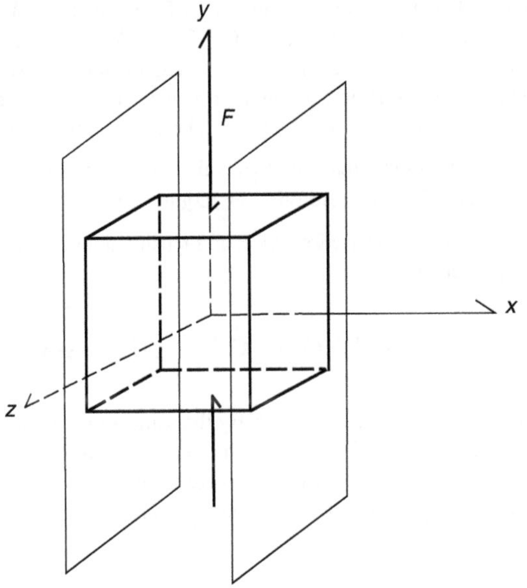

FIGURE 5.17 Stresses and strains in compression of a confined block.

$$\text{Using } \cancel{\mathsf{E}_{xx}} = \frac{1}{\mathsf{E}}\left[\mathrm{S}_{xx} - \nu\left(\mathrm{S}_{yy} + \cancel{\mathrm{S}_{zz}}\right)\right] \text{ furnishes}$$

$$\mathrm{S}_{xx} = \nu \mathrm{S}_{yy} = -\nu \frac{F_y}{A_y} = -\frac{25}{3} \text{ lb}$$

$$\text{Next, from } \mathrm{E}_{yy} = \frac{1}{\mathsf{E}}\Bigg[\mathrm{S}_{yy} - \nu\Big(\cancel{\mathrm{S}_{zz}} + \underbrace{\mathrm{S}_{xx}}_{\nu \mathrm{S}_{yy}}\Big)\Bigg] = \frac{5}{3} \times 10^{-6}$$

$$\mathrm{E}_{yy} = \frac{1-\nu^2}{\mathsf{E}}\mathrm{S}_{yy} = -\frac{1-\nu^2}{\mathsf{E}}\frac{F_y}{A_y} = -\frac{2}{9} \times 10^{-5}$$

Similarly,

$$\mathrm{E}_{zz} = \frac{1}{\mathsf{E}}\left[\cancel{\mathrm{S}_{zz}} - \nu(\mathrm{S}_{xx} + \mathrm{S}_{yy})\right] = -\frac{\nu(1+\nu)}{\mathsf{E}}\mathrm{S}_{yy}$$

from which

$$\mathrm{E}_{zz} = +\frac{\nu(1+\nu)}{\mathsf{E}}\left(\frac{F_y}{A_y}\right) = \frac{10^5}{9}$$

EXAMPLE 5.17

The elements of the 9×9 tangent modulus tensor **C** implied by the Lamé form have already been given in Example 3.6. Find the conditions on λ and μ, and likewise on E and ν, which render it positive definite, which is a requirement for *stability* of the material.

The eigenvalues of a matrix of the form $\boldsymbol{A}+b\boldsymbol{I}$ are given by $\lambda_j(\boldsymbol{A})+b$. Consequently, the eigenvalues of the stiffness tensor $\mathbf{C}=2\mu\boldsymbol{I}_9+\lambda\boldsymbol{i}\boldsymbol{i}^T$ are given by $2\mu+\lambda_j(\boldsymbol{i}\boldsymbol{i}^T)$. The tensor $\boldsymbol{i}\boldsymbol{i}^T$ is of rank 1, so that eight of its nine eigenvalues vanish. The remaining eigenvalue is given by the Rayleigh extremum principle (to be presented in Chapter 10) as $max(\boldsymbol{\zeta}^T\,\boldsymbol{i}\boldsymbol{i}^T\,\boldsymbol{\zeta})$ in which $\boldsymbol{\zeta}$ is a 9×1 vector. The maximizing value of $\boldsymbol{\zeta}$ is in fact $\boldsymbol{i}/\sqrt{\boldsymbol{i}^T\boldsymbol{i}}=\boldsymbol{i}/\sqrt{3}^{|\zeta|=1}$. It follows that the eigenvalues of $\mathbf{C}$ are $2\mu+3\lambda$ once and 2μ eight times. Positive definiteness therefore now requires that $\mu>0$, $\lambda>-\frac{2}{3}\mu$.

We determine the stability restrictions on E and ν using the inverse Lamé relation in Example 5.15. In particular

$$VEC(\mathbf{E})=\mathbf{C}^{-1}VEC(\mathbf{S}),\quad \mathbf{C}^{-1}=\frac{1+\nu}{\mathsf{E}}\mathbf{I}_9-\frac{\nu}{\mathsf{E}}\mathbf{i}\mathbf{i}^T$$

and $\mathbf{C}$ is called the elastic compliance tensor. By the same argument as for μ and λ, we conclude that the eigenvalues of $\mathbf{C}^{-1}$ (i.e., the tangent compliance tensor) are $\frac{1+\nu}{\mathsf{E}}$ eight times and $\frac{1+\nu}{\mathsf{E}}-3\frac{\nu}{\mathsf{E}}=\frac{1-2\nu}{\mathsf{E}}$ once. The material is stable if the eigenvalues of $\mathbf{C}$ and hence of $\mathbf{C}^{-1}$ are positive. It follows after some manipulation that $\mathsf{E}>0$ and $-1\leq\nu\leq 1/2$.

5.9 PRINCIPLE OF VIRTUAL WORK IN LINEAR ELASTICITY

In a continuous medium "dynamic" equilibrium is expressed by the local form of the balance of linear momentum. The finite element method makes use of this equation after it is expressed in variational form as the Principle of Virtual Work

$$\oint \delta u_i \rho \ddot{u}_i\, dV+\oint \delta \mathrm{E}_{ij}\mathrm{S}_{ij}\, dV=\oint \delta u_i t_i\, dS \tag{5.41}$$

under the restriction that the body is referred to a nonrotating coordinate system and has a fixed point within this system. Unrestrained and rotating coordinates are addressed in Chapter 12.

The Principle of Virtual Work is now derived and applied to the following boundary conditions. At a point on the boundary the position vector is denoted by s. The boundary S decomposes into portions S_1, S_2, S_3:

On S_1, the displacement $\boldsymbol{u}(\boldsymbol{s})$ is prescribed as $\boldsymbol{u}_o(\boldsymbol{s})$.
On S_2, the traction $\boldsymbol{t}(\boldsymbol{s})$ is prescribed as $\boldsymbol{t}_o(\mathbf{s})$.
On S_3, the traction satisfies $\boldsymbol{t}(\boldsymbol{s})=\boldsymbol{t}_0(\boldsymbol{s})-\boldsymbol{A}(s)\boldsymbol{u}(\boldsymbol{s})-\boldsymbol{B}(\boldsymbol{s})\ddot{\boldsymbol{u}}(\boldsymbol{s})$.

Under the stated conditions the equation for dynamic equilibrium is given by

$$\rho\ddot{u}_i-\frac{\partial \mathrm{S}_{ij}}{\partial x_j}=0 \tag{5.42}$$

Multiplying through the equation with δu_i and integrating over the domain results in

$$\oint \delta u_i \rho \ddot{u}_i\, dV-\oint \delta u_i\frac{\partial \mathrm{S}_{ij}}{\partial x_j}\, dV=0 \tag{5.43}$$

As usual in variational methods, integration by parts is invoked and furnishes

$$\begin{aligned}\oint \delta u_i \frac{\partial \mathrm{S}_{ij}}{\partial x_j}\mathrm{d}V &= \oint \frac{\partial}{\partial x_j}(\delta u_i \mathrm{S}_{ij})\,\mathrm{d}V - \oint \frac{\partial \delta u_i}{\partial x_j}\mathrm{S}_{ij}\,\mathrm{d}V \\ &= \oint n_j(\delta u_i \mathrm{S}_{ij})\,\mathrm{d}S - \oint\left(\frac{1}{2}\left(\frac{\partial \delta u_i}{\partial x_j}+\frac{\partial \delta u_j}{\partial x_i}\right)+\frac{1}{2}\left(\frac{\partial \delta u_i}{\partial x_j}-\frac{\partial \delta u_j}{\partial x_i}\right)\right)\mathrm{S}_{ij}\,\mathrm{d}V \\ &= \oint \delta u_i t_i\,\mathrm{d}S - \oint(\delta \mathrm{E}_{ij}+\delta\omega_{ij})\mathrm{S}_{ij}\,\mathrm{d}V \end{aligned} \tag{5.44}$$

Now, $\delta\omega_{ij}$ is antisymmetric and S_{ij} is symmetric, implying that $\delta\omega_{ij}\mathrm{S}_{ij}=0$ (Example 3.6 of Chapter 3) follows immediately.

The Principle of Virtual Work is rewritten in the form

$$\oint \delta\mathbf{u}^T\rho\ddot{\mathbf{u}}\,\mathrm{d}V + \oint tr(\delta\mathbf{E}_L\mathbf{S})\,\mathrm{d}V = \oint \delta\mathbf{u}^T\mathbf{t}\,\mathrm{d}S \tag{5.45}$$

and is rewritten yet again using Kronecker product notation as

$$\oint \delta\mathbf{u}^T\rho\ddot{\mathbf{u}}\,\mathrm{d}V + \oint \delta\mathbf{e}_L^T\mathbf{s}\,\mathrm{d}V = \oint \delta\mathbf{u}^T\mathbf{t}\,\mathrm{d}S \tag{5.46}$$

in which $\mathbf{s}=VEC(\mathbf{S})$ and $\mathbf{e}_\mathrm{L}=VEC(\mathbf{E}_L)$.

The constitutive relations of linear elasticity are stated using the general form

$$\mathbf{s}=\boldsymbol{\chi}\mathbf{e}_L \tag{5.47}$$

in which $\boldsymbol{\chi}$ is the positive definite *tangent modulus tensor*. And now

$$\oint \delta\mathbf{u}^T\rho\ddot{\mathbf{u}}\,\mathrm{d}V + \oint \delta\mathbf{e}_L^T\boldsymbol{\chi}\mathbf{e}_L\,\mathrm{d}V = \oint \delta\mathbf{u}^T\mathbf{t}\,\mathrm{d}S \tag{5.48}$$

Suppose a sufficiently accurate approximation exists in the form $\mathbf{u}(\mathbf{X},t)=\boldsymbol{\varphi}^T(\mathbf{X})\boldsymbol{\Phi}\boldsymbol{\gamma}(t)$, in which $\boldsymbol{\varphi}^T(\mathbf{X})$ is a $3\times n$ matrix, $\boldsymbol{\Phi}$ is an $n\times n$ constant nonsingular matrix, and $\boldsymbol{\gamma}(t)$ is an $n\times 1$ vector of parameters such as nodal displacements in a finite element model. As will be seen in Chapter 7, application of the strain–displacement relations serves to derive the subordinate approximation for the linear strains as $\mathbf{e}_L\approx\boldsymbol{\beta}^T(\mathbf{X})\boldsymbol{\Phi}\boldsymbol{\gamma}(t)$. The Principle of Virtual Work now implies that

$$\delta\boldsymbol{\gamma}^T\mathbf{M}\ddot{\boldsymbol{\gamma}}+\delta\boldsymbol{\gamma}^T\mathbf{K}\boldsymbol{\gamma}=\delta\boldsymbol{\gamma}^T\mathbf{f} \tag{5.49}$$

in which $\mathbf{M}$ and $\mathbf{K}$ are $n\times n$ positive definite symmetric matrices and $\mathbf{f}$ is an $n\times 1$ vector. If $\boldsymbol{\gamma}(t)$ represents nodal displacements in a finite element model, $\mathbf{M}$ and $\mathbf{K}$ are called the *mass* and *stiffness matrices*, respectively, and $\mathbf{f}$ is called the *consistent force vector*.

We now consider the boundary conditions more carefully. On S_1 the displacement is prescribed and hence $\delta\mathbf{u} = \mathbf{0}$ and $\delta\boldsymbol{\gamma} = \mathbf{0}$, and there is no contribution to the consistent force. On S_2 the traction is prescribed as $\mathbf{t}_0$ and the contribution is $\mathbf{f}_2 = \boldsymbol{\Phi}^T \int \boldsymbol{\varphi}\mathbf{t}_0 \, dS_2$. Finally, there are both compliant and inertial supports on S_3, furnishing

$$\int \delta\mathbf{u}^T\mathbf{t}\,dS_3 = \boldsymbol{\Phi}^T \int \boldsymbol{\varphi}\mathbf{t}_0\,dS_3 - \boldsymbol{\Phi}^T\left[\int \boldsymbol{\varphi}A(s)\boldsymbol{\varphi}^T dS_3\right]\boldsymbol{\Phi}\boldsymbol{\gamma} - \boldsymbol{\Phi}^T\left[\int \boldsymbol{\varphi}B(s)\boldsymbol{\varphi}^T dS_3\right]\boldsymbol{\Phi}\ddot{\boldsymbol{\gamma}} \tag{5.50}$$

Assuming the foregoing approximations and carrying terms with unknowns to the left-hand side furnishes an alternative form of Equation 5.50:

$$\delta\boldsymbol{\gamma}^T(\mathbf{M} + \mathbf{M}_S)\ddot{\boldsymbol{\gamma}} + \delta\boldsymbol{\gamma}^T(\mathbf{K} + \mathbf{K}_S)\boldsymbol{\gamma} = \delta\boldsymbol{\gamma}^T(\mathbf{f}_2 + \mathbf{f}_3)$$

in which

$$\begin{aligned} \mathbf{M}_S &= \boldsymbol{\Phi}^T\left[\int \boldsymbol{\varphi}B(s)\boldsymbol{\varphi}^T dS_3\right]\boldsymbol{\Phi} \\ \mathbf{K}_S &= \boldsymbol{\Phi}^T\left[\int \boldsymbol{\varphi}A(s)\boldsymbol{\varphi}^T dS_3\right]\boldsymbol{\Phi} \\ \mathbf{f}_3 &= \boldsymbol{\Phi}^T \int \boldsymbol{\varphi}\mathbf{t}_0 dS_3 \end{aligned} \tag{5.51}$$

Clearly, $\mathbf{M}_S$ is the $n \times n$ matrix representing the contribution of *inertial boundary conditions* to the mass matrix and the $n \times n$ matrix $\mathbf{K}_S$ represents the contribution to the stiffness matrix from elastic support on the surface.

EXAMPLE 5.18

The equation of static equilibrium in the presence of body forces, such as gravity, is expressed by

$$\frac{\partial \mathrm{S}_{ij}}{\partial X_j} = -b_i(\mathbf{X},t)$$

Without the body forces, the dynamic Principle of Virtual Work is derived as

$$\int \delta \mathrm{E}_{ij}\mathrm{S}_{ij}\,dV + \int \delta u_i \rho \frac{\partial^2 u_i}{\partial t^2} dV = \int \delta u_i t_i \,dS$$

How should the second equation be modified to include body forces?

Solution

Let **b** denote non-inertial body forces, for example, due to gravity or electromagnetic effects. The Equation of the balance of linear momentum is now stated as

$$\int \rho \ddot{\mathbf{u}}\, \mathrm{d}V = \int \mathbf{t}\, \mathrm{d}S - \int \mathbf{b}\, \mathrm{d}V$$

As before $\int \mathbf{t}\, \mathrm{d}S = \int \mathbf{S}^T \mathbf{n}\, \mathrm{d}S = \int (\nabla^T \mathbf{S})^T \mathrm{d}V$. For the equation to hold over arbitrary subvolumes it is necessary that, now using indicial notation,

$$\frac{\partial \mathrm{S}_{\mathrm{ij}}}{\partial X_{\mathrm{j}}} = \rho \ddot{u}_i + b_i$$

It is easily seen that the same steps as were used to derive the Principle of Virtual Work without the body force **b** now lead to a more general principle in which $\rho \ddot{u}_i + b_i$ replaces $\rho \ddot{u}_i$. In particular,

$$\int \delta \mathbf{e}^T \mathbf{s}\, \mathrm{d}V + \int \delta \mathbf{u}^T (\rho \ddot{\mathbf{u}} + \mathbf{b})\, \mathrm{d}V = \int \delta \mathbf{u}^T \mathbf{t}\, \mathrm{d}S$$

6 Thermal and Thermomechanical Response

6.1 BALANCE OF ENERGY AND PRODUCTION OF ENTROPY

6.1.1 BALANCE OF ENERGY

Sometimes called *The First Law of Thermodynamics*, the balance of energy principle is stated as follows: the rate of change of total energy in a body, including internal energy and kinetic energy, is equal to the corresponding rate of work done by external forces on the body together with the rate of heat added to the body. In rate form

$$\dot{\mathfrak{K}} + \dot{\Xi} = \dot{W} + \dot{Q} \tag{6.1}$$

in which

Ξ is the internal energy with density ξ

$$\dot{\Xi} = \int \rho\dot{\xi}\,\mathrm{d}V \tag{6.2a}$$

$\dot{W}$ is the rate of mechanical work, satisfying

$$\dot{W} = \int \dot{\mathbf{u}}^T\mathbf{t}\,\mathrm{d}S \tag{6.2b}$$

$\dot{Q}$ is the rate of heat input, with heat production h and heat flux $\mathbf{q}$, satisfying

$$\dot{Q} = \int \rho h\,\mathrm{d}V - \int \mathbf{n}^T\mathbf{q}\,\mathrm{d}S \tag{6.2c}$$

and $\dot{\mathfrak{K}}$ is the rate of increase in the kinetic energy,

$$\dot{\mathfrak{K}} = \int \rho\dot{\mathbf{u}}^T\frac{\mathrm{d}\dot{\mathbf{u}}}{\mathrm{d}t}\mathrm{d}V \tag{6.2d}$$

It has been tacitly assumed that all work of external forces is done on the boundary S, and that no work is done by body forces.

Invoking the divergence theorem and balance of linear momentum furnishes

$$\int \left[\rho \dot{\xi} + \dot{\mathbf{u}}^T \left[\rho \frac{\mathrm{d}\dot{\mathbf{u}}}{\mathrm{d}t} - \nabla^T \mathbf{T} \right] - tr(\mathbf{TD}) - \rho h + \nabla^T \mathbf{q} \right] \mathrm{d}V = 0 \tag{6.3}$$

The bracketed term inside the integrand vanishes by virtue of the *balance of linear momentum.* The relation holds for arbitrary volumes, from which the local form of balance of energy, referred to the deformed (current) configuration, is obtained as

$$\rho \dot{\xi} = tr(\mathbf{TD}) - \nabla^T \mathbf{q} + \rho h \tag{6.4}$$

T is the Cauchy stress tensor, which is the stress referred to deformed (current) coordinates, and **D** is the stretching-rate tensor (c.f. Chapter 13). To convert Equation 6.4 to undeformed coordinates relations to be explained in Chapter 13 result in

$$\begin{aligned} \int \mathbf{n}^T \mathbf{q} \,\mathrm{d}S &= \int \mathbf{q}^T J \mathbf{F}^{-T} \mathbf{n}_0 \,\mathrm{d}S_0 \\ &= \int \mathbf{q}_0^T \mathbf{n}_0 \,\mathrm{d}S_0, \quad \mathbf{q}_0 = J \mathbf{F}^{-1} \mathbf{q} \end{aligned} \tag{6.5}$$

in which the subscript "0" indicates undeformed coordinates and **F** is the deformation gradient tensor. In undeformed coordinates, Equation 6.3 is rewritten as

$$\int \left[\rho_0 \dot{\xi} - tr(\mathbf{S}\dot{\mathbf{E}}) - \rho_0 h + \nabla_0^T \mathbf{q}_0 \right] \mathrm{d}V_0 = 0 \tag{6.6a}$$

furnishing the *local form of the balance of energy* as

$$\rho_0 \dot{\xi} - tr(\mathbf{S}\dot{\mathbf{E}}) - \rho_0 h + \nabla_0^T \mathbf{q}_0 = 0 \tag{6.6b}$$

Here, **S** is the second Piola–Kirchhoff or *nominal* stress, which is also explained later in Chapter 13 and which is the stress-referred to deformed coordinates.

6.1.2 Entropy Production Inequality

Generalizing the thermodynamics of ideal and nonideal gases, the entropy production inequality for continua is introduced as follows (Callen, 1985).

$$\dot{\mathrm{H}} = \int \rho \dot{\eta} \,\mathrm{d}V \geq \int \frac{h}{\mathrm{T}} \mathrm{d}V - \int \frac{\mathbf{n}^T \mathbf{q}}{\mathrm{T}} \mathrm{d}S \tag{6.7a}$$

in which $\dot{\mathrm{H}}$ is the total entropy, η is the specific entropy per unit mass, and T is the absolute temperature. This relation provides a "framework" for describing the irreversible nature of dissipative processes such as heat flow and plastic deformation. We apply the divergence theorem to the surface integral leading to the local form of the entropy production inequality:

$$\rho \mathrm{T}\dot{\eta} \geq h - \nabla^T \mathbf{q} + \mathbf{q}^T \nabla \mathrm{T}/\mathrm{T} \tag{6.7b}$$

The counterpart of Equation 6.7b in undeformed coordinates is

$$\rho_0 \mathrm{T}\dot{\eta} \geq h - \nabla_0^T \mathbf{q}_0 + \mathbf{q}_0^T \nabla_0 \mathrm{T}/\mathrm{T} \tag{6.7c}$$

6.1.3 Thermodynamic Potentials in Reversible Processes

The balance of energy introduces the internal energy Ξ, which is an extensive variable—its value accumulates over the domain. The mass and volume averages of extensive variables will also be referred to as extensive variables. This contrasts with intensive or pointwise variables such as the stresses and the temperature. Another extensive variable is the entropy H. In reversible elastic systems, the entropy is completely determined by heat input according to

$$\dot{\mathrm{Q}} = \mathrm{T}H \tag{6.8}$$

(In Chapters 16 and 17 we shall address several irreversible effects including plasticity, viscosity, and heat conduction.) By virtue of Equation 6.8, in undeformed coordinates the balance of energy for *reversible processes* may be written as

$$\rho_0 \dot{\xi} = tr(\mathbf{S}\dot{\mathbf{E}}) + \rho_0 \mathrm{T}\dot{\eta} \tag{6.9}$$

As before, $\mathbf{S}$ is the second Piola–Kirchhoff stress and now $\mathbf{E}$ is the Lagrangian strain to be introduced in Chapter 13. We now invoke conditions for the right-hand side of Equation 6.9 to be uniquely integrable, which renders the internal energy dependent only on the current values of the *state variables* consisting of $\mathbf{E}$ and η. For the sake of understanding we may think of T as a thermal stress and η as a thermal strain. Clearly $\dot{\eta} = 0$ if there is no heat input across the surface or generated in the volume. Consequently the entropy is an attractive state variable for representing *adiabatic processes.*

In Callen (1985), a development is given for the stability of thermodynamic equilibrium according to which, under suitable conditions, the strain and the entropy density assume values which maximize the internal energy. Other thermodynamic potentials, depending on alternate state variables, may be introduced by a Lorentz transformation illustrated below. Doing so is attractive if the new state variables are accessible to measurement. For example, the Gibbs free energy (density) is a function of the intensive variables $\mathbf{S}$ and T:

$$\rho_0 g = \rho_0 \xi - tr(\mathbf{S}\mathbf{E}) - \rho_0 \mathrm{T}\eta \tag{6.10a}$$

so that

$$\rho_0 \dot{g} = -tr(\mathbf{E}\dot{\mathbf{S}}) - \rho_0 \eta \dot{\mathrm{T}} \tag{6.10b}$$

Stability of thermodynamic equilibrium requires that **S** and T assume values which minimize g under suitable conditions. This potential is of interest in fluids experiencing adiabatic conditions in which the pressure (stress) is accessible to measurement using, for example, pitot tubes. It is also commonly used in phase changes.

In solid continua the stress is often more difficult to measure than the strain. Accordingly, for solids the Helmholtz free energy (density) f is introduced using

$$\rho_0 f = \rho_0 \xi - \rho_0 \mathrm{T} \eta \tag{6.11a}$$

furnishing

$$\rho_0 \dot{f} = tr(\mathbf{S}\dot{\mathbf{E}}) - \rho_0 \eta \dot{\mathrm{T}} \tag{6.11b}$$

It is evident that f is a function of an intensive variable and an extensive variable. At thermodynamic equilibrium under suitable conditions it exhibits a (stationary) saddle point rather than a maximum or a minimum. Finally, for the sake of completeness, we mention a fourth potential, known as the enthalpy $\rho_0 h = \rho_0 \xi - tr(\mathbf{SE})$, in terms of which local balance of energy now is expressed as

$$\rho_0 \dot{h} = -tr(\mathbf{E}\dot{\mathbf{S}}) + \rho_0 \mathrm{T} \dot{\eta} \tag{6.12}$$

The enthalpy likewise is a function of an extensive variable and an intensive variable and exhibits a saddle point at equilibrium. It is attractive in fluids under adiabatic conditions.

6.2 CLASSICAL COUPLED LINEAR THERMOELASTICITY

The classical theory of coupled thermoelasticity in isotropic media corresponds to the restriction to the linear strain tensor, $\mathbf{E} \approx \mathbf{E}_\mathrm{L}$, and to the stress–strain temperature relation

$$\mathbf{T} = 2\mu \mathbf{E} + \lambda [tr(\mathbf{E}) - \alpha(\mathrm{T} - \mathrm{T}_0)]\mathrm{I} \tag{6.13}$$

Also $\rho \approx \rho_0$. Here, α is the volumetric coefficient of thermal expansion, typically a very small number in metals and elastomers. If the temperature increases without stress being applied, the volume strain increases according to $e_{\mathrm{vol}} = tr(\mathbf{E}) = \alpha(\mathrm{T} - \mathrm{T}_0)$. Thermoelastic processes are assumed in the present context to be reversible, in which event $-\nabla \cdot \mathbf{q} + h = \rho \mathrm{T} \dot{\eta}$. It is also assumed that the specific heat at constant strain, c_e, given by

$$c_e = \mathrm{T} \left. \frac{\partial \eta}{\partial \mathrm{T}} \right|_{\mathbf{E}} \tag{6.14}$$

is constant. The balance of energy may now be restated as

$$\rho_0\dot{\xi} = tr(\mathbf{T}\dot{\mathbf{E}}) + \rho_0 \mathrm{T}\dot{\eta} \tag{6.15}$$

Recall that ξ is a function of the extensive variables $\mathbf{E}$ and η. To convert to $\mathbf{E}$ and T as state variables which are accessible to measurement, we again invoke the Helmholtz free energy $f = \xi - \mathrm{T}\eta$. Since f is an exact differential to ensure path independence, we infer the Maxwell relation

$$-\rho\left(\frac{\partial\eta}{\partial\mathbf{E}}\right)\bigg|_{\mathrm{T}} = \frac{\partial\mathbf{T}}{\partial\mathrm{T}}\bigg|_{\mathbf{E}} \tag{6.16}$$

Returning to the energy balance equation, to express it using T and $\mathbf{E}$ as state variables, we have

$$\begin{aligned}\rho_0\dot{\xi} &= tr\left(\rho_0\frac{\partial\xi}{\partial\mathbf{E}}\bigg|_{\eta}\dot{\mathbf{E}}\right) + \rho_0\frac{\partial\xi}{\partial\eta}\bigg|_{\mathbf{E}}\dot{\eta} \\ &= tr\left(\left[\rho_0\frac{\partial\xi}{\partial\mathbf{E}}\bigg|_{\eta} + \rho_0\frac{\partial\xi}{\partial\eta}\bigg|_{\mathbf{E}}\frac{\partial\eta}{\partial\mathbf{E}}\bigg|_{\mathrm{T}}\right]\dot{\mathbf{E}}\right) + \rho_0\left[\frac{\partial\xi}{\partial\eta}\bigg|_{\mathbf{E}}\frac{\partial\eta}{\partial\mathrm{T}}\bigg|_{\mathbf{E}}\right]\dot{\mathrm{T}}\end{aligned} \tag{6.17}$$

Also note that

$$\mathbf{T} = \frac{\partial\xi}{\partial\mathbf{E}}\bigg|_{\eta}, \quad \mathrm{T} = \frac{\partial\xi}{\partial\eta}\bigg|_{\mathbf{E}}$$

and hence

$$\begin{aligned}\rho_0\dot{\xi} &= tr\left(\left[\rho_0\frac{\partial\xi}{\partial\mathbf{E}}\bigg|_{\eta} + \rho_0\frac{\partial\xi}{\partial\eta}\bigg|_{\mathbf{E}}\frac{\partial\eta}{\partial\mathbf{E}}\bigg|_{\mathrm{T}}\right]\dot{\mathbf{E}}\right) + \rho_0\left[\frac{\partial\xi}{\partial\eta}\bigg|_{\mathbf{E}}\frac{\partial\eta}{\partial\mathrm{T}}\bigg|_{\mathbf{E}}\right]\dot{\mathrm{T}} \\ &= tr\left(\left[\mathbf{T} - \mathrm{T}\frac{\partial\mathbf{T}}{\partial\mathrm{T}}\bigg|_{\mathbf{E}}\right]\dot{\mathbf{E}}\right) + \rho_0\mathrm{T}\frac{\partial\eta}{\partial\mathrm{T}}\bigg|_{\mathbf{E}}\dot{\mathrm{T}}\end{aligned} \tag{6.18}$$

We previously identified the coefficient of specific heat, assumed constant, as $c_e = \mathrm{T}\frac{\partial\eta}{\partial\mathrm{T}}|_{\mathbf{E}}$ so that

$$\rho_0\dot{\xi} = tr\left(\left[\mathbf{T} - \mathrm{T}\frac{\partial\mathbf{T}}{\partial\mathrm{T}}\right]\dot{\mathbf{E}}\right) + \rho_0 c_e\dot{\mathrm{T}} \tag{6.19}$$

The local form of the balance of energy equation now becomes

$$
\begin{aligned}
-\nabla^T \mathbf{q} &= \rho_0 \dot{\xi} - tr(\mathbf{T}\dot{\mathbf{E}}) \\
&= tr\left(\left[\mathbf{T} - \mathrm{T}\frac{\partial \mathbf{T}}{\partial \mathrm{T}}\right]\dot{\mathbf{E}}\right) + \rho_0 c_e \dot{\mathrm{T}} - tr(\mathbf{T}\dot{\mathbf{E}}) \\
&= tr\left(\left[-\mathrm{T}\frac{\partial \mathbf{T}}{\partial \mathrm{T}}\right]\dot{\mathbf{E}}\right) + \rho_0 c_e \dot{\mathrm{T}}
\end{aligned} \tag{6.20}
$$

From Equation 6.13, approximating T as T_0 yields $\mathrm{T}_0 \frac{\partial \mathbf{T}}{\partial \mathrm{T}} \approx -\alpha\lambda \mathrm{T}_0 \mathbf{I}$. Now assuming the isotropic version of Fourier's Law $\mathbf{q} = -k\nabla \mathrm{T}$ in which the conductivity k is assumed to be constant, the *thermal field equation* emerges

$$
-k\nabla^2 T = \alpha\lambda T_0\, tr\left(\dot{\mathbf{E}}\right) + \rho_0 c_e \dot{\mathrm{T}} \tag{6.21}
$$

Equation 6.21 is subjected to variational operations in Section 6.3 to obtain the finite element equation of the thermal field.

The balance of linear momentum together with the stress–strain and strain–displacement relations of linear isotropic thermoelasticity imply that

$$
\frac{\partial}{\partial x_j}\left[2\mu\left[\frac{1}{2}\left(\frac{\partial u_i}{\partial x_j} + \frac{\partial u_j}{\partial x_i}\right)\right] + \lambda\left[\frac{\partial u_k}{\partial x_k} - \alpha(T - T_0)\delta_{ij}\right]\right] = \rho\frac{\partial^2 u_i}{\partial t^2} \tag{6.22}
$$

from which we obtain the mechanical field equation (Navier's equation for thermoelasticity)

$$
\mu\nabla^2 \mathbf{u} + (\lambda + \mu)\nabla\, tr(\mathbf{E}) - \alpha\lambda\nabla \mathrm{T} = \rho\frac{\partial \mathbf{u}}{\partial t^2} \tag{6.23}
$$

The thermal field equation depends on the mechanical field through the term $\alpha\lambda \mathrm{T}_0\, tr(\dot{\mathbf{E}})$. Consequently, if $\mathbf{E}$ is static there is no coupling. Similarly, the mechanical field depends on the thermal field through $\alpha\lambda\nabla \mathrm{T}$, which likely is quite small in, say, metals, if the assumption of reversibility is a reasonable approximation.

We next derive the entropy. Since $c_e = \mathrm{T}\frac{\partial \eta}{\partial T}\big|_{\mathbf{E}}$ is constant, we conclude that it has the form

$$
\rho_0 \eta = \rho_0 \eta_0 + \rho_0 c_e \ln(\mathrm{T}/\mathrm{T}_0) + \rho_0 \eta^*(\mathbf{E}) \tag{6.24}
$$

where $\eta^*(\mathbf{E})$ remains to be determined. Without loss of generality, we take η_0 to vanish. But $\rho_0 \frac{\partial \eta}{\partial \mathbf{E}}\big|_{\mathrm{T}} = \rho_0 \frac{\partial \eta^*}{\partial \mathbf{E}} = -\frac{\partial \mathbf{T}}{\partial T}\big|_{\mathbf{E}} = \alpha\lambda \mathbf{I}$, implying that

$$
\rho_0 \eta = \rho_0 \eta_0 + \rho_0 c_e \ln(\mathrm{T}/\mathrm{T}_0) + \alpha\lambda\, tr(\mathbf{E}) \tag{6.25}
$$

Now consider f, for which the fundamental relation Equation 6.11b implies

$$\left.\frac{\partial f}{\partial \mathrm{T}}\right|_{\mathbf{E}} = -\eta \quad \rho_0 \left.\frac{\partial f}{\partial \mathbf{E}}\right|_{\mathrm{T}} = \mathbf{T} \tag{6.26}$$

Integrating the entropy,

$$\rho_0 f = \rho_0 f_0 - \rho_0 c_e[\mathrm{T}(\ln(\mathrm{T}/\mathrm{T}_0) - 1)] - \alpha\lambda\, tr(\mathbf{E})\mathrm{T} + f^*(\mathrm{E}) \tag{6.27}$$

in which $f^*(\mathbf{E})$ remains to be determined. Integrating the stress,

$$\rho_0 f = \rho_0 f_0 + \mu\, tr\left(\mathbf{E}^2\right) + \tfrac{\lambda}{2} tr^2(\mathbf{E}) - \alpha\lambda\, tr(\mathbf{E})(\mathrm{T} - \mathrm{T}_0) + f^{**}(\mathrm{T}) \tag{6.28}$$

in which $f^{**}(\mathrm{T})$ likewise remains to be determined. Reconciling Equations 6.27 and 6.28 now furnishes

$$\rho_0 f = \rho_0 f_0 + \mu tr\left(\mathbf{E}^2\right) + \tfrac{\lambda}{2} tr^2(\mathbf{E}) - \alpha\lambda\, tr(\mathbf{E})(\mathrm{T} - \mathrm{T}_0) - \rho_0 c_e\left[(\mathrm{T}\ln(\mathrm{T}/\mathrm{T}_0) - 1)\right] \tag{6.29}$$

EXAMPLE 6.1

(i) Invert the constitutive relations of classical thermoelasticity
(ii) Express the linear thermal expansion coefficient α_L, appearing in the uniaxial tension case, as

$$\mathrm{E}_{xx} = \frac{\mathrm{S}_{xx}}{\mathsf{E}} + \alpha_L(\mathrm{T} - \mathrm{T}_0)$$

in terms of the volumetric thermal expansion coefficient.

Solution

Since classical thermoelasticity assumes small strain, there is no need to distinguish between the Cauchy and the second Piola–Kirchhoff stresses.

The thermoelastic version of Hooke's law is

$$\mathrm{S}_{ij} = 2\mu \mathrm{E}_{ij} + \lambda \mathrm{E}_{kk}\delta_{ij} - \lambda\alpha(\mathrm{T} - \mathrm{T}_0)\delta_{ij}$$

Taking the trace throughout furnishes

$$\mathrm{S}_{kk} = (2\mu + 3\lambda)\mathrm{E}_{kk} - 3\lambda\alpha(\mathrm{T} - \mathrm{T}_0)$$

and

$$\mathrm{E}_{kk} = \frac{1}{2\mu + 3\lambda}\{\mathrm{S}_{kk} + 3\lambda\alpha(\mathrm{T} - \mathrm{T}_0)\}$$

Upon substitution,

$$S_{ij} = 2\mu E_{ij} + \lambda \frac{1}{2\mu + 3\lambda}\{S_{kk} + 3\lambda\alpha(T - T_0)\}\delta_{ij} - \lambda\alpha(T - T_0)\delta_{ij}$$

and

$$2\mu E_{ij} = S_{ij} - \lambda \frac{1}{2\mu + 3\lambda} S_{kk}\delta_{ij} - \left(\lambda - \frac{3\lambda^2}{2\mu + 3\lambda}\right)\alpha(T - T_0)\delta_{ij}$$

and the inverse relation is now

$$S_{ij} = 2\mu E_{ij} + \lambda E_{kk}\delta_{ij} - \lambda\alpha(T - T_0)\delta_{ij}$$

$$S_{kk} = (2\mu + 3\lambda)E_{kk} - 3\lambda\alpha(T - T_0)$$

$$E_{kk} = \frac{1}{2\mu + 3\lambda}\{S_{kk} + 3\lambda\alpha(T - T_0)\}$$

$$S_{ij} = 2\mu E_{ij} + \lambda \frac{1}{2\mu + 3\lambda}\{S_{kk} + 3\lambda\alpha(T - T_0)\}\delta_{ij} - \lambda\alpha(T - T_0)\delta_{ij}$$

$$2\mu E_{ij} = S_{ij} - \lambda \frac{1}{2\mu + 3\lambda} S_{kk}\delta_{ij} - \left(\lambda - \frac{3\lambda^2}{2\mu + 3\lambda}\right)\alpha(T - T_0)\delta_{ij}$$

and so, following obvious steps,

$$E_{ij} = \frac{1 + \nu}{E} S_{ij} - \frac{\nu}{E} S_{kk}\delta_{ij} + \lambda\left(\frac{1 - 2\nu}{E}\right)\alpha(T - T_0)\delta_{ij}$$

We next express λ in terms of E and ν.

$$\begin{aligned}\frac{2\mu + 3\lambda}{3} &= \frac{2}{3}\frac{E}{2(1 + \nu)} + \lambda \\ &= \frac{E}{3(1 - 2\nu)} - \frac{E}{3(1 + \nu)}\end{aligned}$$

so that

$$\begin{aligned}\lambda &= \frac{E[1 + \nu - (1 - 2\nu)]}{3(1 + \nu)(1 - 2\nu)} \\ &= \frac{E\nu}{(1 + \nu)(1 - 2\nu)}\end{aligned}$$

The inverse is now obtained as

$$E_{ij} = \frac{1 + \nu}{E} S_{ij} - \frac{\nu}{E} S_{kk}\delta_{ij} + \frac{\nu}{1 + \nu}\alpha(T - T_0)\delta_{ij}$$

In the absence of stress

$$E_{kk} = 3\frac{\nu}{(1 + \nu)}\alpha(T - T_0)$$

The Kronecker Product form of the inverse relation is extended from Chapter 5 as

$$\mathbf{e} = \left(\frac{1+\nu}{\mathsf{E}}\mathbf{I}_9 - \frac{\nu}{\mathsf{E}}\mathbf{i}\mathbf{i}^T\right)\mathbf{s} + \frac{\nu\alpha}{1+\nu}(\mathrm{T}-\mathrm{T}_0)\mathbf{i}$$

in which $\mathbf{e} = VEC(\mathbf{E})$, $\mathbf{s} = VEC(\mathbf{S})$, and $\mathbf{i} = VEC(\mathbf{I})$. It is elementary that

$$\left.\frac{\partial \mathbf{e}}{\partial \mathbf{s}}\right|_{\mathrm{T}} = \left(\frac{1+\nu}{\mathsf{E}}\mathbf{I}_9 - \frac{\nu}{\mathsf{E}}\mathbf{i}\mathbf{i}^T\right) \quad \text{and} \quad \left.\frac{\partial \mathbf{e}}{\partial \mathrm{T}}\right|_{\mathrm{S}} = \frac{\nu\alpha}{1+\nu}\mathbf{i}$$

Finally, the linear thermal expansion coefficient is identified as

$$\alpha_{\mathrm{L}} = \frac{\nu}{(1+\nu)}\alpha$$

To verify this outcome we return to the Lamé form and set the stress to zero.

$$0 = 2\mu \mathrm{E}_{ij} + \lambda \mathrm{E}_{kk}\delta_{ij} - \lambda\alpha(\mathrm{T}-\mathrm{T}_0)\delta_{ij}$$

$$(2\mu + 3\lambda)\mathrm{E}_{kk} = 3\lambda\alpha(\mathrm{T}-\mathrm{T}_0) \rightarrow \mathrm{E}_{kk} = \frac{3\lambda\alpha}{(2\mu+3\lambda)}(\mathrm{T}-\mathrm{T}_0)$$

$$\frac{3\lambda\alpha}{(2\mu+3\lambda)} = \frac{3(1-2\nu)}{\mathsf{E}}\frac{\mathsf{E}\nu}{(1+\nu)(1-2\nu)} \rightarrow 3\frac{\nu}{(1+\nu)}$$

EXAMPLE 6.2

In classical thermoelasticity, derive the specific heat at constant stress c_s, rather than at constant strain.

Solution

In classical thermoelasticity the internal energy satisfies

$$\rho\dot{\xi} = tr\left(\left[\mathbf{S} - \mathrm{T}\left.\frac{\partial \mathbf{S}}{\partial \mathrm{T}}\right|_{\mathrm{E}}\right]\dot{\mathbf{E}}\right) + \rho c_e \dot{\mathrm{T}}$$

$$= tr\left(\left[\mathrm{S}_{ij} + \alpha\lambda \mathrm{T}\delta_{ij}\right]\left[\frac{\partial \mathrm{E}_{ij}}{\partial \mathbf{S}_{pq}}\dot{\mathbf{S}}_{pq} + \frac{\partial \varepsilon_{ij}}{\partial \mathrm{T}}\dot{\mathrm{T}}\right]\right) + \rho c_e \dot{\mathrm{T}}$$

$$= \left[(\mathbf{s} + \alpha\lambda \mathrm{T}\mathbf{i})^T \left.\frac{\partial \mathbf{e}}{\partial \mathrm{T}}\right|_{\mathrm{s}} + \rho c_e\right]\dot{\mathrm{T}}$$

The inverse thermoelastic relations of Example 6.1 are invoked to furnish

$$c_S = \left[c_e + \frac{(\mathbf{s} + \alpha\lambda \mathrm{T}\mathbf{i})^T \frac{\nu\alpha}{1+\nu}\mathbf{i}}{\rho}\right]$$

$$\approx c_e + \frac{\nu\alpha}{1+\nu}\frac{tr(\mathbf{S}) + \mathrm{T}_0\frac{\nu\alpha\mathsf{E}}{(1+\nu)(1-2\nu)}}{\rho}$$

EXAMPLE 6.3

Determine the adiabatic elastic modulus in uniaxial tension.

SOLUTION

The issue of an adiabatic modulus arises in situations in which a test specimen is thermally insulated to prevent heat transfer. Under uniaxial tension the stress–strain–temperature of classical thermoelasticity is written as

$$\frac{\sigma_{xx}}{\mathsf{E}} = \varepsilon_{xx} - \alpha_L(\mathrm{T} - \mathrm{T}_0)$$

in which α_L is the linear coefficient of thermal expansion. Under adiabatic conditions the entropy is constant so that $\dot{\eta} = 0$. Accordingly,

$$\begin{aligned} 0 &= \rho\, \mathrm{d}\eta \\ &= \rho \frac{\partial \eta}{\partial \mathrm{T}} \mathrm{dT} + \rho \frac{\partial \eta}{\partial \mathrm{E}_{xx}} \mathrm{dE}_{xx} \\ &= \frac{\rho c_e}{\mathrm{T}} \mathrm{dT} - \frac{\mathrm{dE}_{xx}}{\mathrm{dT}} \mathrm{dE}_{xx} \end{aligned}$$

Note that $\frac{\rho c_e}{\mathrm{T}} \approx \frac{\rho c_e}{\mathrm{T}_0}, -\frac{\mathrm{dS}_{xx}}{\mathrm{dT}} = \alpha_L \mathsf{E}$. Accordingly, under adiabatic conditions $\mathrm{dT} = -\frac{\alpha_L \mathsf{E}\mathrm{T}_0}{\rho c_e} \mathsf{dE}_{xx}$. Of course E is the isothermal elastic modulus.
Next,

$$\begin{aligned} \mathrm{dS}_{xx} &= \left.\frac{\partial \mathrm{S}_{xx}}{\partial \mathrm{E}_{xx}}\right|_{\mathrm{T}} \mathrm{dE}_{xx} + \left.\frac{\partial \mathrm{S}_{xx}}{\partial \mathrm{T}}\right|_{\varepsilon_{xx}} \mathrm{dT} \\ &= \mathsf{E}\, \mathrm{dE}_{xx} - \alpha_L \mathsf{E}\, \mathrm{dT} \\ &= \mathsf{E}\left(1 + \frac{{\alpha_L}^2 \mathsf{E}\mathrm{T}_0}{\rho c_e}\right) \mathrm{dE}_{xx} \end{aligned}$$

and finally the adiabatic elastic modulus is

$$\mathsf{E}_{\mathrm{adiabatic}} = \mathsf{E}\left(1 + \frac{{\alpha_L}^2 \mathsf{E}\mathrm{T}_0}{\rho c_e}\right)$$

EXAMPLE 6.4

Neglecting mechanical effects, express the thermal equilibrium equation in:

(i) Cylindrical coordinates
(ii) Spherical coordinates

SOLUTION

In the absence of mechanical effects the thermal equilibrium equation is given by

$$-\nabla^T k_T \nabla \mathrm{T} + \rho c_e \dot{\mathrm{T}} = 0$$

(i) Cylindrical coordinates
Referring to Section 2.7 in Chapter 2, we have the expression for the temperature gradient in cylindrical coordinates as

$$\nabla \mathrm{T} = \frac{\partial \mathrm{T}}{\partial r}\mathbf{e}_r + \frac{1}{r}\frac{\partial \mathrm{T}}{\partial \theta}\mathbf{e}_\theta + \frac{\partial \mathrm{T}}{\partial z}\mathbf{e}_z$$

The divergence of the gradient (cf. Equation 2.87), still using cylindrical coordinates, now gives

$$\begin{aligned}\nabla^T k_\mathrm{T} \nabla \mathrm{T} &= k_\mathrm{T}\left[\frac{1}{r}\frac{\partial}{\partial r}\left(r\frac{\partial \mathrm{T}}{\partial r}\right) + \frac{1}{r}\frac{\partial}{\partial \theta}\left(\frac{1}{r}\frac{\partial \mathrm{T}}{\partial \theta}\right) + \frac{\partial}{\partial z}\left(\frac{\partial \mathrm{T}}{\partial z}\right)\right] \\ &= k_\mathrm{T}\left[\frac{1}{r}\frac{\partial \mathrm{T}}{\partial r} + \frac{\partial^2 \mathrm{T}}{\partial r^2} + \frac{1}{r^2}\frac{\partial^2 \mathrm{T}}{\partial \theta^2} + \frac{\partial^2 \mathrm{T}}{\partial z^2}\right]\end{aligned}$$

Hence in the absence of mechanical effects the thermal equilibrium equation is expressed in cylindrical coordinates as

$$k_\mathrm{T}\left[\frac{1}{r}\frac{\partial \mathrm{T}}{\partial r} + \frac{\partial^2 \mathrm{T}}{\partial r^2} + \frac{1}{r^2}\frac{\partial^2 \mathrm{T}}{\partial \theta^2} + \frac{\partial^2 \mathrm{T}}{\partial z^2}\right] = \rho c_e \frac{\partial \mathrm{T}}{\partial t}$$

(ii) Spherical coordinates
Referring to the Appendix in Chapter 2, the expression for the temperature gradient in spherical coordinates is recognized to be

$$\nabla T = \frac{\partial \mathrm{T}}{\partial r}\mathbf{e}_r + \frac{1}{r\cos\phi}\frac{\partial \mathrm{T}}{\partial \theta}\mathbf{e}_\theta + \frac{1}{r}\frac{\partial \mathrm{T}}{\partial \phi}\mathbf{e}_\phi$$

Applying the divergence furnishes

$$\begin{aligned}\nabla^T k_\mathrm{T} \nabla \mathrm{T} &= k_\mathrm{T}\left[\frac{1}{r^2}\frac{\partial}{\partial r}\left(r^2\frac{\partial \mathrm{T}}{\partial r}\right) + \frac{1}{r\cos\phi}\frac{\partial}{\partial \theta}\left(\frac{1}{r\cos\phi}\frac{\partial \mathrm{T}}{\partial \theta}\right) + \frac{1}{r\cos\phi}\frac{\partial}{\partial \phi}\left(\cos\phi\frac{1}{r}\frac{\partial \mathrm{T}}{\partial \phi}\right)\right] \\ &= k_\mathrm{T}\left[\frac{2}{r}\frac{\partial \mathrm{T}}{\partial r} + \frac{\partial^2 \mathrm{T}}{\partial r^2} + \frac{1}{r^2\cos^2\phi}\frac{\partial^2 \mathrm{T}}{\partial \theta^2} - \frac{\tan\phi}{r^2}\frac{\partial \mathrm{T}}{\partial \phi} + \frac{1}{r^2}\frac{\partial^2 \mathrm{T}}{\partial \phi^2}\right]\end{aligned}$$

Hence, in the absence of mechanical effects the thermal equilibrium equation is expressed in spherical coordinates as,

$$k_{\mathrm{T}}\left[\frac{2}{r}\frac{\partial \mathrm{T}}{\partial r}+\frac{\partial^2 \mathrm{T}}{\partial r^2}+\frac{1}{r^2\cos^2\phi}\frac{\partial^2 \mathrm{T}}{\partial \theta^2}-\frac{\tan\phi}{r^2}\frac{\partial \mathrm{T}}{\partial \phi}+\frac{1}{r^2}\frac{\partial^2 \mathrm{T}}{\partial \phi^2}\right]=\rho c_e\frac{\partial \mathrm{T}}{\partial t}$$

6.3 THERMAL AND THERMOMECHANICAL ANALOGS OF THE PRINCIPLE OF VIRTUAL WORK AND ASSOCIATED FINITE ELEMENT EQUATIONS

6.3.1 Conductive Heat Transfer

Neglecting coupling to the mechanical field and assuming the isotropic form of Fourier's Law, the thermal field equation in an isotropic medium experiencing small deformation may be written as

$$-\nabla^T k_{\mathrm{T}}\nabla \mathrm{T}+\rho c_e\dot{\mathrm{T}}=0 \tag{6.30}$$

We now construct a thermal counterpart of the Principle of Virtual Work. Multiplying by $\delta\mathrm{T}$, the variation of the unknown to be determined, and applying integration by parts, we obtain

$$\int \delta\nabla^T\mathrm{T}k_T\nabla\mathrm{T}\,\mathrm{d}V+\int\delta\mathrm{T}\rho c_e\dot{\mathrm{T}}\,\mathrm{d}V=\int\delta\mathrm{T}(-\mathbf{n}^T\mathbf{q})\,\mathrm{d}S \tag{6.31}$$

Clearly T is regarded as the primary variable and the associated secondary variable is $(-\mathbf{n}^T\mathbf{q})$. Suppose that the boundary is decomposed into three segments: $S=S_I+S_{II}+S_{III}$. On S_I the temperature T is prescribed as, say, T_1. It follows that $\delta\mathrm{T}=0$ on S_I. On S_{II} the heat flux $\mathbf{q}$ is prescribed as $\mathbf{q}_1$. Consequently $\delta\mathrm{T}(-\mathbf{n}^T\mathbf{q})\rightarrow\delta\mathrm{T}(-\mathbf{n}^T\mathbf{q}_1)$. On S_{III} the heat flux is dependent on the surface temperature through a heat transfer matrix $\mathbf{h}$: $\mathbf{q}=\mathbf{q}_0-\mathbf{h}(\mathrm{T}-\mathrm{T}_0)$. The right-hand side of Equation 6.30 now becomes

$$-\int_S\delta\mathrm{T}\mathbf{n}^T\mathbf{q}\,\mathrm{d}S=-\int_{S_{II}}\delta\mathrm{T}\mathbf{n}^T\mathbf{q}_0\,\mathrm{d}S-\int_{S_{III}}\delta\mathrm{T}\mathbf{n}^T\mathbf{q}_0\,\mathrm{d}S+\int_{S_{III}}\delta\mathrm{T}\mathbf{n}^T\mathbf{h}(\mathrm{T}-\mathrm{T}_0)\,\mathrm{d}S \tag{6.32}$$

Next T is approximated using an interpolation function of the form

$$\mathrm{T}-\mathrm{T}_0\sim\mathbf{N}_{\mathrm{T}}^T(\mathbf{x})\boldsymbol{\theta}(t),\quad \delta\mathrm{T}\sim\mathbf{N}_{\mathrm{T}}^T(\mathbf{x})\delta\boldsymbol{\theta}(t) \tag{6.33}$$

from which we obtain

$$\nabla\mathrm{T}\sim\mathbf{B}_{\mathrm{T}}^T(\mathbf{x})\boldsymbol{\theta}(t),\quad \delta\nabla\mathrm{T}\sim\mathbf{B}_{\mathrm{T}}^T(\mathbf{x})\delta\boldsymbol{\theta}(t) \tag{6.34}$$

and $\mathbf{B}_{\mathrm{T}}$ is the thermal analog of the strain–displacement matrix $\boldsymbol{\beta}^T\boldsymbol{\Phi}$, to be illustrated in Example 6.5 and presented in detail in Chapter 7. Upon substitution of the interpolation models, the thermal field equation now reduces to the system of ordinary differential equations

$$\mathbf{K_T}\boldsymbol{\theta}(t) + \mathbf{M_T}\dot{\boldsymbol{\theta}}(t) = \mathbf{f_T} \tag{6.35}$$

in which

$$\begin{aligned}
\mathbf{K_T} &= \mathbf{K_{T1}} + \mathbf{K_{T2}} && \textit{Thermal Stiffness Matrix}\\
\mathbf{K_{T1}} &= \int \mathbf{B_T}(\mathbf{x})k_T\mathbf{B}_T^T(\mathbf{x})\mathrm{d}V && \textit{Conductance Matrix}\\
\mathbf{K_{T2}} &= \int_{S_{III}} \mathbf{N_T}(\mathbf{x})\mathbf{n}^T\mathbf{h}\mathbf{N}_T^T(\mathbf{x})\mathrm{d}S && \textit{Surface Conductance Matrix}\\
\mathbf{M_T} &= \int \mathbf{N_T}(\mathbf{x})\rho c_e\mathbf{N}_T^T(\mathbf{x})\mathrm{d}V && \textit{Thermal Mass Matrix; Capacitance Matrix}\\
\mathbf{f_T} &= -\int_{S_{II}} \mathbf{N_T}(\mathbf{x})\mathbf{n}^T\mathbf{q}_0\mathrm{d}S && \textit{Consistent Thermal Force;}\\
&\quad -\int_{S_{III}} \mathbf{N_T}(\mathbf{x})\mathbf{n}^T\mathbf{q}_1\mathrm{d}S && \textit{Consistent Heat Flux}
\end{aligned}$$

6.3.2 Coupled Linear Isotropic Thermoelasticity

The thermal field equation is repeated as

$$-k_T\nabla^2\mathrm{T} = \alpha\lambda\mathrm{T}_0\, tr\left(\dot{\mathbf{E}}\right) + \rho c_e\dot{\mathrm{T}} \tag{6.36}$$

Following the same steps used for conductive heat transfer furnishes the variational principle

$$\int \delta\nabla^T\mathrm{T}k_T\nabla\mathrm{T}\,\mathrm{d}V + \int \delta\mathrm{T}\rho c_e\dot{\mathrm{T}}\,\mathrm{d}V + \int \delta\mathrm{T}\alpha\lambda\mathrm{T}_0\, tr\left(\dot{\mathrm{E}}\right)\mathrm{d}V = -\int \delta\mathrm{T}\mathbf{n}^T\mathbf{q}\,\mathrm{d}S \tag{6.37}$$

which is the thermal analog of the Principle of Virtual Work.

For the mechanical field the Principle of Virtual Work under small strain is recalled as

$$\int tr(\delta\mathbf{ES})\,\mathrm{d}V + \int \delta\mathbf{u}^T\rho\ddot{\mathbf{u}}\,\mathrm{d}V = \int_S \delta\mathbf{u}^T\mathbf{t}(\mathbf{s})\,\mathrm{d}S \tag{6.38}$$

Recall that $\mathbf{S} = 2\mu\mathbf{E} + \lambda[tr(\mathbf{E}) - \alpha(\mathrm{T} - \mathrm{T}_0)]\mathbf{I}$. The Principle of Virtual Work is expanded to give

$$\begin{aligned}
&\int tr\left(\delta\mathbf{E}[2\mu\mathbf{E} + \lambda\, tr(\mathbf{E})\mathbf{I}]\right)\mathrm{d}V - \int tr\left(\delta\mathbf{E}\lambda\alpha(\mathrm{T} - \mathrm{T}_0)\mathbf{I}\right)\mathrm{d}V + \int \delta\mathbf{u}^T\rho\ddot{\mathbf{u}}\,\mathrm{d}V\\
&\quad = \int \delta\mathbf{u}^T\mathbf{t}(\mathbf{s})\,\mathrm{d}S
\end{aligned} \tag{6.39}$$

Now introduce the interpolation models

$$\mathbf{u}(\mathbf{x}) = \mathbf{N}^T(\mathbf{x})\boldsymbol{\gamma}(t) \tag{6.40}$$

from which we may derive the strain–displacement matrices **B(x)** and **b(x)** satisfying

$$VEC(\mathbf{E}) = \mathbf{B}^T(\mathbf{x})\boldsymbol{\gamma}(t), \quad tr(\mathbf{E}) = \mathbf{b}^T(\mathbf{x})\boldsymbol{\gamma}(t) \tag{6.41}$$

Also, **N** is called the shape function matrix. (The operations to obtain Equation 6.41 are illustrated in Example 6.5 and presented in detail in Chapter 7.)

It follows after some manipulation that

$$\mathbf{K}\boldsymbol{\gamma}(t) + \mathbf{M}\ddot{\boldsymbol{\gamma}}(t) - \boldsymbol{\Omega}^T\boldsymbol{\theta}(t) = \mathbf{f} \tag{6.42}$$

in which

$$\begin{aligned}
\mathbf{K} &= \textstyle\int \mathbf{B}^T(\mathbf{x})\boldsymbol{D}\mathbf{B}(\mathbf{x})\,\mathrm{d}V && \textit{Stiffness Matrix} \\
\mathbf{M} &= \textstyle\int \mathbf{N}(\mathbf{x})\rho\mathbf{N}^T(\mathbf{x})\,\mathrm{d}V && \textit{Mass Matrix} \\
\boldsymbol{\Omega} &= \textstyle\int \alpha\lambda\mathbf{b}^T(\mathbf{x})\mathbf{N}_T(\mathbf{x})\,\mathrm{d}V && \textit{Thermoelastic Matrix} \\
\mathbf{f} &= \textstyle\int \mathbf{N}(\mathbf{x})\mathbf{t}_1(\mathbf{s})\,\mathrm{d}S && \textit{Consistent Force Vector}
\end{aligned} \tag{6.43}$$

We assume that the traction **t(s)** is specified everywhere as $\mathbf{t}_1(\mathbf{s})$ on S. Also $\boldsymbol{D}$ is the isothermal tangent modulus tensor.

For the thermal field, assuming that the heat flux **q** is specified as $\mathbf{q}_1$ on the surface, the thermal counterpart Equation 6.37 of the Principle of Virtual Work, together with the interpolation models Equations 6.33 and 6.34 furnish the finite element equation

$$\tfrac{1}{T_0}\mathbf{K}_T\boldsymbol{\theta}(t) + \tfrac{1}{T_0}\mathbf{M}_T\dot{\boldsymbol{\theta}}(t) + \boldsymbol{\Omega}\dot{\boldsymbol{\gamma}}(t) = \tfrac{1}{T_0}\mathbf{f}_T, \quad \mathbf{f}_T = -\textstyle\int \mathbf{N}_T(\mathbf{x})\mathbf{n}^T\mathbf{q}_1\,\mathrm{d}S \tag{6.44}$$

The combined equations for a thermoelastic medium are now written in state (first order) form as

$$\mathbf{W}_1\frac{\mathrm{d}}{\mathrm{d}t}\begin{pmatrix}\dot{\boldsymbol{\gamma}}(t)\\ \boldsymbol{\gamma}(t)\\ \boldsymbol{\theta}(t)\end{pmatrix} + \mathbf{W}_2\begin{pmatrix}\dot{\boldsymbol{\gamma}}(t)\\ \boldsymbol{\gamma}(t)\\ \boldsymbol{\theta}(t)\end{pmatrix} = \begin{pmatrix}\mathbf{f}(t)\\ \mathbf{0}\\ \frac{1}{T_0}\mathbf{f}_T(t)\end{pmatrix} \tag{6.45}$$

$$\mathbf{W}_1 = \begin{bmatrix}\mathbf{M} & \mathbf{0} & \mathbf{0}\\ \mathbf{0} & \mathbf{K} & \mathbf{0}\\ \mathbf{0} & \mathbf{0} & \frac{1}{T_0}\mathbf{M}_T\end{bmatrix}, \quad \mathbf{W}_2 = \begin{bmatrix}\mathbf{0} & \mathbf{K} & -\boldsymbol{\Omega}^T\\ -\mathbf{K} & \mathbf{0} & \mathbf{0}\\ \boldsymbol{\Omega} & \mathbf{0} & \frac{1}{T_0}\mathbf{K}_T\end{bmatrix}$$

Multiplying by $\mathbf{z}^T = \{\dot{\gamma}(t)\ \gamma(t)\ \boldsymbol{\theta}(t)\}$, letting $\mathbf{y}^T = \begin{pmatrix}\mathbf{f}(t)\\ \mathbf{0}\\ \frac{1}{T_0}\mathbf{f}_T(t)\end{pmatrix}$, and performing simple manipulations furnish the equation

$$\frac{\mathrm{d}}{\mathrm{d}t}\left(\frac{1}{2}\mathbf{r}^T\mathbf{W}_1\mathbf{r}\right) = -\mathbf{r}^T\mathbf{y} - \frac{1}{2}\mathbf{r}^T\frac{1}{2}(\mathbf{W}_2 + \mathbf{W}_2^T)\mathbf{r}. \tag{6.46}$$

Note that $\mathbf{W}_1$ is positive definite and symmetric, while $\frac{1}{2}\left(\mathbf{W}_2+\mathbf{W}_2^T\right)$ is positive semi-definite. This implies that the magnitude of $\mathbf{r}$ is nonincreasing if $\mathbf{y}=\mathbf{0}$. Otherwise stated, coupled linear thermoelasticity is *at least* marginally stable, whereas a purely elastic system is strictly marginally stable. Thus thermal conduction has a stabilizing effect which will next be shown to be analogous to viscous dissipation.

EXAMPLE 6.5

Write down the coupled thermal and elastic equations for a one-dimensional thermoelastic member modeled with one element. Suppose that, on the left end of the member, it is built into a rigid thermal reservoir with fixed temperature T_0. Illustrate the analogy between conductive heat transfer and viscous damping in the case of a constant temperature field.

SOLUTION

This problem will be approached by modeling the member as a single one-dimensional thermoelastic finite element, whose left end is built into a rigid thermal reservoir. The element is supposed to have length L. The displacement is modeled using $u(x,t)=\mathbf{N}^T(x)\boldsymbol{\gamma}_{m1}(t)$ in which

$$\mathbf{N}^T(x)=\boldsymbol{\varphi}^T(x)\boldsymbol{\Phi},\quad \boldsymbol{\varphi}^T(x)=(1\quad x)$$

$$\boldsymbol{\Phi}=\begin{bmatrix}1 & 0\\ 1 & L\end{bmatrix}^{-1}=\frac{1}{L}\begin{bmatrix}L & 0\\ -1 & 1\end{bmatrix}$$

The corresponding strain–displacement matrix $\mathbf{B}$ is given by

$$\mathbf{B}^T(x)=\frac{\partial}{\partial x}\mathbf{N}^T(x)=\boldsymbol{\beta}^T(x)\boldsymbol{\Phi},\quad \mathbf{b}^T(x)=\boldsymbol{\beta}^T(x)\boldsymbol{\Phi},\quad \boldsymbol{\beta}^T(x)=(0\quad 1)$$

The corresponding relations for the thermal degree-of-freedom $\mathrm{T}-\mathrm{T}_0$ are given by

$$\mathbf{N}_\mathrm{T}^T(x)=\boldsymbol{\nu}_\mathrm{T}^T(x)\boldsymbol{\Phi}_T,\quad \boldsymbol{\nu}_\mathrm{T}^T(x)=(1\quad x)$$

$$\boldsymbol{\Phi}_\mathrm{T}=\begin{bmatrix}1 & 0\\ 1 & L\end{bmatrix}^{-1}=\frac{1}{L}\begin{bmatrix}L & 0\\ -1 & 1\end{bmatrix}$$

$$\mathbf{B}_\mathrm{T}^T(x)=\frac{\partial}{\partial x}\mathbf{N}_\mathrm{T}^T(x)=\boldsymbol{\beta}_\mathrm{T}^T(x)\boldsymbol{\Phi}_\mathrm{T},\quad \boldsymbol{\beta}_\mathrm{T}^T(x)=(0\quad 1)$$

For one-dimensional problems, $\mathbf{D}\rightarrow \mathsf{E}$. Following the Principle of Virtual Work the stiffness matrix is given by

$$\begin{aligned}\mathbf{K}&=\int \mathbf{B}(x)\mathsf{E}\mathbf{B}^T(x)\,\mathrm{d}V\\ &=\frac{1}{L^2}\begin{bmatrix}L & -1\\ 0 & 1\end{bmatrix}\int_0^L\begin{pmatrix}0\\ 1\end{pmatrix}\mathsf{E}A(0\quad 1)\,\mathrm{d}x\begin{bmatrix}L & 0\\ -1 & 1\end{bmatrix}\\ &=\frac{\mathsf{E}A}{L}\begin{bmatrix}1 & -1\\ -1 & 1\end{bmatrix}\end{aligned}$$

and for the mass matrix

$$
\begin{aligned}
\mathbf{M} &= \int \mathbf{N}(x)\rho\mathbf{N}^T(x)\,\mathrm{d}V \\
&= \frac{1}{L^2}\begin{bmatrix} L & -1 \\ 0 & 1 \end{bmatrix} \int_0^L \begin{pmatrix} 1 \\ x \end{pmatrix} \rho A(1 \quad x)\,\mathrm{d}x \begin{bmatrix} L & 0 \\ -1 & 1 \end{bmatrix} \\
&= \frac{\rho AL}{3}\begin{bmatrix} 1 & 1/2 \\ 1/2 & 1 \end{bmatrix}
\end{aligned}
$$

No heat source is indicated: i.e., $\mathbf{h} = 0$. The thermal stiffness matrix is now

$$
\begin{aligned}
\mathbf{K}_{\mathrm{T}} &= \int \mathbf{B}_{\mathrm{T}}(x)k_{\mathrm{T}}\mathbf{B}_{\mathrm{T}}^T(x)\,\mathrm{d}V \\
&= \frac{1}{L^2}\begin{bmatrix} L & -1 \\ 0 & 1 \end{bmatrix} \int_0^L \begin{pmatrix} 0 \\ 1 \end{pmatrix} k_{\mathrm{T}} A(0 \quad 1)\,\mathrm{d}x \begin{bmatrix} L & 0 \\ -1 & 1 \end{bmatrix} \\
&= \frac{k_{\mathrm{T}}A}{L}\begin{bmatrix} 1 & -1 \\ -1 & 1 \end{bmatrix}
\end{aligned}
$$

The thermal mass matrix is similarly obtained:

$$
\begin{aligned}
\mathbf{M}_{\mathrm{T}} &= \int \mathbf{N}_{\mathrm{T}}(x)\rho c_e \mathbf{N}_{\mathrm{T}}^T(x)\,\mathrm{d}V \\
&= \frac{1}{L^2}\begin{bmatrix} L & -1 \\ 0 & 1 \end{bmatrix} \int_0^L \begin{pmatrix} 1 \\ x \end{pmatrix} \rho c_e A(1 \quad x)\,\mathrm{d}x \begin{bmatrix} L & 0 \\ -1 & 1 \end{bmatrix} \\
&= \frac{\rho c_e AL}{3}\begin{bmatrix} 1 & 1/2 \\ 1/2 & 1 \end{bmatrix}
\end{aligned}
$$

Finally, the thermoelastic matrix is presented.

$$
\begin{aligned}
\mathbf{\Omega} &= \int \alpha\lambda \mathbf{N}_{\mathrm{T}}(x)\mathbf{b}^T(x)\,\mathrm{d}V \\
&= \frac{1}{L^2}\begin{bmatrix} L & -1 \\ 0 & 1 \end{bmatrix} \int_0^L \begin{pmatrix} 1 \\ x \end{pmatrix} \alpha\lambda A(0 \quad 1)\,\mathrm{d}x \begin{bmatrix} L & 0 \\ -1 & 1 \end{bmatrix} \\
&= \frac{\alpha\lambda A}{2}\begin{bmatrix} -1 & 1 \\ -1 & 1 \end{bmatrix}
\end{aligned}
$$

Collecting the foregoing results furnishes the coupled thermal and elastic finite element equations as

$$
\frac{\mathrm{E}A}{L}\begin{bmatrix} 1 & -1 \\ -1 & 1 \end{bmatrix}\begin{pmatrix} u(0,t) \\ u(L,t) \end{pmatrix} + \frac{\rho AL}{3}\begin{bmatrix} 1 & 1/2 \\ 1/2 & 1 \end{bmatrix}\begin{pmatrix} u(0,t) \\ u(L,t) \end{pmatrix}^{\cdot\cdot} - \frac{\alpha\lambda A}{2}\begin{bmatrix} -1 & -1 \\ 1 & 1 \end{bmatrix}\begin{pmatrix} \mathrm{T}(0,t) - \mathrm{T}_0 \\ \mathrm{T}(L,t) - \mathrm{T}_0 \end{pmatrix} = \mathbf{f}
$$

$$
\frac{1}{\mathrm{T}_0}\frac{k_{\mathrm{T}}A}{L}\begin{bmatrix} 1 & -1 \\ -1 & 1 \end{bmatrix}\begin{pmatrix} \mathrm{T}(0,t) - \mathrm{T}_0 \\ \mathrm{T}(L,t) - \mathrm{T}_0 \end{pmatrix} + \frac{1}{\mathrm{T}_0}\frac{\rho c_e AL}{3}\begin{bmatrix} 1 & 1/2 \\ 1/2 & 1 \end{bmatrix}\begin{pmatrix} \mathrm{T}(0,t) - \mathrm{T}_0 \\ \mathrm{T}(L,t) - \mathrm{T}_0 \end{pmatrix}^{\cdot} + \frac{\alpha\lambda A}{2}\begin{bmatrix} -1 & 1 \\ -1 & 1 \end{bmatrix}\begin{pmatrix} u(0,t) \\ u(L,t) \end{pmatrix}^{\cdot} = \frac{1}{\mathrm{T}_0}\mathbf{f}_{\mathrm{T}}
$$

A rigid "thermal bath" on the left is imagined to constrain both $T(0,t) - T_0$ and $u(0,t)$ to vanish. The constraints serve to eliminate the corresponding rows and columns in the two foregoing equations, resulting in

$$\frac{\mathsf{E}A}{L}u(L,t) + \frac{\rho AL}{3}\ddot{u}(L,t) - \frac{\alpha\lambda A}{2}(\mathrm{T}(L,t) - \mathrm{T}_0) = f$$

$$\frac{1}{\mathrm{T}_0}\frac{k_\mathrm{T}A}{L}(\mathrm{T}(L,t) - \mathrm{T}_0) + \frac{1}{\mathrm{T}_0}\frac{\rho c_e AL}{3}(\mathrm{T}(L,t) - \mathrm{T}_0)^{\bullet} + \frac{\alpha\lambda A}{2}\dot{u}(L,t) = \frac{1}{\mathrm{T}_0}f_\mathrm{T}$$

If the displacement field is static there is no thermomechanical coupling through the thermal field.

If $(\mathrm{T}(L,t) - \mathrm{T}_0)^{\bullet} = 0$, then the thermal field equation reduces to

$$\mathrm{T}(L,t) - \mathrm{T}_0 = \frac{\frac{1}{\mathrm{T}_0}f_\mathrm{T} - \frac{\alpha\lambda A}{2}\dot{u}(L,t)}{\frac{1}{\mathrm{T}_0}\frac{k_\mathrm{T}A}{L}}$$

and upon substitution the mechanical field becomes

$$\frac{\mathsf{E}A}{L}u(L,t) + \frac{\rho AL}{3}\ddot{u}(L,t) - \frac{\alpha\lambda A}{2}\left(\frac{\frac{1}{\mathrm{T}_0}f_\mathrm{T} - \frac{\alpha\lambda A}{2}\dot{u}(L,t)}{\frac{1}{\mathrm{T}_0}\frac{k_\mathrm{T}A}{L}}\right)_0 = f$$

Upon reorganization,

$$\frac{\mathsf{E}A}{L}u(L,t) + \frac{\left(\frac{\alpha\lambda A}{2}\right)^2}{\frac{1}{\mathrm{T}_0}\frac{k_\mathrm{T}A}{L}}\dot{u}(L,t) + \frac{\rho AL}{3}\ddot{u}(L,t) = f + \frac{\alpha\lambda L}{2k_\mathrm{T}}f_\mathrm{T}$$

Mathematically the middle term has exactly the same effect as conventional viscous damping. This illustrates that convective heat transfer, by carrying energy away from a site at which it is concentrated, has a stabilizing effect. However, in many materials the thermal expansion coefficient is very small, rendering the effective damping coefficient due to thermal conduction very small.

7 One-Dimensional Elastic Elements

This chapter introduces finite elements for one-dimensional members, including rods, shafts, beams, and beam-columns. It initially presents interpolation models in physical coordinates for the sake of simplicity and brevity. But interpolation in "natural coordinates" is then presented to enable the use of Gaussian quadrature for integration. Use of natural coordinates to an extent reduces the sensitivity of the elements to geometric details in the physical mesh. A number of examples are given including several illustrating the use of natural coordinates.

7.1 INTERPOLATION MODELS FOR ONE-DIMENSIONAL ELEMENTS

7.1.1 Rods

The governing equation for the displacements in rods (also bars, tendons, and shafts) is

$$\mathsf{E}A\frac{\partial^2 u}{\partial x^2} = \rho A\frac{\partial^2 u}{\partial t^2} \tag{7.1}$$

in which $u(x,t)$ denotes the radial displacement, E, A, and ρ are constants, x is the spatial coordinate, and t denotes time. Since the displacement is governed by a second-order partial differential equation, in the spatial domain it requires two (time-dependent) constants of integration. Applied to an element, the two constants can be supplied implicitly using two nodal displacements as functions of time. We now approximate $u(x,t)$ using its values at x_e and x_{e+1}, as shown in Figure 7.1.

The lowest-order *interpolation model* consistent with two integration constants is linear, in the form

$$u(x,t) = \boldsymbol{\varphi}_{m1}^T(x)\boldsymbol{\Phi}_{m1}\boldsymbol{\gamma}_{m1}(t), \quad \boldsymbol{\gamma}_{m1}(t) = \begin{pmatrix} u_e(t) \\ u_{e+1}(t) \end{pmatrix}, \quad \boldsymbol{\varphi}_{m1}^T(x) = (1 \quad x) \tag{7.2}$$

We seek to identify $\boldsymbol{\Phi}_{m1}$ in terms of the nodal values of u. Letting $u_e = u(x_e)$ and $u_{e+1} = u(x_{e+1})$ furnishes

$$u_e(t) = (1\ x_e)\boldsymbol{\Phi}_{m1}\boldsymbol{\gamma}_{m1}(t), \quad u_{e+1}(t) = (1\ x_{e+1})\boldsymbol{\Phi}_{m1}\boldsymbol{\gamma}_{m1}(t) \tag{7.3}$$

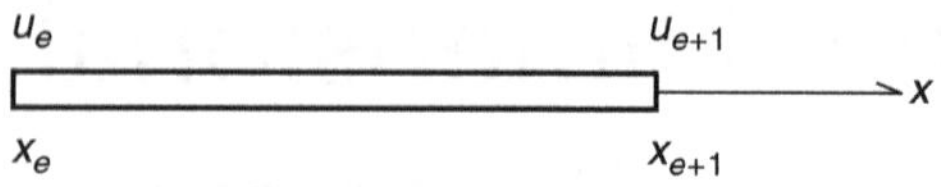

FIGURE 7.1 Rod element.

But since $\boldsymbol{\gamma}_{m1}(t) = \begin{Bmatrix} u_e(t) \\ u_{e+1}(t) \end{Bmatrix}$ we conclude that

$$\begin{aligned} \boldsymbol{\Phi}_{m1} &= \begin{bmatrix} 1 & x_e \\ 1 & x_{e+1} \end{bmatrix}^{-1} \\ &= \frac{1}{l_e}\begin{bmatrix} x_{e+1} & -x_e \\ -1 & 1 \end{bmatrix}, \quad l_e = x_{e+1} - x_e \end{aligned} \tag{7.4}$$

7.1.2 Beams

The equation for a beam, following the classical Euler–Bernoulli theory, is

$$EI\frac{\partial^4 w}{\partial x^4} + \rho A\frac{\partial^2 w}{\partial t^2} = 0 \tag{7.5}$$

in which $w(x,t)$ denotes the transverse (z) displacement of the beam neutral axis, and the constant I is bending moment of area. In the spatial domain there are four constants of integration. In an element the constants can be supplied implicitly by specifying the values of w and $-\partial w/\partial x$ at each of the two-element nodes. Referring to Figure 7.2, we introduce the interpolation model for $w(x,t)$:

$$w(x,t) = \boldsymbol{\varphi}_{b1}^T(x)\boldsymbol{\Phi}_{b1}\boldsymbol{\gamma}_{b1}(t), \quad \boldsymbol{\varphi}_{b1}^T(x) = \begin{pmatrix} 1 & x & x^2 & x^3 \end{pmatrix}, \quad \boldsymbol{\gamma}_{ml}(t) = \begin{pmatrix} w_e \\ -w'_e \\ w_{e+1} \\ -w'_{e+1} \end{pmatrix} \tag{7.6}$$

Enforcing this model at x_e and at x_{e+1} furnishes

$$\boldsymbol{\Phi}_{b1} = \begin{bmatrix} 1 & x_e & x_e^2 & x_e^3 \\ 0 & -1 & -2x_e & -3x_e^2 \\ 1 & x_{e+1} & x_{e+1}^2 & x_{e+1}^3 \\ 0 & -1 & -2x_{e+1} & -3x_{e+1}^2 \end{bmatrix}^{-1} \tag{7.7}$$

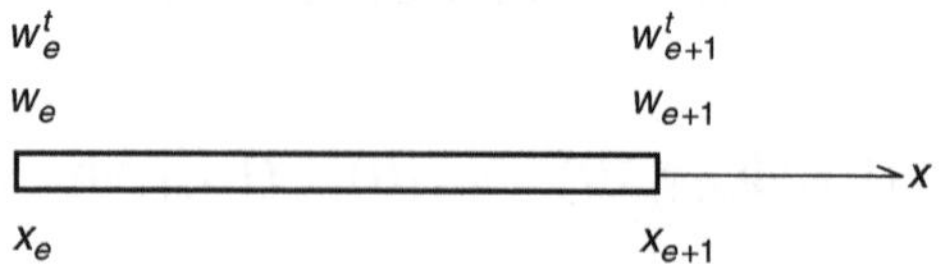

FIGURE 7.2 Beam element.

7.1.3 Beam-Columns

Beam-columns are of interest, among other reasons, to predict buckling according to the Euler criterion. The displacement w is assumed to depend only on x. Also u is viewed as given by

$$u(x,z) = u_0(x) - z\frac{\partial w(x)}{\partial x} \tag{7.8}$$

in which $u_0(x)$ represents the stretching of the neutral axis. It is necessary to know $u_0(x,t)$, $w(x,t)$, and $-\frac{\partial w(x,t)}{\partial x}$ at x_e and x_{e+1}. Combining relations for the rod and the beam element, the interpolation model is now

$$u(x,z,t) = (1 \quad x)\mathbf{\Phi}_{m1}\boldsymbol{\gamma}_{m1} - z(0 \quad 1 \quad 2x \quad 3x^2)\mathbf{\Phi}_{b1}\boldsymbol{\gamma}_{b1} \tag{7.9}$$

$$\boldsymbol{\gamma}_{m1} = \begin{pmatrix} u_{0,e}(t) \\ u_{0,e+1}(t) \end{pmatrix}, \quad \boldsymbol{\gamma}_{b1} = \begin{pmatrix} w_e \\ -w'_e \\ w_{e+1} \\ -w'_{e+1} \end{pmatrix}$$

$$\mathbf{\Phi}_{m1} = \frac{1}{l_e}\begin{bmatrix} x_{e+1} & -x_e \\ -1 & 1 \end{bmatrix}, \quad \mathbf{\Phi}_{b1} = \begin{bmatrix} 1 & x_e & x_e^2 & x_e^3 \\ 0 & -1 & -2x_e & -3x_e^2 \\ 1 & x_{e+1} & x_{e+1}^2 & x_{e+1}^3 \\ 0 & -1 & -2x_{e+1} & -3x_{e+1}^2 \end{bmatrix}^{-1}$$

7.2 STRAIN–DISPLACEMENT RELATIONS IN ONE-DIMENSIONAL ELEMENTS

For the rod, the strain is given by $\varepsilon = E_{11} = \frac{\partial u}{\partial x}$. An estimate for ε implied by the interpolation model Equation 7.3 has the form

$$\varepsilon(x,t) = \boldsymbol{\beta}_{m1}^T(x)\mathbf{\Phi}_{m1}\boldsymbol{\gamma}_{m1}(t) \tag{7.10}$$

$$\boldsymbol{\beta}_{m1}^T = \frac{d\boldsymbol{\varphi}_{m1}^T}{dx} = (0 \quad 1) \tag{7.11}$$

For the beam the corresponding relation is

$$\varepsilon(x,z,t) = -z\frac{\partial^2 w}{\partial x^2} \tag{7.12}$$

from which the consistent approximation is obtained

$$\varepsilon(x,z,t) = -z\beta_{b1}^T(x)\mathbf{\Phi}_{b1}\boldsymbol{\gamma}_{b1}(t) \tag{7.13}$$

$$\boldsymbol{\beta}_{b1}^T = \frac{d\boldsymbol{\varphi}_{b1}^T}{dx} = (0 \quad 1 \quad 2x \quad 3x^2)$$

For the beam-column the strain is given by

$$\varepsilon(x,z,t) = -\boldsymbol{\beta}_{mb1}^T(x)\boldsymbol{\Phi}_{mb1}\boldsymbol{\gamma}_{mb1}(t)$$
$$\boldsymbol{\beta}_{mb1}^T = \begin{pmatrix}\boldsymbol{\beta}_{m1}^T & -z\boldsymbol{\beta}_{b1}^T\end{pmatrix}\begin{bmatrix}\boldsymbol{\Phi}_{m1} & \mathbf{0}\\ \mathbf{0} & \boldsymbol{\Phi}_{mb1}\end{bmatrix}\begin{pmatrix}\boldsymbol{\gamma}_{m1}(t)\\ \boldsymbol{\gamma}_{b1}(t)\end{pmatrix} \tag{7.14}$$

7.3 STRESS–STRAIN RELATIONS IN ONE-DIMENSIONAL ELEMENTS

7.3.1 General

We first recall the stress–strain relations of an isotropic linear elasticity. If **S** is the stress tensor under small deformation, the stress–strain relation for a linearly elastic isotropic solid under small strain is given in Lamé's form by

$$\mathbf{S} = 2\mu\mathbf{E}_L + \lambda\, tr(\mathbf{E}_L)\mathbf{I} \tag{7.15}$$

in which **I** is the identity tensor. The Lamé coefficients are denoted by λ and μ, and are given in terms of the familiar elastic modulus E and Poisson ratio ν as

$$\mu = \frac{\mathsf{E}}{2(1+\nu)}, \quad \lambda = \frac{\nu\mathsf{E}}{(1-2\nu)(1+\nu)} \tag{7.16}$$

Letting $\mathbf{s} = VEC(\mathbf{S})$ and $\mathbf{e} = VEC(\mathbf{E}_L)$, the stress–strain relations are written using Kronecker product operators as

$$\mathbf{s} = \mathbf{De}, \quad \mathbf{D} = 2\mu\mathbf{I}_9 + \lambda\mathbf{ii}^T \tag{7.17}$$

and **D** is the 9×9 tangent modulus tensor introduced in the previous chapters.

7.3.2 One-Dimensional Members

For a beam-column, recalling the foregoing strain–displacement model

$$\mathrm{S}_{11}(x,z,t) = \mathsf{E}E_{11} = -\mathsf{E}z\boldsymbol{\beta}_{mb1}^T(x)\boldsymbol{\Phi}_{mb1}\boldsymbol{\gamma}_{mb1}(t) \tag{7.18}$$

The cases of a rod and a beam are recovered by setting $\boldsymbol{\gamma}_{b1}$ or $\boldsymbol{\gamma}_{m1}$ equal to the zero vectors, respectively.

7.4 ELEMENT STIFFNESS AND MASS MATRICES FROM THE PRINCIPLE OF VIRTUAL WORK

Variational calculus was introduced in Chapter 4 and the Principle of Virtual Work was introduced in Chapter 5. It is repeated here as

$$\int \delta \mathrm{E}_{ij}\mathrm{S}_{ij}\, dV + \int \delta u_i \rho \ddot{u}_i\, dV = \int \delta u_i t_i\, dS \tag{7.19}$$

As before δ represents the variational operator. We assume for present purposes that the displacement, the strain, and the stress satisfy representations of the form

$$\mathbf{u}(\mathbf{x},t) = \boldsymbol{\varphi}^T(\mathbf{x})\boldsymbol{\Phi}\boldsymbol{\gamma}(t), \quad \mathbf{e}(\mathbf{x},t) = \boldsymbol{\beta}^T(\mathbf{x})\boldsymbol{\Phi}\boldsymbol{\gamma}(t), \quad \mathbf{s}(\mathbf{x},t) = \mathbf{De}(\mathbf{x},t) \tag{7.20}$$

in which $\mathbf{e} = VEC(\mathbf{E})$ and $\mathbf{s} = VEC(\mathbf{S})$ are written as one-dimensional arrays. Since small strain is assumed, no distinction is made between the undeformed coordinates $\mathbf{X}$ and the deformed coordinates $\mathbf{x}$.

EXAMPLE 7.1

One-element model for a built-in rod

Suppose that the rod depicted in Figure 7.3 has elastic modulus E, mass density ρ, area A, and length L, and is modeled using a single element. It is built in at $x = 0$. At $x = L$ there is a concentrated mass m to which is attached a spring of stiffness k.

The Principle of Virtual Work reduces to the variational equation

$$\int_0^L \delta \frac{du}{dx} \mathsf{E}A \frac{du}{dx} dx + \int_0^L \delta u \rho A \ddot{u}\, dx = \delta u(L,t)[P - ku(L,t) - m\ddot{u}(L,t)] \tag{7.21}$$

Upon application of the foregoing linear interpolation model for rod elements and enforcement of the constraint $u(0,t) = 0$ at $x = 0$, the stiffness and mass matrices arising from the domain reduce to scalar values as follows: $\mathbf{K} \to \mathsf{E}A/L$, $\mathbf{M} \to \rho AL/3$, $\mathbf{M}_S \to m$, $\mathbf{K}_S \to k$. The governing one-element equation is $\left(\frac{\mathsf{E}A}{L} + k\right)\gamma + \left(\frac{\rho AL}{3} + m\right)\ddot{\gamma} = f$:

EXAMPLE 7.2

One-element model for a built-in rod with a constant distributed surface stress

The configuration of interest is illustrated in Figure 7.4.

(a) To derive the applicable version of the Principle of Virtual Work, **integrate over the domain** and note that

$$\int_0^L \delta u \left[\mathsf{E}A \frac{d^2 u}{dx^2} + p_x(x)\right] dx = 0, \quad p_x(x) = 2\pi r_0 S_{xx}$$

which holds for an arbitrary increment $\delta u(x)$ of u, subject to $\delta u(0) = 0$.

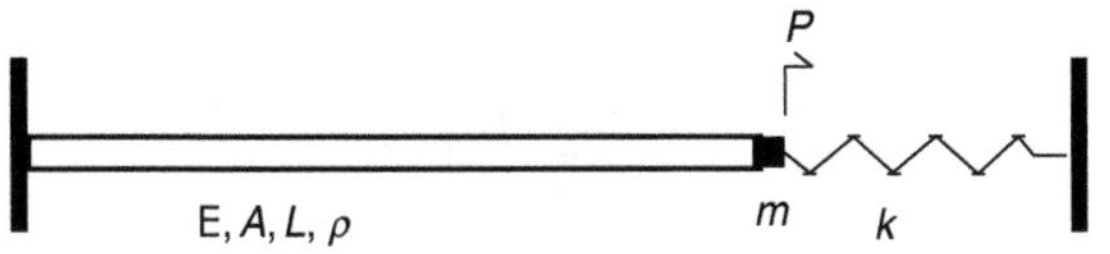

FIGURE 7.3 Rod with inertial and compliant boundary conditions.

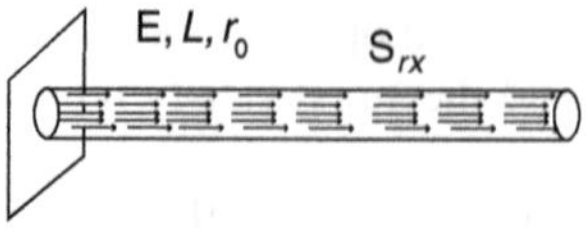

FIGURE 7.4 Rod with constant distributed surface stress.

(b) **Integrate by parts**

$$\int_0^L \frac{d}{dx}\left[\delta u \mathsf{E}A\frac{du}{dx}\right]dx - \int_0^L \frac{d\delta u}{dx}\mathsf{E}A\frac{du}{dx}dx + \int_0^L \delta u p_x(x)\,dx = 0$$

and

$$\int_0^L \frac{d}{dx}\left[\delta u \mathsf{E}A\frac{du}{dx}\right]dx = \delta u \mathsf{E}A\frac{du}{dx}\bigg|_0^L = 0$$

since $\delta u(0) = 0$ and $P(L) = \mathsf{E}A\dfrac{du(L)}{dx} = 0$. Accordingly

$$\int_0^L \frac{d\delta u}{dx}\mathsf{E}A\frac{du}{dx}\,dx = \int_0^L \delta u p_x(x)\,dx$$

which expresses the Principle of Virtual Work in the current case.

(c) We model this member as **one finite element** and use the same approximation for $\dot{u}(x)$ as before, namely:

$$u(x) = \tfrac{x}{L}u(L) \rightarrow \delta u(x) = \tfrac{x}{L}\delta u(L)$$

From before,

$$\delta V = \int_0^L \frac{d\delta u}{dx}\mathsf{E}A\frac{du}{dx}dx$$
$$= \Delta\delta(L)\frac{\mathsf{E}A}{L}u(L)$$

Now, assuming $p_x(x) = p_0$,

$$\delta W = \int_0^L \delta u p_0\,dx$$
$$= \frac{\delta u(L)p_0}{L}\int_0^L x\,dx$$
$$= \tfrac{1}{2}\delta u(L)Lp_0$$

Hence, $\delta V = \delta W$ from which

$$\frac{\mathsf{EA}}{L}u(L) = \frac{1}{2}Lp_0 \quad \text{and also} \quad u(L) = \frac{1}{2}\frac{L^2 p_0}{\mathsf{EA}}$$

Now compare the last result with the exact solution given by

$$u(x) = u'_A x - \frac{p_0}{\mathsf{EA}}\frac{x^2}{2} \quad \text{and} \quad u(L) = \frac{p_0 L^2}{2\mathsf{EA}}$$

The exact displacement function is quadratic, whereas the function assumed in the interpolation model is linear. Despite this difference, the finite element solution gives the exact displacement at $x = L$.

7.4.1 Single-Element Model for Dynamic Response of a Built-in Beam

Next consider a one-element model of a cantilevered beam to which a solid disk is welded at $x = L$. Also attached at L is a linear spring and a torsional spring, the latter having the property that the moment developed is proportional to the (negative of the) slope of the beam. The shear force V_0 and the moment M_0 act at L. The member is illustrated in Figure 7.5.

The Principle of Virtual Work now reduces to the equation

$$\int_0^L \delta w'' \mathsf{E}Iw'' \, dx + \int \delta w \rho A \ddot{w} \, dx$$
$$= \{\delta w(L,t) \ -\delta w'(L,t)\} \begin{Bmatrix} V_0 - kw(L,t) - m\ddot{w}(L,t) \\ M_0 - k_t(-w'(L,t)) - \dfrac{mr^2}{2}(-\ddot{w}'(L,t)) \end{Bmatrix} \tag{7.22}$$

The interpolation model, incorporating the constraints $w(0,t) = -w'(0,t) = 0$ a priori, is

$$w(x,t) = \begin{pmatrix} x^2 & x^3 \end{pmatrix} \begin{bmatrix} L^2 & L^3 \\ -2L & -3L^2 \end{bmatrix}^{-1} \begin{pmatrix} w(L,t) \\ -w'(L,t) \end{pmatrix} \tag{7.23}$$

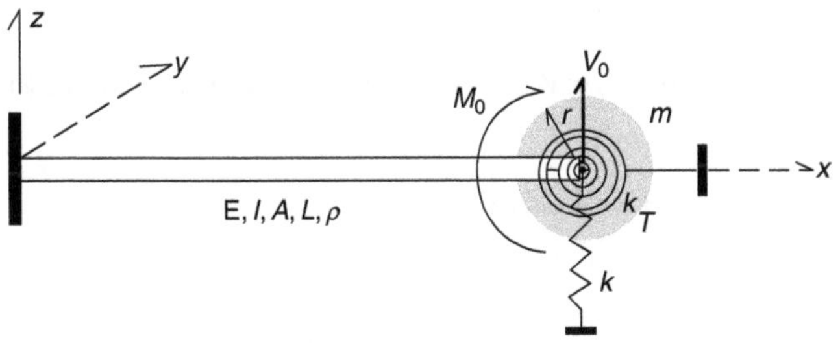

FIGURE 7.5 Beam with translational and rotational inertial and compliant boundary conditions.

Ignoring rotary inertia, the Principle of Virtual Work and the interpolation model Equation 7.23 may now be applied to furnish the domain contributions to the stiffness and mass matrices:

$$\mathbf{K} = \frac{\mathsf{E}I}{L^3}\begin{bmatrix} 12 & 6L \\ 6L & 4L^2 \end{bmatrix}, \quad \mathbf{M} = \rho AL\begin{bmatrix} \frac{13}{35} & \frac{11}{210}L \\ \frac{11}{210}L & \frac{1}{105}L^2 \end{bmatrix} \tag{7.24}$$

The stiffness and mass contributions from the boundary conditions are

$$\mathbf{K}_S = \begin{bmatrix} k & 0 \\ 0 & k_T \end{bmatrix}, \quad \mathbf{M}_S = \begin{bmatrix} m & 0 \\ 0 & \frac{mr^2}{2} \end{bmatrix} \tag{7.25}$$

The governing equation is now

$$(\mathbf{M} + \mathbf{M}_S)\begin{pmatrix} \ddot{w}(L,t) \\ -\ddot{w}'(L,t) \end{pmatrix} + (\mathbf{K} + \mathbf{K}_S)\begin{pmatrix} w(L,t) \\ -w'(L,t) \end{pmatrix} = \begin{pmatrix} V_0 \\ M_0 \end{pmatrix} \tag{7.26}$$

EXAMPLE 7.3

Single-element model for a built-in beam under a uniformly distributed load
Assuming a static response, the Principle of Virtual Work in the present case may be derived from

$$\int_0^L \delta w\left[\mathsf{E}I\frac{\mathrm{d}^4 w}{\mathrm{d}x^4} - p_b(x)\right]\mathrm{d}x = 0, \quad \delta w(0) = 0, \quad -\delta w'(0) = 0$$

in which $p_b = \mathrm{S}_{xx}a$. Integrating by parts twice gives

$$\delta V = \int_0^L \delta w\left[\mathsf{E}I\frac{\mathrm{d}^4 w}{\mathrm{d}x^4}\right]\mathrm{d}x$$

$$= \int_0^L \delta w''\mathsf{E}Iw''\,\mathrm{d}x - V\delta w|_0^L - M(-\delta w')|_0^L$$

Next apply $\delta w(0) = (-\delta w') = 0$; $V(L) = M(L) = 0$. From the interpolation model for $w(x)$,

$$\int_0^L \delta w''\mathsf{E}Iw''\,\mathrm{d}x = \delta V$$

$$= \delta\boldsymbol{\gamma}^T\frac{\mathsf{E}I}{L^3}\begin{bmatrix} 12 & 6L \\ 6L & 4L^2 \end{bmatrix}\boldsymbol{\gamma}$$

in which $\boldsymbol{\gamma} = \begin{Bmatrix} w(L) \\ -w'(L) \end{Bmatrix}$.

We next examine the variation of the work, i.e., $\delta W = \int_0^L \delta w[p_b(x)]\,\mathrm{d}x$, assuming p_b equals the constant value p_{b0}: $p_b = p_{b0}$. From before, enforcing the conditions

$$w(0) = 0,\ w'(0) = 0,\ w(L) = q_1,\ \ -w'(L) = q_2$$

a priori, $w(x)$ is approximated as

$$w(x) = \left[\frac{3x^2}{L^2} - \frac{2x^3}{L^3}\right] w(L) + \left[\frac{x^2}{L} - \frac{x^3}{L^2}\right](-w'(L))$$

and now

$$\begin{aligned}\delta W &= p_{b0}\left[\tfrac{L}{2}\delta w(L) + \tfrac{L^2}{12}(-\delta w'(L))\right] \\ &= \underbrace{\{\delta w(L)\ (-\delta w'(L))\}}_{\delta \boldsymbol{\gamma}^T} \begin{Bmatrix} \dfrac{p_{b0}L}{2} \\ \dfrac{p_{b0}L^2}{12} \end{Bmatrix}\end{aligned}$$

The ensuing finite element equation is

$$\frac{\mathsf{E}I}{L}\begin{bmatrix} 12 & 6L \\ 6L & 4L^2 \end{bmatrix}\begin{Bmatrix} w(L) \\ -w'(L) \end{Bmatrix} = \begin{Bmatrix} \dfrac{p_{b0}L}{2} \\ \dfrac{p_{b0}L^2}{12} \end{Bmatrix}$$

The solution is

$$\begin{aligned}\begin{Bmatrix} w(L) \\ -w'(L) \end{Bmatrix} &= \frac{L^3}{\mathsf{E}I} \frac{\begin{bmatrix} 4L^2 & -6L \\ -6L & 12 \end{bmatrix}}{12L^2}\begin{Bmatrix} \dfrac{L}{2} \\ \dfrac{L^2}{12} \end{Bmatrix} \\ &= \begin{Bmatrix} \dfrac{1}{8}\ \dfrac{p_{b0}L^4}{\mathsf{E}I} \\ -\dfrac{1}{6}\ \dfrac{p_{b0}L^3}{\mathsf{E}I} \end{Bmatrix}\end{aligned}$$

The one-element finite element solution is now compared with the exact solution. The governing equation and the general form of the solution are now

$$\begin{aligned}&\mathsf{E}Iw'''' = p_{b0} \\ &w(x) = \cancel{w_A} + \cancel{w'_A}x + w''_A x^2 + w'''_A x^3 + \frac{p_{b0}}{\mathsf{E}I}\frac{x^4}{24}\end{aligned}$$

Note that the exact solution is one order higher in x (fourth order) than the one-element approximation (third order). Application of the boundary conditions serves to establish that

$$V(L) = 0, \quad w'''(L) = 0, \quad 6w_A''' + \frac{p_{b0}L}{\mathsf{E}I} = 0$$

from which $w_A''' = -\frac{p_{b0}L}{6\mathsf{E}I}$.

$$M(L) = 0, \qquad w''(L) = 0$$

$$2w_A'' + 6w_A''' + \frac{p_{b0}}{\mathsf{E}I}\frac{L^2}{2} = 0, \qquad xw_A'' = \frac{p_{b0}}{\mathsf{E}I}\frac{L^2}{4}$$

The tip displacement is obtained as

$$w(L) = \frac{1}{8}\frac{p_{b0}L^4}{\mathsf{E}I}$$

and it agrees exactly with the one-element finite element solution.

The tip slope is examined next and is obtained as

$$-w'(L) = -\frac{p_{b0}}{\mathsf{E}I}\frac{L^3}{6}$$

which likewise agrees exactly with the one-element finite element solution. Clearly, even though the displacement model is one order lower in x than the exact solution, the one-element model gives the correct results for the displacement and slope at the tip.

EXAMPLE 7.4

Mass and stiffness matrices for a single-element beam-column

Figure 7.7 is now modified to include the axial compressive force P as shown in Figure 7.6. The member is again modeled using one element. Beam-columns are assumed to be described by the classical Euler buckling equation, given in this case by

$$\mathsf{E}Iw^{iv} + Pw'' + \rho A\ddot{w} = 0 \tag{7.27}$$

The Principle of Virtual Work gives rise to the variational equation

$$\int_0^L \delta w''\mathsf{E}Iw''\,\mathrm{d}x - P\int_0^L \delta w'w'\,\mathrm{d}x + \int_0^L \delta w\rho A\ddot{w}\,\mathrm{d}x$$
$$= \{\delta w(L,t) - \delta w'(L,t)\}\left\{\begin{matrix} V_0 - kw(L,t) - m\ddot{w}(L,t) \\ M_0 - k_t(-w'(L,t)) - \frac{mr^2}{2}(-\ddot{w}'(L,t)) \end{matrix}\right\} \tag{7.28}$$

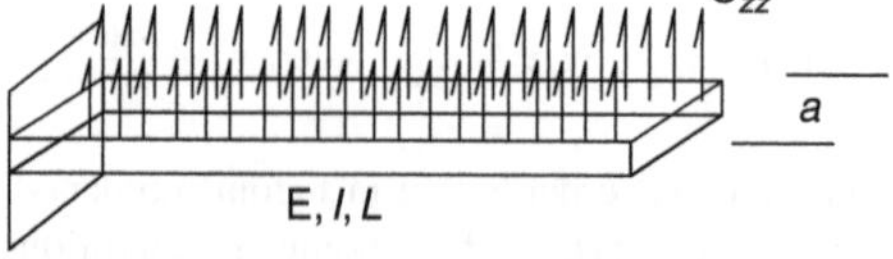

FIGURE 7.6 Cantilevered beam under distributed load.

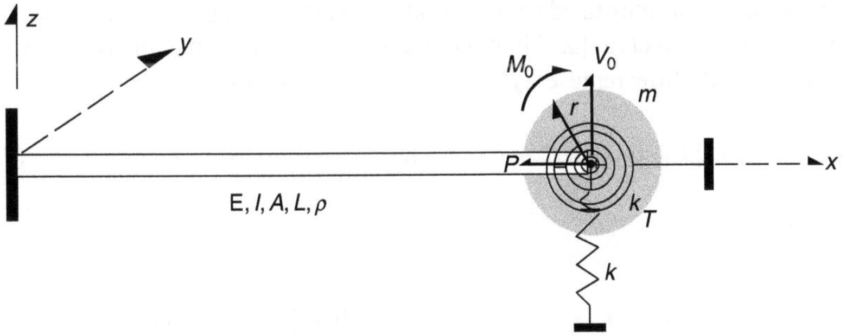

FIGURE 7.7 Built-in beam-column with translational and rotational inertia and compliant boundary conditions.

Upon application of the interpolation model Equation 7.23, the stiffness matrix due to the domain is found after some effort to be

$$\hat{\mathbf{K}} = \frac{EI}{L^3}\begin{bmatrix} 12 & 6L \\ 6L & 4L^2 \end{bmatrix} - \frac{P}{L}\begin{bmatrix} 6/5 & L/10 \\ L/10 & 2L^2/15 \end{bmatrix} \tag{7.29}$$

and the governing equation for the member is now

$$(\mathbf{M} + \mathbf{M}_S)\begin{pmatrix} \ddot{w}(L,t) \\ -\ddot{w}'(L,t) \end{pmatrix} + (\hat{\mathbf{K}} + \mathbf{K}_S)\begin{pmatrix} w(L,t) \\ -w'(L,t) \end{pmatrix} = \begin{pmatrix} V_0 \\ M_0 \end{pmatrix} \tag{7.30}$$

Examination of Equation 7.29 reveals the presence of the negative definite matrix $-\frac{P}{L}\begin{bmatrix} 6/5 & L/10 \\ L/10 & 2L^2/15 \end{bmatrix}$. In static problems, when P becomes large enough to attain the value P_{crit} rendering $\hat{\mathbf{K}}$ singular, buckling is predicted to occur. In a later chapter, we will consider natural frequencies in systems with inertia, at which time we will find that a natural frequency vanishes at P_{crit}.

7.5 INTEGRAL EVALUATION BY GAUSSIAN QUADRATURE: NATURAL COORDINATES

The next step is to formulate stiffness and mass matrices in which there are no constraints, say for the eth element lying between x_e and x_{e+1}. It is helpful to digress briefly to sketch Gaussian quadrature, which underlies the notion of *natural coordinates*. We then return to formulate the matrices of the unconstrained element in terms of both physical and natural coordinates. Gaussian quadrature is optimal in terms of the accuracy gained for a given number of function evaluations. It, also to a certain extent, renders the element matrices insensitive to the initial geometric details of the element.

In the finite element method, computation of element stiffness and mass matrices gives rise to numerous integrations, the accuracy and efficiency of which is critical. Fortunately, a method which is optimal in an important sense, called Gaussian quadrature, has long been known. It is based on converting physical coordinates to natural coordinates as illustrated next. Consider $\int_a^b f(x)\,\mathrm{d}x$. Let $\xi = \frac{1}{b-a}[2x-(a+b)]$. Clearly, ξ maps the interval [*a*,*b*] into the interval [−1,1]. The integral now transforms to $\frac{1}{b-a}\int_{-1}^{1} f(\xi)\,\mathrm{d}\xi$. Now represent $f(\zeta)$ using the power series

$$f(\zeta) = \alpha_0 + \alpha_1\zeta + \alpha_2\zeta^2 + \alpha_3\zeta^3 + \alpha_4\zeta^4 + \alpha_5\zeta^5 + \cdots \tag{7.31}$$

from which

$$\int_{-1}^{1} f(\zeta)\,\mathrm{d}\zeta = 2\alpha_1 + 0 + \frac{2}{3}\alpha_3 + 0 + \frac{2}{5}\alpha_5 + 0 + \cdots \tag{7.32}$$

The advantage of integration procedure on a symmetric interval can be seen in the fact that, with n function evaluations, the integral is evaluated exactly through $(2n-1)$st order.

Consider the first $2n-1$ terms in a power series representation for a function:

$$g(\zeta) = \alpha_1 + \alpha_2\zeta + \cdots + \alpha_{2n}\zeta^{2n-1} \tag{7.33}$$

Now introduce the Gaussian integration formula based on n integration (Gauss) points ξ_i and n weights w_i:

$$\int_{-1}^{1} g(\zeta)\,\mathrm{d}\zeta = \sum_{i=1}^{n} g(\zeta_i)w_i = \alpha_1\sum_{i=1}^{n} w_i + \alpha_2\sum_{i=1}^{n} w_i\zeta_i + \cdots + \alpha_{2n}\sum_{i=1}^{n} w_i\zeta_i^{2n-1} \tag{7.34}$$

Comparison with Equation 7.32 reveals that

$$\sum_{i=1}^{n} w_i = 1, \quad \sum_{i=1}^{n} w_i\xi = 0, \quad \sum_{i=1}^{n} w_i\xi_i^2 = 2/3, \ldots$$
$$\sum_{i=1}^{n} w_i\xi_i^{2n-2} = \frac{2}{2n-1}, \quad \sum_{i=1}^{n} w_i\xi_i^{2n-1} = 0 \tag{7.35}$$

It is necessary to solve for n values ξ_i and n values w_i. These are universal quantities independent of the particular function f. With ξ_i and w_i known *in general*, to integrate a given function $g(\xi)$ *exactly* through ξ^{2n-1} it is necessary to perform n function evaluations, namely to compute $g(\xi_i)$.

As an example we seek two Gauss points and two weights. Now for $n=2$

$$w_1 + w_2 = 2 \tag{7.36a}$$

$$w_1\xi_1 + w_2\xi_2 = 0 \tag{7.36b}$$

$$w_1\xi_1^2 + w_2\xi_2^2 = \tfrac{2}{3} \tag{7.36c}$$

$$w_1\xi_1^3 + w_2\xi_2^3 = 0 \tag{7.36d}$$

From Equations 7.36b and 7.36d, $w_1\xi_1\left[\xi_1^2 - \xi_2^2\right] = 0$, leading to $\xi_2 = -\xi_1$. From Equations 7.36a and 7.36c it now follows that $-\xi_2 = \xi_1 = 1/\sqrt{3}$. Finally, the normalization $w_1 = 1$ implies that $w_2 = 1$.

EXAMPLE 7.5

Modify the rod element to replace the physical coordinate x with the "natural coordinate" ξ in which

$$\xi = ax + b, \quad \xi(x_e) = -1, \quad \xi(x_{e+1}) = +1$$

Rewrite the interpolation model using natural coordinates, and perform the inverse to obtain $\mathbf{\Phi}_{m1}$.

Solution

We seek a, b for which the transformation $\xi = ax + b$ satisfies $\xi(x_e) = -1$ and $\xi(x_{e+1}) = +1$. Elementary manipulation furnishes that $a = \frac{2}{(x_{e+1} - x_e)} = 2/l_e$ and $b = -\frac{x_{e+1} + x_e}{l_e}$, in which l_e is the length of the element. Next $\frac{\partial}{\partial x} = \frac{\partial}{\partial \xi}\frac{\partial \xi}{\partial x}$, so that $\frac{\partial}{\partial x} = a\frac{\partial}{\partial \xi}$ and $\frac{\partial^2}{\partial x^2} = a^2\frac{\partial^2}{\partial \xi^2} = \frac{4}{l_e^2}\frac{\partial^2}{\partial \xi^2}$. Accordingly, the governing equation for the displacements in rods becomes

$$\frac{4EA}{l_e^2}\frac{\partial^2 u}{\partial \xi^2} = \rho A \frac{\partial^2 u}{\partial t^2}$$

For the interpolation model the natural coordinate now is

$$u(\xi(x),t) = \boldsymbol{\varphi}_{m1}^T \mathbf{\Phi}_{m1} \boldsymbol{\Upsilon}_{m1}(t), \quad \boldsymbol{\varphi}_{m1}^T = (1 \ \xi(x)), \quad \boldsymbol{\Upsilon}_{m1}(t) = \begin{pmatrix} u_e(t) \\ u_{e+1}(t) \end{pmatrix}$$

Now letting $u_e = u(\xi(x_e)) = u(-1)$ and $u_{e+1} = u(\xi(x_{e+1})) = u(+1)$ yields

$$\begin{pmatrix} u_e(t) \\ u_{e+1}(t) \end{pmatrix} = \boldsymbol{\Upsilon}_{m1}(t) = \begin{bmatrix} 1 & \xi(x_e) \\ 1 & \xi(x_{e+1}) \end{bmatrix} \mathbf{\Phi}_{m1} \boldsymbol{\Upsilon}_{m1}(t)$$

Consequently,

$$\mathbf{\Phi}_{m1} = \begin{bmatrix} 1 & \xi(x_e) \\ 1 & \xi(x_{e+1}) \end{bmatrix}^{-1} = \begin{bmatrix} 1 & -1 \\ 1 & 1 \end{bmatrix}^{-1} = \frac{1}{2}\begin{bmatrix} 1 & 1 \\ -1 & 1 \end{bmatrix}$$

EXAMPLE 7.6

Rewrite the Euler–Bernoulli equation for the beam using the previous transformation. Rewrite the interpolation model using natural coordinates, and perform the inverse to obtain $\mathbf{\Phi}_{b1}$.

Solution

In the natural coordinate ζ satisfying

$$\xi = a + bx,\ \xi(x_e) = -1,\ \xi(x_{e+1}) = +1,\ a = \frac{2}{(x_{e+1} - x_e)},\ b = \frac{x_{e+1} + x_e}{x_{e+1} - x_e}$$

the beam equation becomes

$$\frac{16\mathrm{E}I}{l_e^4}\frac{\partial^4 w}{\partial \zeta^4} = 0, \quad l_e = x_{e+1} - x_e$$

The interpolation model is

$$w(\xi(x),t) = \boldsymbol{\varphi}_{b1}^T \boldsymbol{\Phi}_{b1} \boldsymbol{\gamma}_{b1}(t)$$

in which

$$\boldsymbol{\varphi}_{b1}^T = \begin{pmatrix} 1 & \xi(x) & \xi^2(x) & \xi^3(x) \end{pmatrix}, \quad \boldsymbol{\gamma}_{b1}(t) = \begin{pmatrix} w_e \\ w'_e/l_e \\ w_{e+1} \\ -2w'_{e+1}/l_2 \end{pmatrix}$$

Enforcing this model at $\xi(x_e) = -1$, and $\xi(x_{e+1}) = +1$ furnishes

$$\boldsymbol{\Phi}_{b1} = \begin{bmatrix} 1 & \xi(x_e) & \xi^2(x_e) & \xi^3(x_e) \\ 0 & -1 & -2\xi(x_e) & -3\xi^2(x_e) \\ 1 & \xi(x_{e+1}) & \xi^2(x_{e+1}) & \xi^3(x_{e+1}) \\ 0 & -1 & -2\xi(x_{e+1}) & -3\xi^2(x_{e+1}) \end{bmatrix}^{-1}$$

$$= \begin{bmatrix} 1 & -1 & 1 & -1 \\ 0 & -1 & 2 & -3 \\ 1 & 1 & 1 & 1 \\ 0 & -1 & -2 & -3 \end{bmatrix}^{-1}$$

$$= \frac{1}{4}\begin{bmatrix} 2 & -1 & 2 & 1 \\ -3 & 1 & 3 & 1 \\ 0 & 1 & 0 & -1 \\ 1 & -1 & -1 & -1 \end{bmatrix}$$

EXAMPLE 7.7

We consider the clamped-free case of a beam-column modeled using a single element. We examine the effect of the compressive load P. A model using a single element gives

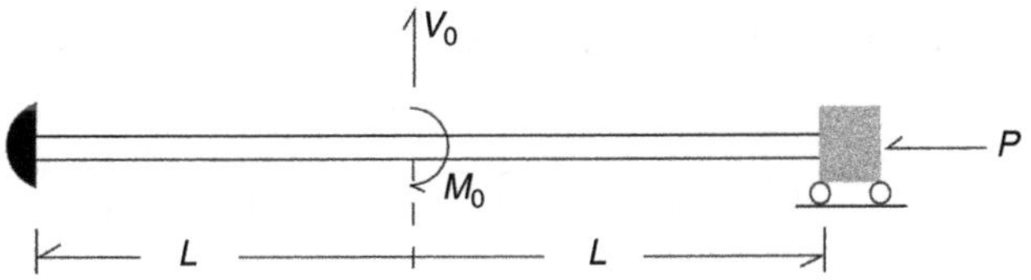

FIGURE 7.8 Buckling of a cantilevered beam-column.

Solution

Enforcing the constraints $w(0) = -w'(0) = 0$ a priori gives the equation

$$\begin{bmatrix} 12 - \frac{6}{5}\xi & (6 - \frac{1}{10}\xi)L \\ (6 - \frac{1}{10}\xi)L & (4 - \frac{2}{15}\xi)L^2 \end{bmatrix} \begin{Bmatrix} w(L) \\ -w'(L) \end{Bmatrix} = \begin{Bmatrix} V_0 \\ M_0 \end{Bmatrix}$$

The generalized displacements become unbounded, and hence *buckling occurs*, at the value of P rendering the stiffness matrix singular (Figure 7.8).

$$\det \begin{bmatrix} 12 - \frac{6}{5}\xi & (6 - \frac{1}{10}\xi)L \\ (6 - \frac{1}{10}\xi)L & (4 - \frac{2}{15}\xi)L^2 \end{bmatrix} = 0, \quad \xi = \frac{PL^2}{EI}$$

from which $\left(12 - \frac{6}{5}\xi\right)\left(4 - \frac{2}{15}\xi\right)L^2 - \left(6 - \frac{1}{10}\xi\right)^2 L^2 = 0.$

On solving the above equation, we have $\xi = 32.18,\ 2.49$. Accordingly, the P values inducing buckling are

$$P_1 = 2.49\frac{EI}{L^2}, \quad P_2 = 32.18\frac{EI}{L^2}$$

The exact solution is taken from Brush and Almroth (1975) as

$$P_1 = \frac{\pi^2}{4}\frac{EI}{L^2} = 2.27\frac{EI}{L^2}, \quad P_2 = \frac{9}{4}\pi^2\frac{EI}{L^2} = 20.42\frac{EI}{L^2}$$

The fundamental critical load for a single element is fairly close to the exact solution, while the second buckling load is about 50% too high. Clearly, a one-element model is not very accurate for buckling. The model becomes more accurate as additional elements are added.

7.6 UNCONSTRAINED ROD ELEMENTS

An element is called unconstrained if none of the points in the element are fixed.

From the Principle of Virtual Work the stiffness matrix satisfies $\mathbf{K} = \int \mathbf{\Phi}^T \boldsymbol{\beta}(\mathbf{x})\mathbf{D}'\boldsymbol{\beta}^T(\mathbf{x})\mathbf{\Phi}\, dV$ which for the foregoing rod element in the physical coordinate becomes

$$\boldsymbol{\varphi}^T(\mathbf{x}) = (1 \quad x), \quad \boldsymbol{\beta}^T(\mathbf{x}) = \frac{\partial \boldsymbol{\varphi}^T(\mathbf{x})}{\partial x} = (0 \quad 1)$$
$$\boldsymbol{\Phi} = \begin{bmatrix} 1 & x_e \\ 1 & x_{e+1} \end{bmatrix}^{-1} = \frac{1}{(x_{e+1} - x_e)} \begin{bmatrix} x_{e+1} & -x_e \\ -1 & 1 \end{bmatrix} \tag{7.37}$$

Also $\mathbf{D}' = \mathsf{E}$, and $\mathrm{d}V = A\,\mathrm{d}x$, in consequence of which

$$\mathbf{K}_e = \int_{x_e}^{x_{e+1}} \frac{1}{(x_{e+1} - x_e)^2} \begin{bmatrix} x_{e+1} & -1 \\ -x_e & 1 \end{bmatrix} \begin{pmatrix} 0 \\ 1 \end{pmatrix} \mathsf{E}A(0 \quad 1) \begin{bmatrix} x_{e+1} & -x_e \\ -1 & 1 \end{bmatrix} \mathrm{d}x$$
$$= \frac{\mathsf{E}A}{l_e^2} \int_{x_e}^{x_{e+1}} \begin{bmatrix} x_{e+1} & -1 \\ -x_e & 1 \end{bmatrix} \begin{bmatrix} 0 & 0 \\ 0 & 1 \end{bmatrix} \begin{bmatrix} x_{e+1} & -x_e \\ -1 & 1 \end{bmatrix} \mathrm{d}x$$
$$= \frac{\mathsf{E}A}{l_e} \begin{bmatrix} 1 & -1 \\ -1 & 1 \end{bmatrix} \tag{7.38}$$

and $l_e = x_{e+1} - x_e$ is the element length.

We now redo the derivation for $\mathbf{K}_e$ using the natural coordinate ξ, in which $\xi = ax + b$, with $-1 = ax_e + b$, $+1 = ax_{e+1} + b$. Interpolation in the natural coordinate now is expressed by

$$\boldsymbol{\varphi}^T = \{1 \quad \xi(x)\}, \quad \boldsymbol{\gamma}(t) = \begin{Bmatrix} u_e(t) \\ u_{e+1}(t) \end{Bmatrix}$$
$$\boldsymbol{\Phi} = \frac{1}{2} \begin{bmatrix} 1 & 1 \\ -1 & 1 \end{bmatrix}, \quad \boldsymbol{\beta}^T = \frac{\partial \boldsymbol{\varphi}^T}{\partial x} = \frac{2}{l_e} \frac{\partial \boldsymbol{\varphi}^T}{\partial \xi} = (0 \quad 1) \tag{7.39}$$
$$\mathbf{K}_e = \int_{-1}^{+1} \frac{1}{4} \begin{bmatrix} 1 & -1 \\ 1 & 1 \end{bmatrix} \left(\frac{2}{l_e}\right)^2 \begin{pmatrix} 0 \\ 1 \end{pmatrix} \mathsf{E}A(0 \quad 1) \begin{bmatrix} 1 & 1 \\ -1 & 1 \end{bmatrix} \mathrm{d}x$$

On substituting $\mathrm{d}x = \frac{l_e}{2}\mathrm{d}\xi$, we obtain that

$$\mathbf{K}_e = \frac{\mathsf{E}A}{2l_e} \int_{-1}^{+1} \begin{bmatrix} 1 & -1 \\ 1 & 1 \end{bmatrix} \begin{bmatrix} 0 & 0 \\ 0 & 1 \end{bmatrix} \begin{bmatrix} 1 & 1 \\ -1 & 1 \end{bmatrix} \mathrm{d}\xi \tag{7.40}$$
$$= \frac{\mathsf{E}A}{l_e} \begin{bmatrix} 1 & -1 \\ -1 & 1 \end{bmatrix} \tag{7.41}$$

Using both the physical and the natural coordinates we next determine the mass matrix $\mathbf{M}_e$ for the segment, in which ρ denotes the mass density. In the physical coordinates, the Principle of Virtual Work applied to the rob element gives the mass matrix as

$$
\begin{aligned}
\mathbf{M} &= \int \rho A \boldsymbol{\Phi}^T \boldsymbol{\varphi}(\mathbf{x})\boldsymbol{\varphi}^T(\mathbf{x})\boldsymbol{\Phi}\,\mathrm{d}x \\
&= \int_{x_e}^{x_{e+1}} \frac{1}{(x_{e+1}-x_e)^2}\begin{bmatrix} x_{e+1} & -1 \\ -x_e & 1 \end{bmatrix}\begin{pmatrix} 1 \\ x \end{pmatrix}\rho A(1 \quad x)\begin{bmatrix} x_{e+1} & -x_e \\ -1 & 1 \end{bmatrix}\mathrm{d}x \\
&= \frac{\rho A}{l_e^2}\begin{bmatrix} x_{e+1} & -1 \\ -x_e & 1 \end{bmatrix}\int_{x_e}^{x_{e+1}}\begin{bmatrix} 1 & x \\ x & x^2 \end{bmatrix}dx\begin{bmatrix} x_{e+1} & -x_e \\ -1 & 1 \end{bmatrix} \\
&= \frac{\rho A}{l_e^2}\begin{bmatrix} x_{e+1} & -1 \\ -x_e & 1 \end{bmatrix}\begin{bmatrix} x_{e+1}-x_e & \frac{1}{2}\left(x_{e+1}^2-x_e^2\right) \\ \frac{1}{2}\left(x_{e+1}^2-x_e^2\right) & \frac{1}{3}\left(x_{e+1}^3-x_e^3\right) \end{bmatrix}\begin{bmatrix} x_{e+1} & -x_e \\ -1 & 1 \end{bmatrix} \\
&= \frac{\rho A}{l_e^2}\begin{bmatrix} \frac{1}{3}(x_{e+1}-x_e)^3 & \frac{1}{6}(x_{e+1}-x_e)^3 \\ \frac{1}{6}(x_{e+1}-x_e)^3 & \frac{1}{3}(x_{e+1}-x_e)^3 \end{bmatrix} \\
&= \frac{\rho A l_e}{3}\begin{bmatrix} 1 & 1/2 \\ 1/2 & 1 \end{bmatrix}
\end{aligned}
\tag{7.42}
$$

In the natural coordinate, we obtain

$$
\begin{aligned}
\mathbf{M}_e &= \int_{-1}^{+1}\frac{1}{4}\begin{bmatrix} 1 & -1 \\ 1 & 1 \end{bmatrix}\begin{pmatrix} 1 \\ \xi \end{pmatrix}\rho A(1 \quad \xi)\begin{bmatrix} 1 & 1 \\ -1 & 1 \end{bmatrix}\frac{l_e}{2}\mathrm{d}\xi \\
&= \frac{\rho A l_e}{8}\begin{bmatrix} 1 & -1 \\ 1 & 1 \end{bmatrix}\int_{-1}^{+1}\begin{bmatrix} 1 & \xi \\ \xi & \xi^2 \end{bmatrix}\mathrm{d}\xi\begin{bmatrix} 1 & 1 \\ -1 & 1 \end{bmatrix} \\
&= \frac{\rho A l_e}{8}\begin{bmatrix} 1 & -1 \\ 1 & 1 \end{bmatrix}\begin{bmatrix} 2 & 0 \\ 0 & 2/3 \end{bmatrix}\begin{bmatrix} 1 & 1 \\ -1 & 1 \end{bmatrix} \\
&= \frac{\rho A l_e}{3}\begin{bmatrix} 1 & 1/2 \\ 1/2 & 1 \end{bmatrix}
\end{aligned}
\tag{7.43}
$$

The terms "bar" and "tendon" are alternate names for rods. Shafts, which support torques by twisting, are described by relations which have the same mathematical form as the relations for rods. The same holds true for one-dimensional thermal or electrical conductors (the latter being described by a potential). The mass and stiffness elements for a shaft are simply

$$
\mathbf{M}_e = \frac{\rho J l_e}{3}\begin{bmatrix} 1 & 1/2 \\ 1/2 & 1 \end{bmatrix}, \quad \mathbf{K}_e = \frac{\mu J}{l_e}\begin{bmatrix} 1 & -1 \\ -1 & 1 \end{bmatrix}
\tag{7.44}
$$

in which $\mu = \frac{\mathsf{E}}{2(1+\nu)}$ is the elastic shear modulus and $J = \pi r_0^4/2$ for a solid circular shaft.

7.7 UNCONSTRAINED ELEMENTS FOR BEAMS AND BEAM-COLUMNS

We next consider the mass matrix for the beam element. It is convenient to introduce the vectors $\boldsymbol{\gamma}_1$, $\boldsymbol{\gamma}_2$ and $\boldsymbol{\gamma}$ as follows:

$$\boldsymbol{\gamma}_1 = \begin{Bmatrix} w_1(x_e,t) \\ -w_1'(x_e,t) \end{Bmatrix}, \quad \boldsymbol{\gamma}_2 = \begin{Bmatrix} w_2(x_{e+1},t) \\ -w_2'(x_{e+1},t) \end{Bmatrix}$$
$$\boldsymbol{\gamma} = \begin{Bmatrix} \boldsymbol{\gamma}_1 \\ \hdashline \boldsymbol{\gamma}_2 \end{Bmatrix} = \begin{Bmatrix} w_1(x_e,t) \\ -w_1'(x_e,t) \\ \hdashline w_2(x_{e+1},t) \\ -w_2'(x_{e+1},t) \end{Bmatrix} \tag{7.45}$$

In physical coordinates the mass matrix of the current element is given by

$$\mathbf{M}_e = \begin{bmatrix} 1 & x_e & x_e^2 & x_e^3 \\ 0 & 1 & 2x_e & 3x_e^2 \\ 1 & x_{e+1} & x_{e+1}^2 & x_{e+1}^3 \\ 0 & 1 & 2x_{e+1} & 3x_e^2 \end{bmatrix}^{-T} \left(\rho A \int_{x_e}^{x_{e+1}} \begin{Bmatrix} 1 \\ x \\ x^2 \\ x^3 \end{Bmatrix} \{1 \quad x \quad x^2 \quad x^3\}\,\mathrm{d}x \right)$$
$$\begin{bmatrix} 1 & x_e & x_e^2 & x_e^3 \\ 0 & 1 & 2x_e & 3x_e^2 \\ 1 & x_{e+1} & x_{e+1}^2 & x_{e+1}^3 \\ 0 & 1 & 2x_{e+1} & 3x_e^2 \end{bmatrix}^{-1} = \begin{bmatrix} \mathbf{M}_{11}^{(e)} & \mathbf{M}_{12}^{(e)} \\ \mathbf{M}_{12}^{(e)T} & \mathbf{M}_{22}^{(2)} \end{bmatrix} \tag{7.46}$$

$$\mathbf{M}_{11}^{(e)} = \frac{\rho A L_e}{210}\begin{bmatrix} 78 & -11L_e \\ -11L_e & 2L_e^2 \end{bmatrix}, \quad \mathbf{M}_{12}^{(e)} = \frac{\rho A L_e}{210}\begin{bmatrix} 27 & \frac{13}{2}L_e \\ -\frac{13}{2}L_e & \frac{3}{2}L_e^2 \end{bmatrix}$$
$$\mathbf{M}_{22}^{(2)} = \frac{\rho A L_e}{210}\begin{bmatrix} 78 & 11L_e \\ 11L_e & 2L_e^2 \end{bmatrix}$$

The stiffness matrix of a beam segment is given in physical coordinates as

$$\mathbf{K}_e = \begin{bmatrix} 1 & x_e & x_e^2 & x_e^3 \\ 0 & -1 & -2x_e & -3x_e^2 \\ 1 & x_{e+1} & x_{e+1}^2 & x_{e+1}^3 \\ 0 & -1 & -2x_{e+1} & -3x_e^2 \end{bmatrix}^{-T} \left(\mathrm{E}I \int_{x_e}^{x_{e+1}} \begin{Bmatrix} 0 \\ 0 \\ 2 \\ 6x \end{Bmatrix} \{0 \quad 0 \quad 2 \quad 6x\}\,\mathrm{d}x \right)$$
$$\times \begin{bmatrix} 1 & x_e & x_e^2 & x_e^3 \\ 0 & -1 & -2x_e & -3x_e^2 \\ 1 & x_{e+1} & x_{e+1}^2 & x_{e+1}^3 \\ 0 & -1 & -2x_{e+1} & -3x_e^2 \end{bmatrix}^{-1}$$
$$= \begin{bmatrix} \mathbf{K}_{11}^{(e)} & \mathbf{K}_{12}^{(e)} \\ \mathbf{K}_{12}^{(e)T} & \mathbf{K}_{22}^{(2)} \end{bmatrix} \tag{7.47}$$

$$\mathbf{K}_{11}^{(e)} = \frac{\mathsf{E}I}{L_e^3}\begin{bmatrix} 12 & -6L_e \\ -6L_e & 4L_e^2 \end{bmatrix}, \quad \mathbf{K}_{12}^{(e)} = \frac{\mathsf{E}I}{L_e^3}\begin{bmatrix} 12 & -6L_e \\ 6L_e & 2L_e^2 \end{bmatrix}, \quad \mathbf{K}_{22}^{(e)} = \frac{\mathsf{E}I}{L_e^3}\begin{bmatrix} 12 & 6L_e \\ 6L_e & 4L_e^2 \end{bmatrix}$$

Finally, if the member serves as a beam-column the stiffness matrix is augmented to furnish the new stiffness matrix $\mathbf{K}_e^{(bc)}$ given by

$$\mathbf{K}_e^{(bc)} = \mathbf{K}_e - \mathbf{K}_e^{(P)}$$

$$\mathbf{K}_e^{(P)} = \begin{bmatrix} 1 & x_e & x_e^2 & x_e^3 \\ 0 & -1 & -2x_e & -3x_e^2 \\ 1 & x_{e+1} & x_{e+1}^2 & x_{e+1}^3 \\ 0 & -1 & -2x_{e+1} & -3x_e^2 \end{bmatrix}^{-T} \left(P \int_{x_e}^{x_{e+1}} \begin{Bmatrix} 0 \\ 1 \\ 2x \\ 3x^2 \end{Bmatrix} \{0 \quad 1 \quad 2x \quad 3x^2\}\, \mathrm{d}x \right)$$

$$\times \begin{bmatrix} 1 & x_e & x_e^2 & x_e^3 \\ 0 & -1 & -2x_e & -3x_e^2 \\ 1 & x_{e+1} & x_{e+1}^2 & x_{e+1}^3 \\ 0 & -1 & -2x_{e+1} & -3x_e^2 \end{bmatrix}^{-1}$$

$$= \begin{bmatrix} \mathbf{K}_{11}^{(eP)} & \mathbf{K}_{12}^{(eP)} \\ \mathbf{K}_{12}^{(eP)T} & \mathbf{K}_{22}^{(eP)} \end{bmatrix} \tag{7.48}$$

$$\mathbf{K}_{11}^{(eP)} = \frac{P}{L_e}\begin{bmatrix} 6/5 & -L_e/10 \\ -L_e/10 & 2L_e^2/15 \end{bmatrix}, \quad \mathbf{K}_{12}^{(eP)} = \frac{P}{L_e}\begin{bmatrix} -6/5 & -L_e/10 \\ L_e/10 & -L_e^2/30 \end{bmatrix}$$

$$\mathbf{K}_{22}^{(eP)} = \frac{P}{L_e}\begin{bmatrix} 6/5 & L_e/10 \\ L_e/10 & 2L_e^2/15 \end{bmatrix}$$

7.8 ASSEMBLAGE AND IMPOSITION OF CONSTRAINTS

7.8.1 Rods

Consider the assemblage consisting of two rod elements: denoted as e and $e+1$, see Figure 7.9a. There are three nodes numbered n, $n+1$, and $n+2$. We first consider assemblage of the stiffness matrices, based on two principles: (a) the forces at the nodes are in equilibrium and (b) the displacements at the nodes are continuous. Principle (a) implies that, in the absence of forces applied externally to the node, at node $n+1$ the force of element $e+1$ on element e is equal and opposite the force of element e on element $e+1$. It is helpful to carefully define global (assemblage level) and local (element level) systems of notation. The global system of forces is shown in (a) while the local system is shown in (b). At the center node

$$P_1^{(e)} - P_2^{(e+1)} = 0 \tag{7.49}$$

since no external load is applied. Also, clearly $P_2^{(e)} = P_n$ and $P_1^{(e+1)} = P_{n+2}$.

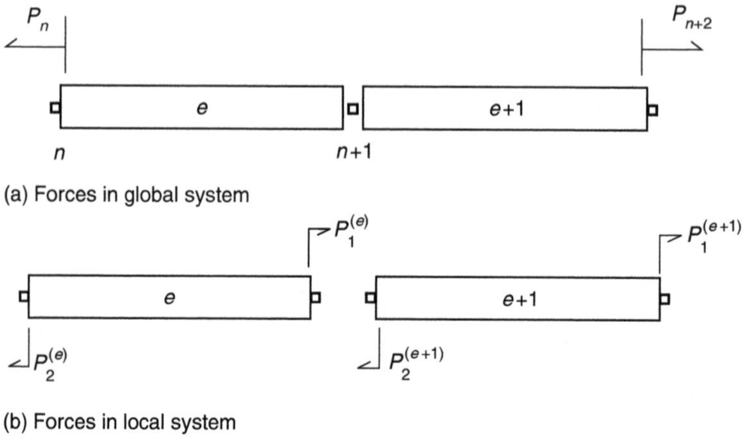

FIGURE 7.9 Assembly of rod elements.

The elements individually satisfy

$$k^{(e)}\begin{bmatrix} 1 & -1 \\ -1 & 1 \end{bmatrix}\begin{pmatrix} u_n \\ u_{n+1} \end{pmatrix} = \begin{pmatrix} -P_2^{(e)} \\ P_1^{(e)} \end{pmatrix}$$
$$k^{(e+1)}\begin{bmatrix} 1 & -1 \\ -1 & 1 \end{bmatrix}\begin{pmatrix} u_{n+1} \\ u_{n+2} \end{pmatrix} = \begin{pmatrix} -P_2^{(e+1)} \\ P_1^{(e+1)} \end{pmatrix} \tag{7.50}$$

and in this case $k^{(e)} = k^{(e+1)} = \mathsf{E}A/L$. These relations may be written as four separate equations:

$$k^{(e)}u_n - k^{(e)}u_{n+1} = -P_2^{(e)} \tag{7.51a}$$

$$-k^{(e)}u_n + k^{(e)}u_{n+1} = P_1^{(e)} \tag{7.51b}$$

$$k^{(e+1)}u_{n+1} - k^{(e+1)}u_{n+2} = -P_2^{(e+1)} \tag{7.51c}$$

$$-k^{(e+1)}u_{n+1} + k^{(e+1)}u_{n+2} = P_1^{(e+1)} \tag{7.51d}$$

Now Equations 7.51b and 7.51c are added and Equation 7.49 is applied to obtain

$$k^{(e)}u_n - k^{(e)}u_{n+1} = -P_2^{(e)} \tag{7.52a}$$

$$-k^{(e)}u_n + \left[k^{(e)} + k^{(e+1)}\right]u_{n+1} - k^{(e+1)}u_{n+2} = 0 \tag{7.52b + 7.52c}$$

$$-k^{(e+1)}u_{n+1} - k^{(e+1)}u_{n+2} = P_1^{(e+1)} \tag{7.52d}$$

and in matrix form

$$\begin{bmatrix} k^{(e)} & -k^{(e)} & 0 \\ -k^{(e)} & k^{(e)}+k^{(e+1)} & -k^{(e+1)} \\ 0 & -k^{(e+1)} & k^{(e+1)} \end{bmatrix} \begin{pmatrix} u_n \\ u_{n+1} \\ u_{n+2} \end{pmatrix} = \begin{pmatrix} -P_n \\ 0 \\ P_{n+1} \end{pmatrix} \tag{7.53}$$

The assembled stiffness matrix shown in Equation 7.53 can be visualized as an overlay of two-element stiffness matrices, referred to global indices, in which there is an intersection of the overlay. The intersection contains the sum of the lowest right hand entry of the upper matrix and the upper left hand entry of the lowest matrix. The overlay structure for a multielement rod is illustrated in Figure 7.10.

The individual element-level stiffness matrices are now rewritten to refer to the global degree-of-freedom numbering system as

$$\mathbf{K}^{(\mathrm{e})} \to \tilde{\mathbf{K}}^{(\mathrm{e})}, \quad \mathbf{K}^{(\mathrm{e+1})} \to \tilde{\mathbf{K}}^{(\mathrm{e+1})}$$
$$\tilde{\mathbf{K}}^{(\mathrm{e})} = k^{(e)} \begin{bmatrix} 1 & -1 & 0 \\ -1 & 1 & 0 \\ 0 & 0 & 0 \end{bmatrix}, \quad \tilde{\mathbf{K}}^{(\mathrm{e+1})} = k^{(e+1)} \begin{bmatrix} 0 & 0 & 0 \\ 0 & 1 & -1 \\ 0 & -1 & 1 \end{bmatrix} \tag{7.54}$$

It is easily recognized that the global stiffness matrix (the assembled stiffness matrix $\mathbf{K}^{(\mathrm{g})}$ of the two-element member) is simply the direct sum of the element stiffness matrices when they are referred to the global degree-of-freedom numbering system: $\mathbf{K}^{(\mathrm{g})} = \tilde{\mathbf{K}}^{(\mathrm{e})} + \tilde{\mathbf{K}}^{(\mathrm{e+1})}$.

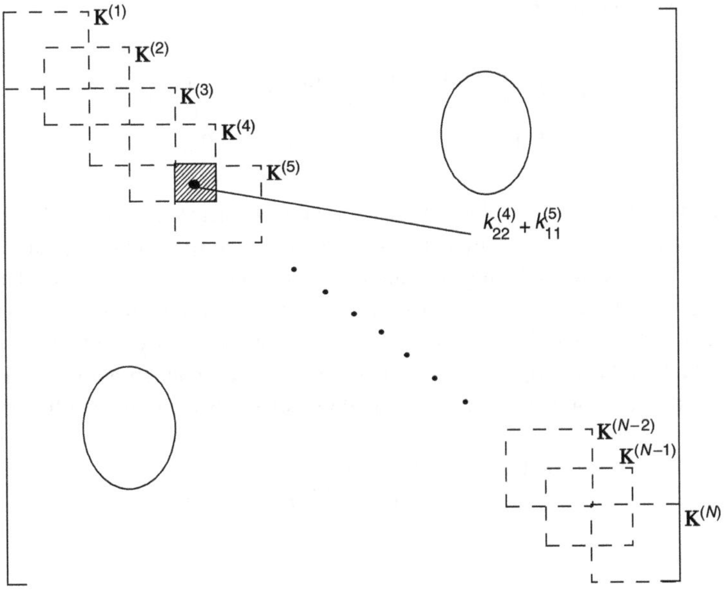

FIGURE 7.10 Assembled beam stiffness matrix.

By extending the two-element example to the multielement rod, the global stiffness matrix is seen to be given by $\mathbf{K}^{(g)} = \sum_e \tilde{\mathbf{K}}^{(e)}$. In the subsequent sections we will see that in the current notation the strain energy in the two elements may be written in the form

$$\begin{aligned} \mathbf{V}^{(e)} &= \tfrac{1}{2}\boldsymbol{\gamma}^T\tilde{\mathbf{K}}^{(e)}\boldsymbol{\gamma}^{(e)} \\ \mathbf{V}^{(e+1)} &= \tfrac{1}{2}\boldsymbol{\gamma}^T\tilde{\mathbf{K}}^{(e+1)}\boldsymbol{\gamma}^{(e)} \\ \boldsymbol{\gamma}^{(e)T} &= (u_n \quad u_{n+1} \quad u_{n+2}) \end{aligned} \tag{7.55}$$

The total strain energy of the two elements simply is the sum: $\frac{1}{2}\boldsymbol{\gamma}^T\mathbf{K}^{(g)}\boldsymbol{\gamma}$.

Finally notice that in the two-rod element assemblage the matrix $\mathbf{K}^{(g)}$ is singular: the sum of the rows is the zero vector as is the sum of the columns. (In fact *prior to the imposition of constraints* the stiffness matrix in any multielement rod is singular.) In this form an attempt to solve the system will give rise to "rigid body motion." To illustrate this reasoning suppose for simplicity that $k^{(e)} = k^{(e+1)}$ in which case equilibrium requires that $P_n = P_{n+2}$. If computations were performed with perfect accuracy, Equation 7.53 would pose no difficulty. However, in performing computations errors arise. For example, suppose P_n is computed as $\hat{P}_n = P_n + \varepsilon_n$ and $\hat{P}_{n+1} = P_{n+1} + \varepsilon_{n+1}$, $P_{n+1} = P_n = P$. Computationally, there is now an unbalanced force $\varepsilon_{n+1} - \varepsilon_n$. In the absence of mass, this in principle implies infinite accelerations, viewed conventionally as rigid body motion. In the finite element method, the problem of rigid body motion can be detected if the output exhibits unrealistically large deformation.

This problem of nonsingularity disappears when the *constraints* of the problem are enforced. In the current example, symmetry implies that $u_{n+1} = 0$. Recalling Equation 7.53 we now have

$$\frac{EA}{L}\begin{bmatrix} 1 & -1 & 0 \\ -1 & 2 & -1 \\ 0 & -1 & 1 \end{bmatrix}\begin{pmatrix} u_n \\ 0 \\ u_{n+2} \end{pmatrix} = \begin{pmatrix} -P_n \\ R \\ P_{n+1} \end{pmatrix} \tag{7.56}$$

in which R is a reaction force which arises to enforce physical symmetry in the presence of numerically generated asymmetry. The equation corresponding to the second equation is useless in predicting the unknowns u_n and u_{n+2} since it introduces the new unknown R. Of course R is a reaction force which arises to enforce the constraint of symmetry. It is possible to "strike out" the second row of the equation and the second column in the matrix, conventionally referred to as "condensation." If the small errors ε_1 and ε_2 are used again, the first and third equations are now rewritten as

$$\frac{EA}{L}\begin{bmatrix} 1 & 0 \\ 0 & 1 \end{bmatrix}\begin{pmatrix} u_n \\ u_{n+1} \end{pmatrix} = \begin{pmatrix} -P + \varepsilon_n \\ P + \varepsilon_{n+1} \end{pmatrix} \tag{7.57}$$

with the solution

$$u_n = [-P + \varepsilon_n]/\tfrac{EA}{L}, \quad u_{n+2} = [P + \varepsilon_{n+1}]/\tfrac{EA}{L} \tag{7.58}$$

To preserve symmetry it is necessary for $u_n + u_{n+1} = 0$. However, the sum is computed as

$$u_n + u_{n+1} = \frac{\varepsilon_n + \varepsilon_{n+1}}{\left(\frac{EA}{L}\right)} \tag{7.59}$$

The reaction force is given by $R = -[\varepsilon_n + \varepsilon_{n+1}]$, and in this case it can be considered as a measure of computational error.

The same assembly arguments apply equally well to the inertial forces as to the elastic forces. This is easily seen if accelerations and mass matrix components are used in Equation 7.53 instead of displacements and stiffness components (i.e., inertial response instead of elastic response). Furthermore, if both inertial and elastic responses are present, the element-level mass matrices are assembled into the global mass matrix in the same manner as in absence of elastic forces.

As will be shown formally in a subsequent section, the total kinetic energy T of the two elements is

$$\begin{aligned} T &= \tfrac{1}{2}\dot{\boldsymbol{\gamma}}^T\left[\tilde{\mathbf{M}}^{(e)} + \tilde{\mathbf{M}}^{(e+1)}\right]\dot{\boldsymbol{\gamma}} \\ \tilde{\mathbf{M}}^{(e)} &= m^{(e)}\begin{bmatrix} 1 & 1/2 \\ 1/2 & 1 \end{bmatrix} \\ m^{(e)} &= \left[\tfrac{1}{3}\rho A l\right]^{(e)} \\ \dot{\boldsymbol{\gamma}}^T &= \{\dot{u}_e \dot{u}_{e+1} \dot{u}_{e+2}\} \end{aligned} \tag{7.60}$$

EXAMPLE 7.8

Write down the assembled mass and stiffness matrices of the following three-element configuration (using rod elements) (Figure 7.11). The elastic modulus is E, the mass density is ρ, and the cross-sectional area is A.

Solution

The stiffness matrix and the mass matrix for a rod element is given by

$$\mathbf{K}^{(e)} = \frac{EA}{L}\begin{bmatrix} 1 & -1 \\ -1 & 1 \end{bmatrix}, \quad \mathbf{M}^{(e)} = \frac{\rho AL}{3}\begin{bmatrix} 1 & 1/2 \\ 1/2 & 1 \end{bmatrix}$$

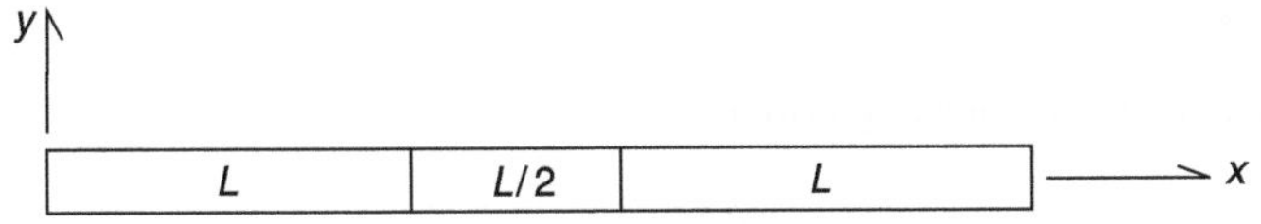

FIGURE 7.11 Three-element model of a rod.

Now, for the first and third elements we have

$$\mathbf{K}^{(1)} = \mathbf{K}^{(3)} = \frac{EA}{L}\begin{bmatrix} 1 & -1 \\ -1 & 1 \end{bmatrix}$$

$$\mathbf{M}^{(1)} = \mathbf{M}^{(3)} = \frac{\rho AL}{3}\begin{bmatrix} 1 & 1/2 \\ 1/2 & 1 \end{bmatrix}$$

But the second element satisfies

$$\mathbf{K}^{(2)} = \frac{EA}{L/2}\begin{bmatrix} 1 & -1 \\ -1 & 1 \end{bmatrix}, \quad \mathbf{M}^{(2)} = \frac{\rho AL/2}{3}\begin{bmatrix} 1 & 1/2 \\ 1/2 & 1 \end{bmatrix}$$

The assembled (global) stiffness matrix is given by

$$\mathbf{K}^{(g)} = \begin{bmatrix} k_{11}^{(1)} & k_{12}^{(1)} & 0 & 0 \\ k_{21}^{(1)} & k_{22}^{(1)} + k_{11}^{(2)} & k_{12}^{(2)} & 0 \\ 0 & k_{21}^{(2)} & k_{22}^{(2)} + k_{11}^{(3)} & k_{12}^{(3)} \\ 0 & 0 & k_{21}^{(3)} & k_{22}^{(3)} \end{bmatrix}$$

Hence

$$\mathbf{K}^{(g)} = \frac{EA}{L}\begin{bmatrix} 1 & -1 & 0 & 0 \\ -1 & 3 & -2 & 0 \\ 0 & -2 & 3 & -1 \\ 0 & 0 & -1 & 1 \end{bmatrix}$$

Continuing, the assembled (global) mass matrix is

$$\mathbf{M}^{(g)} = \frac{\rho AL}{3}\begin{bmatrix} 1 & 1/2 & 0 & 0 \\ 1/2 & 3/2 & 1/4 & 0 \\ 0 & 1/4 & 3/2 & 1/2 \\ 0 & 0 & 1/2 & 1 \end{bmatrix}$$

EXAMPLE 7.9

Show that, for the rod under gravity, a two-element model gives the exact answer for the displacement at $x = L$, as well as a much better approximation to the exact displacement distribution (Figure 7.12).

SOLUTION

The equation for a rod under gravity is

$$EA\frac{\partial^2 u}{\partial x^2} + \rho Ag = 0, x$$

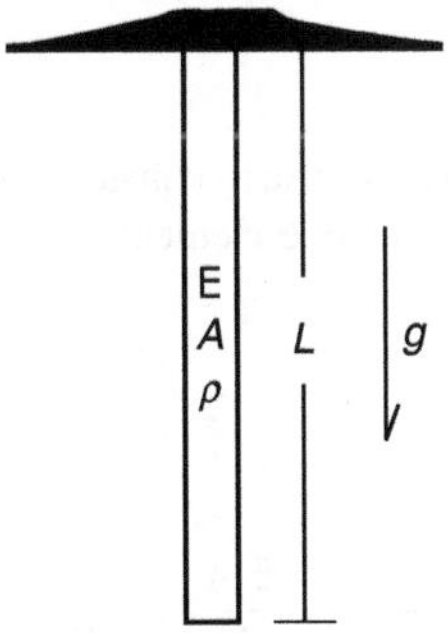

FIGURE 7.12 Rod stretching under gravity.

$$u \text{ positive downward, } u(0) = 0, \; \frac{\partial u(L)}{\partial x} = 0$$

As previously stated, the exact solution is $u(x) = \frac{\rho g}{\mathsf{E}}\left[Lx - \frac{x^2}{2}\right]$, $u(L) = \frac{\rho g}{\mathsf{E}}\frac{L^2}{2}$.

One-element model

Applying variational methods to the foregoing equation furnishes

$$\int_0^L \delta u' \mathsf{E} A u' \,\mathrm{d}x - \int_0^L \delta u \rho A g \,\mathrm{d}x = 0$$

Upon invoking the interpolation model $u(x) = x\frac{u(L)}{L}$, we find

$$\frac{\mathsf{E}A}{L} u(L) = \frac{\rho A g L}{2} \rightarrow u(L) = \frac{\rho g}{\mathsf{E}} \frac{L^2}{2}$$

This result agrees with the exact solution at $x = L$. However, the interpolation model is linear and hence does not agree with the quadratic exact solution. The finite element model predicts that $u(L/2) = \frac{\rho g}{\mathsf{E}}\frac{L}{4}$, while the exact solution at $x = L/2$ is $\frac{3}{8}\frac{\rho g}{\mathsf{E}}L^2$.

Two-element model

Here, the weight of the rod acts (is "lumped") at node "3" and the reaction, which is $-\rho A L g$, acts at node "1." Hence the finite element equation $\mathbf{K}\boldsymbol{\gamma} = \mathbf{f}$ can be written as,

$$\frac{\mathsf{E}A}{(L/2)}\begin{bmatrix} 2 & -1 \\ -1 & 1 \end{bmatrix}\begin{pmatrix} u_2 \\ u_3 \end{pmatrix} = \rho g A \frac{L}{4}\begin{pmatrix} 2 \\ 1 \end{pmatrix} \rightarrow \begin{pmatrix} u_2 \\ u_3 \end{pmatrix} = \frac{\rho g L^2}{\mathsf{E}}\begin{pmatrix} 3/8 \\ 1/2 \end{pmatrix}$$

which is exact at both nodes.

EXAMPLE 7.10

Apply the method of the previous exercise to consider a stepped rod, as shown in Figure 7.13, with each segment modeled as one element.

Two-element solution

The assembled stiffness matrix is

$$\mathbf{K} = \begin{bmatrix} \frac{E_1A_1}{L_1} + \frac{E_2A_2}{L_2} & -\frac{E_2A_2}{L_2} \\ -\frac{E_2A_2}{L_2} & \frac{E_2A_2}{L_2} \end{bmatrix} = \frac{E_2A_2}{L_2}\begin{bmatrix} 1+\alpha & -1 \\ -1 & 1 \end{bmatrix}$$

$$\alpha = \frac{E_1A_1}{L_1} \Big/ \frac{E_2A_2}{L_2}$$

The assembled gravitational terms furnish the consistent gravitational force as

$$\mathbf{f}_g = g\begin{pmatrix} \frac{1}{2}(\rho_1A_1L_1 + \rho_2A_2L_2) \\ \frac{1}{2}\rho_2A_2\frac{L_2}{2} \end{pmatrix} = \rho_2A_2L_2g\begin{pmatrix} \frac{1}{2}(\beta+1) \\ \frac{1}{2} \end{pmatrix}, \quad \beta = \frac{\rho_1A_1L_1}{\rho_2A_2L_2}$$

The solution for the displacements is

$$\begin{pmatrix} u_2 \\ u_3 \end{pmatrix} = \frac{1}{2}\frac{g\rho_2L_2^2}{E_2\alpha}\begin{pmatrix} 2+\beta \\ 2+\alpha+\beta \end{pmatrix}$$

The case of the previous problem is recovered if we set $\alpha=\beta=1$ and $L_2=L/2$.

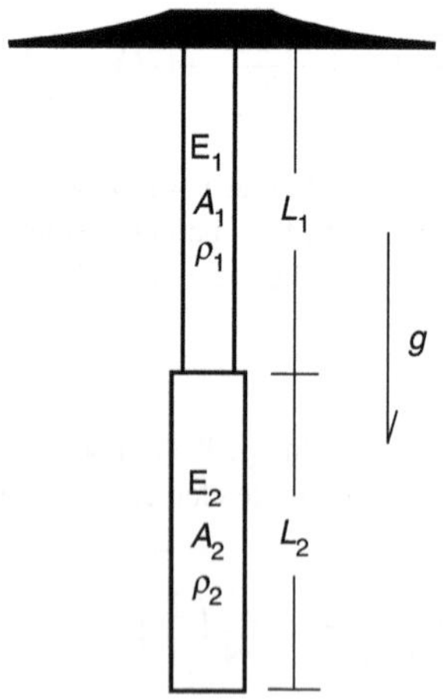

FIGURE 7.13 Two-element rod stretching under gravity.

EXAMPLE 7.11

Using a single finite element in each of the rod segments shown below, obtain expressions for the natural frequencies (Figure 7.14).

SOLUTION

Adding the kinetic and strain energies furnishes

$$T = \frac{1}{2}\{\dot{u}_1 \quad \dot{u}_2\} \cdot \left[\frac{\rho_2 A_2 L_2}{6} \begin{bmatrix} 2 & 1 \\ 1 & 2 \end{bmatrix} \begin{Bmatrix} \dot{u}_1 \\ \dot{u}_2 \end{Bmatrix} \right] + \frac{1}{2}\frac{\rho_1 A_1 L_1}{3} \dot{u}_2^2$$

$$= \frac{1}{2}\{\dot{u}_1 \quad \dot{u}_2\} \cdot \left[\begin{bmatrix} \frac{\rho_2 A_2 L_2}{3} + \frac{\rho_1 A_1 L_1}{3} & \frac{\rho_2 A_2 L_2}{6} \\ \frac{\rho_2 A_2 L_2}{6} & \frac{\rho_2 A_2 L_2}{3} \end{bmatrix} \begin{Bmatrix} \dot{u}_1 \\ \dot{u}_2 \end{Bmatrix} \right]$$

$$V = \frac{1}{2}\{u_1 \quad u_2\} \cdot \left[\frac{\mathrm{E}_2 A_2}{L_2} \begin{bmatrix} 1 & -1 \\ -1 & 1 \end{bmatrix} \begin{Bmatrix} u_1 \\ u_2 \end{Bmatrix} \right] + \frac{1}{2}\frac{\mathrm{E}_1 A_1}{L_1} u_1^2$$

$$= \frac{1}{2}\{u_1 \quad u_2\} \cdot \left[\begin{bmatrix} \frac{\mathrm{E}_2 A_2}{L_2} + \frac{\mathrm{E}_1 A_1}{L_1} & -\frac{\mathrm{E}_2 A_2}{L_2} \\ -\frac{\mathrm{E}_2 A_2}{L_2} & \frac{\mathrm{E}_2 A_2}{L_2} \end{bmatrix} \begin{Bmatrix} u_1 \\ u_2 \end{Bmatrix} \right]$$

Two natural frequencies are obtained from the equation

$$\det \left[\begin{bmatrix} \frac{\mathrm{E}_2 A_2}{L_2} + \frac{\mathrm{E}_1 A_1}{L_1} & -\frac{\mathrm{E}_2 A_2}{L_2} \\ -\frac{\mathrm{E}_2 A_2}{L_2} & \frac{\mathrm{E}_2 A_2}{L_2} \end{bmatrix} - \omega_{n_{1,2}}^2 \begin{bmatrix} \frac{\rho_2 A_2 L_2}{3} + \frac{\rho_1 A_1 L_1}{3} & \frac{\rho_2 A_2 L_2}{6} \\ \frac{\rho_2 A_2 L_2}{6} & \frac{\rho_2 A_2 L_2}{3} \end{bmatrix} \right] = 0$$

Some algebra serves to obtain

$$\omega_{n_{1,2}}^2 = \frac{3\mathrm{E}_2}{\rho_2 L_2^2} \left(\frac{3 + \alpha + \beta \pm \sqrt{(3 + \alpha + \beta)^2 - \alpha(3 + \beta)}}{6 + 2\beta} \right)$$

in which $\alpha = \frac{\mathrm{E}_1 A_1}{L_1} / \frac{\mathrm{E}_2 A_2}{L_2}$ and $\beta = \frac{\rho_1 A_1 L_1}{3} / \frac{\rho_2 A_2 L_2}{3}$.

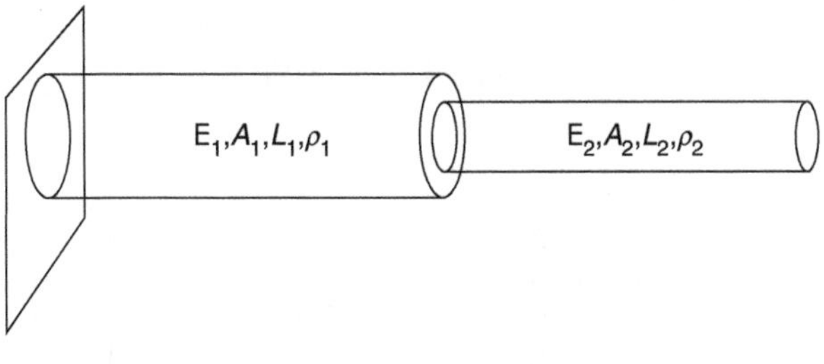

FIGURE 7.14 Natural frequencies of a two-element rod.

7.8.2 Beams

A similar argument applies for beams. The potential energy and stiffness matrix of the *e*th element may be written as

$$V^{(e)} = \tfrac{1}{2}\boldsymbol{\gamma}_b^T \mathbf{K}^{(e)} \boldsymbol{\gamma}_b$$
$$\mathbf{K}^{(e)} = \begin{bmatrix} \mathbf{K}_{11}^{(e)} & \mathbf{K}_{12}^{(e)} \\ \mathbf{K}_{12}^{(e)T} & \mathbf{K}_{22}^{(e)} \end{bmatrix} \tag{7.61}$$
$$\boldsymbol{\gamma}_b^T = \{ w_e \quad -w'_e \quad w_{e+1} \quad -w'_{e+1} \}$$

In a two-element beam model analogous to the foregoing rod model, $V^{(g)} = V^{(e)} + V^{(e+1)}$, implying that

$$\mathbf{K}^{(g)} = \begin{bmatrix} \tilde{\mathbf{K}}_{11}^{(e)} & \tilde{\mathbf{K}}_{12}^{(e)} & \mathbf{0} \\ \tilde{\mathbf{K}}_{12}^{(e)T} & \tilde{\mathbf{K}}_{22}^{(e)} + \tilde{\mathbf{K}}_{11}^{(e+1)} & \tilde{\mathbf{K}}_{12}^{(e+1)} \\ \mathbf{0} & \tilde{\mathbf{K}}_{12}^{(e+1)T} & \tilde{\mathbf{K}}_{22}^{(e+1)} \end{bmatrix} \tag{7.62}$$

Similarly, the global mass matrix for a two-element beam segment is given by

$$\mathbf{M}^{(g)} = \begin{bmatrix} \tilde{\mathbf{M}}_{11}^{(e)} & \tilde{\mathbf{M}}_{12}^{(e)} & \mathbf{0} \\ \tilde{\mathbf{M}}_{12}^{(e)T} & \tilde{\mathbf{M}}_{22}^{(e)} + \tilde{\mathbf{M}}_{11}^{(e+1)} & \mathbf{M}_{12}^{(e+1)} \\ \mathbf{0} & \tilde{\mathbf{M}}_{12}^{(e+1)T} & \tilde{\mathbf{M}}_{22}^{(e+1)} \end{bmatrix} \tag{7.63}$$

EXAMPLE 7.12

For the system indicated below, find the equation governing nodal displacements and slopes (Figure 7.15).

This member is modeled with two beam elements, which is the minimum possible. The global displacement vector, absent the constrained degrees of freedom, is $\gamma_g^T = \{w_1 \quad -w'_1 \quad w_2 \quad -w'_2\}$. The strain energy is the sum of the strain energies of the two elements and the springs:

$$V = V_1 + V_2 + V_3 + V_4$$

in which

$$V_3 = \frac{1}{2}kw_2^2 = \frac{1}{2}\boldsymbol{\gamma}^T \begin{bmatrix} 0 & 0 & 0 & 0 \\ 0 & 0 & 0 & 0 \\ 0 & 0 & k & 0 \\ 0 & 0 & 0 & 0 \end{bmatrix} \boldsymbol{\gamma}_g, \quad V_4 = \frac{1}{2}k_T(-w'_2) = \frac{1}{2}\boldsymbol{\gamma}^T \begin{bmatrix} 0 & 0 & 0 & 0 \\ 0 & 0 & 0 & 0 \\ 0 & 0 & 0 & 0 \\ 0 & 0 & 0 & k_T \end{bmatrix} \boldsymbol{\gamma}$$

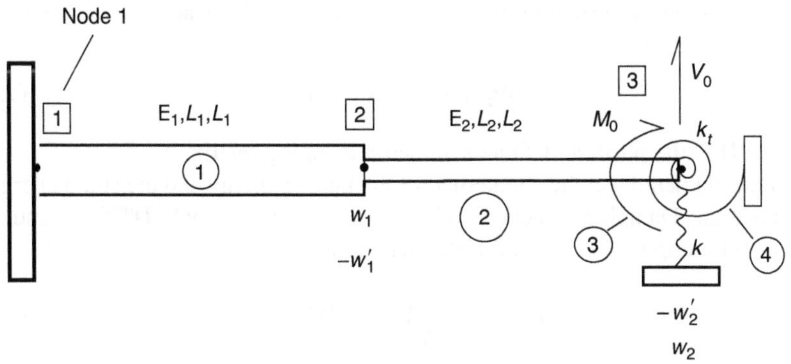

FIGURE 7.15 Two-element beam model with compliant supports.

Also

$$V_1 = \frac{1}{2}\{w_1 \quad -w_1'\}\mathbf{K}_{11}^{(1)}\begin{Bmatrix} w_1 \\ -w_1' \end{Bmatrix} = \frac{1}{2}\boldsymbol{\gamma}_g^T\begin{bmatrix} \mathbf{K}^{(1)} & 0 \\ 0^T & 0 \end{bmatrix}\boldsymbol{\gamma}_g$$

$$V_2 = \frac{1}{2}\boldsymbol{\gamma}_g^T\begin{bmatrix} \mathbf{K}_{11}^{(2)} & \mathbf{K}_{12}^{(2)} \\ \mathbf{K}_{12}^{(2)T} & \mathbf{K}_{22}^{(2)} \end{bmatrix}\boldsymbol{\gamma}_g$$

The total strain energy is obtained by direct addition of the element potential energies

$$V = \frac{1}{2}\boldsymbol{\gamma}_g^T\begin{bmatrix} \mathbf{K}^{(1)} + \mathbf{K}_{\mathrm{I}}^{(2)} & \mathbf{K}_{\mathrm{II}}^{(2)} \\ \mathbf{K}_{\mathrm{II}}^{(2)T} & \mathbf{K}_{\mathrm{III}}^{(2)} + \begin{bmatrix} k & 0 \\ 0 & k_T \end{bmatrix} \end{bmatrix}\boldsymbol{\gamma}_\gamma$$

The Principle of Virtual Work now implies that

$$\begin{bmatrix} 12\left(\frac{\mathsf{E}_1 I_1}{L_1^3}+\frac{\mathsf{E}_2 I_2}{L_2^3}\right) & 6\left(\frac{\mathsf{E}_1 I_1}{L_1^2}-\frac{\mathsf{E}_2 I_2}{L_2^2}\right) & 12\frac{\mathsf{E}_2 I_2}{L_2^3} & -6\frac{\mathsf{E}_2 I_2}{L_2^2} \\ 6\left(\frac{\mathsf{E}_1 I_1}{L_1^2}-\frac{\mathsf{E}_2 I_2}{L_2^2}\right) & 4\left(\frac{\mathsf{E}_1 I_1}{L_1}+\frac{\mathsf{E}_2 I_2}{L_2}\right) & 6\frac{\mathsf{E}_2 I_2}{L_2^2} & 2\frac{\mathsf{E}_2 I_2}{L_2} \\ 12\frac{\mathsf{E}_2 I_2}{L_2^3} & 6\frac{\mathsf{E}_2 I_2}{L_2^2} & 12\frac{\mathsf{E}_2 I_2}{L_2^3}+k & 6\frac{\mathsf{E}_2 I_2}{L_2^2} \\ -6\frac{\mathsf{E}_2 I_2}{L_2^2} & 2\frac{\mathsf{E}_2 I_2}{L_2} & 6\frac{\mathsf{E}_2 I_2}{L_2^2} & 4\frac{\mathsf{E}_2 I_2}{L_2}+k_T \end{bmatrix}\begin{Bmatrix} w_1 \\ -w_1' \\ w_2 \\ -w_2' \end{Bmatrix} = \begin{Bmatrix} 0 \\ 0 \\ V_0 \\ M_0 \end{Bmatrix}$$

The matrix is inverted to solve for the displacements and slopes.

7.9 DAMPING IN RODS AND BEAMS

In addition to mass and stiffness matrices, rods and beams experiencing time-dependent loads are thought to exhibit viscous damping whose effect is represented by a damping matrix. Damping generates a stress proportional to the strain rate.

In linear problems modeled by the finite element method it leads to a vector–matrix equation of the form

$$\mathbf{M}\ddot{\boldsymbol{\gamma}} + \mathbf{D}\dot{\boldsymbol{\gamma}} + \mathbf{K}\boldsymbol{\gamma} = \mathbf{f}(t) \tag{7.64}$$

in which **D** is the positive definite symmetric damping matrix.

At the element level the counterpart of the kinetic energy and the strain energy is the Rayleigh Damping function $\mathrm{D}^{(\mathrm{e})}$ given by $\mathrm{D}^{(\mathrm{e})} = \frac{1}{2}\dot{\boldsymbol{\gamma}}^{(\mathrm{e})^T}\mathbf{D}^{(\mathrm{e})}\dot{\boldsymbol{\gamma}}^{(\mathrm{e})}$, and the "consistent damping force" on the *e*th element is

$$\mathbf{f}_d^{(\mathrm{e})}(t) = \frac{\partial}{\partial\dot{\boldsymbol{\gamma}}}\mathrm{D}^{(\mathrm{e})} = \mathbf{D}^{(\mathrm{e})}\dot{\boldsymbol{\gamma}}^{(\mathrm{e})} \tag{7.65}$$

Just like kinetic and strain energies, the Rayleigh damping function is additive over the elements. Accordingly, if $\hat{\mathbf{D}}^{(\mathrm{e})}$ is the damping matrix of the *e*th element referred to the global node system, the assembled damping matrix is given by

$$\mathbf{D} = \sum_{\mathbf{e}} \hat{\mathbf{D}}^{(\mathrm{e})} \tag{7.66}$$

It should be evident that the global stiffness, mass, and damping matrices have the same bandwidth: a force on one given node depends on the displacements (velocities, accelerations) of the nodes of the elements connected at the given node, thereby determining the bandwidth.

Of course in lightly damped metallic structures the damping properties can be difficult to measure. In finite element practice it is common to assume Rayleigh Damping, in which the damping matrix is assumed to be a linear combination of the mass and stiffness matrices. This has the advantage of enforcing *classical normal modes* in which the damping matrix may be diagonalized by transformations that also diagonalize the mass and stiffness matrices. This topic will appear again in Chapter 9.

7.10 GENERAL DISCUSSION OF ASSEMBLAGE

To explain the basis of the assemblage process, it is convenient to introduce Lagrange's equation. While the Principle of Virtual Work is based on variation of in the spatial domain, Lagrange's equation as stated below also applies variational arguments to the time domain. The equation of interest is

$$\left(\frac{\mathrm{d}}{\mathrm{d}t}\frac{\partial}{\partial\dot{\gamma}_j} - \frac{\partial}{\partial\gamma_j}\right)(T - \mathfrak{B}) = f_j \tag{7.67}$$

Here γ_j is the *j*th entry of the nodal displacement vector in which constraints have been enforced a priori. f_j is the corresponding nonconservative nodal force. T and $\mathfrak{B}$ are the total kinetic energy and total elastic strain energy of the body, expressed in terms of γ_j, $\dot{\gamma}_j$. In particular,

$$T = \frac{1}{2}\int_V \rho\dot{u}_i^2\,\mathrm{d}V, \quad \mathfrak{B} = \frac{1}{2}\int_V \mathrm{S}_{ij}\mathrm{E}_{ij}\,\mathrm{d}V \tag{7.68}$$

But, assuming the body is represented as N elements whose volumes are denoted by V_e,

$$\mathfrak{V} = \frac{1}{2}\sum_{e=1}^{N} \int_{\mathfrak{V}_c} S_{ij}E_{ij}\,dV$$

$$= \frac{1}{2}\sum_{e=1}^{N} \boldsymbol{\gamma}_e^T \mathbf{K}_e \boldsymbol{\gamma}_e \tag{7.69a}$$

in which $\boldsymbol{\gamma}_e$ is the local (element-level) displacement vector, only incorporating the displacements at the nodes of element e. Also $\mathbf{K}_e$ is the corresponding local (element-level) stiffness matrix. For example, for a rod with 10 elements and with the elements numbered from left to right, $\boldsymbol{\gamma}_5^T = \{u_5\ u_6\}$ and $\mathbf{K}_5 = \frac{E_5 A_5}{x_6 - x_5}\begin{bmatrix} 1 & -1 \\ -1 & 1 \end{bmatrix}$.

But, referred to the global displacement vector $\boldsymbol{\gamma}$,

$$\mathfrak{V} = \frac{1}{2}\sum_{e=1}^{N} \boldsymbol{\gamma}^T \tilde{\mathbf{K}}_e \boldsymbol{\gamma}$$

$$= \frac{1}{2}\boldsymbol{\gamma}^T \left(\sum_{e=1}^{N} \tilde{\mathbf{K}}_e\right) \boldsymbol{\gamma}$$

$$= \frac{1}{2}\boldsymbol{\gamma}^T \mathbf{K} \boldsymbol{\gamma}, \quad \mathbf{K} = \left(\sum_{e=1}^{N} \tilde{\mathbf{K}}_e\right) \tag{7.69b}$$

Otherwise stated, the (assembled) stiffness matrix $\mathbf{K}$ is the algebraic sum of the local (element-level) stiffness matrices ($\tilde{\mathbf{K}}_e$) referred to the global nodal displacement vector $\boldsymbol{\gamma}$.

A parallel argument may be applied to the kinetic energy.

$$T = \int_V \rho \dot{u}_i \dot{u}_i\,dV$$

$$= \sum_{e=1}^{N} \left(\int_{V_e} \rho \dot{u}_i \dot{u}_i\,dV \right)$$

$$= \frac{1}{2}\sum_{e=1}^{N} \dot{\boldsymbol{\gamma}}_e^T \mathbf{M}_e \dot{\boldsymbol{\gamma}}_e$$

$$= \frac{1}{2}\sum_{e=1}^{N} \dot{\boldsymbol{\gamma}}^T \tilde{\mathbf{M}}_e \dot{\boldsymbol{\gamma}}$$

$$= \frac{1}{2}\dot{\boldsymbol{\gamma}}^T \left(\sum_{e=1}^{N} \tilde{\mathbf{M}}_e\right) \dot{\boldsymbol{\gamma}}$$

$$= \frac{1}{2}\dot{\boldsymbol{\gamma}}^T \mathbf{M} \dot{\boldsymbol{\gamma}}, \quad \mathbf{M} = \left(\sum_{e=1}^{N} \tilde{\mathbf{M}}_e\right) \tag{7.69c}$$

Applied to the expressions in Equations 7.69a through 7.69c, Lagrange's equation furnishes the expected equation, namely $\mathbf{M}\dot{\boldsymbol{\gamma}} + \mathbf{K}\boldsymbol{\gamma} = \mathbf{f}(t)$.

7.11 GENERAL DISCUSSION ON THE IMPOSITION OF CONSTRAINTS

Constraints serve to remove degrees of freedom. In principle, they should be enforced a priori to reduce the displacement vector to a minimum dimension in which all entries vary *independently*. Alternatively, the functional, whose variation is to be set equal to zero, may be augmented with additional variables (Lagrange multipliers) subject to variation. The variational principles arising from the additional variables enable enforcing the constraints a posteriori. The degrees of freedom, without regard for the constraints, and the Lagrange multipliers are then varied independently.

Of course, enforcing constraints a priori in complex problems can be very difficult, especially if, for example, the constraints are "multipoint." In the finite element method based on the Principle of Virtual Work, despite the mathematical objections to doing so, in most cases the variational principle is formulated without enforcing constraints a priori or without augmenting the functional using Lagrange multipliers or penalty functions. Instead, after the stiffness matrix is formulated the constraints are imposed a posteriori. The author is not aware of any errors resulting from this practice. However, in some classes of constraint problems an augmented functional is used, for example, in incompressible media to be discussed in a subsequent chapter.

There are several classes of constraints. In *simple constraints* one or more displacements vanish at nodes on the boundary, or else assume prescribed fixed values. In *linear multipoint constraints* such as symmetry, several nodes are required to remain on a given line or plane. There are also *internal constraints* such as incompressibility, in which a kinematic requirement is imposed at all nodes throughout the body. As stated before, incompressibility will be addressed in a subsequent chapter.

We first illustrate constraint enforcement when a displacement vanishes at a boundary node, in the case of a static problem. Suppose for simplicity that the constraint occurs at the last entry in the global displacement vector (prior to imposition of constraints). The finite element equation for an n degree-of-freedom system may be written as

$$\mathbf{K}\begin{Bmatrix} \boldsymbol{\gamma}_{n-1} \\ 0 \end{Bmatrix} = \mathbf{f}, \quad \begin{Bmatrix} \mathbf{f}_{n-1} \\ f_n \end{Bmatrix}, \quad \mathbf{K} = \begin{bmatrix} \mathbf{K}_{n-1} & \boldsymbol{\kappa}_n \\ \boldsymbol{\kappa}_n^T & k_{nn} \end{bmatrix}, \quad \mathbf{f} = \begin{Bmatrix} \mathbf{f}_{n-1} \\ f_n \end{Bmatrix} \tag{7.70}$$

The stiffness matrix $\mathbf{K}$ is positive semidefinite and hence singular, since the vector sum of the nodal forces must vanish. If only one nodal displacement is removed the rank of the stiffness matrix is $n-1$, in which case the submatrix $\mathbf{K}_{n-1}$ must be positive definite. The upper block row of Equation 7.70, obtained by "striking out" the nth row and column of $\mathbf{K}$ and nth entry of $\mathbf{f}$, then implies that $\mathbf{K}_{n-1}\boldsymbol{\gamma}_{n-1}=\mathbf{f}_{n-1}$, for which there exists a unique solution. This leaves the reaction force f_n unknown, but the bottom row in Equation 7.70 implies that $f_n=\boldsymbol{\kappa}_n^T\boldsymbol{\gamma}_{n-1}$.

Finally, we illustrate constraint enforcement when a boundary displacement, say the last entry of the global displacement vector, has a prescribed value γ_n. Now the upper block row implies that $\mathbf{K}_{n-1}\boldsymbol{\gamma}_{n-1}=\mathbf{f}_{n-1}-\gamma_n\boldsymbol{\kappa}_n$. Again the matrix of interest is obtained from $\mathbf{K}$ by striking out the nth row and column. However, the force vector is now different from the case of a vanishing displacement. Now there exists a unique solution of the equation $\mathbf{K}_{n-1}\boldsymbol{\gamma}_{n-1}=\mathbf{f}_{n-1}-\gamma_n\boldsymbol{\kappa}_n$.

7.12 INVERSE VARIATIONAL METHOD

Given the functional subject to variation, we derive the underlying differential equation and possible combinations of boundary conditions and constraints. For this purpose we will focus on a simple example. Suppose that a rod satisfies $\delta\Psi=0$, in which Ψ is given by

$$\Psi=\int_0^L \frac{1}{2}\left(\frac{\mathrm{d}u}{\mathrm{d}x}\right)^2 \mathrm{d}x-\frac{Pu(L)}{\mathsf{EA}} \tag{7.71}$$

We again invoke the interpolation model

$$u(x)=\{1 \quad x\}\boldsymbol{\Phi}\begin{pmatrix} u(x_e,t) \\ u(x_{e+1},t) \end{pmatrix} \tag{7.72}$$

For an element $x_e<x<x_{e+1}$, we seek the matrix $\mathbf{K}_e$ such that

$$\mathbf{K}_e\begin{pmatrix} u(x_e,t) \\ u(x_{e+1},t) \end{pmatrix}=\begin{pmatrix} f_e \\ f_{e+1} \end{pmatrix} \tag{7.73}$$

Applying variational operations to the given equation gives

$$\delta\Psi=\int_0^L \left(\frac{\mathrm{d}u}{\mathrm{d}x}\right)\delta\left(\frac{\mathrm{d}u}{\mathrm{d}x}\right)\mathrm{d}x-\frac{P\delta u(L)}{\mathsf{EA}} \tag{7.74}$$

Using integration by parts on the first term furnishes,

$$\delta\Psi=\left[\delta u\cdot\frac{\mathrm{d}u}{\mathrm{d}x}\right]_0^L-\int_0^L \delta u\cdot\frac{\mathrm{d}^2u}{\mathrm{d}x^2}\mathrm{d}x-\frac{P\delta u(L)}{\mathsf{EA}} \tag{7.75}$$

Now $\delta\Psi = 0$ implies that

$$\left[\delta u \cdot \mathsf{E}A \frac{\mathrm{d}u}{\mathrm{d}x}\right]_0^L - \int_0^L \delta u \cdot \mathsf{E}A \frac{\mathrm{d}^2 u}{\mathrm{d}x^2}\mathrm{d}x - P\delta u(L) = 0 \tag{7.76}$$

The domain integral and the endpoint expressions must vanish independently. The endpoint conditions capture possible combinations of boundary conditions and constraints as follows. At $x = 0$, either $u = 0$ or $\mathsf{E}A\frac{\mathrm{d}u}{\mathrm{d}x} = 0$. At $x = L$, $\delta u(L)[\mathsf{E}A\frac{\mathrm{d}u(L)}{\mathrm{d}x} - P] = 0$, from which either $u(L) = 0$ or $\mathsf{E}A\frac{\mathrm{d}u(L)}{\mathrm{d}x} = P$.

From the domain integral we conclude that

$$\int_0^L \delta u\, \mathsf{E}A \frac{\mathrm{d}^2 u}{\mathrm{d}x^2}\mathrm{d}x = 0 \tag{7.77}$$

and the arbitrariness of δu is now appealed to again to conclude that

$$\mathsf{E}A \frac{\mathrm{d}^2 u}{\mathrm{d}x^2} = 0 \tag{7.78}$$

which of course is the well-known differential equation of a rod under static loading.

The expression of the stiffness matrix is, of course,

$$\mathbf{K} = \int \mathbf{\Phi}^T \boldsymbol{\beta}(\mathbf{x})\mathbf{D}'\boldsymbol{\beta}^T(\mathbf{x})\mathbf{\Phi}\,\mathrm{d}V \tag{7.79}$$

For a one-dimensional rod element, as before,

$$\begin{aligned} \boldsymbol{\varphi}^T(\mathbf{x}) &= (1 \quad x), \ \boldsymbol{\beta}^T(\mathbf{x}) = \frac{\partial \boldsymbol{\varphi}^T(\mathbf{x})}{\partial x} = (0 \quad 1), \ \mathbf{\Phi} = \begin{bmatrix} 1 & x_e \\ 1 & x_{e+1} \end{bmatrix}^{-1} \\ &= \frac{1}{(x_{e+1} - x_e)} \begin{bmatrix} x_{e+1} & -x_e \\ -1 & 1 \end{bmatrix} \end{aligned} \tag{7.80}$$

Again $\mathbf{D}' = \mathrm{E}$, and $\mathrm{d}V = A\,\mathrm{d}x$. Now, following now familiar procedures the stiffness matrix is

$$\begin{aligned} \mathbf{K}_e &= \int_{x_e}^{x_{e+1}} \frac{1}{(x_{e+1} - x_e)^2} \begin{bmatrix} x_{e+1} & -1 \\ -x_e & 1 \end{bmatrix} \begin{pmatrix} 0 \\ 1 \end{pmatrix} \mathsf{E}A(0 \quad 1) \begin{bmatrix} x_{e+1} & -x_e \\ -1 & 1 \end{bmatrix} \mathrm{d}x \\ &= \frac{\mathsf{E}A}{l_e^2} \int_{x_e}^{x_{e+1}} \begin{bmatrix} x_{e+1} & -1 \\ -x_e & 1 \end{bmatrix} \begin{bmatrix} 0 & 0 \\ 0 & 1 \end{bmatrix} \begin{bmatrix} x_{e+1} & -x_e \\ -1 & 1 \end{bmatrix} \mathrm{d}x \\ &= \frac{\mathsf{E}A}{l_e} \begin{bmatrix} 1 & -1 \\ -1 & 1 \end{bmatrix} \end{aligned} \tag{7.81}$$

in which $l_e = x_{e+1} - x_e$ is the element length.

EXAMPLE 7.13

Redo this derivation for $\mathbf{K}_e$ in the previous exercise, using the natural coordinate ξ, whereby $\xi = ax + b$, in which a and b are such that $-1 = ax_e + b$, $+1 = ax_{e+1} + b$.

SOLUTION

We show that it is possible to introduce interpolation directly in the natural coordinate ξ. Since $\frac{\partial}{\partial x} = \frac{\partial \xi}{\partial x}\frac{\partial}{\partial \xi} = \frac{2}{\ell_e}\frac{\partial}{\partial \xi}$, the rod equation becomes $\mathsf{E}A\frac{\partial^2 u}{\partial \xi^2} = 0$ and the corresponding interpolation relations are

$$\boldsymbol{\varphi}^T = \{1 \quad \xi(x)\}, \quad \boldsymbol{\gamma}(t) = \begin{Bmatrix} u_e(t) \\ u_{e+1}(t) \end{Bmatrix}$$

$$\boldsymbol{\Phi} = \frac{1}{2}\begin{bmatrix} 1 & 1 \\ -1 & 1 \end{bmatrix}, \quad \boldsymbol{\beta}^T = \frac{\partial \boldsymbol{\varphi}^T}{\partial x} = \frac{2}{l_e}\frac{\partial \boldsymbol{\varphi}^T}{\partial \xi} = (0 \quad 1)$$

Using $dx = \frac{l_e}{2}d\xi$, the stiffness matrix now becomes

$$\mathbf{K}_e = \int_{-1}^{+1} \frac{1}{4}\begin{bmatrix} 1 & -1 \\ 1 & 1 \end{bmatrix}\left(\frac{2}{l_e}\right)^2 \begin{pmatrix} 0 \\ 1 \end{pmatrix} \mathsf{E}A(0 \quad 1)\begin{bmatrix} 1 & 1 \\ -1 & 1 \end{bmatrix}\left(\frac{l_2}{2}d\xi\right)$$

and finally

$$\mathbf{K}_e = \frac{\mathsf{E}A}{l_e}\begin{bmatrix} 1 & -1 \\ -1 & 1 \end{bmatrix}$$

EXAMPLE 7.14

Next regard the nodal displacement vector as a function of t. Find the matrix $\mathbf{M}_e$ such that

$$\mathbf{M}_e \begin{pmatrix} u(x_e, t) \\ u(x_{e+1}, t) \end{pmatrix}^{\bullet\bullet} + \mathbf{K}_e \begin{pmatrix} u(x_e, t) \\ u(x_{e+1}, t) \end{pmatrix} = \begin{pmatrix} f_e \\ f_{e+1} \end{pmatrix}$$

in which ρ is the mass density. Derive $\mathbf{M}_e$ using both physical and natural coordinates.

SOLUTION

First let us derive the mass matrix using the physical coordinates. On substituting $dV = A\,dx$, and letting $l_e = x_{e+1} - x_e$ we have

$$
\begin{aligned}
\mathbf{M}_e &= \int \rho A \boldsymbol{\Phi}^T \boldsymbol{\varphi}(\mathbf{x}) \boldsymbol{\varphi}^T(\mathbf{x}) \boldsymbol{\Phi}\, \mathrm{d}x \\
&= \int_{x_e}^{x_{e+1}} \frac{1}{(x_{e+1}-x_e)^2} \begin{bmatrix} x_{e+1} & -1 \\ -x_e & 1 \end{bmatrix} \begin{pmatrix} 1 \\ x \end{pmatrix} \rho A (1 \quad x) \begin{bmatrix} x_{e+1} & -x_e \\ -1 & 1 \end{bmatrix} \mathrm{d}x \\
&= \frac{\rho A}{l_e^2} \begin{bmatrix} x_{e+1} & -1 \\ -x_e & 1 \end{bmatrix} \int_{x_e}^{x_{e+1}} \begin{bmatrix} 1 & x \\ x & x^2 \end{bmatrix} \mathrm{d}x \begin{bmatrix} x_{e+1} & -x_e \\ -1 & 1 \end{bmatrix} \\
&= \frac{\rho A}{l_e^2} \begin{bmatrix} x_{e+1} & -1 \\ -x_e & 1 \end{bmatrix} \begin{bmatrix} x_{e+1}-x_e & \frac{1}{2}\left(x_{e+1}^2 - x_e^2\right) \\ \frac{1}{2}\left(x_{e+1}^2 - x_e^2\right) & \frac{1}{3}\left(x_{e+1}^3 - x_e^3\right) \end{bmatrix} \begin{bmatrix} x_{e+1} & -x_e \\ -1 & 1 \end{bmatrix} \\
&= \frac{\rho A}{l_e^2} \begin{bmatrix} \frac{1}{3}(x_{e+1}-x_e)^3 & \frac{1}{6}(x_{e+1}-x_e)^3 \\ \frac{1}{6}(x_{e+1}-x_e)^3 & \frac{1}{3}(x_{e+1}-x_e)^3 \end{bmatrix}
\end{aligned}
$$

and this becomes

$$
\mathbf{M}_e = \frac{\rho A l_e}{3} \begin{bmatrix} 1 & 1/2 \\ 1/2 & 1 \end{bmatrix}
$$

For the natural coordinates, following the same procedure evident in the previous exercise gives

$$
\begin{aligned}
\mathbf{M}_e &= \int_{-1}^{+1} \frac{1}{4} \begin{bmatrix} 1 & -1 \\ 1 & 1 \end{bmatrix} \begin{pmatrix} 1 \\ \xi \end{pmatrix} \rho A (1 \quad \xi) \begin{bmatrix} 1 & 1 \\ -1 & 1 \end{bmatrix} \frac{l_e}{2} \mathrm{d}\xi \\
&= \frac{\rho A l_e}{8} \begin{bmatrix} 1 & -1 \\ 1 & 1 \end{bmatrix} \int_{-1}^{+1} \begin{bmatrix} 1 & \xi \\ \xi & \xi^2 \end{bmatrix} \mathrm{d}\xi \begin{bmatrix} 1 & 1 \\ -1 & 1 \end{bmatrix} \\
&= \frac{\rho A l_e}{8} \begin{bmatrix} 1 & -1 \\ 1 & 1 \end{bmatrix} \begin{bmatrix} 2 & 0 \\ 0 & 2/3 \end{bmatrix} \begin{bmatrix} 1 & 1 \\ -1 & 1 \end{bmatrix}
\end{aligned}
$$

and finally

$$
\mathbf{M}_e = \frac{\rho A l_e}{3} \begin{bmatrix} 1 & 1/2 \\ 1/2 & 1 \end{bmatrix}
$$

As has been expected, the stiffness and mass matrices are the same in the rod whether approached using physical or natural coordinates.

8 Two- and Three-Dimensional Elements in Linear Elasticity and Linear Conductive Heat Transfer

8.1 INTERPOLATION MODELS IN TWO DIMENSIONS

8.1.1 Membrane Plate

Consider the unconstrained triangular plate element depicted in Figure 8.1. Suppose that there is no out-of-plane stress (plane stress) or no out-of-plane displacement (plane strain). The displacements $u(x,y,t)$ and $v(x,y,t)$ are to be modeled using the values $u_e(t)$, $v_e(t)$, $u_{e+1}(t)$, $v_{e+1}(t)$, $u_{e+2}(t)$, and $v_{e+2}(t)$. A linear model in x and y suffices for each displacement owing to providing three coefficients to match three nodal values. The interpolation model now is

$$\begin{pmatrix} u(x,y,t) \\ v(x,y,t) \end{pmatrix} = \begin{bmatrix} \boldsymbol{\varphi}_{m2}^T & \mathbf{0}^T \\ \mathbf{0}^T & \boldsymbol{\varphi}_{m2}^T \end{bmatrix} \begin{bmatrix} \boldsymbol{\Phi}_{m2}^T & \mathbf{0}^T \\ \mathbf{0}^T & \boldsymbol{\Phi}_{m2}^T \end{bmatrix} \begin{pmatrix} \boldsymbol{\gamma}_u(t) \\ \boldsymbol{\gamma}_v(t) \end{pmatrix} \tag{8.1}$$

in which

$$\boldsymbol{\gamma}_u(t) = \begin{pmatrix} u_e(t) \\ u_{e+1}(t) \\ u_{e+2}(t) \end{pmatrix}, \quad \boldsymbol{\gamma}_v(t) = \begin{pmatrix} v_e(t) \\ v_{e+1}(t) \\ v_{e+2}(t) \end{pmatrix}, \quad \boldsymbol{\varphi}_{m2} = \begin{pmatrix} 1 \\ x \\ y \end{pmatrix}, \quad \boldsymbol{\Phi}_{m2}^T = \begin{bmatrix} 1 & x_e & y_e \\ 1 & x_{e+1} & y_{e+1} \\ 1 & x_{e+2} & y_{e+2} \end{bmatrix}^{-1}$$

8.1.2 Plate with Bending Stresses Only

In a plate element experiencing bending only, in classical plate theory (e.g., Wang, 1953) the in-plane displacements u and v are expressed by

$$u(x,y,z,t) = -z\frac{\partial w}{\partial x}, \quad v(x,y,z,t) = -z\frac{\partial w}{\partial y} \tag{8.2}$$

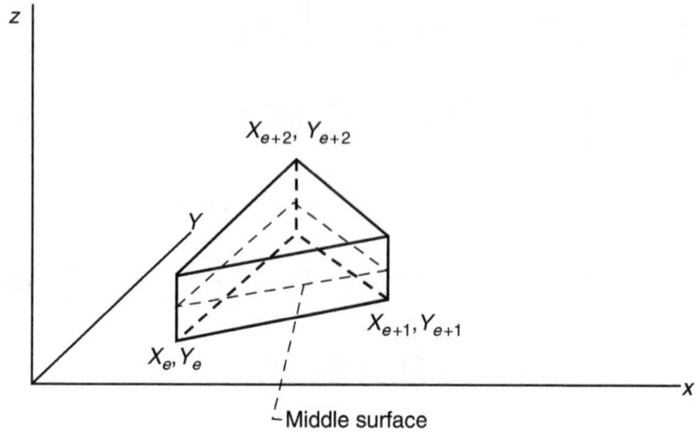

FIGURE 8.1 Triangular plate element.

in which $z = 0$ at the middle (centroidal) plane. The out-of-plane displacement w is assumed to be a function of x and y only. Clearly this model permits no in-plane (membrane) displacements in the middle plane.

An example of an interpolation model is introduced as follows to express $w(x,y)$ throughout the element in terms of the nodal values of w, $-\frac{\partial w}{\partial x}$, and $-\frac{\partial w}{\partial y}$. Clearly w has been assumed to depend only on the in-plane coordinates x and y, and the time t.

$$w(x, y, t) = \boldsymbol{\varphi}_{b2}^T(x, y)\boldsymbol{\Phi}_{b2}\boldsymbol{\gamma}_{b2}(t) \tag{8.3}$$

$$\boldsymbol{\varphi}_{b2}^T(x, y) = \left\{1 \quad x \quad y \quad x^2 \quad xy \quad y^2 \quad x^3 \quad \tfrac{1}{2}(x^2y + y^2x) \quad y^3\right\}$$

$$\boldsymbol{\gamma}_{b2}^T(t) = \Bigg(w_e \quad -\left(\frac{\partial w}{\partial x}\right)_e \quad -\left(\frac{\partial w}{\partial y}\right)_e \quad w_{e+1} \quad -\left(\frac{\partial w}{\partial x}\right)_{e+1}$$
$$-\left(\frac{\partial w}{\partial y}\right)_{e+1} \quad w_{e+2} \quad -\left(\frac{\partial w}{\partial x}\right)_{e+2} \quad -\left(\frac{\partial w}{\partial y}\right)_{e+2} \Bigg)$$

$$\boldsymbol{\Phi}_{b2}^{-1} = \begin{bmatrix} 1 & x_e & y_e & x_e^2 & x_e y_e & y_e^2 & x_e^3 & \frac{1}{2}x_e y_e(x_e + y_e) & y_e^3 \\ 0 & -1 & 0 & -2x_e & -y_e & 0 & -3x_e^2 & -\left(x_e y_e + \frac{1}{2}y_e^2\right) & 0 \\ 0 & 0 & -1 & 0 & -x_e & -2x_e & 0 & -\left(\frac{1}{2}x_e^2 + x_e y_e\right) & -3y_e^2 \\ 1 & x_{e+1} & y_{e+1} & x_{e+1}^2 & x_{e+1}y_{e+1} & y_{e+1}^2 & x_{e+1}^3 & \frac{1}{2}x_{e+1}y_{e+1}(x_{e+1} + y_{e+1}) & y_{e+1}^3 \\ 0 & -1 & 0 & -2x_{e+1} & -y_{e+1} & 0 & -3x_{e+1}^2 & -\left(x_{e+1}y_{e+1} + \frac{1}{2}y_{e+1}^2\right) & 0 \\ 0 & 0 & -1 & 0 & -x_{e+1} & -2y_{e+1} & 0 & -\left(\frac{1}{2}x_{e+1}^2 + x_{e+1}y_{e+1}\right) & -3y_{e+1}^2 \\ 1 & x_{e+2} & y_{e+2} & x_{e+2}^2 & x_{e+2}y_{e+2} & y_{e+2}^2 & x_{e+2}^3 & \frac{1}{2}x_{e+2}y_{e+2}(x_{e+2} + y_{e+2}) & y_{e+2}^3 \\ 0 & -1 & 0 & -2x_{e+2} & -y_{e+2} & 0 & -3x_{e+2}^2 & -\left(x_{e+2}y_{e+2} + \frac{1}{2}y_{e+2}^2\right) & 0 \\ 0 & 0 & -1 & 0 & -x_{e+2} & -2y_{e+2} & 0 & -\left(\frac{1}{2}x_{e+2}^2 + x_{e+2}y_{e+2}\right) & -3y_{e+2}^2 \end{bmatrix}.$$

It follows that

$$\begin{pmatrix} u(x,y,z,t) \\ v(x,y,z,t) \\ w(x,y,z,t) \end{pmatrix} = \begin{pmatrix} -z\dfrac{\partial \boldsymbol{\varphi}_{b2}^T}{\partial x} \\ -z\dfrac{\partial \boldsymbol{\varphi}_{b2}^T}{\partial y} \\ \boldsymbol{\varphi}_{b2}^T \end{pmatrix} \boldsymbol{\Phi}_{b2}\boldsymbol{\gamma}_{b2}(t) \tag{8.4}$$

8.1.3 Plate with Stretching and Bending

Finally, for a plate experiencing both stretching and bending, the displacements are assumed to satisfy

$$\begin{aligned} u(x,y,z,t) &= u_0(x,y,t) - z\frac{\partial w(x,y,t)}{\partial x} \\ v(x,y,z,t) &= v_0(x,y,t) - z\frac{\partial w(x,y,t)}{\partial y} \end{aligned} \tag{8.5}$$

and note that w is a function only of x, y, and t (not z). Here $z = 0$ at the middle surface, while u_0 and v_0 represent the in-plane displacements. Using the nodal values of u_0, v_0, and w, a combined interpolation model is obtained as

$$\begin{aligned} \begin{pmatrix} u(x,y,z,t) \\ v(x,y,z,t) \\ w(x,y,z,t) \end{pmatrix} &= \begin{pmatrix} u_0(x,y,t) \\ v_0(x,y,t) \\ 0 \end{pmatrix} + \begin{pmatrix} -z\dfrac{\partial \boldsymbol{\varphi}_{b2}^T}{\partial x} \\ -z\dfrac{\partial \boldsymbol{\varphi}_{b2}^T}{\partial y} \\ \boldsymbol{\varphi}_{b2}^T \end{pmatrix} \boldsymbol{\Phi}_{b2}\boldsymbol{\gamma}_{b2}(t) \\ &= \begin{bmatrix} \boldsymbol{\varphi}_{m2}^T & \mathbf{0}^T & -z\dfrac{\partial \boldsymbol{\varphi}_{b2}^T}{\partial x} \\ \mathbf{0}^T & \boldsymbol{\varphi}_{m2}^T & -z\dfrac{\partial \boldsymbol{\varphi}_{b2}^T}{\partial y} \\ \mathbf{0}^T & \mathbf{0}^T & \boldsymbol{\varphi}_{b2}^T \end{bmatrix} \begin{bmatrix} \boldsymbol{\Phi}_{m2} & \mathbf{0} & \mathbf{0} \\ \mathbf{0} & \boldsymbol{\Phi}_{m2} & \mathbf{0} \\ \mathbf{0} & \mathbf{0} & \boldsymbol{\Phi}_{b2} \end{bmatrix} \begin{pmatrix} \boldsymbol{\gamma}_u(t) \\ \boldsymbol{\gamma}_v(t) \\ \boldsymbol{\gamma}_w(t) \end{pmatrix} \end{aligned} \tag{8.6}$$

8.1.4 Temperature Field in Two Dimensions

In the two-dimensional triangular element illustrated in Figure 8.1, the linear interpolation model for the temperature is

$$\mathrm{T} - \mathrm{T}_0 = \boldsymbol{\varphi}_{m2}^T \boldsymbol{\Phi}_{m2}\boldsymbol{\theta}_2, \quad \boldsymbol{\theta}_2^T = (\mathrm{T}_e - \mathrm{T}_0 \quad \mathrm{T}_{e+1} - \mathrm{T}_0 \quad \mathrm{T}_{e+2} - \mathrm{T}_0) \tag{8.7}$$

8.1.5 Axisymmetric Elements

An axisymmetric element is displayed in Figure 8.2. It is applicable to bodies which are axisymmetric and are submitted to axisymmetric loads such as all-around pressure. The radial displacement is now denoted by u and the axial displacement is denoted by w. The tangential displacement v vanishes, while radial and axial displacements are independent of θ. Also u and w depend on r, z, and t.

There are two cases which require distinct interpolation models. In the first case none of the nodes are on the axis of revolution ($r = 0$), while in the second case one or two nodes are in fact on the axis. In the first case the linear interpolation model is given by

$$\begin{pmatrix} u(r,z,t) \\ w(r,z,t) \end{pmatrix} = \begin{bmatrix} \boldsymbol{\varphi}_{a1}^T & \mathbf{0}^T \\ \mathbf{0}^T & \boldsymbol{\varphi}_{a1}^T \end{bmatrix} \begin{bmatrix} \boldsymbol{\Phi}_{a1} & \mathbf{0} \\ \mathbf{0} & \boldsymbol{\Phi}_{a1} \end{bmatrix} \begin{pmatrix} \boldsymbol{\gamma}_{ua1}(t) \\ \boldsymbol{\gamma}_{wa1}(t) \end{pmatrix} \tag{8.8}$$

$$\boldsymbol{\varphi}_{a1}^T = (1 \quad r \quad z), \quad \boldsymbol{\Phi}_{a1} = \begin{bmatrix} 1 & r_e & z_e \\ 1 & r_{e+1} & z_{e+1} \\ 1 & r_{e+2} & z_{e+2} \end{bmatrix}^{-1}, \quad \boldsymbol{\gamma}_{ua1} = \begin{pmatrix} u_e \\ u_{e+1} \\ u_{e+2} \end{pmatrix}, \quad \boldsymbol{\gamma}_{wa1} = \begin{pmatrix} w_e \\ w_{e+1} \\ w_{e+2} \end{pmatrix}$$

Now suppose that there are nodes on the axis, and note that the radial displacements are constrained to vanish on the axis. We will see later that, to attain an integrable kernal in the stiffness matrix, it is necessary to enforce the symmetry constraints a priori in the displacement interpolation model. In particular, suppose that node e is on the axis with nodes $e+1$ and $e+2$ defined counterclockwise at the other vertices. A linear interpolation model enforcing the axisymmetry constraint a priori is now

$$\begin{pmatrix} u(r,z,t) \\ w(r,z,t) \end{pmatrix} = \begin{bmatrix} \boldsymbol{\varphi}_{a2}^T & \mathbf{0}^T \\ \mathbf{0}^T & \boldsymbol{\varphi}_{a2}^T \end{bmatrix} \begin{bmatrix} \boldsymbol{\Phi}_{a2} & \mathbf{0} \\ \mathbf{0} & \boldsymbol{\Phi}_{a2} \end{bmatrix} \begin{pmatrix} \boldsymbol{\gamma}_{ua2}(t) \\ \boldsymbol{\gamma}_{wa2}(t) \end{pmatrix} \tag{8.9}$$

$$\boldsymbol{\varphi}_{a2}^T = (r \quad z - z_e), \quad \boldsymbol{\Phi}_{a2} = \begin{bmatrix} r_{e+1} & z_{e+1} - z_e \\ r_{e+2} & z_{e+2} - z_e \end{bmatrix}^{-1}$$

For later purposes note that that ratio $(z - z_e)/r$ is indeterminate as a point in the element is moved toward the node on the axis of revolution.

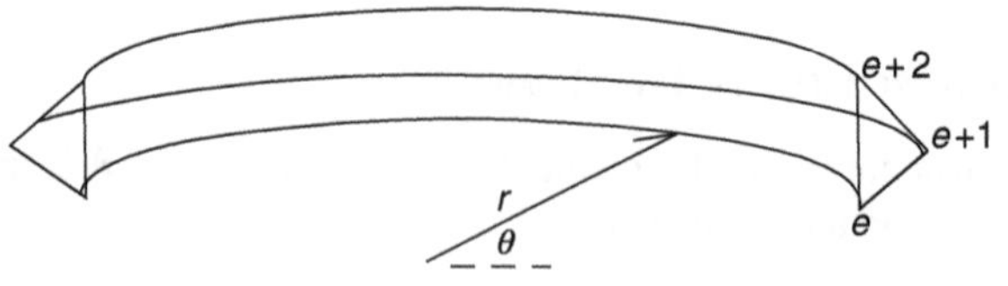

FIGURE 8.2 Axisymmetric element.

A similar formulation can be used if two nodes are on the axis of symmetry, so that the u displacement in the element is modeled using only one nodal displacement, with a coefficient vanishing at each of the nodes on the axis of revolution.

8.2 INTERPOLATION MODELS IN THREE DIMENSIONS

We next consider the tetrahedron illustrated in Figure 8.3. A linear interpolation model for the temperature may be expressed as

$$\mathrm{T} - \mathrm{T}_0 = \boldsymbol{\varphi}_{3T}^T \boldsymbol{\Phi}_{3T} \boldsymbol{\theta}_3 \tag{8.10}$$

$$\boldsymbol{\varphi}_{3T}^T = (1 \quad x \quad y \quad z)$$

$$\boldsymbol{\Phi}_{3T} = \begin{bmatrix} 1 & x_e & y_e & z_e \\ 1 & x_{e+1} & y_{e+1} & z_{e+1} \\ 1 & x_{e+2} & y_{e+2} & z_{e+2} \\ 1 & x_{e+3} & y_{e+3} & z_{e+3} \end{bmatrix}^{-1}$$

$$\boldsymbol{\theta}_{3T}^T = \{\mathrm{T}_e - \mathrm{T}_0 \quad \mathrm{T}_{e+1} - \mathrm{T}_0 \quad \mathrm{T}_{e+2} - \mathrm{T}_0 \quad \mathrm{T}_{e+3} - \mathrm{T}_0\}$$

For elasticity with displacements u, v, and w, the corresponding interpolation model is

$$\begin{pmatrix} u(x,y,z,t) \\ v(x,y,z,t) \\ w(x,y,z,t) \end{pmatrix} = \begin{bmatrix} \boldsymbol{\varphi}_3^T & \mathbf{0}^T & \mathbf{0}^T \\ \mathbf{0}^T & \boldsymbol{\varphi}_3^T & \mathbf{0}^T \\ \mathbf{0}^T & \mathbf{0}^T & \boldsymbol{\varphi}_3^T \end{bmatrix} \begin{bmatrix} \boldsymbol{\Phi}_3 & \mathbf{0} & \mathbf{0} \\ \mathbf{0} & \boldsymbol{\Phi}_3 & \mathbf{0} \\ \mathbf{0} & \mathbf{0} & \boldsymbol{\Phi}_3 \end{bmatrix} \begin{pmatrix} \boldsymbol{\gamma}_u(t) \\ \boldsymbol{\gamma}_v(t) \\ \boldsymbol{\gamma}_w(t) \end{pmatrix} \tag{8.11}$$

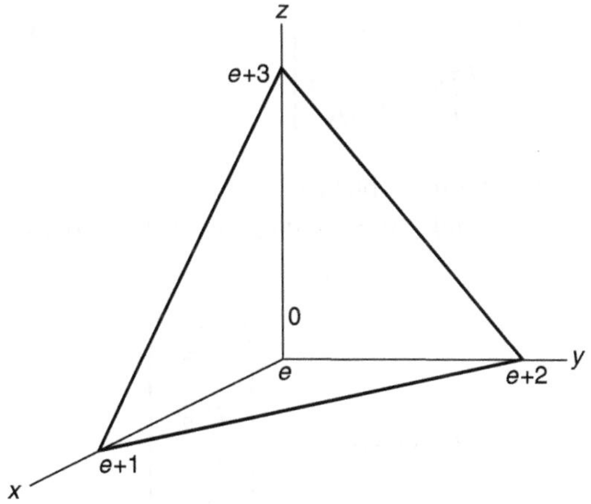

FIGURE 8.3 Tetrahedral element.

8.3 STRAIN–DISPLACEMENT RELATIONS AND THERMAL ANALOGS

8.3.1 Strain–Displacement Relations: Two Dimensions

The (linear) strain tensor for two-dimensional deformation is given by

$$\mathbf{E}(x,y) = \begin{bmatrix} E_{xx} & E_{xy} \\ E_{xy} & E_{yy} \end{bmatrix} = \begin{bmatrix} \dfrac{\partial u}{\partial x} & \dfrac{1}{2}\left(\dfrac{\partial u}{\partial y} + \dfrac{\partial v}{\partial x}\right) \\ \dfrac{1}{2}\left(\dfrac{\partial u}{\partial y} + \dfrac{\partial v}{\partial x}\right) & \dfrac{\partial v}{\partial y} \end{bmatrix} \tag{8.12}$$

In Chapter 5 we encountered the two important cases of plane stress and plane strain. In the latter case, E_{zz} vanishes and S_{zz} is not needed to achieve a solution. In the former case S_{zz} vanishes and E_{zz} is not needed for solution.

As opposed to *VEC* notation, traditional finite element notation (cf. Zienkiewicz and Taylor, 1989) introduces the strain vector $\varepsilon'^T = \{E_{xx} \quad E_{yy} \quad E_{xy}\}$. (Strictly speaking ε' is more properly called an array since it does not have the transformation properties of vectors (cf. Chapter 2).) Upon applying the interpolation model we obtain

$$\boldsymbol{\varepsilon}' = \begin{pmatrix} E_{xx} \\ E_{yy} \\ E_{xy} \end{pmatrix} = \boldsymbol{\beta}_{m2}^T \hat{\boldsymbol{\Phi}}_{m2} \begin{pmatrix} \boldsymbol{\gamma}_{u2} \\ \boldsymbol{\gamma}_{v2} \end{pmatrix} \tag{8.13}$$

in which $\boldsymbol{\beta}_{m2}^T \hat{\boldsymbol{\Phi}}_{m2}$ is called the strain–displacement matrix and

$$\boldsymbol{\beta}_{m2}^T = \begin{bmatrix} \dfrac{\partial \boldsymbol{\varphi}_{m2}^T}{\partial x} & \mathbf{0}^T \\ \mathbf{0}^T & \dfrac{\partial \boldsymbol{\varphi}_{m2}^T}{\partial y} \\ \dfrac{1}{2}\dfrac{\partial \boldsymbol{\varphi}_{m2}^T}{\partial y} & \dfrac{1}{2}\dfrac{\partial \boldsymbol{\varphi}_{m2}^T}{\partial x} \end{bmatrix}, \quad \hat{\boldsymbol{\Phi}}_{m2} = \begin{bmatrix} \boldsymbol{\Phi}_{m2} & \mathbf{0} \\ \mathbf{0} & \boldsymbol{\Phi}_{m2} \end{bmatrix}$$

Hereafter, the prime will not be displayed.

For a plate with bending stresses only, a vector (array) of the strains is displayed as

$$\boldsymbol{\varepsilon}(x,y,z,t) = -z \begin{pmatrix} \dfrac{\partial^2 w}{\partial x^2} \\ \dfrac{\partial^2 w}{\partial y^2} \\ \dfrac{\partial^2 w}{\partial x \partial y} \end{pmatrix} \tag{8.14}$$

from which

$$\boldsymbol{\varepsilon}'(x,y,z,t) = -\mathbf{z}\boldsymbol{\beta}_{b2}^{T}(x,y)\boldsymbol{\Phi}_{b2}\boldsymbol{\gamma}_{b2}(t), \quad \boldsymbol{\beta}_{b2} = \begin{pmatrix} \dfrac{\partial^2 \boldsymbol{\varphi}_{b2}^T}{\partial x^2} \\ \dfrac{\partial^2 \boldsymbol{\varphi}_{b2}^T}{\partial y^2} \\ \dfrac{\partial^2 \boldsymbol{\varphi}_{b2}^T}{\partial x \partial y} \end{pmatrix} \tag{8.15}$$

Note that $\boldsymbol{\varepsilon} \neq VEC(\mathbf{E})$ and instead it represents an ad hoc traditional notation in FEA.

For a plate experiencing both membrane and bending stresses, the foregoing relations can be combined to furnish

$$\boldsymbol{\varepsilon}'(x,y,z,t) = \boldsymbol{\beta}_{mb2}^{T}(x,y,z)\boldsymbol{\Phi}_{mb2}\boldsymbol{\gamma}_{mb2}(t) \tag{8.16}$$

$$\boldsymbol{\beta}_{mb2}^{T} = \begin{pmatrix} \boldsymbol{\beta}_{m2}^{T} & -z\boldsymbol{\beta}_{b2}^{T} \end{pmatrix}, \quad \boldsymbol{\Phi}_{mb2} = \begin{bmatrix} \boldsymbol{\Phi}_{m2} & \mathbf{0} \\ \mathbf{0} & \boldsymbol{\Phi}_{b2} \end{bmatrix}, \quad \boldsymbol{\gamma}_{mb2}(t) = \begin{pmatrix} \boldsymbol{\gamma}_{m2}(t) \\ \boldsymbol{\gamma}_{b2}(t) \end{pmatrix}$$

8.3.2 Axisymmetric Element

For the previously considered toroidal element with a triangular cross section, it is necessary to consider two cases. If there are no nodes on the axis of revolution, then application of the strain–displacement relations to the axisymmetric interpolation model furnishes

$$\boldsymbol{\varepsilon}(r,z,t) = \begin{pmatrix} \dfrac{\partial u}{\partial r} \\ \dfrac{u}{r} \\ \dfrac{\partial w}{\partial z} \\ \dfrac{1}{2}\left(\dfrac{\partial u}{\partial z} + \dfrac{\partial w}{\partial r}\right) \end{pmatrix} = \boldsymbol{\beta}_{a1}^{T} \begin{bmatrix} \boldsymbol{\Phi}_{a1} & \mathbf{0} \\ \mathbf{0} & \boldsymbol{\Phi}_{a1} \end{bmatrix} \begin{pmatrix} \boldsymbol{\gamma}_{ua1}(t) \\ \boldsymbol{\gamma}_{wa1}(t) \end{pmatrix} \tag{8.17}$$

$$\boldsymbol{\beta}_{a1}^{T} = \begin{bmatrix} 0 & 1 & 0 & 0 & 0 & 0 \\ \frac{1}{2} & 1 & \frac{z}{r} & 0 & 0 & 0 \\ 0 & 0 & 0 & 0 & 0 & 1 \\ 0 & 0 & \frac{1}{2} & 0 & \frac{1}{2} & 0 \end{bmatrix}$$

If element e is now located on the axis of revolution, we obtain

$$\boldsymbol{\varepsilon}(r,z,t) = \boldsymbol{\beta}_{a2}^{T} \begin{bmatrix} \boldsymbol{\Phi}_{a2} & \mathbf{0} \\ \mathbf{0} & \boldsymbol{\Phi}_{a1} \end{bmatrix} \begin{pmatrix} \boldsymbol{\gamma}_{ua2}(t) \\ \boldsymbol{\gamma}_{wa1}(t) \end{pmatrix} \tag{8.18}$$

$$\boldsymbol{\beta}_{a2}^T = \begin{bmatrix} 1 & 0 & 0 & 0 & 0 \\ 1 & \frac{z-z_e}{r} & 0 & 0 & 0 \\ 0 & 0 & 0 & 0 & 1 \\ 0 & \frac{1}{2} & 0 & \frac{1}{2} & 0 \end{bmatrix}$$

We now can see the reason for the special interpolation model. If z_e were not present in the foregoing matrix, the quantity z/r would result. Of course, this quantity approaches infinity on a path approaching the node on the axis of revolution. The quantity and its square would appear in the kernal of the integral in the stiffness matrix, rendering the kernal *nonintegrable*. However, with the use of z_e, a path to node *e* produces a path-dependent finite value of $(z - z_e)/r$ in the limit, for which reason the kernal is integrable. The interpolation model enforces the *axisymmetry constraint* a priori, in order to achieve integrability. However, wherever possible the Finite Element Method enforces constraints a posteriori, which is to say that assembled finite element equation is initially obtained without accommodating constraints, and then constraints are used to remove rows and columns from the stiffness matrix.

8.3.3 Thermal Analog for Two-Dimensional and Axisymmetric Elements

The thermal analog of the strain is the temperature gradient. Application of the interpolation model for the temperature furnishes the relation

$$\nabla \mathrm{T} = \boldsymbol{\beta}_{T2}^T \boldsymbol{\Phi}_{T2} \boldsymbol{\theta}_2, \quad \boldsymbol{\beta}_{T2}^T = \begin{bmatrix} 0 & 1 & 0 \\ 0 & 0 & 1 \end{bmatrix} \tag{8.19}$$

There is no need in the axisymmetric case to enforce constraints.

8.3.4 Three-Dimensional Elements

Recalling the tetrahedral element in the previous section, the strain–displacement relation for isotropic linearly elastic materials may be written as

$$\boldsymbol{\varepsilon} = \begin{pmatrix} \mathrm{E}_{xx} \\ \mathrm{E}_{yy} \\ \mathrm{E}_{zz} \\ \mathrm{E}_{xy} \\ \mathrm{E}_{yz} \\ \mathrm{E}_{zx} \end{pmatrix} = \begin{pmatrix} \frac{\partial u}{\partial x} \\ \frac{\partial v}{\partial y} \\ \frac{\partial w}{\partial z} \\ \frac{1}{2}\left(\frac{\partial u}{\partial y} + \frac{\partial v}{\partial x}\right) \\ \frac{1}{2}\left(\frac{\partial v}{\partial z} + \frac{\partial w}{\partial y}\right) \\ \frac{1}{2}\left(\frac{\partial w}{\partial x} + \frac{\partial u}{\partial z}\right) \end{pmatrix} \tag{8.20}$$

$$= \boldsymbol{\beta}_3^T \begin{bmatrix} \boldsymbol{\Phi}_3 & \mathbf{0} & \mathbf{0} \\ \mathbf{0} & \boldsymbol{\Phi}_3 & \mathbf{0} \\ \mathbf{0} & \mathbf{0} & \boldsymbol{\Phi}_3 \end{bmatrix} \begin{pmatrix} \boldsymbol{\gamma}_{u3} \\ \boldsymbol{\gamma}_{v3} \\ \boldsymbol{\gamma}_{w3} \end{pmatrix}$$

in which

$$\boldsymbol{\beta}_3^T = \begin{bmatrix} 0 & 1 & 0 & 0 & 0 & 0 & 0 & 0 & 0 & 0 & 0 & 0 \\ 0 & 0 & 0 & 0 & 0 & 0 & 1 & 0 & 0 & 0 & 0 & 0 \\ 0 & 0 & 0 & 0 & 0 & 0 & 0 & 0 & 0 & 0 & 0 & 1 \\ 0 & 0 & \frac{1}{2} & 0 & 0 & \frac{1}{2} & 0 & 0 & 0 & 0 & 0 & 0 \\ 0 & 0 & 0 & 0 & 0 & 0 & 0 & \frac{1}{2} & 0 & 0 & \frac{1}{2} & 0 \\ 0 & 0 & 0 & \frac{1}{2} & 0 & 0 & 0 & 0 & 0 & \frac{1}{2} & 0 & 0 \end{bmatrix}$$

8.3.5 Thermal Analog in Three Dimensions

Again referring to the tetrahedral element, the relation for the temperature gradient is immediately seen to be

$$\nabla \mathrm{T} = \boldsymbol{\beta}_{3T}^T \boldsymbol{\Phi}_{3T} \boldsymbol{\theta}_3, \quad \boldsymbol{\beta}_{3T}^T = \begin{bmatrix} 0 & 1 & 0 & 0 & 0 & 0 & 0 & 0 & 0 & 0 & 0 & 0 \\ 0 & 0 & 0 & 0 & 0 & 0 & 1 & 0 & 0 & 0 & 0 & 0 \\ 0 & 0 & 0 & 0 & 0 & 0 & 0 & 0 & 0 & 0 & 0 & 1 \end{bmatrix} \tag{8.21}$$

8.4 STRESS–STRAIN RELATIONS

8.4.1 Two-Dimensional Elements

8.4.1.1 Membrane Response

In two-dimensional elements, we have previously distinguished the cases of plane stress and plane strain. In plane stress, the stress–strain relations reduce to

$$\begin{aligned} \mathrm{E}_{xx} &= \tfrac{1}{\mathsf{E}}\left[\mathrm{S}_{xx} - \nu \mathrm{S}_{yy}\right] \\ \mathrm{E}_{yy} &= \tfrac{1}{\mathsf{E}}\left[\mathrm{S}_{yy} - \nu \mathrm{S}_{xx}\right] \\ \mathrm{E}_{xy} &= \tfrac{1+\nu}{\mathsf{E}} \mathrm{S}_{xy} \end{aligned} \tag{8.22}$$

The case of plane strain is retrieved by using $\mathsf{E}^* = \frac{\mathsf{E}}{1-\nu^2}$ and $\nu^* = \frac{\nu}{1-\nu}$ in place of E and ν. In traditional finite element notation Equation 8.22 may be written as

$$\begin{pmatrix} S_{xx} \\ S_{yy} \\ S_{xy} \end{pmatrix} = \mathbf{D}_{m21} \begin{pmatrix} E_{xx} \\ E_{yy} \\ E_{xy} \end{pmatrix} \tag{8.23}$$

in which

$$\mathbf{D}_{m21} = \mathsf{E} \begin{bmatrix} 1 & -\nu & 0 \\ -\nu & 1 & 0 \\ 0 & 0 & 1+\nu \end{bmatrix}^{-1} = \frac{\mathsf{E}}{1-\nu^2} \begin{bmatrix} 1 & \nu & 0 \\ \nu & 1 & 0 \\ 0 & 0 & 1+\nu \end{bmatrix} \tag{8.24}$$

may be called the tangent modulus *matrix* under plane stress (note that, unlike the previous definition, it is not based on the *VEC* operator and is not a tensor). The corresponding matrix for plane strain is denoted by $\mathbf{D}_{m22}$.

However, we shall see that a slightly different quantity from $\mathbf{D}_{m22}$ is needed in plane stress. The Principle of Virtual Work uses the strain energy density given by $\frac{1}{2}S_{ij}E_{ij}$. Elementary manipulation serves to prove that

$$\tfrac{1}{2}S_{ij}E_{ij} = \tfrac{1}{2}\begin{pmatrix} S_{xx} & S_{yy} & S_{zz} \end{pmatrix} \begin{bmatrix} 1 & 0 & 0 \\ 0 & 1 & 0 \\ 0 & 0 & 2 \end{bmatrix} \begin{pmatrix} E_{xx} \\ E_{yy} \\ E_{xy} \end{pmatrix} \tag{8.25}$$

Accordingly, in the Principle of Virtual Work the tangent modulus matrix in plane stress is replaced by

$$\mathbf{D}'_{mb2} = \frac{\mathsf{E}}{1-\nu^2} \begin{bmatrix} 1 & \nu & 0 \\ \nu & 1 & 0 \\ 0 & 0 & 2(1+\nu) \end{bmatrix} \tag{8.26}$$

and similarly for plane strain. (The peculiarity represented by the "2" in the lower right-hand diagonal entry is an artifact of traditional finite element notation and does not appear if *VEC* notation is used.) The stresses are now given in terms of nodal displacements by

$$\begin{pmatrix} S_{xx}(x,y,t) \\ S_{yy}(x,y,t) \\ S_{xy}(x,y,t) \end{pmatrix} = \mathbf{D}'_{m2i} \begin{pmatrix} E_{xx} \\ E_{yy} \\ E_{xy} \end{pmatrix} = \mathbf{D}'_{m2i}\boldsymbol{\beta}^T_{m2}\hat{\boldsymbol{\Phi}}_{m2} \begin{pmatrix} \boldsymbol{\gamma}_{u2} \\ \boldsymbol{\gamma}_{v2} \end{pmatrix}, \quad i = \begin{cases} 1 & \textit{plane stress} \\ 2 & \textit{plane strain} \end{cases} \tag{8.27}$$

8.4.1.2 Two-Dimensional Members: Bending Response of Thin Plates

Thin plates experiencing only bending are assumed to be in a state of plane stress. The tangent modulus matrix is again given by Equation 8.26, and now an approximation for the stress is obtained as

$$\begin{pmatrix} S_{xx} \\ S_{yy} \\ S_{xy} \end{pmatrix} = \mathbf{D}_{m21} \begin{pmatrix} E_{xx} \\ E_{yy} \\ E_{xy} \end{pmatrix} - z\mathbf{D}_{m21}\boldsymbol{\beta}^T_{b2}\hat{\boldsymbol{\Phi}}_{b2}\boldsymbol{\gamma}_{b2}(t) \tag{8.28}$$

8.4.1.3 Element for Plate with Membrane and Bending Response

Plane stress is likewise applicable to the combined case, and consequently the stresses are modeled as

$$\begin{pmatrix} S_{xx} \\ S_{yy} \\ S_{xy} \end{pmatrix} = \mathbf{D}_{m21} \begin{pmatrix} E_{xx} \\ E_{yy} \\ E_{xy} \end{pmatrix} = \mathbf{D}_{m21} \boldsymbol{\beta}^T_{mb2}(x, y, z) \hat{\boldsymbol{\Phi}}_{mb2} \boldsymbol{\gamma}_{mb2}(t) \tag{8.29}$$

8.4.2 Axisymmetric Element

For the purpose of determining a stress model consistent with the underlying interpolation model, it is sufficient to consider the case in which none of the nodes of the element are located on the axis of revolution.

$$\begin{pmatrix} S_{rr} \\ S_{\theta\theta} \\ S_{zz} \\ S_{rz} \end{pmatrix} = \mathbf{D}_a \begin{pmatrix} E_{rr} \\ E_{\theta\theta} \\ E_{zz} \\ E_{rz} \end{pmatrix} = \mathbf{D}_a \boldsymbol{\beta}^T_{a1} \begin{bmatrix} \hat{\boldsymbol{\Phi}}_{a1} & \mathbf{0} \\ \mathbf{0} & \hat{\boldsymbol{\Phi}}_{a1} \end{bmatrix} \begin{pmatrix} \boldsymbol{\gamma}_{ua1}(t) \\ \boldsymbol{\gamma}_{wa1}(t) \end{pmatrix} \tag{8.30}$$

in which the tangent modulus matrix is given by

$$\mathbf{D}_a = \mathsf{E} \begin{bmatrix} 1 & -\nu & -\nu & 0 \\ -\nu & 1 & -\nu & 0 \\ -\nu & -\nu & 1 & 0 \\ 0 & 0 & 0 & 1+\nu \end{bmatrix}^{-1}$$

$$= \frac{\mathsf{E}}{(1-2\nu)(1+\nu)} \begin{bmatrix} 1-\nu & \nu & \nu & 0 \\ \nu & 1-\nu & \nu & 0 \\ \nu & \nu & 1-\nu & 0 \\ 0 & 0 & 0 & 1-2\nu \end{bmatrix}$$

For use in the Principle of Virtual Work, $\mathbf{D}_a$ is modified to furnish $\mathbf{D}'_a$ given by

$$\mathbf{D}'_a = \frac{\mathsf{E}}{(1-2\nu)(1+\nu)} \begin{bmatrix} 1-\nu & \nu & \nu & 0 \\ \nu & 1-\nu & \nu & 0 \\ \nu & \nu & 1-\nu & 0 \\ 0 & 0 & 0 & 2(1-2\nu) \end{bmatrix} \tag{8.31}$$

8.4.3 Three-Dimensional Element

All six stresses and strains are now present. Using traditional finite element notation we write

$$\begin{pmatrix} S_{xx} \\ S_{yy} \\ S_{zz} \\ S_{xy} \\ S_{yz} \\ S_{zx} \end{pmatrix} = \mathbf{D}_3 \begin{pmatrix} E_{xx} \\ E_{yy} \\ E_{zz} \\ E_{xy} \\ E_{yz} \\ E_{zx} \end{pmatrix} = \mathbf{D}_3 \boldsymbol{\beta}_3^T \begin{bmatrix} \boldsymbol{\Phi}_3 & \mathbf{0} & \mathbf{0} \\ \mathbf{0} & \boldsymbol{\Phi}_3 & \mathbf{0} \\ \mathbf{0} & \mathbf{0} & \boldsymbol{\Phi}_3 \end{bmatrix} \begin{pmatrix} \boldsymbol{\gamma}_{u3} \\ \boldsymbol{\gamma}_{v3} \\ \boldsymbol{\gamma}_{w3} \end{pmatrix} \tag{8.32}$$

$$\mathbf{D}_3 = \mathsf{E} \begin{bmatrix} 1 & -\nu & -\nu & 0 & 0 & 0 \\ -\nu & 1 & -\nu & 0 & 0 & 0 \\ -\nu & -\nu & 1 & 0 & 0 & 0 \\ 0 & 0 & 0 & 1+\nu & 0 & 0 \\ 0 & 0 & 0 & 0 & 1+\nu & 0 \\ 0 & 0 & 0 & 0 & 0 & 1+\nu \end{bmatrix}^{-1}$$

$$= \frac{\mathsf{E}}{(1-2\nu)(1+\nu)} \begin{bmatrix} 1-\nu & \nu & \nu & 0 & 0 & 0 \\ \nu & 1-\nu & \nu & 0 & 0 & 0 \\ \nu & \nu & 1-\nu & 0 & 0 & 0 \\ 0 & 0 & 0 & 1-2\nu & 0 & 0 \\ 0 & 0 & 0 & 0 & 1-2\nu & 0 \\ 0 & 0 & 0 & 0 & 0 & 1-2\nu \end{bmatrix}$$

and for the Principle of Virtual Work, the associated matrix is

$$\mathbf{D}_3' = \frac{\mathsf{E}}{(1-2\nu)(1+\nu)} \times \begin{bmatrix} 1-\nu & \nu & \nu & 0 & 0 & 0 \\ \nu & 1-\nu & \nu & 0 & 0 & 0 \\ \nu & \nu & 1-\nu & 0 & 0 & 0 \\ 0 & 0 & 0 & 2(1-2\nu) & 0 & 0 \\ 0 & 0 & 0 & 0 & 2(1-2\nu) & 0 \\ 0 & 0 & 0 & 0 & 0 & 2(1-2\nu) \end{bmatrix} \tag{8.33}$$

8.4.4 Elements for Conductive Heat Transfer

Assuming the isotropic version of the Fourier Law, the heat flux vector, which may be considered the thermal analog of the stress, is obtained using

$$\mathbf{q} = -k \begin{cases} \boldsymbol{\beta}_{T1}^T \boldsymbol{\Phi}_{T1} \boldsymbol{\theta}_1, & 1-D \\ \boldsymbol{\beta}_{T2}^T \boldsymbol{\Phi}_{T2} \boldsymbol{\theta}_2, & 2-D \\ \boldsymbol{\beta}_{T3}^T \boldsymbol{\Phi}_{T3} \boldsymbol{\theta}_3, & 3-D \end{cases} \tag{8.34}$$

8.5 STIFFNESS AND MASS MATRICES AND THEIR THERMAL ANALOGS

Elements of variational calculus were discussed in Chapter 3 and the Principle of Virtual Work was introduced in Chapter 5. It is repeated here as

$$\int \delta \mathrm{E}_{ij}\mathrm{S}_{ij}\,\mathrm{d}V + \int \delta u_i \rho \ddot{u}_i\,\mathrm{d}V = \int \delta u_i t_i\,\mathrm{d}S \tag{8.35}$$

As before we assume that the displacement, the strain, and the stress may to satisfactory accuracy be approximated using expressions of the form

$$\mathbf{u}(\mathbf{x},t) = \boldsymbol{\varphi}^T(\mathbf{x})\boldsymbol{\Phi}\boldsymbol{\gamma}(t), \quad \mathbf{E} = \boldsymbol{\beta}^T(\mathbf{x})\boldsymbol{\Phi}\boldsymbol{\gamma}(t), \quad \mathbf{S} = \mathbf{DE} \tag{8.36}$$

in which **E** and **S** are written as one-dimensional arrays in accordance with traditional finite element notation, and of course **t** is the traction vector. Also, for use of traditional finite element notation in the Principle of Virtual Work, it is necessary to make use of $\mathbf{D}'$, which introduces the factor 2 into the entries corresponding to shear. We suppose that the boundary is decomposed into four segments: $S = S_I + S_{II} + S_{III} + S_{IV}$. On S_I, **u** is prescribed, in which event $\delta\mathbf{u}$ vanishes. On S_{II} the traction **t** is prescribed as $\mathbf{t}_0$. On S_{III} there is an elastic foundation described by $\mathbf{t} = \mathbf{t}_0 - \mathbf{A}(\mathbf{x})\mathbf{u}$, in which $\mathbf{A}(\mathbf{x})$ is a known matrix function of **x**. On S_{IV} there are inertial boundary conditions by virtue of which $\mathbf{t} = \mathbf{t}_0 - \mathbf{B}\ddot{\mathbf{u}}$. The right-hand term now becomes

$$\begin{aligned}\int \delta u_i t_i\,\mathrm{d}S = \delta\boldsymbol{\gamma}^T &\int_{S_{II}+S_{III}+S_{IV}} \boldsymbol{\Phi}^T\boldsymbol{\varphi}(\mathbf{x})\mathbf{t}_0\,\mathrm{d}S \\ - \delta\boldsymbol{\gamma}^T &\int_{S_{III}} \boldsymbol{\Phi}^T\boldsymbol{\varphi}(\mathbf{x})\mathbf{A}\boldsymbol{\varphi}^T(\mathbf{x})\boldsymbol{\Phi}\,\mathrm{d}S\,\boldsymbol{\gamma}(t) \\ - \delta\boldsymbol{\gamma}^T &\int_{S_{IV}} \boldsymbol{\Phi}^T\boldsymbol{\varphi}(\mathbf{x})\mathbf{B}\boldsymbol{\varphi}^T(\mathbf{x})\boldsymbol{\Phi}\,\mathrm{d}S\,\ddot{\boldsymbol{\gamma}}(t)\end{aligned} \tag{8.37}$$

The leftmost term in Equation 8.35 becomes

$$\begin{aligned}\int \delta \mathrm{E}_{ij}\mathrm{S}_{ij}\,\mathrm{d}V &= \delta\boldsymbol{\gamma}^T\mathbf{K}\boldsymbol{\gamma}(t), \quad \mathbf{K} = \int \boldsymbol{\Phi}^T\boldsymbol{\beta}(\mathbf{x})\mathbf{D}'\boldsymbol{\beta}^T(\mathbf{x})\boldsymbol{\Phi}\,\mathrm{d}V \\ \int \delta u_i \rho \ddot{u}_i\,\mathrm{d}V &= \delta\boldsymbol{\gamma}^T\mathbf{M}\ddot{\boldsymbol{\gamma}}(t), \quad \mathbf{M} = \int \rho\boldsymbol{\Phi}^T\boldsymbol{\varphi}(\mathbf{x})\boldsymbol{\varphi}^T(\mathbf{x})\boldsymbol{\Phi}\,\mathrm{d}V\end{aligned} \tag{8.38}$$

in which **K** is of course the stiffness matrix and **M** the mass matrix. Canceling the arbitrary variation and bringing terms with unknowns to the left-hand side furnish the equation

$$(\mathbf{K} + \mathbf{K}_S)\boldsymbol{\gamma}(t) + (\mathbf{M} + \mathbf{M}_S)\ddot{\boldsymbol{\gamma}}(t) = \mathbf{f}(t) \tag{8.39}$$

$$\mathbf{f} = \int_{S_{II}+S_{III}} \boldsymbol{\Phi}^T \boldsymbol{\varphi}(\mathbf{x}) \boldsymbol{\tau}_0 \, dS$$

$$\mathbf{K}_S = \int_{S_{III}} \boldsymbol{\Phi}^T \boldsymbol{\varphi}(\mathbf{x}) \mathbf{A} \boldsymbol{\varphi}^T(\mathbf{x}) \boldsymbol{\Phi} \, dS$$

$$\mathbf{M}_S = \int_{S_{IV}} \boldsymbol{\Phi}^T \boldsymbol{\varphi}(\mathbf{x}) \mathbf{B} \boldsymbol{\varphi}^T(\mathbf{x}) \boldsymbol{\Phi} \, dS$$

Clearly, elastic supports on S_{III} furnish a boundary contribution to the stiffness matrix, while mass on the boundary segment S_{IV} furnishes a contribution to the mass matrix.

8.6 THERMAL COUNTERPART OF THE PRINCIPLE OF VIRTUAL WORK

For current purposes we focus on the equation of conductive heat transfer as

$$k\nabla^2 \mathrm{T} = \rho c_e \frac{\partial \mathrm{T}}{\partial t} \tag{8.40}$$

Multiplying by the variation of $T - T_0$, integrating by parts and applying the divergence theorem furnishes

$$\int \delta \nabla^T \mathrm{T} k \nabla \mathrm{T} \, dV + \int \delta \mathrm{T} \rho c_e \frac{\partial \mathrm{T}}{\partial t} dV = \int \delta \mathrm{T} \mathbf{n}^T \mathbf{q} \, dS \tag{8.41}$$

Now suppose that the interpolation models for temperature in the current element furnish a relation of the form

$$\mathrm{T} - \mathrm{T}_0 = \boldsymbol{\varphi}_T^T(\mathbf{x}) \boldsymbol{\Phi}_T \boldsymbol{\theta}(t), \quad \nabla \mathrm{T} = \boldsymbol{\beta}_T^T(\mathbf{x}) \boldsymbol{\Phi}_T \boldsymbol{\theta}(t), \quad \mathbf{q} = -k_T^T(\mathbf{x}) \boldsymbol{\Phi}_T \boldsymbol{\theta}(t) \tag{8.42}$$

The left-hand terms in Equation 8.41 may now be written as

$$\begin{aligned} &\int \delta \nabla^T \mathrm{T} k \nabla \mathrm{T} \, dV \rightarrow \delta \boldsymbol{\theta}^T(t) \mathbf{K}_T \boldsymbol{\theta}(t), \quad \mathbf{K}_T = \int k \boldsymbol{\Phi}_T^T \boldsymbol{\beta}_T \boldsymbol{\beta}_T^T \boldsymbol{\Phi}_T \, dV \\ &\int \delta \mathrm{T} \rho c_e \frac{\partial \mathrm{T}}{\partial t} dV = \delta \boldsymbol{\theta}^T(t) \mathbf{M}_T \dot{\boldsymbol{\theta}}(t), \quad \mathbf{M}_T = \int k \boldsymbol{\Phi}_T^T \boldsymbol{\varphi}_T \boldsymbol{\varphi}_T^T \boldsymbol{\Phi}_T \, dV \end{aligned} \tag{8.43}$$

$\mathbf{K}_T$ and $\mathbf{M}_T$ may be called the thermal stiffness (or conductance) matrix and thermal mass (or capacitance) matrix, respectively.

Next, suppose that the boundary S has four zones: $S = S_I + S_{II} + S_{III} + S_{IV}$. On S_I the temperature is prescribed as T_1, from which we conclude that $\delta \mathrm{T} = 0$. On S_{II} the heat flux is prescribed as $\mathbf{n}^T \mathbf{q}_1$. On S_{III}, the heat flux satisfies $\mathbf{n}^T \mathbf{q} = \mathbf{n}^T \mathbf{q}_1 - h_1(\mathrm{T} - \mathrm{T}_0)$, while on S_{IV}, $\mathbf{n}^T \mathbf{q} = \mathbf{n}^T \mathbf{q}_1 - h_2 \, d\mathrm{T}/dt$. The governing finite element equation is now

$$[\mathbf{M}_T + \mathbf{M}_{TS}]\dot{\boldsymbol{\theta}}(t) + [\mathbf{K}_T + \mathbf{K}_{TS}]\boldsymbol{\theta}(t) = \mathbf{f}_T(t)$$

$$\mathbf{M}_{TS} = \boldsymbol{\Phi}_{TT}^T \int_{S_{IV}} \boldsymbol{\varphi}_T h_2 \boldsymbol{\varphi}_T^T \, \mathrm{d}S \, \boldsymbol{\Phi}_T, \quad \mathbf{K}_{TS} = \boldsymbol{\Phi}_T^T \int_{S_{III}} \boldsymbol{\varphi}_T h_1 \boldsymbol{\varphi}_T^T \, \mathrm{d}S \, \boldsymbol{\Phi}_T \tag{8.44}$$

$$\mathbf{f}_T(t) = \boldsymbol{\Phi}_T^T \int_{\Omega} \boldsymbol{\varphi}_T \mathbf{n}^T q_1 \mathrm{d}S, \quad \Omega = S_{II} + S_{III} + S_{IV}$$

8.7 CONVERSION TO NATURAL COORDINATES IN TWO AND THREE DIMENSIONS

The notion of natural coordinates is applicable in two and three dimensions. It requires transforming the undeformed coordinates of the physical element to a reference element with suitable symmetry properties. As an illustration consider the quadrilateral element shown below.

We seek transformations $\zeta_k(X_j)$ and their inverses $X_j(\zeta_k)$ such that the nodes of the physical element are mapped to the values $\left(\zeta_1^{(e)}, \zeta_2^{(e)}\right) = (-1,-1)$; $\left(\zeta_1^{(e+1)}, \zeta_2^{(e+1)}\right) = (1,-1)$; $\left(\zeta_1^{(e+2)}, \zeta_2^{(e+2)}\right) = (1,1)$; $\left(\zeta_1^{(e+3)}, \zeta_2^{(e+3)}\right) = (1,-1)$ in the transformed element. The coordinates $\zeta_k(X_j)$ are of course the *natural coordinates.* In this instance we also require that straight lines remain straight lines. The element in the transformed coordinates is depicted as follows. The mapping is achieved by the functions

$$X_1 = \{1 \quad \zeta_1 \quad \zeta_2 \quad \zeta_1\zeta_2\}\Phi_\zeta \begin{Bmatrix} X_1^{(e)} \\ X_1^{(e+1)} \\ X_1^{(e+2)} \\ X_1^{(e+3)} \end{Bmatrix} \quad X_2 = \{1 \quad \zeta_1 \quad \zeta_2 \quad \zeta_1\zeta_2\}\Phi_\zeta \begin{Bmatrix} X_2^{(e)} \\ X_2^{(e+1)} \\ X_2^{(e+2)} \\ X_2^{(e+3)} \end{Bmatrix}$$

$$\Phi_\zeta = \begin{bmatrix} 1 & -1 & -1 & 1 \\ 1 & 1 & -1 & -1 \\ 1 & 1 & 1 & 1 \\ 1 & -1 & 1 & 1 \end{bmatrix}^{-1} = \frac{1}{4}\begin{bmatrix} 1 & 1 & 1 & 1 \\ -1 & 1 & 1 & -1 \\ -1 & -1 & 1 & 1 \\ 1 & -1 & 1 & -1 \end{bmatrix} \tag{8.45}$$

Along the sides $\zeta_1 = -1, 1$, X_1 and X_2 are linear functions of ζ_2. Accordingly they are linear functions of each other and hence the top and bottom faces of the square in Figure 8.5 map into lines between the nodal values at the endpoints in Figure 8.4. A similar observation holds regarding the right and left faces.

Clearly the transformation relations between the physical and the natural coordinates are reminiscent of interpolation models introduced heretofore for deformed coordinates in terms of undeformed coordinates. If the transformation model involves the same order of polynomial as the interpolation model, the element in the natural coordinates is said to be *isoparametric.*

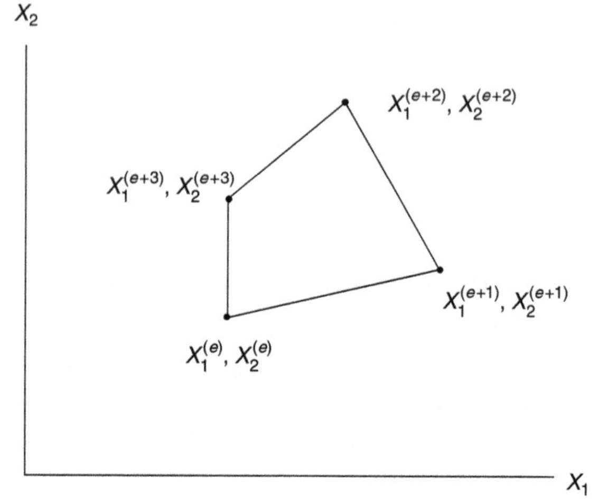

FIGURE 8.4 Two-dimensional element in physical coordinates.

The Jacobian matrix for the transformation is defined here as by

$$\mathbf{J} = \frac{\partial \mathbf{X}}{\partial \boldsymbol{\zeta}} = \begin{bmatrix} \dfrac{\partial X_1}{\partial \zeta_1} & \dfrac{\partial X_1}{\partial \zeta_2} \\ \dfrac{\partial X_2}{\partial \zeta_1} & \dfrac{\partial X_2}{\partial \zeta_2} \end{bmatrix} \tag{8.46}$$

in which $\mathbf{X} = \begin{Bmatrix} X_1(\zeta_1,\zeta_2) \\ X_2(\zeta_1,\zeta_2) \end{Bmatrix}$ and $\boldsymbol{\zeta} = \begin{Bmatrix} \zeta_1 \\ \zeta_2 \end{Bmatrix}$. (Sometimes $\mathbf{J}$ is defined as $(d\boldsymbol{\zeta}/d\mathbf{X})^T$.) The reader may find our definition surprising since, as first glance, ζ denotes the coordinates being introduced by the transformation. However, we take the view that

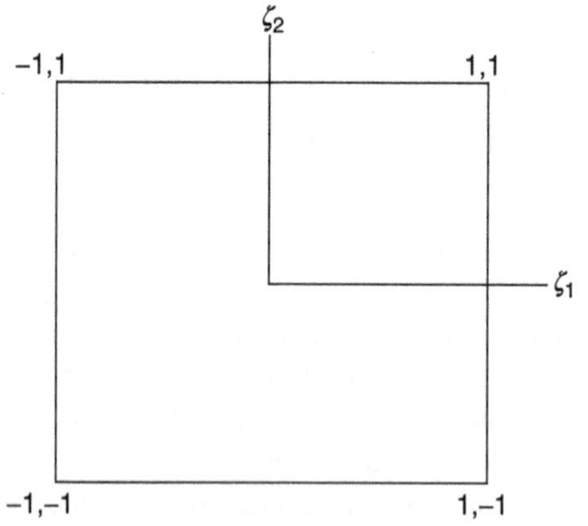

FIGURE 8.5 Two-dimensional element in natural coordinates.

the natural coordinates now represent the reference configuration, since the final matrices and vectors will be expressed in terms of the natural coordinates.)

The Jacobian matrix is reminiscent of the deformation gradient tensor $\mathbf{F}$. The transformation $\zeta_k(X_j)$ is invertible and the inverse transformation $X_j(\zeta_k)$ exists if $\mathbf{J}$ and hence its inverse are nonsingular. The volume (area) in the physical element dA and in the transformed element dA_ζ are related by $dA_\zeta = \det(\mathbf{J}^{-1})\, dA$, and clearly a singular inverse Jacobian matrix would imply mapping the physical element onto a zero-volume element in natural coordinates. The determinant of the Jacobian matrix is denoted as J: $J = \det(\mathbf{J})$.

Assuming nonsingularity, the inverse of the Jacobian matrix is readily verified using the Chain Rule of calculus to be given by

$$\mathbf{J}^{-1} = \begin{bmatrix} \frac{\partial \zeta_1}{\partial X_1} & \frac{\partial \zeta_1}{\partial X_2} \\ \frac{\partial \zeta_2}{\partial X_1} & \frac{\partial \zeta_2}{\partial X_2} \end{bmatrix} \tag{8.47}$$

We now consider referring the Principle of Virtual Work to natural coordinates. The inertial term becomes

$$\int \delta\mathbf{u}^T \ddot{\mathbf{u}}\,\rho\, dV = \int \delta\mathbf{u}^T \ddot{\mathbf{u}}\,\rho J\, dV_\zeta \tag{8.48}$$

Now suppose that the displacement vector in element e is approximated using an interpolation model in the natural coordinates:

$$\mathbf{u}(\mathbf{X},t) \to \mathbf{u}(\boldsymbol{\zeta},t) \approx \boldsymbol{\varphi}_\zeta^T(\boldsymbol{\zeta})\boldsymbol{\Phi}_{\zeta e}\boldsymbol{\gamma}_{\zeta e} \tag{8.49}$$

The inertial term in the eth element is now

$$\begin{aligned} \int \delta\mathbf{u}^T \ddot{\mathbf{u}}\,\rho J\, dV_\zeta &= \delta\boldsymbol{\gamma}_{\zeta e}^T \mathbf{M}_{\zeta e} \ddot{\boldsymbol{\gamma}}_{\zeta e} \\ \mathbf{M}_{\zeta e} &= \boldsymbol{\Phi}_\zeta^T \left[\int \boldsymbol{\varphi}_{\zeta e} \boldsymbol{\varphi}_{\zeta e}^T \rho J\, dV_\zeta \right] \boldsymbol{\Phi}_\zeta \end{aligned} \tag{8.50}$$

We next consider the consistent force, assuming that the traction vector is specified at all points on the exterior boundary. In Chapter 13, we will encounter a relation between the deformed surface area element dS_d and the corresponding undeformed element: namely $dS_d = \mu\, dS$ in which $\mu = \det(\mathbf{F})\sqrt{\mathbf{n}_0^T \mathbf{C}^{-1} \mathbf{n}_0}$ and $\mathbf{n}_0$ is the surface normal vector of the undeformed element. We may now write

$$dS = \mu_\zeta\, dS_\zeta, \quad \mu_\zeta = \det(\mathbf{J})\sqrt{\mathbf{n}_\zeta^T \mathbf{J}^{-1} \mathbf{J}^{-T} \mathbf{n}_\zeta} \tag{8.51}$$

in which dS_ζ is the area of the element in the natural coordinate system and $\mathbf{n}_\zeta$ is the corresponding unit normal vector. We now determine the corresponding consistent force.

The term in the Principle of Virtual Work which represents the *virtual external work of the traction* is transformed according to

$$\int \delta\mathbf{u}^T(\mathbf{X},t)\mathbf{t}(\mathbf{X},t)\,dS = \int \delta\mathbf{u}^T(\boldsymbol{\zeta},t)\mathbf{t}(\boldsymbol{\zeta},t)\mu_\zeta(\boldsymbol{\zeta},t)\,dS_\zeta$$

$$= \delta\boldsymbol{\gamma}_{\zeta e}^T\mathbf{f}_{\zeta e} \tag{8.52}$$

$$\mathbf{f}_{\zeta e} = \boldsymbol{\Phi}_{\zeta e}\int \boldsymbol{\varphi}_\zeta(\boldsymbol{\zeta})\mathbf{t}(\boldsymbol{\zeta},t)\mu_\zeta(\boldsymbol{\zeta},t)\,dS_\zeta$$

Finally we address the term representing *virtual internal work of the stress*, and is hereafter called the stiffness term. First note that, in linear elasticity, owing to the total symmetry of the tangent modulus tensor $\mathbf{C}(c_{ijkl})$,

$$\begin{aligned}
\delta E_{ji}S_{ij} &= \delta E_{ij}c_{ijkl}E_{kl}\\
&= \frac{\partial\delta u_j}{\partial X_i}c_{ijkl}\frac{\partial u_l}{\partial X_k}\\
&= \frac{\partial\delta u_j}{\partial\zeta_m}\left[\frac{\partial\zeta_m}{\partial X_i}c_{ijkl}\frac{\partial\zeta_n}{\partial X_k}\right]\frac{\partial u_l}{\partial\zeta_n}\\
&= tr\left(\left\{\frac{\partial\delta\mathbf{u}}{\partial\boldsymbol{\zeta}}\right\}^T\mathbf{C}_\zeta\frac{\partial\delta\mathbf{u}}{\partial\boldsymbol{\zeta}}\right)
\end{aligned} \tag{8.53}$$

in which the fourth-order tangent modulus referred to natural coordinates is expressed as

$$[\mathbf{C}_\zeta]_{mjln} = \left[\mathbf{J}^{-T}\mathbf{C}\mathbf{J}^{-1}\right]_{mjln} = \frac{\partial\zeta_m}{\partial X_i}c_{ijkl}\frac{\partial\zeta_n}{\partial X_k}$$

The stiffness term referred to natural coordinates may now be identified.

$$\int tr(\delta\mathbf{E}\mathbf{S})\,dV = \int tr\left(\left\{\frac{\partial\delta\mathbf{u}}{\partial\boldsymbol{\zeta}}\right\}^T\mathbf{C}_\zeta\frac{\partial\delta\mathbf{u}}{\partial\boldsymbol{\zeta}}\right)J\,dV_\zeta \tag{8.54}$$

The interpolation model (Equation 8.49) implies a subsidiary model of the form

$$VEC\left(\frac{\mathbf{du}(\boldsymbol{\zeta},t)}{d\boldsymbol{\zeta}}\right) = \boldsymbol{\beta}_\zeta^T(\boldsymbol{\zeta})\boldsymbol{\Phi}_{\zeta e}\boldsymbol{\gamma}_{\zeta e} \tag{8.55}$$

Accordingly, the stiffness term assumes the form

$$\int tr(\delta \mathbf{ES})\,\mathrm{d}V \rightarrow \delta\boldsymbol{\gamma}_{\zeta e}^{T}\mathbf{K}_{\zeta e}\boldsymbol{\gamma}_{\zeta_e}$$
$$\mathbf{K}_{\zeta e} = \boldsymbol{\Phi}_{\zeta e}^{T}\int \boldsymbol{\beta}_{\zeta}(\boldsymbol{\zeta})\left(\mathbf{J}^{-T}(\boldsymbol{\zeta})\mathbf{C}\mathbf{J}^{-1}(\boldsymbol{\zeta})\right)(\boldsymbol{\zeta})\boldsymbol{\beta}_{\zeta}^{T}(\boldsymbol{\zeta})J(\zeta)\,\mathrm{d}V_{\zeta} \tag{8.56}$$

The ensuing element level finite element equation is now stated as

$$\mathbf{M}_{\zeta e}\ddot{\boldsymbol{\gamma}}_{\zeta e} + \mathbf{K}_{\gamma e}\boldsymbol{\gamma}_{\zeta e} = \mathbf{f}_{\zeta e} \tag{8.57}$$

The transformation to natural coordinates preserves the kinetic energy and the potential energy of the element, in consequence of which assemblage of element matrices to obtain global matrices proceeds by direct addition in the manner introduced in Chapter 7.

Similar arguments to the above furnish the transformations for the thermal stiffness matrix.

$$\int \delta\nabla^{T}\mathrm{T}k\nabla^{T}\mathrm{T}\,\mathrm{d}A = \int (\nabla_{\varsigma}\mathrm{T}\delta\mathrm{T})k\mathbf{J}^{-T}\mathbf{J}^{-1}(\nabla_{\varsigma}\mathrm{T})J\,\mathrm{d}A_{\zeta}$$
$$= \delta\boldsymbol{\theta}_{\zeta e}^{T}\mathbf{K}_{T\zeta e}\boldsymbol{\theta}_{\zeta e} \tag{8.58}$$
$$\mathbf{K}_{T\zeta e} = \boldsymbol{\Phi}_{T\zeta}^{T}\int k\boldsymbol{\beta}_{T\zeta}\mathbf{J}^{-T}\mathbf{J}^{-1}\boldsymbol{\beta}_{T\zeta}^{T}J\,\mathrm{d}V_{\zeta}\,\boldsymbol{\Phi}_{T\zeta}$$

The transformations for the thermal mass matrix and the consistent thermal force are parallel to the mechanical field counterparts.

EXAMPLE 8.1

Find the Jacobian matrix and its determinant for the transformation shown below (Figure 8.6).

SOLUTION

The transformation is achieved using

$$x = \{1 \quad \varsigma \quad \eta \quad \eta\varsigma\}\boldsymbol{\Phi}\begin{Bmatrix} x_1 \\ x_2 \\ x_3 \\ x_4 \end{Bmatrix}, \quad y = \{1 \quad \varsigma \quad \eta \quad \eta\varsigma\}\boldsymbol{\Phi}\begin{Bmatrix} y_1 \\ y_2 \\ y_3 \\ y_4 \end{Bmatrix}$$

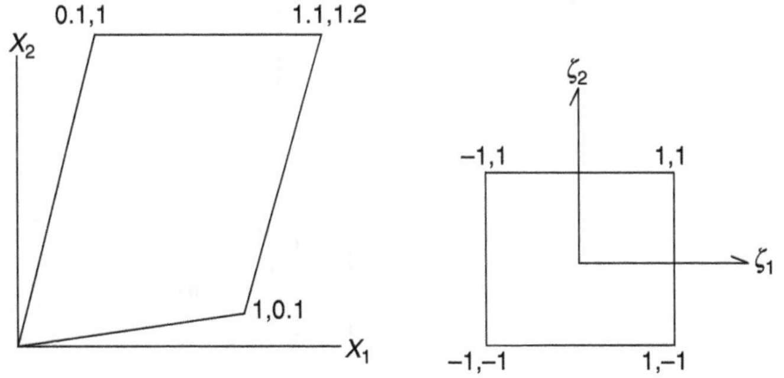

FIGURE 8.6 Figure for determinant of a Jacobian.

The matrix $\mathbf{\Phi}$ is given above in Equation 8.45. The Jacobian matrix is obtained as

$$\mathbf{J}=\begin{bmatrix} \{0\ \ 1\ \ 0\ \ \zeta_2\}\mathbf{\Phi}\begin{Bmatrix} x_1^{(1)} \\ x_1^{(2)} \\ x_1^{(3)} \\ x_1^{(4)} \end{Bmatrix} & \{0\ \ 0\ \ 1\ \ \zeta_1\}\mathbf{\Phi}\begin{Bmatrix} x_1^{(1)} \\ x_1^{(2)} \\ x_1^{(3)} \\ x_1^{(4)} \end{Bmatrix} \\ \{0\ \ 1\ \ 0\ \ \varsigma_1\}\mathbf{\Phi}\begin{Bmatrix} x_2^{(1)} \\ x_2^{(2)} \\ x_2^{(3)} \\ x_2^{(4)} \end{Bmatrix} & \{0\ \ 0\ \ 1\ \ \varsigma_1\}\mathbf{\Phi}\begin{Bmatrix} x_2^{(1)} \\ x_2^{(2)} \\ x_2^{(3)} \\ x_2^{(4)} \end{Bmatrix} \end{bmatrix}$$

The determinant J is recognized as

$$\begin{aligned} J=&\left(\{0\ \ 0\ \ 1\ \ \varsigma_1\}\mathbf{\Phi}\begin{Bmatrix} x_2^{(1)} \\ x_2^{(2)} \\ x_2^{(3)} \\ x_2^{(4)} \end{Bmatrix}\right)\left(\{0\ \ 1\ \ 0\ \ \zeta_2\}\mathbf{\Phi}\begin{Bmatrix} x_1^{(1)} \\ x_1^{(2)} \\ x_1^{(3)} \\ x_1^{(4)} \end{Bmatrix}\right) \\ &-\left(\{0\ \ 0\ \ 1\ \ \varsigma_1\}\mathbf{\Phi}\begin{Bmatrix} x_1^{(1)} \\ x_1^{(2)} \\ x_1^{(3)} \\ x_1^{(4)} \end{Bmatrix}\right)\left(\{0\ \ 1\ \ 0\ \ \zeta_2\}\mathbf{\Phi}\begin{Bmatrix} x_2^{(1)} \\ x_2^{(2)} \\ x_2^{(3)} \\ x_2^{(4)} \end{Bmatrix}\right) \\ &=\left\{x_1^{(1)}\ \ x_1^{(2)}\ \ x_1^{(3)}\ \ x_1^{(4)}\right\}\mathbf{\Phi}^T\left[\begin{Bmatrix} 0 \\ 1 \\ 0 \\ \zeta_2 \end{Bmatrix}\{0\ \ 0\ \ 1\ \ \varsigma_1\}-\begin{Bmatrix} 0 \\ 0 \\ 1 \\ \varsigma_1 \end{Bmatrix}\{0\ \ 1\ \ 0\ \ \varsigma_2\}\right]\mathbf{\Phi}\begin{Bmatrix} x_2^{(1)} \\ x_2^{(2)} \\ x_2^{(3)} \\ x_2^{(4)} \end{Bmatrix} \\ &=\left\{x_1^{(1)}\ \ x_1^{(2)}\ \ x_1^{(3)}\ \ x_1^{(4)}\right\}\mathbf{\Phi}^T\begin{bmatrix} 0 & 0 & 0 & 0 \\ 0 & 0 & 1 & \zeta_1 \\ 0 & -1 & 0 & -\zeta_2 \\ 0 & -\zeta_1 & \zeta_2 & 0 \end{bmatrix}\mathbf{\Phi}\begin{Bmatrix} x_2^{(1)} \\ x_2^{(2)} \\ x_2^{(3)} \\ x_2^{(4)} \end{Bmatrix} \end{aligned}$$

and finally

$$\left\{x_1^{(1)}\ x_1^{(2)}\ x_1^{(3)}\ x_1^{(4)}\right\}=\{0\ \ 1\ \ 1.1\ \ 0.1\};\quad \begin{Bmatrix} x_2^{(1)} \\ x_2^{(2)} \\ x_2^{(3)} \\ x_2^{(4)} \end{Bmatrix}=\begin{Bmatrix} 0 \\ 0.1 \\ 1.2 \\ 1 \end{Bmatrix},\quad \mathbf{\Phi}=\frac{1}{4}\begin{bmatrix} 1 & 1 & 1 & 1 \\ -1 & 1 & 1 & -1 \\ -1 & -1 & 1 & 1 \\ 1 & -1 & 1 & -1 \end{bmatrix}$$

It is left to the reader to compute the value of J using the relations in the preceding lines.

8.8 ASSEMBLY OF TWO- AND THREE-DIMENSIONAL ELEMENTS

We next consider assembly of stiffness matrices for physical elements in two dimensions. Assembly in three dimensions follows the same procedures. Consider the model depicted below (Figure 8.7), consisting of four rectangular elements, denoted as element e, $e+1$, $e+2$, $e+3$. The nodes are also numbered in the global system. Locally, the nodes in an element are numbered in a counterclockwise scheme starting from the lower left-hand corner. Suppose there is one degree of freedom per node (e.g., x-displacement), and one corresponding force.

In the local numbering system, the force on the center node induces displacements according to

$$\begin{aligned}
f_{e,3} &= k_{3,1}^{(e)}u_{e,1} + k_{3,2}^{(e)}u_{e,2} + k_{3,3}^{(e)}u_{e,3} + k_{3,4}^{(e)}u_{e,4}\\
f_{e+1,4} &= k_{4,1}^{(e+1)}u_{e+1,1} + k_{4,2}^{(e+1)}u_{e+1,2} + k_{4,3}^{(e+1)}u_{e+1,3} + k_{4,4}^{(e+1)}u_{e+1,4}\\
f_{e+2,1} &= k_{1,1}^{(e+2)}u_{e+2,1} + k_{1,2}^{(e+2)}u_{e+2,2} + k_{1,3}^{(e+2)}u_{e+2,3} + k_{1,4}^{(e+2)}u_{e+2,4}\\
f_{e+3,2} &= k_{2,1}^{(e+3)}u_{e+3,1} + k_{2,2}^{(e+3)}u_{e+3,2} + k_{2,3}^{(e+3)}u_{e+3,3} + k_{2,4}^{(e+3)}u_{e+3,4}
\end{aligned} \tag{8.59}$$

The conversion from local to global coordinates is expressed by

$$\begin{array}{llll}
u_{e,1} \rightarrow u_1 & u_{e,2} \rightarrow u_2 & u_{e,3} \rightarrow u_5 & u_{e,4} \rightarrow u_6\\
u_{e+1,1} \rightarrow u_2 & u_{e+1,2} \rightarrow u_3 & u_{e+1,3} \rightarrow u_4 & u_{e+1,4} \rightarrow u_5\\
u_{e+2,1} \rightarrow u_5 & u_{e+2,2} \rightarrow u_4 & u_{e+2,3} \rightarrow u_9 & u_{e+2,4} \rightarrow u_8\\
u_{e+3,1} \rightarrow u_6 & u_{e+3,2} \rightarrow u_6 & u_{e+3,3} \rightarrow u_8 & u_{e+3,4} \rightarrow u_7
\end{array} \tag{8.60}$$

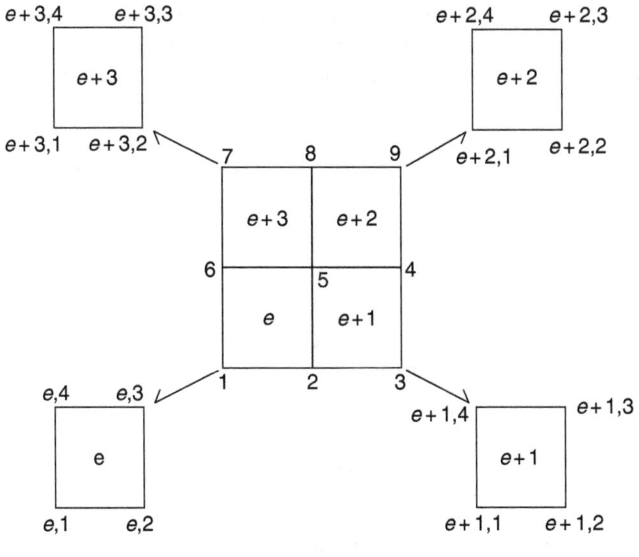

FIGURE 8.7 Two-dimensional assembly process.

Adding the forces of the elements on the center node gives

$$
\begin{aligned}
f_5 &= k_{3,1}^{(e)} u_1 + \left[k_{3,2}^{(e)} + k_{4,1}^{(e+1)}\right] u_2 + k_{4,2}^{(e+1)} u_3 + \left[k_{4,3}^{(e+1)} + k_{1,2}^{(e+2)}\right] u_4 \\
&+ \left[k_{3,3}^{(e)} + k_{4,4}^{(e+1)} + k_{1,1}^{(e+2)} + k_{2,2}^{(e+3)}\right] u_5 + \left[k_{3,4}^{(e)} + k_{2,2}^{(e+3)}\right] u_6 \\
&+ k_{2,4}^{(e+3)} u_7 + \left[k_{1,4}^{(e+2)} + k_{2,3}^{(e+3)}\right] u_8 + k_{1,3}^{(e+2)} u_9
\end{aligned} \tag{8.61}
$$

Taking advantage of the symmetry of the stiffness matrix, this implies that the fifth row of the stiffness matrix is

$$
\begin{aligned}
\kappa_5^T = \{ & k_{3,1}^{(e)} \ \left[k_{3,2}^{(e)} + k_{4,1}^{(e+1)}\right] \ k_{4,2}^{(e+1)} \ \left[k_{4,3}^{(e+1)} + k_{1,2}^{(e+2)}\right] \\
& \left[k_{3,3}^{(e)} + k_{4,4}^{(e+1)} + k_{1,1}^{(e+2)} + k_{2,2}^{(e+3)}\right] \cdots symmetry \cdots \}
\end{aligned} \tag{8.62}
$$

The process can be repeated for all of the nodes, leading to the assembled stiffness matrix. The process is essentially the same for solid and axisymmetric elements.

Of course, a conceptually easier way is to add the kinetic and strain energies of the individual elements, referred to the global numbering system for degrees of freedom, as explained in Chapter 7.

EXAMPLE 8.2

Assemble the stiffness coefficients associated with node n below, assuming plane stress elements. The modulus is E and the Poisson's ratio is ν. $\mathbf{K}^{(1)}$, $\mathbf{K}^{(2)}$, and $\mathbf{K}^{(3)}$ denote the stiffness matrices of the elements (Figure 8.8).

SOLUTION

Now suppose there are two degrees of freedom per node (x- and y-displacements), and two corresponding forces. Since the elements are three-noded, and each node has two degrees of freedom, the element stiffness matrices will be 6×6 and the force vectors will be 6×1.

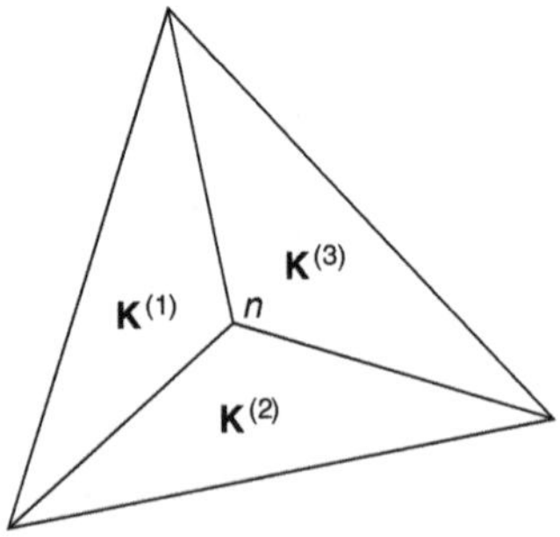

FIGURE 8.8 Assemblage of triangular elements.

The element level equation for element (1) is stated as

$$\begin{bmatrix} k_{11}^{(1)} & k_{12}^{(1)} & k_{13}^{(1)} & k_{14}^{(1)} & k_{15}^{(1)} & k_{16}^{(1)} \\ k_{21}^{(1)} & k_{22}^{(1)} & k_{23}^{(1)} & k_{24}^{(1)} & k_{25}^{(1)} & k_{26}^{(1)} \\ k_{31}^{(1)} & k_{32}^{(1)} & k_{33}^{(1)} & k_{34}^{(1)} & k_{35}^{(1)} & k_{36}^{(1)} \\ k_{41}^{(1)} & k_{42}^{(1)} & k_{43}^{(1)} & k_{44}^{(1)} & k_{45}^{(1)} & k_{46}^{(1)} \\ k_{51}^{(1)} & k_{52}^{(1)} & k_{53}^{(1)} & k_{54}^{(1)} & k_{55}^{(1)} & k_{56}^{(1)} \\ k_{61}^{(1)} & k_{62}^{(1)} & k_{63}^{(1)} & k_{64}^{(1)} & k_{65}^{(1)} & k_{66}^{(1)} \end{bmatrix} \begin{pmatrix} u_{1,1} \\ u_{1,2} \\ u_{1,3} \\ v_{1,1} \\ v_{1,2} \\ v_{1,3} \end{pmatrix} = \begin{pmatrix} f_{x1,1} \\ f_{x1,2} \\ f_{x1,3} \\ f_{y1,1} \\ f_{y1,2} \\ f_{y1,3} \end{pmatrix}$$

and similarly for two remaining elements.

Following the assemblage procedure, in the local system the force on the center node "n" induces displacements according to

$$f_{x1,2} = k_{21}^{(1)}u_{1,1} + k_{22}^{(1)}u_{1,2} + k_{23}^{(1)}u_{1,3} + k_{24}^{(1)}v_{1,1} + k_{25}^{(1)}v_{1,2} + k_{26}^{(1)}v_{1,3}$$

$$f_{y1,2} = k_{51}^{(1)}u_{1,1} + k_{52}^{(1)}u_{1,2} + k_{53}^{(1)}u_{1,3} + k_{54}^{(1)}v_{1,1} + k_{55}^{(1)}v_{1,2} + k_{56}^{(1)}v_{1,3}$$

$$f_{x2,2} = k_{21}^{(2)}u_{2,1} + k_{22}^{(2)}u_{2,2} + k_{23}^{(2)}u_{2,3} + k_{24}^{(2)}v_{2,1} + k_{25}^{(2)}v_{2,2} + k_{26}^{(2)}v_{2,3}$$

$$f_{y2,2} = k_{51}^{(2)}u_{2,1} + k_{52}^{(2)}u_{2,2} + k_{53}^{(2)}u_{2,3} + k_{54}^{(2)}v_{2,1} + k_{55}^{(2)}v_{2,2} + k_{56}^{(2)}v_{2,3}$$

$$f_{x3,2} = k_{21}^{(3)}u_{3,1} + k_{22}^{(3)}u_{3,2} + k_{23}^{(3)}u_{3,3} + k_{24}^{(3)}v_{3,1} + k_{25}^{(3)}v_{3,2} + k_{26}^{(3)}v_{3,3}$$

$$f_{y3,2} = k_{51}^{(3)}u_{3,1} + k_{52}^{(3)}u_{3,2} + k_{53}^{(3)}u_{3,3} + k_{54}^{(3)}v_{3,1} + k_{55}^{(3)}v_{3,2} + k_{56}^{(3)}v_{3,3}$$

Conversion to the global numbering scheme for degrees of freedom is expressed by

$$\begin{array}{llllll} u_{1,1} \to u_{n+1} & u_{1,2} \to u_n & u_{1,3} \to u_{n+3} & v_{1,1} \to v_{n+1} & v_{1,2} \to v_n & v_{1,3} \to v_{n+3} \\ u_{2,1} \to u_{n+1} & u_{2,2} \to u_n & u_{2,3} \to u_{n+2} & v_{2,1} \to v_{n+1} & v_{2,2} \to v_n & v_{2,3} \to v_{n+2} \\ u_{3,1} \to u_{n+2} & u_{3,2} \to u_n & u_{3,3} \to u_{n+3} & v_{3,1} \to v_{n+2} & v_{3,2} \to v_n & v_{3,3} \to v_{n+3} \end{array}$$

Now adding the forces of the elements on the center node gives

$$\begin{aligned} f_{nx} = {} & \left[k_{22}^{(1)} + k_{22}^{(2)} + k_{22}^{(3)}\right]u_n + \left[k_{21}^{(1)} + k_{21}^{(2)}\right]u_{n+1} + \left[k_{23}^{(2)} + k_{21}^{(3)}\right]u_{n+2} + \left[k_{23}^{(1)} + k_{23}^{(3)}\right]u_{n+3} \\ & + \left[k_{25}^{(1)} + k_{25}^{(2)} + k_{25}^{(3)}\right]v_n + \left[k_{24}^{(1)} + k_{24}^{(2)}\right]v_{n+1} + \left[k_{26}^{(2)} + k_{24}^{(3)}\right]v_{n+2} + \left[k_{26}^{(1)} + k_{26}^{(3)}\right]v_{n+3} \end{aligned}$$

$$\begin{aligned} f_{ny} = {} & \left[k_{52}^{(1)} + k_{52}^{(2)} + k_{52}^{(3)}\right]u_n + \left[k_{51}^{(1)} + k_{51}^{(2)}\right]u_{n+1} + \left[k_{53}^{(2)} + k_{51}^{(3)}\right]u_{n+2} + \left[k_{53}^{(1)} + k_{53}^{(3)}\right]u_{n+3} \\ & + \left[k_{55}^{(1)} + k_{55}^{(2)} + k_{55}^{(3)}\right]v_n + \left[k_{54}^{(1)} + k_{54}^{(2)}\right]v_{n+1} + \left[k_{56}^{(2)} + k_{54}^{(3)}\right]v_{n+2} + \left[k_{56}^{(1)} + k_{56}^{(3)}\right]v_{n+3} \end{aligned}$$

The finite element equation for the three element configuration is

$$\begin{bmatrix} \mathbf{k}_{n,x}^T \\ \mathbf{k}_{n+1,x}^T \\ \mathbf{k}_{n+2,x}^T \\ \mathbf{k}_{n+3,x}^T \\ \mathbf{k}_{n,y}^T \\ \mathbf{k}_{n+1,y}^T \\ \mathbf{k}_{n+2,y}^T \\ \mathbf{k}_{n+3,y}^T \end{bmatrix} \begin{pmatrix} u_n \\ u_{n+1} \\ u_{n+2} \\ u_{n+3} \\ v_n \\ v_{n+1} \\ v_{n+2} \\ v_{n+3} \end{pmatrix} = \begin{pmatrix} f_{nx} \\ f_{n+1x} \\ f_{n+2x} \\ f_{n+3x} \\ f_{ny} \\ f_{n+1y} \\ f_{n+2y} \\ f_{n+3y} \end{pmatrix}$$

The stiffness coefficients associated with the center node are as follows:

$$\mathbf{k}_{n,x} = \begin{pmatrix} \left[k_{22}^{(1)} + k_{22}^{(2)} + k_{22}^{(3)}\right] \\ \left[k_{21}^{(1)} + k_{21}^{(2)}\right] \\ \left[k_{23}^{(2)} + k_{21}^{(3)}\right] \\ \left[k_{23}^{(1)} + k_{23}^{(3)}\right] \\ \left[k_{25}^{(1)} + k_{25}^{(2)} + k_{25}^{(3)}\right] \\ \left[k_{24}^{(1)} + k_{24}^{(2)}\right] \\ \left[k_{26}^{(2)} + k_{24}^{(3)}\right] \\ \left[k_{26}^{(1)} + k_{26}^{(3)}\right] \end{pmatrix}, \quad \mathbf{k}_{n,y} = \begin{pmatrix} \left[k_{52}^{(1)} + k_{52}^{(2)} + k_{52}^{(3)}\right] \\ \left[k_{51}^{(1)} + k_{51}^{(2)}\right] \\ \left[k_{53}^{(2)} + k_{51}^{(3)}\right] \\ \left[k_{53}^{(1)} + k_{53}^{(3)}\right] \\ \left[k_{55}^{(1)} + k_{55}^{(2)} + k_{55}^{(3)}\right] \\ \left[k_{54}^{(1)} + k_{54}^{(2)}\right] \\ \left[k_{56}^{(2)} + k_{54}^{(3)}\right] \\ \left[k_{56}^{(1)} + k_{56}^{(3)}\right] \end{pmatrix}$$

The remaining rows of the stiffness matrix may similarly be obtained, and the entries of the mass matrix may be obtained by a similar process.

9 Solution Methods for Linear Problems: I

9.1 NUMERICAL METHODS IN FEA

9.1.1 Solving the Finite Element Equations: Static Problems

We consider numerical solution of the linear system $\mathbf{K}\boldsymbol{\gamma}=\mathbf{f}$, in which $\mathbf{K}$ is the positive definite and symmetric stiffness matrix. In many problems it has a large dimension, but is also banded. The matrix may be "triangularized" to yield the form $\mathbf{K}=\mathbf{L}\mathbf{L}^T$, in which $\mathbf{L}$ is a lower triangular nonsingular matrix (zeroes in all entries above the diagonal). We may introduce $\mathbf{z}=\mathbf{L}^T\boldsymbol{\gamma}$ and obtain $\mathbf{z}$ by solving $\mathbf{L}\mathbf{z}=\mathbf{f}$. Next $\boldsymbol{\gamma}$ can be computed by solving $\mathbf{L}^T\boldsymbol{\gamma}=\mathbf{z}$. We now see that $\mathbf{L}\mathbf{z}=\mathbf{f}$ can be conveniently solved by *forward substitution*. $\mathbf{L}\mathbf{z}=\mathbf{f}$ may be expanded as

$$\begin{bmatrix} l_{11} & 0 & . & . & . & 0 \\ l_{21} & l_{22} & . & . & . & . \\ l_{31} & l_{32} & l_{33} & . & . & . \\ . & . & . & . & . & . \\ . & . & . & . & . & 0 \\ l_{n1} & l_{n2} & . & . & . & l_{nn} \end{bmatrix} \begin{pmatrix} z_1 \\ z_2 \\ z_3 \\ . \\ . \\ z_n \end{pmatrix} = \begin{pmatrix} f_1 \\ f_2 \\ f_3 \\ . \\ . \\ f_n \end{pmatrix} \tag{9.1}$$

Assuming that the diagonal entries are not too small, this equation can be solved, starting from the upper left entry, using simple arithmetic: $z_1=f_1/l_{11}$, $z_2=[f_2-l_{21}z_1]/l_{22}$, $z_3=[f_3-l_{31}z_1-l_{32}z_2]/l_{33},\ldots$.

Next the equation $\mathbf{L}^T\boldsymbol{\gamma}=\mathbf{z}$ can be solved by *back substitution*. The equation is expanded as

$$\begin{bmatrix} l_{11} & l_{12} & . & . & . & l_{1n} \\ 0 & l_{22} & . & . & . & . \\ 0 & 0 & . & . & . & . \\ . & . & . & l_{n-2,n-2} & l_{n-2,n-1} & l_{n-2,n} \\ . & . & . & 0 & l_{n-1,n-1} & l_{n-1,n} \\ 0 & l_{n2} & . & 0 & 0 & l_{nn} \end{bmatrix} \begin{pmatrix} \gamma_1 \\ \gamma_2 \\ \gamma_3 \\ . \\ . \\ \gamma_n \end{pmatrix} = \begin{pmatrix} f_1 \\ . \\ . \\ f_{n-2} \\ f_{n-1} \\ f_n \end{pmatrix} \tag{9.2}$$

Starting from the lower right-hand entry, solution can be achieved by simple arithmetic as

$$\gamma_n = f_n/l_{nn},\ \gamma_{n-1} = [f_{n-1} - l_{n-1,1}\gamma_n]/l_{n-1,n-1}$$
$$\gamma_{n-2} = [f_{n-2} - l_{n-2,n}\gamma_n - l_{n-2,n-1}\gamma_{n-1}]/l_{n-2,n-2},\ldots$$

In both procedures, only one unknown is encountered in each step (row).

9.1.2 Matrix Triangularization and Solution of Linear Systems

We next consider how to triangularize $\mathbf{K}_j$. Suppose that the upper left hand $(j-1)\times(j-1)$ block $\mathbf{K}_{j-1}$ has been triangularized to furnish $\mathbf{K}_{j-1} = \mathbf{L}_{j-1}\mathbf{L}_{j-1}^T$. To determine whether the $j\times j$ block $\mathbf{K}_j$ can be triangularized, we seek $\boldsymbol{\lambda}_j$ and l_{jj} satisfying

$$\mathbf{K}_j = \begin{bmatrix} \mathbf{K}_{j-1} & \boldsymbol{\kappa}_j \\ \boldsymbol{\kappa}_j^T & k_{jj} \end{bmatrix} = \begin{bmatrix} \mathbf{L}_{j-1} & \mathbf{0} \\ \boldsymbol{\lambda}_j^T & l_{jj} \end{bmatrix} \begin{bmatrix} \mathbf{L}_{j-1}^T & \boldsymbol{\lambda}_j \\ \mathbf{0}^T & l_{jj} \end{bmatrix} \tag{9.3}$$

in which $\boldsymbol{\kappa}_j$ is a $(j-1)\times 1$ array of the first $j-1$ entries of the jth column of $\mathbf{K}_j$. Simple manipulation suffices to furnish $\boldsymbol{\kappa}_j$ and l_{jj}.

$$\begin{aligned} \boldsymbol{\kappa}_j &= \mathbf{L}_{j-1}\boldsymbol{\lambda}_j \\ l_{jj} &= \sqrt{k_{jj} - \boldsymbol{\lambda}_j^T\boldsymbol{\lambda}_j} \end{aligned} \tag{9.4}$$

Note that $\boldsymbol{\lambda}_j$ can be conveniently computed using forward substitution. Also, note that $l_{jj} = \sqrt{k_{jj} - \boldsymbol{\kappa}_j^T\mathbf{K}_{j-1}^{-1}\boldsymbol{\kappa}_j}$. The fact that $\mathbf{K}_j > 0$ implies that l_{jj} is real. The triangularization process proceeds to the $(j+1)$st block and from there to the complete stiffness matrix.

As an illustration, consider

$$\mathbf{A}_3 = \begin{bmatrix} 1 & \frac{1}{2} & \frac{1}{3} \\ \frac{1}{2} & \frac{1}{3} & \frac{1}{4} \\ \frac{1}{3} & \frac{1}{4} & \frac{1}{5} \end{bmatrix} \tag{9.5}$$

Clearly $\mathbf{L}_1 = \mathbf{L}_1^T \to 1$. For the second block

$$\begin{bmatrix} 1 & 0 \\ \lambda_2 & l_{22} \end{bmatrix} \begin{bmatrix} 1 & \lambda_2 \\ 0 & l_{22} \end{bmatrix} = \begin{bmatrix} 1 & \frac{1}{2} \\ \frac{1}{2} & \frac{1}{3} \end{bmatrix} \tag{9.6}$$

from which $\lambda_2 = 1/2$ and $l_{22} = \sqrt{1/3 - (1/2)^2} = 1/\sqrt{12}$. And so

$$\mathbf{L}_2 = \begin{bmatrix} 1 & 0 \\ \frac{1}{2} & \frac{1}{\sqrt{12}} \end{bmatrix} \tag{9.7}$$

We now proceed to the full matrix:

$$\begin{bmatrix} 1 & \frac{1}{2} & \frac{1}{3} \\ \frac{1}{2} & \frac{1}{3} & \frac{1}{4} \\ \frac{1}{3} & \frac{1}{4} & \frac{1}{5} \end{bmatrix} = \mathbf{L}_3\mathbf{L}_3^T$$

$$= \begin{bmatrix} 1 & 0 & 0 \\ \frac{1}{2} & \frac{1}{\sqrt{12}} & 0 \\ l_{31} & l_{32} & l_{33} \end{bmatrix} \begin{bmatrix} 1 & \frac{1}{2} & l_{31} \\ 0 & \frac{1}{\sqrt{12}} & l_{32} \\ 0 & 0 & l_{33} \end{bmatrix} \tag{9.8}$$

$$= \begin{bmatrix} 1 & \frac{1}{2} & l_{31} \\ \frac{1}{2} & \frac{1}{3} & l_{31}/2 + l_{32}/\sqrt{12} \\ l_{31} & l_{31}/2 + l_{32}/\sqrt{12} & l_{31}^2 + l_{32}^2 + l_{33}^2 \end{bmatrix}$$

We conclude that $l_{31} = 1/3$, $l_{32} = 1/\sqrt{12}$, $l_{33}^2 = 1/5 - 1/9 - 1/12 = 7/180$.

The finite element problems so far considered are direct: They involve known tractions and unknown displacements and can be solved uniquely owing to the positive definiteness of the stiffness matrix. In Chapter 10, the solution method is extended to a type of inverse problem in which tractions and displacements are both specified for some degrees of freedom on the boundary nodes, while neither is specified for other degrees of freedom.

EXAMPLE 9.1

Verify that the triangular factors $\mathbf{L}_3$ and $\mathbf{L}_3^T$ for $\mathbf{A}_3$ in Equation 9.5 are correct.

Solution

From Equation 9.5

$$\mathbf{A}_3 = \begin{bmatrix} 1 & \frac{1}{2} & \frac{1}{3} \\ \frac{1}{2} & \frac{1}{3} & \frac{1}{4} \\ \frac{1}{3} & \frac{1}{4} & \frac{1}{5} \end{bmatrix}, \quad \mathbf{L}_3 = \begin{bmatrix} 1 & 0 & 0 \\ \frac{1}{2} & \frac{1}{\sqrt{12}} & 0 \\ \frac{1}{3} & \frac{1}{\sqrt{12}} & \frac{1}{\sqrt{180}} \end{bmatrix}$$

Now

$$\mathbf{L}_3\mathbf{L}_3^T = \begin{bmatrix} 1 & 0 & 0 \\ \frac{1}{2} & \frac{1}{\sqrt{12}} & 0 \\ \frac{1}{3} & \frac{1}{\sqrt{12}} & \frac{1}{\sqrt{180}} \end{bmatrix} \begin{bmatrix} 1 & \frac{1}{2} & \frac{1}{3} \\ 0 & \frac{1}{\sqrt{12}} & \frac{1}{\sqrt{12}} \\ 0 & 0 & \frac{1}{\sqrt{180}} \end{bmatrix}$$

$$= \begin{bmatrix} 1 & \frac{1}{2} & \frac{1}{3} \\ \frac{1}{2} & \frac{1}{3} & \frac{1}{4} \\ \frac{1}{3} & \frac{1}{4} & \frac{1}{5} \end{bmatrix}$$

as expected.

EXAMPLE 9.2

Invoking $\mathbf{A}_3$ in Equation 9.5, use forward substitution followed by back substitution to solve

$$\mathbf{A}_3\boldsymbol{\gamma} = \begin{pmatrix} 1 \\ 1 \\ 1 \end{pmatrix}$$

SOLUTION

Introducing $\mathbf{L}_3^T\boldsymbol{\gamma} = \mathbf{z}$ and recalling $\mathbf{L}_3$ from Equation 9.5, the foregoing equation becomes

$$\begin{bmatrix} 1 & 0 & 0 \\ \frac{1}{2} & \frac{1}{\sqrt{12}} & 0 \\ \frac{1}{3} & \frac{1}{\sqrt{12}} & \frac{1}{180} \end{bmatrix} \begin{pmatrix} z_1 \\ z_2 \\ z_3 \end{pmatrix} = \begin{pmatrix} 1 \\ 1 \\ 1 \end{pmatrix}$$

Using forward substitution

$$z_1 = 1, \quad z_2 = \sqrt{12}\left(1 - \tfrac{1}{2}z_1\right) = \sqrt{3}, \quad z_3 = \sqrt{180}\left(1 - \tfrac{1}{3}z_1 - \tfrac{1}{\sqrt{12}}z_2\right) = \sqrt{5}$$

Hence

$$\mathbf{z} = \begin{pmatrix} 1 \\ \sqrt{3} \\ \sqrt{5} \end{pmatrix}$$

Next

$$\begin{bmatrix} 1 & \frac{1}{2} & \frac{1}{3} \\ 0 & \frac{1}{\sqrt{12}} & \frac{1}{\sqrt{12}} \\ 0 & 0 & \frac{1}{180} \end{bmatrix} \begin{pmatrix} \gamma_1 \\ \gamma_2 \\ \gamma_3 \end{pmatrix} = \begin{pmatrix} 1 \\ \sqrt{3} \\ \sqrt{5} \end{pmatrix}$$

and using back substitution

$$\gamma_3 = \sqrt{180}\sqrt{5} = 30, \quad \gamma_2 = \sqrt{12}\left(\sqrt{3} - \tfrac{1}{\sqrt{12}}\gamma_3\right) = -24, \quad \gamma_1 = 1 - \tfrac{1}{2}\gamma_2 - \tfrac{1}{3}\gamma_3 = 3$$

We conclude that

$$\boldsymbol{\gamma} = \begin{pmatrix} 30 \\ -24 \\ 3 \end{pmatrix}$$

EXAMPLE 9.3

Triangularize the matrix

$$\mathbf{K} = \begin{bmatrix} 36 & 30 & 18 \\ 30 & 41 & 23 \\ 18 & 23 & 14 \end{bmatrix}$$

SOLUTION

Triangularizing the upper left-hand 2×2 block $\mathbf{K}_2$ in the form $\mathbf{K}_2 = \mathbf{L}_2\mathbf{L}_2^T$ gives

$$\begin{bmatrix} 36 & 30 \\ 30 & 41 \end{bmatrix} = \begin{bmatrix} 6 & 0 \\ \lambda_2 & l_{22} \end{bmatrix} \begin{bmatrix} 6 & \lambda_2 \\ 0 & l_{22} \end{bmatrix}$$

from which

$$\lambda_2 = 5, \quad l_{22} = \sqrt{41 - \lambda_2^2} = 4$$

Consequently, the 2×2 matrix $\mathbf{K}_2$ triangularizes to

$$\begin{bmatrix} 36 & 30 \\ 30 & 41 \end{bmatrix} = \begin{bmatrix} 6 & 0 \\ 5 & 4 \end{bmatrix} \begin{bmatrix} 6 & 5 \\ 0 & 4 \end{bmatrix}$$

Now extend the procedure to the 3×3 matrix $\mathbf{K}$:

$$\begin{bmatrix} 36 & 30 & 18 \\ 30 & 41 & 23 \\ 18 & 23 & 14 \end{bmatrix} = \begin{bmatrix} 6 & 0 & 0 \\ 5 & 4 & 0 \\ \lambda_{31} & \lambda_{32} & l_{33} \end{bmatrix} \begin{bmatrix} 6 & 5 & \lambda_{31} \\ 0 & 4 & \lambda_{32} \\ 0 & 0 & l_{33} \end{bmatrix}$$

After simple manipulation we obtain

$$\lambda_{31} = 18/6 = 3, \quad \lambda_{32} = \tfrac{1}{4}(23 - 5\lambda_{31}) = 2, \quad l_{33} = \sqrt{(14 - \lambda_{31}^2 - \lambda_{32}^2)} = 1$$

Accordingly, the triangular factor of $\mathbf{K}$ is

$$\mathbf{L} = \begin{bmatrix} 6 & 0 & 0 \\ 5 & 4 & 0 \\ 3 & 2 & 1 \end{bmatrix}$$

EXAMPLE 9.4

For the linear system

$$\begin{bmatrix} 36 & 30 & 24 \\ 30 & 41 & 32 \\ 24 & 32 & 27 \end{bmatrix} \begin{pmatrix} \gamma_1 \\ \gamma_2 \\ \gamma_3 \end{pmatrix} = \begin{pmatrix} 1 \\ 2 \\ 3 \end{pmatrix}$$

triangularize the matrix and solve for γ_1, γ_2, γ_3.

SOLUTION

Triangularizing the upper left-hand 2×2 block $\mathbf{A}_2$ in the form $\mathbf{A}_2 = \mathbf{L}_2\mathbf{L}_2^T$ gives

$$\begin{bmatrix} 36 & 30 \\ 30 & 41 \end{bmatrix} = \begin{bmatrix} 6 & 0 \\ \lambda_2 & l_{22} \end{bmatrix} \begin{bmatrix} 6 & \lambda_2 \\ 0 & l_{22} \end{bmatrix}$$

from which

$$\lambda_2 = 5, \quad l_{22} = \sqrt{41 - \lambda_2^2} = 4$$

Accordingly $\mathbf{A}_2$ triangularizes as follows:

$$\begin{bmatrix} 36 & 30 \\ 30 & 41 \end{bmatrix} = \begin{bmatrix} 6 & 0 \\ 5 & 4 \end{bmatrix} \begin{bmatrix} 6 & 5 \\ 0 & 4 \end{bmatrix}$$

For the 3×3 matrix $\mathbf{A}$,

$$\begin{bmatrix} 36 & 30 & 24 \\ 30 & 41 & 32 \\ 24 & 32 & 27 \end{bmatrix} = \begin{bmatrix} 6 & 0 & 0 \\ 5 & 4 & 0 \\ \lambda_{31} & \lambda_{32} & l_{33} \end{bmatrix} \begin{bmatrix} 6 & 5 & \lambda_{31} \\ 0 & 4 & \lambda_{32} \\ 0 & 0 & l_{33} \end{bmatrix}$$

from which

$$\lambda_{31} = 24/6 = 4, \quad \lambda_{32} = \tfrac{1}{4}(32 - 5\lambda_{31}) = 3, \quad l_{33} = \sqrt{(27 - \lambda_{31}^2 - \lambda_{32}^2)} = \sqrt{2}$$

The triangular factor of $\mathbf{L}$ of $\mathbf{A}$ is now

$$\mathbf{L} = \begin{bmatrix} 6 & 0 & 0 \\ 5 & 4 & 0 \\ 4 & 3 & \sqrt{2} \end{bmatrix}$$

On introducing $\mathbf{L}^T\boldsymbol{\gamma} = \mathbf{z}$, we encounter

$$\begin{bmatrix} 6 & 0 & 0 \\ 5 & 4 & 0 \\ 4 & 3 & \sqrt{2} \end{bmatrix} \begin{pmatrix} z_1 \\ z_2 \\ z_3 \end{pmatrix} = \begin{pmatrix} 1 \\ 2 \\ 3 \end{pmatrix}$$

Forward substitution results in

$$z_1 = 1/6, \quad z_2 = 1/4(2 - 5z_1) = 7/24, \quad z_3 = 1/\sqrt{2}(3 - 4z_1 - 3z_2) = 35\sqrt{2}/48$$

and also

$$\mathbf{z} = \begin{pmatrix} 1/6 \\ 7/24 \\ 35\sqrt{2}/48 \end{pmatrix}$$

Next

$$\begin{bmatrix} 6 & 5 & 4 \\ 0 & 4 & 3 \\ 0 & 0 & \sqrt{2} \end{bmatrix} \begin{pmatrix} \gamma_1 \\ \gamma_2 \\ \gamma_3 \end{pmatrix} = \begin{pmatrix} 1/6 \\ 7/24 \\ 35\sqrt{2}/48 \end{pmatrix}$$

and using backward substitution gives

$$\gamma_3 = 35\sqrt{2}/48 \cdot 1/\sqrt{2} = 35/48, \quad \gamma_2 = 1/4(7/24 - 3\gamma_3) = -91/192,$$
$$\gamma_1 = 1/6(1/6 - 5\gamma_2 - 4\gamma_3) = -73/1152$$

We conclude that

$$\boldsymbol{\gamma} = \begin{pmatrix} -73/1152 \\ -91/192 \\ 35/48 \end{pmatrix}$$

9.1.3 Triangularization of Asymmetric Matrices

The foregoing triangularization is applicable to positive definite symmetric matrices. Asymmetric matrices arise in a number of finite element problems, including problems with incompressibility, unsteady rotation, or thermomechanical coupling. If the matrix is nonsingular, it may be decomposed into the product of a lower triangular and an upper triangular matrix, followed by forward and back substitution.

$$\mathbf{K} = \mathbf{LU} \tag{9.9}$$

(It is also possible to triangularize a singular matrix, but $\mathbf{L}$ will then be singular, preventing the use of forward substitution.)

Assuming the $(j-1)$st diagonal block has been triangularized, we consider whether the jth block admits the decomposition

$$\mathbf{K}_j = \begin{bmatrix} \mathbf{K}_{j-1} & \boldsymbol{\kappa}_{1j} \\ \boldsymbol{\kappa}_{2j}^T & k_{jj} \end{bmatrix} = \begin{bmatrix} \mathbf{L}_{j-1} & \mathbf{0} \\ \boldsymbol{\lambda}_j^T & l_{jj} \end{bmatrix} \begin{bmatrix} \mathbf{U}_{j-1} & u_j \\ \mathbf{0}^T & u_{jj} \end{bmatrix}$$
$$= \begin{bmatrix} \mathbf{L}_{j-1}\mathbf{U}_{j-1} & \mathbf{L}_{j-1}\mathbf{u}_j \\ \boldsymbol{\lambda}_j^T\mathbf{U}_{j-1} & \boldsymbol{\lambda}_j^T\mathbf{u}_j + u_{jj}l_{jj} \end{bmatrix} \tag{9.10}$$

Now $\mathbf{u}_j$ is obtained by forward substitution using $\mathbf{L}_{j-1}\mathbf{u}_j = \boldsymbol{\kappa}_{1j}$, and $\boldsymbol{\lambda}_j$ is obtained by back substitution using $\mathbf{U}_{j-1}^T\,\boldsymbol{\lambda}_j = \boldsymbol{\kappa}_{2j}$. Finally, $u_{jj}l_{jj} = k_{jj} - \boldsymbol{\lambda}_j^T\mathbf{u}_j$, for which purpose u_{jj} may be arbitrarily set to unity. After the triangularization process is completed, an equation of the form $\mathbf{Ku} = \mathbf{f}$ can now be solved by forward substitution applied to $\mathbf{Lz} = \mathbf{f}$, followed by back substitution applied to $\mathbf{Uu} = \mathbf{z}$.

EXAMPLE 9.5

Triangularize the asymmetric matrix

$$\begin{bmatrix} 2\mu A/L & 0 & -A \\ 0 & 4\mu AL/Y^2 & -2\mu A/L \\ A & 2\mu A/L & 0 \end{bmatrix}$$

(This matrix will be seen later in Chapter 11 concerning incompressible materials.)

Solution

Triangularizing the upper left-hand 2×2 block $\mathbf{B}_2$ in the form $\mathbf{B}_2 = \mathbf{L}_2\mathbf{U}_2$ gives

$$\begin{bmatrix} 2\mu A/L & 0 \\ 0 & 4\mu AL/Y^2 \end{bmatrix} = \begin{bmatrix} l_{11} & 0 \\ l_{21} & l_{22} \end{bmatrix} \begin{bmatrix} u_{11} & u_{12} \\ 0 & u_{22} \end{bmatrix}$$

On setting u_{11} and u_{22} as unity,

$$l_{11} = 2\mu A/L, \quad l_{21} = 0, \quad u_{12} = 0, \quad l_{22} = 4\mu AL/Y^2$$

Now consider the 3×3 matrix $\mathbf{B}$:

$$\begin{bmatrix} 2\mu A/L & 0 & -A \\ 0 & 4\mu AL/Y^2 & -2\mu A/L \\ A & 2\mu A/L & 0 \end{bmatrix} = \begin{bmatrix} 2\mu A/L & 0 & 0 \\ 0 & 4\mu AL/Y^2 & 0 \\ l_{31} & l_{32} & l_{33} \end{bmatrix} \begin{bmatrix} 1 & 0 & u_{13} \\ 0 & 1 & u_{23} \\ 0 & 0 & u_{33} \end{bmatrix}$$

On setting $u_{33} = 1$, simple manipulation furnishes

$$l_{31} = A, \quad l_{32} = 2\mu A/L, \quad u_{13} = -L/2\mu, \quad u_{23} = -Y^2/2L^2, \quad l_{33} = \left(AL/2\mu + \mu AY^2/L^3\right)$$

Finally

$$\mathbf{B} = \mathbf{LU} = \begin{bmatrix} 2\mu A/L & 0 & 0 \\ 0 & 4\mu AL/Y^2 & 0 \\ A & 2\mu A/L & (AL/2\mu + \mu AY^2/L^3) \end{bmatrix} \begin{bmatrix} 1 & 0 & -L/2\mu \\ 0 & 1 & -Y^2/2L^2 \\ 0 & 0 & 1 \end{bmatrix}$$

The decomposition is not unique since it is based on setting the diagonal entries of $\mathbf{U}$ to unity.

9.2 TIME INTEGRATION: STABILITY AND ACCURACY

Much insight may be gained from considering the model equation

$$\frac{dy}{dt} = -\lambda y \tag{9.11}$$

in which λ is complex. If $\mathrm{Re}(\lambda) > 0$, for the initial value $y(0) = y_0$ the solution is $y(t) = y_0 \exp(-\lambda t)$, and clearly $y(t) \to 0$. In this event the system is called *asymptotically stable.*

We now ask whether *numerical integration schemes* to integrate Equation 9.11 have stability properties corresponding to asymptotic stability. In other words, does the numerical solution decay when the exact solution decays, and diverge when the exact solution diverges? For this purpose we consider the trapezoidal rule, the properties of which will be discussed in Section 9.3. Consider time steps of duration h, and suppose that the solution has been calculated through the nth time step. We seek to compute the solution at the $(n+1)$st time step. The trapezoidal rule (see Section 9.3) is given by

$$\frac{\mathrm{d}y}{\mathrm{d}t} \approx \frac{y_{n+1} - y_n}{h}, \qquad -\lambda y \approx -\tfrac{\lambda}{2}[y_{n+1} + y_n] \tag{9.12}$$

Consequently,

$$\begin{aligned} y_{n+1} &= \frac{1 - \lambda h/2}{1 + \lambda h/2} y_n \\ &= \left[\frac{1 - \lambda h/2}{1 + \lambda h/2}\right]^n y_0 \end{aligned} \tag{9.13}$$

Clearly, $y_{n+1} \to 0$ if $\left|\frac{1-\lambda h/2}{1+\lambda h/2}\right| < 1$, and $y_{n+1} \to \infty$ if $\left|\frac{1-\lambda h/2}{1+\lambda h/2}\right| > 1$, in which $|\cdot|$ implies the magnitude. If the first inequality is satisfied the numerical method is called *A-stable* (Dahlquist and Bjork, 1974). We next write $\lambda = \lambda_r + i\lambda_i$, and now A-stability requires that

$$\frac{\left(1 - \frac{\lambda_r h}{2}\right)^2 + \left(\frac{\lambda_i h}{2}\right)^2}{\left(1 + \frac{\lambda_r h}{2}\right)^2 + \left(\frac{\lambda_i h}{2}\right)^2} < 1 \tag{9.14}$$

A-stability obtains if $\lambda_r > 0$, which is precisely the condition for asymptotic stability.

Next consider the matrix–vector system arising in the finite element method.

$$\mathbf{M}\ddot{\boldsymbol{\gamma}} + \mathbf{D}\dot{\boldsymbol{\gamma}} + \mathbf{K}\boldsymbol{\gamma} = \mathbf{0}, \quad \boldsymbol{\gamma}(0) = \boldsymbol{\gamma}_0, \quad \dot{\boldsymbol{\gamma}}(0) = \dot{\boldsymbol{\gamma}}_0 \tag{9.15}$$

in which **M**, **D**, and **K** are positive definite. Elementary manipulation serves to establish that

$$\frac{\mathrm{d}}{\mathrm{d}t}\left[\frac{1}{2}\dot{\boldsymbol{\gamma}}^T\mathbf{M}\dot{\boldsymbol{\gamma}} + \frac{1}{2}\boldsymbol{\gamma}^T\mathbf{K}\boldsymbol{\gamma}\right] = -\dot{\boldsymbol{\gamma}}^T\mathbf{D}\dot{\boldsymbol{\gamma}} < 0 \tag{9.16}$$

It follows that $\dot{\boldsymbol{\gamma}} \to \mathbf{0}$ and $\boldsymbol{\gamma} \to \mathbf{0}$. We conclude that the system is asymptotically stable.

Introducing the vector $\mathbf{p} = \dot{\boldsymbol{\gamma}}$, the n-dimensional second-order system is written in *state form* as the $(2n)$-dimensional first-order system of ordinary differential equations:

$$\begin{bmatrix} \mathbf{M} & \mathbf{0} \\ \mathbf{0} & \mathbf{I} \end{bmatrix} \begin{pmatrix} \mathbf{p} \\ \boldsymbol{\gamma} \end{pmatrix}^{\bullet} + \begin{bmatrix} \mathbf{D} & \mathbf{K} \\ -\mathbf{I} & \mathbf{0} \end{bmatrix} \begin{pmatrix} \mathbf{p} \\ \boldsymbol{\gamma} \end{pmatrix} = \begin{pmatrix} \mathbf{f} \\ \mathbf{0} \end{pmatrix} \tag{9.17}$$

We next apply the trapezoidal rule to the system:

$$\begin{bmatrix} \mathbf{M} & \mathbf{0} \\ \mathbf{0} & \mathbf{I} \end{bmatrix} \begin{pmatrix} \frac{1}{h}(\mathbf{p}_{n+1} - \mathbf{p}_n) \\ \frac{1}{h}(\boldsymbol{\gamma}_{n+1} - \boldsymbol{\gamma}_n) \end{pmatrix} + \begin{bmatrix} \mathbf{D} & \mathbf{K} \\ -\mathbf{I} & \mathbf{0} \end{bmatrix} \begin{pmatrix} \frac{1}{2}(\mathbf{p}_{n+1} + \mathbf{p}_n) \\ \frac{1}{2}(\boldsymbol{\gamma}_{n+1} + \boldsymbol{\gamma}_n) \end{pmatrix} = \begin{pmatrix} \frac{1}{2}(\mathbf{f}_{n+1} + \mathbf{f}_n) \\ \mathbf{0} \end{pmatrix} \tag{9.18}$$

From the equation in the lower row, $\mathbf{p}_{n+1} = \frac{2}{h}[\boldsymbol{\gamma}_{n+1} - \boldsymbol{\gamma}_n] - \mathbf{p}_n$. Eliminating $\mathbf{p}_{n+1}$ in the upper row furnishes a formula underlying the classical Newmark method:

$$\begin{aligned} \mathbf{K_D}\boldsymbol{\gamma}_{n+1} &= \mathbf{r}_{n+1}, \quad \mathbf{K_D} = \left[\mathbf{M} + \tfrac{h}{2}\mathbf{D} + \tfrac{h^2}{4}\mathbf{K}\right] \\ \mathbf{r}_{n+1} &= \left[\mathbf{M} + \tfrac{h}{2}\mathbf{D} - \tfrac{h^2}{4}\mathbf{K}\right]\mathbf{y}_n + \left[\mathbf{M} + \tfrac{h}{2}\mathbf{D}\right]\tfrac{h}{2}\mathbf{p}_n + \tfrac{h^2}{4}(\mathbf{f}_{n+1} + \mathbf{f}_n) \end{aligned} \tag{9.19}$$

and $\mathbf{K_D}$ may be called the dynamic stiffness matrix. Equation 9.19 may be solved by tiangularization of $\mathbf{K_D}$, followed by forward and backward substitution.

9.3 PROPERTIES OF THE TRAPEZOIDAL RULE

We consider the accuracy of the trapezoidal rule, and by extension of Newmark's method. To fix important notions consider the model equation

$$\frac{dy}{dx} = f(y) \tag{9.20}$$

Suppose this equation is approximated as

$$\alpha y_{n+1} + \beta y_n + h[\gamma f_{n+1} + \delta f_n] = 0 \tag{9.21}$$

We now use the Taylor series to express y_{n+1} and f_{n+1} in terms of y_n and f_n. Noting that $y'_n = f_n$ and $y''_n = f'_n$, we obtain

$$\begin{aligned} 0 = \alpha&\left[y_n + y'_n h + y''_n h^2/2\right] \\ &+ \beta y_n + h\gamma\left[y'_n + y''_n h\right] + h\delta y'_n \end{aligned} \tag{9.22}$$

For exact agreement through h^2, the coefficients must satisfy

$$\alpha + \beta = 0, \quad \alpha + \gamma + \delta = 0, \quad \alpha/2 + \gamma = 0 \tag{9.23}$$

We also introduce the convenient normalization $\gamma + \delta = 1$. Simple manipulation serves to derive that $\alpha = -1$, $\beta = 1$, $\gamma = 1/2$, $\delta = 1/2$, furnishing

$$\frac{y_{n+1} - y_n}{h} = \frac{1}{2}[f(y_{n+1}) + f(y_n)] \tag{9.24}$$

which may be recognized as the trapezoidal rule.

In fact the trapezoidal rule is unique and optimal in having the following three characteristics:

(a) It is a *one-step method*, using only the values at the beginning of the current time step.
(b) It is *second-order accurate*—it agrees exactly with the Taylor series through h^2.
(c) Applied to $dy/dt + \lambda y = 0$ with initial condition $y(0) = y_0$, it is A-stable, i.e., numerically stable whenever system described by the equation is asymptotically stable.

EXAMPLE 9.6

For the model equation $dy/dx = f(y)$, develop a *two-step* numerical quadrature formula:

$$(\alpha y_{n+1} + \beta y_n + \gamma y_{n-1}) + h[\delta f(y_{n+1}) + \varepsilon f(y_n) + \zeta f(y_{n-1})] = 0$$

What is the order of the integration method (highest power in h with exact agreement with the Taylor series)?

Solution

Expressing y_{n+1}, y_{n-1}, $f(y_{n+1})$, and $f(y_{n-1})$ using the Taylor expansion gives

$$y_{n+1} = y_n + hy'_n + \frac{h^2}{2!}y''_n + \frac{h^3}{3!}y'''_n + \frac{h^4}{4!}y^{iv}_n + \cdots$$

$$y_{n-1} = y_n - hy'_n + \frac{h^2}{2!}y''_n - \frac{h^3}{3!}y'''_n + \frac{h^4}{4!}y^{iv}_n - \cdots$$

$$f(y_{n+1}) = f_n + hf'_n + \frac{h^2}{2!}f''_n + \frac{h^3}{3!}f'''_n + \cdots$$

$$f(y_{n-1}) = f_n - hf'_n + \frac{h^2}{2!}f''_n - \frac{h^3}{3!}f'''_n + \cdots$$

But $f_n = y'_n$ also implies

$$f(y_{n+1}) = y'_n + hy''_n + \frac{h^2}{2!}y'''_n + \frac{h^3}{3!}y^{iv}_n + \cdots$$
$$f(y_{n-1}) = y'_n - hy''_n + \frac{h^2}{2!}y'''_n - \frac{h^3}{3!}y^{iv}_n + \cdots$$

On substitution into the quadrature formula, we have

$$\begin{aligned}&\alpha\left[y_n + hy'_n + \frac{h^2}{2!}y''_n + \frac{h^3}{3!}y'''_n + \frac{h^4}{4!}y^{iv}_n\right] + \beta y_n \\ &+ \gamma\left[y_n - hy'_n + \frac{h^2}{2!}y''_n - \frac{h^3}{3!}y'''_n + \frac{h^4}{4!}y^{iv}_n\right] \\ &+ h\left[\begin{array}{l}\delta\left\{y'_n + hy''_n + \frac{h^2}{2!}y'''_n + \frac{h^3}{3!}y^{iv}_n\right\} + \varepsilon y'_n \\ +\zeta\left\{y'_n - hy''_n + \frac{h^2}{2!}y'''_n - \frac{h^3}{3!}y^{iv}_n\right\}\end{array}\right] = 0 + 0(h^5)\end{aligned}$$

For exact agreement through h^4, the coefficients must satisfy

$$\alpha + \beta + \gamma = 0, \quad \alpha - \gamma + \delta + \varepsilon + \zeta = 0, \quad \alpha/2 + \gamma/2 + \delta - \zeta = 0$$
$$\alpha/6 - \gamma/6 + \delta/2 + \zeta/2 = 0, \quad \alpha/24 + \gamma/24 + \delta/6 - \zeta/6 = 0$$

We now introduce the convenient normalization $\delta + \varepsilon + \zeta = 1$. Simple manipulation serves to derive

$$\alpha = -1/2, \quad \beta = 0, \quad \gamma = 1/2, \quad \delta = \zeta = 1/6, \quad \varepsilon = 2/3$$

The quadrature formula is now stated as

$$[-1/2y_{n+1} + 0 + 1/2y_{n-1}] + h[1/6f(y_{n+1}) + 2/3f(y_n) + 1/6f(y_{n-1})] = 0$$

On rearranging

$$\frac{y_{n+1} - y_{n-1}}{h} = \tfrac{1}{3}[f(y_{n-1}) + 4f(y_n) + f(y_{n-1})]$$

which is a the *two-step* numerical integration model. Since the relations are exact through h^4, this is a fourth-order accurate and its order of integration is four. However, it is not A-stable, as discussed in Chapter 10.

EXAMPLE 9.7

In the damped linear mechanical system

$$\mathbf{M}\ddot{\mathbf{y}} + \mathbf{D}\dot{\mathbf{y}} + \mathbf{K}\mathbf{y} = \mathbf{f}(t)$$

suppose that $\boldsymbol{\gamma}(t) = \boldsymbol{\gamma}_n$ at the nth time step. Derive $\mathbf{K}_D$ and $\mathbf{r}_{n+1}$ such that $\boldsymbol{\gamma}$ at the $(n+1)$ st time step satisfies

$$\mathbf{K}_D\boldsymbol{\gamma}_{n+1} = \mathbf{r}_{n+1}$$

Solution

Introducing the state form relation $\mathbf{p} = \dot{\boldsymbol{\gamma}}$ in the equation, we have

$$\begin{bmatrix} \mathbf{M} & \mathbf{0} \\ \mathbf{0} & \mathbf{I} \end{bmatrix} \begin{pmatrix} \mathbf{p} \\ \boldsymbol{\gamma} \end{pmatrix}^{\bullet} + \begin{bmatrix} \mathbf{D} & \mathbf{K} \\ -\mathbf{I} & \mathbf{0} \end{bmatrix} \begin{pmatrix} \mathbf{p} \\ \boldsymbol{\gamma} \end{pmatrix} = \begin{pmatrix} \mathbf{f} \\ \mathbf{0} \end{pmatrix}$$

This expression may be rewritten using the trapezoidal rule as

$$\begin{bmatrix} \mathbf{M} & \mathbf{0} \\ \mathbf{0} & \mathbf{I} \end{bmatrix} \begin{pmatrix} \frac{1}{h}(\mathbf{p}_{n+1} - \mathbf{p}_n) \\ \frac{1}{h}(\boldsymbol{\gamma}_{n+1} - \boldsymbol{\gamma}_n) \end{pmatrix} + \begin{bmatrix} \mathbf{D} & \mathbf{K} \\ -\mathbf{I} & \mathbf{0} \end{bmatrix} \begin{pmatrix} \frac{1}{2}(\mathbf{p}_{n+1} + \mathbf{p}_n) \\ \frac{1}{2}(\boldsymbol{\gamma}_{n+1} + \boldsymbol{\gamma}_n) \end{pmatrix} = \begin{pmatrix} \frac{1}{2}(\mathbf{f}_{n+1} + \mathbf{f}_n) \\ \mathbf{0} \end{pmatrix}$$

Now the lower row implies

$$\mathbf{p}_{n+1} = \frac{2}{h}(\boldsymbol{\gamma}_{n+1} - \boldsymbol{\gamma}_n) - \mathbf{p}_n$$

Now consider the upper row:

$$\frac{1}{h}\mathbf{M}(\mathbf{p}_{n+1} - \mathbf{p}_n) + \frac{1}{2}\mathbf{D}(\mathbf{p}_{n+1} + \mathbf{p}_n) + \frac{1}{2}\mathbf{K}(\boldsymbol{\gamma}_{n+1} + \boldsymbol{\gamma}_n) = \frac{1}{2}(\mathbf{f}_{n+1} + \mathbf{f}_n)$$

Eliminating $\mathbf{p}_{n+1}$ results in

$$\frac{1}{h}\mathbf{M}\left[\frac{2}{h}(\boldsymbol{\gamma}_{n+1} - \boldsymbol{\gamma}_n) - 2\mathbf{p}_n\right] + \frac{1}{2}\mathbf{D}\frac{2}{h}(\boldsymbol{\gamma}_{n+1} - \boldsymbol{\gamma}_n) + \frac{1}{2}\mathbf{K}(\boldsymbol{\gamma}_{n+1} + \boldsymbol{\gamma}_n) = \frac{1}{2}(\mathbf{f}_{n+1} + \mathbf{f}_n)$$

On multiplying throughout by $h^2/2$ and rearranging,

$$\left[\mathbf{M} + \frac{h}{2}\mathbf{D} + \frac{h^2}{4}\mathbf{K}\right]\boldsymbol{\gamma}_{n+1} = \left[\mathbf{M} + \frac{h}{2}\mathbf{D} - \frac{h^2}{4}\mathbf{K}\right]\boldsymbol{\gamma}_n + h\mathbf{M}\mathbf{p}_n + \frac{h^2}{4}(\mathbf{f}_{n+1} + \mathbf{f}_n)$$

Comparing the above equation with the given equation $\mathbf{K}_D\boldsymbol{\gamma}_{n+1} = \mathbf{r}_{n+1}$, it follows that

$$\mathbf{K}_D = \left[\mathbf{M} + \frac{h}{2}\mathbf{D} + \frac{h^2}{4}\mathbf{K}\right]$$

$$\mathbf{r}_{n+1} = \left[\mathbf{M} + \frac{h}{2}\mathbf{D} - \frac{h^2}{4}\mathbf{K}\right]\boldsymbol{\gamma}_n + h\mathbf{M}\mathbf{p}_n + \frac{h^2}{4}(\mathbf{f}_{n+1} + \mathbf{f}_n)$$

A higher-order time integration scheme is presented in Chapter 10.

9.4 INTEGRAL EVALUATION BY GAUSSIAN QUADRATURE

There are many integrations in the finite element method, the accuracy and efficiency of which is critical. Fortunately, a method which is optimal in an important sense, called Gaussian quadrature, has long been known (it was previously introduced in Chapter 7). It is based on converting physical coordinates to intrinsic "natural" coordinates. Consider $\int_a^b f(x)\,\mathrm{d}x$. Let $\xi = \frac{1}{b-a}[2x-(a+b)]$. Clearly, ξ maps the interval $[a,b]$ into the interval $[-1,1]$. The integral now becomes $\frac{1}{b-a}\int_{-1}^{1} f(\xi)\,\mathrm{d}\xi$. Now consider the power series

$$f(\xi) = \alpha_0 + \alpha_1\xi + \alpha_2\xi^2 + \alpha_3\xi^3 + \alpha_4\xi^4 + \alpha_5\xi^5 + \cdots \tag{9.25}$$

from which

$$\int_{-1}^{1} f(\xi)\,\mathrm{d}\xi = 2\alpha_1 + 0 + \tfrac{2}{3}\alpha_3 + 0 + \tfrac{2}{5}\alpha_5 + 0 + \cdots \tag{9.26}$$

The advantages for integration on a symmetric interval, evident in the zeroes in Equation 9.26, can be seen in the fact that, with n function evaluations, an integral can be evaluated *exactly* through $(2n-1)$st order in the Taylor series.

Consider the first $2n-1$ terms in a power series representation for a function:

$$g(\xi) = \alpha_1 + \alpha_2\xi + \cdots + \alpha_{2n}\xi^{2n-1} \tag{9.27}$$

Assume that n integration (Gauss) points ξ_i and n weights are used as follows:

$$\int_{-1}^{1} g(\xi)\,\mathrm{d}\xi = \sum_{i=1}^{n} g(\xi_i)w_i = \alpha_1\sum_{i=1}^{n} w_i + \alpha_2\sum_{i=1}^{n} w_i\xi_i + \cdots + \alpha_{2n}\sum_{i=1}^{n} w_i\xi_i^{2n-1} \tag{9.28}$$

Comparison with Equation 9.26 implies that

$$\sum_{i=1}^{n} w_i = 2, \quad \sum_{i=1}^{n} w_i\xi_i = 0, \quad \sum_{i=1}^{n} w_i\xi_i^2 = 2/3, \ldots$$
$$\sum_{i=1}^{n} w_i\xi_i^{2n-2} = \frac{2}{2n-1}, \quad \sum_{i=1}^{n} w_i\xi_i^{2n-1} = 0 \tag{9.29}$$

It is necessary to solve for n integration points ξ_i and n weights w_i. *These are universal quantities.* Thereafter, to integrate a given function $g(\xi)$ exactly through ξ^{2n-1}, it is necessary to perform n function evaluations to compute $g(\xi_i)$.

As an example, we seek two Gauss points and two weights ($n=2$). After simple manipulation

$$\begin{aligned} w_1 + w_2 = 2 \quad \text{(a)}, \quad w_1\xi_1 + w_2\xi_2 = 0 \quad \text{(b)} \\ w_1\xi_1^2 + w_2\xi_2^2 = \tfrac{2}{3} \quad \text{(c)}, \quad w_1\xi_1^3 + w_2\xi_2^3 = 0 \quad \text{(d)} \end{aligned} \tag{9.30}$$

From Equations 9.30b and 9.30d, $w_1\xi_1\left[\xi_1^2 - \xi_2^2\right] = 0$, leading to $\xi_2 = -\xi_1$. From Equations 9.30a and 9.30c it now follows that $-\xi_2 = \xi_1 = 1/\sqrt{3}$. Finally, the normalization $w_1 = 1$ implies that $w_2 = 1$.

EXAMPLE 9.8

Find the integration (Gauss) points and weights for $n = 3$.

SOLUTION

Owing to complexity the steps in the solution are given in detail. On substituting $n = 3$ in Equation 9.29

$$w_1 + w_2 + w_3 = 2 \tag{9.31}$$

$$w_1\xi_1 + w_2\xi_2 + w_3\xi_3 = 0 \tag{9.32}$$

$$w_1\xi_1^2 + w_2\xi_2^2 + w_3\xi_3^2 = 2/3 \tag{9.33}$$

$$w_1\xi_1^3 + w_2\xi_2^3 + w_3\xi_3^3 = 0 \tag{9.34}$$

$$w_1\xi_1^4 + w_2\xi_2^4 + w_3\xi_3^4 = 2/5 \tag{9.35}$$

$$w_1\xi_1^5 + w_2\xi_2^5 + w_3\xi_3^5 = 0 \tag{9.36}$$

Multiplying Equations 9.32 and 9.34 by ξ_2^2 furnishes

$$w_1\xi_1\xi_2^2 + w_2\xi_2^3 + w_3\xi_2^2\xi_3 = 0 \tag{9.37}$$

$$w_1\xi_1^3\xi_2^2 + w_2\xi_2^5 + w_3\xi_2^2\xi_3^3 = 0 \tag{9.38}$$

Now, on subtracting Equation 9.34 from Equation 9.37, and Equation 9.36 from Equation 9.38,

$$w_1\xi_1\left(\xi_2^2 - \xi_1^2\right) + w_3\xi_3\left(\xi_2^2 - \xi_3^2\right) = 0 \tag{9.39}$$

$$w_1\xi_1^3\left(\xi_2^2 - \xi_1^2\right) + w_3\xi_3^3\left(\xi_2^2 - \xi_3^2\right) = 0 \tag{9.40}$$

On multiplying Equation 9.39 by ξ_2^2,

$$w_1\xi_1^3\left(\xi_2^2 - \xi_1^2\right) + w_3\xi_1^2\xi_3\left(\xi_2^2 - \xi_3^2\right) = 0 \tag{9.41}$$

Equation 9.40 is subtracted from Equation 9.41 to furnish

$$w_3\xi_3\left(\xi_1^2 - \xi_3^2\right)\left(\xi_2^2 - \xi_3^2\right) = 0 \tag{9.42}$$

Assuming that the integration points are equally spaced, Equation 9.42 implies

$$\xi_1 = -\xi_3 \tag{9.43}$$

On substituting Equation 9.43 in Equation 9.39, we conclude that

$$w_1 = w_3 \tag{9.44}$$

On substituting Equations 9.43 and 9.44 in Equation 9.32, we learn that

$$w_2 \xi_2 = 0$$

and also

$$\xi_2 = 0 \tag{9.45}$$

Next substituting Equations 9.43 through 9.45 in Equation 9.33 and 9.35 leads to

$$w_1 \xi_1^2 = 1/3 \tag{9.46}$$

$$w_1 \xi_1^4 = 1/5 \tag{9.47}$$

Equations 9.43, 9.46, and 9.47 serve to derive that

$$\xi_1 = -\sqrt{3/5}, \quad \xi_3 = \sqrt{3/5}$$

Simple manipulations furnish the remaining unknowns as

$$w_1 = 5/9 = w_3, \quad w_2 = 8/9$$

The results are summarized as

$$\begin{array}{lll} w_1 = 5/9, & w_2 = 8/9, & w_3 = 5/9 \\ \xi_1 = -\sqrt{3/5}, & \xi_2 = 0, & \xi_3 = \sqrt{3/5} \end{array}$$

9.5 MODAL ANALYSIS BY FEA

9.5.1 MODAL DECOMPOSITION

In the absence of damping the finite element equation for a linear mechanical system, which is unforced but has nonzero initial values, is described by

$$\mathbf{M}\ddot{\boldsymbol{\gamma}} + \mathbf{K}\boldsymbol{\gamma} = \mathbf{0}, \quad \boldsymbol{\gamma}(0) = \boldsymbol{\gamma}_0, \quad \dot{\boldsymbol{\gamma}}(0) = \dot{\boldsymbol{\gamma}}_0 \tag{9.48}$$

Assume a solution of the form $\boldsymbol{\gamma} = \hat{\boldsymbol{\gamma}} \exp(\lambda t)$: it furnishes upon substitution

$$\left[\mathbf{K} + \lambda^2 \mathbf{M}\right]\hat{\boldsymbol{\gamma}} = \mathbf{0} \tag{9.49}$$

The jth eigenvalue λ_j is obtained by solving $\det(\mathbf{K} + \lambda_j^2 \mathbf{M}) = 0$, and a corresponding eigenvector vector $\boldsymbol{\gamma}_j$ may likewise be computed (see example below). For the sake of generality suppose that λ_j and $\boldsymbol{\gamma}_j$ are complex. Let $\boldsymbol{\gamma}_j^H$ denote the complex conjugate (Hermitian) transpose of $\boldsymbol{\gamma}_j$. Now λ_j^2 satisfies $\lambda_j^2 = -\frac{\boldsymbol{\gamma}_j^H\mathbf{K}\boldsymbol{\gamma}_j}{\boldsymbol{\gamma}_j^H\mathbf{M}\boldsymbol{\gamma}_j}$. Since $\mathbf{M}$ and $\mathbf{K}$ are real and positive definite, it follows that λ_j is pure imaginary: $\lambda_j = i\omega_j$. Also, without loss of generality, we may take $\boldsymbol{\gamma}_j$ to be real and orthogonal with respect to both $\mathbf{M}$ and $\mathbf{K}$.

EXAMPLE 9.9

We seek the modes of the system governed by the equation

$$\begin{bmatrix} 2 & 0 \\ 0 & 1 \end{bmatrix}\begin{pmatrix} \gamma_1 \\ \gamma_2 \end{pmatrix}^{\bullet\bullet} + \frac{k}{m}\begin{bmatrix} 2 & -1 \\ -1 & 1 \end{bmatrix}\begin{pmatrix} \gamma_1 \\ \gamma_2 \end{pmatrix} = \begin{pmatrix} 0 \\ 0 \end{pmatrix}$$

Let $\zeta^2 = \omega_j^2/\omega_0^2$, $\omega_0^2 = k/m$. For the determinant to vanish, $1 - \zeta_\pm^2 = \pm 1/\sqrt{2}$. Using $1 - \zeta_+^2 = 1/\sqrt{2}$, the first eigenvector satisfies

$$\begin{bmatrix} \sqrt{2} & -1 \\ -1 & 1/\sqrt{2} \end{bmatrix}\begin{pmatrix} \gamma_1^{(1)} \\ \gamma_2^{(1)} \end{pmatrix} = \begin{pmatrix} 0 \\ 0 \end{pmatrix}, \quad \left[\gamma_1^{(1)}\right]^2 + \left[\gamma_2^{(1)}\right]^2 = 1$$

implying that $\gamma_1^{(1)} = 1/\sqrt{3}$, $\gamma_2^{(1)} = \sqrt{2}/\sqrt{3}$. The corresponding procedures for the second eigenvalue furnish that $\gamma_1^{(2)} = 1/\sqrt{3}$, $\gamma_2^{(2)} = -\sqrt{2}/\sqrt{3}$. It is readily verified that

$$\begin{aligned} \boldsymbol{\gamma}^{(2)T}\mathbf{M}\boldsymbol{\gamma}^{(1)} &= \boldsymbol{\gamma}^{(1)T}\mathbf{M}\boldsymbol{\gamma}^{(2)} = 0, \quad \mu_1 = 4/3, \quad \mu_2 = 4/3 \\ \boldsymbol{\gamma}^{(2)T}\mathbf{K}\boldsymbol{\gamma}^{(1)} &= \boldsymbol{\gamma}^{(1)T}\mathbf{K}\boldsymbol{\gamma}^{(2)} = 0, \quad \kappa_1 = \tfrac{4}{3}\left[1 - 1/\sqrt{2}\right], \quad \kappa_2 = \tfrac{4}{3}\left[1 + 1/\sqrt{2}\right] \end{aligned}$$

Returning to the general development, the *modal matrix* $\mathbf{X}$ is now defined as

$$\mathbf{X} = [\boldsymbol{\gamma}_1 \quad \boldsymbol{\gamma}_2 \quad \boldsymbol{\gamma}_3 \quad \cdots \quad \boldsymbol{\gamma}_n] \tag{9.50}$$

Since the *jk*th entry of $\mathbf{X}^T\mathbf{M}\mathbf{X}$ and $\mathbf{X}^T\mathbf{K}\mathbf{X}$ are $\boldsymbol{\gamma}_j^T\mathbf{M}\boldsymbol{\gamma}_k$ and $\boldsymbol{\gamma}_j^T\mathbf{K}\boldsymbol{\gamma}_k$, respectively, it follows that

$$\mathbf{X}^T\mathbf{M}\mathbf{X} = \begin{bmatrix} \mu_1 & 0 & . & . & . \\ 0 & \mu_2 & . & . & . \\ . & . & . & . & . \\ . & . & . & . & . \\ . & . & . & . & \mu_n \end{bmatrix}, \quad \mathbf{X}^T\mathbf{K}\mathbf{X} = \begin{bmatrix} \kappa_1 & 0 & . & . & . \\ 0 & \kappa_2 & . & . & . \\ . & . & . & . & . \\ . & . & . & . & . \\ . & . & . & . & \kappa_n \end{bmatrix} \tag{9.51}$$

The modal matrix is said to be *orthogonal with respect to both* $\mathbf{M}$ *and* $\mathbf{K}$, but it is not simply orthogonal since $\mathbf{X}^{-1} \neq \mathbf{X}^T$.

The governing equation is now rewritten as

$$\mathbf{X}^T\mathbf{M}\mathbf{X}\ddot{\boldsymbol{\xi}} + \mathbf{X}^T\mathbf{K}\mathbf{X}\boldsymbol{\xi} = \mathbf{g}, \quad \boldsymbol{\xi} = \mathbf{X}^{-1}\boldsymbol{\gamma}, \quad \mathbf{g} = \mathbf{X}^T\mathbf{f} \tag{9.52}$$

implying the (uncoupled) modes

$$\mu_j\ddot{\xi}_j + \kappa_j\xi_j = g_j(t) \tag{9.53}$$

Suppose that $g_j(t) = g_{j0}(t)\sin\omega t$. Neglecting transients, the steady state solution for the jth mode is

$$\xi_j = \frac{g_{j0}}{\kappa_j - \omega^2\mu_j}\sin(\omega t) \tag{9.54}$$

It is evident that, if $\omega^2 \sim \omega_j^2 = \kappa_j/\mu_j$ (resonance), the response amplitude for the jth mode is much greater than for the other modes, so that the structural motion under this excitation frequency illustrates the mode. For this reason the modes can easily be animated.

EXAMPLE 9.10

For the system shown below (Figure 9.1),

(a) Find the eigenvalues
(b) Verify that the rows of $\mathbf{K} - \omega_{n1}^2\mathbf{M}$ and of $\mathbf{K} - \omega_{n2}^2\mathbf{M}$ are linearly dependent
(c) Find the eigenvectors
(d) Verify their orthogonality with respect to the mass and stiffness matrices
(e) Find the modal matrix
(f) Find the modal stiffnesses and modal dampers
(g) Find the modal masses
(h) Verify that the natural frequencies in the modal equations are the same as the system eigenvalues
(i) Find the modal coordinates
(j) Find the modal forces
(k) Find the solutions for the two modes under the conditions

$$\mathbf{f}(t) = \mathbf{0}, \quad \mathbf{x}(0) = \mathbf{0}, \quad \dot{\mathbf{x}}(0) = \dot{\mathbf{x}}_0$$

(l) Transform the modal solutions back to the physical coordinates

SOLUTION

The governing equation is

$$m\begin{bmatrix} 3 & 0 \\ 0 & 2 \end{bmatrix}\begin{Bmatrix} \ddot{x}_1 \\ \ddot{x}_2 \end{Bmatrix} + c\begin{bmatrix} 3 & -2 \\ -2 & 2 \end{bmatrix}\begin{Bmatrix} \dot{x}_1 \\ \dot{x}_2 \end{Bmatrix} + k\begin{bmatrix} 3 & -2 \\ -2 & 2 \end{bmatrix}\begin{Bmatrix} x_1 \\ x_2 \end{Bmatrix} = \begin{Bmatrix} F_1 \\ F_2 \end{Bmatrix}$$

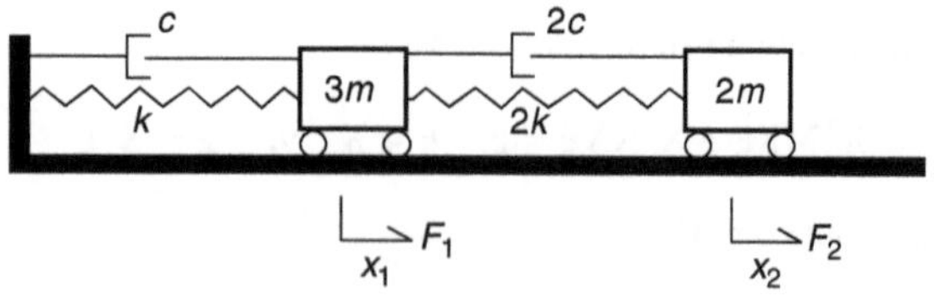

FIGURE 9.1 Two-degree-of-freedom vibrating system.

The natural frequencies are obtained from:

$$\det\left[\begin{bmatrix} 3 & -2 \\ -2 & 2 \end{bmatrix} - \frac{\omega^2}{\omega_0^2}\begin{bmatrix} 3 & 0 \\ 0 & 2 \end{bmatrix}\right] = 0, \quad \omega_0^2 = k/m$$

resulting in the eigenvalues $\omega_{n_{1,2}}^2 = \left(1 \pm \sqrt{2/3}\right)\omega_0^2$.

If the first eigenvalue is used, the two rows become $\{-3\sqrt{2/3}\ \ -2\}$ and $\{-2\ \ -2\sqrt{2/3}\}$. The second row is $\sqrt{3/2} \times$ first row, so that the two rows are linearly dependent.

The rows corresponding to the second eigenvalue are $\{3\sqrt{2/3}\ \ -2\}$ and $\{-2\ \ 2\sqrt{2/3}\}$. The second row is $-\sqrt{3/2} \times$ first row.

We now seek the eigenvectors. Using the first eigenvalue gives

$$3\left(1 - \left(1 + \sqrt{2/3}\right)\right)x_1^{(1)} - 2x_2^{(1)} = 0 \rightarrow x_2^{(1)} = -\sqrt{\tfrac{3}{2}}x_1^{(1)}$$

$$1 = \left(x_1^{(1)}\right)^2 + \left(x_2^{(1)}\right)^2 \rightarrow \mathbf{x}^{(1)} = \frac{1}{\sqrt{5}}\begin{Bmatrix} \sqrt{2} \\ -\sqrt{3} \end{Bmatrix}$$

Similarly, $\mathbf{x}^{(2)} = \frac{1}{\sqrt{5}}\begin{Bmatrix} \sqrt{2} \\ \sqrt{3} \end{Bmatrix}$.

The eigenvectors are orthogonal with respect to **M** and **K** since

$$\frac{1}{5}\left\{\sqrt{2}\sqrt{3}\right\}\begin{bmatrix} 3 & 0 \\ 0 & 2 \end{bmatrix}\begin{Bmatrix} \sqrt{2} \\ -\sqrt{3} \end{Bmatrix} = \frac{1}{5}(6 - 6) = 0$$

$$\frac{1}{5}\left\{\sqrt{2}\sqrt{3}\right\}\begin{bmatrix} 3 & -2 \\ -2 & 2 \end{bmatrix}\begin{Bmatrix} \sqrt{2} \\ -\sqrt{3} \end{Bmatrix} = \frac{1}{5}(6 - 2\sqrt{6} + 2\sqrt{6} - 6) = 0.$$

The modal matrix is $\mathbf{X} = \frac{1}{\sqrt{5}}\begin{bmatrix} \sqrt{2} & \sqrt{2} \\ \sqrt{3} & -\sqrt{3} \end{bmatrix}$. The modal masses satisfy

$$\mu_1 = m\frac{1}{5}\{\sqrt{2}\ \ \sqrt{3}\}\begin{bmatrix} 3 & 0 \\ 0 & 2 \end{bmatrix}\begin{Bmatrix} \sqrt{2} \\ \sqrt{3} \end{Bmatrix} = \frac{12}{5}m$$

$$\mu_2 = m\frac{1}{5}\{\sqrt{2}\ \ -\sqrt{3}\}\begin{bmatrix} 3 & 0 \\ 0 & 2 \end{bmatrix}\begin{Bmatrix} \sqrt{2} \\ -\sqrt{3} \end{Bmatrix} = \frac{12}{5}m$$

Similarly, the modal stiffnesses are $\kappa_1 = \frac{12-4\sqrt{6}}{5}k$ and $\kappa_2 = \frac{12+4\sqrt{6}}{5}k$, and the modal dampers are $\delta_1 = \frac{12-4\sqrt{6}}{5}c$ and $\delta_2 = \frac{12+4\sqrt{6}}{5}c$. Finally, the modal damping factors are $\zeta_{1,2} = \frac{\delta_1}{2\sqrt{\kappa_1\mu_1}}, \frac{\delta_2}{2\sqrt{\kappa_2\mu_2}}$.

The Modal Equations imply the natural frequencies

$$\frac{\kappa_1}{\mu_1} = \frac{k}{m}\frac{\frac{12-4\sqrt{6}}{5}}{\frac{12}{5}} = \frac{k}{m}\left(1 - \sqrt{2/3}\right) = \omega_{n2}^2,$$

$$\frac{\kappa_2}{\mu_2} = \frac{k}{m}\frac{\frac{12+4\sqrt{6}}{5}}{\frac{12}{5}} = \frac{k}{m}\left(1+\sqrt{2/3}\right) = \omega_{n1}^2$$

in exact agreement with the relations obtained from the original (coupled) system equation.

Since $\mathbf{X}^{-1} = \frac{\sqrt{5}}{2\sqrt{6}}\begin{bmatrix}\sqrt{3} & \sqrt{2}\\ \sqrt{3} & -\sqrt{2}\end{bmatrix}$, the modal coordinates satisfy

$$\begin{Bmatrix} y_1 \\ y_2 \end{Bmatrix} = \mathbf{X}^{-1}\begin{Bmatrix} x_1 \\ x_2 \end{Bmatrix} = \frac{\sqrt{5}}{2\sqrt{6}}\begin{Bmatrix} \sqrt{3}x_1 + \sqrt{2}x_2 \\ \sqrt{3}x_1 - \sqrt{2}x_2 \end{Bmatrix}$$

The modal forces are obtained as

$$\begin{Bmatrix} g_1 \\ g_2 \end{Bmatrix} = \mathbf{X}^{T}\begin{Bmatrix} f_1 \\ f_2 \end{Bmatrix} = \frac{1}{\sqrt{5}}\begin{Bmatrix} \sqrt{2}f_1 + \sqrt{3}f_2 \\ \sqrt{2}f_1 - \sqrt{3}f_2 \end{Bmatrix}$$

We now seek the solutions of the modal equations. Initial conditions in terms of modal coordinates are given by

$$\begin{Bmatrix} y_1(0) \\ y_2(0) \end{Bmatrix} = \begin{Bmatrix} 0 \\ 0 \end{Bmatrix} \quad \begin{Bmatrix} \dot{y}_1(0) \\ \dot{y}_2(0) \end{Bmatrix} = \frac{\sqrt{5}}{2\sqrt{6}}\begin{Bmatrix} \sqrt{3}\dot{x}_1(0) + \sqrt{2}\dot{x}_2(0) \\ \sqrt{3}\dot{x}_1(0) - \sqrt{2}\dot{x}_2(0) \end{Bmatrix}$$

and the two modal forces vanish: $\begin{Bmatrix} g_1 \\ g_2 \end{Bmatrix} = \begin{Bmatrix} 0 \\ 0 \end{Bmatrix}$. Introducing $\omega_{d1,2} = \omega_{n1,2}\sqrt{1-\xi_{1,2}^2}$, the foregoing reduces to two initial single-degree-of-freedom initial value problems with solutions

$$y_1(t) = \exp(-\zeta_1\omega_{n1}t)\frac{\dot{y}_1(0)}{\omega_{d1}}\sin(\omega_{d1}t)$$

$$y_2(t) = \exp(-\zeta_2\omega_{n2}t)\frac{\dot{y}_2(0)}{\omega_{d2}}\sin(\omega_{d2}t)$$

The solutions in physical coordinates are recovered using

$$\begin{Bmatrix} x_1 \\ x_2 \end{Bmatrix} = \mathbf{X}\begin{Bmatrix} y_1 \\ y_2 \end{Bmatrix} = \frac{1}{\sqrt{5}}\begin{Bmatrix} \sqrt{2}[y_1(t) + y_2(t)] \\ \sqrt{3}[y_1(t) - y_2(t)] \end{Bmatrix}$$

EXAMPLE 9.11

Figure 9.2 shows a clamped–clamped beam, modeled as two finite elements. Show that the finite element equations decompose into two uncoupled modes: One representing symmetric response and the other representing antisymmetric response.

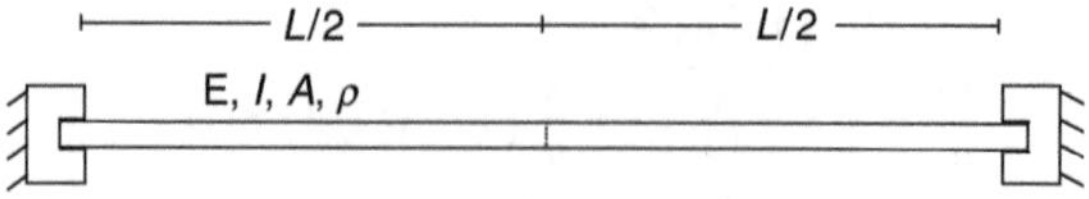

FIGURE 9.2 Clamped–clamped beam modeled as two elements.

Solution

The mass and stiffness matrices of a beam element with no constraints are given by

$$\mathbf{M}^{(e)} = \begin{bmatrix} \mathbf{M}_{11}^{(e)} & \mathbf{M}_{12}^{(e)} \\ \mathbf{M}_{12}^{T(e)} & \mathbf{M}_{22}^{(e)} \end{bmatrix}, \quad \mathbf{K}^{(e)} = \begin{bmatrix} \mathbf{K}_{11}^{(e)} & \mathbf{K}_{12}^{(e)} \\ \mathbf{K}_{12}^{T(e)} & \mathbf{K}_{22}^{(e)} \end{bmatrix}$$

$$\mathbf{M}_{11}^{(e)} = \frac{\rho A L_e}{420}\begin{bmatrix} 156 & -22L_e \\ -22L_e & 4L_e^2 \end{bmatrix}, \quad \mathbf{M}_{12}^{(e)} = \frac{\rho A L_e}{420}\begin{bmatrix} 54 & 13L_e \\ -13L_e & -3L_e^2 \end{bmatrix}, \quad \mathbf{M}_{22}^{(e)} = \frac{\rho A L_e}{420}\begin{bmatrix} 156 & 22L_e \\ 22L_e & 4L_e^2 \end{bmatrix}$$

$$\mathbf{K}_{11}^{(e)} = \frac{\mathsf{E}I}{L_e^3}\begin{bmatrix} 12 & -6L_e \\ -6L_e & 4L_e^2 \end{bmatrix}, \quad \mathbf{K}_{12}^{(e)} = \frac{\mathsf{E}I}{L_e^3}\begin{bmatrix} 12 & -6L_e \\ 6L_e & 2L_e^2 \end{bmatrix}, \quad \mathbf{K}_{22}^{(e)} = \frac{\mathsf{E}I}{L_e^3}\begin{bmatrix} 12 & 6L_e \\ 6L_e & 4L_e^2 \end{bmatrix}$$

Using $L_e = L/2$, assembling the stiffness and mass matrices, and imposing the clamped constraints results in

$$\begin{bmatrix} 192\mathsf{E}I/L^3 & 0 \\ 0 & 16\mathsf{E}I/L \end{bmatrix}\begin{Bmatrix} w_2 \\ -w_2' \end{Bmatrix} + \begin{bmatrix} 13\rho AL/35 & 0 \\ 0 & \rho AL^3/420 \end{bmatrix}\begin{Bmatrix} \ddot{w}_2 \\ -\ddot{w}_2' \end{Bmatrix} = \begin{Bmatrix} 0 \\ 0 \end{Bmatrix}$$

in which w_2 is the transverse displacement of the mid-node. This equation represents two separate single-degree-of-freedom systems, with the natural frequencies

$$\omega_{n1} = \sqrt{\frac{35 \times 192}{13}}\sqrt{\frac{\mathsf{E}}{\rho}}\sqrt{\frac{I}{A}}\frac{1}{L^2}, \quad \omega_{n2} = 16 \times 420\sqrt{\frac{\mathsf{E}}{\rho}}\sqrt{\frac{I}{A}}\frac{1}{L^2}$$

To consider the question of symmetry and antisymmetry, consider the case in which the right-hand side satisfies $M = 0$, but $V = \begin{cases} 0, & t < 0 \\ V_0, & t \geq 0 \end{cases}$. This is to say that there is a shear force imposed at $t = 0$ at the midpoint but no bending moment. If initially $(w_2(L/2,0) = \dot{w}_2(L/2,0) = 0,\ -w_2'(L/2,0) = -\dot{w}_2'(L/2,0) = 0)$, then $-w'(L/2,t) = 0$ and the deformation is symmetric with natural frequency ω_{n1}. On the other hand, if $V = 0$, but $V = \begin{cases} 0, & t < 0 \\ M_0, & t \geq 0 \end{cases}$, then $w_2(L/2,\ t) = 0$ and the deformation is antisymmetric with natural frequency ω_{n2}.

EXAMPLE 9.12

As an example of eigenvalue determination, consider

$$\mathbf{M} = \begin{bmatrix} 1 & 0 \\ 0 & 1 \end{bmatrix}, \quad \mathbf{K} = \begin{bmatrix} k_{11} & k_{12} \\ k_{12} & k_{22} \end{bmatrix}$$

Now det $[\mathbf{K} + \lambda^2\mathbf{I}] = 0$ reduces to

$$(\lambda^2)^2 + [k_{11} + k_{22}]\lambda^2 + [k_{11}k_{22} + k_{12}^2] = 0$$

with the roots

$$\lambda^2_{+,-} = \frac{1}{2}\left[-[k_{11}+k_{22}] \pm \sqrt{[k_{11}+k_{22}]^2 - 4\left[k_{11}k_{22}+k_{12}^2\right]}\right]$$
$$= \frac{1}{2}\left[-[k_{11}+k_{22}] \pm \sqrt{[k_{11}-k_{22}]^2}\right]$$

so that both λ^2_+ and λ^2_- are negative (since k_{11} and k_{22} are positive).

Returning to the general development, we now consider eigenvectors. The eigenvalue equations for the *i*th and *j*th eigenvectors are written as

$$\left[\mathbf{K}+\omega_j^2\mathbf{M}\right]\boldsymbol{\gamma}_j = \mathbf{0}, \quad \left[\mathbf{K}+\omega_k^2\mathbf{M}\right]\boldsymbol{\gamma}_k = \mathbf{0} \tag{9.55}$$

It is easily seen that the eigenvectors have arbitrary magnitudes, and for convenience we assume that they have unit magnitude: $\boldsymbol{\gamma}_j^T\boldsymbol{\gamma}_j = 1$. Simple manipulation furnishes that

$$\boldsymbol{\gamma}_k^T\mathbf{K}\boldsymbol{\gamma}_j - \boldsymbol{\gamma}_j^T\mathbf{K}\boldsymbol{\gamma}_k - \left[\omega_j^2\boldsymbol{\gamma}_k^T\mathbf{M}\boldsymbol{\gamma}_j - \omega_k^2\boldsymbol{\gamma}_j^T\mathbf{M}\boldsymbol{\gamma}_k\right] = 0 \tag{9.56}$$

Symmetry of **K** and **M** imply that

$$\boldsymbol{\gamma}_k^T\mathbf{K}\boldsymbol{\gamma}_j - \boldsymbol{\gamma}_j^T\mathbf{K}\boldsymbol{\gamma}_k = 0, \quad \left[\omega_j^2\boldsymbol{\gamma}_k^T\mathbf{M}\boldsymbol{\gamma}_j - \omega_k^2\boldsymbol{\gamma}_j^T\mathbf{M}\boldsymbol{\gamma}_k\right] = \left(\omega_j^2-\omega_k^2\right)\boldsymbol{\gamma}_k^T\mathbf{M}\boldsymbol{\gamma}_j = 0 \tag{9.57}$$

Assuming for convenience that the eigenvalues are all distinct, it follows that

$$\boldsymbol{\gamma}_j^T\mathbf{M}\boldsymbol{\gamma}_k = 0, \quad \boldsymbol{\gamma}_j^T\mathbf{K}\boldsymbol{\gamma}_k = 0, \quad j \neq k \tag{9.58}$$

The eigenvectors are thus said to be orthogonal with respect to **M** and **K**. The quantities $\mu_j = \boldsymbol{\gamma}_j^T\mathbf{M}\boldsymbol{\gamma}_j$ and $\kappa_j = \boldsymbol{\gamma}_j^T\mathbf{K}\boldsymbol{\gamma}_j$ are called the (*j*th) modal mass and (*j*th) modal stiffness.

EXAMPLE 9.13

(a) Find the modal masses μ_1 and μ_2 and the modal stiffnesses κ_1 and κ_2 of the system

$$3\begin{bmatrix}1 & 0\\ 0 & 2\end{bmatrix}\begin{pmatrix}\ddot{\gamma}_1\\ \gamma_2\end{pmatrix} + 27\begin{bmatrix}1 & -1\\ -1 & 2\end{bmatrix}\begin{pmatrix}\gamma_1\\ \gamma_2\end{pmatrix} = \begin{pmatrix}10\\ 20\end{pmatrix}\sin(10t)$$

(b) Determine the steady state response of the system (i.e., particular solution to the equation).

Solution

Consider the homogeneous equation

$$\mathbf{M}\begin{pmatrix}\ddot{\gamma}_1\\ \ddot{\gamma}_2\end{pmatrix} + \mathbf{K}\begin{pmatrix}\gamma_1\\ \gamma_2\end{pmatrix} = \mathbf{0}$$

$$\mathbf{M} = 3\begin{bmatrix} 1 & 0 \\ 0 & 2 \end{bmatrix}\begin{pmatrix} \ddot{\gamma}_1 \\ \ddot{\gamma}_2 \end{pmatrix}, \quad \mathbf{K} = 27\begin{bmatrix} 1 & -1 \\ -1 & 2 \end{bmatrix}\begin{pmatrix} \gamma_1 \\ \gamma_2 \end{pmatrix}$$

For det$[\mathbf{K} - \zeta^2\mathbf{M}]$ to vanish $(9 - \zeta_{\pm}^2) = \pm 9/\sqrt{2}$. For $(9 - \zeta_{+}^2) = +9/\sqrt{2}$, the eigenvector is obtained from

$$\begin{bmatrix} 27/\sqrt{2} & -27 \\ -27 & 54/\sqrt{2} \end{bmatrix}\begin{pmatrix} \gamma_1^{(1)} \\ \gamma_2^{(1)} \end{pmatrix} = \begin{pmatrix} 0 \\ 0 \end{pmatrix}, \quad \left[\gamma_1^{(1)}\right]^2 + \left[\gamma_2^{(1)}\right]^2 = 1$$

Simple manipulation furnishes

$$\gamma_1^{(1)} = \sqrt{2/3} \quad \text{and} \quad \gamma_2^{(1)} = \sqrt{1/3}$$

The corresponding procedure for $(9 - \zeta_{-}^2) = -9/\sqrt{2}$ furnishes

$$\gamma_1^{(2)} = \sqrt{2/3} \quad \text{and} \quad \gamma_2^{(2)} = -\sqrt{1/3}$$

The modal matrix $\mathbf{X}$ is given by

$$\mathbf{X} = \begin{bmatrix} \boldsymbol{\gamma}^{(1)} & \boldsymbol{\gamma}^{(2)} \end{bmatrix} = \begin{bmatrix} \sqrt{2/3} & \sqrt{2/3} \\ \sqrt{1/3} & -\sqrt{1/3} \end{bmatrix}$$

Hence, the modal masses are given by

$$\mathbf{X}^T\mathbf{M}\mathbf{X} = \begin{bmatrix} \sqrt{2/3} & \sqrt{1/3} \\ \sqrt{2/3} & -\sqrt{1/3} \end{bmatrix} 3\begin{bmatrix} 1 & 0 \\ 0 & 2 \end{bmatrix}\begin{bmatrix} \sqrt{2/3} & \sqrt{2/3} \\ \sqrt{1/3} & -\sqrt{1/3} \end{bmatrix} = \begin{bmatrix} 4 & 0 \\ 0 & 4 \end{bmatrix}$$

so that $\mu_1 = \mu_2 = 4$. The modal stiffnesses are obtained using

$$\mathbf{X}^T\mathbf{K}\mathbf{X} = \begin{bmatrix} \sqrt{2/3} & \sqrt{1/3} \\ \sqrt{2/3} & -\sqrt{1/3} \end{bmatrix} 27\begin{bmatrix} 1 & -1 \\ -1 & 2 \end{bmatrix}\begin{bmatrix} \sqrt{2/3} & \sqrt{2/3} \\ \sqrt{1/3} & -\sqrt{1/3} \end{bmatrix}$$
$$= \begin{bmatrix} 36 - 18\sqrt{2} & 0 \\ 0 & 36 + 18\sqrt{2} \end{bmatrix}$$

and so $\kappa_1 = 36 - 18\sqrt{2}$ and $\kappa_2 = 36 + 18\sqrt{2}$. The steady state response of the system satisfies

$$\xi_j = \frac{g_{j0}}{\kappa_j - \omega^2\mu_j}\sin(\omega t)$$

Since $\mathbf{g}_0 \sin(\omega t) = \mathbf{g}(t) = \begin{pmatrix} 10 \\ 20 \end{pmatrix}\sin(10t)$, simple substitution furnishes

$$\xi_1 = \frac{10\sin(10t)}{36 - 18\sqrt{2} - 100(4)} \quad \text{and} \quad \xi_2 = \frac{20\sin(10t)}{36 + 18\sqrt{2} - 100(4)}$$

from which the steady state solution results as

$$\xi(t) = \begin{pmatrix} \dfrac{1}{182 - 9\sqrt{2}} \\ \dfrac{1}{182 + 9\sqrt{2}} \end{pmatrix} 5\sin(10t)$$

EXAMPLE 9.14

Express the following equations in state form, apply the trapezoidal rule, and triangularize the ensuing dynamic stiffness matrix:

$$\mathbf{M}\ddot{\boldsymbol{\gamma}} + \mathbf{K}\boldsymbol{\gamma} - \boldsymbol{\Sigma}\boldsymbol{\pi} = \mathbf{f}, \quad \boldsymbol{\Sigma}^T\boldsymbol{\gamma} = \mathbf{0}$$

(Equations in this form will be seen to arise in finite element models of incompressible elastic bodies.)

Solution

The foregoing equation is expressed in state form as follows:

$$\begin{bmatrix} \mathbf{M} & \mathbf{0} & \mathbf{0} \\ \mathbf{0}^T & \mathbf{K} & \mathbf{0} \\ \mathbf{0}^T & \mathbf{0}^T & \mathbf{0} \end{bmatrix} \begin{pmatrix} \dot{\boldsymbol{\gamma}}(t) \\ \boldsymbol{\gamma}(t) \\ \boldsymbol{\pi}(t) \end{pmatrix}^{\bullet} + \begin{bmatrix} \mathbf{M} & \mathbf{K} & -\boldsymbol{\Sigma} \\ -\mathbf{K} & \mathbf{0} & \mathbf{0} \\ \mathbf{0}^T & \boldsymbol{\Sigma}^T & \mathbf{0} \end{bmatrix} \begin{pmatrix} \dot{\boldsymbol{\gamma}}(t) \\ \boldsymbol{\gamma}(t) \\ \boldsymbol{\pi}(t) \end{pmatrix} = \begin{pmatrix} \mathbf{f}(t) \\ \mathbf{0} \\ \mathbf{0} \end{pmatrix}$$

This expression is rewritten using the trapezoidal rule and $\mathbf{p} = \dot{\boldsymbol{\gamma}}$ as

$$\begin{bmatrix} \mathbf{M} & \mathbf{0} & \mathbf{0} \\ \mathbf{0}^T & \mathbf{K} & \mathbf{0} \\ \mathbf{0}^T & \mathbf{0}^T & \mathbf{0} \end{bmatrix} \begin{pmatrix} \frac{1}{h}(\mathbf{p}_{n+1} - \mathbf{p}_n) \\ \frac{1}{h}(\boldsymbol{\gamma}_{n+1} - \boldsymbol{\gamma}_n) \\ \frac{1}{h}(\boldsymbol{\pi}_{n+1} - \boldsymbol{\pi}_n) \end{pmatrix} + \begin{bmatrix} \mathbf{M} & \mathbf{K} & -\boldsymbol{\Sigma} \\ -\mathbf{K} & \mathbf{0} & \mathbf{0} \\ \mathbf{0}^T & \boldsymbol{\Sigma}^T & \mathbf{0} \end{bmatrix} \begin{pmatrix} \frac{1}{2}(\mathbf{p}_{n+1} + \mathbf{p}_n) \\ \frac{1}{2}(\boldsymbol{\gamma}_{n+1} + \boldsymbol{\gamma}_n) \\ \frac{1}{2}(\boldsymbol{\pi}_{n+1} + \boldsymbol{\pi}_n) \end{pmatrix}$$

$$= \begin{pmatrix} \frac{1}{2}(\mathbf{f}_{n+1} + \mathbf{f}_n) \\ \mathbf{0} \\ \mathbf{0} \end{pmatrix}$$

The second row implies that $\mathbf{p}_{n+1} = \frac{2}{h}(\boldsymbol{\gamma}_{n+1} - \boldsymbol{\gamma}_n) - \mathbf{p}_n$, enabling the first row can be rewritten as

$$\frac{1}{h}\mathbf{M}\left[\frac{2}{h}(\boldsymbol{\gamma}_{n+1} - \boldsymbol{\gamma}_n) - 2\mathbf{p}_n\right] + \frac{1}{2}\mathbf{K}(\boldsymbol{\gamma}_{n+1} + \boldsymbol{\gamma}_n) - \frac{1}{2}\boldsymbol{\Sigma}(\boldsymbol{\pi}_{n+1} + \boldsymbol{\pi}_n) = \frac{1}{2}(\mathbf{f}_{n+1} + \mathbf{f}_n)$$

Multiplying throughout by $h^2/2$ and rearranging gives

$$\left[\mathbf{M} + \frac{h^2}{4}\mathbf{K}\right]\boldsymbol{\gamma}_{n+1} - \frac{h^2}{4}\boldsymbol{\Sigma}\boldsymbol{\pi}_{n+1} = \frac{h^2}{4}[\mathbf{f}_{n+1} + \mathbf{f}_n - \mathbf{K}\boldsymbol{\gamma}_n + \boldsymbol{\Sigma}\boldsymbol{\pi}_n] + \mathbf{M}[\boldsymbol{\gamma}_n + h\mathbf{p}_n] \quad (9.59)$$

The third row, after multiplying by $h^2/2$, is now

$$\frac{h^2}{4}\boldsymbol{\Sigma}^T\boldsymbol{\gamma}_{n+1} = -\frac{h^2}{4}\boldsymbol{\Sigma}^T\boldsymbol{\gamma}_n \tag{9.60}$$

Equations 9.59 and 9.60 are written in matrix–vector notation as

$$\begin{bmatrix} \mathbf{M}+\frac{h^2}{4}\mathbf{K} & -\frac{h^2}{4}\boldsymbol{\Sigma} \\ \frac{h^2}{4}\boldsymbol{\Sigma}^T & \mathbf{0} \end{bmatrix}\begin{pmatrix} \boldsymbol{\gamma}_{n+1} \\ \boldsymbol{\pi}_{n+1} \end{pmatrix} = \begin{pmatrix} \mathbf{g}_{n+1} \\ -\frac{h^2}{4}\boldsymbol{\Sigma}^T\boldsymbol{\gamma}_n \end{pmatrix}$$

in which

$$\mathbf{g}_{n+1} = \frac{h^2}{4}[\mathbf{f}_{n+1}+\mathbf{f}_n-\mathbf{K}\boldsymbol{\gamma}_n+\boldsymbol{\Sigma}\boldsymbol{\pi}_n]+\mathbf{M}[\boldsymbol{\gamma}_n+h\mathbf{p}_n]$$

The *dynamic stiffness matrix* emerges as

$$\mathbf{K}_D = \begin{bmatrix} \mathbf{M}+\frac{h^2}{4}\mathbf{K} & -\frac{h^2}{4}\boldsymbol{\Sigma} \\ \frac{h^2}{4}\boldsymbol{\Sigma}^T & \mathbf{0} \end{bmatrix}$$

Now $\mathbf{K}_D$ is decomposed into a product of a lower triangular and an upper triangular matrix $\mathbf{K}_D = \mathbf{LU}$ as

$$\begin{bmatrix} \mathbf{M}+\frac{h^2}{4}\mathbf{K} & -\frac{h^2}{4}\boldsymbol{\Sigma} \\ \frac{h^2}{4}\boldsymbol{\Sigma}^T & \mathbf{0} \end{bmatrix} = \begin{bmatrix} \mathbf{L}_{11} & \mathbf{0} \\ \mathbf{L}_{21} & \mathbf{L}_{22} \end{bmatrix}\begin{bmatrix} \mathbf{U}_{11} & \mathbf{U}_{12} \\ \mathbf{0} & \mathbf{U}_{22} \end{bmatrix}$$

On setting $\mathbf{U}_{11} = \mathbf{L}_{11}^T$ and $\mathbf{L}_{22} = \mathbf{I}$, we find that

$$\mathbf{L}_{11}\mathbf{L}_{11}^T = \mathbf{M}+\frac{h^2}{4}\mathbf{K}$$

which can be triangularized to find $\mathbf{L}_{11}$ since, $\mathbf{M}$ and $\mathbf{K}$ are positive definite. In addition

$$\mathbf{U}_{12} = -\frac{h^2}{4}\mathbf{L}_{11}^{-1}\boldsymbol{\Sigma}, \quad \mathbf{L}_{21} = \frac{h^2}{4}\boldsymbol{\Sigma}^T\mathbf{L}_{11}^{-T}, \quad \mathbf{U}_{22} = -\mathbf{L}_{21}\mathbf{U}_{12} = \frac{h^4}{16}\boldsymbol{\Sigma}^T\mathbf{L}_{11}^{-T}\mathbf{L}_{11}^{-1}\boldsymbol{\Sigma}$$

and finally

$$\mathbf{K}_D = \mathbf{LU} = \begin{bmatrix} \mathbf{L}_{11} & \mathbf{0} \\ \frac{h^2}{4}\boldsymbol{\Sigma}^T\mathbf{L}_{11}^{-T} & \mathbf{I} \end{bmatrix}\begin{bmatrix} \mathbf{L}_{11}^T & -\frac{h^2}{4}\mathbf{L}_{11}^{-1}\boldsymbol{\Sigma} \\ \mathbf{0}^T & \frac{h^4}{16}\boldsymbol{\Sigma}^T\mathbf{L}_{11}^{-T}\mathbf{L}_{11}^{-1}\boldsymbol{\Sigma} \end{bmatrix}$$

EXAMPLE 9.15

Derive general expressions for the modal decomposition of the two-degree-of-freedom system

$$\begin{bmatrix} k_{11} & k_{12} \\ k_{12} & k_{22} \end{bmatrix}\begin{Bmatrix} x_1 \\ x_2 \end{Bmatrix} + \begin{bmatrix} m_{11} & m_{12} \\ m_{12} & m_{22} \end{bmatrix}\begin{Bmatrix} \ddot{x}_1 \\ \ddot{x}_2 \end{Bmatrix} = \begin{Bmatrix} 0 \\ 0 \end{Bmatrix}$$

Solution

(a) *Eigenvalues*

$$\begin{aligned} 0 &= \det\begin{bmatrix} k_{11}-\omega^2 m_{11} & k_{12}-\omega^2 m_{12} \\ k_{12}-\omega^2 m_{12} & k_{22}-\omega^2 m_{22} \end{bmatrix} \\ &= \left(k_{11}-\omega^2 m_{11}\right)\left(k_{22}-\omega^2 m_{22}\right) - \left(k_{12}-\omega^2 m_{12}\right)\left(k_{12}-\omega^2 m_{12}\right) \\ &= \left(k_{11}k_{22}-k_{12}^2\right) - \omega^2(m_{11}k_{22}+k_{11}m_{22}-2k_{12}m_{12}) + \left(\omega^2\right)^2\left(m_{11}m_{22}-m_{12}^2\right) \end{aligned}$$

Letting

$$\Delta_{kk} = \left(k_{11}k_{22}-k_{12}^2\right),\ \Delta_{km} = (m_{11}k_{22}+k_{11}m_{22}-2k_{12}m_{12}),\ \Delta_{mm} = \left(m_{11}m_{22}-m_{12}^2\right)$$

the eigenvalues are obtained as

$$\left(\omega^2\right)_{1.2} = \frac{1}{2\Delta_m}\Delta_{mk} \pm \sqrt{(\Delta_{mk})^2 - 4\Delta_k}$$

We know from other considerations that $(\Delta_{mk})^2 - 4\Delta_k > 0$.

(b) *Eigenvectors*

The eigenvector paired with the eigenvalue ω_j satisfies

$$\left(k_{11}-\omega_j^2 m_{11}\right)x_1^{(j)} + \left(k_{12}-\omega_j^2 m_{12}\right)x_2^{(j)} = 0,\quad \left(x_1^{(j)}\right)^2 + \left(x_2^{(j)}\right)^2 = 1$$

from which

$$x_1^{(j)} = \frac{1}{\sqrt{1+\left(\dfrac{k_{11}-\omega_j^2 m_{11}}{k_{12}-\omega_j^2 m_{12}}\right)^2}},\quad x_2^{(j)} = \frac{-\dfrac{k_{11}-\omega_j^2 m_{11}}{k_{12}-\omega_j^2 m_{12}}}{\sqrt{1+\left(\dfrac{k_{11}-\omega_j^2 m_{11}}{k_{12}-\omega_j^2 m_{12}}\right)^2}}$$

(c) *Modal matrix*

$$\mathbf{X} = \begin{bmatrix} x_{11} & x_{12} \\ x_{21} & x_{22} \end{bmatrix} = \begin{bmatrix} \dfrac{1}{\sqrt{1+\left(\frac{k_{11}-\omega_1^2 m_{11}}{k_{12}-\omega_1^2 m_{12}}\right)^2}} & \dfrac{1}{\sqrt{1+\left(\frac{k_{11}-\omega_2^2 m_{11}}{k_{12}-\omega_2^2 m_{12}}\right)^2}} \\ \dfrac{-\frac{k_{11}-\omega_1^2 m_{11}}{k_{12}-\omega_1^2 m_{12}}}{\sqrt{1+\left(\frac{k_{11}-\omega_1^2 m_{11}}{k_{12}-\omega_1^2 m_{12}}\right)^2}} & \dfrac{-\frac{k_{11}-\omega_2^2 m_{11}}{k_{12}-\omega_2^2 m_{12}}}{\sqrt{1+\left(\frac{k_{11}-\omega_2^2 m_{11}}{k_{12}-\omega_2^2 m_{12}}\right)^2}} \end{bmatrix}$$

(d) *Modal masses and stiffnesses*

$$\mathbf{X}^T\mathbf{M}\mathbf{X} = \begin{bmatrix} \mu_1 & 0 \\ 0 & \mu_2 \end{bmatrix}, \quad \mathbf{X}^T\mathbf{K}\mathbf{X} = \begin{bmatrix} \kappa_1 & 0 \\ 0 & \kappa_2 \end{bmatrix}$$

$$\mu_1 = x_{11}^2 m_{11} + 2x_{11}x_{21}m_{12} + x_{21}^2 m_{22}, \qquad \kappa_1 = x_{11}^2 k_{11} + 2x_{11}x_{21}k_{12} + x_{21}^2 k_{22}$$
$$\mu_2 = x_{12}^2 m_{11} + 2x_{12}x_{22}m_{12} + x_{22}^2 m_{22}, \qquad \kappa_2 = x_{12}^2 k_{11} + 2x_{12}x_{22}m_{12} + x_{22}^2 m_{22}$$

(e) *Modal coordinates*

The modal coordinate vector $\mathbf{y}$ is obtained from the physical coordinate vector $\mathbf{x}$ using

$$\begin{Bmatrix} y_1 \\ y_2 \end{Bmatrix} = \begin{bmatrix} x_{11} & x_{12} \\ x_{21} & x_{22} \end{bmatrix}^{-1} \begin{Bmatrix} x_1 \\ x_2 \end{Bmatrix}$$

Carrying out the manipulations gives

$$\begin{Bmatrix} x_1 \\ x_2 \end{Bmatrix} = \frac{\begin{bmatrix} x_{22} & -x_{12} \\ -x_{21} & x_{11} \end{bmatrix}}{x_{11}x_{22} - x_{21}x_{12}} \begin{Bmatrix} y_1 \\ y_2 \end{Bmatrix}$$
$$= \frac{1}{x_{11}x_{22} - x_{21}x_{12}} \begin{Bmatrix} x_{22}y_1 - x_{12}y_2 \\ -x_{21}y_1 + x_{11}y_2 \end{Bmatrix}$$

(f) *Modal force and physical force*

The modal force vector $\mathbf{g}$ is now obtained from the physical force vector $\mathbf{f}$:

$$\begin{Bmatrix} g_1 \\ g_2 \end{Bmatrix} = \mathbf{X}^T \begin{Bmatrix} f_1 \\ f_2 \end{Bmatrix}$$
$$= \begin{Bmatrix} x_{11}f_1 + x_{21}f_2 \\ x_{12}f_1 + x_{22}f_2 \end{Bmatrix}$$

(g) *Modes*

Using the foregoing relations, the system is expressed as modes, that is to say, as two uncoupled second-order single-degree-of-freedom systems.

$$\mu_1\ddot{y}_1 + \kappa_1 y_1 = g_1(t)$$
$$\mu_2\ddot{y}_2 + \kappa_2 y_2 = g_2(t)$$

EXAMPLE 9.16

Consider the effect of the mesh on the eigenvalues of the simple system illustrated (Figure 9.3).

SOLUTION

Formulating and assembling the stiffness and mass matrices and imposing the constraints at the two ends results in a single-degree-of-freedom system governed by

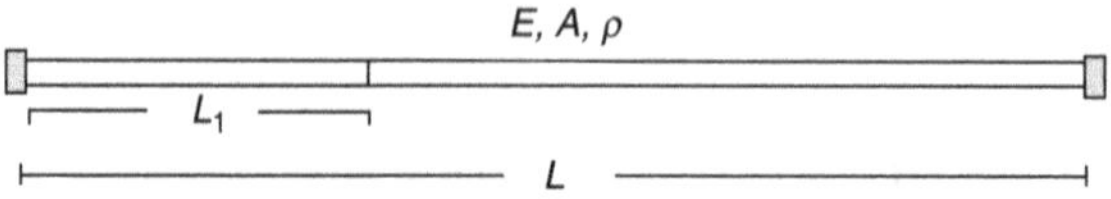

FIGURE 9.3 Two-element model of a clamped–clamped rod.

$$\mathrm{E}A\left(\frac{1}{L_1}+\frac{1}{L-L_1}\right)u(L_1,t)+\frac{\rho AL}{3}\ddot{u}(L_1,t)=0$$

The eigenvalue is given by

$$\omega_n=\frac{\sqrt{3}}{\sqrt{\alpha(1-\alpha)}}\sqrt{\frac{\mathrm{E}}{\rho}\frac{1}{L}},\quad \alpha=\frac{L_1}{L}$$

If $\alpha=1/2$, $\omega_n=\frac{2\sqrt{3}}{L}\sqrt{\frac{\mathrm{E}}{\rho}}=\frac{3.42}{L}\sqrt{\frac{\mathrm{E}}{\rho}}$. However, If $\alpha=1/4$, $\omega_n=\frac{4}{L}\sqrt{\frac{\mathrm{E}}{\rho}}$. These two results are significantly different from each other and from the exact solution, which is $\omega_n=\frac{\pi}{L}\sqrt{\frac{\mathrm{E}}{\rho}}$.

However, accurate results may be obtained for this minimum eigenvalue by using a sufficient number of elements.

9.5.2 Comments on Eigenstructure Computation in Large Finite Element Systems

Jacobi and Subspace Iterations (cf. Bathe, 1996) are prominent among the many methods which have been formulated to compute the eigenvalues and eigenvectors of a large finite element system. The first finds all eigenvalues of the system, and then finds the eigenvectors using the eigenvalues. The second selectively finds the lowest eigenvalues and corresponding eigenvectors.

One difficulty with using the finite element method for eigenstructures is that the mesh makes a contribution to the values obtained. This is a relatively minor problem for the lowest modes, but becomes progressively more bothersome as the mode number increases.

In Chapter 10, we describe an easily visualized method, which we call the *hypercircle* method (Nicholson and Lin, 2006). It represents a kind of steepest descent optimization method. For present purposes, we describe the minimum properties of eigenvalues and eigenvectors. For finite element equation of a linear elastic system $\mathbf{M}\ddot{\boldsymbol{\gamma}}+\mathbf{K}\boldsymbol{\gamma}=\mathbf{0}$ with positive definite symmetric matrices $\mathbf{K}$ and $\mathbf{M}$, the minimum eigenvalue and corresponding eigenvector (called an eigenpair) satisfy the Rayleigh quotient relation

$$\lambda_n=\min_j\lambda_j=\frac{\mathbf{x}_n^T\mathbf{K}\mathbf{x}_n}{\mathbf{x}_n^T\mathbf{M}\mathbf{x}_n}=\min_{\mathbf{x}^T\mathbf{x}=1}\frac{\mathbf{x}^T\mathbf{K}\mathbf{x}}{\mathbf{x}^T\mathbf{M}\mathbf{x}} \tag{9.61}$$

Since the vector iterates in the minimization process are constrained to have unit magnitude, all such vectors are related to each other through a proper orthogonal matrix. If $\mathbf{x}^{(0)}$ is the initial iterate, then the νth iterate may be expressed as $\mathbf{x}^{(\nu)} = \mathbf{Q}_\nu \mathbf{x}^{(0)}$. It is readily seen that this reduces the iteration process to determining a sequence of rotation matrices which progressively reduce the magnitude of the Rayleigh quotient.

However, once an eigenvalue and an eigenvector have been determined, the matrices must be altered in such a way that the subsequent minimization steps do not return to the same pair. If the eigenpair is actually removed the process is called *deflation*. In contrast in Chapter 10, a *replacement* process is described in which the lowest eigenvalue is replaced with a higher value without altering the eigenvectors.

10 Solution Methods for Linear Problems: II

10.1 INTRODUCTION

Chapter 9 presented a number of conventional numerical methods in the linear FEA. This chapter introduces three comparatively advanced topics. The first is solution on an inverse problem, while the second addresses use of fourth-order time integration method, and the third presents an optimization-based method for computing eigenvalues and eigenvectors.

10.2 SOLUTION METHOD FOR AN INVERSE PROBLEM

10.2.1 INVERSE PROBLEM IN ELASTICITY

Many practical applications involving response of elastic bodies give rise to *inverse problems*. In FEA, for a given mesh and set of physical properties, even though a well-posed *direct* problem generally possesses a unique solution in classical linear elasticity, a corresponding inverse problem (based on the same stiffness matrix) may not. Furthermore, even when the inverse problem possesses a unique solution when modeled "exactly" using the classical linear theory of elasticity, an unfortunate choice of a mesh may cause the finite element version of the inverse problem to fail to do so. The current formulation addresses a particular example of an inverse problem and exploits a matrix *nonsingularity criterion* for assuring that the finite element model possesses a unique solution. A numerical test is applied to verify satisfaction of the criterion. The test is based on the linear independence of the rows of a nonsingular matrix. If the nonsingularity condition is violated, the mesh can be modified and the nonsingularity condition applied again.

In a linearly elastic body which has known physical properties and which is experiencing small strains under statically applied loads, the finite element equation may be written as

$$\mathbf{K}\begin{Bmatrix}\mathbf{u}_1\\ \mathbf{u}_2\\ \mathbf{u}_i\end{Bmatrix}=\begin{Bmatrix}\mathbf{f}_1\\ \mathbf{f}_2\\ \mathbf{0}\end{Bmatrix},\quad \mathbf{K}=\begin{bmatrix}\mathbf{K}_{11} & \mathbf{K}_{12} & \mathbf{K}_{13}\\ \mathbf{K}_{21} & \mathbf{K}_{22} & \mathbf{K}_{23}\\ \mathbf{K}_{31} & \mathbf{K}_{32} & \mathbf{K}_{33}\end{bmatrix} \tag{10.1}$$

Here the $n_i \times 1$ vector $\mathbf{u}_i$ denotes the displacement degrees of freedom at *interior nodes*, while the $n_1 \times 1$ vector and the $n_2 \times 1$ vector $\mathbf{u}_2$ denote displacement degrees of freedom at two different sets of *boundary nodes*. It is assumed that external forces are applied only to boundary nodes, with the $n_1 \times 1$ force vector $\mathbf{f}_1$ and the $n_2 \times 1$ force vector $\mathbf{f}_2$ referred to the two sets. All matrices and vectors are real. Furthermore, the finite element stiffness matrix $\mathbf{K}$ is positive definite, written $\mathbf{K} > 0$.

The system has n degrees of freedom in which $n = n_1 + n_2 + n_i$. It is assumed that Equation 10.1 reflects degrees of freedom remaining after any *simple constraints* on the body have been applied. Here, in a simple constraint a displacement degree of freedom is specified and the corresponding *reaction force* is an unknown. (Direct problems only exhibit simple constraints.) This contrasts with *complex* (overspecified) *constraints* appearing in inverse problems in which *both* the displacement *and* force are prescribed for a degree of freedom.

In finite element models of static problems in linear elasticity, the $n \times n$ stiffness matrix $\mathbf{K}$ is symmetric and positive definite (after simple constraints have been enforced). In the direct problem, which is considered "well posed," the force vectors $\mathbf{f}_1$ and $\mathbf{f}_2$ are prescribed, and the corresponding displacement vectors $\mathbf{u}_1$ and $\mathbf{u}_2$ are unknowns to be determined. Accordingly, at each boundary degree of freedom only one quantity (the force) is specified. Positive definiteness of $\mathbf{K}$ implies that the solution of the direct problem *exists and is unique*.

In contrast, in the particular type of inverse problem (e.g., Dennis et al., 2004) of interest here, the first node set is "overspecified" in that displacements and tractions are specified at the *same* nodal degrees of freedom (complex constraints). Correspondingly, the second node set is "underspecified" in that *neither* displacements nor tractions are specified at the degrees of freedom. *As shown subsequently there is no assurance that a unique solution exists for the finite element inverse problem even when it does in the direct problem.*

10.2.2 Existence of a Unique Solution

We first employ an example to demonstrate that in inverse problems a unique solution may not exist even though the solution of the corresponding direct problem does.

EXAMPLE 10.1

Let $\mathbf{K}_{11}$ and $\mathbf{K}_{22}$ be two $n \times n$ positive definite symmetric matrices, and let $\mathbf{K}_{12}$ denote a *singular* $n \times n$ matrix. Next, introduce the matrix $\mathbf{H}$ given by

$$\mathbf{H} = \begin{bmatrix} \mathbf{K}_{11} & \sqrt{\mathbf{K}_{11}}\mathbf{K}_{12} \\ \mathbf{K}_{12}^T\sqrt{\mathbf{K}_{11}} & \mathbf{K}_{12}^T\mathbf{K}_{12} + \mathbf{K}_{22} \end{bmatrix} = \begin{bmatrix} \sqrt{\mathbf{K}_{11}} & \mathbf{0} \\ \mathbf{K}_{12}^T & \sqrt{\mathbf{K}_{22}} \end{bmatrix} \begin{bmatrix} \sqrt{\mathbf{K}_{11}} & \mathbf{K}_{12} \\ \mathbf{0}^T & \sqrt{\mathbf{K}_{22}} \end{bmatrix} \tag{10.2}$$

Of course, under the stated conditions $\mathbf{H}$ is positive definite and symmetric, written as $\mathbf{H} > 0$.

Direct Problem: Letting $\begin{Bmatrix} \mathbf{f}_{k1} \\ \mathbf{f}_{k2} \end{Bmatrix}$ be a *known* $2n \times 2n$ vector, positive definiteness of $\mathbf{H}$ implies that there exists a unique solution for the unknown vector $\begin{Bmatrix} \mathbf{u}_{u1} \\ \mathbf{u}_{u2} \end{Bmatrix}$ satisfying $\mathbf{H}\begin{Bmatrix} \mathbf{u}_{u1} \\ \mathbf{u}_{u2} \end{Bmatrix} = \begin{Bmatrix} \mathbf{f}_{k1} \\ \mathbf{f}_{k2} \end{Bmatrix}$.

Inverse Problem: Now suppose that the right-hand side contains the vector is $\begin{Bmatrix} \mathbf{f}_{k1} \\ \mathbf{f}_{u2} \end{Bmatrix}$ in which $\mathbf{f}_{k1}$ is known but $\mathbf{f}_{u2}$ is unknown. Also, suppose that the left-hand side now contains the vector $\begin{Bmatrix} \mathbf{u}_{k1} \\ \mathbf{u}_{u2} \end{Bmatrix}$ in which $\mathbf{u}_{k1}$ is known but $\mathbf{u}_{u2}$ is unknown. We are now confronted with the inverse problem

$$\begin{bmatrix} \mathbf{K}_{11} & \sqrt{\mathbf{K}}_{11}\mathbf{K}_{12} \\ \mathbf{K}_{12}^T\sqrt{\mathbf{K}}_{11} & \mathbf{K}_{12}^T\mathbf{K}_{12} + \mathbf{K}_{22} \end{bmatrix} \begin{Bmatrix} \mathbf{u}_{k1} \\ \mathbf{u}_{u2} \end{Bmatrix} = \begin{Bmatrix} \mathbf{f}_{k1} \\ \mathbf{f}_{u2} \end{Bmatrix} \tag{10.3}$$

The upper block row implies that

$$\sqrt{\mathbf{K}}_{11}\mathbf{K}_{12}\mathbf{u}_{u2} = \mathbf{f}_{k1} - \mathbf{K}_{11}\mathbf{u}_{k1} \tag{10.4}$$

But $\mathbf{K}_{12}$ and hence $\sqrt{\mathbf{K}}_{11}\mathbf{K}_{12}$ are singular. Hence, either there is no unique solution for $\mathbf{u}_{u2}$ or there are many solutions. The second block row does not mitigate this difficulty since it introduces the additional unknown vector $\mathbf{f}_{u2}$. Again the inverse problem involving $\mathbf{H}$ does not possess a unique solution although the direct problem does.

Returning to the general development, a sufficient condition is now presented for the existence and uniqueness of the solution to a finite element model of the inverse problem of interest. However, as will be demonstrated, it is quite possible that the stiffness matrix in one mesh will not satisfy the sufficient condition, while the stiffness matrix in another mesh for the same physical problem will do so.

From the third row in Equation 10.1, we have

$$\mathbf{K}_{31}\mathbf{u}_1 + \mathbf{K}_{32}\mathbf{u}_2 + \mathbf{K}_{33}\mathbf{u}_i = 0 \tag{10.5}$$

and, recalling that $\mathbf{K}_{33} > 0$,

$$\mathbf{u}_3 = -\mathbf{K}_{33}^{-1}(\mathbf{K}_{31}\mathbf{u}_1 + \mathbf{K}_{32}\mathbf{u}_2) \tag{10.6}$$

Upon substitution, the upper two rows provide the equations

$$\left[\mathbf{K}_{11} - \mathbf{K}_{13}\mathbf{K}_{33}^{-1}\mathbf{K}_{31}\right]\mathbf{u}_1 + \left[\mathbf{K}_{12} - \mathbf{K}_{13}\mathbf{K}_{33}^{-1}\mathbf{K}_{32}\right]\mathbf{u}_2 = \mathbf{f}_1 \tag{10.7}$$

$$\left[\mathbf{K}_{21} - \mathbf{K}_{23}\mathbf{K}_{33}^{-1}\mathbf{K}_{31}\right]\mathbf{u}_1 + \left[\mathbf{K}_{22} - \mathbf{K}_{23}\mathbf{K}_{33}^{-1}\mathbf{K}_{32}\right]\mathbf{u}_2 = \mathbf{f}_2 \tag{10.8}$$

Positive definiteness of $\mathbf{K}$ implies that its principal minors are positive definite. It follows that

$$\mathbf{K}_{11} - \mathbf{K}_{13}\mathbf{K}_{33}^{-1}\mathbf{K}_{31} > 0$$
$$\mathbf{K}_{22} - \mathbf{K}_{23}\mathbf{K}_{33}^{-1}\mathbf{K}_{32} > 0$$

and

$$\begin{bmatrix} \mathbf{K}_{11} - \mathbf{K}_{13}\mathbf{K}_{33}^{-1}\mathbf{K}_{31} & \mathbf{K}_{12} - \mathbf{K}_{13}\mathbf{K}_{33}^{-1}\mathbf{K}_{32} \\ \mathbf{K}_{11} - \mathbf{K}_{13}\mathbf{K}_{33}^{-1}\mathbf{K}_{31} & \mathbf{K}_{22} - \mathbf{K}_{23}\mathbf{K}_{33}^{-1}\mathbf{K}_{32} \end{bmatrix} > 0 \tag{10.9}$$

Now suppose that $\mathbf{u}_1 = \mathbf{u}_k$ and $\mathbf{f}_1 = \mathbf{f}_k$ are known, while $\mathbf{u}_2 = \mathbf{u}_u$ and $\mathbf{f}_2 = \mathbf{f}_u$ are unknown, thereby introducing an *inverse problem* of the type of interest here. Equations 10.7 and 10.8 are now rewritten as

$$\left[\mathbf{K}_{11} - \mathbf{K}_{13}\mathbf{K}_{33}^{-1}\mathbf{K}_{31}\right]\mathbf{u}_k + \left[\mathbf{K}_{12} - \mathbf{K}_{13}\mathbf{K}_{33}^{-1}\mathbf{K}_{32}\right]\mathbf{u}_u = \mathbf{f}_k \tag{10.10}$$

$$\left[\mathbf{K}_{21} - \mathbf{K}_{23}\mathbf{K}_{33}^{-1}\mathbf{K}_{31}\right]\mathbf{u}_k + \left[\mathbf{K}_{22} - \mathbf{K}_{23}\mathbf{K}_{33}^{-1}\mathbf{K}_{32}\right]\mathbf{u}_u = \mathbf{f}_u \tag{10.11}$$

Equation 10.10 immediately furnishes

$$\left[\mathbf{K}_{12} - \mathbf{K}_{13}\mathbf{K}_{33}^{-1}\mathbf{K}_{32}\right]\mathbf{u}_u = \mathbf{f}_k - \left[\mathbf{K}_{11} - \mathbf{K}_{13}\mathbf{K}_{33}^{-1}\mathbf{K}_{31}\right]\mathbf{u}_k \tag{10.12}$$

For the solution of the inverse problem expressed by Equations 10.10 and 10.11 to exist and be unique it is necessary and sufficient that $\mathbf{K}_{12} - \mathbf{K}_{13}\mathbf{K}_{33}^{-1}\mathbf{K}_{32}$ be nonsingular, regardless of $\mathbf{u}_k$ and $\mathbf{f}_k$. Once $\mathbf{u}_u$ is obtained by solving Equation 10.12, $\mathbf{f}_u$ is immediately found from Equation 10.11.

EXAMPLE 10.2

Inverse problems in two-element cantilevered beams

Figure 10.1 depicts a cantilevered beam modeled by two elements. The elastic modulus E and the bending moment of area I are the same in the two elements, but their lengths L_1 and L_2 differ. The (unclamped) nodes are denoted as 1 and 2.

The vertical (z) displacement and slope are denoted by w, $-w'$, the shear force by V, and the bending moment by M. The finite element equation for the two-element beam configuration is given by

$$\mathrm{E}I\begin{bmatrix} \begin{bmatrix} \left(\frac{12}{L_1^3}+\frac{12}{L_2^3}\right) & \left(\frac{6}{L_1^2}-\frac{6}{L_2^2}\right) \\ \left(\frac{6}{L_1^2}-\frac{6}{L_2^2}\right) & \left(\frac{4}{L_1}+\frac{4}{L_2}\right) \end{bmatrix} & \begin{bmatrix} -\frac{12}{L_2^3} & -\frac{6}{L_2^2} \\ \frac{6}{L_2^2} & \frac{4}{L_2} \end{bmatrix} \\ \begin{bmatrix} -\frac{12}{L_2^3} & \frac{6}{L_2^2} \\ -\frac{6}{L_2^2} & \frac{4}{L_2} \end{bmatrix} & \begin{bmatrix} \frac{12}{L_2^3} & \frac{6}{L_2^2} \\ \frac{6}{L_2^2} & \frac{4}{L_2} \end{bmatrix} \end{bmatrix} \begin{Bmatrix} w(L_1) \\ -w'(L_1) \\ w(L_1+L_2) \\ -w'(L_1+L_2) \end{Bmatrix} = \begin{Bmatrix} V(L_1) \\ M(L_1) \\ V(L_1+L_2) \\ M(L_1+L_2) \end{Bmatrix} \tag{10.13}$$

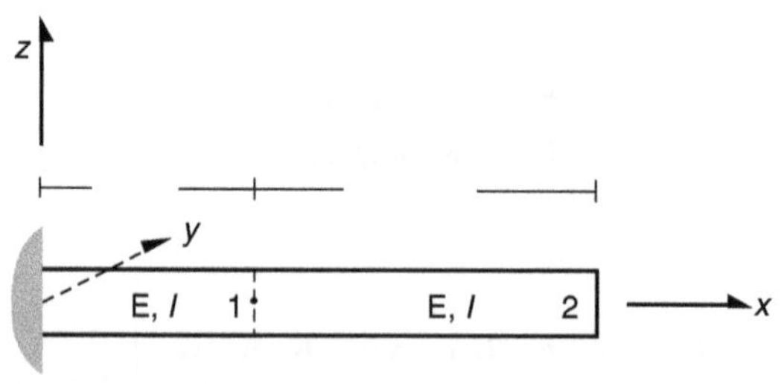

FIGURE 10.1 Inverse problem for a two-element beam problem.

Four distinct inverse problems are now considered.

Case I: $w(L_1)$, $V(L_1)$, $w(L_1+L_2)$, $V(L_1+L_2)$ are prescribed.

The solution exists and is unique if $\begin{bmatrix} 6/L_1^2 - 6/L_2^2 & -6/L_2^2 \\ 6/L_2^2 & 6/L_2^2 \end{bmatrix}$ *is nonsingular. In fact, it is* with determinant equalling $36/L_1^2L_2^2$.

Case II: $w(L_1)$, $V(L_1)$, $-w'(L_1+L_2)$, $M(L_1+L_2)$ are prescribed.

The matrix of interest is $\begin{bmatrix} 6/L_1^2 - 6/L_2^2 & 6/L_2^2 \\ -6/L_2^2 & 6/L_2^2 \end{bmatrix}$, *which is the transpose of the matrix in the first case and has the same nonvanishing determinant.*

Case III: $-w'(L_1)$, $M(L_1)$, $-w'(L_1+L_2)$, $M(L_1+L_2)$ are prescribed.

The matrix is $\begin{bmatrix} 12/L_1^3 + 12/L_2^3 & -6/L_2^2 \\ -6/L_2^2 & 4/L_2 \end{bmatrix}$ *arises which is nonsingular with determinant* equalling $48/L_1^3L_2 + 12/L_2^4$.

Case IV: $-w'(L_1)$, $M(L_1)$, $w(L_1+L_2)$, $V(L_1+L_2)$ are prescribed.

The matrix $\begin{bmatrix} 12/L_1^3 + 12/L_2^3 & -12/L_2^3 \\ -6/L_2^2 & 6/L_2^2 \end{bmatrix}$ *is nonsingular, with determinant* $72/L_1^3L_2^2$.

10.2.3 Nonsingularity Test

We next introduce a test for nonsingularity of $\mathbf{K}_{12} - \mathbf{K}_{13}\mathbf{K}_{33}^{-1}\mathbf{K}_{32}$. Of course it is necessary that $n_1 = n_2$, since otherwise this matrix is not square and hence is singular. The first result, which is easily proved, is that $\mathbf{K}_{12} - \mathbf{K}_{13}\mathbf{K}_{22}^{-1}\mathbf{K}_{32}$ is nonsingular if, and only if, $\begin{bmatrix} \mathbf{K}_{12} & \mathbf{K}_{23} \\ \mathbf{K}_{32} & \mathbf{K}_{33} \end{bmatrix}$ is nonsingular.

Next a test of nonsingularity of $\begin{bmatrix} \mathbf{K}_{12} & \mathbf{K}_{23} \\ \mathbf{K}_{32} & \mathbf{K}_{33} \end{bmatrix}$ is given. An $n \times n$ matrix $\mathbf{A}$ may be written in the form

$$\mathbf{A} = \begin{bmatrix} \mathbf{a}_1^T \\ \mathbf{a}_2^T \\ \cdot \\ \mathbf{a}_j^T \\ \cdot \\ \mathbf{a}_{n-1}^T \\ \mathbf{a}_n^T \end{bmatrix} \tag{10.14}$$

in which the ith row of the matrix $\mathbf{A}$ is written as the row vector $\mathbf{a}_j^T$. Using Gram-Schmidt orthogonalization, we may construct a set of *orthonormal base vectors* $\mathbf{e}_i$ as follows. The n base vectors $\mathbf{e}_j$ are given by

$$
\begin{aligned}
& & \mathbf{e}_1 &= \mathbf{a}_1/|\mathbf{a}_1| \\
\hat{\mathbf{e}}_2 &= \mathbf{a}_2 - (\mathbf{a}_2^T\mathbf{e}_1)\mathbf{e}_1, & \mathbf{e}_2 &= \hat{\mathbf{e}}_2/|\hat{\mathbf{e}}_2| \\
&\vdots & &\vdots \\
\hat{\mathbf{e}}_{n-1} &= \mathbf{a}_{n-1} - \sum_{j=1}^{n-1}(\mathbf{a}_{n-1}^T\mathbf{e}_j)\mathbf{e}_j, & \mathbf{e}_{n-1} &= \hat{\mathbf{e}}_{n-1}/|\hat{\mathbf{e}}_{n-1}| \\
\hat{\mathbf{e}}_n &= \mathbf{a}_n - \sum_{j=1}^{n-1}(\mathbf{a}_n^T\mathbf{e}_j)\mathbf{e}_j, & \mathbf{e}_n &= \hat{\mathbf{e}}_n/|\hat{\mathbf{e}}_n|
\end{aligned}
\tag{10.15}
$$

If $\mathbf{A}$ is nonsingular, the jth row vector cannot be a linear combination of the foregoing $j-1$ row vectors. Accordingly, the jth row vector $\mathbf{a}_j$ exists in the j-dimensional subspace spanned by orthonormal base vectors $\mathbf{e}_1, \mathbf{e}_2, \ldots, \mathbf{e}_j$. For the moment, suppose instead that the matrix $\mathbf{A}$ is singular with unit rank deficiency, and that $\mathbf{a}_n^T$ is a linear combination of the foregoing row vectors and *hence exists in an $n-1$-dimensional subspace spanned by the base vectors* $\mathbf{e}_1, \mathbf{e}_2, \ldots, \mathbf{e}_j, \ldots, \mathbf{e}_{n-1}$. But, if $\hat{\mathbf{e}}_n$ simultaneously (i) lies in the $n-1$-dimensional subspace and (ii) is orthogonal to the base vectors of the subspace, it must equal the null vector: $\hat{\mathbf{e}}_n = \mathbf{0}$.

Accordingly, the condition for the matrix $\mathbf{A}$ to be nonsingular is

$$
\hat{\mathbf{e}}_k \neq 0, \quad k = 1, 2, 3, \ldots, n \tag{10.16}
$$

Conversely, if any of the vectors $\hat{\mathbf{e}}_k$ vanish, $\mathbf{A}$ is singular.

Simple examples are now introduced to illustrate the application and performance of the nonsingularity test expressed in Equation 10.16.

EXAMPLE 10.3

We first consider the matrix

$$
\mathbf{A} = \begin{bmatrix} 1 & 2 \\ 2 & 4+\varepsilon \end{bmatrix} \tag{10.17}
$$

in which $\varepsilon \ll 1$. Following the operations in Equation 10.15 the base vectors are found to be

$$
\begin{aligned}
\mathbf{e}_1 &= \frac{1}{\sqrt{5}}\begin{Bmatrix} 1 \\ 2 \end{Bmatrix} \\
\hat{\mathbf{e}}_2 &= \begin{Bmatrix} 2 \\ 4 \end{Bmatrix} - \left[\frac{1}{\sqrt{5}}\{2 \quad 4+\varepsilon\}\begin{Bmatrix} 1 \\ 2 \end{Bmatrix}\right]\frac{1}{\sqrt{5}}\begin{Bmatrix} 1 \\ 2 \end{Bmatrix} \\
&= -\frac{1}{5}\begin{Bmatrix} \varepsilon \\ 4\varepsilon \end{Bmatrix}
\end{aligned}
\tag{10.18}
$$

In fact, this matrix $\mathbf{A}$ is nonsingular if $\varepsilon > 0$ and becomes singular when $\varepsilon \downarrow 0$. But also note that, in the current test, $\hat{\mathbf{e}}_2 \neq \mathbf{0}$ if $\varepsilon > 0$ and $\hat{\mathbf{e}}_2 \rightarrow \mathbf{0}$ as $\varepsilon \downarrow 0$, consistent with Equation 10.16.

EXAMPLE 10.4

Next, consider

$$\mathbf{A} = \begin{bmatrix} 1 & 2 & 3 \\ 2 & 4 & 6+\varepsilon_1 \\ 3 & 6 & 9+\varepsilon_2 \end{bmatrix} \tag{10.19}$$

We first seek to determine whether the current test correctly identifies values of ε_1 and ε_2 for which $\mathbf{A}$ is singular. We first check $\hat{\mathbf{e}}_2$.

$$\mathbf{e}_1 = \frac{1}{\sqrt{14}}\begin{Bmatrix} 1 \\ 2 \\ 3 \end{Bmatrix}$$

$$\hat{\mathbf{e}}_2 = \begin{Bmatrix} 2 \\ 4 \\ 6+\varepsilon_1 \end{Bmatrix} - \{2 \quad 4 \quad 6+\varepsilon_1\}\frac{1}{\sqrt{14}}\begin{Bmatrix} 1 \\ 2 \\ 3 \end{Bmatrix}\frac{1}{\sqrt{14}}\begin{Bmatrix} 1 \\ 2 \\ 3 \end{Bmatrix}$$

$$= \frac{1}{14}\begin{Bmatrix} -3 \\ -6 \\ 5 \end{Bmatrix}\varepsilon_1 \tag{10.20}$$

It is evident that $\mathbf{A}$ is singular if $\varepsilon_1 = 0$. The current procedure furnishes that $\hat{\mathbf{e}}_2 \neq \mathbf{0}$ if $\varepsilon_1 > 0$, but $\hat{\mathbf{e}}_2 \to \mathbf{0}$ as $\varepsilon_1 \downarrow 0$. Assume for the moment that $\varepsilon_1 > 0$, so that

$$\mathbf{e}_2 = \frac{1}{\sqrt{70}}\begin{Bmatrix} -3 \\ -6 \\ 5 \end{Bmatrix} \tag{10.21}$$

Note that ε_1 does not appear in $\mathbf{e}_2$.

We next determine the third base vector, assuming $\varepsilon_1 \neq 0$.

$$\hat{\mathbf{e}}_3 = \begin{Bmatrix} 3 \\ 6 \\ 9+\varepsilon_2 \end{Bmatrix} - \left[\{3 \quad 6 \quad 9+\varepsilon_2\}\frac{1}{\sqrt{14}}\begin{Bmatrix} 1 \\ 2 \\ 3 \end{Bmatrix}\right]\frac{1}{\sqrt{14}}\begin{Bmatrix} 1 \\ 2 \\ 3 \end{Bmatrix}$$

$$- \left[\{3 \quad 6 \quad 9+\varepsilon_2\}\frac{1}{\sqrt{70}}\begin{Bmatrix} -3 \\ -6 \\ 5 \end{Bmatrix}\right]\frac{1}{\sqrt{70}}\begin{Bmatrix} -3 \\ -6 \\ 5 \end{Bmatrix} = \begin{Bmatrix} 0 \\ 0 \\ 0 \end{Bmatrix} \tag{10.22}$$

The test in Equation 10.16 indicates that the matrix $\mathbf{A}$ is singular *regardless of* ε_2. That this result is *correct* is easily seen by recognizing that the second column in $\mathbf{A}$ is proportional to the first column regardless of ε_2 (or ε_1).

10.3 ACCELERATED EIGENSTRUCTURE COMPUTATION IN FEA

10.3.1 INTRODUCTION

Calculation of lowest modes is a common task in finite element modeling of large systems. The primary established method in FEA is Subspace Iteration (Bathe, 1996). Here, an optimization algorithm (Nicholson and Lin, 2006) is described which exploits the fact that (i) the eigenvectors terminate on a unit hypersphere and (ii) that the minimum eigenvalue and associated eigenvector satisfy the Rayleigh minimum principle. At the current estimate for the minimizing eigenvector, a direction on the hypersphere is found analytically in which the Rayleigh quotient experiences "steepest descent." Kronecker product algebra is instrumental in the derivation. The current eigenvector and a unit vector representing the direction of steepest descent define a plane intersecting the hypersphere along a unit hypercircle. An analytical solution is found for the vector minimizing the Rayleigh quotient on the hypercircle, constituting a hypercircle counterpart of a "line search." At this last vector a new steepest descent vector is determined and the process is repeated. In a numerical example, the algorithm converges very rapidly. Also introduced is a counterpart of "deflation" to replace the smallest eigenvalue while leaving the eigenvectors and the dimension of the matrix unchanged. The deflation procedure currently is based on the assumption that none of the eigenvalues in the lowest modes are repeated.

10.3.2 PROBLEM STATEMENT

For dynamic response of a linear elastic system, the finite element equation is conventionally written as

$$\hat{\mathbf{M}}\ddot{\hat{\mathbf{x}}} + \hat{\mathbf{K}}\hat{\mathbf{x}} = \hat{\mathbf{f}}(t) \tag{10.23}$$

in which the matrices $\hat{\mathbf{M}}$ and $\hat{\mathbf{K}}$ are $n \times n$, real, positive definite, and symmetric, while $\hat{\mathbf{x}}$ and $\hat{\mathbf{f}}(t)$ are real $n \times 1$ vectors. For convenience the foregoing system is rewritten as

$$\ddot{\mathbf{x}} + \mathbf{K}\mathbf{x} = \mathbf{f}(t) \tag{10.24}$$

in which $\mathbf{K} = \mathbf{M}^{-1/2}\hat{\mathbf{K}}\mathbf{M}^{-1/2}$ and $\mathbf{x} = \mathbf{M}^{1/2}\hat{\mathbf{x}}$. The modes of the system are determined by the eigenvalues λ_j and the real orthonormal eigenvectors $\mathbf{x}_j$, with the assumed magnitude ordering $\lambda_1 > \lambda_2 > \cdots > \lambda_n$. The primary interest is in the low-frequency modes, say with λ_j having magnitudes less than a user-specified threshold value.

The foundation of the method being introduced is minimization of the Rayleigh quotient (e.g., Dahlquist and Bjork, 1976). Namely, the minimum eigenvalue λ_n and the corresponding minimizing eigenvector $\mathbf{x}_n$ satisfy

$$\begin{aligned} \lambda_n(\mathbf{K}) &= \min_{1 \le j \le n} \lambda_j(\mathbf{K}) \\ &= \mathbf{x}_n^T \mathbf{K}\mathbf{x}_n \\ &= \min_{\mathbf{x}^T\mathbf{x}=1} (\mathbf{x}^T\mathbf{K}\mathbf{x}) \end{aligned} \tag{10.25}$$

To enforce unit magnitude of $\mathbf{n} = \mathbf{x}/\sqrt{\mathbf{x}^T\mathbf{x}}$ a priori, we replace $\mathbf{n}$ with

$$\mathbf{n} = \mathbf{Q}\boldsymbol{\upsilon} \tag{10.26}$$

in which $\boldsymbol{\upsilon}$ is a unit vector chosen to give the initial estimate of the minimizing eigenvector. Also $\mathbf{Q}$ is an orthogonal transformation rotating $\boldsymbol{\upsilon}$ to $\mathbf{n}$, which of course is also of unit magnitude. Clearly, the tip of $\mathbf{n}$ lies on a unit *hypersphere* (a sphere in n-space). As $\mathbf{Q}$ changes during minimization, the tip of $\mathbf{n}$ prescribes a path on the hypersphere. Otherwise stated, the task is to determine the hyperspherical path which is optimal in a meaningful sense and which terminates at the tip of the desired eigenvector.

10.3.3 Hypersphere Path of Steepest Descent

We now determine the differential $\mathrm{d}\mathbf{Q}$ corresponding to the most rapid rate of decrease of $\Phi_1 = \boldsymbol{\upsilon}^T\mathbf{Q}^T\mathbf{K}\mathbf{Q}\boldsymbol{\upsilon}$. Letting $\mathbf{q}$ denote $VEC(\mathbf{Q})$, Kronecker Product Algebra (presented in Chapter 3) furnishes the following relations:

$$\begin{aligned}
\mathrm{d}\Phi_1(\mathbf{Q}) &= \boldsymbol{\upsilon}^T\,\mathrm{d}\mathbf{Q}^T\mathbf{K}\mathbf{Q}\boldsymbol{\upsilon} + \boldsymbol{\upsilon}^T\mathbf{Q}^T\mathbf{K}\,\mathrm{d}\mathbf{Q}\boldsymbol{\upsilon} \\
&= \boldsymbol{\upsilon}^T \otimes \boldsymbol{\upsilon}^T\,VEC(\mathrm{d}\mathbf{Q}^T\,\mathbf{K}\mathbf{Q} + \mathbf{Q}^T\mathbf{K}\,\mathrm{d}\mathbf{Q}) \\
&= \boldsymbol{\upsilon}^T \otimes \boldsymbol{\upsilon}^T[(\mathbf{Q}^T\mathbf{K}) \otimes \mathbf{I}\mathbf{U} + \mathbf{I} \otimes (\mathbf{Q}^T\mathbf{K})]\,\mathrm{d}\mathbf{q} \\
&= [(\boldsymbol{\upsilon}^T\mathbf{Q}^T\mathbf{K}) \otimes \boldsymbol{\upsilon}^T\mathbf{U} + \boldsymbol{\upsilon}^T \otimes (\boldsymbol{\upsilon}^T\mathbf{Q}^T\mathbf{K})]\,\mathrm{d}\mathbf{q} \\
&= \mathbf{m}^T\,\mathrm{d}\mathbf{q}
\end{aligned} \tag{10.27}$$

in which

$$\mathrm{d}\mathbf{q} = VEC(\mathrm{d}\mathbf{Q}) \quad \mathbf{m}^T = (\mathbf{I} + \mathbf{U})\boldsymbol{\upsilon}^T \otimes (\mathbf{n}^T\mathbf{K}), \quad \mathbf{n} = \mathbf{Q}\boldsymbol{\upsilon}$$

However, we may also write

$$\mathrm{d}\Phi_1 = \tau(\mathbf{Q}\mathbf{M}^T\,\mathrm{d}\mathbf{Q}) \tag{10.28}$$

in which $\mathbf{M} = IVEC(\mathbf{m})$ and τ denotes the trace. Invariance of the trace under similarity transformations implies

$$\mathrm{d}\Phi_1 = \tau(\mathbf{M}^T\,\mathrm{d}\mathbf{Q}\,\mathbf{Q}^T) \tag{10.29}$$

Note that the matrix $\mathrm{d}\mathbf{Q}\,\mathbf{Q}^T$ is *antisymmetric.*

Let $\mathbf{B} = \mathbf{Q}\mathbf{M}^T$ and let $\mathbf{C} = \frac{1}{2}(\mathbf{B} - \mathbf{B}^T)$. The trace vanishes for the product of a symmetric and an antisymmetric matrix:

$$\begin{aligned}
\mathrm{d}\Phi_1 &= \tau(\mathbf{B}\,\mathrm{d}\mathbf{Q}\,\mathbf{Q}^T) \\
&= \tau(\mathbf{C}\,\mathrm{d}\mathbf{Q}\,\mathbf{Q}^T)
\end{aligned} \tag{10.30}$$

Note that $\mathbf{C}^2$ is negative definite if the antisymmetric matrix $\mathbf{C}$ is nonsingular, so that steepest descent is attained if $\mathrm{d}\mathbf{Q}\mathbf{Q}^T$ is proportional to $\mathbf{C}$. Accordingly, we seek the path defined by the relation

$$\mathrm{d}\mathbf{Q} = \mathrm{d}\Lambda \frac{\mathbf{C}(\mathbf{Q})\mathbf{Q}}{\tau^{1/2}(-\mathbf{C}^2(\mathbf{Q}))} \tag{10.31}$$

and $\mathrm{d}\Lambda$ may be viewed as a differential arc length since

$$\mathrm{d}\Lambda = \tau^{1/2}(\mathrm{d}\mathbf{Q}\,\mathrm{d}\mathbf{Q}^T) \tag{10.32}$$

Away from the minimum, the procedure *coerces* Φ to decrease since

$$\mathrm{d}\Phi = -\mathrm{d}\Lambda\tau^{1/2}(-\mathbf{C}^2(\mathbf{Q})) \leq 0 \tag{10.33}$$

Of course the current position vector after j iterations is given by $\mathbf{n}_j = \mathbf{Q}_j\boldsymbol{\upsilon}$, in which $\mathbf{Q}_j$ denotes the current value of $\mathbf{Q}$. The method for identifying an orthogonal matrix $\mathbf{Q}$ (i.e., $\mathbf{Q}_j$) rotating $\boldsymbol{\upsilon}$ to $\mathbf{n}(\mathbf{n}_j)$ is described in Nicholson and Lin (2006).

10.3.4 Hypercircle Search

Suppose that j iterations have occurred leading to the current vector $\mathbf{n}_j = \mathbf{Q}_j\boldsymbol{\upsilon}$. We introduce the incremental position vector $\mathrm{d}\mathbf{n}_j = \mathrm{d}\mathbf{Q}_j\,\mathbf{n}_j$. It follows that to first order in the increments

$$\begin{aligned}\mathbf{n}_j^T\,\mathrm{d}\mathbf{n}_j &= \boldsymbol{\upsilon}^T\mathbf{Q}_j^T\,\mathrm{d}\mathbf{Q}_j\,\boldsymbol{\upsilon}\\ &= 0\end{aligned} \tag{10.34}$$

since $\mathbf{Q}^T\mathrm{d}\mathbf{Q}$ is antisymmetric. It also follows that the unit vector $\mathbf{r}(\mathbf{n}_j)$ given by

$$\mathbf{r}(\mathbf{n}_j) = \frac{\mathrm{d}\mathbf{n}_j}{\sqrt{\mathrm{d}\mathbf{n}_j^T\,\mathrm{d}\mathbf{n}_j}} \tag{10.35}$$

is orthogonal to $\mathbf{n}_j$ and hence is tangent to the hypersphere. We say that the *steepest descent* from position $\mathbf{n}_j$ occurs in the direction represented by the unit vector $\mathbf{r}(\mathbf{n}_j)$. The vectors $\mathbf{n}_j$ and $\mathbf{r}(\mathbf{n}_j)$ define a hyperplane whose intersection with the hypersphere is a hypercircle, as illustrated in Figure 10.2. Any vector $\mathbf{p}$ terminating on the hypercircle may be expressed in terms of $\mathbf{n}_j$, $\mathbf{r}(\mathbf{n}_j)$, and an angle η, illustrated in Figure 10.3. Likewise, on the hypercircle Φ may be expressed as a simple function of $\mathbf{n}_j$, $\mathbf{r}(\mathbf{n}_j)$, and η. The hypercircle search simply consists of determining the angle η_0 which minimizes Φ_1 on the hypercircle. It will now be seen that determination of η_0 reduces to a simple algebraic problem with an analytical solution. Finally, if we designate the value of $\mathbf{Q}$ minimizing the Rayleigh quotient as $\mathbf{Q}_{\min}$, the minimizing eigenvector is obtained immediately as $\mathbf{n}_{\min} = \mathbf{Q}_{\min}\boldsymbol{\upsilon}$, and of course, the minimum eigenvalue satisfies $\lambda_{\min} = \Phi(\mathbf{Q}_{\min})$.

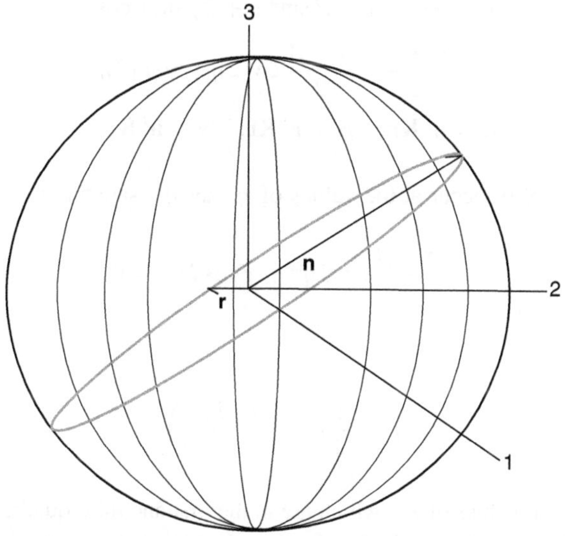

FIGURE 10.2 Hypersphere and hyperplane determined by **n** and **r**.

Minimizing Φ on the hypercircle reduces to determining an angle which can be expressed analytically in terms of an inverse tangent. The vector $\mathbf{p}(\eta)$ illustrated below in Figure 10.3 is given by

$$\mathbf{p}(\eta) = \cos\eta\mathbf{n} + \sin\eta\mathbf{r} \tag{10.36}$$

On the hypercircle the function Φ is expressed in terms of the angle η by

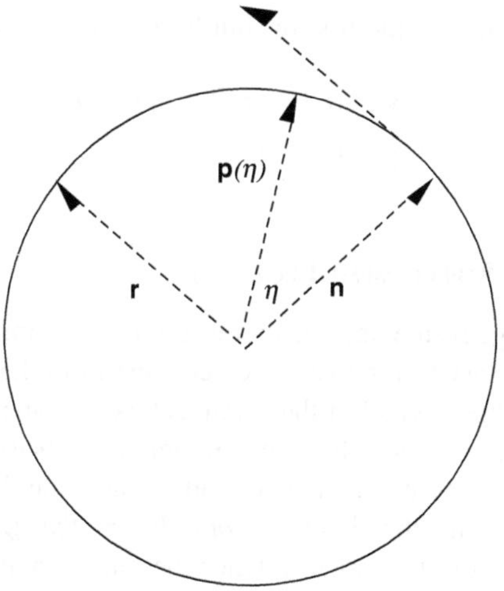

FIGURE 10.3 Vector terminating on the hypercircle.

$$
\begin{aligned}
\Phi &= \alpha \cos^2 \eta + \beta \sin^2 \eta + 2\gamma \sin \eta \cos \eta \\
&= \frac{\alpha+\beta}{2} + \frac{\alpha-\beta}{2} \cos 2\eta + \gamma \sin 2\eta \\
\alpha &= \mathbf{n}^T\mathbf{K}\mathbf{n}, \quad \beta = \mathbf{r}^T\mathbf{K}\mathbf{r}, \quad \gamma = \mathbf{n}^T\mathbf{K}\mathbf{r}
\end{aligned} \tag{10.37}
$$

The extrema of Φ occur at the values of η, say η^*, satisfying

$$
-\frac{\alpha-\beta}{2} \sin 2\eta^* + \gamma \cos 2\eta^* = 0 \tag{10.38}
$$

from which

$$
\eta^* = \frac{1}{2} \tan^{-1}\left(\frac{2\gamma}{\alpha-\beta}\right) \tag{10.39}
$$

There are two such values of η^*, which are in the first and third quadrants if $\frac{2\gamma}{\alpha-\beta} > 0$, but the second or fourth quadrants if $\frac{2\gamma}{\alpha-\beta} < 0$. One value minimizes Φ_1 on the hypercircle, and the second maximizes Φ_1. For the present extremum to be a minimum, η^* must satisfy

$$
-\frac{\alpha-\beta}{2} \cos 2\eta^* - \gamma \sin 2\eta^* > 0 \tag{10.40}
$$

Simple manipulation furnishes

$$
-\gamma\left[\frac{1}{\gamma^2}\left(\frac{\alpha-\beta}{2}\right)^2 + 1\right] \sin 2\eta^* > 0 \tag{10.41}
$$

Consequently, η^* is found in the first or fourth quadrants as follows:

$$
\begin{aligned}
&\text{if } \gamma \geq 0, \quad -\pi/2 \leq \eta^* \leq 0 \\
&\text{if } \gamma < 0, \quad 0 < \eta^* \leq \pi/2
\end{aligned} \tag{10.42}
$$

10.3.5 Eigenvalue Replacement Procedure

Once the minimum eigenvalue and the corresponding (minimizing) eigenvector are determined, it is necessary to remove the eigenvalue from the matrix while leaving the remaining eigenvalues and all of the eigenvectors unchanged, a process we refer to as *eigenvalue replacement*. The scheme presented below replaces the most recently computed minimum eigenvalue with magnitude less than a threshold value with a value above the threshold, *without altering the eigenvectors* or reducing the dimensions of (deflating) the matrix. Once the minimum eigenvalue is replaced, the minimization process is repeated to compute the next largest eigenvalue and corresponding eigenvector, and continues until an eigenvalue is obtained whose magnitude equals or exceeds a user-specified threshold value. Currently, the foregoing deflation procedure is based on the restriction that none of the eigenvalues of the modes of interest has multiplicity greater than unity.

Suppose λ_n and $\mathbf{n}_n$ have been computed. We may construct a set of vectors $\mathbf{p}_{n-1}^{(n-1)}, \mathbf{p}_{n-2}^{(n-1)}, \ldots, \mathbf{p}_1^{(n-1)}$ which are orthonormal to each other and to $\mathbf{n}_n$. They are likewise orthogonal to $\mathbf{Kn}_n$ since it coincides in direction with $\mathbf{n}_n$. In particular, $\mathbf{p}_{n-1}^{(n-1)}, \mathbf{p}_{n-2}^{(n-1)}, \ldots, \mathbf{p}_1^{(n-1)}$ are obtained sequentially using the Gram-Schmidt scheme

$$\hat{\mathbf{p}}_j^{(n-1)} = \mathbf{q}_j^{(n-1)} - \sum_{k=1}^{n-j-1} (\mathbf{q}_j^{(n-1)T}\mathbf{p}_{j+k}^{(n-1)})\mathbf{p}_{j+k}^{(n-1)} - (\mathbf{q}_j^{(n-1)T}\mathbf{n}_n)\mathbf{n}_n$$
$$j = n-1, n-2, \ldots, 1, \quad \mathbf{p}_j^{(n-1)} = \hat{\mathbf{p}}_j^{(n-1)} \Big/ \sqrt{\hat{\mathbf{p}}_j^{(n-1)T}\hat{\mathbf{p}}_j^{(n-1)}} \tag{10.43}$$

in which $\mathbf{q}_j^{(n-1)}$ are "judiciously chosen" "trial" vectors. The matrix $\mathbf{R}_n^{(n-1)}$ given by

$$\mathbf{R}_n^{(n-1)} = \begin{bmatrix}\mathbf{R}_{n-1}^{(n-1)} & n_n\end{bmatrix}, \quad \mathbf{R}_{n-1}^{(n-1)} = \begin{bmatrix}\mathbf{p}_1^{(n-1)} & \mathbf{p}_2^{(n-1)} & \cdots & \mathbf{p}_{n-2}^{(n-1)} & \mathbf{p}_{n-1}^{(n-1)}\end{bmatrix} \tag{10.44}$$

is orthogonal and gives rise to the singularity transformation

$$\mathbf{R}_n^{(n-1)T}\mathbf{K}\mathbf{R}_n^{(n-1)} = \begin{bmatrix}\mathbf{K}_{n-1}^* & \mathbf{0} \\ \mathbf{0}^T & \lambda_n\end{bmatrix} \tag{10.45}$$

In Equation 10.45 note that (i) $[\mathbf{R}_n^{(n-1)T}\mathbf{K}\mathbf{R}_n^{(n-1)}]_{jn} = \mathbf{p}_j^{(n-1)T}(\mathbf{Kn}_n) = 0$, (ii) $[\mathbf{R}_n^{(n-1)T}\mathbf{K}\mathbf{R}_n^{(n-1)}]_{nj} = \mathbf{n}_n^T\mathbf{K}\mathbf{p}_j^{(n-1)} = \mathbf{p}_j^{(n-1)T}(\mathbf{Kn}_n) = 0$, and (iii) $\left[\mathbf{K}_{n-1}^*\right]_{ij} = \mathbf{p}_i^{(n-1)^T}\mathbf{K}\mathbf{p}_j^{(n-1)}$.

The eigenvalues of $\mathbf{K}_{n-1}^*$ are $\lambda_{n-1}, \lambda_{n-2}, \ldots, \lambda_1$, which are the largest $n-1$ eigenvalues of $\mathbf{K}$. Of course, the eigenvectors of $\mathbf{K}_{n-1}^*$ are also eigenvectors of $\mathbf{K}$. Suppose the largest eigenvalue of interest is no greater in magnitude than a threshold value λ_{th}, corresponding to the mode with the highest natural frequency of interest. We may introduce the matrix

$$\mathbf{K}^{(n-1)} = \mathbf{R}_n^{(n-1)}\begin{bmatrix}\mathbf{K}_{n-1}^* & \mathbf{0} \\ \mathbf{0}^T & \lambda_{th}\end{bmatrix}\mathbf{R}_n^{(n-1)T} \tag{10.46}$$

Now the eigenvalues of $\mathbf{K}^{(n-1)}$ are $\lambda_{n-1}, \lambda_{n-2}, \ldots, \lambda_{th}, \ldots, \lambda_1$. It is next demonstrated that $\mathbf{K}^{(n-1)}$ has the same eigenvectors as $\mathbf{K}$.

Since $\mathbf{K}_{n-1}^*$ is symmetric and positive definite there exists an orthogonal matrix $\mathbf{V}_{n-1}$ (which need not be computed) such that

$$\mathbf{K}^{(n-1)} = \mathbf{R}_n^{(n-1)}\begin{bmatrix}\mathbf{V}_{n-1} & 0 \\ \mathbf{0}^T & 1\end{bmatrix}\begin{bmatrix}\mathbf{\Lambda}\left(\mathbf{K}_{n-1}^*\right) & 0 \\ \mathbf{0}^T & \lambda_n + \lambda_{th}\end{bmatrix}\begin{bmatrix}\mathbf{V}_{n-1}^T & 0 \\ \mathbf{0}^T & 1\end{bmatrix}\mathbf{R}_n^{(n-1)T} \tag{10.47}$$

in which, assuming obvious eigenvalue ordering,

$$\boldsymbol{\Lambda}\left(\mathbf{K}_{n-1}^{*}\right)=\begin{bmatrix}\lambda_1 & 0 & . & . & . \\ 0 & \lambda_2 & 0 & . & . \\ . & . & . & . & . \\ . & . & . & . & 0 \\ . & . & . & 0 & \lambda_{n-1}\end{bmatrix} \tag{10.48}$$

It follows that

$$\begin{bmatrix}\boldsymbol{\Lambda}\left(\mathbf{K}_{n-1}^{*}\right) & 0 \\ \mathbf{0}^T & \lambda_{\text{th}}\end{bmatrix}=\left[\begin{bmatrix}\mathbf{V}_{n-1}^T & 0 \\ \mathbf{0}^T & 1\end{bmatrix}\mathbf{R}_n^{(n-1)T}\right]\mathbf{K}^{(n-1)}\left[\mathbf{R}_n^{(n-1)}\begin{bmatrix}\mathbf{V}_{n-1} & 0 \\ \mathbf{0}^T & 1\end{bmatrix}\right] \tag{10.49}$$

However, $\left[\mathbf{R}_n^{(n-1)}\begin{bmatrix}\mathbf{V}_{n-1} & 0 \\ \mathbf{0}^T & 1\end{bmatrix}\right]$ is also recognized as an orthogonal matrix which diagonalizes $\mathbf{K}$:

$$\begin{bmatrix}\boldsymbol{\Lambda}\left(\mathbf{K}_{n-1}^{*}\right) & \mathbf{0} \\ \mathbf{0}^T & \lambda_n\end{bmatrix}=\left[\begin{bmatrix}\mathbf{V}_{n-1}^T & \mathbf{0} \\ \mathbf{0}^T & 1\end{bmatrix}\mathbf{R}_n^{(n-1)T}\right]\mathbf{K}\left[\mathbf{R}_n^{(n-1)}\begin{bmatrix}\mathbf{V}_{n-1} & \mathbf{0} \\ \mathbf{0}^T & 1\end{bmatrix}\right] \tag{10.50}$$

But matrices which are diagonalized by the same similarity transformation, in particular $\mathbf{K}$ and $\mathbf{K}^{(n-1)}$, *have the same eigenvectors*, Nicholson and Lin (1996).

The replacement process continues as follows. The next largest eigenvalue is λ_{n-1} with corresponding eigenvector $\mathbf{n}_{n-1}$, both of which are computed by minimizing the Rayleigh quotient applied to $\mathbf{K}^{(n-1)}$. A new set of unit vectors $\mathbf{p}_j^{(n-2)}$, $j=n-2, n-1, \ldots, 1$ is generated, which are orthogonal to each other as well as $\mathbf{n}_{n-1}$ and $\mathbf{n}_n$. Extending the steps shown above furnishes a matrix of the form

$$\mathbf{R}_n^{(n-2)T}\mathbf{K}^{(n-2)}\mathbf{R}_n^{(n-2)}=\begin{bmatrix}K_{n-2}^{*} & \mathbf{0} & \mathbf{0} \\ \mathbf{0}^T & \lambda_{n-1}+\lambda_{\text{th}} & \mathbf{0} \\ \mathbf{0}^T & 0 & \lambda_n+\lambda_{\text{th}}\end{bmatrix} \tag{10.51}$$

The deflation process is repeated to furnish the desired eigenvalues (with magnitude not exceeding λ_{th}) in order of increasing magnitude, and to furnish the corresponding eigenvectors.

10.3.6 Example: Minimum Eigenvalue of the 3 × 3 Hilbert Matrix

We illustrate the hyperspherical method by applying it to determine the minimum eigenvalue $\lambda_{\min}$ and corresponding eigenvector $\mathbf{n}_{\min}$ of the 3×3 Hilbert matrix

$$\mathbf{H}_3=\begin{bmatrix}1 & 1/2 & 1/3 \\ 1/2 & 1/3 & 1/4 \\ 1/3 & 1/4 & 1/5\end{bmatrix} \tag{10.52}$$

Hilbert matrices are notorious for being ill-conditioned even though positive definite and symmetric. A simple program has been written in high precision to compute λ_3 and $\mathbf{n}_3$. The initial vector is assumed using the diagonal terms of $\mathbf{H}_3$:

$$v = \frac{1}{\sqrt{(1/2)^2 + (1/3)^2 + (1/5)^2}} \begin{Bmatrix} 1/2 \\ 1/2 \\ 1/5 \end{Bmatrix} \quad (10.53)$$

The procedure generates and makes use of two vectors in each step after the first step. To use the hypercircle method in the first step, a second vector $\boldsymbol{\vartheta}$ must be introduced at the outset. We used the vector

$$\boldsymbol{\vartheta} = \begin{Bmatrix} 1 \\ 0 \\ 0 \end{Bmatrix} \quad (10.54)$$

The computed iterates for the lowest eigenvalues and the relative errors are shown in Table 10.1 below.

Even though the errors are very high in the initial estimate, convergence still occurs rapidly and furnishes extremely accurate values. Convergence appears to be much more rapid than in a linear convergence scheme. (Subspace Iteration exhibits linear convergence.) The eigenvectors likewise converge very rapidly.

10.4 FOURTH-ORDER TIME INTEGRATION

10.4.1 Introduction

In FEA of elastic systems with light viscous damping, the widely used Newmark method is the second order, one step, and A-stable. A systematic presentation of established time integration methods for FEA is given in Zienkiewicz and Taylor (1989), covering the well-known methods of Newmark, Houbolt, Wilson, Hibler, and others. The Newmark method is a reformulation of the classical Trapezoidal

TABLE 10.1
Convergence of Iterations for Lowest Eigenvalue

Computed Minimum Eigenvalue	Percent Error (%)
1.3422136422136	49845.800104181800000
0.1126113206822	4090.437599027870000
0.0047836190059	78.005699785540800
0.0026907039866	0.125165789790772
0.0026873457448	0.000200532865081
0.0026873403644	0.000000321273041
0.0026873403558	0.000000000515009
0.0026873403558	0.000000000001485

Rule for systems with accelerations. Even though it is A-stable, *time step sizes are severely limited by considerations of accuracy*. Also, "numerical damping," for example in the Wilson-Theta method, has been introduced to attenuate higher order modes. It does so at a modest cost in accuracy (near-second order).

In this section a fourth-order counterpart of the Newmark method is described which extends the fourth-order three-step Adams–Moulton (AM) method to systems with acceleration, referred to below as AMX. (AM refers to first-order systems while AMX refers to second-order systems.) The AMX method is three-step. No three-step method can be A-stable, by virtue of a classical theorem of Dahlquist (cf. Dahlquist and Bjork, 1974; Gear, 1971). However if, after every time step (or set of several steps) numerically unstable higher order modes are filtered from the response using, for example, the Wavelet Packet transform (Kaplan, 2002), the stability-based restrictions on the time step size can be comparable to the restrictions ensuing from accuracy (Nicholson and Lin, 2006). Also, a modification of the AMX method is given to incorporate numerical damping, rendering the modification near-fourth order. AMX gives rise to a linear system involving a dynamic stiffness matrix. Solution using triangularization followed by forward and back substitution is seen to require exactly the same computational effort as the Newmark method.

10.4.2 Error Growth in the Newmark Method

The errors and error growth in the Newmark method have been extensively described by Nicholson and Lin (2005) for free and forced response, and results are quoted here on undamped free response.

10.4.2.1 Undamped Free Vibration

To illustrate the error properties of the Newmark method, suppose that the highest mode to be computed is the Jth mode with natural frequency ω_J, and suppose there is no damping. The Newmark method does not produce any magnitude error, but there is a phase angle error since the Newmark method evidently approximates $\omega_J h/2$ as $\tan^{-1}(\omega_J h/2)$. The phase error in a step is given by

$$\varepsilon_h \approx \frac{\pi^2}{3}\left(\frac{h}{T_J}\right)^2 \tag{10.55}$$

in which T_J is the time period of the model. For illustration we choose h to attain a stepwise relative error of $\frac{1}{300}$, in which instance the number of time steps per period is found to be $\frac{T_J}{h} = 10\pi \approx 32$.

Furthermore, we determine the effect of stepwise error on *cumulative error*. The exact solution after M time steps is $y_{n+1} = \exp(M\lambda h)y_0 = [\exp(\lambda h)]^{M+1}y_0$. Recalling Equation 10.55, the numerical solution is $\tilde{y}_{n+1} = [(1+\varepsilon_h)\exp(\lambda h)]^{M+1}$ $y_0 = (1+\varepsilon_h)^{M+1}y_{n+1}$. The cumulative error after $M+1$ steps is $\varepsilon_T = \frac{\tilde{y}_{n+1} - y_{n+1}}{y_{n+1}} = (1+\varepsilon_h)^{M+1} - 1$. The number of time steps after N time periods is given by $M+1 = \frac{T_J}{h}N = \frac{\pi N}{\sqrt{3\varepsilon_h}}$. Now $(1+\varepsilon_h)^{\frac{\pi N}{\sqrt{3\varepsilon_h}}} = 1+\varepsilon_T$, and $\frac{\pi N}{\sqrt{3\varepsilon_h}}\ln(1+\varepsilon_h) = \ln(1+\varepsilon_T)$.

We assume that both relative errors are much less than unity in magnitude, and take the first nonvanishing terms in the Taylor expansion of the natural logarithms. Now $\frac{\pi N}{\sqrt{3}}\sqrt{\varepsilon_h} \approx \varepsilon_T$, so that $\varepsilon_h \approx \frac{3}{\pi^2 N^2}\varepsilon_T^2$. Otherwise stated, the time step is to be selected to satisfy $\frac{T_1}{h} = \frac{\pi^2}{3}\frac{N}{\varepsilon_T}$. As an example, suppose that $N = 10$ and $\varepsilon_T = 1/10$. This implies the number of time steps per period is 333.

10.4.3 Adams–Moulton Formula

Consider the differential equation

$$\frac{dy}{dt} = f(y), \quad y(0) = y_0 \tag{10.56}$$

in which $f(y)$ is a known function. A three-step integration formula has the form (Gear, 1971)

$$\alpha_0 y_{n+1} + \alpha_1 y_n + \alpha_2 y_{n-1} + \alpha_3 y_{n-2} + h\beta_0 f_{n+1} + h\beta_1 f_n + h\beta_2 f_{n-1} + h\beta_3 f_{n-2} = 0 \tag{10.57}$$

in which the coefficients α_j and β_j are to be determined from accuracy and stability considerations. Elements of the derivations are reproduced here from, for example, Gear (1971), to set the stage for a stability and accuracy analysis and for the introduction of damping. Taylor expansion if y_{n+j} and f_{n+j}, $j = 1, 0, -1, -2$ through fifth order furnishes the following relations:

$$\beta_0 + \beta_1 + \beta_2 + \beta_3 = 1 \tag{10.58a}$$

$$\alpha_0 + \alpha_1 + \alpha_2 + \alpha_3 = 0 : \mathrm{d}(h^{(0)}) \tag{10.58b}$$

$$\alpha_1 + 2\alpha_2 + 3\alpha_3 - \beta_0 - \beta_1 - \beta_2 - \beta_3 = 0 : \mathrm{d}(h) \tag{10.58c}$$

$$\frac{7}{12}\alpha_0 + \frac{1}{12}\alpha_1 - \frac{17}{12}\alpha_2 - \frac{47}{12}\alpha_3 + \beta_1 + 2\beta_2 + 3\beta_3 = 0 : \mathrm{d}(h^2) \tag{10.58d}$$

$$-\frac{5}{12}\alpha_1 + \frac{1}{6}\alpha_2 + \frac{11}{4}\alpha_3 + \frac{7}{12}\beta_0 + \frac{1}{12}\beta_1 - \frac{17}{12}\beta_2 - \frac{47}{12}\beta_3 = 0 : \mathrm{d}(h^3) \tag{10.58e}$$

$$\frac{49}{144}\alpha_0 + \frac{13}{144}\alpha_1 - \frac{23}{144}\alpha_2 + \frac{157}{144}\alpha_3 + \frac{5}{12}\beta_1 - \frac{1}{6}\beta_2 - \frac{11}{4}\beta_3 = 0 : \mathrm{d}(h^4) \tag{10.58f}$$

The error coefficient (coefficient of h^5) is given by

$$\begin{aligned} C_5 = {} & \frac{7}{48}\alpha_0 + \frac{5}{48}\alpha_1 - \frac{1}{48}\alpha_2 + \frac{11}{48}\alpha_3 - \frac{49}{144}\beta_0 - \frac{13}{144}\beta_1 \\ & + \frac{23}{144}\beta_2 - \frac{157}{144}\beta_3 \end{aligned} \tag{10.59}$$

Equations 10.58a through 10.58f represent six equations in eight unknowns. Accordingly, $\alpha_0, \alpha_1, \alpha_2, \alpha_3$, β_1, and β_2 may be expressed in terms of β_0 and β_3. Simple calculation serves to obtain

$$\begin{Bmatrix} \alpha_1 \\ \alpha_2 \\ \alpha_3 \\ \beta_1 \\ \beta_2 \end{Bmatrix} = \begin{Bmatrix} -3/4 \\ 3/4 \\ 1/12 \\ 1/2 \\ 1/2 \end{Bmatrix} + \beta_0 \begin{Bmatrix} 9/2 \\ -3/2 \\ -1/2 \\ 1 \\ -2 \end{Bmatrix} + \beta_3 \begin{Bmatrix} 3/2 \\ -9/2 \\ 5/2 \\ -2 \\ 1 \end{Bmatrix} \tag{10.60}$$

$$\text{and} \quad \alpha_0 = -1/12 - 5/2\beta_0 + 1/2\,\beta_3$$

There are several types of errors in a multistep method. One is primary error associated with the primary eigenvalue λ_J, which in the current context determines magnitude error and phase error. The primary eigenvalue should match the exact eigenvalues in the physical mode. In addition there are "extraneous errors," also called "parasitic errors," which are associated with secondary eigenvalues arising in Equation 10.57 and are purely numerical in origin. It is reasonable to choose β_0 and β_3 to minimize the magnitudes of the two extraneous eigenvalues *at vanishing values of* h. With this choice, a finite time step is required before the extraneous eigenvalues (as a function of the time step) approach unity in magnitude and thereby imply instability. The secondary eigenvalues vanish at $h = 0$ if β_0 and β_3 are chosen such that $\beta_0 = \frac{3}{8}$ and $\beta_3 = \frac{1}{24}$. In fact, doing so yields the classical fourth-order three-step Adams–Moulton (AM) formula

$$y_{n+1} - y_n + \lambda h\left[\tfrac{3}{8}y_{n+1} + \tfrac{19}{24}y_n - \tfrac{5}{24}y_{n-1} + \tfrac{1}{24}y_{n-2}\right] = 0 \tag{10.61}$$

and more generally

$$y_{n+1} - y_n - h\left[\tfrac{3}{8}f_{n+1} + \tfrac{19}{24}f_n - \tfrac{5}{24}f_{n-1} + \tfrac{1}{24}f_{n-2}\right] = 0 \tag{10.62}$$

AM is next seen to preserve magnitude and compute phase angle with significantly larger time steps than Newmark for comparable levels of accuracy. However, recall that Newmark is A-stable, while AM exhibits numerical stability only up to a critical ratio (to be discussed) of the time step to the period. AM is not suitable if high-frequency modes are present—a likely occurrence in large finite element systems. As stated previously, one possible way to remove high-frequency modes is filtering using, for example, the Wavelet Packet transform.

10.4.4 Stepwise and Cumulative Error in the Adams–Moulton Method

We consider errors in free vibration of the undamped Jth mode. Again there is no magnitude error, but as in Newmark there is a phase error. Nicholson and Lin (2005) reported that the stepwise error estimate

$$\varepsilon_h \approx \frac{\pi^4}{45}\left(\frac{h}{T_J}\right)^4 \approx 2.165\left(\frac{h}{T_J}\right)^4 \tag{10.63}$$

For comparison with Newmark we again choose the stepwise relative phase angle error to be $1/300$. The result is $T_J/h \approx 4$, or four time steps per period. *This contrasts with 32 time steps per period in the Newmark method.*

The AM method shows an even greater advantage when cumulative error is considered. From Equation 10.63 the error measure $\varepsilon_h = (h/T_J)^4$ is representative of the relative phase error per step in the AM method applied to the transient solution. The number of time steps in N periods satisfies $M+1 = NT_J/h = N/(\varepsilon_h)^{1/4}$. Now $(1+\varepsilon_h)^{N/(\varepsilon_h)^{1/4}} = 1+\varepsilon_T$. Use of the natural logarithm and the first term in the Taylor series leads to $N(\varepsilon_h)^{3/4} = \varepsilon_T$, implying that the number of time steps per period is given by $\frac{T_J}{h} = \left(\frac{N}{\varepsilon_T}\right)^{1/3}$. As before, consider $N = 10$ and $\varepsilon_T = 1/10$. The number of time steps per period is now 4.7, *much lower* than 333 for Newmark.

10.4.5 Stability Limit on Time Step in the Adams–Moulton Method

For a finite time step h the eigenvalues of the AM formula are obtained from (Gear, 1971)

$$\left(1+\tfrac{3}{8}\lambda_J h\right)\eta^3 - \left(1-\tfrac{19}{24}\lambda_J h\right)\eta^2 - \tfrac{5}{24}\lambda_J h\eta + \tfrac{1}{24}\lambda_J h = 0 \tag{10.64}$$

We already know that the primary eigenvalue is $\eta = \exp(-\lambda_J h)$ to fourth order in h. This permits approximating Equation 10.64 as

$$(\eta - \exp(-\lambda_J h))\left[\eta^2 + \gamma_1\eta + \gamma_2\right] = 0$$

$$\gamma_1 = \frac{\left(\exp(-2\lambda_J h)\dfrac{\lambda_J h}{24} + \exp(-\lambda_J h)\dfrac{5\lambda_J h}{24}\right)}{\left(1+\frac{3}{8}\lambda_J h\right)}, \quad \gamma_2 = \frac{-\exp(-\lambda_J h)\dfrac{\lambda_J h}{24}}{\left(1+\frac{3}{8}\lambda_J h\right)} \tag{10.65}$$

Numerical stability requires that the extraneous eigenvalues be interior to the unit circle in the complex plane. They are given by

$$\eta_1, \eta_2 = -\gamma_1 \pm \sqrt{\gamma_1^2 - 4\gamma_2} \tag{10.66}$$

Clearly, the complex values and magnitudes can be readily computed; they are presented and displayed in the following paragraphs.

We concern ourselves with underdamped media, to which end we write

$$\lambda_J h = \omega_J h\left(\zeta_J + i\sqrt{1-\zeta_J^2}\right),\ i = \sqrt{-1},\ \mathrm{Re}(\lambda_J h) = \omega_J h\zeta_J,\ \mathrm{Im}(\lambda_J h) = \omega_J h\sqrt{1-\zeta_J^2} \tag{10.67}$$

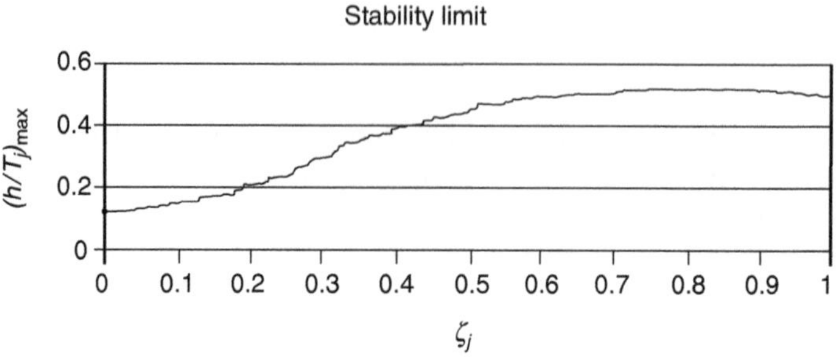

FIGURE 10.4 Maximum time step ratio vs. damping factor.

Following the language of vibration theory, ζ_J is called the damping factor and ω_J the *undamped* natural frequency (of the *J*th mode). The undamped period is $T_J = \frac{2\pi}{\omega_J}$. Upon writing $\lambda_J h = \frac{2\pi h}{T}\left(\zeta_J + i\sqrt{1-\zeta_J^2}\right)$, the goal becomes numerical evaluation of the critical time step *relative to the undamped period*, denoted $(h/T_J)_{\max}$, as a function of ζ_J. The results are depicted in Figure 10.4. Computations are performed over the "underdamped" range $0 < \zeta_J < 1$.

Two important observations may be made from Figure 10.4.

- The numerical stability limits on the time step are comparable to the previously obtained accuracy limits. For example, a critical ratio of 0.2 corresponds to five time steps in a period, while a phase error of 1/300 in a step involves four time steps per period.
- Subcritical damping (i.e., $\zeta_J < 1$), whether numerical or viscous, may triple the critical time step for stability compared to the undamped case. But, for a system whose frequencies span a large spectrum, in the high-frequency modes numerical instability associated with extraneous roots *is not necessarily suppressed by subcritical damping*, whether viscous or "numerical."

Of course, in large finite element systems there are many modes. Suppose the time step is chosen for accuracy in the lower modes but is otherwise limited from below by considerations of computational effort.

- Above the modes of interest, say the lowest 10, in a large finite element system there likely are modes for which the time step ratio violates the AM stability criteria if the time step is chosen for moderate computational effort.
- Newmark is A-stable. Even so high-frequency modes are still calculated inaccurately in consequence of the high value of h/T_K, $K \gg J$. Fortunately, in lightly damped systems, at a slight cost in accuracy it is possible to employ numerical damping to attenuate higher-order modes to remove the inaccuracy in the Newmark method.
- But, as shown above, numerical damping does not appear to be capable of removing the numerical instability in the AM method. Instead, it appears necessary to remove the high-frequency content from the output by another means, such as filtering.

10.4.6 Introducing Numerical Damping into the Adams–Moulton Method

Of course it may not be wise to insist that, in the absence of (physical) damping (i.e., "marginal stability"), the magnitude be preserved in each time step, even in the lower modes. There is a risk that, owing to round-off or truncation errors, the magnitude will grow slightly in a time step. Accordingly, in lightly damped elastic systems there is good reason to introduce "numerical damping" (cf. Zienkiewicz and Taylor, 1989), to ensure that no such growth can occur. A formulation for numerical damping in AM is given below, to ensure that magnitude decay in one time step occurs in fourth order.

In terms of the primary eigenvalue the AM formula is equivalent to the Pade approximation (Nicholson and Lin, 2005)

$$\left(1+ax+bx^2\right) \approx \left(1-x+\tfrac{1}{2}x^2-\tfrac{1}{6}x^3+\tfrac{1}{24}x^4\right)\left(1+cx+dx^2\right) \qquad (10.68)$$

in which the four coefficients a, b, c, d serve to enforce accuracy through fourth order. Now exact agreement through h^3 rather than h^4 requires (i) $h^{(1)}$: $a = c-1$, (ii) $h^{(2)}: b = d - c + \frac{1}{2}$, (iii) $h^{(3)}: 0 = -d + \frac{1}{2}c - \frac{1}{6}$. We next require that the remainder at fourth order equal χh^4, in which event (iv) $\chi = \frac{1}{2}d - \frac{1}{6}c + \frac{1}{24}$. As will be seen, the parameter χ controls magnitude decay and is selected to represent "numerical damping." The coefficients a, b, c, d are now expressed in terms of χ as

$$\text{(i) } a = -\tfrac{1}{2} + 12\chi,\ \text{(ii) } b = \tfrac{1}{12} - 6\chi,\ \text{(iii) } c = \tfrac{1}{2} + 12\chi,\ \text{(iv) } d = \tfrac{1}{12} + 6\chi \qquad (10.69)$$

We determine the effect of χ on magnitude under undamped oscillation. The amplitude decreases in a given step if $\left|\dfrac{1-\left(\frac{1}{2}-12\chi\right)i\omega h+\left(\frac{1}{12}-6\chi\right)(i\omega h)^2}{1+\left(\frac{1}{2}+12\chi\right)i\omega h+\left(\frac{1}{12}+6\chi\right)(i\omega h)^2}\right| < 1$, which is equivalent to

$$\begin{aligned} 0 &< A + B \\ A &= \left(\tfrac{1}{2}+12\chi\right)^2(\omega h)^2 - \left(\tfrac{1}{2}-12\chi\right)^2(\omega h)^2 \\ &= (\omega h)^2 24\chi \\ B &= \left[1-\left(\tfrac{1}{12}+6\chi\right)(\omega h)^2\right]^2 - \left[1-\left(\tfrac{1}{12}-6\chi\right)(\omega h)^2\right]^2 \\ &= \left[2-\tfrac{1}{6}(\omega h)^2\right]\left[-12\chi(\omega h)^2\right] \end{aligned} \qquad (10.70)$$

Consequently, the magnitude decreases if

$$0 < A + B = (\omega h)^2[24\chi - 24\chi] + 2\chi(\omega h)^4 = 2\chi(\omega h)^4 \qquad (10.71)$$

The numerical value of the magnitude decreases in each time step by Δmag given by

$$\Delta mag = \frac{A+B}{\left|1+\left(\frac{1}{2}+12\chi\right)i\omega h+\left(\left(\frac{1}{12}+6\chi\right)(i\omega h)^2\right)\right|} \approx 2\chi(\omega h)^4 + d\left(h^5\right) \quad (10.72)$$

in which the Taylor approximation has again been used. The coefficients in the numerical integration formula, Equation 10.58a through 10.58f, are now modified to satisfy

$$\beta_0+\beta_1+\beta_2+\beta_3=1 \quad (10.73a)$$

$$\alpha_0+\alpha_1+\alpha_2+\alpha_3=0 : d(h^{(0)}) \quad (10.73b)$$

$$\alpha_1+2\alpha_2+3\alpha_3-\beta_0-\beta_1-\beta_2-\beta_3=0 : d(h) \quad (10.73c)$$

$$\tfrac{7}{12}\alpha_0+\tfrac{1}{12}\alpha_1-\tfrac{17}{12}\alpha_2-\tfrac{47}{12}\alpha_3+\beta_1+2\beta_2+3\beta_3=0 : d(h^2) \quad (10.73d)$$

$$-\tfrac{5}{12}\alpha_1+\tfrac{1}{6}\alpha_2+\tfrac{11}{4}\alpha_3+\tfrac{7}{12}\beta_0+\tfrac{1}{12}\beta_1-\tfrac{17}{12}\beta_2-\tfrac{47}{12}\beta_3=0 : d(h^3) \quad (10.73e)$$

$$\tfrac{49}{144}\alpha_0+\tfrac{13}{144}\alpha_1-\tfrac{23}{144}\alpha_2+\tfrac{157}{144}\alpha_3+\tfrac{5}{12}\beta_1-\tfrac{1}{6}\beta_2-\tfrac{11}{4}\beta_3=\chi : d(h^4) \quad (10.73f)$$

from which

$$\begin{Bmatrix}\alpha_1\\ \alpha_2\\ \alpha_3\\ \beta_1\\ \beta_2\end{Bmatrix} = \begin{Bmatrix}-3/4\\ 3/4\\ 1/12\\ 1/2\\ 1/2\end{Bmatrix} + \beta_0\begin{Bmatrix}9/2\\ -3/2\\ -1/2\\ 1\\ -2\end{Bmatrix} + \beta_3\begin{Bmatrix}3/2\\ -9/2\\ 5/2\\ -2\\ 1\end{Bmatrix} + \begin{Bmatrix}-3\\ -3\\ 3\\ 6\\ -6\end{Bmatrix}\chi \quad (10.74)$$

and

$$\alpha_0 = -1/12 - 5/2\beta_0 + \tfrac{1}{2}\beta_3 + 3\chi$$

Again choosing the AM values $\beta_0 = 3/8$, $\beta_3 = 1/24$, the coefficients are now

$$\alpha_0=-1+3\chi,\ \alpha_1=1-3\chi,\ \alpha_2=-3\chi,\ \alpha_3=3\chi,\ \beta_1=\tfrac{19}{24}+6\chi,\ \beta_2=-\tfrac{5}{24}-6\chi \quad (10.75)$$

The desired numerical integration formula, representing the AM formula, modified to incorporate numerical damping, now emerges as

$$\frac{y_{n+1}-y_n+\frac{3\chi}{1-3\chi}(y_{n-1}-y_{n-2})}{h} = \frac{1}{1-3\chi}\left[\frac{3}{8}f_{n+1}+\left(\frac{19}{24}+6\chi\right)f_n - \left(\frac{5}{24}+6\chi\right)f_{n-1}+\frac{1}{24}f_{n-2}\right] \quad (10.76)$$

Clearly, if χ is set equal to zero, the AM formula (Equation 10.61) is recovered.

To illustrate attractive values of χ, suppose that the Jth mode is the highest mode for which accurate values are sought. Recall that amplitude reduction is given by $2\chi(\omega_j h)^4 = 2\chi\left(\frac{2\pi h}{T_j}\right)^4 \approx 3200\chi\left(\frac{h}{T_j}\right)^4$. To attain low attenuation of this mode, for example, for a magnitude reduction of 1/10000 per time step at 10 steps per period, we obtain $\chi \approx 3 \times 10^{-4}$.

10.4.7 AMX: Adams–Moulton Method Applied to Systems with Acceleration

For an elastic system with viscous damping, the conventional finite element equation is written as

$$\mathbf{M}\ddot{\mathbf{x}} + \mathbf{D}\dot{\mathbf{x}} + \mathbf{K}\mathbf{x} = \mathbf{f}(t) \tag{10.77}$$

in which, as usual, **M**, **D**, and **K** are the $n \times n$ mass, damping, and stiffness matrices, assumed positive definite and symmetric, **x** is the $n \times 1$ global displacement vector, and **f**(t) is the $n \times 1$ force vector, which is prescribed as a function of time.

To attain a counterpart of the Newmark formula, Equation 10.77 is expressed in *state form* as

$$\begin{bmatrix} \mathbf{M} & \mathbf{0} \\ \mathbf{0} & \mathbf{I} \end{bmatrix} \begin{Bmatrix} \dot{\mathbf{x}} \\ \mathbf{x} \end{Bmatrix}^{\cdot} + \begin{bmatrix} \mathbf{D} & \mathbf{K} \\ -\mathbf{I} & \mathbf{0} \end{bmatrix} \begin{Bmatrix} \dot{\mathbf{x}} \\ \mathbf{x} \end{Bmatrix} = \begin{Bmatrix} \mathbf{f} \\ \mathbf{0} \end{Bmatrix} \tag{10.78}$$

Equation 10.76 implies the substitutions

$$\begin{aligned}
\begin{Bmatrix} \dot{\mathbf{x}} \\ \mathbf{x} \end{Bmatrix}^{\cdot} &\approx \frac{1}{h} \begin{Bmatrix} \dot{\mathbf{x}}_{n+1} - \dot{\mathbf{x}}_n + \dfrac{3\chi}{1-3\chi}(\dot{\mathbf{x}}_{n-1} - \dot{\mathbf{x}}_{n-2}) \\ \mathbf{x}_{n+1} - \mathbf{x}_n + \dfrac{3\chi}{1-3\chi}(\mathbf{x}_{n-1} - \mathbf{x}_{n-2}) \end{Bmatrix} \\
\begin{Bmatrix} \dot{\mathbf{x}} \\ \mathbf{x} \end{Bmatrix} &\approx \begin{Bmatrix} \frac{3}{8}\dot{\mathbf{x}}_{n+1} + \left(\frac{19}{24} + 6\chi\right)\dot{\mathbf{x}}_n - \left(\frac{5}{24} + 6\chi\right)\dot{\mathbf{x}}_{n-1} + \frac{1}{24}\dot{\mathbf{x}}_{n-2} \\ \frac{3}{8}\mathbf{x}_{n+1} + \left(\frac{19}{24} + 6\chi\right)\mathbf{x}_n - \left(\frac{5}{24} + 6\chi\right)\mathbf{x}_{n-1} + \frac{1}{24}\mathbf{x}_{n-2} \end{Bmatrix} \\
\begin{Bmatrix} \mathbf{f} \\ \mathbf{0} \end{Bmatrix} &\approx \frac{1}{1-3\chi} \begin{Bmatrix} \frac{3}{8}\mathbf{f}_{n+1} + \left(\frac{19}{24} + 6\chi\right)\mathbf{f}_n - \left(\frac{5}{24} + 6\chi\right)\mathbf{f}_{n-1} + \frac{1}{24}\mathbf{f}_{n-2} \\ \mathbf{0} \end{Bmatrix}
\end{aligned} \tag{10.79}$$

Obvious reorganization furnishes

$$\frac{1}{h}\begin{bmatrix} \mathbf{M} & \mathbf{0} \\ \mathbf{0} & \mathbf{I} \end{bmatrix} \begin{Bmatrix} \dot{\mathbf{x}}_{n+1} \\ \mathbf{x}_{n+1} \end{Bmatrix} + \frac{\frac{3}{8}}{1-3\chi} \begin{bmatrix} \mathbf{D} & \mathbf{K} \\ -\mathbf{I} & \mathbf{0} \end{bmatrix} \begin{Bmatrix} \dot{\mathbf{x}}_{n+1} \\ \mathbf{x}_{n+1} \end{Bmatrix} = \begin{Bmatrix} \mathbf{r}_{n+1} \\ \mathbf{s}_{n+1} \end{Bmatrix} \tag{10.80}$$

in which

$$\begin{Bmatrix} \mathbf{r}_{n+1} \\ \mathbf{s}_{n+1} \end{Bmatrix} = \frac{1}{h}\begin{bmatrix} \mathbf{M} & \mathbf{0} \\ \mathbf{0} & \mathbf{I} \end{bmatrix} \begin{Bmatrix} \dot{\mathbf{x}}_n - \dfrac{3\chi}{1-3\chi}(\dot{\mathbf{x}}_{n-1} - \dot{\mathbf{x}}_{n-2}) \\ \mathbf{x}_n - \dfrac{3\chi}{1-3\chi}(\mathbf{x}_{n-1} - \mathbf{x}_{n-2}) \end{Bmatrix}$$
$$+ \frac{1}{1-3\chi}\begin{bmatrix} \mathbf{D} & \mathbf{K} \\ -\mathbf{I} & \mathbf{0} \end{bmatrix} \begin{Bmatrix} -\left(\frac{19}{24}+6\chi\right)\dot{\mathbf{x}}_n + \left(\frac{5}{24}+6\chi\right)\dot{\mathbf{x}}_{n-1} - \frac{1}{24}\dot{\mathbf{x}}_{n-2} \\ -\left(\frac{19}{24}+6\chi\right)\mathbf{x}_n + \left(\frac{5}{24}+6\chi\right)\mathbf{x}_{n-1} - \frac{1}{24}\mathbf{x}_{n-2} \end{Bmatrix}$$
$$+ \frac{1}{1-3\chi}\begin{Bmatrix} \frac{3}{8}\mathbf{f}_{n+1} + \left(\frac{19}{24}+6\chi\right)\mathbf{f}_n - \left(\frac{5}{24}+6\chi\right)\mathbf{f}_{n-1} + \frac{1}{24}\mathbf{f}_{n-2} \\ \mathbf{0} \end{Bmatrix} \tag{10.81}$$

Of course $\begin{Bmatrix} \mathbf{r}_{n+1} \\ \mathbf{s}_{n+1} \end{Bmatrix}$ is known from the solutions at the previous time steps. Use of the second block row in Equation 10.80 provides the identification $\dot{\mathbf{x}}_{n+1} = \frac{8}{3}(1-3\chi)\left(\frac{1}{h}\mathbf{x}_{n+1} - \mathbf{s}_{n+1}\right)$. Upon substitution in the first block row the desired fourth-order counterpart AMX of the Newmark formula, modified to incorporate numerical damping, is now

$$\begin{aligned} \mathbf{K}_D\mathbf{x}_{n+1} &= \frac{1}{(1-3\chi)}\frac{3h^2}{8}\left[\mathbf{r}_{n+1} + \left[\frac{1}{h}\mathbf{M}\frac{8}{3}(1-3\chi) + \frac{8}{3}\mathbf{D}\right]\mathbf{s}_{n+1}\right] \\ \mathbf{K}_D &= \mathbf{M} + \frac{3}{8}\frac{h}{1-3\chi}\mathbf{D} + \left(\frac{3}{8}\frac{h}{1-3\chi}\right)^2\mathbf{K} \end{aligned} \tag{10.82}$$

Solution of the linear system Equation 10.82 may be accomplished by conventional finite element procedures consisting of triangularization of the positive definite, symmetric, banded dynamic stiffness matrix $\mathbf{K}_D$, together with forward and backward substitution. In fact, the effort to do so is *exactly the same* as in the Newmark method.

10.4.8 Comments on Filtering to Remove High-Order Modes

By being confined to one step, the Newmark and other early methods (Zienkiewicz and Taylor, 1989) give rise only to the primary eigenvalue and avoid numerical instability ensuing from extraneous eigenvalues. However, high-frequency modes may still be computed inaccurately since h/T_K becomes large when $K > J$. To "attenuate" inaccuracy in high-frequency modes in one-step methods, numerical damping has often been introduced at a slight loss of accuracy, for example the Wilson-theta method. We refer to such modified Newmark methods as *near-second order*.

The AMX method attains high accuracy in the lower modes with appreciably larger and/or fewer time steps than in the Newmark method. However, in systems with a broad eigenvalue spectrum, it appears wise to introduce a technique to remove the potentially unstable high-frequency modes. Recall that subcritical damping, viscous or numerical, cannot be relied upon to obviate the numerical instability associated with extraneous eigenvalues in AM. Fortunately, even in the presence of transients, in the last two decades high-frequency mode removal has become possible using filters based on the fast and discrete versions of the Wavelet transform (cf. Kaplan, 2002; Bettayeb et al., 2004).

The classical Fourier transform introduces one parameter (the frequency) and maps the time domain onto the frequency domain. The Fourier components are globally regular functions such as sinusoids. The Wavelet transform has two parameters, which are scale and time, scale being similar to the reciprocal of the frequency. Wavelets are locally regular functions.

In simple wavelet filters, the scale function represents a "low pass filter" covering the lower half of the frequency spectrum, while the wavelet function represents a "high pass filter" covering the upper half of the spectrum (Bettayeb et al., 2004). In the current context the discrete signal of interest corresponds to displacement values computed during the last few time periods, assuming filtering was applied prior to these periods. Doing so sets the present values of the modal amplitudes to zero in the upper half of the spectrum. The reduced signal is then reconstructed in the time domain. The filter is now again applied to the reconstructed signal to partition it into low pass and high pass segments. The process is continued until the reconstructed signal has a spectrum which (a) is in the stable range, and (b) contains the highest natural frequency (mode) for which accurate computations are sought (assuming (b) is compatible with (a)).

11 Additional Topics in Linear Thermoelastic Systems

Topics addressed in this chapter include linear conductive heat transfer, linear thermoelasticity, incompressible elastic materials, elastic torsion, and buckling.

11.1 TRANSIENT CONDUCTIVE HEAT TRANSFER IN LINEAR MEDIA

11.1.1 Finite Element Equation

The governing equation for conductive heat transfer without heat sources in an isotropic medium is

$$k\nabla^2 \mathrm{T} = \rho c_e \dot{\mathrm{T}} \tag{11.1}$$

in which T is the (absolute) temperature, k is the thermal conductivity, and c_e is the coefficient of specific heat at constant strain. We invoke the interpolation model $\mathrm{T}(t) - \mathrm{T}_0 = \boldsymbol{\varphi}_T^T(x)\boldsymbol{\Phi}_T\boldsymbol{\theta}(t)$ in which $\boldsymbol{\theta}(t)$ is the vector of nodal temperatures (minus T_0), while $\boldsymbol{\varphi}_T^T(x)$ and $\boldsymbol{\Phi}_T$ are the thermal counterparts of $\boldsymbol{\varphi}^T(x)$ and $\boldsymbol{\Phi}$ in mechanical fields. Also application of the gradient leads to a relation of the form $\nabla\mathrm{T} = \boldsymbol{\beta}_T^T\,\boldsymbol{\Phi}_T\,\boldsymbol{\theta}(t)$, and the finite element equation assumes the form

$$\mathbf{K}_T\boldsymbol{\theta} + \mathbf{M}_T\dot{\boldsymbol{\theta}} = -\boldsymbol{q}(\mathrm{T}) \tag{11.2}$$

$$\mathbf{K}_T = \int \boldsymbol{\Phi}_T^T\boldsymbol{\beta}_T k \boldsymbol{\beta}_T^T\boldsymbol{\Phi}_T\,\mathrm{d}V, \quad \mathbf{M}_T = \int \boldsymbol{\Phi}_T^T\boldsymbol{\varphi}_T\rho c_e\boldsymbol{\varphi}_T^T\boldsymbol{\Phi}_T^T\,\mathrm{d}V$$

This equation is parabolic (first order in the time rates), and implies that the temperature changes occur immediately at all points in the domain, but at smaller initial rates away from where the heat is added. This contrasts with the hyperbolic (second order in time rates) solid mechanics equations, in which information propagates into the unperturbed medium as finite velocity waves, and in which oscillatory response occurs in response to a perturbation.

11.1.2 Direct Integration by the Trapezoidal Rule

Equation 11.1 is already in state form since it is first order, and the trapezoidal rule can be applied directly.

$$\mathbf{M}_T \frac{\boldsymbol{\theta}_{n+1} - \boldsymbol{\theta}_n}{h} + \mathbf{K}_T \frac{\boldsymbol{\theta}_{n+1} + \boldsymbol{\theta}_n}{2} = -\frac{\mathbf{q}_{n+1} + \mathbf{q}_n}{2} \tag{11.3}$$

from which

$$\mathbf{K}_{DT}\boldsymbol{\theta}_{n+1} = \mathbf{r}_{n+1} \tag{11.4}$$

$$\mathbf{K}_{DT} = \mathbf{M}_T + \tfrac{h}{2}\mathbf{K}_T \mathbf{r}_{n+1} = \mathbf{M}_T\boldsymbol{\theta}_n - \tfrac{h}{2}\mathbf{K}_T\boldsymbol{\theta}_n - \tfrac{h}{2}(\mathbf{q}_{n+1} + \mathbf{q}_n)$$

For the assumed conditions the dynamic thermal stiffness matrix is positive definite and for the current time step the foregoing equation can be solved in the same manner as in the static counterpart, namely triangularization followed by forward substitution.

11.1.3 Modal Analysis in Linear Thermoelasticity

Modes are not of much interest in thermal problems since the modes are not oscillatory or useful to visualize. However, the foregoing equation can still be decomposed into independent single degree of freedom systems. First we note that the thermal system is asymptotically stable. In particular, suppose the inhomogeneous term vanishes and that $\boldsymbol{\theta}$ at $t = 0$ does not vanish. Multiplying the foregoing equation by $\boldsymbol{\theta}^T$ and elementary manipulation furnishes that

$$\frac{\mathrm{d}}{\mathrm{d}t}\left(\frac{\boldsymbol{\theta}^T\mathbf{M}_T\boldsymbol{\theta}}{2}\right) = -\boldsymbol{\theta}^T\mathbf{K}_T\boldsymbol{\theta} < 0 \tag{11.5}$$

Clearly the product $\boldsymbol{\theta}^T\mathbf{M}_T\boldsymbol{\theta}$ decreases continuously. But it vanishes only if $\boldsymbol{\theta}$ vanishes.

Next, to examine the modes assume a solution of the form $\boldsymbol{\theta}(t) = \boldsymbol{\theta}_{0j}\exp(\lambda_j t)$. The eigenvectors $\boldsymbol{\theta}_{0j}$ satisfy

$$\boldsymbol{\theta}_{0j}^T\mathbf{M}_T\boldsymbol{\theta}_{0k} = \begin{cases} \mu_{Tj}, & j = k \\ 0, & j \neq k \end{cases}, \quad \boldsymbol{\theta}_{0j}^T\mathbf{K}_T\boldsymbol{\theta}_{0k} = \begin{cases} \kappa_{Tj}, & j = k \\ 0, & j \neq k \end{cases} \tag{11.6}$$

and we call μ_{Tj} and κ_{Tj} the jth modal thermal mass and jth modal thermal stiffness. We may also form the modal matrix $\boldsymbol{\Theta} = [\boldsymbol{\theta}_{01} \cdots \boldsymbol{\theta}_{0n}]$, and again

$$\boldsymbol{\Theta}^T\mathbf{M}_T\boldsymbol{\Theta} = \begin{bmatrix} \mu_{Tj} & 0 & . & . & 0 \\ 0 & \mu_{Tj} & . & . & . \\ . & . & . & . & . \\ . & . & . & . & . \\ 0 & . & . & . & . \end{bmatrix}, \quad \boldsymbol{\Theta}^T\mathbf{K}_T\boldsymbol{\Theta} = \begin{bmatrix} \kappa_{Tj} & 0 & . & . & 0 \\ 0 & \kappa_{Tj} & . & . & . \\ . & . & . & . & . \\ . & . & . & . & . \\ 0 & . & . & . & . \end{bmatrix} \tag{11.7}$$

Let $\xi = \boldsymbol{\Theta}^{-1}\boldsymbol{\theta}$ and $\mathbf{g}(t) = \boldsymbol{\Theta}^T\mathbf{q}(t)$. Pre- and post-multiplying Equation 11.2 with $\boldsymbol{\Theta}^T$ and $\boldsymbol{\Theta}$, respectively, furnishes the decoupled equations

$$\mu_{Tj}\dot{\xi}_j + \kappa_{Tj}\xi_j = g_j \tag{11.8}$$

Supposing for convenience that g_j is a constant, the general solution is of the form

$$\xi_j(t) = \xi_{j0}\exp\left(-\frac{\kappa_{Tj}}{\mu_{Tj}}t\right) + \int_0^t \exp\left(-\frac{\kappa_{Tj}}{\mu_{Tj}}(t-\tau)\right)g_j(\tau)\,\mathrm{d}\tau \tag{11.9}$$

illustrating the monotonically decreasing nature of the free ($\mathbf{g} = \mathbf{0}$) response. Now there are n uncoupled single degrees of freedom.

11.2 COUPLED LINEAR THERMOELASTICITY

11.2.1 Finite Element Equation

The classical theory of coupled thermoelasticity accommodates the fact that the thermal and mechanical fields are coupled. For isotropic materials, assuming that temperature only affects the volume of an element, the stress–strain relation is

$$\mathrm{S}_{ij} = 2\mu\mathrm{E}_{ij} + \lambda(\mathrm{E}_{kk} - \alpha(\mathrm{T} - \mathrm{T}_0))\delta_{ij} \tag{11.10}$$

in which α denotes the volumetric thermal expansion coefficient. The equilibrium equation is repeated as $\frac{\partial \mathrm{S}_{ij}}{\partial x_i} = \rho\ddot{u}_j$. The Principle of Virtual Work (Chapter 5) implies that

$$\int \delta\mathrm{E}_{ij}\left[2\mu\mathrm{E}_{ij} + \lambda\mathrm{E}_{kk}\delta_{ij}\right]\mathrm{d}V_0 + \int \delta u_i\rho\ddot{u}_i\,\mathrm{d}V_0 - \alpha\lambda\int \delta\mathrm{E}_{ij}(\mathrm{T} - \mathrm{T}_0)\delta_{ij}\,\mathrm{d}V_0 = \int \delta\delta u_j t_j\,\mathrm{d}S_0 \tag{11.11}$$

in which, as before, t_j refers to the traction vector.

Now consider the interpolation models

$$\mathbf{u}(\mathbf{x},t) = \mathbf{N}^T(\mathbf{x})\boldsymbol{\gamma}(t),\ \mathrm{E}_{ij} \rightarrow \mathbf{e} = \mathbf{B}^T(\mathbf{x})\boldsymbol{\gamma}(t),\ \mathrm{T} - \mathrm{T}_0 = \boldsymbol{\nu}^T(\mathbf{x})\boldsymbol{\theta}(t),\ \nabla\mathrm{T} = \mathbf{B}_T^T(\mathbf{x})\boldsymbol{\theta}(\mathbf{x}) \tag{11.12}$$

in which $\mathbf{e}$ is the strain written in conventional finite element notation as a column vector. As before, $\mathbf{N}$ is the shape function matrix, its thermal counterpart is $\boldsymbol{\nu}(\mathbf{x})$, $\mathbf{B}$ is the strain–displacement matrix, and $\mathbf{B}_T$ is its thermal counterpart. Now familiar procedures furnish the finite element equation

$$\mathbf{M}\ddot{\boldsymbol{\gamma}}(t) + \mathbf{K}\boldsymbol{\gamma}(t) - \boldsymbol{\Sigma}\boldsymbol{\theta}(t) = \mathbf{f}(t), \quad \boldsymbol{\Sigma} = \alpha\lambda\int \mathbf{B}\boldsymbol{\nu}^T\,\mathrm{d}V_0 \tag{11.13}$$

The quantity $\boldsymbol{\Sigma}$ may be called the *thermomechanical stiffness matrix*. If there are n_m displacement degrees of freedom and n_t temperature degrees of freedom, the quantities appearing in the equation have dimensions according to

$$\mathbf{M}, \mathbf{K} : n_m \times n_m, \quad \boldsymbol{\gamma}(t), \mathbf{f}(t) : n_m \times 1, \quad \boldsymbol{\Sigma} : n_m \times n_t, \quad \boldsymbol{\theta}(t) : n_t \times 1$$

We next address the thermal field. The energy balance equation, including coupling to mechanical effects, is given by

$$k\nabla^2 \mathrm{T} = \rho c_e \dot{\mathrm{T}} + \alpha\lambda \mathrm{T}_0\, tr(\dot{\mathbf{E}}) \tag{11.14}$$

Application of the usual variational methods and interpolation models implies that

$$\mathbf{K}_T\boldsymbol{\theta}(t) + \mathbf{M}_T\dot{\boldsymbol{\theta}}(t) + T_0\boldsymbol{\Sigma}^T\dot{\boldsymbol{\gamma}}(t) = -\mathbf{q}, \quad \mathbf{q} = \int \nu\mathbf{n}\cdot\mathbf{q}\,\mathrm{d}S \tag{11.15}$$

Now consider the special case in which T is constant. Then, at the global level, $\boldsymbol{\theta}(t) = -\mathrm{T}_0\mathbf{K}_T^{-1}\Sigma^T\dot{\boldsymbol{\gamma}}(t)$. The thermal field is thus eliminated at the global level, giving the new governing equation as

$$\mathbf{M}\ddot{\boldsymbol{\gamma}}(t) + \mathrm{T}_0\boldsymbol{\Sigma}\mathbf{K}_T^{-1}\boldsymbol{\Sigma}^T\dot{\boldsymbol{\gamma}}(t) + \mathbf{K}\boldsymbol{\gamma}(t) = \mathbf{f}(t) \tag{11.16}$$

Conductive heat transfer is thereby seen to be analogous to damping. The thermomechanical system is now asymptotically stable (positive effective damping) rather than asymptotically marginally stable (no effective damping).

We next express the global equations in *state form* as

$$\mathbf{Q}_1\dot{\mathbf{z}} + \mathbf{Q}_2\mathbf{z} = \mathbf{f} \tag{11.17}$$

in which

$$\mathbf{Q}_1 = \begin{bmatrix} \mathbf{M} & \mathbf{0} & \mathbf{0} \\ \mathbf{0} & \mathbf{K} & \mathbf{0} \\ \mathbf{0} & \mathbf{0} & \mathbf{M}_T/\mathrm{T}_0 \end{bmatrix}, \quad \mathbf{z} = \begin{pmatrix} \dot{\boldsymbol{\gamma}} \\ \boldsymbol{\gamma} \\ \boldsymbol{\theta} \end{pmatrix}$$

$$\mathbf{Q}_2 = \begin{bmatrix} \mathbf{0} & \mathbf{K} & -\boldsymbol{\Sigma} \\ -\mathbf{K} & \mathbf{0} & \mathbf{0} \\ \boldsymbol{\Sigma}^{\mathrm{T}} & \mathbf{0} & \mathbf{K}_T/\mathrm{T}_0 \end{bmatrix}, \quad f = \begin{pmatrix} \mathbf{f} \\ \mathbf{0} \\ -\mathbf{q}/\mathrm{T}_0 \end{pmatrix}$$

Clearly Equation 11.17 can be integrated numerically using the trapezoidal rule:

$$\left[\mathbf{Q}_1 + \tfrac{h}{2}\mathbf{Q}_2\right]\mathbf{z}_{n+1} = \left[\mathbf{Q}_1 - \tfrac{h}{2}\mathbf{Q}_2\right]\mathbf{z}_n + \tfrac{1}{2}[\mathbf{f}_{n+1} + \mathbf{f}_n] \tag{11.18}$$

We consider asymptotic stability, for which purpose it is sufficient to take $\mathbf{f} = \mathbf{0}$, $\mathbf{z}(0) = \mathbf{z}_0$. Upon pre-multiplying Equation 11.17 by $\mathbf{z}^T$, we obtain

$$\frac{\mathrm{d}}{\mathrm{d}t}\left(\tfrac{1}{2}\mathbf{z}^T\mathbf{Q}_1\mathbf{z}\right) = -\mathbf{z}^T\mathbf{Q}_2\mathbf{z}$$
$$= -\mathbf{z}^T\tfrac{1}{2}\left[\mathbf{Q}_2 + \mathbf{Q}_2^T\right]\mathbf{z}$$
$$= -\boldsymbol{\theta}^T\mathbf{K}_T\boldsymbol{\theta} \tag{11.19}$$

and $\mathbf{z}$ must be real. Assuming that $\boldsymbol{\theta} \neq \mathbf{0}$, it follows that $\mathbf{z} \downarrow \mathbf{0}$, and hence the system is asymptotically stable.

EXAMPLE 11.1

Find the exact solution for a circular rod of length L, radius r, mass density ρ, specific heat c_e, conductivity k, and cross-sectional area $A=\pi r^2$. The initial temperature is T_0, and the rod is built into a large wall at fixed temperature T_0 (see Figure 11.1). However, at time $t=0$, the temperature T_1 is imposed at $x=L$. Compare the exact solution to the one- and two-element solutions. Note that for a one-element model

$$\frac{kA}{L}\theta(L,t) + \frac{\rho c_e AL}{3}\dot{\theta}(L,t) = -q(L)$$

Solution

(i) *Exact solution*

The governing equation for conductive heat transfer in one dimension is given by

$$k\frac{\partial^2 \mathrm{T}}{\partial x^2} = \rho c_e \frac{\partial \mathrm{T}}{\partial t}$$

We seek to solve this equation using "Separation of Variables." If $\alpha = k/\rho c_e$.

$$\frac{\partial^2 \mathrm{T}}{\partial x^2} = \frac{1}{\alpha}\frac{\partial \mathrm{T}}{\partial t}$$

Now let $\mathrm{T}^\# = \mathrm{T} - [\mathrm{T}_0 + x(\mathrm{T}_1 - \mathrm{T}_0)/L]$. Now $\frac{\partial^2 \mathrm{T}^\#}{\partial x^2} = \frac{1}{\alpha}\frac{\partial \mathrm{T}^\#}{\partial t}$, with $\mathrm{T}^\#(0) = \mathrm{T}^\#(L) = 0$.

Next assume the spatial-temporal decomposition $\mathrm{T}^\#(x,t) = X(x)\hat{\mathrm{T}}(t)$, with the consequence that

$$\frac{X''}{X} = \frac{1}{\alpha}\frac{\dot{\hat{\mathrm{T}}}}{\hat{\mathrm{T}}} = -\lambda_j^2, \quad \lambda_j \text{ the } j\text{th eigenvalue}$$

The function $X(x)$ corresponding to λ_j is now denoted as $X_j(x)$ and similarly for $\mathrm{T}_j(t)$. The two functions satisfy

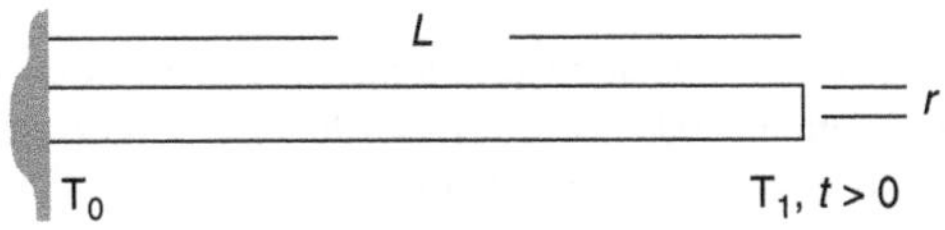

FIGURE 11.1 One-element thermal conductor.

$$X_j'' + \lambda_j^2 X_j = 0 \rightarrow X_j = A_j \cos(\lambda_j x) + B_j \sin(\lambda_j x)$$
$$\mathrm{T}_j' + \alpha\lambda_j^2 \mathrm{T}_j = 0 \rightarrow \mathrm{T}_j = C_j \exp\left(-\alpha\lambda_j^2 t\right)$$

in which A_j, B_j, and C_j remain to be determined. The eigenvalues satisfy $\sin(\lambda_j L) = 0$ which implies that $\lambda_j = j\pi/L$.

With $D_j = A_j C_j$ and $E_j = B_j C_j$, the solution assumes the form

$$\mathrm{T}(x,t) = \sum_j \left(D_j \cos(j\pi x/L) + E_j \sin(j\pi x/L)\right) \exp\left(-\alpha j^2 \pi^2 t/L\right) + \mathrm{T}_0 + \frac{x(\mathrm{T}_1 - \mathrm{T}_0)}{L}$$

The initial condition $\mathrm{T}(x,0) = \mathrm{T}_0$ and standard application of the orthogonality properties of the eigenfunctions (e.g., Hildebrand, 1976) $\cos(j\pi x/L)$ and $\sin(j\pi x/L)$ permit determination of the coefficients D_j and E_j.

(ii) *Finite element solution*

For a thermal element with natural coordinates $\xi = -1$ at $x = x_e$, and $\xi = +1$ at $x = x_{e+1}$,

$$\boldsymbol{\phi}_T^T = (1 \quad \xi), \quad \boldsymbol{\Phi}_T = \begin{bmatrix} 1 & -1 \\ 1 & 1 \end{bmatrix}^{-1} = \frac{1}{2}\begin{bmatrix} 1 & 1 \\ -1 & 1 \end{bmatrix}$$

Also $\mathrm{d}x = \frac{L}{2}\mathrm{d}\xi$, and $\frac{\mathrm{d}}{\mathrm{d}x} = \frac{2}{L}\frac{\mathrm{d}}{\mathrm{d}\xi}$. Consequently, the thermal stiffness (conductance) matrix is given by

$$\begin{aligned}
\mathbf{K}_T &= \int \boldsymbol{\Phi}_T^T \boldsymbol{\beta}_T k \boldsymbol{\beta}_T^T \boldsymbol{\Phi}_T \, \mathrm{d}V \\
&= \frac{1}{4}\begin{bmatrix} 1 & -1 \\ 1 & 1 \end{bmatrix} \int_{-1}^{+1} \frac{2}{L}\begin{pmatrix} 0 \\ 1 \end{pmatrix} k \frac{2}{L}(0 \quad 1) A \frac{L}{2} \mathrm{d}\xi \begin{bmatrix} 1 & 1 \\ -1 & 1 \end{bmatrix} \\
&= \frac{kA}{L}\begin{bmatrix} 1 & -1 \\ -1 & 1 \end{bmatrix}
\end{aligned}$$

Continuing, the thermal mass (capacitance) matrix is now

$$\begin{aligned}
\mathbf{M}_T &= \int \boldsymbol{\Phi}_T^T \boldsymbol{\varphi}_T \rho c_e \boldsymbol{\varphi}_T^T \boldsymbol{\Phi}_T \, \mathrm{d}V \\
&= \frac{\rho A c_e L}{8}\begin{bmatrix} 1 & -1 \\ 1 & 1 \end{bmatrix} \int_{-1}^{+1} \begin{pmatrix} 1 \\ \xi \end{pmatrix} (1 \quad \xi)\, \mathrm{d}\xi \begin{bmatrix} 1 & 1 \\ -1 & 1 \end{bmatrix} \\
&= \frac{\rho A c_e L}{3}\begin{bmatrix} 1 & 1/2 \\ 1/2 & 1 \end{bmatrix}
\end{aligned}$$

(iia) *One-element solution: conductor built on the right-hand side (rhs)*

Here, the interpolation model is $\mathrm{T} - \mathrm{T}_0 = x\theta(L,t)/L$. Consequently, the thermal stiffness and mass matrices degenerate to scalars:

$$\mathbf{K}_T = \frac{kA}{L}, \quad \mathbf{M}_T = \frac{\rho A c_e L}{3}$$

Substitution into Equation 11.2 furnishes the one-element equation

$$\frac{kA}{L}\theta(L,t) + \frac{\rho c_e AL}{3}\dot{\theta}(L,t) = -q(L)$$

This equation is to be solved for $\theta(L,t)$ with an assumed value of $q(L)$ adjusted to enforce the constraint $T(L) = T_1$.

(iib) *Two-element solution*

Now the length of each element is $L/2$, and so

$$\mathbf{K}_T = \frac{kA}{(L/2)}\begin{bmatrix} 1 & -1 \\ -1 & 1 \end{bmatrix}, \quad \mathbf{M}_T = \frac{\rho A c_e(L/2)}{3}\begin{bmatrix} 1 & 1/2 \\ 1/2 & 1 \end{bmatrix}$$

After simple manipulation the assembled global finite element equation emerges as

$$\frac{kA}{(L/2)}\begin{bmatrix} 1 & -1 & 0 \\ -1 & 2 & -1 \\ 0 & -1 & 1 \end{bmatrix}\begin{pmatrix} \theta(0,t) \\ \theta(L/2,t) \\ \theta(L,t) \end{pmatrix} + \frac{\rho A c_e(L/2)}{3}\begin{bmatrix} 1 & 1/2 & 0 \\ 1/2 & 2 & 1/2 \\ 0 & 1/2 & 1 \end{bmatrix}\begin{pmatrix} \theta(0,t) \\ \theta(L/2,t) \\ \theta(L,t) \end{pmatrix}^{\cdot} = \begin{pmatrix} q(L) \\ 0 \\ -q(L) \end{pmatrix}$$

But $\theta(0,t) = 0$, and in consequence

$$\frac{kA}{(L/2)}\begin{bmatrix} 2 & -1 \\ -1 & 1 \end{bmatrix}\begin{pmatrix} \theta(L/2,t) \\ \theta(L,t) \end{pmatrix} + \frac{\rho A c_e(L/2)}{3}\begin{bmatrix} 2 & 1/2 \\ 1/2 & 1 \end{bmatrix}\begin{pmatrix} \theta(L/2,t) \\ \theta(L,t) \end{pmatrix}^{\cdot} = \begin{pmatrix} 0 \\ -q(L) \end{pmatrix}$$

as expected.

11.2.2 Thermoelasticity in an Elastic Conductor

Consider a thermoelastic rod which is built into a large rigid, nonconducting temperature reservoir at $x = 0$. The force f_0 and heat flux $-q_0$ are prescribed at $x = L$. A single element is used to model the rod. Now

$$u(x,t) = x\gamma(t)/L, \quad E(x,t) = \gamma(t)/L, \quad T - T_0 = x\theta(t)/L, \quad dT/dx = \theta(t)/L \quad (11.20)$$

The thermoelastic stiffness matrix becomes $\mathbf{\Sigma} = \alpha\lambda \int \mathbf{B}\nu^T dV \to \Sigma = \alpha\lambda A/2$. The governing equations are now

$$\begin{aligned} \frac{\rho AL}{3}\ddot{\gamma} + \frac{EA}{L}\gamma - \frac{1}{2}\alpha\lambda A\theta &= f_0 \\ \frac{1}{T_0}\frac{\rho c_e AL}{3}\dot{\theta} + \frac{1}{T_0}\frac{kA}{L}\theta + \frac{1}{2}\alpha\lambda A\dot{\gamma} &= -q_0 \end{aligned} \quad (11.21)$$

EXAMPLE 11.2

State the equations of a thermoelastic medium in state form.

Solution

The equations of a thermoelastic medium are:

$$\mathbf{M}\ddot{\boldsymbol{\gamma}}(t) + \mathbf{K}\boldsymbol{\gamma}(t) - \boldsymbol{\Sigma}\boldsymbol{\theta}(t) = \mathbf{f}(t)$$

$$\frac{1}{\mathrm{T}_0}\mathbf{K}_T\boldsymbol{\theta}(t) + \frac{1}{\mathrm{T}_0}\mathbf{M}_T\dot{\boldsymbol{\theta}}(t) + \boldsymbol{\Sigma}^{\mathrm{T}}\dot{\boldsymbol{\gamma}}(t) = -\frac{1}{\mathrm{T}_0}\mathbf{q}$$

$$\mathbf{M} \rightarrow \frac{\rho AL}{3}, \quad \mathbf{K} \rightarrow \frac{\mathrm{E}A}{L}, \quad \mathbf{K}_T \rightarrow \frac{kA}{L}, \quad \mathbf{M}_T \rightarrow \frac{\rho c_e AL}{3}, \quad \boldsymbol{\Sigma} = \frac{\lambda\alpha A}{2}$$

On converting state form using $\mathbf{p} = \dot{\boldsymbol{\gamma}}$, we have

$$\begin{bmatrix} \mathbf{M} & \mathbf{0} & \mathbf{0} \\ \mathbf{0} & \mathbf{K} & \mathbf{0} \\ \mathbf{0} & \mathbf{0} & \mathbf{M}_T/\mathrm{T}_0 \end{bmatrix} \begin{pmatrix} \mathbf{p} \\ \boldsymbol{\gamma} \\ \boldsymbol{\theta} \end{pmatrix}^{\cdot} + \begin{bmatrix} \mathbf{0} & \mathbf{K} & -\boldsymbol{\Sigma} \\ -\mathbf{K} & \mathbf{0} & \mathbf{0} \\ \boldsymbol{\Sigma}^{\mathrm{T}} & \mathbf{0} & \mathbf{K}_T/\mathrm{T}_0 \end{bmatrix} \begin{pmatrix} \mathbf{p} \\ \boldsymbol{\gamma} \\ \boldsymbol{\theta} \end{pmatrix} = \begin{pmatrix} \mathbf{f} \\ \mathbf{0} \\ -\mathbf{q}/\mathrm{T}_0 \end{pmatrix}$$

If the trapezoidal rule with timestep h is applied, an algebraic equation arises in which appears the dynamic thermoelastic stiffness matrix

$$\mathbf{K}_{DT} = \begin{bmatrix} \mathbf{M} & \frac{h}{2}\mathbf{K} & -\frac{h}{2}\boldsymbol{\Sigma} \\ -\frac{h}{2}\mathbf{K} & \mathbf{K} & \mathbf{0} \\ \frac{h}{2}\boldsymbol{\Sigma}^T & \mathbf{0}^T & \mathbf{K}_T \end{bmatrix}$$

Some manipulation serves to verify that $\mathbf{K}_{DT}$ satisfies the triangularization

$$\mathbf{K}_{DT} = \begin{bmatrix} \mathbf{M}^{1/2} & \mathbf{0}^T \\ -\frac{h}{2}\begin{bmatrix} \mathbf{K}\mathbf{M}^{-1/2} \\ -\boldsymbol{\Sigma}^T\mathbf{M}^{-1/2} \end{bmatrix} \begin{bmatrix} \mathbf{K} & \mathbf{0} \\ \mathbf{0}^T & \mathbf{K}_T \end{bmatrix} + \frac{h^2}{4}\begin{bmatrix} \mathbf{K}\mathbf{M}^{-1/2} \\ \boldsymbol{\Sigma}^T\mathbf{M}^{-1/2} \end{bmatrix} & \begin{bmatrix} \mathbf{M}^{-1/2}\mathbf{K} & -\mathbf{M}^{-1/2}\boldsymbol{\Sigma} \end{bmatrix} \end{bmatrix}$$

$$\times \begin{bmatrix} \mathbf{M}^{1/2} & \frac{h}{2}\begin{bmatrix} \mathbf{M}^{-1/2}\mathbf{K} & -\mathbf{M}^{-1/2}\boldsymbol{\Sigma} \end{bmatrix} \\ \mathbf{0}^T & \mathbf{I} \end{bmatrix}$$

11.3 INCOMPRESSIBLE ELASTIC MEDIA

Rubber and polymer-based plastics, as well as biological tissue, frequently exemplify the *internal constraint of incompressibility*.

For a compressible elastic material, the isotropic stress S_{kk} and the volume (dilatational) strain E_{kk} are related by $\mathrm{S}_{kk} = 3\kappa\mathrm{E}_{kk}$, in which $\kappa = \mathrm{E}/[3(1-2\nu)]$ is recognized as the bulk modulus. Clearly, as $\nu \rightarrow 1/2$, the (hydrostatic) pressure $p = -\mathrm{S}_{kk}/3$ needed to attain a finite compressive volume strain ($\mathrm{E}_{kk} < 0$) becomes infinite using elastic relations in their current form. However, we will see that this difficulty is avoided by correctly incorporating incompressibility through a limiting process as $\nu \rightarrow 1/2$.

Consider the case of plane strain implying that $\mathrm{E}_{zz} = 0$. The tangent modulus matrix $\mathbf{D}$ is now readily found as

$$\begin{pmatrix} S_{xx} \\ S_{yy} \\ S_{zz} \end{pmatrix} = \boldsymbol{D} \begin{pmatrix} E_{xx} \\ E_{yy} \\ E_{zz} \end{pmatrix}, \quad \boldsymbol{D} = \frac{\mathsf{E}}{(1+\nu)(1-2\nu)} \begin{bmatrix} 1-\nu^2 & -\nu(1+\nu) & 0 \\ -\nu(1+\nu) & 1-\nu^2 & 0 \\ 0 & 0 & 1-2\nu \end{bmatrix} \tag{11.22}$$

Clearly, in the current form $\boldsymbol{D}$ becomes unbounded as $\nu \rightarrow 1/2$. Further, suppose that for a material through to be nearly incompressible ν is estimated as 0.495 while the correct value is 0.49. It might be supposed that the estimated value is a good approximation for the correct value. However, for the correct value $(1-2\nu)^{-1} = 50$, but for the estimated value $(1-2\nu)^{-1} = 100$, implying 100% error.

The problem of unbounded magnitude is addressed as follows. In an incompressible material a pressure field $p(\mathbf{x})$ arises which serves to enforce the incompressibility constraint. Since the trace of the strains vanishes everywhere, the strains are not sufficient to determine the stresses. However, the strains together with the pressure are sufficient. In FEA, a general interpolation model is used at the outset for the displacement field. Another interpolation model must be introduced for the pressure field. Owing to the fact that pressures are stress variables, the order of the interpolation should be one degree lower than for the displacement interpolation model (Hughes, 2000). The Principle of Virtual Work is now expressed in terms of the displacements and pressure, and an adjoining equation is introduced to enforce the incompressibility constraint a posteriori. The pressure may be shown to serve as a Lagrange multiplier, in which event the displacement vector and the pressure are *varied independently*.

In incompressible materials, to preserve finite stresses we suppose that the second Lamé coefficient satisfies $\lambda \rightarrow \infty$ as $tr(\mathbf{E}) \rightarrow 0$, in such a way that the product is an indeterminate quantity denoted by p:

$$\lambda \, tr(\mathbf{E}) \rightarrow -p(\mathbf{x},t) \tag{11.23}$$

The Lamé form of the elastic constitutive relations is replaced by

$$S_{ij} = 2\mu E_{ij} - p\delta_{ij} \tag{11.24}$$

together with the incompressibility constraint $E_{ij}\delta_{ij} = 0$. There now are two independent principal strains and the pressure with which to determine the three principal stresses.

In a compressible elastic material the strain energy function w satisfies $S_{ij} = \frac{\partial w}{\partial E_{ij}}$, and the domain term in the Principle of Virtual Work may be rewritten as $\int \delta E_{ij} S_{ij} \, dV = \int \delta w \, dV$. The elastic strain energy is given by $w = \mu E_{ij} E_{ij} + \frac{\lambda}{2} E_{kk}^2$. For reasons to be seen shortly, we introduce the augmented strain energy function

$$w' = \mu E_{ij} E_{ij} - pE_{kk} \tag{11.25}$$

and assume the variational principle

$$\int \delta w' \, dV_0 + \int \delta \mathbf{u}^T \rho \ddot{\mathbf{u}} \, dV_0 = \int \delta \mathbf{u}^T \mathbf{t} \, dS_0 \tag{11.26}$$

Now considering $\mathbf{u}$ and p to vary independently, the integrand of the first term becomes $\delta w' = \delta E_{ij}[2\mu E_{ij} - p\delta_{ij}] - \delta p E_{kk}$, furnishing two variational relations

$$\int tr(\delta\mathbf{ES})\,\mathrm{d}V_0 + \int \delta\mathbf{u}^T\rho\ddot{\mathbf{u}}\,\mathrm{d}V_0 = \int \delta\mathbf{u}^T\mathbf{t}\,\mathrm{d}S_0 \tag{11.27a}$$

$$\int \delta p \mathrm{E}_{kk}\,\mathrm{d}V_0 = 0 \tag{11.27b}$$

The first relation may be recognized as a counterpart the Principle of Virtual Work (variational principle for the displacement field), and the second equation serves to enforce the internal constraint of incompressibility (variational principle for the pressure field).

Next introduce the interpolation models

$$\begin{aligned} \mathbf{u} &= \mathbf{N}^T(\mathbf{x})\boldsymbol{\gamma}(t), \quad \mathbf{e} = \mathbf{B}^T(\mathbf{x})\boldsymbol{\gamma}(t) \\ \mathrm{E}_{kk} &= \mathbf{b}^T(\mathbf{x})\boldsymbol{\gamma}(t), \quad p(\mathbf{x},t) = \boldsymbol{\xi}^T(\mathbf{x})\boldsymbol{\pi}(t) \end{aligned} \tag{11.28}$$

Substitution serves to derive that

$$\begin{aligned} &\mathbf{M}\ddot{\boldsymbol{\gamma}}(t) + \mathbf{K}\boldsymbol{\gamma}(t) - \boldsymbol{\Sigma}\boldsymbol{\pi}(t) = \mathbf{f}(t) \\ &\boldsymbol{\Sigma} = \int \mathbf{b}\boldsymbol{\xi}^T\mathrm{d}V_0, \quad \boldsymbol{\Sigma}^T\boldsymbol{\gamma}(t) = \mathbf{0} \end{aligned} \tag{11.29}$$

Assuming these equations apply at the global level, *state form* is expressed as

$$\begin{bmatrix} \mathbf{M} & \mathbf{0} & \mathbf{0} \\ \mathbf{0} & \mathbf{K} & \mathbf{0} \\ \mathbf{0}^T & \mathbf{0}^T & \mathbf{0} \end{bmatrix} \frac{\mathrm{d}}{\mathrm{d}t}\begin{pmatrix} \dot{\boldsymbol{\gamma}}(t) \\ \boldsymbol{\gamma}(t) \\ \boldsymbol{\pi}(t) \end{pmatrix} + \begin{bmatrix} \mathbf{M} & \mathbf{K} & -\boldsymbol{\Sigma} \\ -\mathbf{K} & \mathbf{0} & \mathbf{0} \\ \boldsymbol{\Sigma}^T & \mathbf{0}^T & \mathbf{0} \end{bmatrix}\begin{pmatrix} \dot{\boldsymbol{\gamma}}(t) \\ \boldsymbol{\gamma}(t) \\ \boldsymbol{\pi}(t) \end{pmatrix} = \begin{pmatrix} \mathbf{f}(t) \\ \mathbf{0} \\ \mathbf{0} \end{pmatrix} \tag{11.30}$$

The second matrix is antisymmetric except for the upper left-hand diagonal term. Further, the system exhibits marginal asymptotic stability; namely, if $\mathbf{f}(t) = \mathbf{0}$ while $\dot{\boldsymbol{\gamma}}(0)$, $\boldsymbol{\gamma}(0)$, and $\boldsymbol{\pi}(0)$ do not all vanish, then

$$\frac{\mathrm{d}}{\mathrm{d}t}\left[\frac{1}{2}\begin{pmatrix} \dot{\boldsymbol{\gamma}}^T(t) & \boldsymbol{\gamma}^T(t) & \boldsymbol{\pi}^T(t) \end{pmatrix}\begin{bmatrix} \mathbf{M} & \mathbf{0} & \mathbf{0} \\ \mathbf{0} & \mathbf{K} & \mathbf{0} \\ \mathbf{0}^T & \mathbf{0}^T & \mathbf{0} \end{bmatrix}\begin{pmatrix} \dot{\boldsymbol{\gamma}}(t) \\ \boldsymbol{\gamma}(t) \\ \boldsymbol{\pi}(t) \end{pmatrix}\right] = 0 \tag{11.31}$$

EXAMPLE 11.3

Put the following equations in state form, apply the trapezoidal rule, and triangularize the ensuing dynamic stiffness matrix assuming that the triangular factors of **M** and **K** are known.

$$\mathbf{M}\ddot{\boldsymbol{\gamma}} + \mathbf{K}\boldsymbol{\gamma} - \boldsymbol{\Sigma}\boldsymbol{\pi} = \mathbf{f}, \quad \boldsymbol{\Sigma}^T\boldsymbol{\gamma} = \mathbf{0}$$

Solution

The given equations are expressed in state form as follows:

$$\begin{bmatrix} \mathbf{M} & \mathbf{0} & \mathbf{0} \\ \mathbf{0} & \mathbf{K} & \mathbf{0} \\ \mathbf{0} & \mathbf{0} & \mathbf{0} \end{bmatrix}\begin{pmatrix} \dot{\boldsymbol{\gamma}}(t) \\ \boldsymbol{\gamma}(t) \\ \boldsymbol{\pi}(t) \end{pmatrix}^{\cdot} + \begin{bmatrix} \mathbf{M} & \mathbf{K} & -\boldsymbol{\Sigma} \\ -\mathbf{K} & \mathbf{0} & \mathbf{0} \\ \mathbf{0} & \boldsymbol{\Sigma}^T & \mathbf{0} \end{bmatrix}\begin{pmatrix} \dot{\boldsymbol{\gamma}}(t) \\ \boldsymbol{\gamma}(t) \\ \boldsymbol{\pi}(t) \end{pmatrix} = \begin{pmatrix} \mathbf{f}(t) \\ \mathbf{0} \\ \mathbf{0} \end{pmatrix}$$

Using the trapezoidal rule and $\mathbf{p} = \dot{\boldsymbol{\gamma}}$ gives

$$\begin{bmatrix} \mathbf{M} & \mathbf{0} & \mathbf{0} \\ \mathbf{0} & \mathbf{K} & \mathbf{0} \\ \mathbf{0} & \mathbf{0} & \mathbf{0} \end{bmatrix} \begin{pmatrix} \frac{1}{h}(\mathbf{p}_{n+1} - \mathbf{p}_n) \\ \frac{1}{h}(\boldsymbol{\gamma}_{n+1} - \boldsymbol{\gamma}_n) \\ \frac{1}{h}(\boldsymbol{\pi}_{n+1} - \boldsymbol{\pi}_n) \end{pmatrix} + \begin{bmatrix} \mathbf{M} & \mathbf{K} & -\boldsymbol{\Sigma} \\ -\mathbf{K} & \mathbf{0} & \mathbf{0} \\ \mathbf{0} & \boldsymbol{\Sigma}^T & \mathbf{0} \end{bmatrix} \begin{pmatrix} \frac{1}{2}(\mathbf{p}_{n+1} + \mathbf{p}_n) \\ \frac{1}{2}(\boldsymbol{\gamma}_{n+1} + \boldsymbol{\gamma}_n) \\ \frac{1}{2}(\boldsymbol{\pi}_{n+1} + \boldsymbol{\pi}_n) \end{pmatrix} = \begin{pmatrix} \frac{1}{2}(\mathbf{f}_{n+1} + \mathbf{f}_n) \\ \mathbf{0} \\ \mathbf{0} \end{pmatrix}$$

The second row implies that $\mathbf{p}_{n+1} = \frac{2}{h}(\boldsymbol{\gamma}_{n+1} - \boldsymbol{\gamma}_n) - \mathbf{p}_n$. Thence the first row becomes

$$\frac{1}{h}\mathbf{M}\left[\frac{2}{h}(\boldsymbol{\gamma}_{n+1} - \boldsymbol{\gamma}_n) - 2\mathbf{p}_n\right] + \frac{1}{2}\mathbf{K}(\boldsymbol{\gamma}_{n+1} + \boldsymbol{\gamma}_n) - \frac{1}{2}\boldsymbol{\Sigma}(\boldsymbol{\pi}_{n+1} + \boldsymbol{\pi}_n) = \frac{1}{2}(\mathbf{f}_{n+1} + \mathbf{f}_n)$$

Multiplying throughout by $h^2/2$ and rearranging results in

$$\left[\mathbf{M} + \frac{h^2}{4}\mathbf{K}\right]\boldsymbol{\gamma}_{n+1} - \frac{h^2}{4}\boldsymbol{\Sigma}\boldsymbol{\pi}_{n+1} = \frac{h^2}{4}[\mathbf{f}_{n+1} + \mathbf{f}_n - \mathbf{K}\boldsymbol{\gamma}_n + \boldsymbol{\Sigma}\boldsymbol{\pi}_n] + \mathbf{M}[\boldsymbol{\gamma}_n + h\mathbf{p}_n] \quad (11.32a)$$

The third row, after multiplying by $h^2/2$, becomes

$$\frac{h^2}{4}\boldsymbol{\Sigma}^T\boldsymbol{\gamma}_{n+1} = -\frac{h^2}{4}\boldsymbol{\Sigma}^T\boldsymbol{\gamma}_n \quad (11.32b)$$

The equations are restated in vector–matrix notation as

$$\begin{bmatrix} \mathbf{M} + \frac{h^2}{4}\mathbf{K} & -\frac{h^2}{4}\boldsymbol{\Sigma} \\ \frac{h^2}{4}\boldsymbol{\Sigma}^T & \mathbf{0} \end{bmatrix} \begin{pmatrix} \boldsymbol{\gamma}_{n+1} \\ \boldsymbol{\pi}_{n+1} \end{pmatrix} = \begin{pmatrix} \mathbf{g}_{n+1} \\ -\frac{h^2}{4}\boldsymbol{\Sigma}^T\boldsymbol{\gamma}_n \end{pmatrix}$$

in which

$$\mathbf{g}_{n+1} = \frac{h^2}{4}[\mathbf{f}_{n+1} + \mathbf{f}_n - \mathbf{K}\boldsymbol{\gamma}_n + \boldsymbol{\Sigma}\boldsymbol{\pi}_n] + \mathbf{M}[\boldsymbol{\gamma}_n + h\mathbf{p}_n]$$

The dynamic stiffness matrix now is

$$\mathbf{K}_D = \begin{bmatrix} \mathbf{M} + \frac{h^2}{4}\mathbf{K} & -\frac{h^2}{4}\boldsymbol{\Sigma} \\ \frac{h^2}{4}\boldsymbol{\Sigma}^T & \mathbf{0} \end{bmatrix}$$

Next, $\mathbf{K}_D$ is decomposed into a product of a lower triangular and an upper triangular matrix. We first write

$$\begin{bmatrix} \mathbf{M} + \frac{h^2}{4}\mathbf{K} & -\frac{h^2}{4}\boldsymbol{\Sigma} \\ \frac{h^2}{4}\boldsymbol{\Sigma}^T & \mathbf{0} \end{bmatrix} = \begin{bmatrix} \mathbf{L}_{11} & \mathbf{0} \\ \mathbf{L}_{21} & \mathbf{L}_{22} \end{bmatrix} \begin{bmatrix} \mathbf{U}_{11} & \mathbf{U}_{12} \\ \mathbf{0} & \mathbf{U}_{22} \end{bmatrix}$$

Setting $\mathbf{U}_{11} = \mathbf{L}_{11}^T$, and $\mathbf{L}_{22} = \mathbf{I}$ gives

$$\mathbf{L}_{11}\mathbf{L}_{11}^T = \mathbf{M} + \frac{h^2}{4}\mathbf{K}$$

which may be triangularized to find $\mathbf{L}_{11}$ since, the triangular factors of $\mathbf{M}$ and $\mathbf{K}$ are known. Further

$$\mathbf{U}_{12} = -\frac{h^2}{4}\mathbf{L}_{11}^{-1}\mathbf{\Sigma}$$

$$\mathbf{L}_{21} = \frac{h^2}{4}\mathbf{\Sigma}^T\mathbf{L}_{11}^{-T}$$

$$\mathbf{U}_{22} = -\mathbf{L}_{21}\mathbf{U}_{12} = \frac{h^4}{16}\mathbf{\Sigma}^T\mathbf{L}_{11}^{-T}\mathbf{L}_{11}^{-1}\mathbf{\Sigma}$$

Accordingly, the triangularization is expressed as

$$\mathbf{K}_D = \mathbf{LU} = \begin{bmatrix} \mathbf{L}_{11} & \mathbf{0} \\ \frac{h^2}{4}\mathbf{\Sigma}^T\mathbf{L}_{11}^{-T} & \mathbf{I} \end{bmatrix}\begin{bmatrix} \mathbf{L}_{11}^T & -\frac{h^2}{4}\mathbf{L}_{11}^{-1}\mathbf{\Sigma} \\ \mathbf{0} & \frac{h^4}{16}\mathbf{\Sigma}^T\mathbf{L}_{11}^{-T}\mathbf{L}_{11}^{-1}\mathbf{\Sigma} \end{bmatrix}$$

EXAMPLE 11.4

In an element of an incompressible square rod of cross-sectional area A, it is necessary to consider the displacements v and w. Suppose the length is L, the lateral dimension is Y, and the interpolation models are linear for the displacements (u linear in x, with v, w linear in y) and constant for the pressure. Show that the finite element equation assumes the form

$$\begin{bmatrix} 2\mu A/L & 0 & -A \\ 0 & 4\mu AL/Y^2 & -2AL/Y \\ A & 2AL/Y & 0 \end{bmatrix}\begin{pmatrix} u(L) \\ v(Y) \\ p \end{pmatrix} = \begin{pmatrix} f \\ 0 \\ 0 \end{pmatrix}$$

and that this implies the relation $3\mu\frac{u(L)}{L} = f$.

Solution

The interpolation models are given by

$$u = \frac{x}{L}u(L), \quad v = w = \frac{y}{Y}v(Y)$$

Now consider the virtual work term for virtual strain energy.

$$\begin{aligned}\int \delta\varepsilon_{ij}2\mu\varepsilon_{ij}\,\mathrm{d}V &= 2\mu AL\delta\varepsilon_{ij}\varepsilon_{ij} \\ &= 2\mu AL\left[\delta\varepsilon_{xx}\varepsilon_{xx} + \delta\varepsilon_{yy}\varepsilon_{yy} + \delta\varepsilon_{zz}\varepsilon_{zz}\right] \\ &= 2\mu AL\left[\frac{\delta u(L)}{L}\frac{u(L)}{L} + 2\frac{\delta v(Y)}{Y}\frac{v(Y)}{Y}\right] \\ &= 2\mu AL(\delta u(L) \quad \delta v(Y))\begin{bmatrix} 1/L^2 & 0 \\ 0 & 2/Y^2 \end{bmatrix}\begin{pmatrix} u(L) \\ v(Y) \end{pmatrix}\end{aligned}$$

But $\int \delta\varepsilon_{ij}2\mu\varepsilon_{ij}\,\mathrm{d}V = \delta\boldsymbol{\gamma}^T\mathbf{K}\boldsymbol{\gamma}$. Accordingly, the foregoing equation implies

$$\boldsymbol{\gamma} = \begin{pmatrix} u(L) \\ v(Y) \end{pmatrix}, \quad \mathbf{K} = \begin{bmatrix} 2\mu A/L & 0 \\ 0 & 4\mu AL/Y^2 \end{bmatrix}$$

Next

$$\int \delta\varepsilon_{kk} p \, \mathrm{d}V = ALp\left[\frac{\delta u(L)}{L} + 2\frac{\delta v(Y)}{Y}\right] = (\delta u(L), \quad \delta v(Y))AL\begin{pmatrix} u(L) \\ v(Y) \end{pmatrix} p$$

But, $\int \delta\varepsilon_{kk} p \, \mathrm{d}V = \delta\boldsymbol{\gamma}^T \boldsymbol{\Sigma}\boldsymbol{\pi}$. Also, if we assume that p is a constant, we have

$$\boldsymbol{\pi} = p, \quad \boldsymbol{\Sigma} = \begin{pmatrix} A \\ 2AL/Y \end{pmatrix}$$

After neglecting the inertia term in Equation 11.29, the equations of interest become

$$\mathbf{K}\boldsymbol{\gamma} - \boldsymbol{\Sigma}\boldsymbol{\pi} = \mathbf{f}, \quad \boldsymbol{\Sigma}^T\boldsymbol{\gamma} = \mathbf{0}$$

Use of state form furnishes

$$\begin{bmatrix} \mathbf{K} & -\boldsymbol{\Sigma} \\ \boldsymbol{\Sigma}^T & \mathbf{0} \end{bmatrix}\begin{pmatrix} \boldsymbol{\gamma} \\ \boldsymbol{\pi} \end{pmatrix} = \begin{pmatrix} \mathbf{f} \\ \mathbf{0} \end{pmatrix}$$

Accordingly

$$\begin{bmatrix} 2\mu A/L & 0 & -A \\ 0 & 4\mu AL/Y^2 & -2AL/Y \\ A & 2AL/Y & 0 \end{bmatrix}\begin{pmatrix} u(L) \\ v(Y) \\ p \end{pmatrix} = \begin{pmatrix} f \\ 0 \\ 0 \end{pmatrix}$$

The foregoing matrix product is equivalent to the three relations

$$\frac{2\mu A}{L}u(L) - Ap = f \tag{11.33a}$$

$$\frac{4\mu AL}{Y^2}v(Y) - \frac{2AL}{Y}p = 0 \tag{11.33b}$$

$$Au(L) + \frac{2AL}{Y}v(Y) = 0 \tag{11.34}$$

Multiplying Equation 11.34 by $\frac{2\mu}{Y}$ gives rise to

$$\frac{2\mu A}{Y}u(L) + \frac{4\mu AL}{Y^2}v(Y) = 0 \tag{11.35}$$

Next subtract Equation 11.35 from Equation 11.33b.

$$u(L) + \frac{2AL}{Y}p = 0, \quad p = -\frac{\mu}{L}u(L)$$

Substituting the last expression in Equation 11.34 now produces the relation

$$3\mu A\frac{u(L)}{L} = f$$

This agrees exactly with the uniaxial result $\mathsf{E}A\frac{u(L)}{L} = f$ in the incompressible case since $\nu = 1/2$ and $\mu = \mathsf{E}/2(1+\nu)$.

11.4 TORSION OF PRISMATIC BARS

11.4.1 Basic Relations

Figure 11.2 illustrates a member experiencing torsion. The member in this case is cylindrical with length L and radius r_0. The base is fixed and a torque is applied at the top surface which causes the member to twist. The twist at height z is $\theta(z)$, and at height L it is θ_0.

Ordinarily, in finite element problems so far considered, the displacement is the basic unknown. It is approximated by an interpolation model, from which an approximation for the strain tensor is obtained. Then an approximation for the stress tensor is obtained using the stress–strain relations. The nodal displacements are then solved by an equilibrium principle, in the form of the Principle of Virtual Work. In the current problem an alternative path is followed, in which the stress tensor, or more precisely a stress potential, is the unknown. The strains are determined from the stresses. However, for arbitrary stresses satisfying equilibrium the strain field may not be compatible, enabling it to determine a displacement field which is unique to within a rigid body translation and rotation. The compatibility condition (cf. Chapter 5) is now enforced, furnishing a partial differential equation known as the Poisson equation. A variational argument is applied to furnish a finite element expression for the *torsional constant* of the section.

For the member before twist, consider points X and Y at angle ϕ and at radial position r. Clearly, $X = r\cos\phi$ and $Y = r\sin\phi$. Twist induces a rotation through angle $\theta(z)$ but it does not affect the radial position of a given point. Now $x = r\cos(\phi + \theta)$, $y = r\sin(\phi + \theta)$. Use of double angle formulae furnishes the displacements, and restriction to small angles θ furnishes, to first order,

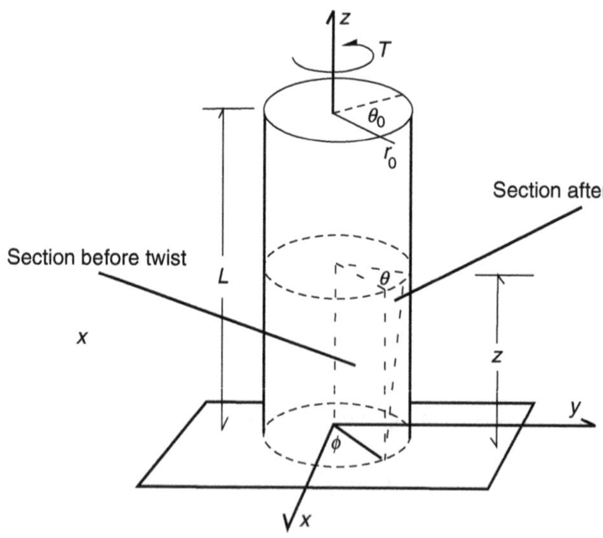

FIGURE 11.2 Twist of a prismatic rod.

$$u = -Y\theta, \quad v = X\theta \tag{11.36}$$

It is also assumed that torsion does not increase the length of the member, which is attained by requiring that w only depends on X and Y. The quantity $w(X,Y)$ is called the *warping function*.

Elementary manipulation serves to verify that all strains vanish except E_{xz} and E_{yz}, for which

$$E_{xz} = \frac{1}{2}\left[\frac{\partial w}{\partial x} - y\frac{\partial \theta}{\partial z}\right], \quad E_{yz} = \frac{1}{2}\left[\frac{\partial w}{\partial y} + x\frac{\partial \theta}{\partial z}\right] \tag{11.37}$$

Equilibrium requires that

$$\frac{\partial S_{xz}}{\partial x} + \frac{\partial S_{yz}}{\partial y} = 0 \tag{11.38}$$

with all other stresses vanishing owing to the assumed displacement field. The equilibrium relation may be identically satisfied by a potential function ψ for which

$$S_{xz} = \frac{\partial \psi}{\partial y}, \quad S_{yz} = -\frac{\partial \psi}{\partial x} \tag{11.39}$$

It remains to satisfy the compatibility condition, to ensure that the strain field arises from a displacement field which is unique to within a rigid body translation and rotation. (Compatibility is automatically satisfied if the displacements are considered the unknowns and are approximated by a continuous interpolation model. Here the stresses are the unknowns.) From the stress–strain relation

$$E_{xz} = \frac{1}{2\mu}S_{xz} = \frac{1}{2\mu}\frac{\partial \psi}{\partial y}, \quad E_{yz} = \frac{1}{2\mu}S_{yz} = -\frac{1}{2\mu}\frac{\partial \psi}{\partial x} \tag{11.40}$$

Compatibility (integrability) now requires that $\frac{\partial^2 w}{\partial x \partial y} = \frac{\partial^2 w}{\partial y \partial x}$, from which

$$-\frac{\partial}{\partial y}\left[\frac{1}{2\mu}\frac{\partial \psi}{\partial y} + \frac{1}{2}y\frac{d\theta}{dz}\right] + \frac{\partial}{\partial x}\left[-\frac{1}{2\mu}\frac{\partial \psi}{\partial x} - \frac{1}{2}x\frac{d\theta}{dz}\right] = 0 \tag{11.41}$$

furnishing Poisson's equation for the potential function ψ:

$$\frac{\partial^2 \psi}{\partial x^2} + \frac{\partial^2 \psi}{\partial y^2} = -2\mu\frac{d\theta}{dz} \tag{11.42}$$

For boundary conditions, it is assumed that the lateral boundaries of the member are traction free. Now the assumed displacement fields already imply that $t_x = 0$ and $t_y = 0$ on the lateral boundary S. For traction t_z to vanish requires that

$$t_z = n_x S_{xz} + n_y S_{yz} = 0 \quad \text{on } S \tag{11.43}$$

Upon examining Figure 11.3 it may be seen that $n_x = \mathrm{d}y/\mathrm{d}s$ and $n_y = -\mathrm{d}x/\mathrm{d}s$, with s the arc length along the boundary at z. Consequently,

$$\begin{aligned} t_z &= \frac{\mathrm{d}y}{\mathrm{d}s} S_{xz} - \frac{\mathrm{d}x}{\mathrm{d}s} S_{yz} \\ &= \frac{\mathrm{d}y}{\mathrm{d}s}\frac{\partial\psi}{\partial y} + \frac{\mathrm{d}x}{\mathrm{d}s}\frac{\partial\psi}{\partial x} \\ &= \frac{\mathrm{d}\psi}{\mathrm{d}s} \end{aligned} \tag{11.44}$$

Now $\frac{\mathrm{d}\psi}{\mathrm{d}s} = 0$ on S, and therefore ψ is a constant on S, and it may, in general, be taken as zero. Since ψ is elsewhere the unknown, the vanishing values on the exterior boundary are analogous to constraints in conventional displacement-based problems.

We next consider the total torque T on the member. Figure 11.4 depicts the cross section at z in which the torque $\mathrm{d}T$ on the element at x and y is given by

$$\begin{aligned} \mathrm{d}T &= xS_{yz}\,\mathrm{d}x\,\mathrm{d}y - yS_{xz}\,\mathrm{d}x\,\mathrm{d}y \\ &= \left[-x\frac{\mathrm{d}\psi}{\mathrm{d}x} - y\frac{\mathrm{d}\psi}{\mathrm{d}y}\right]\mathrm{d}x\,\mathrm{d}y \end{aligned} \tag{11.45}$$

Integration furnishes

$$\begin{aligned} T &= -\iint\left[x\frac{\mathrm{d}\psi}{\mathrm{d}x} + y\frac{\mathrm{d}\psi}{\mathrm{d}y}\right]\mathrm{d}x\,\mathrm{d}y \\ &= -\iint\left[\left[\frac{\mathrm{d}(x\psi)}{\mathrm{d}x} + \frac{\mathrm{d}(y\psi)}{\mathrm{d}y}\right] - \psi\left[\frac{\mathrm{d}x}{\mathrm{d}x} + \frac{\mathrm{d}y}{\mathrm{d}y}\right]\right]\mathrm{d}x\,\mathrm{d}y \\ &= -\int\nabla\cdot\begin{pmatrix} x\psi \\ y\psi \end{pmatrix}\mathrm{d}x\,\mathrm{d}y + 2\int\psi\,\mathrm{d}x\,\mathrm{d}y \end{aligned} \tag{11.46}$$

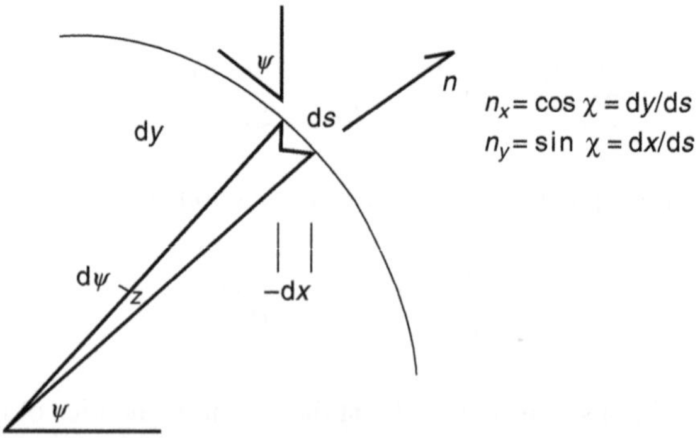

FIGURE 11.3 Illustration of geometric relation.

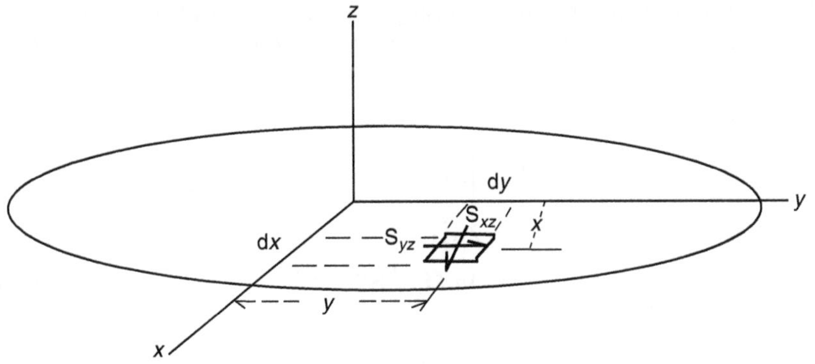

FIGURE 11.4 Evaluation of twisting moment.

Application of the divergence theorem to the first term leads to $\int \psi[xn_x + yn_y]\,ds$, which vanishes since ψ vanishes on S. Finally

$$T = 2\int \psi\,dx\,dy \tag{11.47}$$

We now apply variational methods to the Poisson equation, considering the stress potential function ψ to be the unknown.

$$\int \delta\psi[\nabla \cdot \nabla\psi + 2\mu\theta']\,dx\,dy = 0 \tag{11.48}$$

in which $\theta' = d\theta/dx$. Integration by parts, use of the divergence theorem and imposition of the "constraint" $\psi = 0$ on S furnishes

$$\int (\nabla\delta\psi) \cdot \nabla\psi\,dx\,dy = \int \delta\psi 2\mu\theta'\,dx\,dy \tag{11.49}$$

The integrals are to be evaluated over a set of small elements. In the eth element approximate ψ as $\boldsymbol{v}_\psi^T(x,y)\boldsymbol{\Psi}_e\boldsymbol{\eta}_e$ in which $\boldsymbol{v}_\varphi$ is a vector with dimension (number of rows) equal to the number of nonvanishing nodal values of ψ. The gradient $\nabla\psi$ has a corresponding (derived) interpolation model $\nabla\psi = \boldsymbol{\beta}_\psi^T(x,y)\boldsymbol{\Psi}_e\boldsymbol{\eta}_e$, in which $\boldsymbol{\beta}_\varphi$ is a matrix. The finite element counterpart of the Poisson equation at the element level now is written as

$$\mathbf{K}_\psi^{(e)}\boldsymbol{\eta}_e = 2\mu\theta'\mathbf{f}_\psi^{(e)} \tag{11.50}$$

$$\mathbf{K}_\psi^{(e)} = \boldsymbol{\Psi}_e^T \int \boldsymbol{\beta}_\psi \boldsymbol{\beta}_\psi^T\,dx\,dy\,\boldsymbol{\Psi}_e, \quad \mathbf{f}_\psi^{(e)} = \boldsymbol{\Psi}_e^T \int v_\psi(x,y)\,dx\,dy$$

and the stiffness matrix should be nonsingular since the constraint $\psi = 0$ on S has already been used. It follows that, globally, $\boldsymbol{\eta}_g = 2\mu\theta' \mathbf{K}_\psi^{(g)-1} \mathbf{f}_\psi^{(g)}$. Next, the torque satisfies

$$\begin{aligned} T &= \int 2\psi \, \mathrm{d}x \, \mathrm{d}y \\ &= 2\boldsymbol{\eta}_g^T \mathbf{f}_\psi^{(g)} \\ &= 4\mu\theta' \mathbf{f}_\psi^{(g)T} \mathbf{K}_\psi^{(g)-1} \mathbf{f}_\psi^{(g)} \end{aligned} \tag{11.51}$$

In the theory of torsion it is common to introduce the torsional constant J for which $T = \mu J \theta'$. It follows immediately that $J = 4\mathbf{f}_\psi{}^{(g)T} \, \mathbf{K}_\psi{}^{(g)-1} \, \mathbf{f}_\psi{}^{(g)}$.

EXAMPLE 11.5

Figure 11.5 depicts a single triangular element in the cross section of a shaft experiencing torsion. Assuming the interpolation model

$$\psi(x,y) = (1 \quad x \quad y) \begin{bmatrix} 1 & x_1 & y_1 \\ 1 & x_2 & y_2 \\ 1 & x_3 & y_3 \end{bmatrix}^{-1} \begin{pmatrix} \psi_1 \\ \psi_2 \\ \psi_3 \end{pmatrix}$$

find $\mathbf{K}_\psi$, $\mathbf{f}_\psi$, and the torsional constant J for this element.

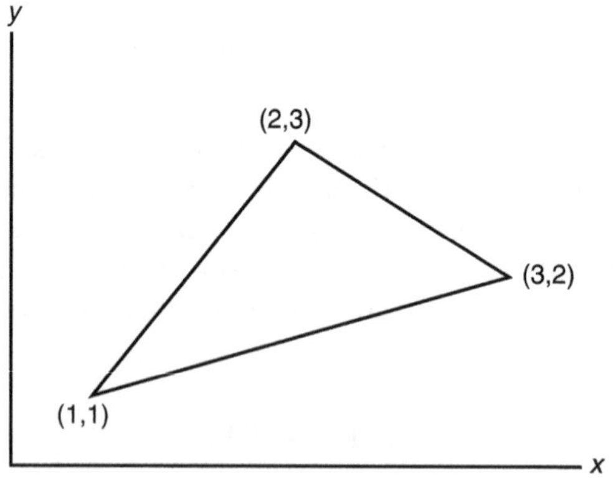

FIGURE 11.5 Triangular element in shaft cross section.

Solution

Here

$$v_T^T(x,y) = (1 \quad x \quad y)$$

$$\boldsymbol{\Psi} = \begin{bmatrix} 1 & x_1 & y_1 \\ 1 & x_2 & y_2 \\ 1 & x_3 & y_3 \end{bmatrix}^{-1} = \begin{bmatrix} 1 & 1 & 1 \\ 1 & 3 & 2 \\ 1 & 2 & 3 \end{bmatrix}^{-1} = \frac{1}{3}\begin{bmatrix} 5 & -1 & -1 \\ -1 & 2 & -1 \\ -1 & -1 & 2 \end{bmatrix}$$

$$\boldsymbol{\eta} = \begin{pmatrix} \psi_1 \\ \psi_2 \\ \psi_3 \end{pmatrix}, \quad \boldsymbol{\beta}_T^T = \nabla v_T^T(x,y) = \begin{bmatrix} 0 & 1 & 0 \\ 0 & 0 & 1 \end{bmatrix}$$

Also

$$\mathbf{K}_T = \boldsymbol{\Psi}^T \int \boldsymbol{\beta}_T \boldsymbol{\beta}_T^T \, \mathrm{d}x \, \mathrm{d}y \, \boldsymbol{\Psi}$$

$$= \frac{1}{9}\begin{bmatrix} 5 & -1 & -1 \\ -1 & 2 & -1 \\ -1 & -1 & 2 \end{bmatrix}\begin{bmatrix} 0 & 0 & 0 \\ 0 & 1 & 0 \\ 0 & 0 & 1 \end{bmatrix}\int \mathrm{d}x \, \mathrm{d}y \begin{bmatrix} 5 & -1 & -1 \\ -1 & 2 & -1 \\ -1 & -1 & 2 \end{bmatrix}$$

But

$$\int \mathrm{d}x \, \mathrm{d}y = \frac{1}{2}\det\begin{vmatrix} 1 & x_1 & y_1 \\ 1 & x_2 & y_2 \\ 1 & x_3 & y_3 \end{vmatrix} = \text{Area of the triangle} = \frac{3}{2}$$

Consequently,

$$\mathbf{K}_T = \frac{1}{6}\begin{bmatrix} 2 & -1 & -1 \\ -1 & 5 & -4 \\ -1 & -4 & 5 \end{bmatrix}$$

Further,

$$\mathbf{f}_\psi = \boldsymbol{\Psi}^T \int v_\psi(x,y) \, \mathrm{d}x \, \mathrm{d}y = \frac{1}{3}\begin{bmatrix} 5 & -1 & -1 \\ -1 & 2 & -1 \\ -1 & -1 & 2 \end{bmatrix}\int \begin{pmatrix} 1 \\ x \\ y \end{pmatrix} \mathrm{d}x \, \mathrm{d}y$$

and

$$\int x \, \mathrm{d}x \, \mathrm{d}y = \int (x - x_0 + x_0) \, \mathrm{d}x \, \mathrm{d}y$$

in which (x_0,y_0) is the centroid of the triangle and is given by

$$x_0 = \frac{x_1 + x_2 + x_3}{3} = 2, \quad y_0 = \frac{y_1 + y_2 + y_3}{3} = 2$$

Continuing,

$$\int (x - x_0)\,\mathrm{d}x_0\,\mathrm{d}y_0 = 0, \text{ and } \int \mathrm{d}x_0\,\mathrm{d}y_0 = \text{Area of the triangle} = \frac{3}{2}$$

Finally $\int x\,\mathrm{d}x\,\mathrm{d}y = 3$, $\int y\,\mathrm{d}x\,\mathrm{d}y = 3$, and

$$\mathbf{f}_T = \frac{1}{2}\begin{pmatrix}1\\1\\1\end{pmatrix}$$

The torsional constant J for the element is now found to be given by

$$\begin{aligned} J &= 4\mathbf{f}_\psi^T\mathbf{K}_\psi^{-1}\mathbf{f}_\psi \\ &= 0.6 \end{aligned}$$

EXAMPLE 11.6

Find the torsional constant in the circular shaft shown below using four axisymmetric elements, as depicted in Figure 11.6.

SOLUTION

Since the configuration is axisymmetric and the constraints $\psi_2 = \psi_3 = \psi_4 = \psi_5 = 0$ are enforced a priori, the linear interpolation model in each element is the same and may be taken as

$$\psi(r) = \frac{r_0 - r}{r_0}\psi_1$$

We conclude that

$$v_\varphi \to r_0 - r, \quad \mathbf{\Psi}_\varphi = \frac{1}{r_0}, \quad \mathbf{\eta}_\varphi = \psi_1, \quad \nabla = \frac{\mathrm{d}}{\mathrm{d}r}, \quad \mathbf{\beta}_\varphi = -1.$$

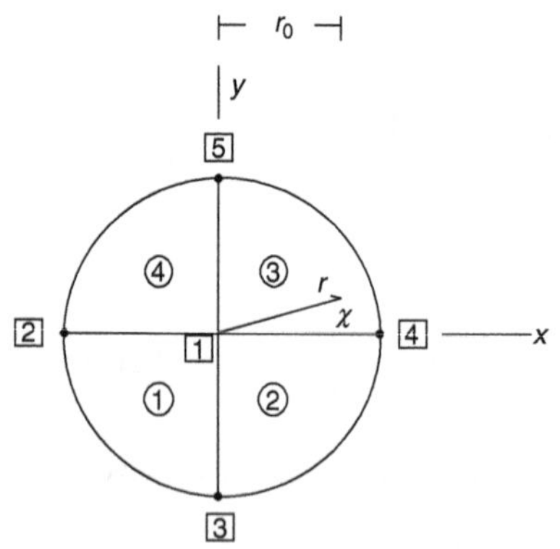

FIGURE 11.6 Torsional constant of a circular shaft.

Simple manipulation yields $\mathbf{K}_{\varphi}^{(g)} = \pi$ and $\mathbf{f}_{\varphi}^{(g)} = \frac{1}{r_0}\int (r_0 - r) r\,\mathrm{d}r\,\mathrm{d}\chi = \frac{\pi r_0^2}{3}$, from which

$$J = 4\mathbf{f}_{\varphi}^{(g)T}\mathbf{K}_{\varphi}^{(g)-1}\mathbf{f}_{\varphi}^{(g)} \\ = \tfrac{4}{9}\pi r_0^4$$

The exact answer is $J = \frac{1}{2}\pi r_0^4$, so that the one degree of freedom finite element model is accurate to within 10%.

11.5 BUCKLING OF ELASTIC BEAMS AND PLATES

11.5.1 Euler Buckling of Beam Columns

11.5.1.1 Static Buckling

Under in-plane compressive loads, the resistance of a thin member (beam or plate) can be reduced progressively, culminating in *buckling*. There are two equilibrium states that the member potentially can sustain—compression only, or compression with bending. The member will "snap" to the second state if it involves less "potential energy" than the first state. The notions explaining buckling are addressed in detail in subsequent sections. For present purposes we focus on beams and plates, using classical equations in which, by retaining lowest order "linearized" corrections for geometric nonlinearity, in-plane compressive forces appear.

For the beam shown in Figure 11.7, the classical Euler buckling equation is

$$\mathsf{E}Iw^{iv} + Pw'' + \rho A\ddot{w} = 0 \tag{11.52}$$

and P is the axial compressive force. The interpolation model for $w(x)$ has the form $w(x) = \boldsymbol{\varphi}^T(x)\boldsymbol{\Phi}\boldsymbol{\gamma}$. Following the usual variational procedures (integration by parts) furnishes

$$\int \delta w \rho A\ddot{w}\,\mathrm{d}x \rightarrow \delta\boldsymbol{\gamma}^T\mathbf{M}\ddot{\boldsymbol{\gamma}}, \quad \mathbf{M} = \boldsymbol{\Phi}^T\int \boldsymbol{\phi}(x)\rho A\boldsymbol{\phi}^T(x)\,\mathrm{d}V\,\boldsymbol{\Phi} \tag{11.53}$$

$$\int \delta w\left[\mathsf{E}Iw^{iv} + Pw''\right]\mathrm{d}x = \int \delta w''\mathsf{E}Iw''\,\mathrm{d}x - \int \delta w'Pw'\,\mathrm{d}x \\ -[(\delta w)(-Pw' - \mathsf{E}Iw''')]_0^L - [(-\delta w')(-\mathsf{E}Iw'')]_0^L$$

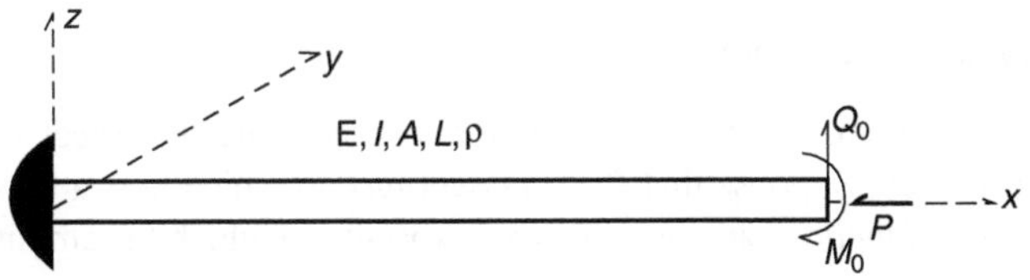

FIGURE 11.7 Euler buckling of a beam column.

At $x = 0$, both δw and $-\delta w'$ vanish. Note that the shear force V and the bending moment M are identified as $V = -\mathsf{E}Iw'''$ and $M = -\mathsf{E}Iw''$. The "effective shear force" Q is defined as $Q = -Pw' - \mathsf{E}Iw'''$.

For the specific case illustrated in Figure 11.7, for a one-element model we may use the interpolation model

$$w(x) = \{x^2 \quad x^3\} \begin{bmatrix} L^2 & L^3 \\ -2L & -3L^2 \end{bmatrix}^{-1} \boldsymbol{\gamma}(t), \quad \boldsymbol{\gamma}(t) = \begin{pmatrix} w(L) \\ -w'(L) \end{pmatrix} \tag{11.54}$$

As seen in Example 11.7, the mass matrix is shown after some algebra to be

$$\mathbf{M} = \rho AL\mathbf{K}_0, \quad \hat{\mathbf{K}}_0 = \begin{bmatrix} \frac{13}{35} & \frac{11}{210}L \\ \frac{11}{210}L & \frac{1}{105}L^2 \end{bmatrix} \tag{11.55}$$

Similarly,

$$\begin{aligned} \int \delta w' P w' \, \mathrm{d}x &= \delta\boldsymbol{\gamma}^T \frac{P}{L} \hat{\mathbf{K}}_1 \boldsymbol{\gamma}, \quad \hat{\mathbf{K}}_1 = \begin{bmatrix} \frac{6}{5} & \frac{1}{10}L \\ \frac{1}{10}L & \frac{2}{15}L^2 \end{bmatrix} \\ \int \delta w'' \mathsf{E}I w'' \, \mathrm{d}x &= \delta\boldsymbol{\gamma}^T \frac{\mathsf{E}I}{L^3} \hat{\mathbf{K}}_2 \boldsymbol{\gamma}, \quad \hat{\mathbf{K}}_2 = \begin{bmatrix} 12 & 6L \\ 6L & 4L^2 \end{bmatrix} \end{aligned} \tag{11.56}$$

The governing equation is written in finite element form as

$$\left[\frac{\mathsf{E}I}{L^3}\hat{\mathbf{K}}_2 - \frac{P}{L}\hat{\mathbf{K}}_1\right]\boldsymbol{\gamma} + \rho AL K_0 \ddot{\boldsymbol{\gamma}} = \mathbf{f}, \quad \mathbf{f} = \begin{pmatrix} Q_0 \\ M_0 \end{pmatrix} \tag{11.57}$$

In a static problem, $\ddot{\boldsymbol{\gamma}} = \mathbf{0}$ in which case the solution has the form

$$\boldsymbol{\gamma} = \frac{\mathrm{cof}\left(\hat{\mathbf{K}}_2 - \frac{PL^2}{\mathsf{E}I}\hat{\mathbf{K}}_1\right)}{\det\left(\hat{\mathbf{K}}_2 - \frac{PL^2}{\mathsf{E}I}\hat{\mathbf{K}}_1\right)}\mathbf{f} \tag{11.58}$$

in which cof(·) denotes the cofactor matrix, and clearly $\boldsymbol{\gamma} \to \infty$ for values of $\frac{PL^2}{\mathsf{E}I}$ which render $\det\left(\hat{\mathbf{K}}_2 - \frac{PL^2}{\mathsf{E}I}\hat{\mathbf{K}}_1\right) = 0$.

11.5.1.2 Dynamic Buckling

In a dynamic problem, it may be of interest to determine the effect of P on the resonance frequency. Suppose that $\mathbf{f}(t) = \mathbf{f}_0 \exp(i\omega t)$, in which $\mathbf{f}_0$ is a known vector. The displacement function satisfies $\boldsymbol{\gamma}(t) = \boldsymbol{\gamma}_0 \exp(i\omega t)$, in which the amplitude vector $\boldsymbol{\gamma}_0$ satisfies

$$\left[\frac{EI}{L^3}\hat{\mathbf{K}}_2 - \frac{P}{L}\hat{\mathbf{K}}_1 - \omega^2\rho AL\hat{\mathbf{K}}_0\right]\boldsymbol{\gamma}_0 = \mathbf{f}_0 \tag{11.59}$$

Resonance occurs at a frequency ω_0 for which

$$\det\left[\frac{EI}{L^3}\hat{\mathbf{K}}_2 - \frac{P}{L}\hat{\mathbf{K}}_1 - \omega_0^2\rho AL\hat{\mathbf{K}}_0\right] = 0 \tag{11.60}$$

Clearly, ω_0^2 is an eigenvalue (often the minimum) of the matrix $\frac{1}{\rho AL}\hat{\mathbf{K}}_0^{-1/2}$ $\left[\frac{EI}{L^3}\hat{\mathbf{K}}_2 - \frac{P}{L}\hat{\mathbf{K}}_1\right]\hat{\mathbf{K}}_0^{-1/2}$. The resonance frequency ω_0^2 is reduced by the presence of P and *vanishes precisely at the critical value of P.*

EXAMPLE 11.7

Derive the matrices $\hat{\mathbf{K}}_0$, $\hat{\mathbf{K}}_1$, $\hat{\mathbf{K}}_2$ for a cantilevered beam modeled as one element.

Solution

Enforcing the clamped constraints a priori, the interpolation model is

$$\boldsymbol{\varphi}^T(x) = (x^2 \quad x^3), \quad \boldsymbol{\Phi} = \begin{bmatrix} L^2 & L^3 \\ -2L & -3L^2 \end{bmatrix}^{-1} = \frac{1}{L^3}\begin{bmatrix} 3L & L^2 \\ -2 & -L \end{bmatrix}$$

The mass matrix satisfies

$$\mathbf{M} = \boldsymbol{\Phi}^T \int \boldsymbol{\varphi}(x)\rho A\boldsymbol{\varphi}^T(x)\,dx\,\boldsymbol{\Phi}$$

$$= \frac{1}{L^6}\begin{bmatrix} 3L & -2 \\ L^2 & -L \end{bmatrix}\int_0^L \begin{pmatrix} x^2 \\ x^3 \end{pmatrix}\rho A(x^2 \quad x^3)\,dx \begin{bmatrix} 3L & L^2 \\ -2 & -L \end{bmatrix}$$

$$= \rho AL\begin{bmatrix} \frac{13}{35} & \frac{11}{210}L \\ \frac{11}{210}L & \frac{1}{105}L^2 \end{bmatrix}$$

But $\mathbf{M} = \rho AL\hat{\mathbf{K}}_0$, and so

$$\hat{\mathbf{K}}_0 = \begin{bmatrix} \frac{13}{35} & \frac{11}{210}L \\ \frac{11}{210}L & \frac{1}{105}L^2 \end{bmatrix}$$

Continuing,

$$w' = \boldsymbol{\beta}^T\boldsymbol{\Phi}\boldsymbol{\gamma}, \quad \boldsymbol{\beta}^T = \partial/\partial x\,\boldsymbol{\varphi}^T(x) = (2x \quad 3x^2)$$

$$\int \delta w' P w' \,\mathrm{d}x = \delta\boldsymbol{\gamma}^T \boldsymbol{\Phi}^T \int \boldsymbol{\beta} P \boldsymbol{\beta}^T \,\mathrm{d}x \,\boldsymbol{\Phi}\boldsymbol{\gamma}$$

$$= \delta\boldsymbol{\gamma}^T \frac{P}{L^6} \begin{bmatrix} 3L & -2 \\ L^2 & -L \end{bmatrix} \int_0^L \begin{pmatrix} 2x \\ 3x^2 \end{pmatrix} \begin{pmatrix} 2x & 3x^2 \end{pmatrix} \mathrm{d}x \begin{bmatrix} 3L & L^2 \\ -2 & -L \end{bmatrix} \boldsymbol{\gamma}$$

$$= \delta\boldsymbol{\gamma}^T \frac{P}{L} \begin{bmatrix} \frac{6}{5} & \frac{1}{10}L \\ \frac{1}{10}L & \frac{2}{15}L^2 \end{bmatrix} \boldsymbol{\gamma}$$

But since $\int \delta w' P w' \,\mathrm{d}x = \delta\boldsymbol{\gamma}^T \frac{P}{L} \hat{\mathbf{K}}_1 \boldsymbol{\gamma}$, we conclude that

$$\hat{\mathbf{K}}_1 = \begin{bmatrix} \frac{6}{5} & \frac{1}{10}L \\ \frac{1}{10}L & \frac{2}{15}L^2 \end{bmatrix}$$

Finally,

$$\int \delta w'' \mathsf{E}I w'' \,\mathrm{d}x = \delta\boldsymbol{\gamma}^T \boldsymbol{\Phi}^T \int_0^L \begin{pmatrix} 2 \\ 6x \end{pmatrix} \mathsf{E}I \begin{pmatrix} 2 & 6x \end{pmatrix} \mathrm{d}x \,\boldsymbol{\Phi}\boldsymbol{\gamma}$$

$$= \delta\boldsymbol{\gamma}^T \frac{\mathsf{E}I}{L^6} \begin{bmatrix} 3L & -2 \\ L^2 & -L \end{bmatrix} \begin{bmatrix} 4L & 6L^2 \\ 6L^2 & 12L^3 \end{bmatrix} \begin{bmatrix} 3L & L^2 \\ -2 & -L \end{bmatrix} \boldsymbol{\gamma}$$

$$= \delta\boldsymbol{\gamma}^T \frac{\mathsf{E}I}{L^3} \begin{bmatrix} 12 & 6L \\ 6L & 4L^2 \end{bmatrix} \boldsymbol{\gamma}$$

But $\int \delta w'' \mathsf{E}I w'' \,\mathrm{d}x = \delta\boldsymbol{\gamma}^T \frac{\mathsf{E}I}{L^3} \hat{\mathbf{K}}_2 \boldsymbol{\gamma}$, from which

$$\hat{\mathbf{K}}_2 = \begin{bmatrix} 12 & 6L \\ 6L & 4L^2 \end{bmatrix}$$

EXAMPLE 11.8

Determine the two buckling load of a cantilevered beam modeled as one element.

SOLUTION

The two critical values can be obtained using $\det\left[\hat{\mathbf{K}}_2 - \frac{PL^2}{\mathsf{E}I}\hat{\mathbf{K}}_1\right] = 0$, leading to

$$\det \begin{bmatrix} 12 - \frac{6}{5}\xi & (6 - \frac{1}{10}\xi)L \\ (6 - \frac{1}{10}\xi)L & (4 - \frac{2}{15}\xi)L^2 \end{bmatrix} = 0, \quad \xi = \frac{PL^2}{\mathsf{E}I}$$

and so

$$\left(12 - \tfrac{6}{5}\xi\right)\left(4 - \tfrac{2}{15}\xi\right)L^2 - \left(6 - \tfrac{1}{10}\xi\right)^2 L^2 = 0$$

On solving the above equation, we have $\xi = 2.49$, 32.18, with the corresponding critical values

$$P = 2.49\frac{\mathsf{E}I}{L^2},\ 32.18\frac{\mathsf{E}I}{L^2}$$

The lower value of the coefficient of $\mathsf{E}I/L$ is very close to the exact value of $\pi^2/4$, but the higher value is not very close to the exact value, given by $9\pi^2/4 = 22.2$.

EXAMPLE 11.9

Interpretation of Buckling Modes

Next consider static buckling of a clamped–clamped beam as shown in Figure 11.8.

If we impose clamped constraints from both the left and right equations, the first and second elements contribute the following stiffness matrices, using notation introduced in Chapter 7.

$$\mathbf{K}_{22}^{(1)} = \frac{8\mathsf{E}I}{L^3}\begin{bmatrix} 12 & 3L \\ 3L & L^2 \end{bmatrix},\quad \mathbf{K}_{11}^{(2)} = \frac{\mathsf{E}I}{l_e^3}\begin{bmatrix} 12 & -3L \\ -3L & L^2 \end{bmatrix}$$

$$-\mathbf{K}_{22}^{(1)} = -\frac{2P}{L}\begin{bmatrix} 6/5 & L/20 \\ L/20 & L^2/30 \end{bmatrix},\quad -\mathbf{K}_{11}^{(1)} = -\frac{2P}{L}\begin{bmatrix} 6/5 & -L/20 \\ -L/20 & L^2/30 \end{bmatrix}$$

The assembled stiffness matrix is now obtained by direct addition:

$$\begin{aligned}\mathbf{K} &= \mathbf{K}_{22}^{(1)} + \mathbf{K}_{11}^{(2)} - \left(\mathbf{K}_{22}^{(1)} + \mathbf{K}_{11}^{(2)}\right) \\ &= \frac{8\mathsf{E}I}{L^3}\begin{bmatrix} 24 & 0 \\ 0 & 2L^2 \end{bmatrix} - \frac{2P}{L}\begin{bmatrix} 12/5 & 0 \\ 0 & L^2/30 \end{bmatrix}\end{aligned}$$

and the load–deflection relation is

$$\left[\frac{8\mathsf{E}I}{L^3}\begin{bmatrix} 24 & 0 \\ 0 & L^2 \end{bmatrix} - \frac{2P}{L}\begin{bmatrix} 12/5 & 0 \\ 0 & L^2/30 \end{bmatrix}\right]\begin{Bmatrix} w_2 \\ -w_2' \end{Bmatrix} = \begin{Bmatrix} V_2 \\ M_2 \end{Bmatrix}$$

The two critical buckling loads can be obained analytically since the two buckling modes are uncoupled.

$$\frac{P_1L^2}{\mathsf{E}I} = 40 \sim 4\pi^2,\quad \frac{P_2L^2}{\mathsf{E}I} = 120 \sim 12\pi^2$$

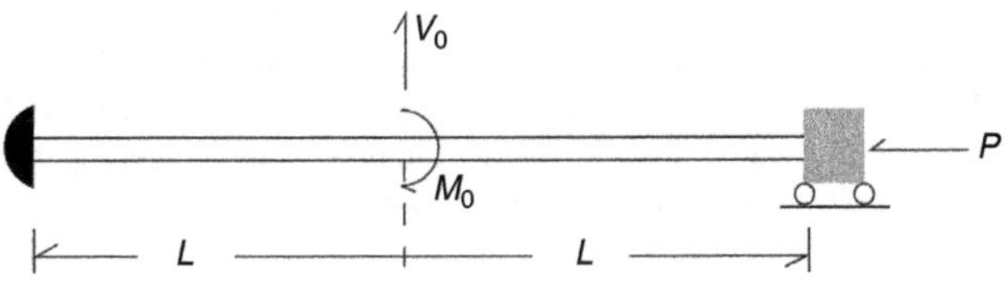

FIGURE 11.8 Buckling of a clamped–clamped beam.

Note that the two element values of the critical load correspond to the *symmetric* and *antisymmetric* behaviors. (i) In particular, if $M_0=0$ there is only one critical load and it is P_1. Furthermore, $w_2'=0$ and $w(L/2-\zeta)=w(L/2+\zeta)$. Alternatively stated, the deformation is *symmetric* about $L/2$. (ii) However, if $V_0=0$, there is only one critical load and it is P_2. In this case, $w_2=0$ and $w(L/2-\zeta)=-w(L/2+\zeta)$. This of course represents *antisymmetry*.

Returning to the main development, but referring to Figure 11.8, we now compare the finite element approach with the exact method, assuming static conditions. Consider the symmetric case. Let $w(x)=w_c(x)+w_p(x)$, in which $w_c(x)$ is the characteristic solution and $w_p(x)$ is the particular solution reflecting the perturbation. From the Euler buckling Equation 11.52 $w_c(x)$ has a general solution of the form $w_c(x)=\alpha+\beta x+\gamma\cos\kappa x+\delta\sin\kappa x$, in which $\kappa=\sqrt{\frac{P}{EI}}$. Now $w=-w'=0$ at $x=0$, $-w'(L)=0$ and $EIw'''\left(\frac{L}{2}\right)=V_1$, expressed as the conditions

$$\begin{aligned}1\alpha+0\beta+1\gamma+0\delta&=-w_p(0)\\0\alpha+1\beta+0\gamma+\kappa\delta&=-w_p'(0)\\0\alpha+1\beta-\gamma\kappa\sin(\kappa L/2)+\delta\kappa\cos(\kappa L/2)&=-w_p'(L/2)\\0\alpha+0\beta+\gamma\kappa^3\sin(\kappa L/2)-\delta\kappa^3\cos(\kappa L/2)&=-EIw_p'''(L/2)+V_1\end{aligned}\tag{11.61}$$

or, in matrix–vector notation,

$$\mathbf{Bz}=\begin{pmatrix}-w_p(0)\\-w_p'(0)\\-w_p'(L/2)\\-EIw_p'''(L/2)+V_1\end{pmatrix},\quad \mathbf{B}=\begin{bmatrix}1&0&1&0\\0&1&0&\kappa\\0&1&-\kappa\sin(\kappa L/2)&\kappa\cos(\kappa L/2)\\0&0&\kappa^3\sin(\kappa L/2)&-\kappa^3\cos(\kappa L/2)\end{bmatrix},\quad \mathbf{z}=\begin{pmatrix}\alpha\\\beta\\\gamma\\\delta\end{pmatrix}\tag{11.62}$$

For the solution to "blow up" it is necessary for the matrix $\mathbf{B}$ to be singular, which occurs if the corresponding *homogeneous* problem in fact possesses a unique solution. Accordingly, we seek conditions under which there exists a nonvanishing vector $\mathbf{z}$ for which $\mathbf{Bz}=\mathbf{0}$. Direct elimination of α and β furnishes $\alpha=-\gamma$ and $\beta=-\kappa\delta$. The remaining coefficients must satisfy

$$\begin{bmatrix}-\sin(\kappa L/2)&\cos(\kappa L/2)-1\\\sin(\kappa L/2)&\cos(\kappa L/2)\end{bmatrix}\begin{pmatrix}\gamma\\\delta\end{pmatrix}=\begin{pmatrix}0\\0\end{pmatrix}\tag{11.63}$$

A nonvanishing solution is possible only if the determinant vanishes, which reduces to the requirement $\sin\kappa L=0$. This equation has many solutions for $\kappa L/2$, including $\kappa L/2=0$. The lowest nontrivial solution is $\kappa L/2=\pi$, from which $P_{crit}=4\pi^2\ EI/L^2=39.88\ EI/L^2$. Clearly, the symmetric solution in the foregoing two-element model ($P_{crit}=40\ EI/L^2$) gives a very accurate result.

For the antisymmetric case the corresponding result is that $\tan \kappa L/2 = \kappa L/2$. The lowest meaningful root of this equation is $\kappa L = 4.49$ (Brush and Almroth, 1975), giving $P_{\text{crit}} = 80.76\ \mathsf{E}I/L^2$. Clearly, unlike the symmetric solution, the axisymmetric solution from the two-element model ($P_{\text{crit}} = 120\ \mathsf{E}I/L^2$) is not very accurate.

To this point it has been implicitly assumed that the beam column is initially perfectly straight. This assumption can lead to overestimates of the critical buckling load. Now consider that there is a known initial distribution $w_0(x)$. The governing equation is now

$$\frac{d^2}{dx^2}\mathsf{E}I\frac{d^2}{dx^2}(w - w_0) + P\frac{d^2}{dx^2}(w - w_0) = 0 \tag{11.64}$$

or equivalently

$$\frac{d^2}{dx^2}\mathsf{E}I\frac{d^2}{dx^2}w + P\frac{d^2}{dx^2}w = \frac{d^2}{dx^2}\mathsf{E}I\frac{d^2}{dx^2}w_0 + P\frac{d^2}{dx^2}w_0 \tag{11.65}$$

Now crookedness is modeled as a perturbation. Similarly, if the cross-sectional properties of the beam column exhibit a small amount of variation, say $\mathsf{E}I(x) = \mathsf{E}I_0[1 + \vartheta \sin(\pi x/L)]$, the effect of the variation may likewise be modeled as a perturbation.

EXAMPLE 11.10

In the clamped–clamped beam column use four equal length elements to determine how much improvement, if any, occurs in the symmetric and antisymmetric cases.

Solution

Consider the right half of the configuration, which has two beams of length $L/2$. Denoting $L/4$ as $\hat{L}$, we assume an interpolation model $w(x) = \boldsymbol{\varphi}^T(x)\boldsymbol{\Phi}\boldsymbol{\gamma}$ in which

$$\boldsymbol{\varphi}^T(x) = (1 \quad x \quad x^2 \quad x^3)$$

$$\boldsymbol{\Phi} = \begin{bmatrix} 1 & 0 & 0 & 0 \\ 0 & -1 & 0 & 0 \\ 1 & \hat{L} & \hat{L}^2 & \hat{L}^3 \\ 0 & -1 & -2\hat{L} & -3\hat{L}^2 \end{bmatrix}^{-1} = \frac{1}{\hat{L}^3}\begin{bmatrix} \hat{L}^3 & 0 & 0 & 0 \\ 0 & -\hat{L}^3 & 0 & 0 \\ -3\hat{L} & 2\hat{L}^2 & 3\hat{L} & \hat{L}^2 \\ 2 & -\hat{L} & -2 & -\hat{L} \end{bmatrix}, \quad \boldsymbol{\gamma} = \begin{pmatrix} w_e \\ -w'_e \\ w_{e+1} \\ -w'_{e+1} \end{pmatrix}$$

The interpolation model also involves the relations

$$w' = \boldsymbol{\beta}^T\boldsymbol{\Phi}\boldsymbol{\gamma}, \quad \boldsymbol{\beta}^T = \frac{\partial \boldsymbol{\varphi}^T(x)}{\partial x} = (0 \quad 1 \quad 2x \quad 3x^2), \quad w'' = (0 \quad 0 \quad 2 \quad 6x)\boldsymbol{\Phi}\boldsymbol{\gamma}$$

Now

$$\hat{\mathbf{K}}_1 = \boldsymbol{\Phi}^T \int \beta L \beta^T \, dx \, \Phi$$

$$= \frac{1}{\hat{L}^5}\begin{bmatrix} \hat{L}^3 & 0 & -3\hat{L} & 2 \\ 0 & -\hat{L}^3 & 2\hat{L}^2 & -\hat{L} \\ 0 & 0 & 3\hat{L} & -2 \\ 0 & 0 & \hat{L}^2 & -\hat{L} \end{bmatrix} \int_0^L \begin{bmatrix} 0 & 0 & 0 & 0 \\ 0 & 1 & 2x & 3x^2 \\ 0 & 2x & 4x^2 & 6x^3 \\ 0 & 3x^2 & 6x^3 & 9x^4 \end{bmatrix} dx \begin{bmatrix} \hat{L}^3 & 0 & 0 & 0 \\ 0 & -\hat{L}^3 & 0 & 0 \\ -3\hat{L} & 2\hat{L}^2 & 3\hat{L} & \hat{L}^2 \\ 2 & -\hat{L} & -2 & -\hat{L} \end{bmatrix}$$

$$= \begin{bmatrix} \frac{6}{5} & -\frac{1}{10}\hat{L} & -\frac{6}{5} & -\frac{1}{10}\hat{L} \\ -\frac{1}{10}\hat{L} & \frac{2}{15}\hat{L}^2 & \frac{1}{10}\hat{L} & -\frac{1}{30}\hat{L}^2 \\ -\frac{6}{5} & \frac{1}{10}\hat{L} & \frac{6}{5} & \frac{1}{10}\hat{L} \\ -\frac{1}{10}\hat{L} & -\frac{1}{30}\hat{L}^2 & \frac{1}{10}\hat{L} & \frac{2}{15}\hat{L}^2 \end{bmatrix}$$

Continuing,

$$\hat{\mathbf{K}}_2 = \frac{1}{\hat{L}^5}\begin{bmatrix} \hat{L}^3 & 0 & -3\hat{L} & 2 \\ 0 & -\hat{L}^3 & 2\hat{L}^2 & -\hat{L} \\ 0 & 0 & 3\hat{L} & -2 \\ 0 & 0 & \hat{L}^2 & -\hat{L} \end{bmatrix} \int_0^L \begin{pmatrix} 0 \\ 0 \\ 2 \\ 6x \end{pmatrix} (0 \quad 0 \quad 2 \quad 6x)\, dx \begin{bmatrix} \hat{L}^3 & 0 & 0 & 0 \\ 0 & -\hat{L}^3 & 0 & 0 \\ -3\hat{L} & 2\hat{L}^2 & 3\hat{L} & \hat{L}^2 \\ 2 & -\hat{L} & -2 & -\hat{L} \end{bmatrix}$$

$$= \begin{bmatrix} 12 & -6\hat{L} & -12 & -6\hat{L} \\ -6\hat{L} & 4\hat{L}^2 & 6\hat{L} & 2\hat{L}^2 \\ -12 & 6\hat{L} & 12 & 6\hat{L} \\ -6\hat{L} & 2\hat{L}^2 & 6\hat{L} & 4\hat{L}^2 \end{bmatrix}$$

Since the element length is $L/4$, the element matrices become

$$[\mathbf{K}_1]_1 = [\mathbf{K}_1]_2 = \begin{bmatrix} \frac{3}{5} & -\frac{1}{40}L & -\frac{3}{5} & -\frac{1}{40}L \\ -\frac{1}{40}L & \frac{1}{120}L^2 & \frac{1}{40}L & -\frac{1}{480}L^2 \\ -\frac{3}{5} & \frac{1}{40}L & \frac{3}{5} & \frac{1}{40}L \\ -\frac{1}{40}L & -\frac{1}{480}L^2 & \frac{1}{40}L & \frac{1}{120}L^2 \end{bmatrix}, \quad [\mathbf{K}_2]_1 = [\mathbf{K}_2]_2 = \begin{bmatrix} 12 & -3L/2 & -12 & -3L/2 \\ -3L/2 & L^2/4 & 3L/2 & L^2/8 \\ -12 & 3L/2 & 12 & 3L/2 \\ -3L/2 & L^2/8 & 3L/2 & L^2/4 \end{bmatrix}$$

The assembled matrices, after enforcing the constraints $w(0) = w'(0) = 0$ are

$$\hat{\mathbf{K}}_1 = \begin{bmatrix} \frac{1}{5} & 0 & -\frac{6}{5} & -\frac{1}{40}L \\ 0 & \frac{1}{60}L^2 & \frac{1}{40}L & -\frac{1}{480}L^2 \\ -\frac{6}{5} & \frac{1}{40}L & \frac{6}{5} & \frac{1}{40}L \\ -\frac{1}{40}L & -\frac{1}{480}L^2 & \frac{1}{40}L & \frac{1}{120}L^2 \end{bmatrix}, \quad \hat{\mathbf{K}}_2 = \begin{bmatrix} 24 & 0 & -12 & -3L/2 \\ 0 & L^2/2 & 3L/2 & L^2/8 \\ -12 & 3L/2 & 12 & 3L/2 \\ -3L/2 & L^2/8 & 3L/2 & L^2/4 \end{bmatrix}$$

The governing equation is written in finite element form as

$$\left[\frac{\mathsf{E}I}{(L/4)^3}\begin{bmatrix} 24 & 0 & -12 & -3L/2 \\ 0 & L^2/2 & 3L/2 & L^2/8 \\ -12 & 3L/2 & 12 & 3L/2 \\ -3L/2 & L^2/8 & 3L/2 & L^2/4 \end{bmatrix} - \frac{P}{(L/4)}\begin{bmatrix} \frac{12}{5} & 0 & -\frac{6}{5} & -\frac{1}{40}L \\ 0 & \frac{1}{240}L^2 & \frac{1}{40}L & -\frac{1}{480}L^2 \\ -\frac{6}{5} & \frac{1}{40}L & \frac{6}{5} & \frac{1}{40}L \\ -\frac{1}{40}L & -\frac{1}{480}L^2 & \frac{1}{40}L & \frac{1}{120}L^2 \end{bmatrix}\right]\boldsymbol{\gamma}_3 = \mathbf{f}_3$$

$$\boldsymbol{\gamma}_3 = \begin{pmatrix} w(L/4) \\ -w'(L/4) \\ w(L/2) \\ -w'(L/2) \end{pmatrix}, \quad \mathbf{f}_3 = \begin{pmatrix} 0 \\ 0 \\ V_1 \\ M_1 \end{pmatrix}$$

Symmetric Case: $M_0 = 0$; $w'(L) = 0$

The governing equation reduces to

$$\left[\frac{\mathsf{E}I}{(L/4)^3}\begin{bmatrix} 24 & 0 & -12 \\ 0 & L^2/2 & 3L/2 \\ -12 & 3L/2 & 12 \end{bmatrix} - \frac{P}{(L/4)}\begin{bmatrix} \frac{12}{5} & 0 & -\frac{6}{5} \\ 0 & \frac{1}{60}L^2 & \frac{1}{40}L \\ -\frac{6}{5} & \frac{1}{40}L & \frac{6}{5} \end{bmatrix}\right]\begin{pmatrix} w(L/4) \\ -w'(L/4) \\ w(L/2) \end{pmatrix} = \begin{pmatrix} 0 \\ 0 \\ V_1 \end{pmatrix}$$

Buckling occurs if

$$\det\begin{bmatrix} 24 - \frac{12}{5}\xi & 0 & -12 + \frac{6}{5}\xi \\ 0 & (2 - \frac{1}{15}\xi)L^2/4 & (3 - \frac{1}{20}\xi)L/2 \\ -12 + \frac{6}{5}\xi & (3 - \frac{1}{20}\xi)L/2 & 12 - \frac{6}{5}\xi \end{bmatrix} = 0, \quad \xi = \frac{P/(L/4)}{\mathsf{E}I/(L/4)^3} = 16\frac{PL^2}{\mathsf{E}I}$$

The roots of the above equation are $\xi = 10$, 2.486, 32.1807, independently of L. The lowest nontrivial root is $\xi = 2.486$ and so

$$P_1 = 39.736\frac{\mathsf{E}I}{L^2}$$

which is very close to the exact solution ($4\pi^2\ \mathsf{E}I/L^2$).

Antisymmetric Case: $V_0 = 0$; $w(L) = 0$

Following the same steps as before, the governing equation is found to be

$$\left[\frac{\mathsf{E}I}{(L/4)^3}\begin{bmatrix} 24 & 0 & -3/2 \\ 0 & L/2 & \frac{1}{8}L \\ -3/2 & \frac{1}{8}L & L \end{bmatrix} - \frac{P}{(L/4)}\begin{bmatrix} \frac{12}{5} & 0 & -\frac{1}{40} \\ 0 & \frac{1}{60}L & -\frac{1}{480}L \\ -\frac{1}{40} & -\frac{1}{480}L & \frac{1}{120}L \end{bmatrix}\right]\begin{pmatrix} w(L/4) \\ -Lw'(L/4) \\ Lw'(L/2) \end{pmatrix} = \begin{pmatrix} 0 \\ 0 \\ M_1 \end{pmatrix}$$

The critical loads are obtained from

$$\det\begin{bmatrix} 24-\frac{12}{5}\xi & 0 & \frac{1}{2}\left(-3+\frac{1}{20}\xi\right) \\ 0 & \left(2-\frac{1}{15}\xi\right)L/4 & \left(\frac{1}{2}+\frac{1}{120}\xi\right)L/4 \\ \frac{1}{2}\left(-3+\frac{1}{20}\xi\right)L & \left(\frac{1}{2}+\frac{1}{120}\xi\right)L/4 & \left(1-\frac{1}{30}\xi\right)L/4 \end{bmatrix}=0$$

The roots are found with a little effort to be $\xi = 5.18$, 18.78, 49.38, giving the lowest nontrivial buckling load as

$$P_1 = 82.88\frac{EI}{L^2}$$

which is very close to the exact solution (80.76 EI/L^2).

Clearly, a four-element model gives a much better approximation than the two-element model in the antisymmetric case. The percentage of error decreases from 48.6% to 2.6%.

11.5.2 Euler Buckling of Plates

The governing equation for an isotropic plate element subject to in-plane loads is (Wang, 1953)

$$\frac{Eh^2}{12(1-\nu^2)}\nabla^4 w + P_x\frac{\partial^2 w}{\partial x^2} + P_y\frac{\partial^2 w}{\partial y^2} + P_{xy}\frac{\partial^2 w}{\partial x\partial y} = 0 \quad (11.66)$$

in which the loads are illustrated in Figure 11.9. The variational methods in Chapter 4 furnish

$$\int \delta w\nabla^4 w\,dA = \int tr(\delta\mathbf{W}\mathbf{W})\,dA + \int \delta w(\mathbf{n}\cdot\nabla)\nabla^2 w\,dS - \int \delta\nabla w\cdot\nabla(\mathbf{n}\cdot\nabla w)\,dS \quad (11.67)$$

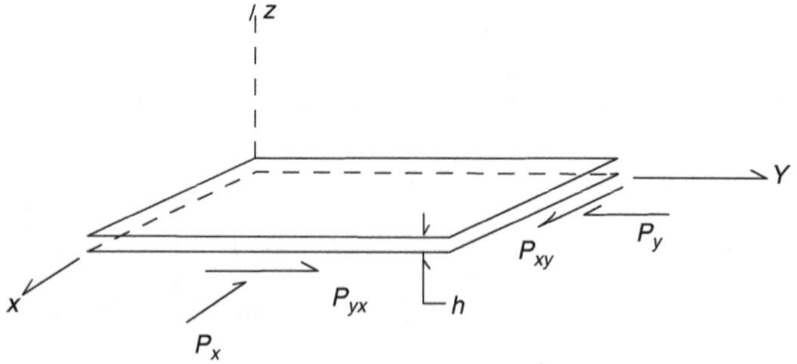

FIGURE 11.9 Plate element with in-plane compressive loads.

in which $\mathbf{W} = \nabla\nabla^T w$ (a matrix!). In addition

$$\int \delta w \left[P_x \frac{\partial^2 w}{\partial x^2} + P_y \frac{\partial^2 w}{\partial y^2} + P_{xy} \frac{\partial^2 w}{\partial x \partial y} \right] \mathrm{d}A$$
$$= \int \delta w \mathbf{n}^T \mathbf{p} \, \mathrm{d}S - \int (\nabla \delta w)^T \mathbf{P} (\nabla w) \, \mathrm{d}A \qquad (11.68)$$

$$\mathbf{p} = \begin{pmatrix} \left[P_x \frac{\partial \mathrm{w}}{\partial \mathrm{x}} + \frac{1}{2} P_{xy} \frac{\partial \mathrm{w}}{\partial \mathrm{y}} \right] \\ \left[\frac{1}{2} P_{xy} \frac{\partial \mathrm{w}}{\partial \mathrm{x}} + P_y \frac{\partial \mathrm{w}}{\partial \mathrm{y}} \right] \end{pmatrix}, \quad \mathbf{P} = \begin{bmatrix} P_x & \frac{1}{2} P_{xy} \\ \frac{1}{2} P_{xy} & P_y \end{bmatrix}$$

We recall the interpolation model $w(x,y) = \boldsymbol{\varphi}_{b2}^T \boldsymbol{\Phi}_{b2} \boldsymbol{\gamma}_{b2}$, from which we may obtain the relations of the form

$$\nabla w = \begin{pmatrix} w_x \\ w_y \end{pmatrix} = \boldsymbol{\beta}_{1b2}^T \boldsymbol{\Phi}_{b2} \boldsymbol{\gamma}_{b2}, \quad VEC(\mathbf{W}) = \boldsymbol{\beta}_{2b2}^T \boldsymbol{\Phi}_{b2} \boldsymbol{\gamma}_{b2} \qquad (11.69a)$$

$$\boldsymbol{\beta}_{1b2}^T = \begin{bmatrix} \frac{\partial}{\partial x} \boldsymbol{\varphi}_{b2}^T \\ \frac{\partial}{\partial y} \boldsymbol{\varphi}_{b2}^T \end{bmatrix}, \quad \boldsymbol{\beta}_{2b2}^T = \begin{bmatrix} \frac{\partial^2 \boldsymbol{\varphi}_{b2}^T}{\partial x^2} \\ \frac{\partial^2 \boldsymbol{\varphi}_{b2}^T}{\partial x \partial y} \\ \frac{\partial^2 \boldsymbol{\varphi}_{b2}^T}{\partial x \partial y} \\ \frac{\partial^2 \boldsymbol{\varphi}_{b2}^T}{\partial x^2} \end{bmatrix} \qquad (11.69b)$$

We also assume that the *secondary variables* $\frac{\mathrm{E}h^2}{12(1-\nu^2)} (\mathbf{n} \cdot \nabla) \nabla^2 w$, $\frac{\mathrm{E}h^2}{12(1-\nu^2)} (\mathbf{n} \cdot \nabla) \nabla w$, and $\mathbf{p}$ are prescribed on S. Doing so serves to derive

$$[\mathbf{K}_{b21} - \mathbf{K}_{b22}] \boldsymbol{\gamma}_{b2} = \mathbf{f} \qquad (11.70)$$

$$\mathbf{K}_{b21} = \frac{\mathrm{E}h^2}{12(1-\nu^2)} \boldsymbol{\Phi}_{b2}^T \int \boldsymbol{\beta}_{2b2} \boldsymbol{\beta}_{2b2}^T \, \mathrm{d}A \, \boldsymbol{\Phi}_{b2}, \quad \mathbf{K}_{b22} = \boldsymbol{\Phi}_{b2}^T \int \boldsymbol{\beta}_{1b2} \mathbf{P} \boldsymbol{\beta}_{1b2}^T \, \mathrm{d}V \, \boldsymbol{\Phi}_{b2}$$

and $\mathbf{f}$ reflects the secondary variables prescribed on S.

As illustrated in Figure 11.10 we now consider a three-dimensional loading space in which P_x, P_y, and P_{xy} correspond to the axes in terms of which we seek to determine a surface of critical values at which buckling occurs. In this space a straight line emanating from the origin represents a *proportional loading* path. Let the load intensity λ denote the distance to a given point on this line. By analogy with spherical coordinates, there exist two angles θ and ϕ such that

$$P_x = \lambda \cos\theta \cos\phi, \quad P_y = \lambda \sin\theta \cos\phi, \quad P_{xy} = \lambda \sin\phi \qquad (11.71)$$

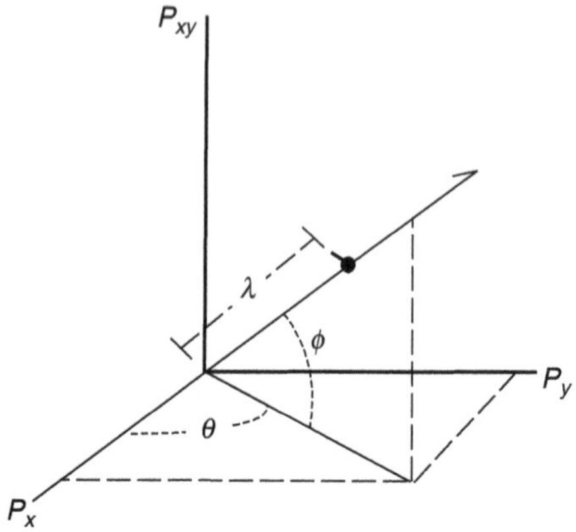

FIGURE 11.10 Loading space for plate buckling.

Now

$$\begin{aligned} \mathbf{K}_{b22} &= \lambda \hat{\mathbf{K}}_{b22}(\theta,\varphi) \\ \hat{\mathbf{K}}_{b22}(\theta,\varphi) &= \boldsymbol{\Phi}_{b2}^T \int \boldsymbol{\beta}_{1b2} \hat{\mathbf{P}}(\theta,\varphi) \boldsymbol{\beta}_{1b2}^T \, dV \, \boldsymbol{\Phi}_{b2} \\ \hat{\mathbf{P}}(\theta,\phi) &= \begin{bmatrix} \cos\theta\cos\phi & \sin\phi \\ \sin\phi & \sin\theta\cos\phi \end{bmatrix} \end{aligned} \tag{11.72}$$

For each pair (θ,ϕ), buckling occurs at a critical load intensity $\lambda_{\text{crit}}(\theta,\phi)$, satisfying

$$\det\left[\mathbf{K}_{b21} - \lambda_{\text{crit}}(\theta,\phi)\hat{\mathbf{K}}_{b22}\right] = 0 \tag{11.73}$$

A surface of critical load intensities $\lambda_{\text{crit}}(\theta,\phi)$ can be drawn in the loading space of Figure 11.10 by evaluating $\lambda_{\text{crit}}(\theta,\phi)$ over all values of (θ,ϕ) and discarding values which are negative.

Recalling Example 3.8 in Chapter 3, we may write

$$\mathbf{K}_{b22} = IVEC\left(\left[\boldsymbol{\Phi}_{b2}^T \otimes \boldsymbol{\Phi}_{b2}^T \int \boldsymbol{\beta}_{1b2}^T(\mathbf{x}) \otimes \boldsymbol{\beta}_{1b2}^T(\mathbf{x}) \, dV\right] VEC(\hat{\mathbf{P}})\right) \tag{11.74}$$

Accordingly, in computing the part of the stiffness matrix incorporating the in-plane load, volume integration need only be performed once, independently of λ, θ, and φ.

11.6 INTRODUCTION TO CONTACT PROBLEMS

11.6.1 GAP

In many practical problems the information required to develop a finite element model, for example, the geometry of a member and the properties of its constituent materials, can be determined with little uncertainty or ambiguity. However, often the loads experienced by the member are not easily identified. This is especially true if loads are transmitted to

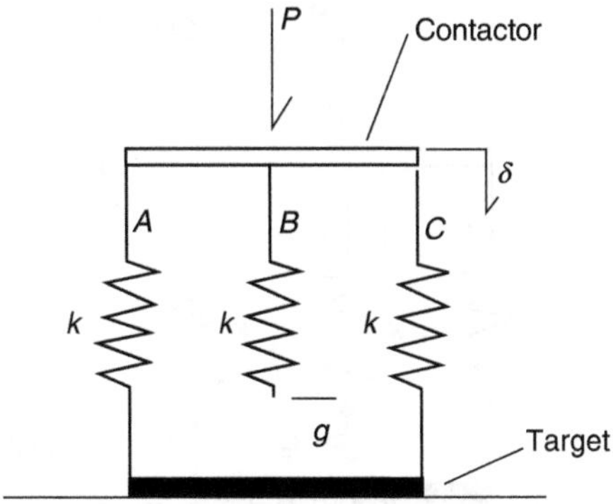

FIGURE 11.11 Simple contact problem.

the member along an interface with a second member. This class of problems has been called contact problems, and they are arguably the most common boundary conditions encountered in "the real world." The finite element community has devoted and continues to devote a great deal of effort to contact problems, culminating in gap and interface elements for contact. Here, we provide a simple introduction to gap elements.

First consider the three spring configuration in Figure 11.11. All springs are of stiffness k. Springs A and C extend from the top plate, called the *contactor*, to the bottom plate, called the *target*. The bottom of spring B is initially remote from the target by a gap g. The exact stiffness of this configuration is bilinear:

$$k_c = \begin{cases} 2k, & \delta < g \\ 3k, & \delta \geq g \end{cases} \tag{11.75}$$

From the viewpoint of the finite element method Figure 11.11 poses the following difficulty. If a node is set at the lowest point on Spring B and at the point directly below it on the target, these nodes are not initially connected, but may later be connected in the physical problem after contact is established. Further, it is necessary to satisfy the *nonpenetration constraint,* which may be stated as an inequality, whereby the middle spring does not move through the target after contact is established. If the nodes are considered unconnected in the finite element model, there is nothing to enforce the nonpenetration constraint. If, however, the nodes are considered connected, the nonpenetration constraint can be satisfied but the stiffness is artificially high.

This difficulty is overcome in an approximate sense by a bilinear contact element. In particular, we introduce a new spring k_g as shown in Figure 11.12.

The stiffness of the middle spring pair (B in series with the contact spring) is now denoted as k_m and is given by

$$k_m = \left[\frac{1}{k} + \frac{1}{k_g}\right]^{-1} \tag{11.76}$$

It is desirable for the middle spring to be soft when the gap is open ($g > \delta$), and to be stiff when the gap is closed ($g < \delta$). For the purpose of illustration we make the selection

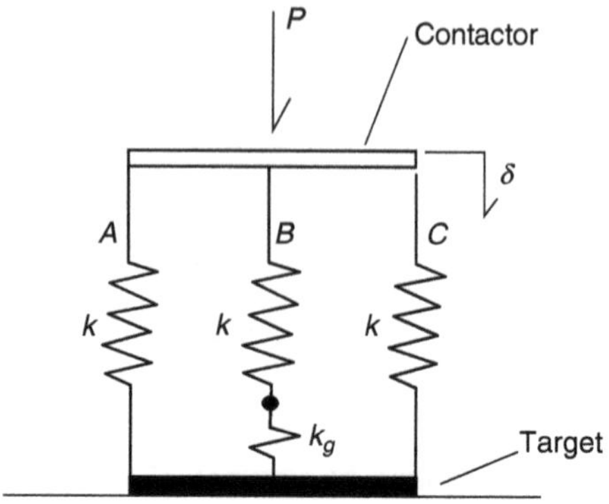

FIGURE 11.12 Spring representing contact element.

$$k_g = \begin{cases} k/100, & g > \delta \\ 100k, & g \leq \delta \end{cases} \tag{11.77}$$

Elementary algebra suffices to demonstrate that

$$k_c \approx \begin{cases} 2k + 0.01k, & g > \delta \\ 2k + 0.99k, & g \leq \delta \end{cases} \tag{11.78}$$

Consequently, the model with the contact element is too stiff by 0.5% when the gap is open and is too soft by 0.33% when the gap is closed (contact). One observation from this example is that the stiffness of the gap element should be related to the stiffnesses of the contactor and the target in the vicinity of the contact point.

In finite element modeling, "slave" nodes placed at points "i" on the target and "master" nodes placed at points "j" on the contactor are not initially connected, but may later be connected in the physical problem when contact is established. If, however, in the finite element model the nodes are always connected by the gap element, penetration is prevented, and the bilinear spring serves to overcome the difficulty of overstiffness before closure and understiffness after closure by rendering the effective stiffnesses close to the correct stiffnesses.

11.6.2 Point-to-Point Contact

Of course, in the more general case, it is not known what points, say on the target, will come into contact with the contactor, and there is no guarantee that target nodes will come into contact with contactor nodes. The gap elements can be used to account for the unknown contact area, as follows. Figure 11.13 shows a contactor and a target on which are indicated candidate contact areas dS_c and dS_t, containing nodes $c_1, c_2, \ldots, c_n, t_1, t_2, \ldots, t_n$. The candidate contact areas must contain all points for which there is a possibility of establishing contact.

The gap (i.e., the distance in the undeformed configuration) from the ith node of the contactor to the jth node of the target is denoted by g_{ij}. In point-to-point contact,

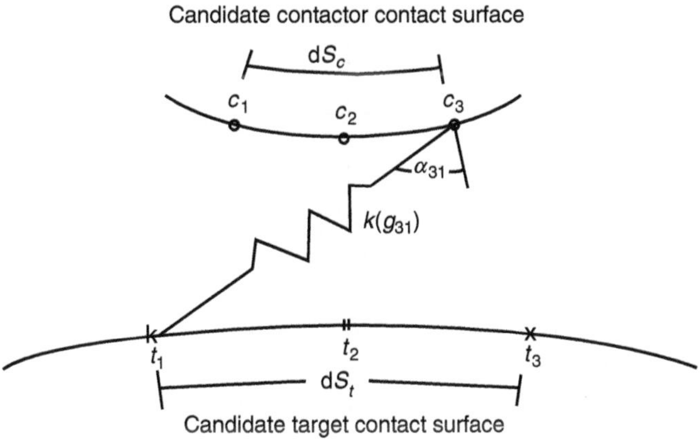

FIGURE 11.13 Point-to-point contact.

in preselected candidate contact zones, each node on the contactor is connected to each node of the target by a spring with a bilinear stiffness. (Clearly, this element may miss the edge of the contact zone when the edge does not occur at a node.) The angle between the spring and the normal at the contactor node is α_{ij}, while the angle between the spring and the normal to the target is α_{ji}. Under load, the ith contactor node experiences displacement u_{ij} in the direction of the jth target node, and the jth target node experiences displacement u_{ji}. For illustration, the spring connecting the ith contactor node with the jth target node has stiffness k_{ij} given by

$$k_{ij} = \begin{cases} k_{ij\text{lower}}, & \delta_{ij} < g_{ij} \\ k_{ij\text{upper}}, & \delta_{ij} \geq g_{ij} \end{cases} \tag{11.79}$$

in which $\delta_{ij} = u_{ij} + u_{ji}$ is the relative displacement. The force in the spring connecting the ith contactor node and jth target node is $f_{ij} = k_{ij}(g_{ij})\delta_{ij}$ (no summation). The total normal force experienced by the ith contactor node is $f_i = \sum f_{ij}\cos(\alpha_{ij})$.

As an example of how the spring stiffness might be taken to depend on the gap to achieve a continuous approximation to a bilinear function, consider the expression

$$k_{ij}(g_{ij} - \delta_{ij}) = k_0\left[\varepsilon + (1 - 2\varepsilon)\frac{2}{\pi}\tan^{-1}\left(\frac{\alpha}{2}\left[\left|\sqrt{(g_{ij} - \delta_{ij} - \gamma)^2} - (g_{ij} - \delta_{ij} - \gamma)\right|\right]\right)\right] \tag{11.80}$$

in which γ, α, and ε are positive parameters selected as follows. When $g_{ij} - \delta_{ij} - \gamma > 0$, k_{ij} attains the lower shelf value $k_0\varepsilon$, and we assume that $\varepsilon \ll 1$. If $g_{ij} - \delta_{ij} - \gamma < 0$, k_{ij} approaches the upper shelf value $k_0(1 - \varepsilon)$. We choose γ to be a small value to attain a narrow transition range from the lower to the upper shelf values. In the range $0 < g_{ij} - \delta_{ij} < \gamma$ there is a rapid but continuous transition from the lower shelf (soft) value to the upper shelf (stiff) value. If we now choose α such that $\alpha\gamma = 1$, k_{ij} equals $k_0/2(+O(\varepsilon))$ when the gap initially closes ($g_{ij} = \delta_{ij}$). The spring characteristic is illustrated in Figure 11.14.

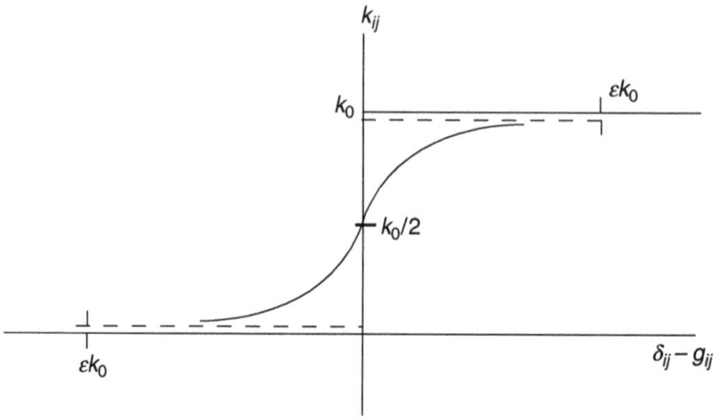

FIGURE 11.14 Illustration of a gap stiffness function.

The total normal force on a contactor node is the sum of the individual contact element forces. For example, for the jth contactor node,

$$f_{nj} = \sum_{i}^{N_c} k_{ij}(g_{ij} - \delta_{ij})\delta_{ij}\cos(\alpha_{ij}) \tag{11.81}$$

Clearly, significant forces are exerted only by the contact elements which are "closed."

11.6.3 Point-to-Surface Contact

We now briefly consider point-to-surface contact, illustrated in Figure 11.15 using a triangular element. Here target node t_3 is connected via a triangular element to contactor nodes c_1 and c_2. The stiffness matrix of the element can be written as $k([g_1 - \delta_1], [g_2 - \delta_2])\hat{\mathbf{K}}$, in which $g_1 - \delta_1$ is the gap between nodes t_1 and c_1, and $\hat{\mathbf{K}}$

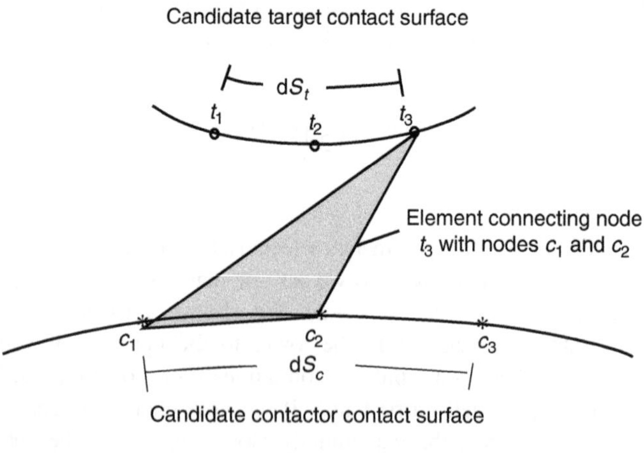

FIGURE 11.15 Element for point-to-surface contact.

is the geometric part of the stiffness matrix of a triangular elastic element. The stiffness matrix of the element may be made a function of both gaps. Total force normal to the target node is the sum of the forces exerted by the contact elements to the candidate contactor nodes.

In some finite element codes schemes such as illustrated in Figure 11.15 are also used to approximate the tangential force in the case of friction. Namely, an 'elastic friction' force is assumed in which the tangential tractions are assumed to be proportional to the normal traction through a friction coefficient. Elastic friction models do not appear to consider sliding and may be considered bonded contact. Advanced models address sliding contact and incorporate friction laws not based on the Coulomb model.

EXAMPLE 11.11

1. Consider a finite element model for a set of springs, illustrated below (Figure 11.16). A load moves the left-hand plate toward the fixed right-hand plate.
   (a) What is the load–deflection curve of the configuration?
   (b) For a finite element model, suppose a bilinear spring is supplied to bridge the gap H. What is the load–deflection curve of the finite element model?
   (c) Identify a k_g value for which the load–deflection behavior of the finite element model is close to the actual configuration.
   (d) Why is the new spring needed in the finite element model?
2. Suppose a contact element is added in the foregoing problem, in which the stiffness (spring rate) satisfies

$$k_{ij}(g_{ij}-\delta_{ij})=k_0\left[\varepsilon+(1-2\varepsilon)\frac{2}{\pi}\tan^{-1}\left(\frac{\alpha}{2}\left[\left|\sqrt{(g_{ij}-\delta_{ij}-\gamma)^2}-(g_{ij}-\delta_{ij}-\gamma)\right|\right]\right)\right]$$

with $\alpha\gamma=1$, $k_L=k/100$, and $k_u=100k$. Compute the stiffness/k for the configuration as a function of the deflection.

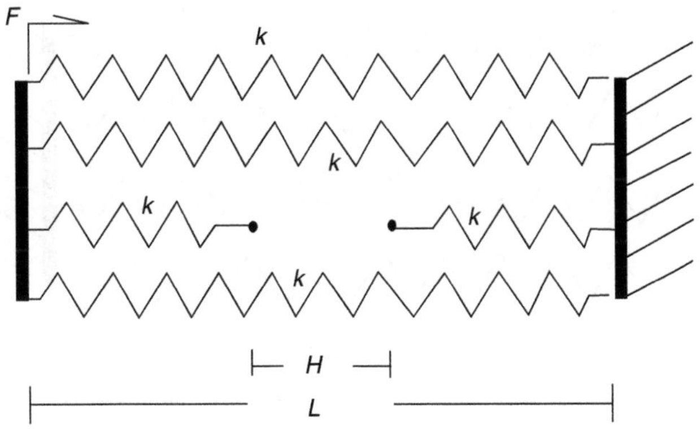

FIGURE 11.16 Set of springs with gap.

Solution

1. The exact stiffness of this configuration is

$$k_c = \begin{cases} 3k, & \delta < H, \text{ gap is open} \\ 3.5k, & \delta \geq H, \text{ gap is closed} \end{cases}$$

in which when $\delta \geq H k_c$ is calculated as follows:

$$k_c = 3k + \left(\frac{1}{k} + \frac{1}{k}\right)^{-1} = 3.5k$$

The exact load–deflection curve is plotted as in Figure 11.17

Now let us introduce a gap element of stiffness k_g as illustrated in Figure 11.18.

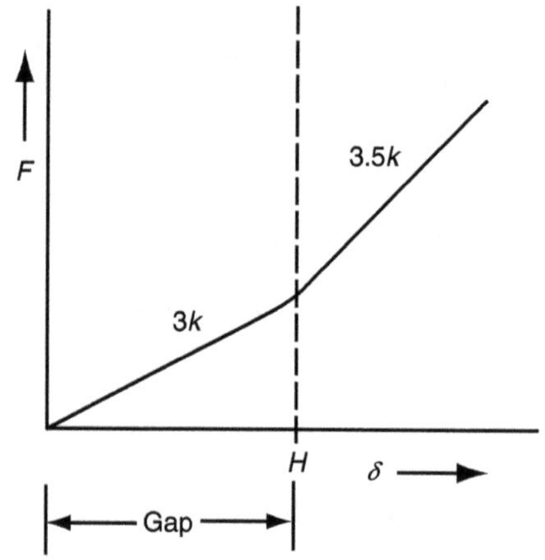

FIGURE 11.17 Load–deflection curve of the actual configuration.

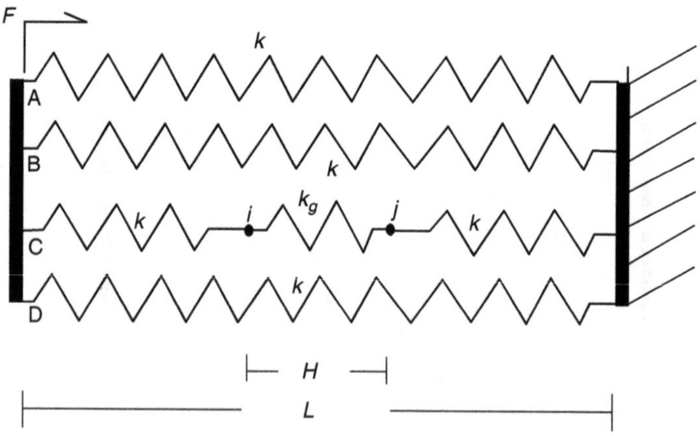

FIGURE 11.18 Illustration of gap element.

It is desirable for the spring C to be soft when the gap is open ($\delta < H$), and to be stiff when the gap is closed ($\delta > H$). Suppose

$$k_g = \begin{cases} k/100, & \delta < H \\ 100k, & \delta \geq H \end{cases}$$

When the gap is open, i.e., $\delta < H$, the stiffness of the configuration is given by

$$k_c = 3k + \left(\frac{1}{k} + \frac{1}{k_g} + \frac{1}{k}\right)^{-1} = 3k + 0.0098k$$

And when the gap is closed, i.e., $\delta > H$, the stiffness of the configuration is given by

$$k_c = 3k + \left(\frac{1}{k} + \frac{1}{k_g} + \frac{1}{k}\right)^{-1} = 3k + 0.4975k$$

Accordingly

$$k_c = \begin{cases} 3k + 0.0098k, & \delta < H \\ 3k + 0.4975k, & \delta \geq H \end{cases}$$

The model with the gap element is overly stiff by 0.33% when the gap is open, and is overly soft by 0.07% when the gap is closed. The exact and approximate values are illustrated as in Figure 11.19.

Clearly, the assumed value of k_g gives a close approximation to the actual configuration.

2. *Case (1)*: *gap is open*: When the gap is open, $g > \delta + \gamma$, i.e., $g - \delta - \gamma > 0$. Hence we have

$$\left[\left|\sqrt{(g_{ij} - \delta_{ij} - \gamma)^2} - (g_{ij} - \delta_{ij} - \gamma)\right|\right] = 0$$

$$\Rightarrow k_{ij}(g_{ij} - \delta_{ij}) = k_L = k_0\varepsilon$$

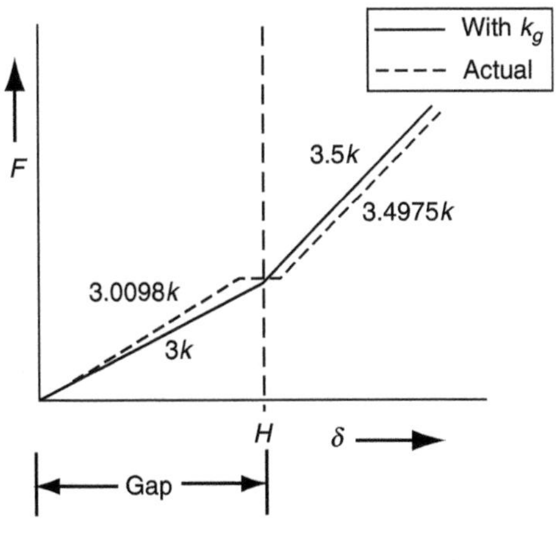

FIGURE 11.19 Load–deflection curve of the actual and the finite element configuration.

Case (2) gap is closed: When the gap is closed, $g < \delta + \gamma$, i.e., $-(g - \delta - \gamma) > 0$. Now

$$k_{ij}(g_{ij} - \delta_{ij}) = k_0 \left[\varepsilon + (1 - 2\varepsilon) \frac{2}{\pi} \tan^{-1} \left(\frac{\alpha}{2} 2(\delta + \gamma - g) \right) \right]$$

Now $\tan^{-1}(\alpha(\delta + \gamma - g))$ asymptotically approaches $\pi/2$, for which

$$k_{ij}(g_{ij} - \delta_{ij}) = k_U = k_0(1 - \varepsilon)$$

Case (3) gap just closes: When the gap just closes, $g < \delta + \gamma$, i.e., $g - \delta - \gamma < 0$. Now,

$$\tan^{-1}\left(\frac{\alpha}{2} 2(\delta + \gamma - g) \right) = \tan^{-1}(\alpha(\delta + \gamma - g))$$

On substituting $\alpha\gamma = 1$, we obtain

$$k_{ij}(g_{ij} - \delta_{ij}) = k = k_0 \left[\varepsilon + (1 - 2\varepsilon) \frac{2}{\pi} \tan^{-1}(\alpha(\delta - g) + 1) \right]$$

Since $\delta = g$ under the stated condition

$$k_{ij}(g_{ij} - \delta_{ij}) = k = k_0 \left(\varepsilon + \frac{1 - \varepsilon}{2} \right) \sim \frac{k_0}{2}$$

The results can be summarized as

$$k_{ij}(g_{ij} - \delta_{ij}) = \begin{cases} k_0\varepsilon, & \text{gap is open} \\ k_0/2, & \text{gap just closes} \\ k_0, & \text{gap is closed} \end{cases}$$

12 Rotating and Unrestrained Elastic Bodies

12.1 FINITE ELEMENTS IN ROTATION

12.1.1 ANGULAR VELOCITY AND ANGULAR ACCELERATION VECTORS

We consider a vector **b** which is referred to an instantaneous coordinate system with base vectors $\mathbf{e}_x$, $\mathbf{e}_y$, and $\mathbf{e}_z$. Suppose that in a time increment dt there is rotation of the x–y plane clockwise about the z-axis through a small angle $-d\psi$, generating a new vector $\mathbf{b}'$, and giving rise to the new coordinates x', y', z' in which z' coincides with z. This rotation is depicted in Figure 12.1. *Note: A negative rotation of the coordinate system is equivalent to a positive rotation of the vector about the instantaneously fixed coordinate system.* Prior to the rotation, the vector may be expressed as $\mathbf{b}^T = \{b_x \; b_y \; b_z\}$. For the moment we assume that b_x, b_y, and b_z are constants.

The rotation gives rise to $\mathbf{b}'$ which satisfies

$$\mathbf{b}' = \begin{bmatrix} \cos(-d\psi) & \sin(-d\psi) & 0 \\ -\sin(-d\psi) & \cos(-d\psi) & 0 \\ 0 & 0 & 1 \end{bmatrix} \mathbf{b} \sim \left[\begin{bmatrix} 1 & 0 & 0 \\ 0 & 1 & 0 \\ 0 & 0 & 1 \end{bmatrix} + d\psi \begin{bmatrix} 0 & -1 & 0 \\ 1 & 0 & 0 \\ 0 & 0 & 0 \end{bmatrix} \right] \mathbf{b} \tag{12.1}$$

from which

$$\mathbf{b}' = \mathbf{b} + d\psi \begin{bmatrix} 0 & -1 & 0 \\ 1 & 0 & 0 \\ 0 & 0 & 0 \end{bmatrix} \mathbf{b}$$

Next, suppose that the y'–z' plane is rotated about the x-axis through the incremental angle $-d\theta$, corresponding to a positive rotation of the vector with respect to the instantaneously fixed $x'y'z'$ coordinate system (cf. Figure 12.2). We now have the vector $\mathbf{b}''$ and the directions x'', y'', z'', in which $\mathbf{b}''$ is given by

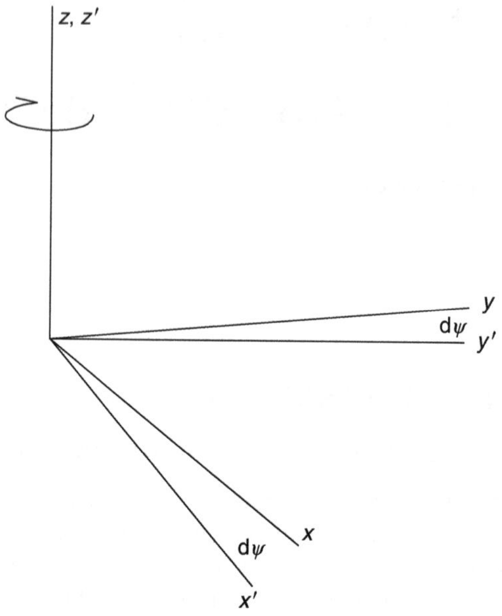

FIGURE 12.1 Differential clockwise rotation about the z-axis.

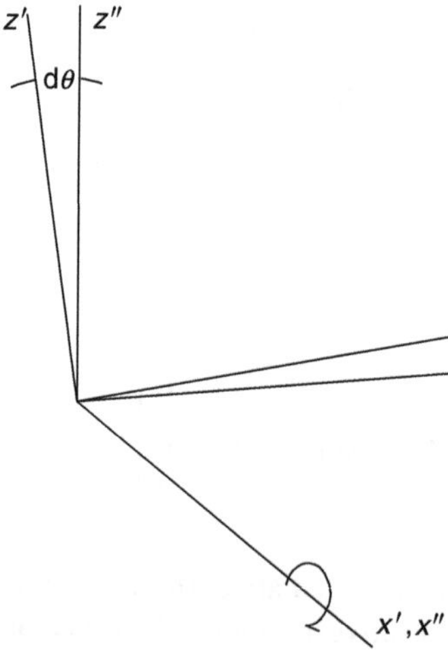

FIGURE 12.2 Differential rotation about the x'-axis.

$$\mathbf{b}'' = \begin{bmatrix} 1 & 0 & 0 \\ 0 & \cos(-\mathrm{d}\theta) & \sin(-\mathrm{d}\theta) \\ 0 & -\sin(-\mathrm{d}\theta) & \cos(-\mathrm{d}\theta) \end{bmatrix} \mathbf{b}' \approx \begin{bmatrix} 1 & 0 & 0 \\ 0 & 1 & -\mathrm{d}\theta \\ 0 & \mathrm{d}\theta & 1 \end{bmatrix} \mathbf{b}' \tag{12.2}$$

But eliminating $\mathbf{b}'$ furnishes

$$\mathbf{b}'' = \left[\begin{bmatrix} 1 & 0 & 0 \\ 0 & 1 & 0 \\ 0 & 0 & 1 \end{bmatrix} + \mathrm{d}\theta \begin{bmatrix} 0 & 0 & 0 \\ 0 & 0 & -1 \\ 0 & 1 & 0 \end{bmatrix}\right]\left[\begin{bmatrix} 1 & 0 & 0 \\ 0 & 1 & 0 \\ 0 & 0 & 1 \end{bmatrix} + \mathrm{d}\psi \begin{bmatrix} 0 & -1 & 0 \\ 1 & 0 & 0 \\ 0 & 0 & 0 \end{bmatrix}\right]\mathbf{b}$$

$$\sim \mathbf{b} + \left[\mathrm{d}\theta \begin{bmatrix} 0 & 0 & 0 \\ 0 & 0 & -1 \\ 0 & 1 & 0 \end{bmatrix} + \mathrm{d}\psi \begin{bmatrix} 0 & -1 & 0 \\ 1 & 0 & 0 \\ 0 & 0 & 0 \end{bmatrix}\right]\mathbf{b} \tag{12.3}$$

Finally, referring to Figure 12.3, the $z''-x''$ plane is rotated around the y''-axis clockwise through an angle $-\mathrm{d}\phi$, furnishing the new axes x''', y''', z''', and the new vector $\mathbf{b}'''$.

The vector $\mathbf{b}'''$ is given by

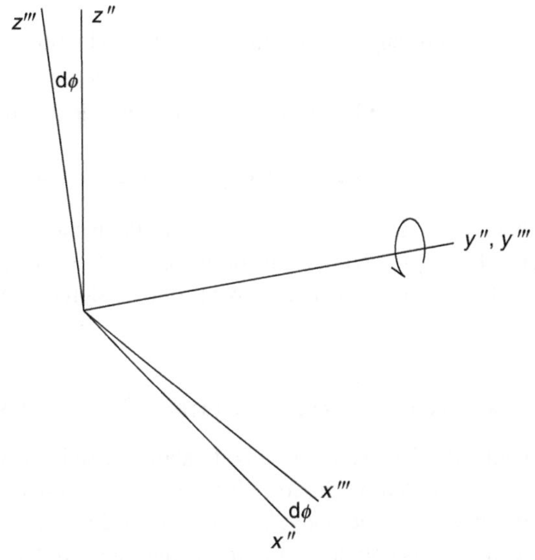

FIGURE 12.3 Differential clockwise rotation about the y''-axis.

$$\mathbf{b}''' = \begin{bmatrix} \cos(-d\phi) & 0 & -\sin(-d\phi) \\ 0 & 1 & 0 \\ \sin(-d\phi) & 0 & \cos(-d\phi) \end{bmatrix} \mathbf{b}'' \approx \left[\begin{bmatrix} 1 & 0 & 0 \\ 0 & 1 & 0 \\ 0 & 0 & 1 \end{bmatrix} + d\phi \begin{bmatrix} 0 & 0 & 1 \\ 0 & 0 & 0 \\ -1 & 0 & 0 \end{bmatrix} \right] \mathbf{b}''$$

$$\approx \mathbf{b} + \left[d\phi \begin{bmatrix} 0 & 0 & 1 \\ 0 & 0 & 0 \\ -1 & 0 & 0 \end{bmatrix} + d\theta \begin{bmatrix} 0 & 0 & 0 \\ 0 & 0 & -1 \\ 0 & 1 & 0 \end{bmatrix} + d\psi \begin{bmatrix} 0 & -1 & 0 \\ 1 & 0 & 0 \\ 0 & 0 & 0 \end{bmatrix} \right] \mathbf{b}$$

$$\approx \mathbf{b} + \begin{bmatrix} 0 & -d\psi & d\phi \\ d\psi & 0 & -d\theta \\ -d\phi & d\theta & 0 \end{bmatrix} \mathbf{b} \tag{12.4}$$

The derivative of $\mathbf{b}$ is equated with $\frac{\mathbf{b}''' - \mathbf{b}}{dt}$, and now

$$\begin{aligned} \frac{d\mathbf{b}}{dt} &= \begin{bmatrix} 0 & -\dot{\psi} & \dot{\phi} \\ \dot{\psi} & 0 & -\dot{\theta} \\ -\dot{\phi} & \dot{\theta} & 0 \end{bmatrix} \mathbf{b} \\ &= (\dot{\phi} b_z - \dot{\psi} b_y)\mathbf{e}_x + (\dot{\psi} b_x - \dot{\theta} b_z)\mathbf{e}_y + (\dot{\theta} b_y - \dot{\phi} b_x)\mathbf{e}_z \\ &= \boldsymbol{\omega} \times \mathbf{b} \end{aligned} \tag{12.5}$$

in which the instantaneous angular position vector is identified as $\boldsymbol{\theta} = \begin{Bmatrix} \theta \\ \phi \\ \psi \end{Bmatrix} = \theta \mathbf{e}_x + \phi \mathbf{e}_y + \psi \mathbf{e}_z$, and the angular velocity vector is identified as $\boldsymbol{\omega} = \frac{d\boldsymbol{\theta}}{dt} = \begin{Bmatrix} \dot{\theta} \\ \dot{\phi} \\ \dot{\psi} \end{Bmatrix} = \dot{\theta}\mathbf{e}_x + \dot{\phi}\mathbf{e}_y + \dot{\psi}\mathbf{e}_z$. The angular velocity vector describes the rate at which the vector $\mathbf{b}$ rotates in a counterclockwise sense about the instantaneous axes. In the appendix, the vector $\boldsymbol{\theta}$ is expressed in terms of angles used in spherical coordinates.

The angular position vector presented above is one version of what are called the Euler angles.

12.1.2 Velocity and Acceleration in Rotating Coordinates

More generally, again consider rotation of a body about a fixed axis in which at least one point in the body is located on the axis. The coordinate system is embedded in the fixed point and rotates with the body. The undeformed position vector $\mathbf{X}'$ in the rotated system is related to its counterpart $\mathbf{X}$ in the unrotated system by $\mathbf{X}' = \mathbf{Q}(t)\mathbf{X}$, in which $\mathbf{Q}$ is a proper orthogonal tensor. If the deformed body is viewed in the same coordinate system, the counterpart for the deformed position is $\mathbf{x}' = \mathbf{Q}(t)\mathbf{x}$. The displacement likewise satisfies $\mathbf{u}' = \mathbf{Q}(t)\mathbf{u}$. Time differentiation gives

$$\frac{\mathrm{d}\mathbf{u}'(\mathbf{X},t)}{\mathrm{d}t} = \mathbf{Q}(t)\frac{\mathrm{d}\mathbf{u}(\mathbf{X},t)}{\mathrm{d}t} + \dot{\mathbf{Q}}(t)\mathbf{u}(\mathbf{X},t)$$
$$= \mathbf{Q}(t)\frac{\mathrm{d}\mathbf{u}(\mathbf{X},t)}{\mathrm{d}t} + \dot{\mathbf{Q}}(t)\mathbf{Q}^T(t)\mathbf{u}'(\mathbf{X},t). \tag{12.6}$$

We make the identification

$$\frac{\partial \mathbf{u}'(\mathbf{X},t)}{\partial t} = \mathbf{Q}(t)\frac{\mathrm{d}\mathbf{u}(\mathbf{X},t)}{\mathrm{d}t} \tag{12.7}$$

with the interpretation that $\frac{\partial \mathbf{u}'(\mathbf{X},t)}{\partial t}$ denotes the derivative of $\mathbf{u}'$ with the rotation tensor $\mathbf{Q}(t)$ held fixed at its current position, and vanishes if $\mathbf{u}$ is unchanging relative to the rotating system. In addition, since $\mathbf{0} = \frac{\mathrm{d}}{\mathrm{d}t}\mathbf{Q}\mathbf{Q}^T = \dot{\mathbf{Q}}\mathbf{Q}^T + \mathbf{Q}\dot{\mathbf{Q}}^T$, $\boldsymbol{\Omega} = \dot{\mathbf{Q}}\mathbf{Q}^T$ is antisymmetric. Finally, from Chapter 3 for the 3×3 antisymmetric tensor $\boldsymbol{\Omega}$ and for any 3×1 vector $\mathbf{b}$ there exists a 3×1 vector $\boldsymbol{\omega}$ satisfying $\boldsymbol{\Omega}\mathbf{b} = \boldsymbol{\omega} \times \mathbf{b}$. Of course $\boldsymbol{\omega}$ is the *angular velocity vector* of dynamics, and it is referred to the instantaneous (rotating) coordinate system; its time derivative $\frac{\partial \boldsymbol{\omega}}{\partial t}$ is the angular acceleration vector, hereafter denoted by $\boldsymbol{\alpha}$ and referred to the instantaneous coordinate system. Corresponding to $\boldsymbol{\alpha}$ there is an antisymmetric tensor $\boldsymbol{A} = \partial\boldsymbol{\Omega}/\partial t$ such that $\boldsymbol{A}\mathbf{b} = \boldsymbol{\alpha} \times \mathbf{b}$.

Now with the prime on $\mathbf{u}$ no longer displayed, the total time derivative of the displacement vector, i.e., the velocity vector $\mathbf{v}$, may be expressed in the rotating system as

$$\mathbf{v} = \frac{\mathrm{d}\mathbf{u}}{\mathrm{d}t} = \left(\frac{\partial}{\partial t} + \boldsymbol{\omega} \times\right)\mathbf{u} = \frac{\partial \mathbf{u}}{\partial t} + \boldsymbol{\omega} \times \mathbf{u} \tag{12.8}$$

The acceleration vector is similarly expressed as

$$\mathbf{a} = \mathrm{d}\mathbf{v}/\mathrm{d}t$$
$$= (\partial/\partial t + \boldsymbol{\omega}\ \times)(\partial\mathbf{u}/\partial t + \omega \times \mathbf{u})$$
$$= \partial^2\mathbf{u}/\partial t^2 + 2\boldsymbol{\omega} \times \partial\mathbf{u}/\partial t + \boldsymbol{\omega} \times \boldsymbol{\omega} \times \mathbf{u} + \boldsymbol{\alpha} \times \mathbf{u} \tag{12.9}$$

The four right-hand terms in Equation 12.9 are, respectively, called the *translational*, *Coriolis*, *centrifugal*, and *angular accelerations*, respectively.

In the Principle of Virtual Work, for reasons to be explained in a subsequent sections let $\delta\hat{\mathbf{u}}$ denote the variation of $\mathbf{u}$ *with the coordinate system held fixed at its instantaneous position.* The corresponding inertial term becomes $\int \delta\hat{\mathbf{u}}^T\rho\frac{\mathrm{d}^2}{\mathrm{d}t^2}(\mathbf{u}' + \mathbf{X}')\mathrm{d}V$. Now assuming the usual interpolation model $\mathbf{u}'(\mathbf{X}',t) = \varphi^T(\mathbf{X}')\boldsymbol{\Phi}\gamma(t)$, the inertial term becomes

$$\int \delta\hat{\mathbf{u}}^T\rho\frac{\mathrm{d}^2\mathbf{u}'}{\mathrm{d}t^2}\mathrm{d}V = \delta\boldsymbol{\gamma}^T\left[\mathbf{M}\frac{\mathrm{d}^2\boldsymbol{\gamma}}{\mathrm{d}t^2} + \mathbf{G}_1\frac{\mathrm{d}\boldsymbol{\gamma}}{\mathrm{d}t} + (\mathbf{G}_2 + \boldsymbol{A})\boldsymbol{\gamma}\right] \tag{12.10}$$

$$\mathbf{M} = \mathbf{\Phi}^T \int \rho \boldsymbol{\varphi} \boldsymbol{\varphi}^T \, \mathrm{d}V \, \mathbf{\Phi}, \qquad \mathbf{G}_1 = \mathbf{\Phi}^T \int \rho \boldsymbol{\varphi} \mathbf{\Omega} \boldsymbol{\varphi}^T \, \mathrm{d}V \, \mathbf{\Phi}$$

$$\mathbf{G}_2 = \mathbf{\Phi}^T \int \rho \boldsymbol{\varphi} \mathbf{\Omega}^2 \boldsymbol{\varphi}^T \, \mathrm{d}V \, \mathbf{\Phi}, \quad A = \mathbf{\Phi}^T \int \rho \boldsymbol{\varphi} \mathbf{A} \boldsymbol{\varphi}^T \, \mathrm{d}V \, \mathbf{\Phi}$$

The matrix $\mathbf{M}$ is the conventional positive definite and symmetric mass matrix in a nonrotating system. The Coriolis matrix $\mathbf{G}_1$ is antisymmetric, the centrifugal matrix $\mathbf{G}_2$ is negative definite, and the angular acceleration matrix A is antisymmetric. ($\mathbf{G}_2$ involves the square of an antisymmetric matrix. Since an antisymmetric matrix has pure imaginary eigenvalues [Chapter 2], its square is symmetric with negative real eigenvalues and hence is negative definite.)

There is also a rigid body force term due to centrifugal and angular accelerations.

$$\int \delta \hat{\mathbf{u}}^T \rho \frac{\mathrm{d}^2}{\mathrm{d}t^2} \mathbf{X}' \, \mathrm{d}V = \delta \boldsymbol{\gamma}^T \mathbf{f}_{\mathrm{rot}}, \quad \mathbf{f}_{\mathrm{rot}} = \mathbf{\Phi}^T \int \rho \boldsymbol{\varphi} \left[\mathbf{\Omega}^2 + A \right] \mathbf{X}' \, \mathrm{d}V \tag{12.11}$$

Consolidating the terms, the ensuing governing finite element equation is now

$$\mathbf{M} \frac{\mathrm{d}^2 \boldsymbol{\gamma}}{\mathrm{d}t^2} + \mathbf{G}_1 \frac{\mathrm{d}\boldsymbol{\gamma}}{\mathrm{d}t} + [\mathbf{K} + \mathbf{G}_2 + A] \boldsymbol{\gamma} = \mathbf{f} - \mathbf{f}_{\mathrm{rot}} \tag{12.12}$$

EXAMPLE 12.1

Compare a one-element model of a rotating rod to the exact model. The rod is depicted as follows (Figure 12.4).

We consider the member a rod and accordingly neglect effects in direction transverse to the motion. The radial displacement is denoted as $u(r)$. The acceleration is obtained as

$$\mathbf{u} = u \mathbf{e}_r, \quad \boldsymbol{\omega} = \omega \mathbf{e}_z, \quad \boldsymbol{\alpha} = \alpha \mathbf{e}_z$$

$$\ddot{\mathbf{u}} = \frac{\partial^2 u}{\partial t^2} \mathbf{e}_r + 2\omega \mathbf{e}_z \times \frac{\partial u}{\partial t} \mathbf{e}_r + \omega \mathbf{e}_z \times \omega \mathbf{e}_z \times u \mathbf{e}_r + \alpha \mathbf{e}_z \times u \mathbf{e}_r$$

The Coriolis and angular accelerations terms are aligned with $\mathbf{e}_\theta$ and are therefore not of interest in a rod. The centrifugal acceleration points back radially toward the origin. The acceleration vector therefore reduces to

$$\ddot{\mathbf{u}} = \left[\frac{\partial^2 u}{\partial t^2} - \omega^2 u \right] \mathbf{e}_r$$

We also obtain the rigid body contribution $\ddot{\mathbf{r}}_0 = -\omega^2 r_0 \mathbf{e}_r$. Enforcing the constraint at the shaft a priori, we introduce the interpolation model $u(r_0,t) = \frac{r_0}{L} \gamma(t)$. The Principle of Virtual Work (using $\delta \hat{\mathbf{u}} \to \delta u$) now implies

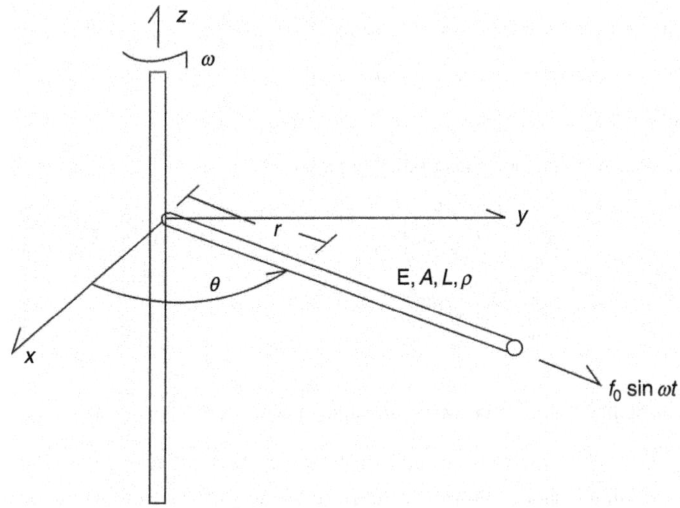

FIGURE 12.4 Steadily rotating elastic rod.

$$\int_0^L \delta u \rho \left[\frac{\partial^2 u}{\partial t^2} - \omega^2 r_0 - \omega^2 u\right] A\, dr_0 = \delta\gamma\left[\frac{\rho AL}{3}\ddot{\gamma} - \frac{\rho AL}{3}\omega^2\gamma - \frac{\rho AL^2}{3}\omega\right]$$

The stiffness term is found as

$$\int_0^L \delta \mathrm{E}_{rr} \mathrm{S}_{rr} A\, \mathrm{d}r_0 = \delta\gamma \frac{\mathrm{E}A}{L}\gamma$$

The one-element finite element equation is now

$$\left[\frac{\mathrm{E}A}{L} - \frac{\rho AL}{3}\omega^2\right]\gamma + \frac{\rho AL}{3}\ddot{\gamma} = \frac{\rho AL^2}{3}\omega$$

In the steady state $\ddot{\gamma} = 0$. Under steady state rotation, $\gamma \to \infty$ if $\omega \to \omega_{cr} = \sqrt{\frac{\mathrm{E}A/L}{\rho AL/3}} = \sqrt{\frac{3}{L}}\sqrt{\frac{\mathrm{E}}{\rho}}$, and ω_{cr} is called the critical speed. Note that the critical speed is nothing but the natural frequency of the nonrotating rod.

To assess accuracy the one-element finite element model is now compared to the exact solution.

The exact governing differential equation is expressed as

$$\mathrm{E}A\frac{\mathrm{d}^2 u}{\mathrm{d}r_0^2} + \rho\omega^2 u = -\rho\omega^2 r_0$$

The particular solution is $u_p = -r_0$, in which event the characteristic solution satisfies

$$\mathrm{E}A\frac{\mathrm{d}^2 u_c}{\mathrm{d}r_0^2} + \rho\omega^2 u_c = 0$$

The characteristic solution has the form $u_c = \alpha \sin\left(\sqrt{\frac{\rho\omega^2}{\mathsf{E}}} r_0\right) + \beta \cos\left(\sqrt{\frac{\rho\omega^2}{\mathsf{E}}} r_0\right)$, and the boundary conditions are $u(0) = 0$, $\mathsf{E}A\frac{du(L)}{dr_0} = 0$, giving $\beta = 0$, $\alpha = L\Big/\sqrt{\frac{\rho\omega^2}{\mathsf{E}}}\cos\left(\sqrt{\frac{\rho\omega^2}{\mathsf{E}}}L\right)$.

Finally, the displacement function emerges as

$$u(r_0;\omega^2) = \frac{L\sin\left(\sqrt{\frac{\rho\omega^2}{\mathsf{E}}} r_0\right)}{\sqrt{\frac{\rho\omega^2}{\mathsf{E}}}\cos\left(\sqrt{\frac{\rho\omega^2}{\mathsf{E}}}L\right)} - r_0, \qquad \gamma = L\left[\frac{\tan\left(\sqrt{\frac{\rho\omega^2}{\mathsf{E}}}L\right)}{\sqrt{\frac{\rho\omega^2}{\mathsf{E}}}} - 1\right]$$

The nodal displacement γ becomes unbounded when $\sqrt{\frac{\rho\omega^2}{\mathsf{E}}}L = \frac{\pi}{2}$, and hence when $\omega = \frac{\pi}{2L}\sqrt{\frac{\mathsf{E}}{\rho}}$, which is the exact natural frequency of the nonrotating rod. The critical speed in the single element finite element model is within 10% of the exact value.

EXAMPLE 12.2

Unsteady rotation of a beam column about an axis

Consider a thin *beam column* that is rotating unsteadily around a shaft. Its thin (local z) direction points in the direction of the motion, giving rise to Coriolis effects in bending. Derive the ensuing one-element model. Note that a beam column couples extension and bending.

SOLUTION

Enforcing the clamped constraints at the axis of revolution, the interpolation model for a beam column (Figure 12.5) is given by $u(x,y,t) = u_0(x,t) - y\frac{\partial v(x,t)}{\partial x}$, with

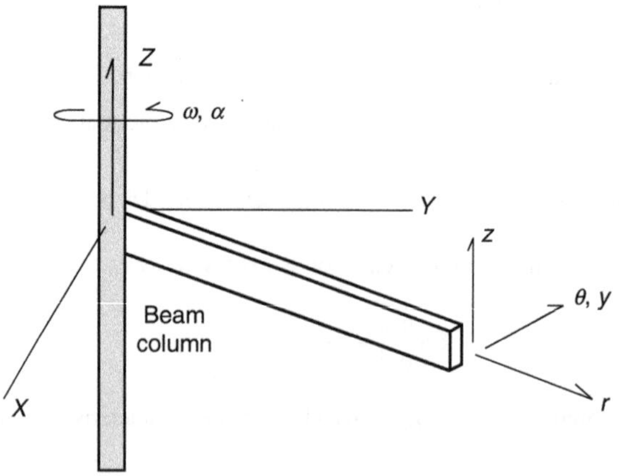

FIGURE 12.5 Beam column in unsteady rotation about an axis.

$$u_0(x,t) = \frac{x}{L}u_0(L,t), \quad v = \boldsymbol{\varphi}_b^T \boldsymbol{\Phi}_b \boldsymbol{\gamma}_b$$

$$\boldsymbol{\varphi}_b^T = (x^2 \quad x^3), \quad \boldsymbol{\Phi}_b = \begin{bmatrix} L^2 & L^3 \\ -2L & -3L^2 \end{bmatrix}^{-1} = \frac{1}{L^4}\begin{bmatrix} 3L^2 & L^3 \\ -2L & -L^2 \end{bmatrix}$$

$$\boldsymbol{\gamma}_b = \begin{pmatrix} v(L) \\ -v'(L) \end{pmatrix}, \quad v'(x,t) = \frac{\partial v(x,t)}{\partial x} = \boldsymbol{\beta}_b^T \boldsymbol{\Phi}_b \boldsymbol{\gamma}_b, \quad \boldsymbol{\beta}_b^T = (2x \quad 3x^2)$$

The stiffness matrix not affected by the motion and is given by (cf. Chapter 7)

$$\int \delta\varepsilon_{ij}\sigma_{ij}\,\mathrm{d}V = \delta\boldsymbol{\gamma}^T \mathbf{K}\boldsymbol{\gamma}, \quad \mathbf{K} = \begin{bmatrix} \mathbf{K}_m & \mathbf{0} \\ \mathbf{0} & \mathbf{K}_b \end{bmatrix}$$

$$\mathbf{K}_m = \frac{\mathsf{E}A}{L}, \quad \mathbf{K}_b = \frac{\mathsf{E}I}{L^3}\begin{bmatrix} 12 & 6 \\ 6 & 4 \end{bmatrix}, \quad \boldsymbol{\gamma} = \begin{pmatrix} u_0(L,t) \\ \boldsymbol{\gamma}_b \end{pmatrix}$$

The inertial terms now are stated as

$$\frac{\mathrm{d}^2\mathbf{u}}{\mathrm{d}t^2} = \frac{\partial^2\mathbf{u}}{\partial t^2} + 2\boldsymbol{\omega} \times \frac{\partial \mathbf{u}}{\partial t} + \boldsymbol{\omega} \times \boldsymbol{\omega} \times \mathbf{u} + \boldsymbol{\alpha} \times \mathbf{u}$$

$$= \int (\delta u_0 - y\delta v' \delta v)\begin{pmatrix} \dfrac{\partial^2(u_0 - yv')}{\partial t^2} - \omega^2(u_0 - yv') - 2\omega\dfrac{\partial v}{\partial t} - \alpha v \\ 2\omega\dfrac{\partial(u_0 - yv')}{\partial t} + \alpha(u_0 - yv') + \dfrac{\partial^2 v}{\partial t^2} - \omega^2 v \end{pmatrix}\rho\,\mathrm{d}V$$

$$= \int \delta u_0\left(\frac{\partial^2 u_0}{\partial t^2} - \omega^2 u_0 - 2\omega\frac{\partial v}{\partial t} - \alpha v\right)\rho\,\mathrm{d}V + \int y\delta v'\left(y\frac{\partial^2 v'}{\partial t^2} - \omega^2 yv'\right)\rho\,\mathrm{d}V$$

$$+ \int \delta v\left(2\omega\frac{\partial u_0}{\partial t} + \alpha u_0 + \frac{\partial^2 v}{\partial t^2} - \omega^2 v\right)\rho\,\mathrm{d}V$$

Performing the indicated manipulations results in the following relations:

$$\int \delta u_0 \frac{\partial^2 u_0}{\partial t^2}\rho\,\mathrm{d}V = \delta u_0(L,t)\int_0^L \frac{x}{L}\frac{x}{L}\rho A\,\mathrm{d}x \frac{\partial^2 u_0(L,t)}{\partial t^2}$$

$$= \delta u_0(L,t)\left(\frac{\rho AL}{3}\right)\frac{\partial^2 u_0(L,t)}{\partial t^2}$$

Similarly, $\int \delta u_0(-\omega^2)u_0\rho\,\mathrm{d}V = \delta u_0(L,t)\left(-\omega^2\frac{\rho AL}{3}\right)u_0(L,t)$

$$\int \delta u_0(-2\omega)\frac{\partial v}{\partial t}\rho\,\mathrm{d}V = \delta u_0(L,t)\int_0^L \frac{x}{L}(-2\omega)\boldsymbol{\phi}_b^T \boldsymbol{\Phi}_b \rho A\,\mathrm{d}x\,\dot{\boldsymbol{\gamma}}_b$$

$$= \delta u_0(L,t)(-2\omega\rho A)\left(\frac{1}{4} \quad \frac{L}{5}\right)\begin{bmatrix} 3L^2 & L^3 \\ -2L & -L^2 \end{bmatrix}\dot{\boldsymbol{\gamma}}_b$$

$$= \delta u_0(L,t)\left(-2\omega\frac{\rho AL}{20}(7 \quad L)\right)\dot{\boldsymbol{\gamma}}_b$$

$$\int \delta u_0(-\alpha v)\rho\, dV = \delta u_0(L,t)\int_0^L \frac{x}{L}(-\alpha)\boldsymbol{\phi}_b^T\boldsymbol{\Phi}_b\rho A\, dx\, \boldsymbol{\gamma}_b$$

$$= \delta u_0(L,t)\left(-\alpha\frac{\rho A}{L}\right)\left(\frac{1}{4}\ \frac{L}{5}\right)\begin{bmatrix} 3L^2 & L^3 \\ -2L & -L^2 \end{bmatrix}\boldsymbol{\gamma}_b$$

$$= \delta u_0(L,t)\left(-\alpha\frac{\rho AL}{20}(7\quad L)\right)\boldsymbol{\gamma}_b$$

$$\int y^2\delta v'\frac{\partial^2 v'}{\partial t^2}\rho\, dV = \delta\boldsymbol{\gamma}_b^T\boldsymbol{\Phi}_b^T\int \boldsymbol{\beta}_b\rho I\boldsymbol{\beta}_b^T\, dx\, \boldsymbol{\Phi}_b\ddot{\boldsymbol{\gamma}}_b$$

$$= \delta\boldsymbol{\gamma}_b^{\mathrm{T}}\boldsymbol{\Phi}_b^T\int_0^L \rho I\begin{bmatrix} 4x^2 & 6x^3 \\ 6x^3 & 9x^4 \end{bmatrix} dx\, \boldsymbol{\Phi}_b\ddot{\boldsymbol{\gamma}}_b$$

$$= \delta\boldsymbol{\gamma}_b^T\frac{\rho I}{L}\begin{bmatrix} \frac{6}{5} & \frac{1}{10}L \\ \frac{1}{10}L & \frac{2}{15}L^2 \end{bmatrix}\ddot{\boldsymbol{\gamma}}_b$$

The matrix in the preceding expression is recognized as proportional to $\hat{\mathbf{K}}_1$ encountered in Chapter 10 in conjunction with buckling.
Continuing,

$$\int \delta v\alpha u_0\rho\, dV = \delta\boldsymbol{\gamma}_b^T\alpha\rho AL\begin{pmatrix} \frac{7}{20} \\ \frac{1}{20}L \end{pmatrix}u_0(L,t)$$

$$\int \delta v\frac{\partial^2 v}{\partial t^2}\rho\, dV = \delta\boldsymbol{\gamma}_b^T\boldsymbol{\Phi}_b^T\int \boldsymbol{\varphi}_b\rho A\boldsymbol{\varphi}_b^T\, dx\, \boldsymbol{\Phi}_b\ddot{\boldsymbol{\gamma}}_b$$

$$= \delta\boldsymbol{\gamma}_b^T\boldsymbol{\Phi}_b^T\int_0^L \rho A\begin{bmatrix} x^4 & x^5 \\ x^5 & x^6 \end{bmatrix} dx\, \boldsymbol{\Phi}_b\ddot{\boldsymbol{\gamma}}_b$$

$$= \delta\boldsymbol{\gamma}_b^T\frac{\rho A}{L^8}\begin{bmatrix} 3L^2 & -2L \\ L^3 & -L^2 \end{bmatrix}\begin{bmatrix} \frac{1}{5}L^5 & \frac{1}{6}L^6 \\ \frac{1}{6}L^6 & \frac{1}{7}L^7 \end{bmatrix}\begin{bmatrix} 3L^2 & L^3 \\ -2L & -L^2 \end{bmatrix}\ddot{\boldsymbol{\gamma}}_b$$

$$= \delta\boldsymbol{\gamma}_b^T\rho AL\begin{bmatrix} \frac{13}{35} & \frac{11}{210}L \\ \frac{11}{210}L & \frac{1}{105}L^2 \end{bmatrix}\ddot{\boldsymbol{\gamma}}_b$$

$$\int y^2\delta v'(-\omega^2)v'\rho\, dV = \delta\boldsymbol{\gamma}_b^T(-\omega^2)\frac{\rho I}{L}\begin{bmatrix} \frac{6}{5} & \frac{1}{10}L \\ \frac{1}{10}L & \frac{2}{15}L^2 \end{bmatrix}\ddot{\boldsymbol{\gamma}}_b$$

$$\int \delta v(2\omega)\frac{\partial u_0}{\partial t}\rho\,\mathrm{d}V = \delta\boldsymbol{\gamma}_b^T\boldsymbol{\Phi}_b^T\int(2\omega)\boldsymbol{\varphi}_b\frac{x}{L}\rho A\,\mathrm{d}x\,\dot{u}_0(L,t)$$

$$= \delta\boldsymbol{\gamma}_b^T\frac{2\omega\rho A}{L}\begin{bmatrix}3L^2 & -2L\\ L^3 & -L^2\end{bmatrix}\begin{pmatrix}\frac{1}{4}L^4\\ \frac{1}{5}L^5\end{pmatrix}\dot{u}_0(L,t)$$

$$= \delta\boldsymbol{\gamma}_b^T 2\omega\rho AL\begin{pmatrix}\frac{7}{20}\\ \frac{1}{20}L\end{pmatrix}\dot{u}_0(L,t)$$

Finally,

$$\int \delta v(-\omega^2)v\rho\,\mathrm{d}V = \delta\boldsymbol{\gamma}_b^T(-\omega^2)\rho AL\begin{bmatrix}\frac{13}{35} & \frac{11}{210}L\\ \frac{11}{210}L & \frac{1}{105}L^2\end{bmatrix}\ddot{\boldsymbol{\gamma}}_b$$

Upon combining $\gamma_m = u_0$ and $\boldsymbol{\gamma}_b$ into the vector $\boldsymbol{\gamma}$, the inertial terms in the Principle of Virtual Work are now stated as

$$\int \delta u_0\frac{\partial^2 u_0}{\partial t^2}\rho\,\mathrm{d}V = \delta\boldsymbol{\gamma}^T\left(\frac{\rho AL}{3}\right)\begin{bmatrix}1 & 0 & 0\\ 0 & 0 & 0\\ 0 & 0 & 0\end{bmatrix}\ddot{\boldsymbol{\gamma}}$$

$$\int \delta u_0(-\omega^2)u_0\rho\,\mathrm{d}V = \delta\boldsymbol{\gamma}^T\left(-\omega^2\frac{\rho AL}{3}\right)\begin{bmatrix}1 & 0 & 0\\ 0 & 0 & 0\\ 0 & 0 & 0\end{bmatrix}\boldsymbol{\gamma}$$

$$\int \delta u_0(-2\omega)\frac{\partial v}{\partial t}\rho\,\mathrm{d}V = \delta\boldsymbol{\gamma}^T\left(-2\omega\frac{\rho AL}{20}\right)\begin{bmatrix}0 & 7 & L\\ 0 & 0 & 0\\ 0 & 0 & 0\end{bmatrix}\dot{\boldsymbol{\gamma}}$$

$$\int \delta u_0(-\alpha v)\rho\,\mathrm{d}V = \delta\boldsymbol{\gamma}^T\left(-\alpha\frac{\rho AL}{20}\right)\begin{bmatrix}0 & 7 & L\\ 0 & 0 & 0\\ 0 & 0 & 0\end{bmatrix}\boldsymbol{\gamma}$$

$$\int y^2\delta v'\frac{\partial^2 v'}{\partial t^2}\rho\,\mathrm{d}V = \delta\boldsymbol{\gamma}^T\frac{\rho I}{L}\begin{bmatrix}0 & 0 & 0\\ 0 & \frac{6}{5} & \frac{1}{10}L\\ 0 & \frac{1}{10}L & \frac{2}{15}L^2\end{bmatrix}\ddot{\boldsymbol{\gamma}}$$

$$\int y^2\delta v'(-\omega^2)v'\rho\,\mathrm{d}V = \delta\boldsymbol{\gamma}^T\left(-\omega^2\frac{\rho I}{L}\right)\begin{bmatrix}0 & 0 & 0\\ 0 & \frac{6}{5} & \frac{1}{10}L\\ 0 & \frac{1}{10}L & \frac{2}{15}L^2\end{bmatrix}\boldsymbol{\gamma}$$

$$\int \delta v(2\omega)\frac{\partial u_0}{\partial t}\rho\,\mathrm{d}V = \delta\boldsymbol{\gamma}^T\left(2\omega\frac{\rho AL}{20}\right)\begin{bmatrix} 0 & 0 & 0 \\ 7 & 0 & 0 \\ L & 0 & 0 \end{bmatrix}\dot{\boldsymbol{\gamma}}$$

$$\int \delta v\alpha u_0\rho\,\mathrm{d}V = \delta\boldsymbol{\gamma}^T\left(\alpha\frac{\rho AL}{20}\right)\begin{bmatrix} 0 & 0 & 0 \\ 7 & 0 & 0 \\ L & 0 & 0 \end{bmatrix}\boldsymbol{\gamma}$$

$$\int \delta v\frac{\partial^2 v}{\partial t^2}\rho\,\mathrm{d}V = \delta\boldsymbol{\gamma}^T\rho AL\begin{bmatrix} 0 & 0 & 0 \\ 0 & \frac{13}{35} & \frac{11}{210}L \\ 0 & \frac{11}{210}L & \frac{1}{105}L^2 \end{bmatrix}\ddot{\boldsymbol{\gamma}}$$

$$\int \delta v(-\omega^2)v\rho\,\mathrm{d}V = \delta\boldsymbol{\gamma}^T(-\omega^2\rho AL)\begin{bmatrix} 0 & 0 & 0 \\ 0 & \frac{13}{35} & \frac{11}{210}L \\ 0 & \frac{11}{210}L & \frac{1}{105}L^2 \end{bmatrix}\boldsymbol{\gamma}$$

and the stiffness term becomes

$$\int \delta\varepsilon_{ij}\sigma_{ij}\,\mathrm{d}V = \delta\boldsymbol{\gamma}^T\begin{bmatrix} \frac{\mathsf{E}A}{L} & 0 & 0 \\ 0 & 12\frac{\mathsf{E}I}{L^3} & 6\frac{\mathsf{E}I}{L^3} \\ 0 & 6\frac{\mathsf{E}I}{L^3} & 4\frac{\mathsf{E}I}{L^3} \end{bmatrix}\boldsymbol{\gamma}$$

The one-element model for the rotating beam column emerges as

$$\begin{bmatrix} \frac{\rho AL}{3} & 0 & 0 \\ 0 & \frac{13}{35}\rho AL + \frac{6}{5}\frac{\rho I}{L} & \frac{11}{210}\rho AL^2 + \frac{1}{10}\rho I \\ 0 & \frac{11}{210}\rho AL^2 + \frac{1}{10}\rho I & \frac{1}{105}\rho AL^3 + \frac{2}{15}\rho IL \end{bmatrix}\ddot{\boldsymbol{\gamma}} + \left(2\omega\frac{\rho AL}{20}\right)\begin{bmatrix} 0 & -7 & -L \\ 7 & 0 & 0 \\ L & 0 & 0 \end{bmatrix}\dot{\boldsymbol{\gamma}}$$

$$+\begin{bmatrix} \frac{\mathsf{E}A}{L} - \omega^2\frac{\rho AL}{3} & -\frac{7}{20}\alpha\rho AL & -\frac{1}{20}\alpha\rho AL^2 \\ \frac{7}{20}\alpha\rho AL & 12\frac{\mathsf{E}I}{L^3} - \omega^2\left(\frac{13}{35}\rho AL + \frac{6}{5}\frac{\rho I}{L}\right) & 6\frac{\mathsf{E}I}{L^3} - \omega^2\left(\frac{11}{210}\rho AL^2 + \frac{1}{10}\rho I\right) \\ \frac{1}{20}\alpha\rho AL^2 & 6\frac{\mathsf{E}I}{L^3} - \omega^2\left(\frac{11}{210}\rho AL^2 + \frac{1}{10}\rho I\right) & 4\frac{\mathsf{E}I}{L^3} - \omega^2\left(\frac{1}{105}\rho AL^3 + \frac{2}{15}\rho IL\right) \end{bmatrix}\boldsymbol{\gamma} = \begin{pmatrix} f_u \\ f_v \\ M_v \end{pmatrix}$$

We observe that unsteady rotation of a beam column gives rise to Coriolis and angular accelerations effects *serving to couple the membrane with the bending response*.

Returning to the main development, in two-dimensional steady rotation of an undamped elastic medium without Coriolis effects, the governing equation reduces to

$$\mathbf{M}\ddot{\boldsymbol{\gamma}} + (\mathbf{K} - \omega^2\mathbf{M})\boldsymbol{\gamma} = \mathbf{f} - \mathbf{f}_\omega \tag{12.13}$$

We consider the eigenvalues and eigenvectors implied by this equation. The eigenvalues ω_j^2 and eigenvectors $\mathbf{x}_j$ satisfy the generalized eigenvalue equation

$$\begin{aligned}\mathbf{0} &= \left[(\mathbf{K} - \omega^2\mathbf{M}) - \omega_j^2\mathbf{M}\right]\mathbf{x}_j \\ &= \left(\mathbf{K} - \left(\omega^2 + \omega_j^2\right)\mathbf{M}\right)\mathbf{x}_j\end{aligned} \tag{12.14}$$

Now $\omega^2 + \omega_j^2 = \omega_{nj}^2$ in which ω_{nj} are the natural frequencies found in the nonrotating system with the same stiffness and mass matrix. It follows that the eigenvalues of the rotating system satisfy

$$\omega_j^2 = \omega_{nj}^2 - \omega^2 \tag{12.15}$$

However, the eigenvectors $\mathbf{x}_j$ simultaneously diagonalize $\mathbf{K}$ and $\mathbf{M}$ regardless of ω. We conclude that *the eigenvectors are not affected by rotation.*

On the other hand, if the Coriolis and angular acceleration matrices need to be taken into account, the eigenvalues may in general have nonzero real and imaginary parts.

12.2 CRITICAL SPEEDS IN SHAFT ROTOR SYSTEMS

There is instability in elastic bodies under steady rotation, associated with the notion of the critical speed, already touched upon Example 12.1. Critical speeds are illustrated through several additional examples.

EXAMPLE 12.3

Find the critical speeds in a beam rotating about its axis, in the symmetric and antisymmetric cases.

Case 1: Transverse load as shown in Figure 12.6

SOLUTION

The member under study is called a *shaft* in power generation applications. But here we refer to it as a *beam* since we are concerned with the transverse displacements induced by rotation. The force F_r rotates with the beam. An imperfection (mass unbalance) in the beam will cause it to bend away from the axis, and the bent configuration will

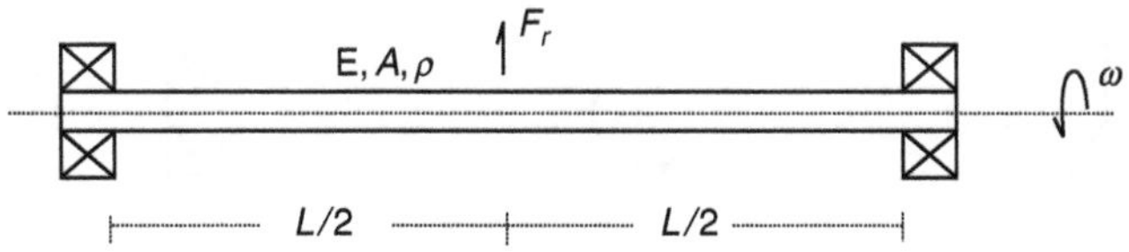

FIGURE 12.6 Rotating shaft with a transverse load.

likewise rotate with the shaft. The more the beam bends the greater the centrifugal forces, thereby tending to cause the beam to bend even further.

Using two elements and imposing clamped constraints on the right and left bearings gives the stiffness in terms in the Principle of Virtual Work as

$$\left(\mathbf{K}_{22}^{(1)}+\mathbf{K}_{11}^{(2)}\right)\begin{Bmatrix} w_2 \\ -w_2' \end{Bmatrix}=\begin{bmatrix} 192\mathrm{E}I/L^3 & 0 \\ 0 & 16\mathrm{E}I/L \end{bmatrix}\begin{Bmatrix} w_2 \\ -w_2' \end{Bmatrix}$$

in which the subscript 2 denotes the central node. (The bearings are considered rigid.)

The inertial terms of the beam are similarly found. Under the assumed steady rotation and assuming that the deformed profile is constant in the radial (outward) direction, the Coriolis and angular acceleration matrices vanish, leaving

$$\left(\mathbf{M}_{22}^{(1)}+\mathbf{M}_{11}^{(2)}\right)\begin{Bmatrix} \ddot{w}_2 \\ -\ddot{w}_2' \end{Bmatrix}-\omega^2\left(\mathbf{M}_{22}^{(1)}+\mathbf{M}_{11}^{(2)}\right)\begin{Bmatrix} w_2 \\ -w_2' \end{Bmatrix}$$
$$=\begin{bmatrix} \frac{13}{35}\rho AL & 0 \\ 0 & \frac{1}{420}\rho AL^3 \end{bmatrix}\left\{\begin{Bmatrix} \ddot{w}_2 \\ -\ddot{w}_2' \end{Bmatrix}-\omega^2\begin{Bmatrix} w_2 \\ -w_2' \end{Bmatrix}\right\}$$

The ensuing finite element equation is therefore

$$\begin{bmatrix} \frac{13}{35}\rho AL & 0 \\ 0 & \frac{1}{420}\rho AL^2 \end{bmatrix}\begin{Bmatrix} \ddot{w}_2 \\ -\ddot{w}_2' \end{Bmatrix}+\left(\begin{bmatrix} 192\mathrm{E}I/L^3 & 0 \\ 0 & 16\mathrm{E}I/L \end{bmatrix}\right.$$
$$\left.-\,\omega^2\begin{bmatrix} \frac{13}{35}\rho AL & 0 \\ 0 & \frac{1}{420}\rho AL^2 \end{bmatrix}\right)\begin{Bmatrix} w_2 \\ -w_2' \end{Bmatrix}=\begin{Bmatrix} F_r \\ 0 \end{Bmatrix} \tag{12.16}$$

Of course the equations represented by the rows of Equation 12.16 are uncoupled, leading to the two critical speeds

$$\omega_{c1}=\sqrt{\frac{192\mathrm{E}I/L^3}{13\rho AL/35}},\quad \omega_{c2}=\sqrt{\frac{16\mathrm{E}I/L}{\rho AL^3/420}}$$

If $\omega=\omega_{c1}$, the symmetric bending profile (about the midpoint) grows in a unstable fashion.

This happens in the current case since the force F_r induces a symmetric deformation.

In either of these cases the stiffness matrix $\mathbf{K}$ is singular.

However, only low amplitude motion will happen in the antisymmetric mode unless there is a moment at the midpoint, as discussed in Case 2.

Case 2: Suppose instead that at the midpoint there is a moment which rotates with the shaft (Figure 12.7).

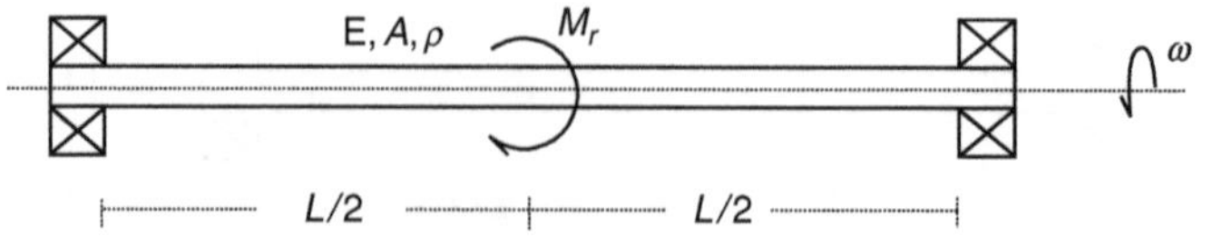

FIGURE 12.7 Rotating beam with bending moment.

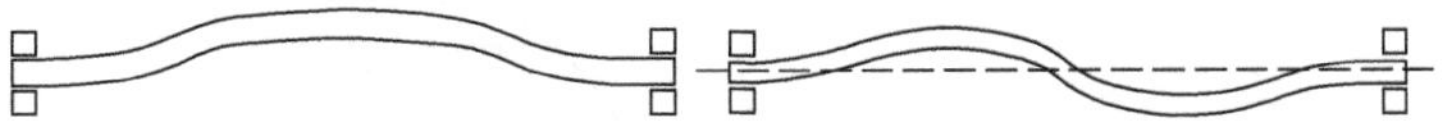

FIGURE 12.8 Symmetric and antisymmetric deformations.

Now the response is antisymmetric. If $\omega = \omega_{c2}$, an unstable antisymmetric deformation occurs. However, in this case if $\omega = \omega_{c1}$, there is low amplitude response since the applied moment only induces an antisymmetric motion.

The symmetric and antisymmetric deformations are illustrated in Figure 12.8.

EXAMPLE 12.4

Shaft rotor systems neglecting the mass of the shaft

Power transmission shafts use rotors to store kinetic energy. We seek the critical speeds in the shaft rotor system depicted in Figure 12.9 below.

Suppose the shaft is balanced such that its center of mass is located at its geometric center. The rotor does not induce bending if it is balanced side-to-side as well as top-to-bottom. Top-to-bottom symmetry and side-to-side symmetry are shown below (Figure 12.10).

In top-to-bottom balance, each half has the same mass, and the centers of mass are located at the geometric centers of the semicircular disks.

In side-to-side balance, the right half disk has its center of mass on the midpoint along the axis, and similarly for the left-hand side (Figure 12.11).

Now suppose that the rotor is not balanced. For simplicity suppose that the centers of mass of the top and bottom of the right and left halves *lie in the same plane*, which of course rotate with the shaft. The *unbalance* can be modeled as follows (Figure 12.12).

Here e_1 and h_1 represent the horizontal and vertical offsets of the center of mass of the upper right portion of the disk from the geometric center of the rotor, and similarly for e_2, e_3, e_4, h_2, h_3, h_4.

The disk imposes a radial force and a radial moment on the shaft at its midpoint. Taking account of the deformation of the shaft, the kinetic energy of the disk is

$$\begin{aligned}T_d &= \frac{1}{2}\left(\frac{mr^2}{4} + \frac{ma^2}{12}\right)\omega^2 \\ &\quad + \frac{1}{2}\frac{m}{4}\omega^2\left[(w_2 + h_1 - (-w_2')e_1)^2 + (w_2 - h_2 - (-w_2')e_2)^2\right. \\ &\quad \left. + (w_2 + h_3 + (-w_2')e_3)^2 + (w_2 - h_4 + (-w_2')e_4)^2\right]\end{aligned}$$

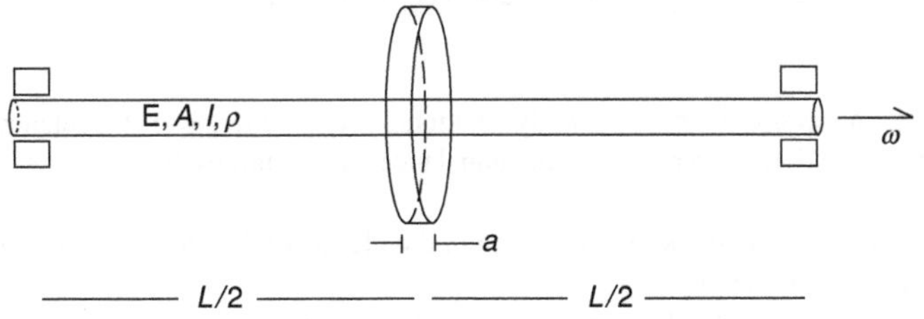

FIGURE 12.9 Simple shaft rotor system.

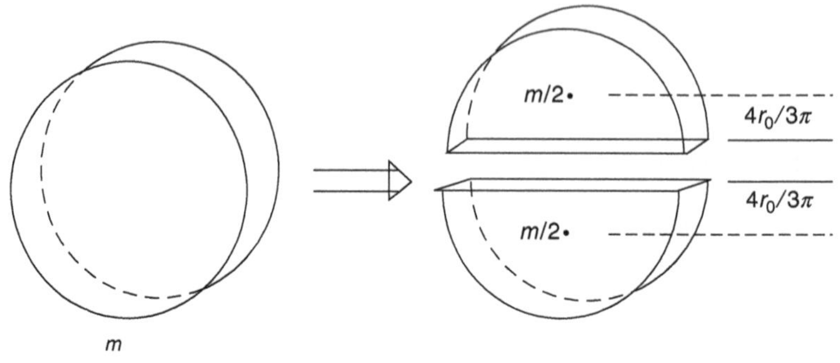

FIGURE 12.10 Top-to-bottom balance; rotor has mass m and radius r.

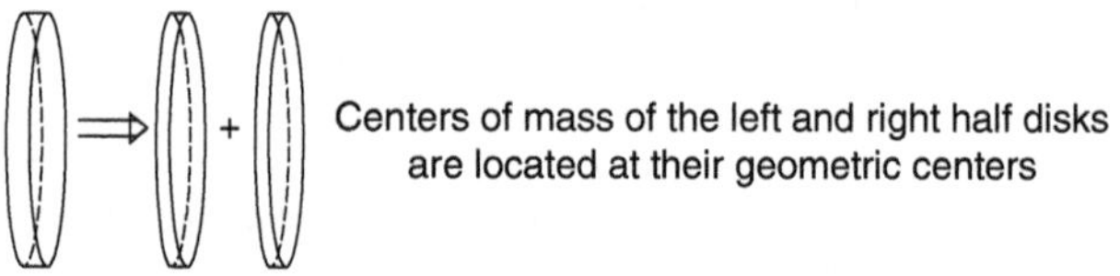

FIGURE 12.11 Side-to-side balance.

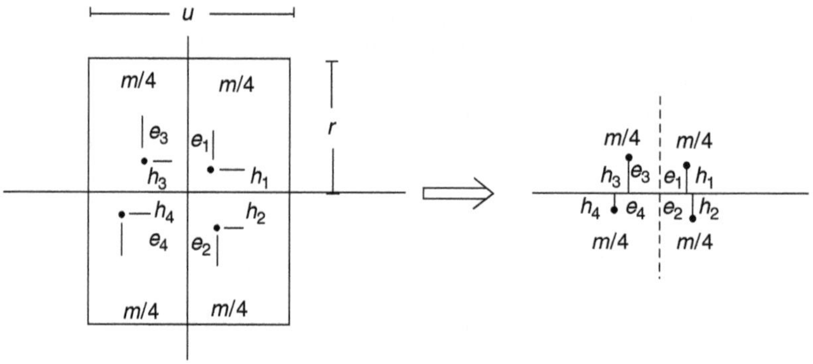

FIGURE 12.12 Rotor with top-to-bottom and side-to-side unbalance.

The first term represents the rigid body rotational kinetic energy of the balanced rotor while the second term represents the unbalance contributions to the (translational) kinetic energy.

Neglecting the kinetic energy of the beam, the elastic strain energy is the same as in Cases 1 and 2 of Example 12.3.

The two-element finite element model is expressed as

$$\left(\begin{bmatrix} 192\mathrm{E}I/L^3 & 0 \\ 0 & 16\mathrm{E}I/L \end{bmatrix} - m\omega^2 \begin{bmatrix} 1 & -\dfrac{e_1+e_2-e_3-e_4}{4} \\ -\dfrac{e_1+e_2-e_3-e_4}{4} & \dfrac{e_1^2+e_2^2+e_3^2+e_4^2}{4} \end{bmatrix}\right)\begin{Bmatrix} w_2 \\ -w_2' \end{Bmatrix}$$
$$= \frac{m\omega^2}{4}\begin{Bmatrix} h_1 - h_2 + h_3 - h_4 \\ -h_1e_1 + h_2e_2 + h_3e_3 - h_4e_4 \end{Bmatrix}$$

The right-hand side represents a force and a moment induced by rigid body motion. If the rotors are perfectly balanced side-to-side ($e_1 = e_2 = e_3 = e_4 = a/2$) the off-diagonal entries of the second matrix vanish. If the rotor is perfectly balanced top-to-bottom, then $h_1 = h_2 = h_3 = h_4 = 0$ and the rigid body contribution vanishes.

The critical speeds are obtained by solving

$$\det\left(\begin{bmatrix} 192\mathrm{E}I/L^3 & 0 \\ 0 & 16\mathrm{E}I/L \end{bmatrix} - m\omega^2 \begin{bmatrix} 1 & -\dfrac{e_1+e_2-e_3-e_4}{4} \\ -\dfrac{e_1+e_2-e_3-e_4}{4} & \dfrac{e_1^2+e_2^2+e_3^2+e_4^2}{4} \end{bmatrix}\right) = 0$$

once numerical values have been established for the offsets.

Example 12.4 is now extended to include the effects of the shaft (beam) kinetic energy. Using the previously developed mass matrix for the beam elements (cf. Chapter 11), we have the two-element finite element equation, and again the critical speeds are obtained by finding the determinant of the matrix on the left-hand side of the equation.

$$\left(\begin{bmatrix} 192\mathrm{E}I/L^3 & 0 \\ 0 & 16\mathrm{E}I/L \end{bmatrix} - m\omega^2 \begin{bmatrix} 1+\frac{13}{35}\frac{\rho AL}{m} & \dfrac{e_1+e_2-e_3-e_4}{4} \\ \dfrac{e_1+e_2-e_3-e_4}{4} & \dfrac{e_1^2+e_2^2-e_3^2-e_4^2}{4}+\frac{1}{420}\frac{\rho AL^2}{m} \end{bmatrix}\right)\begin{Bmatrix} w_2 \\ -w_2' \end{Bmatrix}$$
$$= -\frac{m\omega^2}{4}\begin{Bmatrix} h_1 - h_2 + h_3 - h_4 \\ h_1e_1 + h_2e_2 - h_3e_3 - h_4e_4 \end{Bmatrix}$$

Computation of the two critical speeds is straightforward and is left to the reader as an exercise.

12.3 FINITE ELEMENT ANALYSIS FOR UNCONSTRAINED ELASTIC BODIES

12.3.1 BODY AXES

We next consider the three-dimensional response of an elastic body which has no fixed points, as in spacecraft, for example as depicted in Figure 12.13. It is assumed that the tractions are prescribed on the *undeformed* surface of the elastic body, and that they rotate and translate with the body. The response is thereby referred to the "body axes," i.e., axes embedded in the corresponding rigid body. An example of body axes is the principal axes of the moment of inertia tensor in the body in Figure 12.13, assuming it is rigid. Otherwise stated, it is assumed that the points of the rigid body coincide with the points of the elastic body in its undeformed configuration.

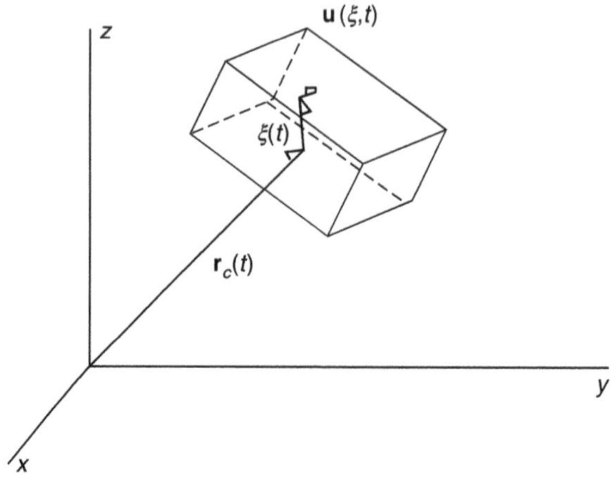

FIGURE 12.13 Three-dimensional unconstrained 3-D element.

The position vector $\mathbf{r}$ of a point in the rotating and translating elastic body may be decomposed as

$$\mathbf{r} = \mathbf{r}_c + \boldsymbol{\xi} + \mathbf{u} \tag{12.17}$$

in which $\mathbf{r}_c$ is the position vector to the center of mass, $\boldsymbol{\xi}$ is the relative position vector from the center of mass to the undeformed position of the current point, in the body system, and $\mathbf{u}$ is the displacement from the undeformed to the deformed position, likewise in the body system.

It is necessary to develop expressions for the variations of $\mathbf{r}_c$, $\boldsymbol{\xi}$, and $\mathbf{u}$. To this end we regard $\mathbf{r}_c$ as being referred to the absolute coordinate system (*xyz*), in which case we need only write $\delta\mathbf{r}_c$.

Next $\boldsymbol{\xi}$ is fixed in the rotating and translating system, so that its variation comes purely from the coordinate system. In analogy with $\dot{\boldsymbol{\xi}} = \boldsymbol{\omega} \times \boldsymbol{\xi}$ with $\boldsymbol{\omega} = \frac{\partial\boldsymbol{\theta}}{\partial t}$, we now have $\delta\boldsymbol{\xi} = \delta\boldsymbol{\theta} \times \boldsymbol{\xi}$. Finally, the variation of the displacement vector has contributions from the coordinate system as well as from the fact that it changes relative to the rotating system. In analogy with $\frac{d\mathbf{u}}{dt} = \frac{\partial\mathbf{u}}{\partial t} + \boldsymbol{\omega} \times \mathbf{u}$, the variation of $\mathbf{u}$ is expressed as $\delta\mathbf{u} = \delta\hat{\mathbf{u}} + \delta\boldsymbol{\theta} \times \mathbf{u}$, in which $\delta\hat{\mathbf{u}}$ is the variation of $\mathbf{u}$ with the coordinate system instantaneously constrained not to rotate.

In order to conduct FEA of an elastic body undergoing unconstrained motion (rigid body translation and rotation), it is necessary to compute the motion of the body axes, which is achieved by integrating the Euler equations presented next.

12.3.2 Euler Equations of a Rigid Body

We first restrict attention to rigid bodies for which $\mathbf{u} = \mathbf{0}$. The position vector now has the decomposition

$$\mathbf{r} = \mathbf{r}_c + \boldsymbol{\xi}, \quad \delta\boldsymbol{\xi} = \delta\boldsymbol{\theta} \times \boldsymbol{\xi} \tag{12.18}$$

Neglecting body forces, equilibrium (referred to the translating/rotating frame) is expressed as $\frac{\partial S_{ij}}{\partial \xi_j} - \rho[\ddot{r}_{ci} + \ddot{\xi}_i] = 0$. Variation with respect to $\mathbf{r}_c$, namely $\int \delta r_{ci}\left[\frac{\partial S_{ij}}{\partial \xi_j} - \rho[\ddot{r}_{ci} + \ddot{\xi}_i]\right] dV = 0$, results in

$$\int \delta r_{ci}\left[\frac{\partial \mathrm{S}_{ij}}{\partial \varsigma_j}\right]\mathrm{d}V = \delta \mathrm{r}_{ci}\int [n_j\mathrm{S}_{ij}]\,\mathrm{d}S$$
$$= \delta \mathrm{r}_{ci}\int t_i\,\mathrm{d}S \tag{12.19}$$

and we note that $\mathbf{F} = \int \mathbf{t}\,\mathrm{d}S$ is the total force exerted by the surface tractions. For the inertial terms

$$\int \delta r_{ci}[\rho[\ddot{r}_{ci} + \ddot{\varsigma}_i]]\,\mathrm{d}V = \delta \mathrm{r}_{ci}\underbrace{\int \rho\,\mathrm{d}V}_{m}\,\ddot{r}_{ci} + \delta \mathrm{r}_{ci}\int \rho\ddot{\xi}_i\,\mathrm{d}V \tag{12.20}$$

and so $\delta \mathbf{r}_c^T[\mathbf{F} - m\ddot{\mathbf{r}}_c] = 0$ since the definition of the center of mass implies that $\int \rho\ddot{\xi}_i\,\mathrm{d}V = 0$, and m is the total mass. The first Euler equation may now be stated as

$$\mathbf{F} = m\ddot{\mathbf{r}}_c \tag{12.21}$$

We next consider the effect of the variation $\delta\boldsymbol{\xi}$,

$$0 = \int \delta\xi_i\left(\left[\frac{\partial \mathrm{S}_{ij}}{\partial \xi_j} - \rho[\ddot{r}_{ci} + \ddot{\xi}]\right]\right)\mathrm{d}V$$
$$= \int \in_{ipq}\delta\theta_p\xi_q\left[\frac{\partial \mathrm{S}_{ij}}{\partial \xi_j} - \rho[\ddot{r}_{ci} + \ddot{\varsigma}_i]\right]\mathrm{d}V \tag{12.22}$$

and $\in_{ipq}$ is recognized as the (third order) permutation tensor (cf. Chapter 3). Observe that

$$\int \in_{ipq}\delta\theta_p\varsigma_q\ddot{r}_{ci}\rho\,\mathrm{d}V = \in_{ipq}\delta\theta_p\left[\int \xi_q\rho\,\mathrm{d}V\right]\ddot{r}_{ci} = 0 \tag{12.23}$$

Next,

$$\int \in_{ipq}\delta\theta_p\varsigma_q\left[\frac{\partial \mathrm{S}_{ij}}{\partial \xi_j}\right]\mathrm{d}V = \int\left[\frac{\partial}{\partial \xi_j}[\in_{ipq}\delta\theta_p\varsigma_q\mathrm{S}_{ij}] - \frac{\partial}{\partial \xi_j}[\in_{ipq}\delta\theta_p\xi_q]\mathrm{S}_{ij}\right]\mathrm{d}V$$
$$= \int [n_j[\in_{ipq}\delta\theta_p\xi_q\mathrm{S}_{ij}]]\,\mathrm{d}S - \int\left[\in_{ipq}\delta\theta_p\frac{\partial \xi_q}{\partial \xi_j}\right]\mathrm{S}_{ij}\,\mathrm{d}V$$
$$= \delta\theta_p\int \in_{pqi}\xi_q t_i\,\mathrm{d}V - \int [\in_{ipq}\delta\theta_p\delta_{qj}]\mathrm{S}_{ij}\,\mathrm{d}V \tag{12.24}$$

We observe that $\mathbf{M}_c = \int \boldsymbol{\xi}\times\mathbf{t}\,\mathrm{d}S$ is the total moment of the tractions about the center of mass. Also, $\int[\in_{ipq}\delta\theta_p\delta_{qj}]\mathrm{S}_{ij}\,\mathrm{d}V = 0$ since $a_{ij} = \in_{ipq}\delta\theta_p\delta_{qj}$ is antisymmetric.

It remains to consider the inertial terms $\int \in_{ipq}\delta\theta_p\varsigma_q[-\rho[\ddot{r}_{ci} + \ddot{\xi}_i]]\,\mathrm{d}V$. First

$$\int \in_{ipq}\delta\theta_p\xi_q\ddot{r}_{ci}\rho\,\mathrm{d}V = \in_{ipq}\delta\theta_p\int \ddot{\xi}_q\rho\,\mathrm{d}V = 0 \tag{12.25}$$

Continuing,

$$\begin{aligned}\int \in_{ipq}\delta\theta_p\xi_q\ddot{\xi}_i\,\mathrm{d}V &= \delta\boldsymbol{\theta}^T\int \rho\boldsymbol{\xi}\times\ddot{\boldsymbol{\xi}}\,\mathrm{d}S\\ &= \delta\boldsymbol{\theta}^T\int \rho\boldsymbol{\xi}\times[\boldsymbol{\omega}\times\boldsymbol{\omega}\times\boldsymbol{\xi}+\boldsymbol{\alpha}\times\boldsymbol{\xi}]\,\mathrm{d}V\\ &= \delta\boldsymbol{\theta}^T\left[\int[-(\boldsymbol{\omega}\times\boldsymbol{\xi})\times(\boldsymbol{\xi}\times\boldsymbol{\omega})]\rho\,\mathrm{d}V+\int(-\boldsymbol{\xi}\times\boldsymbol{\xi}\times\boldsymbol{\alpha})\rho\,\mathrm{d}V\right]\\ &= \delta\boldsymbol{\theta}^T[\mathbf{J}\boldsymbol{\alpha}+\boldsymbol{\omega}\times\mathbf{J}\boldsymbol{\omega}]\end{aligned} \tag{12.26}$$

in which $\mathbf{J}=\int\mathbf{Z}^T\mathbf{Z}\rho\,\mathrm{d}V$ is the positive definite (moment of) inertia tensor, and $\mathbf{Z}$ is the the antisymmetric tensor satisfying $\mathbf{Zb}=\boldsymbol{\xi}\times\mathbf{b}$ for any 3×1 vector $\mathbf{b}$.

The second Euler equation now arises as

$$\mathbf{M}=\mathbf{J}\boldsymbol{\alpha}+\boldsymbol{\omega}\times\mathbf{J}\boldsymbol{\omega} \tag{12.27}$$

If we write $\boldsymbol{\xi}^T=\{\xi_1\quad\xi_2\quad\xi_3\}$ the inertia tensor is found with routine effort to be given by

$$\mathbf{J}=\int\begin{bmatrix}\xi_2^2+\xi_3^2 & -\xi_1\xi_2 & -\xi_1\xi_3\\ -\xi_1\xi_2 & \xi_3^2+\xi_1^2 & -\xi_2\xi_3\\ -\xi_3\xi_1 & -\xi_2\xi_3 & \xi_1^2+\xi_2^2\end{bmatrix}\rho\,\mathrm{d}V \tag{12.28}$$

12.3.3 Variational Equations of an Unconstrained Elastic Body

The relations below for an elastic body are developed on the assumption that the reference undeformed configuration in the current coordinate system coincides with the rigid body. The balance of linear momentum expressed in the body system is

$$\frac{\partial S_{ij}}{\partial\xi_j}=\rho[\ddot{r}_{ci}+\ddot{\xi}_i+\ddot{u}_i] \tag{12.29}$$

Recall that $\delta\mathbf{r}=\delta\mathbf{r_c}+\delta\boldsymbol{\theta}\times(\boldsymbol{\xi}+\mathbf{u})+\delta\mathbf{u}'$, in which $\delta\mathbf{u}'$ is the variation of $\mathbf{u}$ with the axes instantaneously held fixed. The quantities $\mathbf{r}_c$, $\boldsymbol{\upsilon}$, and $\mathbf{u}'$ *may be varied independently* since there is no constraint relating them. For $\delta\mathbf{r}_c$, the variational statement is

$$\int\delta r_{ci}\left[\frac{\partial S_{ij}}{\partial\xi_j}-\rho[\ddot{r}_{ci}+\ddot{\xi}_i+\ddot{u}_i]\right]\mathrm{d}V=0 \tag{12.30}$$

From the previous section $\int\delta r_{ci}\left[\frac{\partial S_{ij}}{\partial\xi_j}-\rho[\ddot{r}_{ci}+\ddot{\xi}_i]\right]\mathrm{d}V=0$. The term involving the divergence of the stress becomes the traction term on application of the Divergence

Theorem. But recall that the tractions are considered to be specified on the undeformed body, which renders the tractions the same as in the rigid body. We conclude that $\int \delta r_{ci}\rho\ddot{u}_i\,\mathrm{d}V=0$ and also $\int \rho\ddot{u}_i\,\mathrm{d}V=0$. Assuming vanishing initial values of $\mathbf{u}$ and $\dot{\mathbf{u}}$, it follows that $\int \rho\mathbf{u}\,\mathrm{d}V=\mathbf{0}$ with the consequence that the *center of mass in the elastic body coincides with that in the rigid body*. It likewise follows that in finite element modeling we may impose the constraint $\mathbf{u}(\mathbf{0},t)=\mathbf{0}$ at $\boldsymbol{\xi}=\mathbf{0}$. Of course the constraint is applied in the body axis system.

Secondly, consider the effect of $\delta\boldsymbol{\theta}$, namely $\mathbf{0}=\int \in_{ipq}\delta\theta_p(\xi_q+u_q)\left[\frac{\partial S_{ij}}{\partial \xi_j}-\rho[\ddot{r}_{ci}+\ddot{\xi}_i+\ddot{u}_i]\right]\mathrm{d}V$.

The assumption that the tractions are prescribed in the undeformed configuration but referred to the current coordinate system implies that the moments are the same in the rigid body and undeformed configuration of the elastic body, and hence that $\int \in_{ipq}\delta\theta_p u_q \frac{\partial S_{ij}}{\partial \xi_j}\mathrm{d}V=0$. The implication is that

$$\int \in_{ipq}\delta\theta_p(\xi_q+u_q)\rho[\ddot{\xi}_i+\ddot{u}_i]\,\mathrm{d}V=\int \in_{ipq}\delta\theta_p\xi_q\rho\ddot{\xi}_i\,\mathrm{d}V \tag{12.31}$$

Assuming small displacements and thereby neglecting products of $\mathbf{u}$ and its time derivatives results in the relation

$$\int \boldsymbol{\xi}\times\rho\ddot{\mathbf{u}}\,\mathrm{d}V+\int \mathbf{u}\times\rho\ddot{\boldsymbol{\xi}}\,\mathrm{d}V=\mathbf{0} \tag{12.32}$$

Upon expansion of the accelerations,

$$\begin{aligned}\mathbf{0}&=\int \boldsymbol{\xi}\times\rho\ddot{\mathbf{u}}\,\mathrm{d}V+\int \mathbf{u}\times\rho\ddot{\boldsymbol{\xi}}\,\mathrm{d}V\\&=\int \mathbf{u}\times(\boldsymbol{\omega}\times\boldsymbol{\omega}+\boldsymbol{\alpha})\times\boldsymbol{\xi}\rho\,\mathrm{d}V+\int \boldsymbol{\xi}\times(\boldsymbol{\omega}\times\boldsymbol{\omega}+\boldsymbol{\alpha})\times\mathbf{u}\rho\,\mathrm{d}V\\&\quad+\int \boldsymbol{\xi}\times\frac{\partial^2\mathbf{u}}{\partial t^2}\rho\,\mathrm{d}V+\int \boldsymbol{\xi}\times 2\boldsymbol{\omega}\times\frac{\partial\mathbf{u}}{\partial t}\rho\,\mathrm{d}V\end{aligned} \tag{12.33}$$

But $\int \boldsymbol{\xi}\times(\boldsymbol{\omega}\times\boldsymbol{\omega}+\boldsymbol{\alpha})\times\mathbf{u}\rho\,\mathrm{d}V=-\int \mathbf{u}\times(\boldsymbol{\omega}\times\boldsymbol{\omega}+\boldsymbol{\alpha})\times\boldsymbol{\xi}\rho\,\mathrm{d}V$ has the consequence that

$$\int \boldsymbol{\xi}\times\frac{\partial^2\mathbf{u}}{\partial t^2}\rho\,\mathrm{d}V-2\boldsymbol{\omega}\times\int \boldsymbol{\xi}\times\frac{\partial\mathbf{u}}{\partial t}\rho\,\mathrm{d}V=\mathbf{0} \tag{12.34}$$

Now the partial time derivative at fixed body axes in general has no relation to the angular velocity vector $\boldsymbol{\omega}$ describing the rotation of the body axes. We conclude that $\frac{\partial}{\partial t}\int \boldsymbol{\xi}\times\mathbf{u}\rho\,\mathrm{d}V=\mathbf{0}$, and with suitable initial conditions that

$$\int \boldsymbol{\xi}\times\mathbf{u}\rho\,\mathrm{d}V=\mathbf{0} \tag{12.35}$$

The relations $\int \rho \mathbf{u}\,dV = \mathbf{0}$ and $\int \boldsymbol{\xi} \times \rho \mathbf{u}\,dV = \mathbf{0}$ have classically been derived by Truesdell and Noll (1965) by assuming that the deformed configuration minimizes kinetic energy.

EXAMPLE 12.5

Write down the proof that, for an unconstrained elastic body, if $\int \boldsymbol{\xi} \times \mathbf{u}\rho\,dV = \mathbf{0}$

$$\int \rho \frac{d^2\boldsymbol{\xi}}{dt^2} \times \mathbf{u}\,dV = \mathbf{0}, \quad \int \rho \boldsymbol{\xi} \times \frac{d^2\mathbf{u}}{dt^2}\,dV = \mathbf{0}$$

Comment: In Chapter 13 we will encounter the Reynolds Transport Theorem which states that $\frac{d}{dt}\int \rho \mathbf{b}\,dV = \int \rho \frac{d\mathbf{b}}{dt}\,dV$, for any vector $\mathbf{b}$. It is used in the solution below.

Solution

First $\int \frac{d^2\boldsymbol{\xi}}{dt^2} \times \rho \mathbf{u}\,dV = [(\boldsymbol{\omega} \times \boldsymbol{\omega}) + \boldsymbol{\alpha}] \times \left(\int \boldsymbol{\xi} \times \rho \mathbf{u}\,dV\right) = \mathbf{0}$. Secondly, $\mathbf{0} = \frac{d}{dt}\int \boldsymbol{\xi} \times \rho \mathbf{u}\,dV = \boldsymbol{\omega} \times \int \boldsymbol{\xi} \times \rho \mathbf{u}\,dV + \int \boldsymbol{\xi} \times \rho \frac{d\mathbf{u}}{dt}\,dV$. The first right-hand term vanishes by virtue of the problem statement, and we conclude that $\int \boldsymbol{\xi} \times \rho \frac{d\mathbf{u}}{dt}\,dV = \mathbf{0}$. Finally, $\mathbf{0} = \frac{d^2}{dt^2}\int \boldsymbol{\xi} \times \rho \mathbf{u}\,dV = (\boldsymbol{\omega} \times \boldsymbol{\omega} + \boldsymbol{\alpha}) \times \int \boldsymbol{\xi} \times \rho \mathbf{u}\,dV + \boldsymbol{\omega} \times \int \boldsymbol{\xi} \times \rho \frac{d\mathbf{u}}{dt}\,dV + \int \boldsymbol{\xi} \times \rho \frac{d^2}{dt^2}\mathbf{u}\,dV$. The first two right-hand terms vanish, in consequence of which $\int \boldsymbol{\xi} \times \rho \frac{d^2}{dt^2}\mathbf{u}\,dV = \mathbf{0}$.

Returning to the general presentation, the rotational relation $\int \boldsymbol{\xi} \times \rho \mathbf{u}\,dV = \mathbf{0}$ together with $\mathbf{u}(0, t) = \mathbf{0}$ represents a set of constraints that prevent the center of mass of the elastic body from displacing from that the rigid body, and also prevents the body axes in the elastic body from rotating relative to the body axes of the rigid body.

The rotational constraint is global while the center-of-mass constraint can be enforced at a point. To examine the treatment of the rotational constraint we assume the global interpolation model (referred to the body axes) in the form $u_k = \varphi_{kl}(\boldsymbol{\xi})\boldsymbol{\Phi}_{mn}\gamma_n(t)$. The constraint is now expressed in the form

$$\boldsymbol{\Upsilon}\boldsymbol{\gamma} = \mathbf{0}, \quad \boldsymbol{\Upsilon} = \int \in_{ijk} \xi_j \varphi_{kl} \boldsymbol{\Phi}_{mn} \rho\,dV \tag{12.36}$$

But Equation 12.36 may be restated as

$$[\boldsymbol{\Upsilon}_{3,n-3} \quad \boldsymbol{\Upsilon}_{3,3}]\begin{Bmatrix} \boldsymbol{\gamma}_{n-3} \\ \boldsymbol{\gamma}_3 \end{Bmatrix} = \mathbf{0}, \quad \text{with } \boldsymbol{\gamma}_3 = -\boldsymbol{\Upsilon}_{3,3}^{-1}\boldsymbol{\Upsilon}_{3,n-3}\boldsymbol{\gamma}_{n-3} \tag{12.37}$$

provided that $\boldsymbol{\Upsilon}_{3,3}$ is nonsingular. We therefore use a global constraint to remove three degrees of freedom.

For illustration consider a finite element model in which

$$\begin{bmatrix} \mathbf{K}_{n-3,n-3} & \mathbf{K}_{n-3,3} \\ \mathbf{K}_{n-3,3} & \mathbf{K}_{3,3} \end{bmatrix}\begin{Bmatrix} \boldsymbol{\gamma}_{n-3} \\ \boldsymbol{\gamma}_3 \end{Bmatrix} = \begin{Bmatrix} f_{n-3} \\ f_3 \end{Bmatrix} \quad \text{and} \quad \boldsymbol{\gamma}_3 = -\boldsymbol{\Upsilon}_{3,3}^{-1}\boldsymbol{\Upsilon}_{3,n-3}\boldsymbol{\gamma}_{n-3} \tag{12.38}$$

It appears that f_3 must be an as-yet-unknown reaction force arising to enforce the rotational constraint. If the constraint is used to eliminate $\boldsymbol{\gamma}_3$ in terms of $\boldsymbol{\gamma}_{n-3}$, we encounter $\boldsymbol{\gamma}_{n-3}$.

The finite element equation to be solved is now

$$[\mathbf{K}_{n-3,n-3} - \mathbf{K}_{n-3,3}\mathbf{Y}_{3,3}^{-1}\mathbf{Y}_{3,n-3}]\boldsymbol{\gamma}_{n-3} = \boldsymbol{f}_{n-3} \tag{12.39}$$

and note that the resulting stiffness matrix is generally nonsymmetric. If the reaction forces enforcing the constraints are of interest, they can be computed using

$$\boldsymbol{f}_3 = [\mathbf{K}_{n-3,3} - \mathbf{K}_{3,3}\mathbf{Y}_{3,3}^{-1}\mathbf{Y}_{3,n-3}]\boldsymbol{\gamma}_{n-3} \tag{12.40}$$

12.3.4 Principle of Virtual Work in Body Coordinates

To consider the effect of $\delta\mathbf{u}'$, recall that $\int \delta u_i'\left[\frac{\partial S_{ij}}{\partial \xi_j} - \rho[\ddot{r}_{ci} + \ddot{\xi}_i + \ddot{u}_i]\right]\mathrm{d}V = 0$. First note that $\int \delta u_i'\rho\,\mathrm{d}V\,\ddot{r}_{ci} = 0$ since $\int \delta u_i'\rho\,\mathrm{d}V = 0$. The remaining terms are formally the same as for a body with a fixed point, and consequently the variational equations reduce to

$$\mathbf{M}\frac{\mathrm{d}^2\boldsymbol{\gamma}}{\mathrm{d}t^2} + \mathbf{G}_1\frac{\mathrm{d}\boldsymbol{\gamma}}{\mathrm{d}t} + [\mathbf{K} + \mathbf{G}_2 + A]\boldsymbol{\gamma} = \mathbf{f} - \mathbf{f}_{\mathrm{rot}}$$

$$\mathbf{G}_1 = \boldsymbol{\Phi}^T\int \rho\boldsymbol{\varphi}\boldsymbol{\Omega}\boldsymbol{\varphi}^T\,\mathrm{d}V\,\boldsymbol{\Phi},\quad \mathbf{G}_2 = \boldsymbol{\Phi}^T\int \rho\boldsymbol{\varphi}\boldsymbol{\Omega}^2\boldsymbol{\varphi}^T\,\mathrm{d}V\,\boldsymbol{\Phi}, \tag{12.41}$$

$$A = \boldsymbol{\Phi}^T\int \rho\boldsymbol{\varphi}\mathbf{A}\boldsymbol{\varphi}^T\,\mathrm{d}V\,\boldsymbol{\Phi},\quad \mathbf{f}_{\mathrm{rot}} = \boldsymbol{\Phi}^T\int \rho\boldsymbol{\varphi}[\boldsymbol{\Omega}^2 + \mathrm{A}]X'\,]\,\mathrm{d}V$$

The terms in Equation 12.41 appeared in the first section of this chapter and their names and meanings were given.

12.3.5 Numerical Determination of the Current Position of the Body Axes

For a complete finite element solution, it is necessary to be able to compute the current position of the body axes using the Euler equations, although this is not actually accomplished using finite element equations. Recall that it has been assumed that the tractions, and hence the total force $\mathbf{F}(t)$ and moment $\mathbf{M}(t)$, are prescribed as functions of time on the undeformed configuration of the elastic body, referred to rotating axes. The current position of the center of mass in the absolute coordinate system is obtained with little effort by integrating $m\ddot{\mathbf{r}}_c = \mathbf{F}(t)$. The current value of the angular velocity is obtained by integrating the equations

$$\begin{aligned}\frac{\mathrm{d}\boldsymbol{\omega}}{\mathrm{d}t} &= \mathbf{J}^{-1}[\mathbf{M}(t) - \boldsymbol{\omega}\times\mathbf{J}\boldsymbol{\omega}]\\ \frac{\mathrm{d}\mathbf{Q}}{\mathrm{d}t} &= \boldsymbol{\Omega}(t)\mathbf{Q}(t)\end{aligned} \tag{12.42}$$

the first of which is nonlinear. Recall that $\boldsymbol{\omega}$ is the axial vector of $\mathbf{Q}$. Suppose the time step in numerical time integration is h, that the time after n steps is denoted by $t_n = nh$, and that the solution has been computed at and before t_n. Application of the Trapezoidal Rule and some manipulation serve to derive the integration formula

$$\left[\mathbf{I}+\frac{h}{2}\mathbf{J}^{-1}\boldsymbol{\Omega}(t_{n+1}\mathbf{J})\right]\boldsymbol{\omega}(t_{n+1}) = \left(\left[\mathbf{I}-\frac{h}{2}\mathbf{J}^{-1}\boldsymbol{\Omega}(t_n)\mathbf{J}\right]\boldsymbol{\omega}(t_n)+\frac{h}{2}\mathbf{J}^{-1}[\mathbf{M}(t_{n+1})+\mathbf{M}(t_n)]\right)$$

$$\mathbf{Q}(t_{n+1}) = \left[\mathbf{I}+\frac{h}{2}\boldsymbol{\Omega}(t_{n+1})\right]^{-1}\left[\mathbf{I}-\frac{h}{2}\boldsymbol{\Omega}(t_n)\right]\mathbf{Q}(t_n) \tag{12.43}$$

12.4 APPENDIX: ANGULAR VELOCITY VECTOR IN SPHERICAL COORDINATES

This appendix gives an interpretation of the angular velocity vector in terms of the angles appearing in spherical coordinates, as depicted in Figure 12.14. We assume that the spherical coordinate system coincides with the body axes.

The xyz axes are a rectilinear inertial (nonchanging) system. The base vector $\mathbf{e}_r$ is collinear with the position vector $\mathbf{r}$. The second vector $\mathbf{e}_\xi$ is perpendicular to r and lies in a plane parallel to the x–y plane. The vector $\mathbf{e}_\zeta$ completes the right-handed $r\xi\zeta$ system and points toward the z-axis.

The base vectors of the body system may be derived as

$$\begin{aligned} \mathbf{e}_r &= \cos\zeta(\cos\xi e_x + \sin\xi e_t) + \sin\zeta e_Z \\ \mathbf{e}_\xi &= -\sin\xi e_x + \cos\xi e_t \\ \mathbf{e}_\zeta &= -\sin\zeta(\cos\xi e_x + \sin\xi e_t) + \cos\zeta e_Z \end{aligned} \tag{12A.1}$$

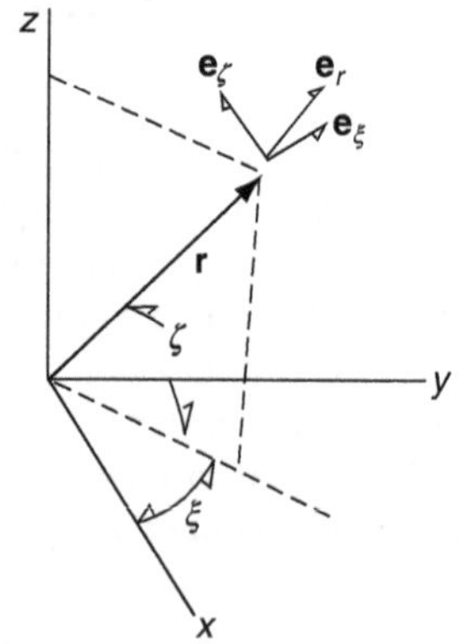

FIGURE 12.14 Angles used in spherical coordinates.

Let **b** denote the fixed length vector given by $\mathbf{b} = a\mathbf{e}_x + b\mathbf{e}_y + c\mathbf{e}_d$. It may be rotated to the spherical system as $\mathbf{b}' = a\mathbf{e}_r + b\mathbf{e}_\xi + c\mathbf{e}_\zeta$, which upon making the substitutions according to Equation 12A.1 gives

$$\mathbf{b}' = (\cos\xi\cos\zeta a - \sin\xi a - \sin\zeta\cos\xi c)\mathbf{e}_x + (\sin\xi\cos\zeta a + \cos\xi b - \sin\xi\sin\zeta c)\mathbf{e}_y + (\sin\zeta a + \cos\zeta c)\mathbf{e}_z \qquad (12A.2)$$

$$= \mathbf{Q}\begin{Bmatrix} a \\ b \\ c \end{Bmatrix} = \begin{Bmatrix} a' \\ b' \\ c' \end{Bmatrix}, \quad \mathbf{Q} = \begin{bmatrix} \cos\xi\cos\zeta & -\sin\xi & -\sin\zeta\cos\xi \\ \sin\xi\cos\zeta & \cos\xi & -\sin\xi\sin\zeta \\ \sin\zeta & 0 & \cos\zeta \end{bmatrix}$$

Also

$$\dot{\mathbf{Q}} = \begin{bmatrix} -\sin\xi\cos\zeta & -\cos\xi & \sin\zeta\sin\xi \\ \cos\xi\cos\zeta & -\sin\xi & -\cos\xi\sin\zeta \\ 0 & 0 & 0 \end{bmatrix}\dot{\xi} + \begin{bmatrix} -\cos\xi\sin\zeta & 0 & -\cos\zeta\cos\xi \\ -\sin\zeta\sin\xi & 0 & -\sin\xi\cos\zeta \\ \cos\zeta & 0 & -\sin\zeta \end{bmatrix}\dot{\zeta} \qquad (12A.3)$$

But note also that

$$\begin{bmatrix} -\sin\xi\cos\zeta & -\cos\xi & \sin\zeta\sin\xi \\ \cos\xi\cos\zeta & -\sin\xi & -\cos\xi\sin\zeta \\ 0 & 0 & 0 \end{bmatrix}\begin{bmatrix} \cos\xi\cos\zeta & \sin\xi\cos\zeta & \sin\zeta \\ -\sin\xi & \cos\xi & 0 \\ -\sin\zeta\cos\xi & -\sin\xi\sin\zeta & \cos\zeta \end{bmatrix} = \begin{bmatrix} 0 & -1 & 0 \\ 1 & 0 & 0 \\ 0 & 0 & 0 \end{bmatrix}$$

$$\begin{bmatrix} -\cos\xi\sin\zeta & 0 & -\cos\zeta\cos\xi \\ -\sin\zeta\sin\xi & 0 & -\sin\xi\cos\zeta \\ \cos\zeta & 0 & -\sin\zeta \end{bmatrix}\begin{bmatrix} \cos\xi\cos\zeta & \sin\xi\cos\zeta & \sin\zeta \\ -\sin\xi & \cos\xi & 0 \\ -\sin\zeta\cos\xi & -\sin\xi\sin\zeta & \cos\zeta \end{bmatrix} = \begin{bmatrix} 0 & 0 & -\cos\xi \\ 0 & 0 & -\sin\xi \\ \cos\xi & \sin\xi & 0 \end{bmatrix}$$

Now we regard the angles ξ, ζ to be time dependent, in which case $\dot{\mathbf{b}}' = \dot{\mathbf{Q}}\mathbf{b} = \boldsymbol{\Omega}\mathbf{b}'$, $\boldsymbol{\Omega} = \dot{\mathbf{Q}}\mathbf{Q}^T$.

After some manipulation

$$\boldsymbol{\Omega} = \begin{bmatrix} 0 & -\dot{\xi} & -\cos\xi\dot{\zeta} \\ \dot{\xi} & 0 & -\sin\xi\dot{\zeta} \\ \cos\xi\dot{\zeta} & \sin\xi\dot{\zeta} & 0 \end{bmatrix} \qquad (12A.4)$$

The corresponding angular velocity vector is $\boldsymbol{\omega}^T = \{\sin\xi\dot{\zeta} \quad -\cos\xi\dot{\zeta} \quad \dot{\xi}\}$. Of course it represents the counterclockwise rotation of the body about its instantaneous position. Referring to the angles introduced in Section 12.1, we now make the identifications

$$\begin{Bmatrix} \dot{\theta} \\ \dot{\phi} \\ \dot{\psi} \end{Bmatrix} = \begin{Bmatrix} \sin\xi\dot{\zeta} \\ -\cos\xi\dot{\zeta} \\ \dot{\xi} \end{Bmatrix} \tag{12A.5}$$

13 Aspects of Nonlinear Continuum Thermomechanics

13.1 INTRODUCTION

The first 12 topics of the current monograph have been concerned with linear FEA. The underlying notions of linear solid mechanics and conductive heat transfer were addressed in Chapter 5 and 6. Here, we briefly present several nonlinear topics from continuum thermomechanics, which will enable treating nonlinear finite element techniques thereafter. A more complete account is given, for example, in Chandrasekharaiah and Debnath (1994).

13.2 NONLINEAR KINEMATICS OF DEFORMATION

13.2.1 Deformation Gradient Tensor

Displacement: In FEA for finite deformation, it is necessary to carefully distinguish between the current (or "deformed") configuration (i.e., at the current time or load step) and a reference configuration which is usually considered strain-free. Here, both configurations are referred to the same orthogonal coordinate system characterized by the base vectors $\mathbf{e}_1$, $\mathbf{e}_2$, $\mathbf{e}_3$ (see Figure 2.1). Consider a body with volume V and surface S in the current configuration. The particle P occupies a position represented by the position vector $\mathbf{x}$, and experiences (empirical) temperature T. In the corresponding undeformed configuration, the position of P is described by $\mathbf{X}$, and the temperature has the value T_0 independent of $\mathbf{X}$. It is now assumed that $\mathbf{x}$ is a function of $\mathbf{X}$ and t, and T is also a function of $\mathbf{X}$ and t. The relations are written as $\mathbf{x}(\mathbf{X},t)$ and $T(\mathbf{X},t)$ and it is assumed that $\mathbf{x}$ and T are continuously differentiable in $\mathbf{X}$ and t through whatever order needed in the subsequent development (Figure 13.1).

Displacement Vector: The vector $\mathbf{u}(\mathbf{X},t)$ in Equation 13.1 represents the displacement from positions $\mathbf{X}$ to $\mathbf{x}$:

$$\mathbf{u}(\mathbf{X},t) = \mathbf{x} - \mathbf{X} \tag{13.1}$$

Now consider two close points P and Q in the undeformed configuration. The vector difference $\mathbf{X}_P - \mathbf{X}_Q$ is represented as a differential $d\mathbf{X}$ with squared length

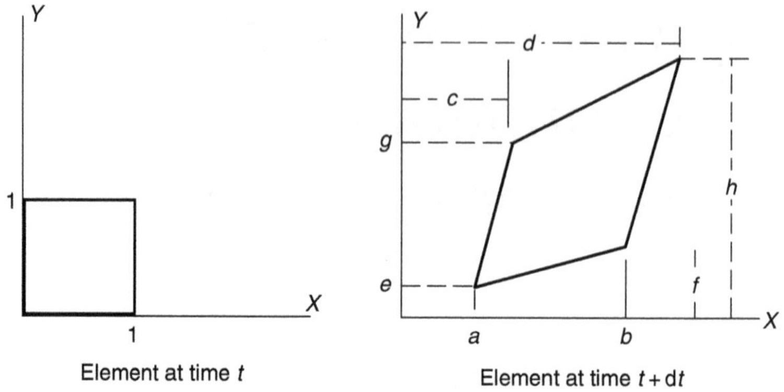

FIGURE 13.1 Position vectors in deformed and undeformed configurations.

$dS^2 = d\mathbf{X}^T d\mathbf{X}$. The corresponding quantity in the deformed configuration is $d\mathbf{x}$, with $ds^2 = d\mathbf{x}^T d\mathbf{x}$ (Figure 13.2).

Deformation Gradient Tensor: The deformation gradient tensor $\mathbf{F}$ is introduced as

$$d\mathbf{x} = \mathbf{F}\, d\mathbf{X}, \quad \mathbf{F} = \frac{\partial \mathbf{x}}{\partial \mathbf{X}} \tag{13.2}$$

$\mathbf{F}$ satisfies the polar decomposition theorem

$$\mathbf{F} = \mathbf{U}\boldsymbol{\Sigma}\mathbf{V}^T \tag{13.3}$$

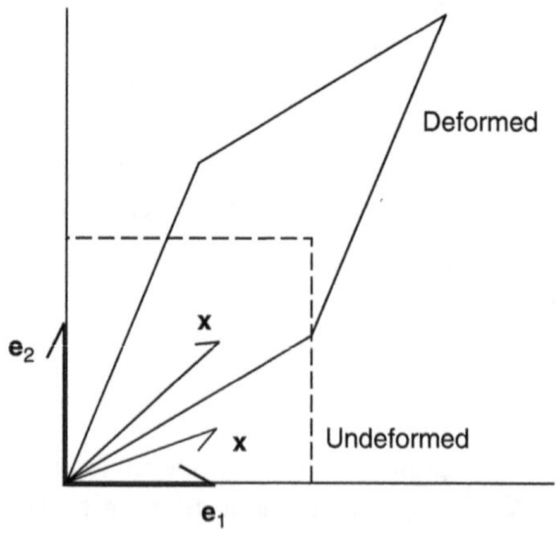

FIGURE 13.2 Deformed and undeformed distances between adjacent points.

in which $\mathbf{U}$ and $\mathbf{V}$ are orthogonal and $\boldsymbol{\Sigma}$ is a positive definite diagonal tensor whose entries λ_j, the eigenvalues of $\sqrt{\mathbf{F}^T\mathbf{F}}$, are called the *principal stretches*.

$$\boldsymbol{\Sigma} = \begin{bmatrix} \lambda_1 & 0 & 0 \\ 0 & \lambda_2 & 0 \\ 0 & 0 & \lambda_3 \end{bmatrix} \tag{13.4}$$

From Equation 13.3 $\mathbf{F}$ may be visualized as representing a rotation, followed by a stretch, followed by a second rotation.

13.2.2 Lagrangian Strain Tensor

The deformation-induced change in squared length is given by

$$\mathrm{d}s^2 - \mathrm{d}S^2 = \mathrm{d}\mathbf{X}^T\mathbf{E}\,\mathrm{d}\mathbf{X}, \quad \mathbf{E} = \tfrac{1}{2}[\mathbf{F}^T\mathbf{F} - \mathbf{I}] \tag{13.5}$$

in which $\mathbf{E}$ denotes the *Lagrangian strain tensor*. Also of interest is the *right Cauchy–Green strain* $\mathbf{C} = \mathbf{F}^T\mathbf{F} = 2\mathbf{E} + \mathbf{I}$. Note that $\mathbf{F} = \mathbf{I} + \partial\mathbf{u}/\partial\mathbf{X}$. If quadratic terms in $\partial\mathbf{u}/\partial\mathbf{X}$ are neglected the *linear strain tensor* $\mathbf{E}_L$ introduced in Chapter 5 is recovered as

$$\mathbf{E}_L = \frac{1}{2}\left[\frac{\partial\mathbf{u}}{\partial\mathbf{X}} + \left(\frac{\partial\mathbf{u}}{\partial\mathbf{X}}\right)^T\right] \tag{13.6}$$

Upon application of Equation 13.3, $\mathbf{E}$ is rewritten as

$$\mathbf{E} = \mathbf{V}\left[\tfrac{1}{2}(\boldsymbol{\Sigma}^2 - \mathbf{I})\mathbf{V}^T\right] \tag{13.7}$$

Under pure rigid body translation and rotation $\mathbf{x} = \mathbf{b}(t) + \mathbf{QX}$, we obtain $\mathbf{F} = \mathbf{Q}$ and $\mathbf{E} = \frac{1}{2}[\mathbf{Q}^T\mathbf{Q} - \mathbf{I}] = \mathbf{0}$, implying that the Lagrangian strain identically vanishes under rigid body motion, unlike the linear strain (see Chapter 5).

EXAMPLE 13.1

$\mathbf{F}$, $\mathbf{E}$, and $\mathbf{u}$ in cylindrical coordinates

Cylindrical Coordinates: In cylindrical coordinates, the position vector in the undeformed (reference) configuration is given by $\mathbf{R} = R\mathbf{e}_R + Z\mathbf{e}_Z$, with $\mathbf{e}_R = \cos\Theta\mathbf{e}_x + \sin\Theta\mathbf{e}_y$ and $\mathbf{e}_\Theta = -\sin\Theta\mathbf{e}_x + \cos\Theta\mathbf{e}_y$. In the deformed (current) configuration, the position vector is given by $\mathbf{r} = r\mathbf{e}_r + z\mathbf{e}_z$, $\mathbf{e}_r = \cos\theta\mathbf{e}_x + \sin\theta\mathbf{e}_y$ and $\mathbf{e}_\theta = -\sin\theta\mathbf{e}_x + \cos\theta\mathbf{e}_y$. We first seek $\mathbf{F}$.

$$
\begin{aligned}
\mathbf{dr} &= \mathrm{d}r\mathbf{e}_r + r\,\mathrm{d}\theta\,\mathbf{e}_\theta + \mathrm{d}z\,\mathbf{e}_z \\
&= \left(\frac{\partial r}{\partial R}\mathrm{d}R + \frac{1}{R}\frac{\partial r}{\partial \Theta}R\,\mathrm{d}\Theta + \frac{\partial r}{\partial Z}\mathrm{d}Z\right)\mathbf{e}_r \\
&\quad + \left(r\frac{\partial \theta}{\partial R}\mathrm{d}R + \frac{r}{R}\frac{\partial \theta}{\partial \Theta}R\,\mathrm{d}\Theta + r\frac{\partial \theta}{\partial Z}\mathrm{d}Z\right)\mathbf{e}_\theta \\
&\quad + \left(\frac{\mathrm{d}z}{\mathrm{d}R}\mathrm{d}R + \frac{1}{R}\frac{\mathrm{d}z}{\mathrm{d}\Theta}R\,\mathrm{d}\Theta + \frac{\partial z}{\partial Z}\mathrm{d}Z\right)\mathbf{e}_z \\
&= \mathbf{F}'\begin{Bmatrix}\mathrm{d}R \\ R\,\mathrm{d}\Theta \\ \mathrm{d}Z\end{Bmatrix}', \quad \mathbf{F}' = \begin{bmatrix}\frac{\partial r}{\partial R} & \frac{1}{R}\frac{\partial r}{\partial \Theta} & \frac{\partial r}{\partial Z} \\ r\frac{\partial \theta}{\partial R} & \frac{r}{R}\frac{\partial \theta}{\partial \Theta} & r\frac{\partial \theta}{\partial Z} \\ \frac{\mathrm{d}z}{\mathrm{d}R} & \frac{1}{R}\frac{\mathrm{d}z}{\mathrm{d}\Theta} & \frac{\partial z}{\partial Z}\end{bmatrix}
\end{aligned}
$$

in which the prime calls attention to the fact that the base vectors are still $\mathbf{e}_r$, $\mathbf{e}_\theta$, $\mathbf{e}_z$. Rotation of the $\mathbf{e}_r$, $\mathbf{e}_\theta$, $\mathbf{e}_z$ system to coincide with the base $\mathbf{e}_R$, $\mathbf{e}_\Theta$, $\mathbf{e}_Z$ system gives rise to the orthogonal tensor

$$
\mathbf{Q}^T = \begin{bmatrix}\cos(\theta-\Theta) & -\sin(\theta-\Theta) & 0 \\ \sin(\theta-\Theta) & \cos(\theta-\Theta) & 0 \\ 0 & 0 & 1\end{bmatrix}
$$

We conclude that the deformation gradient tensor is given by $\mathbf{F} = \mathbf{Q}^T\mathbf{F}'$.

The displacement vector is now

$$
\mathbf{u} = [r\cos(\theta-\Theta) - R]\mathbf{e}_R + r\sin\theta\mathbf{e}_\Theta + (z-Z)\mathbf{e}_Z
$$

We now apply the chain rule to $\mathrm{d}s^2$ in cylindrical coordinates to obtain the right Cauchy–Green strain tensor $\mathbf{C}$.

$$
\begin{aligned}
\mathrm{d}s^2 &= \mathbf{dr}\cdot\mathbf{dr} = \mathrm{d}r^2 + r^2\,\mathrm{d}\theta^2 + \mathrm{d}z^2 \\
&= \{\mathrm{d}R \quad R\,\mathrm{d}\Theta \quad \mathrm{d}Z\}\mathbf{C}\begin{pmatrix}\mathrm{d}R \\ R\,\mathrm{d}\Theta \\ \mathrm{d}Z\end{pmatrix}
\end{aligned}
$$

in which

$$
\begin{aligned}
c_{RR} &= \left(\frac{\mathrm{d}r}{\mathrm{d}R}\right)^2 + \left(r\frac{\mathrm{d}\theta}{\mathrm{d}R}\right)^2 + \left(\frac{\mathrm{d}z}{\mathrm{d}R}\right)^2, & e_{RR} &= \frac{1}{2}(c_{RR}-1) \\
c_{\Theta\Theta} &= \left(\frac{1}{R}\frac{\mathrm{d}r}{\mathrm{d}\Theta}\right)^2 + \left(\frac{r}{R}\frac{\mathrm{d}\theta}{\mathrm{d}\Theta}\right)^2 + \left(\frac{1}{R}\frac{\mathrm{d}z}{\mathrm{d}\Theta}\right)^2, & e_{\Theta\Theta} &= \frac{1}{2}(c_{\Theta\Theta}-1) \\
c_{ZZ} &= \left(\frac{\mathrm{d}r}{\mathrm{d}Z}\right)^2 + \left(r\frac{\mathrm{d}\theta}{\mathrm{d}Z}\right)^2 + \left(\frac{\mathrm{d}z}{\mathrm{d}Z}\right)^2, & e_{ZZ} &= \frac{1}{2}(c_{ZZ}-1) \\
c_{R\Theta} &= \left(\frac{\mathrm{d}r}{\mathrm{d}R}\right)\left(r\frac{\mathrm{d}\theta}{\mathrm{d}R}\right) + \left(r\frac{\mathrm{d}\theta}{\mathrm{d}R}\right)\left(\frac{r}{R}\frac{\mathrm{d}\theta}{\mathrm{d}\Theta}\right) + \left(\frac{\mathrm{d}z}{\mathrm{d}R}\right)\left(\frac{1}{R}\frac{\mathrm{d}z}{\mathrm{d}\Theta}\right), & e_{R\Theta} &= \frac{1}{2}c_{R\Theta}
\end{aligned}
$$

$$c_{\Theta Z} = \left(\frac{1}{R}\frac{\mathrm{d}r}{\mathrm{d}\Theta}\right)\left(\frac{\mathrm{d}r}{\mathrm{d}Z}\right) + \left(\frac{r}{R}\frac{\mathrm{d}\theta}{\mathrm{d}\Theta}\right)\left(r\frac{\mathrm{d}\theta}{\mathrm{d}Z}\right) + \left(\frac{1}{R}\frac{\mathrm{d}z}{\mathrm{d}\Theta}\right)\left(\frac{\mathrm{d}z}{\mathrm{d}Z}\right), \quad e_{\Theta Z} = \frac{1}{2}c_{\Theta Z}$$

$$c_{ZR} = \left(\frac{\mathrm{d}r}{\mathrm{d}Z}\right)\left(\frac{\mathrm{d}r}{\mathrm{d}R}\right) + \left(r\frac{\mathrm{d}\theta}{\mathrm{d}Z}\right)\left(r\frac{\mathrm{d}\theta}{\mathrm{d}R}\right) + \left(\frac{\mathrm{d}z}{\mathrm{d}Z}\right)\left(\frac{\mathrm{d}z}{\mathrm{d}R}\right), \quad e_{ZR} = \frac{1}{2}c_{ZR}$$

Of course the Lagrangian strain tensor is obtained from $\mathbf{E} = \frac{1}{2}(\mathbf{C} - \mathbf{I})$, with $\mathbf{Q}$ not appearing.

13.2.3 Velocity Gradient Tensor, Deformation Rate Tensor, and Spin Tensor

We now introduce the particle velocity $\mathbf{v} = \partial\mathbf{x}/\partial t$ and assume that it is an explicit function of $\mathbf{x}(t)$ and t. The *velocity gradient tensor* $\mathbf{L}$ is introduced using $\mathrm{d}\mathbf{v} = \mathbf{L}\,\mathrm{d}\mathbf{x}$, from which

$$\begin{aligned} \mathbf{L} &= \frac{\mathrm{d}\mathbf{v}}{\mathrm{d}\mathbf{x}} \\ &= \frac{\mathrm{d}\mathbf{v}}{\mathrm{d}\mathbf{X}}\frac{\mathrm{d}\mathbf{X}}{\mathrm{d}\mathbf{x}} \\ &= \dot{\mathbf{F}}\mathbf{F}^{-1} \end{aligned} \tag{13.8}$$

Its symmetric part, called the *deformation rate tensor*, is

$$\mathbf{D} = \tfrac{1}{2}\left[\mathbf{L} + \mathbf{L}^T\right] \tag{13.9}$$

It may be regarded as a strain rate referred to the current configuration. The corresponding strain rate referred to the undeformed configuration is the Lagrangian strain rate:

$$\begin{aligned} \dot{\mathbf{E}} &= \tfrac{1}{2}\left[\mathbf{F}^T\dot{\mathbf{F}} + \dot{\mathbf{F}}^T\mathbf{F}\right] \\ &= \mathbf{F}^T\left\{\tfrac{1}{2}\left[\dot{\mathbf{F}}\mathbf{F}^{-1} + \mathbf{F}^{-T}\dot{\mathbf{F}}^T\right]\right\}\mathbf{F} \\ &= \mathbf{F}^T\mathbf{D}\mathbf{F} \end{aligned} \tag{13.10}$$

The antisymmetric portion of $\mathbf{L}$ is called the *spin tensor* $\mathbf{W}$:

$$\mathbf{W} = \tfrac{1}{2}[\mathbf{L} - \mathbf{L}^T] \tag{13.11}$$

Suppose the deformation consists only of a time-dependent rigid body motion expressed by

$$\mathbf{x}(t) = \mathbf{Q}(t)\mathbf{X} + \mathbf{b}(t), \quad \mathbf{Q}^T(t)\mathbf{Q}(t) = \mathbf{I} \tag{13.12}$$

Clearly, $\mathbf{F}=\mathbf{Q}$ and $\mathbf{E}=\mathbf{0}$, $\mathbf{D}=\mathbf{0}$ and $\mathbf{L}=\mathbf{W}=\dot{\mathbf{Q}}\mathbf{Q}^T$, and recall from Chapter 12 that $\dot{\mathbf{Q}}\mathbf{Q}^T$ is antisymmetric.

EXAMPLE 13.2

$\mathbf{v}$, $\mathbf{L}$, $\mathbf{D}$, and $\mathbf{W}$ *in cylindrical coordinates*

The velocity vector in cylindrical coordinates is

$$\begin{aligned}\mathbf{v} &= \frac{\partial r}{\partial t}\mathbf{e}_r + r\frac{\partial \theta}{\partial t}\mathbf{e}_\theta + \frac{\partial z}{\partial t}\mathbf{e}_\phi \\ &= v_r\dot{\mathbf{e}}_r + v_\theta\mathbf{e}_\theta + v_\phi\mathbf{e}_\phi\end{aligned}$$

Observe that

$$\begin{aligned}\mathrm{d}\mathbf{v} &= \mathrm{d}v_r\,\mathbf{e}_r + \mathrm{d}v_\theta\,\mathbf{e}_\theta + \mathrm{d}v_z\,\mathbf{e}_z + v_r\,\mathrm{d}\mathbf{e}_r + v_\theta\,\mathrm{d}\mathbf{e}_\theta \\ &= \left[\mathrm{d}v_r - \frac{v_\theta}{r}r\,\mathrm{d}\theta\right]\mathbf{e}_r + \left[\mathrm{d}v_\theta + \frac{v_r}{r}r\,\mathrm{d}\theta\right]\mathbf{e}_\theta + \mathrm{d}v_z\,\mathbf{e}_z\end{aligned}$$

Now converting to matrix–vector notation,

$$\mathrm{d}\mathbf{v} = \begin{pmatrix} \frac{\mathrm{d}v_r}{\mathrm{d}r}\mathrm{d}r + \frac{1}{r}\frac{\mathrm{d}v_r}{\mathrm{d}\theta}r\,\mathrm{d}\theta + \frac{\mathrm{d}v_r}{\mathrm{d}z}\mathrm{d}z - \frac{v_\theta}{r}r\,\mathrm{d}\theta \\ \frac{\mathrm{d}v_\theta}{\mathrm{d}r}\mathrm{d}r + \frac{1}{r}\frac{\mathrm{d}v_\theta}{\mathrm{d}\theta}r\,\mathrm{d}\theta + \frac{\mathrm{d}v_\theta}{\mathrm{d}z}\mathrm{d}z - \frac{v_r}{r}r\,\mathrm{d}\theta \\ \frac{\mathrm{d}v_z}{\mathrm{d}r}\mathrm{d}r + \frac{1}{r}\frac{\mathrm{d}v_z}{\mathrm{d}\theta}r\,\mathrm{d}\theta + \frac{\mathrm{d}v_z}{\mathrm{d}z}\mathrm{d}z \end{pmatrix}$$

$$= \mathbf{L}\begin{pmatrix}\mathrm{d}r \\ r\mathrm{d}\theta \\ \mathrm{d}z\end{pmatrix}, \quad \mathbf{L} = \begin{bmatrix} \frac{\mathrm{d}v_r}{\mathrm{d}r} & \frac{1}{r}\frac{\mathrm{d}v_r}{\mathrm{d}\theta} - \frac{v_\theta}{r} & \frac{\mathrm{d}v_r}{\mathrm{d}z} \\ \frac{\mathrm{d}v_\theta}{\mathrm{d}r} & \frac{1}{r}\frac{\mathrm{d}v_\theta}{\mathrm{d}\theta} + \frac{v_r}{r} & \frac{\mathrm{d}v_\theta}{\mathrm{d}z} \\ \frac{\mathrm{d}v_z}{\mathrm{d}r} & \frac{1}{r}\frac{\mathrm{d}v_z}{\mathrm{d}\theta} & \frac{\mathrm{d}v_z}{\mathrm{d}z} \end{bmatrix}$$

Of course $\mathbf{D}$ and $\mathbf{W}$ are obtained as the symmetric and antisymmetric portions of $\mathbf{L}$.

13.2.4 Differential Volume Element

The volume implied by the differential position vector $\mathrm{d}\mathbf{R}$ is given by the vector triple product

$$\begin{aligned}\mathrm{d}V_0 &= \mathrm{d}\mathbf{X}_1 \cdot \mathrm{d}\mathbf{X}_2 \times \mathrm{d}\mathbf{X}_3 = \mathrm{d}\mathbf{X}_1\,\mathrm{d}\mathbf{X}_2\,\mathrm{d}\mathbf{X}_3 \\ \mathrm{d}\mathbf{X}_1 &= \mathrm{d}X_1\mathbf{e}_1, \quad \mathrm{d}\mathbf{X}_2 = \mathrm{d}X_2\mathbf{e}_2, \quad \mathrm{d}\mathbf{X}_3 = \mathrm{d}X_3\,\mathbf{e}_3\end{aligned} \tag{13.13}$$

The vectors $\mathrm{d}\mathbf{X}_i$ deform into $\mathrm{d}\mathbf{x}_j = \frac{\mathrm{d}x_j}{\mathrm{d}X_i}\mathbf{e}_j\,\mathrm{d}\mathbf{X}_i$. The deformed volume is now readily verified as

$$\begin{aligned}\mathrm{d}V &= \mathrm{d}\mathbf{x}_1 \cdot \mathrm{d}\mathbf{x}_2 \times \mathrm{d}\mathbf{x}_3 \\ &= J\mathrm{d}V_0, \quad J = \det(\mathbf{F}) = \det^{\frac{1}{2}}(\mathbf{C})\end{aligned} \tag{13.14}$$

and J as before is called the *Jacobian*.

The time derivative of J is prominent in incremental formulations in continuum mechanics. Recalling that $J = \sqrt{I_3}$, we have

$$\begin{aligned}\frac{\mathrm{d}}{\mathrm{d}t}J &= \frac{1}{2J}\frac{\mathrm{d}I_3}{\mathrm{d}t} \\ &= \tfrac{J}{2}tr(\mathbf{C}^{-1}\dot{\mathbf{C}}) \\ &= \tfrac{J}{2}[tr(\mathbf{F}^{-1}\dot{\mathbf{F}}) + tr(\mathbf{F}^{-1}\mathbf{F}^{-T}\dot{\mathbf{F}}^T\mathbf{F})] \\ &= J\,tr\left(\frac{\mathbf{F}^{-1}\dot{\mathbf{F}} + \mathbf{F}^{-T}\dot{\mathbf{F}}^T}{2}\right) \\ &= J\,tr(\mathbf{D})\end{aligned} \tag{13.15}$$

EXAMPLE 13.3

Relation of $\mathbf{D}$ *to* $\dot{\mathbf{E}}$

We now express $\mathbf{D}$ and its trace in terms of $\dot{\mathbf{E}}$, $\mathbf{E}$, and $\mathbf{F}$. First differentiate $\mathbf{E}$ to find

$$\dot{\mathbf{E}} = \tfrac{1}{2}(\mathbf{F}^T\dot{\mathbf{F}} + \dot{\mathbf{F}}^T\mathbf{F})$$

from which

$$\begin{aligned}\mathbf{F}^{-T}\dot{\mathbf{E}}\mathbf{F}^{-1} &= \tfrac{1}{2}(\dot{\mathbf{F}}\mathbf{F}^{-1} + \mathbf{F}^{-T}\dot{\mathbf{F}}^T) \\ &= \mathbf{D}\end{aligned}$$

Next,

$$\begin{aligned}tr(\mathbf{D}) &= tr(\mathbf{F}^{-T}\dot{\mathbf{E}}\mathbf{F}^{-1}) \\ &= tr(\dot{\mathbf{E}}\mathbf{F}^{-1}\mathbf{F}^{-T}) \\ &= tr(\dot{\mathbf{E}}(\mathbf{I} + 2\mathbf{E})^{-1})\end{aligned}$$

13.2.5 Differential Surface Element

Let $\mathrm{d}S$ denote a surface element in the deformed configuration, with exterior unit normal $\mathbf{n}$ illustrated in Figure 13.3. The counterparts from the reference configuration are $\mathrm{d}S_0$ and $\mathbf{n}_0$. A surface element $\mathrm{d}S$ obeys Nanson's theorem (cf. Chandrasekharaiah and Debnath, 1994)

$$\mathbf{n}\,\mathrm{d}S = J\mathbf{F}^{-T}\mathbf{n}_0\,\mathrm{d}S_0 \tag{13.16}$$

Taking the magnitude of $\mathbf{n}\,\mathrm{d}S$, we conclude that

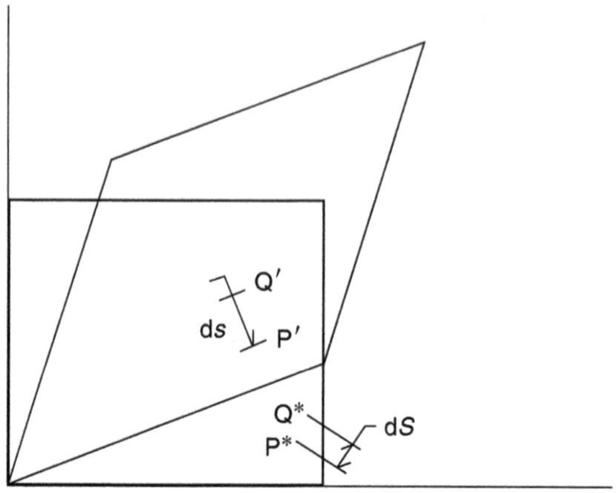

FIGURE 13.3 Differential length changes.

$$dS = J\sqrt{\mathbf{n}_0^T\mathbf{C}^{-1}\mathbf{n}_0}\,dS_0, \quad \mathbf{n} = \frac{\mathbf{F}^{-T}\mathbf{n}_0}{\sqrt{\mathbf{n}_0^T\mathbf{C}^{-1}\mathbf{n}_0}} \tag{13.17}$$

Of course during deformation the surface normal changes direction (cf. Figure 13.4), a fact which is important, for example, in contact problems. For later use, in incremental variational methods we consider the differential $\frac{d\mathbf{n}}{dt}$ and $d(\mathbf{n}\,dS)$:

$$\frac{d}{dt}[\mathbf{n}\,dS] = \frac{dJ}{dt}\mathbf{F}^{-T}\mathbf{n}_0\,dS + J\frac{d\mathbf{F}^{-T}}{dt}\mathbf{n}_0\,dS_0 \tag{13.18}$$

But, recalling Equation 13.15,

$$\begin{aligned}\frac{dJ}{dt} &= J\,tr(\mathbf{D})\\ \frac{dJ}{dt}\mathbf{F}^{-T}\mathbf{n}_0\,dS_0 &= tr(\mathbf{D})J\mathbf{F}^{-T}\mathbf{n}_0\,dS_0\\ &= tr(\mathbf{D})\mathbf{n}\,dS\end{aligned} \tag{13.19}$$

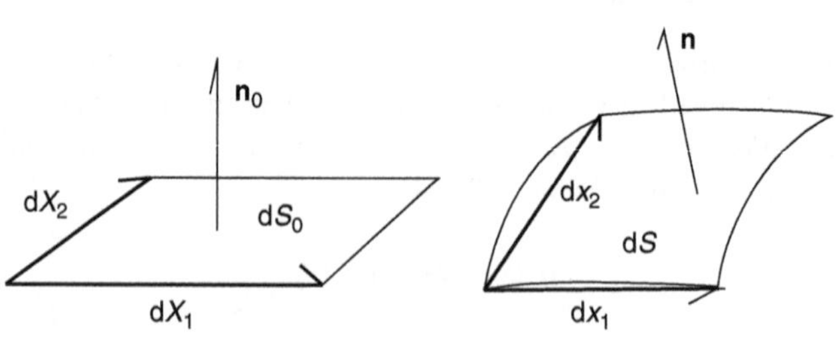

FIGURE 13.4 Undeformed and deformed surface patches.

Also, since $d(\mathbf{F}^T\mathbf{F}^{-T}) = \mathbf{0}$,

$$\begin{aligned}\frac{d\mathbf{F}^{-T}}{dt} &= -\mathbf{F}^{-T}\frac{d\mathbf{F}^T}{dt}\mathbf{F}^{-T} \\ &= -\mathbf{L}^T\mathbf{F}^T\end{aligned} \tag{13.20}$$

Finally, we have

$$\frac{d[\mathbf{n}\,dS]}{dt} = [tr(\mathbf{D})\mathbf{I} - \mathbf{L}^T]\mathbf{n}\,dS \tag{13.21}$$

EXAMPLE 13.4

Referring to Figure 13.5, determine **u**, **F**, J, and **E** as functions of **X** and **Y**: use $H=1$, $W=1$, $a=0.1$, $b=0.1$, $c=0.3$, $d=0.2$, $e=0.2$, $f=0.1$. Assume a unit thickness in the Z-direction in both the deformed and undeformed configurations.

Solution

A deformation model which captures the fact that straight sides remain straight is given in the form assumed in the form

$$x = \alpha X + \beta Y + \gamma XY, \quad y = \delta X + \varepsilon Y + \zeta XY$$

It is necessary to determine α, β, γ, δ, ε, and ζ from the coordinates of the vertices in the deformed configuration.

$$\begin{aligned}
&\text{At } (X, Y) = (W, 0): && W + a = \alpha W, && b = \delta W \\
&\text{At } (X, Y) = (0, H): && e = \beta H, && H + f = \varepsilon H \\
&\text{At } (X, Y) = (W, H): && W + c = \alpha W + \beta H + \gamma WH, && H + d = \delta W + \varepsilon H + \zeta WH
\end{aligned}$$

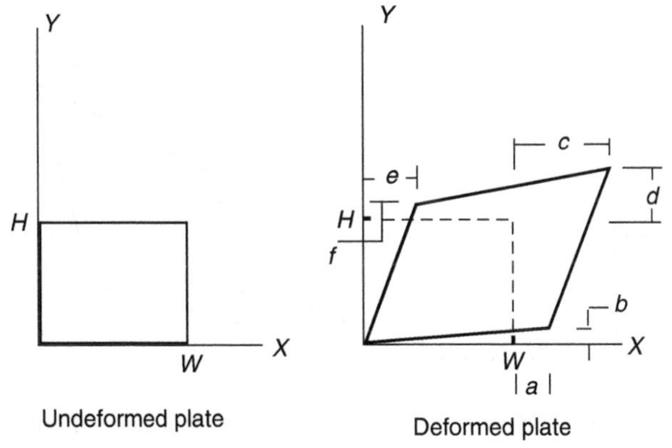

FIGURE 13.5 Plate elements in undeformed and deformed states.

After elementary manipulations,

$$\alpha = 1 + \frac{a}{W} = 1.1, \quad \beta = \frac{e}{H} = 0.2, \qquad \delta = \frac{b}{W} = 0.1$$

$$\varepsilon = 1 + \frac{f}{H} = 1.2, \quad \gamma = \frac{W + c - (W + a) - e}{WH} = 0.4, \quad \zeta = \frac{H + d - b - (H + f)}{WH} = 0$$

The 2×2 deformation gradient tensor **F** and its determinant now are:

$$\mathbf{F} = \begin{bmatrix} \alpha + \gamma Y & \beta + \gamma X \\ \delta + \zeta Y & \varepsilon + \zeta X \end{bmatrix} = \begin{bmatrix} 1.1 + 0.4Y & 0.2 + 0.4X \\ 0.1 & 1.2 \end{bmatrix}$$

$$J = 1.3 + 0.48Y - 0.04X$$

The displacement vector is

$$\mathbf{u} = \begin{pmatrix} x - X \\ y - Y \end{pmatrix} = \begin{pmatrix} (\alpha - 1)X + \beta Y + \gamma XY \\ \delta X + (\varepsilon - 1)Y + \zeta XY \end{pmatrix} = \begin{pmatrix} 0.1X + 0.2Y + 0.4XY \\ 0.1X + 0.2Y \end{pmatrix}$$

The Lagrangian strain $\mathbf{E} = \frac{1}{2}[\mathbf{F}^T\mathbf{F} - \mathbf{I}]$ is found to be

$$\mathbf{E} = \frac{1}{2}\begin{bmatrix} 1.1 + 0.4Y & 0.1 \\ 0.2 + 0.4X & 1.2 \end{bmatrix}\begin{bmatrix} 1.1 + 0.4Y & 0.2 + 0.4X \\ 0.1 & 1.2 \end{bmatrix} - \frac{1}{2}\begin{bmatrix} 1 & 0 \\ 0 & 1 \end{bmatrix}$$

$$= \begin{bmatrix} 0.11 + 0.44Y + 0.08Y^2 & 0.17 + 0.22X + 0.04Y + 0.08XY \\ 0.17 + 0.22X + 0.04Y + 0.08XY & 0.24 + 0.08X + 0.08X^2 \end{bmatrix}$$

EXAMPLE 13.5

Figure 13.6 shows a square element at time t and at $t + \mathrm{d}t$. Estimate **L**, **D**, and **W** at time t. Use $a = 0.1\,\mathrm{d}t$, $b = 1 + 0.2\,\mathrm{d}t$, $c = 0.2\,\mathrm{d}t$, $d = 1 + 0.4\,\mathrm{d}t$, $e = 0.05\,\mathrm{d}t$, $f = 0.1\,\mathrm{d}t$,

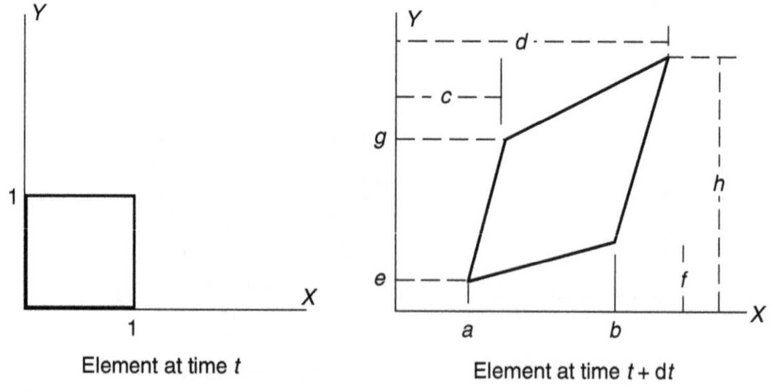

FIGURE 13.6 Element experiencing rigid body motion and deformation.

$g = 1 - 0.1\,\mathrm{d}t$, $h = 1 + 0.5\,\mathrm{d}t$. Assume a unit thickness in the Z-direction in both the deformed and undeformed configurations.

Solution

First represent the deformed position vectors in terms of the undeformed position vectors using eight coefficients to be determined using the given geometry. In particular,

$$x = \alpha + \beta X + \gamma Y + \delta XY, \qquad y = \varepsilon + \zeta X + \eta Y + \theta XY$$

This expression likewise captures the fact that straight sides remain straight. Following procedures analogous to Example 13.3, we find

$$\begin{aligned} \alpha &= 0.1\,\mathrm{d}t, & \varepsilon &= 0.05\,\mathrm{d}t \\ \beta &= 1 + 0.2\,\mathrm{d}t, & \zeta &= 0.1\,\mathrm{d}t \\ \gamma &= 0.2\,\mathrm{d}t, & \eta &= 1 - 0.1\,\mathrm{d}t \\ \delta &= -0.5\,\mathrm{d}t, & \theta &= -0.05\,\mathrm{d}t \end{aligned}$$

The velocities may be estimated using $v_x \approx \frac{x-X}{\mathrm{d}t}$ and $v_y \approx \frac{y-Y}{\mathrm{d}t}$, from which

$$\begin{aligned} v_x &= 0.1 + 0.2X + 0.2Y - 0.5XY \\ v_y &= 0.05 + 0.1X - 0.1Y - 0.05XY \end{aligned}$$

The tensors **L**, **D**, and **W** are now readily found as

$$\mathbf{L} = \begin{bmatrix} 0.2 - 0.5Y & 0.2 - 0.5X \\ 0.1 - 0.05Y & -0.1 - 0.05X \end{bmatrix}$$

$$\mathbf{D} = \begin{bmatrix} 0.2 - 0.5Y & 0.15 - 0.25X - 0.025Y \\ 0.15 - 0.25X - 0.025Y & -0.1 - 0.05X \end{bmatrix}$$

$$\mathbf{W} = \begin{bmatrix} 0 & -0.4 - 0.25X + 0.025Y \\ 0.4 + 0.25X - 0.025Y & 0 \end{bmatrix}$$

13.3 MECHANICAL EQUILIBRIUM AND THE PRINCIPLE OF VIRTUAL WORK

13.3.1 Traction Vector and Stress Tensors

A stress tensor was previously introduced in Chapter 5. However, it was in the context of small deformation, in which case it was not indicated how the stress changes when referred to the undeformed as opposed to the current (deformed) configuration. The distinction is central to the development below, leading to the Cauchy stress tensor (current configuration) and the first and second Piola–Kirchoff stress tensor.

Cauchy Stress: We consider a differential tetrahedron enclosing the point x in the deformed configuration, as illustrated in Figure 13.7. The area of the inclined face

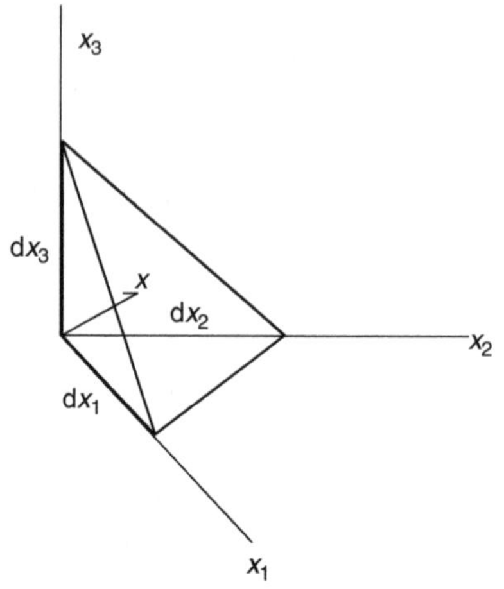

FIGURE 13.7 Differential tetrahedron.

is dS, and dS_i is the area of the face whose exterior normal vector is $-\mathbf{e}_i$. Simple vector analysis serves to derive that $n_i = \mathrm{d}S_i/\mathrm{d}S$, see Example 2.5. Now referring to Figure 13.8, let d$\mathbf{F}$ denote the force on the surface element dS, and let d$\mathbf{F}^{(i)}$ denote the force on area dS_i. As the tetrahedron shrinks to a point, the contribution of volume forces such as inertia decays faster than surface forces. Balance of forces requires that $\mathrm{d}F_j = \sum_i \mathrm{d}F_j^{(i)}$.

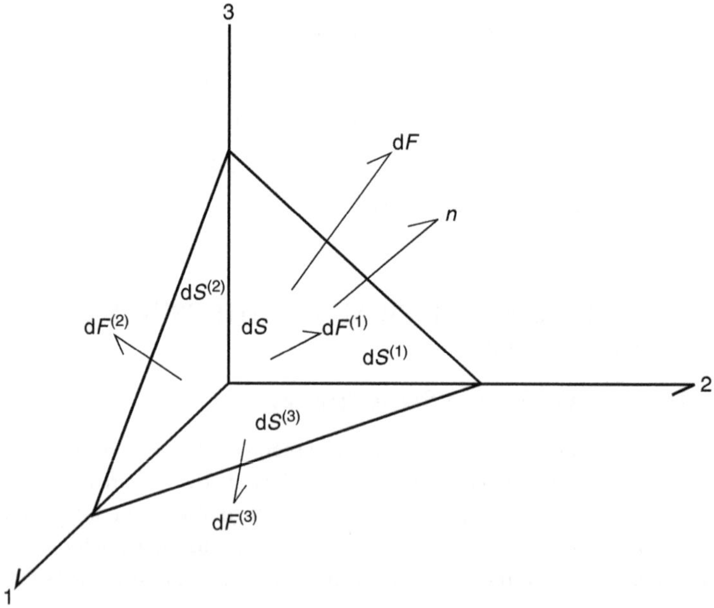

FIGURE 13.8 Forces applied to a differential tetrahedron.

The traction vector acting on the inclined face is defined by

$$\mathbf{t} = \frac{\mathrm{d}\mathbf{F}}{\mathrm{d}S} \tag{13.22}$$

from which

$$\begin{aligned} t_j &= \sum_i \frac{\mathrm{d}F_j^{(i)}}{\mathrm{d}S_i}\frac{\mathrm{d}S_i}{\mathrm{d}S} \\ &= \mathrm{T}_{ij}n_i \end{aligned} \tag{13.23}$$

in which

$$\mathrm{T}_{ij} = \frac{\mathrm{d}P_j^{(i)}}{\mathrm{d}S_i} \tag{13.24}$$

It is readily seen that T_{ij} can be interpreted as the intensity of the force acting in the j direction on the facet pointing in the $-i$ direction, and is recognized as ijth entry of the Cauchy stress $\mathbf{T}$. In matrix–vector notation the stress–traction relation is written as

$$\mathbf{t} = \mathbf{T}^T\mathbf{n} \tag{13.25}$$

In Section 13.3.2 it will be seen that $\mathbf{T}$ is symmetric by virtue of the balance of angular momentum. Equation 13.24 implies that $\mathbf{T}^T$ is a tensor, from which it follows that $\mathbf{T}$ is a tensor.

In traditional depictions the stresses on the back faces are represented by arrows pointing in negative directions. However, this depiction can be confusing—the arrows actually represent the directions of the traction components. Consider the one-dimensional member in Figure 13.9. The traction vector $t\mathbf{e}_1$ acts at $x = L$, while the traction vector $-t\mathbf{e}_1$ at $x = 0$. At $x = L$, the corresponding stress is $t_{11} = t\mathbf{e}_1 \cdot \mathbf{e}_1 = t$. At $x = 0$ the stress is given by $(-t\mathbf{e}_1) \cdot (-\mathbf{e}_1) = t$. Clearly, the stress at both ends, and in fact throughout the member, is positive (tensile).

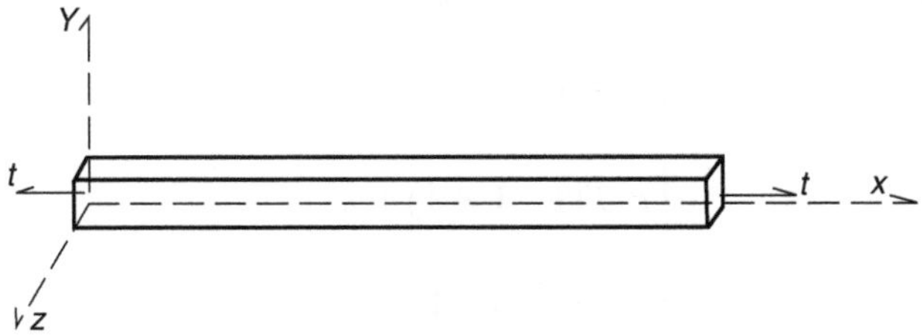

FIGURE 13.9 Tractions on a bar experiencing uniaxial tension.

We will see later that the stress tensor is symmetric by virtue of the balance of angular momentum.

First Piola–Kirchhoff Stress Tensor: Transformation to undeformed coordinates is now considered. From the transformation properties of a surface element, we have

$$\begin{aligned}\mathbf{t}\,\mathrm{d}S &= \mathbf{T}^T\mathbf{n}\,\mathrm{d}S \\ &= \mathbf{T}^T J\mathbf{F}^{-T}\mathbf{n}_0\,\mathrm{d}S_0 \\ &= \bar{\mathbf{S}}^T\mathbf{n}_0\,\mathrm{d}S_0, \quad \bar{\mathbf{S}} = J\mathbf{F}^{-1}\mathbf{T}\end{aligned} \tag{13.26}$$

$\bar{\mathbf{S}}$ is known as the first Piola–Kirchhoff stress tensor and it *is not symmetric*.

Second Piola–Kirchhoff Stress: We next derive the stress tensor which is conjugate to the Lagrangian strain rate, i.e., gives the correct amount of work per unit undeformed volume. At a segment $\mathrm{d}S$ at $\mathbf{x}$ on the deformed boundary, assuming static conditions the rate of work $\mathrm{d}\dot{W}$ of the tractions is

$$\begin{aligned}\mathrm{d}\dot{W} &= \mathrm{d}\mathbf{F}^T\dot{\mathbf{u}} \\ &= \mathbf{t}^T\dot{\mathbf{u}}\,\mathrm{d}S\end{aligned} \tag{13.27}$$

Over the surface S, shifting to tensor-indicial notation and invoking the divergence theorem, we find

$$\begin{aligned}\dot{W} &= \int \mathbf{t}^T\dot{\mathbf{u}}\,\mathrm{d}S \\ &= \int \mathrm{T}_{ij}n_i\dot{u}_j\,\mathrm{d}S \\ &= \int \frac{\partial \dot{u}_j}{\partial x_i}\mathrm{T}_{ij}\,\mathrm{d}V + \int \dot{u}_j\frac{\partial \mathrm{T}_{ij}}{\partial x_i}\,\mathrm{d}V\end{aligned} \tag{13.28}$$

We will shortly see that static equilibrium implies that $\frac{\partial \mathrm{T}_{ij}}{\partial x_i} = 0$, which will enable us to conclude that

$$\dot{W} = \int \frac{\partial \dot{u}_j}{\partial x_i}T_{ij}\,\mathrm{d}V = \int tr(\mathbf{TL})\,\mathrm{d}V = \int tr(\mathbf{TD})\,\mathrm{d}V \tag{13.29}$$

To convert to undeformed coordinates, note that

$$\begin{aligned}\dot{W} &= \int tr(\mathbf{TD})\,\mathrm{d}V \\ &= \int tr(J\mathbf{T}\mathbf{F}^{-1}\dot{\mathbf{E}}\mathbf{F}^{-T})\,\mathrm{d}V_0 \\ &= \int tr(J\mathbf{F}^{-T}\mathbf{T}\mathbf{F}^{-1}\dot{\mathbf{E}})\,\mathrm{d}V_0 \\ &= \int tr(\mathbf{S}\dot{\mathbf{E}})\,\mathrm{d}V, \qquad \mathbf{S} = J\mathbf{F}^{-T}\mathbf{T}\mathbf{F}^{-1}\end{aligned} \tag{13.30}$$

and the tensor **S** is called the second *Piola–Kirchhoff stress tensor*. It is symmetric if **T** is symmetric, which we will shortly see to be the case.

EXAMPLE 13.6

At point (0,0,0) the tractions $\mathbf{t}_1$, $\mathbf{t}_2$, $\mathbf{t}_3$ act on planes with normal vectors $\mathbf{n}_1$, $\mathbf{n}_2$, and $\mathbf{n}_3$. Find the Cauchy stress tensor **T**. Given:

$$\mathbf{n}_1 = \tfrac{1}{\sqrt{3}}[\mathbf{e}_1 + \mathbf{e}_2 + \mathbf{e}_3], \qquad \boldsymbol{\tau}_1 = \tfrac{1}{\sqrt{3}}[6\mathbf{e}_1 + 9\mathbf{e}_2 + 12\mathbf{e}_3]$$

$$\mathbf{n}_2 = \tfrac{1}{\sqrt{3}}[\mathbf{e}_1 + \mathbf{e}_2 - \mathbf{e}_3], \qquad \boldsymbol{\tau}_2 = \tfrac{1}{\sqrt{3}}[0\mathbf{e}_1 + 1\mathbf{e}_2 + 3\mathbf{e}_3]$$

$$\mathbf{n}_3 = \tfrac{1}{\sqrt{3}}[\mathbf{e}_1 - \mathbf{e}_2 - \mathbf{e}_3], \qquad \boldsymbol{\tau}_3 = -\tfrac{1}{\sqrt{3}}[4\mathbf{e}_1 + 5\mathbf{e}_2 + 6\mathbf{e}_3]$$

SOLUTION

This problem requires application of the stress–traction relation. Now

$$\frac{1}{\sqrt{3}}\begin{pmatrix}6\\9\\12\end{pmatrix} = \begin{bmatrix}T_{11} & T_{12} & T_{13}\\ T_{21} & T_{22} & T_{23}\\ T_{31} & T_{32} & T_{33}\end{bmatrix}\frac{1}{\sqrt{3}}\begin{pmatrix}1\\1\\1\end{pmatrix}$$

which gives

$$\mathrm{T}_{11} + \mathrm{T}_{12} + \mathrm{T}_{13} = 6 \tag{13.31}$$

$$\mathrm{T}_{21} + \mathrm{T}_{22} + \mathrm{T}_{23} = 9 \tag{13.32}$$

$$\mathrm{T}_{31} + \mathrm{T}_{32} + \mathrm{T}_{33} = 12 \tag{13.33}$$

Similarly,

$$\frac{1}{\sqrt{3}}\begin{pmatrix}0\\1\\2\end{pmatrix} = \begin{bmatrix}T_{11} & T_{12} & T_{13}\\ T_{21} & T_{22} & T_{23}\\ T_{31} & T_{32} & T_{33}\end{bmatrix}\frac{1}{\sqrt{3}}\begin{pmatrix}1\\1\\-1\end{pmatrix}$$

from which

$$\mathrm{T}_{11} + \mathrm{T}_{12} - \mathrm{T}_{12} = 0 \tag{13.34}$$

$$\mathrm{T}_{21} + \mathrm{T}_{22} - \mathrm{T}_{23} = 1 \tag{13.35}$$

$$\mathrm{T}_{31} + \mathrm{T}_{32} - \mathrm{T}_{33} = 2 \tag{13.36}$$

$$\frac{1}{\sqrt{3}}\begin{pmatrix}-4\\-5\\-6\end{pmatrix} = \begin{bmatrix}T_{11} & T_{12} & T_{13}\\ T_{21} & T_{22} & T_{23}\\ T_{31} & T_{32} & T_{33}\end{bmatrix}\frac{1}{\sqrt{3}}\begin{pmatrix}1\\-1\\-1\end{pmatrix}$$

and now

$$T_{11} - T_{12} - T_{13} = -4 \tag{13.37}$$

$$T_{21} - T_{22} - T_{23} = -5 \tag{13.38}$$

$$T_{31} - T_{32} - T_{33} = -6 \tag{13.39}$$

It is elementary to attain the solution, which is

$$\mathbf{T} = \begin{bmatrix} 1 & 2 & 3 \\ 2 & 3 & 4 \\ 3 & 4 & 5 \end{bmatrix}$$

13.3.2 Stress Flux

Consider two deformations $\mathbf{x}_1$ and $\mathbf{x}_2$ differing only be a rigid body motion:

$$\mathbf{x}_2 = \mathbf{V}(t)\mathbf{x}_1 + \mathbf{b}(t) \tag{13.40}$$

in which $\mathbf{V}(t)$ is an arbitrary orthonormal tensor. A tensor $\mathbf{A}(\mathbf{x})$ is objective (cf. Eringen, 1962) if

$$\mathbf{A}(\mathbf{x}_2) = \mathbf{V}\mathbf{A}(\mathbf{x}_1)\mathbf{V}^T \tag{13.41}$$

If a tensor is objective, the differences seen by observers at $\mathbf{x}_1$ and $\mathbf{x}_2$ are accounted for by the transformations relating the two associated motions.

It turns out that the matrix of time derivatives of the Cauchy stress, $\dot{\mathbf{T}}$, is not objective, while the deformation rate tensor $\mathbf{D}$ is. Accordingly, if $\boldsymbol{\xi}$ is a fourth-order tensor, a constitutive equation of the form $\dot{\mathbf{T}} = \boldsymbol{\xi}\mathbf{D}$ would be senseless. Instead, the time derivative of $\mathbf{T}$ is replaced with an *objective stress flux*, as explained below. First note that

$$\begin{aligned}
\mathbf{F}_2 &= \mathbf{V}\mathbf{F}_1 \\
\mathbf{L}_2 &= \dot{\mathbf{F}}_2\mathbf{F}_2^{-1} \\
&= \left[\dot{\mathbf{V}}\mathbf{V}^T\mathbf{V}\mathbf{F}_1 + \mathbf{V}\dot{\mathbf{F}}_1\right]\mathbf{F}_1^{-1}\mathbf{V}^T \\
&= \boldsymbol{\Omega} + \mathbf{V}\mathbf{L}_1\mathbf{V}^T, \quad \boldsymbol{\Omega} = \dot{\mathbf{V}}\mathbf{V}^T
\end{aligned} \tag{13.42}$$

The tensor $\boldsymbol{\Omega}$ is antisymmetric since $\mathrm{d}\mathbf{I}/\mathrm{d}t = \mathbf{0} = \boldsymbol{\Omega} + \boldsymbol{\Omega}^T$. Clearly, the tensors $\mathbf{F}$ and $\mathbf{L}$ are not objective.

We seek a stress flux affording the simplest conversion from deformed to undeformed coordinates. To this end we examine the time derivative of the second Piola–Kirchhoff stress.

$$\begin{aligned}
\frac{\mathrm{d}\mathbf{S}}{\mathrm{d}t} &= \frac{\mathrm{d}}{\mathrm{d}t}[J\mathbf{F}^{-T}\mathbf{T}\mathbf{F}^{-1}] \\
&= J tr(\mathbf{D})\mathbf{F}^{-T}\mathbf{T}\mathbf{F}^{-1} + J\dot{\mathbf{F}}^{-T}\mathbf{T}\mathbf{F}^{-1} + J\mathbf{F}^{-T}\dot{\mathbf{T}}\mathbf{F}^{-1} + J\mathbf{F}^{-T}\mathbf{T}\dot{\mathbf{F}}^{-1}
\end{aligned} \tag{13.43}$$

But $(\mathbf{F}^{-1}\dot{\mathbf{F}}) = \mathbf{0}$ so that $\dot{\mathbf{F}}^{-1} = -\mathbf{F}^{-1}\dot{\mathbf{F}}\mathbf{F}^{-1}$ and $\dot{\mathbf{F}}^{-T} = -\mathbf{F}^{-T}\dot{\mathbf{F}}^{T}\mathbf{F}^{-T}$. Continuing,

$$\frac{\mathrm{d}\mathbf{S}}{\mathrm{d}t} = J\mathbf{F}^T\mathring{\mathbf{T}}\mathbf{F}^{-1} \tag{13.44}$$

in which

$$\mathring{\mathbf{T}} = \dot{\mathbf{T}} + tr(\mathbf{D})\mathbf{T} - \mathbf{L}\mathbf{T} - \mathbf{T}\mathbf{L}^T \tag{13.45}$$

is known as the *Truesdell stress flux*. Under pure rotation $\mathbf{F} = \mathbf{Q}$ and $\dot{\mathbf{S}} = \mathbf{J}\mathbf{Q}\mathring{\mathbf{T}}\mathbf{Q}^T$.

To prove the objectivity of $\mathring{\mathbf{T}}$, note that

$$\begin{aligned}
\mathring{\mathbf{T}}_2 &= \dot{\mathbf{T}}_2 + \mathbf{T}_2\, tr(\mathbf{D}_2) - \mathbf{L}_2\mathbf{T}_2 - \mathbf{T}_2\mathbf{L}_2^T \\
&= \left[\mathbf{V}\mathbf{T}_1\mathbf{V}^T\right]^{\bullet} + \mathbf{V}\mathbf{T}_1\mathbf{V}^T tr(\mathbf{D}_1) - \left[\mathbf{V}\mathbf{L}_1\mathbf{V}^T + \boldsymbol{\Omega}\right]\mathbf{V}\mathbf{T}_1\mathbf{V}^T \\
&\quad - \mathbf{V}\mathbf{T}_1^T\mathbf{V}^T\left[\mathbf{V}\mathbf{L}_1{}^T\mathbf{V}^T - \boldsymbol{\Omega}\right] \\
&= \mathbf{V}\dot{\mathbf{T}}_1\mathbf{V}^T + \boldsymbol{\Omega}\mathbf{V}\mathbf{T}_1\mathbf{V}^T - \mathbf{V}\mathbf{T}_1\mathbf{V}^T\boldsymbol{\Omega} + \mathbf{V}\mathbf{T}_1\mathbf{V}^T\, tr(\mathbf{D}_1) \\
&\quad - \mathbf{V}\mathbf{L}_1\mathbf{T}_1\mathbf{V}^T - \mathbf{V}\mathbf{T}_1\mathbf{L}_1{}^T\mathbf{V}^T - \boldsymbol{\Omega}\mathbf{V}\mathbf{T}_1\mathbf{V}^T + \mathbf{V}\mathbf{T}_1\mathbf{V}^T\boldsymbol{\Omega} \\
&= \mathbf{V}\mathring{\mathbf{T}}_1\mathbf{V}^T
\end{aligned} \tag{13.46}$$

as desired.

The choice of stress flux is not unique. For example, the stress flux given by

$$\begin{aligned}
\mathring{\mathbf{T}} &= \mathring{\mathbf{T}} - \mathbf{T}\, tr(\mathbf{D}) \\
&= \dot{\mathbf{T}} - \mathbf{L}\mathbf{T} - \mathbf{T}\mathbf{L}^T
\end{aligned} \tag{13.47a}$$

is also objective, as is the widely used Jaumann stress flux

$$\overset{\triangle}{\mathbf{T}} = \dot{\mathbf{T}} + \mathbf{T}\mathbf{W} - \mathbf{W}\mathbf{T}, \quad \mathbf{W} = \tfrac{1}{2}(\mathbf{L} - \mathbf{L}^T) \tag{13.47b}$$

EXAMPLE 13.7

Relate the Jaumann stress flux to the Truesdell stress flux and to the rate of the second Piola–Kirchhoff stress

SOLUTION

Expanding the velocity gradient tensor in the Truesdell stress flux gives

$$\begin{aligned}
\mathring{\mathbf{T}} &= \dot{\mathbf{T}} + \mathbf{T}\, tr\mathbf{D} - \mathbf{L}\mathbf{T} - \mathbf{T}\mathbf{L}^T \\
&= \dot{\mathbf{T}} + \mathbf{T}\, tr\mathbf{D} - (\mathbf{D} + \mathbf{W})\mathbf{T} - \mathbf{T}(\mathbf{D} - \mathbf{W})
\end{aligned}$$

$$= \mathbf{T}\,tr\mathbf{D} - \mathbf{DT} - \mathbf{TD} + (\dot{\mathbf{T}} + \mathbf{TW} - \mathbf{WT})$$

$$= \mathbf{T}\,tr\mathbf{D} - \mathbf{DT} - \mathbf{TD} + \overset{\triangle}{\mathbf{T}}$$

Secondly,

$$J\mathbf{F}^{-1}\overset{\circ}{\mathbf{T}}\mathbf{F}^{-T} = \dot{\mathbf{S}} - J\mathbf{S}\,tr\left(\dot{\mathbf{E}}(2\mathbf{E}+\mathbf{I})^{-1}\right) - \{(2\mathbf{E}+\mathbf{I})^{-1}\dot{\mathbf{E}}\mathbf{S} + \mathbf{S}\dot{\mathbf{E}}(2\mathbf{E}+\mathbf{I})^{-1}\}$$

We will discuss this result further in conjunction with incremental stress–strain relations. It is clear that, if the Jaumann stress flux is used, conversion to undeformed coordinates introduces increments of strain *not being proportional* to increments of stress.

13.3.3 Balance of Mass, Linear Momentum, and Angular Momentum

Balance of Mass: Balance of mass requires that the total mass of an isolated body not change:

$$\frac{\mathrm{d}}{\mathrm{d}t}\int \rho\,\mathrm{d}V = 0 \tag{13.48}$$

in which $\rho(\mathbf{x},t)$ is the mass density. Since $\mathrm{d}V = J\,\mathrm{d}V_0$, it follows that $\rho J = \rho_0$.

Reynolds Transport Theorem: This useful principle is a consequence of balance of mass. Let $\mathbf{w}(\mathbf{x},t)$ denote a vector-valued function. Conversion of the volume integral from deformed to undeformed coordinates is simply achieved as $\int_V \rho\mathbf{w}(\mathbf{x},t)\,\mathrm{d}V = \int_{V_0} \rho_0\mathbf{w}(\mathbf{x},t)\,\mathrm{d}V_0$. The *Reynolds Transport Theorem* follows as

$$\frac{\mathrm{d}}{\mathrm{d}t}\int \rho\mathbf{w}(\mathbf{x},t)\,\mathrm{d}V = \int\left(\rho\left[\frac{\mathrm{d}}{\mathrm{d}t}\mathbf{w}(\mathbf{x},t)\right] + \frac{1}{J}\left(\frac{\mathrm{d}}{\mathrm{d}t}(\rho J)\right)\mathbf{w}(\mathbf{x},t)\right)\mathrm{d}V$$

$$= \int \rho\frac{\mathrm{d}}{\mathrm{d}t}\mathbf{w}(\mathbf{x},t)\,\mathrm{d}V \tag{13.49}$$

Balance of Linear Momentum: In a fixed coordinate system, balance of linear momentum requires that the total force on a body with volume V and surface S be equal to the rate of change of linear momentum. Assuming that all force is applied on the exterior surface, the equation of interest is

$$\mathbf{F} = \int \mathbf{t}\,\mathrm{d}S = \frac{\mathrm{d}}{\mathrm{d}t}\int \rho\frac{\mathrm{d}\mathbf{u}}{\mathrm{d}t}\mathrm{d}V \tag{13.50}$$

Invoking the Reynolds Transport Theorem yields

$$\int \mathbf{t}\,\mathrm{d}S = \int \rho\frac{\mathrm{d}^2\mathbf{u}}{\mathrm{d}t^2}\mathrm{d}V \tag{13.51}$$

In current coordinates, application of the divergence theorem (Equation 3.18) furnishes the equilibrium equation

$$\begin{aligned}\int \mathbf{t}\,\mathrm{d}S &= \int \mathbf{T}^T\mathbf{n}\,\mathrm{d}S \\ &= \int [\nabla^T\mathbf{T}]^T\,\mathrm{d}V \end{aligned} \tag{13.52}$$

Equation 13.51 now becomes

$$\int \left[\nabla^T\mathbf{T} - \rho\frac{\mathrm{d}^2\mathbf{u}^T}{\mathrm{d}t^2}\right]\mathrm{d}V = \mathbf{0}^T \tag{13.53}$$

Since this equation applies not only to the whole body but to arbitrary subdomains of the body, the argument of the integral in Equation 13.53 must vanish pointwise:

$$\nabla^T\mathbf{T} = \rho\frac{\mathrm{d}^2\mathbf{u}^T}{\mathrm{d}t^2} \tag{13.54}$$

To convert to undeformed coordinates the first Piola–Kirchhoff stress is invoked to furnish

$$\int \bar{\mathbf{S}}^T\mathbf{n}_0\,\mathrm{d}S_0 = \int \rho_0\frac{\mathrm{d}^2\mathbf{u}}{\mathrm{d}t^2}\mathrm{d}V_0 \tag{13.55}$$

and the divergence theorem furnishes

$$\nabla_0^T\bar{\mathbf{S}} = \rho_0\frac{\mathrm{d}^2\mathbf{u}^T}{\mathrm{d}t^2} \tag{13.56}$$

in which ∇_0 denotes the divergence operator referred to undeformed coordinates. This equation will later be the starting point in the formulation of incremental variational principles.

Balance of Angular Momentum: Assuming that only surface forces are present, relative to the origin and a fixed coordinate system the total moment of the traction is equal to the rate of change of angular momentum:

$$\int \mathbf{x}\times\mathbf{t}\,\mathrm{d}S = \frac{\mathrm{d}}{\mathrm{d}t}\int \mathbf{x}\times\rho\dot{\mathbf{x}}\,\mathrm{d}V \tag{13.57}$$

To examine this principle further it is convenient to use tensor-indicial notation. First note using the divergence theorem that

$$
\begin{aligned}
\int \mathbf{x} \times \mathbf{t}\, dS &= \int \varepsilon_{ijk} x_j t_k \, dS \\
&= \int \varepsilon_{ijk} x_j \mathrm{T}_{lk} n_l \, dS \\
&= \int \frac{\partial}{\partial x_l} (\varepsilon_{ijk} x_j \mathrm{T}_{lk}) \, dV \\
&= \int \varepsilon_{ijk} \delta_{jl} \mathrm{T}_{lk} \, dV + \int \varepsilon_{ijk} x_j \frac{\partial}{\partial x_l} \mathrm{T}_{lk} \, dV \\
&= \int \varepsilon_{ijk} \mathrm{T}_{jk} \, dV + \int \varepsilon_{ijk} x_j \frac{\partial}{\partial x_l} \mathrm{T}_{lk} \, dV
\end{aligned} \tag{13.58}
$$

and of course ε_{ijk} is the third-order permutation tensor introduced in Chapter 3.

Continuing,

$$
\begin{aligned}
\frac{d}{dt} \int \mathbf{x} \times \boldsymbol{\rho} \dot{\mathbf{x}} \, dV &= \int \mathbf{x} \times \boldsymbol{\rho} \ddot{\mathbf{x}} \, dV + \int \dot{\mathbf{x}} \times \boldsymbol{\rho} \dot{\mathbf{x}} \, dV \\
&= \int \mathbf{x} \times \boldsymbol{\rho} \ddot{\mathbf{x}} \, dV \\
&= \int \varepsilon_{ijk} x_j \rho \ddot{x}_k \, dV
\end{aligned} \tag{13.59}
$$

Balance of angular momentum may thus be restated as

$$
0 = \int \varepsilon_{ijk} \mathrm{T}_{jk} \, dV + \int \varepsilon_{ijk} x_j \left[\frac{\partial}{\partial x_l} \mathrm{T}_{lk} - \rho \ddot{x}_k \right] dV \tag{13.60}
$$

The second term vanishes by virtue of balance of linear momentum (Equation 13.54), leaving

$$
\varepsilon_{ijk} \mathrm{T}_{jk} = 0 \tag{13.61}
$$

which implies that $\mathbf{T}$ is symmetric: $\mathbf{T} = \mathbf{T}^T$ (Example 2.6). Note also that $\mathbf{S}$ is also symmetric but that $\bar{\mathrm{S}}$ is not symmetric.

13.4 PRINCIPLE OF VIRTUAL WORK UNDER LARGE DEFORMATION

The balance of linear and the balance of angular momentum lead to auxiliary variational principles that are fundamental to the finite element method. Variational methods were introduced in Chapter 4. We recall the balance of linear momentum in rectilinear coordinates as

$$
\frac{\partial}{\partial x_l} \mathrm{T}_{kl} - \rho \ddot{u}_k = 0 \tag{13.62}
$$

and $T_{lk} = T_{kl}$ by virtue of the balance of angular momentum. We have tacitly assumed that $\ddot{x} = \ddot{u}_k$ which is to say that deformed positions are referred to a coordinate system that does not translate or rotate. A variational principle is sought from

$$\int \delta u_k \left[\frac{\partial}{\partial x_l} T_{kl} - \rho \ddot{u}_k \right] dV = 0 \tag{13.63}$$

in which δu_k is an admissible (i.e., consistent with constraints) variation of u_k. We consider the spatial dependence of u_k to be subjected to variation, but not the temporal dependence. For example, if u_k can be represented, at least locally, as

$$\mathbf{u} = \mathbf{N}^T(\mathbf{x})\boldsymbol{\gamma}(t)$$

then

$$\delta u_k = [\mathbf{N}^T(\mathbf{x})]_{kl} \delta \gamma_l(t) \tag{13.64}$$

The second term in the variational equation simply remains as $-\int \delta u_k \rho \ddot{u}_k \, dV$. The first term is integrated by parts once for reasons which will be identified shortly: it becomes $\int \frac{\partial}{\partial x_l} [\delta u_k T_{lk}] \, dV - \int \frac{\partial \delta u_k}{\partial x_l} T_{lk} \, dV$. From the divergence theorem,

$$\begin{aligned} \int \frac{\partial}{\partial x_l} [\delta u_k T_{lk}] \, dV &= \int n_l [\delta u_k T_{lk}] \, dS \\ &= \int \delta u_k t_k \, dS \end{aligned} \tag{13.65}$$

which may be interpreted as the virtual work of the tractions on the exterior boundary. Next, since $\mathbf{T}$ is symmetric,

$$\begin{aligned} \int \frac{\partial \delta u_k}{\partial x_l} T_{lk} \, dV &= \int \delta \ni_{kl} T_{lk} \, dV \\ \ni_{kl} &= \frac{1}{2} \left[\frac{\partial u_k}{\partial x_l} + \frac{\partial u_l}{\partial x_k} \right] \end{aligned} \tag{13.66}$$

and we call $\ni_{kl}$ the Eulerian strain. The term $\int \delta \ni_{kl} T_{lk} \, dV$ may be called the virtual work of the stresses. Next, to evaluate $\int \delta u_k T_k \, dS$ we suppose that the exterior boundary consists of three zones: $S = S_1 + S_2 + S_3$. On S_1 the displacement u_k is prescribed, causing the integral over S_1 to vanish. On S_2 suppose that the traction is prescribed as $\bar{t}_k(\mathbf{s})$. The contribution is $\int_{S_2} \delta u_k \bar{t}_k(\mathbf{s}) \, dS$. Finally on S_3 suppose that $t_k = \bar{t}_k(\mathbf{s}) - [\mathbf{A}(\mathbf{s})]_{kl} u_l(\mathbf{s})$, furnishing $\int_{S_3} \delta u_k \bar{t}_k(\mathbf{s}) \, dS - \int_{S_3} \delta u_k [\mathbf{A}(\mathbf{s})]_{kl} u_l(\mathbf{s}) \, dS$. The various contributions are consolidated into the large deformation form of the principle of virtual work (Zienkiewicz and Taylor, 1989) in current (deformed) coordinates as

$$\int \delta \ni_{kl} \mathrm{T}_{lk}\, \mathrm{d}V + \int \delta u_k \rho \ddot{u}_k\, \mathrm{d}V$$
$$= \int_{S_2} \delta u_k \bar{\mathrm{t}}_k(\mathbf{s})\, \mathrm{d}S + \int_{S_3} \delta u_k \bar{\mathrm{t}}_k(\mathbf{s})\, \mathrm{d}S - \int_{S_3} \delta u_k [\mathbf{A}(\mathbf{s})]_{kl} u_l(\mathbf{s})\, \mathrm{d}S \tag{13.67}$$

Now consider the case in which, as in classical elasticity,

$$\mathrm{T}_{lk} = d_{lkmn} \ni_{mn}, \quad \mathbf{T} = \mathbf{D} \ni \tag{13.68}$$

in which d_{lkmn} are the entries of a fourth-order constant positive definite tangent modulus tensor $\mathbf{D}$. The first term in Equation 13.67 now becomes $\delta \int \frac{1}{2} \ni_{kl} d_{lkmn} \ni_{mn} \mathrm{d}V$, with a positive definite integrand. Achieving this outcome is the motivation behind integrating by parts once. In the language of Chapter 4, u_k is the primary variable and t_k is the secondary variable: on any boundary point either u_k or t_k is typically specified, and t_k may be specified as a function of u_k. Finally, application of the interpolation model (Equation 13.64) and cancellation of $\delta \boldsymbol{\gamma}^T$ furnishes the ordinary differential equation

$$\mathbf{M}\ddot{\boldsymbol{\gamma}} + [\mathbf{K} + \mathbf{H}]\boldsymbol{\gamma} = \mathbf{f}(t) \tag{13.69}$$

in which

$$\mathbf{M} = \int \mathbf{N}(\mathbf{x})\mathbf{N}^T(\mathbf{x})\rho\, \mathrm{d}V, \qquad \mathbf{K} = \int \mathbf{BDB}^T \mathrm{d}V$$
$$\mathbf{H} = \int_{S_3} \mathbf{N}(\mathbf{x})\mathbf{A}\mathbf{N}^T(\mathbf{x})\, \mathrm{d}V, \quad \mathbf{f}(t) = \int_{S_2+S_3} \mathbf{N}(\mathbf{s})\bar{\mathbf{t}}_k(\mathbf{s})\, \mathrm{d}S \tag{13.70}$$

and in which $VEC(\mathbf{T}) = \boldsymbol{\chi} VEC(\ni)$, $\boldsymbol{\chi} = TEN22(\mathbf{D})$. Also $VEC(\ni) = \mathbf{B}^T(\mathbf{x})\boldsymbol{\gamma}$, and $\mathbf{B}$ is derived from the strain–displacement relations, $\mathbf{M}$ is the positive definite *mass matrix*, $\mathbf{K}$ is the positive definite matrix representing the *domain contribution* to the *stiffness matrix*, $\mathbf{H}$ is the *boundary contribution* to the *stiffness matrix*, and $\mathbf{f}$ is the consistent force vector. These notions will be addressed in greater detail in subsequent chapters.

Owing to its importance in nonlinear FEA we go into considerable detail to convert the foregoing large deformation form of the Principle of Virtual Work to undeformed coordinates. First

$$\int \delta u_k \rho \ddot{u}_k\, \mathrm{d}V \rightarrow \int \delta u_k \rho_0 \ddot{u}_k\, \mathrm{d}V_0$$
$$\int_{S_2} \delta u_k \bar{t}_k(\mathbf{s})\, \mathrm{d}S + \int_{S_3} \delta u_k \bar{t}_k(\mathbf{s})\, \mathrm{d}S \rightarrow \int_{S_{20}} \delta u_k \bar{t}_k^0(\mathbf{s}_0)\, \mathrm{d}S_0 + \int_{S_{30}} \delta u_k \bar{t}_k^0(\mathbf{s}_0)\, \mathrm{d}S_0 \tag{13.71}$$

using

$$\bar{t}_k^0 = \mu \bar{t}_k \quad \mu = J\sqrt{\mathbf{n}_0^T \mathbf{C}^{-1} \mathbf{n}_0}$$

The traction relation $\mathbf{t}_0 = \mu\mathbf{t}$ is seen from the fact that $\mathbf{t}_0\,\mathrm{d}S_0 = \mathbf{t}\,\mathrm{d}S = \mathbf{t}\mu\,\mathrm{d}S_0$.

Next,

$$\int_{S_3} \delta u_k[\mathbf{A}(\mathbf{s})]_{kl}u_l(\mathbf{s})\,\mathrm{d}S \rightarrow \int_{S_3} \delta u_k[\mu\mathbf{A}(\mathbf{s})]_{kl}u_l(\mathbf{s}_0)\,\mathrm{d}S_0 \tag{13.72}$$

Some manipulation is required to convert the virtual work of the stresses. Observe that

$$\begin{aligned}
\delta\ni &= \frac{1}{2}\left[\frac{\partial\delta\mathbf{u}}{\partial\mathbf{x}} + \left(\frac{\partial\delta\mathbf{u}}{\partial\mathbf{x}}\right)^T\right] \\
&= \frac{1}{2}\left[\frac{\partial\delta\mathbf{u}}{\partial\mathbf{X}}\frac{\partial\mathbf{X}}{\partial\mathbf{x}} + \left(\frac{\partial\delta\mathbf{u}}{\partial\mathbf{X}}\frac{\partial\mathbf{X}}{\partial\mathbf{x}}\right)^T\right] \\
&= \tfrac{1}{2}[\delta\mathbf{F}\mathbf{F}^{-1} + \mathbf{F}^{-T}\delta\mathbf{F}^T] \\
&= \mathbf{F}^{-T}\left[\tfrac{1}{2}[\mathbf{F}^T\delta\mathbf{F} + \delta\mathbf{F}^T\mathbf{F}]\right]\mathbf{F}^{-1} \\
&= \mathbf{F}^{-T}\delta\mathbf{E}\mathbf{F}^{-1}
\end{aligned} \tag{13.73}$$

Third

$$\begin{aligned}
\int \delta\ni_{kl}\mathrm{T}_{lk}\,\mathrm{d}V &= \int tr(\delta\ni\mathbf{T})\,\mathrm{d}V \\
&= \int tr(\mathbf{F}^{-T}\delta\mathbf{E}\mathbf{F}^{-1}\mathbf{T})J\,\mathrm{d}V_0 \\
&= \int tr(\delta\mathbf{E}J\mathbf{F}^{-1}\mathbf{T}\mathbf{F}^{-T})\,\mathrm{d}V_0 \\
&= \int tr(\delta\mathbf{E}\mathbf{S})\,\mathrm{d}V_0 \\
&= \int \delta\mathrm{E}_{ji}\mathrm{S}_{ij}\,\mathrm{d}V_0
\end{aligned} \tag{13.74}$$

Consolidating the foregoing terms the Principle of Virtual Work in undeformed coordinates is

$$\begin{aligned}
\int \delta\mathrm{E}_{ji}\mathrm{S}_{ij}\,\mathrm{d}V_0 + \int \delta u_k\rho_0\ddot{u}_k\,\mathrm{d}V_0 &= \int_{S_{20}} \delta u_k\bar{t}_k^0(\mathbf{s}_0)\,\mathrm{d}S_0 + \int_{S_{30}} \delta u_k\bar{t}_k^0(\mathbf{s}_0)\,\mathrm{d}S_0 \\
&\quad - \int_{S_{30}} \delta u_k[\mu\mathbf{A}(\mathbf{s}_0)]_{kl}u_l(\mathbf{s}_0)\,\mathrm{d}S_0
\end{aligned} \tag{13.75}$$

EXAMPLE 13.8

Given the Cauchy stress tensor **T**, find the first and second *Piola–Kirchhoff stress* tensors if $\mathbf{x}(t) = \mathbf{Q}(t)\boldsymbol{\Delta}(t)\mathbf{X}$, in which

$$\boldsymbol{\Delta}(t) = \begin{bmatrix} 1+at & 0 & 0 \\ 0 & 1+bt & 0 \\ 0 & 0 & 1+ct \end{bmatrix}$$

Assume **Q** represents a plane rotation about the z-axis.

Solution

We know that the deformation gradient tensor is given by $\mathbf{F} = \frac{\mathrm{d}\mathbf{x}(t)}{\mathrm{d}\mathbf{X}}$. Hence,

$$\mathbf{F} = \mathbf{Q}(t)\boldsymbol{\Delta}(t) = \begin{bmatrix} \cos\theta & \sin\theta & 0 \\ -\sin\theta & \cos\theta & 0 \\ 0 & 0 & 1 \end{bmatrix} \begin{bmatrix} 1+at & 0 & 0 \\ 0 & 1+bt & 0 \\ 0 & 0 & 1+ct \end{bmatrix}$$

$$= \begin{bmatrix} (1+at)\cos\theta & (1+bt)\sin\theta & 0 \\ -(1+at)\sin\theta & (1+bt)\cos\theta & 0 \\ 0 & 0 & (1+ct) \end{bmatrix}$$

and also

$$\mathbf{F}^{-1} = \begin{bmatrix} \dfrac{\cos\theta}{(1+at)} & -\dfrac{\sin\theta}{(1+bt)} & 0 \\ \dfrac{\sin\theta}{(1+at)} & \dfrac{\cos\theta}{(1+bt)} & 0 \\ 0 & 0 & \dfrac{1}{(1+ct)} \end{bmatrix}$$

The Jacobian $J = \det \mathbf{F}$ is given by

$$J = \det(\mathbf{F}) = (1+at)(1+bt)(1+ct)$$

The first Piola–Kirchhoff stress is now given by

$$\bar{\mathbf{S}} = \mathbf{J}\mathbf{F}^{-1}\mathbf{T} = (1+at)(1+bt)(1+ct) \begin{bmatrix} \dfrac{\cos\theta}{(1+at)} & -\dfrac{\sin\theta}{(1+bt)} & 0 \\ \dfrac{\sin\theta}{(1+at)} & \dfrac{\cos\theta}{(1+bt)} & 0 \\ 0 & 0 & \dfrac{1}{(1+ct)} \end{bmatrix} \mathbf{T}$$

and the second Piola–Kirchhoff stress is obtained immediately as

$$\mathbf{S} = \mathbf{J}\mathbf{F}^{-1}\mathbf{T}\mathbf{F}^{-T}$$

$$= (1+at)(1+bt)(1+ct)\begin{bmatrix} \frac{\cos\theta}{(1+at)} & -\frac{\sin\theta}{(1+bt)} & 0 \\ \frac{\sin\theta}{(1+at)} & \frac{\cos\theta}{(1+bt)} & 0 \\ 0 & 0 & \frac{1}{(1+ct)} \end{bmatrix}$$

$$\times \mathbf{T}\begin{bmatrix} \frac{\cos\theta}{(1+at)} & \frac{\sin\theta}{(1+bt)} & 0 \\ -\frac{\sin\theta}{(1+at)} & \frac{\cos\theta}{(1+bt)} & 0 \\ 0 & 0 & \frac{1}{(1+ct)} \end{bmatrix}$$

13.5 NONLINEAR STRESS–STRAIN–TEMPERATURE RELATIONS: THE ISOTHERMAL TANGENT MODULUS TENSOR

13.5.1 Classical Elasticity

Under small deformation, the *fourth-order* tangent modulus tensor D in linear elasticity is defined implicitly by

$$\mathrm{d}\mathbf{T} = \mathsf{D}\,\mathrm{d}\mathbf{E}_L \tag{13.76}$$

in which $\mathbf{E}_L$ is the small strain tensor. In linear isotropic elasticity, the stress–strain relations are written in the Lamé form as

$$\mathbf{T} = 2\mu\mathbf{E}_L + \lambda\, tr(\mathbf{E}_L)\mathbf{I} \tag{13.77}$$

Using Kronecker product notation from Chapter 3, Equation 13.77 may be rewritten as

$$VEC(\mathbf{T}) = \left[2\mu\mathbf{I}\otimes\mathbf{I}\mathbf{E}_L + \lambda\mathbf{i}\mathbf{i}^T\right]VEC(\mathbf{E}_L) \tag{13.78}$$

from which we conclude that

$$\mathsf{D} = ITEN22\left(2\mu\mathbf{I}\otimes\mathbf{I} + \lambda\mathbf{i}\mathbf{i}^T\right) \tag{13.79}$$

13.5.2 Compressible Hyperelastic Materials

In isotropic hyperelasticity, which is descriptive of compressible rubber, the second Piola–Kirchhoff stress is taken to be derivable from a strain energy function w which depends on the principal invariants I_1, I_2, I_3 of the right Cauchy–Green strain tensor, introduced in Chapter 3.

$$\mathbf{S} = \frac{\mathrm{d}w}{\mathrm{d}\mathbf{E}} = 2\frac{\mathrm{d}w}{\mathrm{d}\mathbf{C}}, \quad \mathbf{s}^T = \frac{\mathrm{d}w}{\mathrm{d}\mathbf{e}} = 2\frac{\mathrm{d}w}{\mathrm{d}\mathbf{c}} \tag{13.80a}$$

$$\mathbf{s} = VEC(\mathbf{S}), \quad \mathbf{e} = VEC(\mathbf{E}), \quad \mathbf{c} = VEC(\mathbf{C}) \tag{13.80b}$$

Simple manipulation serves to obtain

$$\mathbf{s} = 2\phi_i \mathbf{n}_i, \quad \phi_i = \frac{\partial w}{\partial I_i}, \quad \mathbf{n}_i^T = \frac{\partial I_i}{\partial \mathbf{c}} \tag{13.81}$$

From Chapter 3, Section 3.6.8,

$$\mathbf{n}_1 = \mathbf{i}, \quad \mathbf{n}_2 = I_1 \mathbf{i} - \mathbf{c}, \quad \mathbf{n}_3 = VEC(\mathbf{C}^{-1}) I_3 \tag{13.82}$$

The tangent modulus tensor $\mathbf{D}_0$ referred to the undeformed configuration is given by

$$\mathrm{d}\mathbf{S} = \mathbf{D}_0 \, \mathrm{d}\mathbf{E}, \quad \mathrm{d}\mathbf{s} = TEN22(\mathbf{D}_0) \, \mathrm{d}\mathbf{e} \tag{13.83}$$

and furthermore

$$TEN22(\mathbf{D}_0) = 4\phi_{ij} \mathbf{n}_i \mathbf{n}_j^T + 4\phi_i \mathbf{A}_i, \quad \mathbf{A}_i = \frac{\mathrm{d}\mathbf{n}_i}{\mathrm{d}\mathbf{c}} \tag{13.84}$$

Finally, recalling Chapter 3,

$$\mathbf{A}_1 = \frac{\mathrm{d}\mathbf{n}_1}{\mathrm{d}\mathbf{c}} = \mathbf{0} \tag{13.85a}$$

$$\begin{aligned} \mathbf{A}_2 &= \frac{\mathrm{d}\mathbf{n}_2}{\mathrm{d}\mathbf{c}} \\ &= \frac{\mathrm{d}}{\mathrm{d}\mathbf{c}} [I_1 \mathbf{i} - \mathbf{c}] \\ &= \mathbf{i}\mathbf{i}^T - \mathbf{I}_9 \end{aligned} \tag{13.85b}$$

$$\begin{aligned} \mathbf{A}_3 &= \frac{\mathrm{d}}{\mathrm{d}\mathbf{c}} \left[VEC(\mathbf{C}^{-1}) I_3 \right] \\ &= \frac{\mathrm{d}}{\mathrm{d}\mathbf{c}} \left[I_2 \mathbf{i} - I_1 \mathbf{c} + VEC(\mathbf{C}^2) \right] \\ &= \mathbf{i}\mathbf{n}_2^T - \mathbf{c}\mathbf{i}^T + \mathbf{C} \oplus \mathbf{C} \\ &= I_1 \left[\mathbf{i}\mathbf{i}^T - \mathbf{I}_9 \right] - \left[\mathbf{i}\mathbf{c}^T + \mathbf{c}\mathbf{i}^T \right] + \mathbf{C} \oplus \mathbf{C} \end{aligned} \tag{13.85c}$$

13.5.3 Incompressible and Near-Incompressible Hyperelastic Materials

Polymeric materials such as natural rubber are often nearly incompressible. For some applications they may be idealized as incompressible. But for applications involving confinement, such as in the corners of seal wells, it may be necessary to accommodate the small degree of compressibility to achieve high accuracy in the stresses. Incompressibility and near-incompressibility represent *internal constraints*. The principal (Eulerian) strains are not independent, and the (Cauchy) stresses are not determined completely by the strains, instead, differences in the principal stresses are determined by differences in principal strains (Oden, 1972).

An additional field is introduced to enforce the internal constraint, and we will see that this internal field may be identified as the hydrostatic pressure (referred to the current configuration).

13.5.3.1 Incompressibility

The constraint of incompressibility is expressed by the relation $J = 1$ and note that

$$\begin{aligned} J &= \det \mathbf{F} \\ &= \sqrt{\det^2(\mathbf{F})} \\ &= \sqrt{\det(\mathbf{F})\det(\mathbf{F}^T)} \\ &= \sqrt{\det^2(\mathbf{F}^T\mathbf{F})} \\ &= \sqrt{I_3} \end{aligned} \tag{13.86}$$

and consequently the constraint of incompressibility may be restated as $I_3 = 1$.

The constraint $I_3 = 1$ may be enforced using a Lagrange multiplier (Oden, 1972) denoted here as $-p$. The multiplier depends on $\mathbf{X}$ and is in fact the additional field just mentioned. Oden (1972) proposed introducing an augmented strain energy function w' similar to

$$w' = w\left(I_1', I_2'\right) - \frac{1}{2}p(J-1), \quad I_1' = \frac{I_1}{I_3^{1/3}}, \; I_2' = \frac{I_2}{I_3^{2/3}} \tag{13.87}$$

in which w is interpreted as the conventional strain energy function but with dependence on $I_3(=1)$ removed. I_1' and I_2' are called the deviatoric invariants. For reasons to be explained in Chapter 15 which presents additional variational principles to address global constraints, this form serves to enforce incompressibility, with $\mathbf{S}$ now given by

$$\begin{aligned} \mathbf{s}^T &= \frac{\partial w'}{\partial \mathbf{e}} \\ &= 2\phi_1'\mathbf{n}_1{}' + 2\phi_2'\mathbf{n}_2' - \tfrac{1}{J}p\mathbf{n}_3 \end{aligned} \tag{13.88}$$

$$\phi_1' = \frac{\partial w}{\partial I_1'}, \quad \phi_2' = \frac{\partial w}{\partial I_2'}, \quad \mathbf{n}_1{}' = \left[\frac{\partial I_1'}{\partial \mathbf{c}}\right]^T, \quad \mathbf{n}_2' = \left[\frac{\partial I_2'}{\partial \mathbf{c}}\right]^T$$

To convert to deformed coordinates, recall that $\mathbf{S} = J\mathbf{F}^{-1}\mathbf{T}\mathbf{F}^{-T}$. An example presented below serves to derive ψ_1' and ψ_2' and also

$$\mathbf{t} = VEC(\mathbf{T}) = 2\psi_1'\mathbf{m}_1' + 2\psi_2'\mathbf{m}_2' - p\mathbf{i} \tag{13.89}$$

The example will also establish that $\mathbf{i}^T\mathbf{m}_1' = 0$ and $\mathbf{i}^T\mathbf{m}_2' = 0$. Note that $\mathbf{t}$ does not denote the traction vector in the present context. We find that $p = -tr(\mathbf{T})/3$ since

$$
\begin{aligned}
tr(\mathbf{T}) &= \mathbf{i}^T\mathbf{t} \\
&= 2\psi_1'\mathbf{i}^T\mathbf{m}_1' + 2\psi_2'\mathbf{i}^T\mathbf{m}_2' - p\mathbf{i}^T\mathbf{i} \\
&= -\mathbf{i}^\mathbf{T}\mathbf{i}p \\
&= -3p
\end{aligned} \tag{13.90}
$$

Evidently the Lagrange multiplier enforcing incompressibility is the "true" hydrostatic pressure.

Finally, the tangent modulus tensor is somewhat more complicated, because d**S** depends on d**E** and dp. We will see in a subsequent chapter that the tangent modulus tensor may be defined as **D*** using

$$
TEN22(\mathbf{D}^*) = \begin{bmatrix} \frac{\mathrm{d}\mathbf{s}}{\mathrm{d}\mathbf{e}} & \frac{\mathrm{d}\mathbf{s}}{\mathrm{d}p} \\ -\left(\frac{\mathrm{d}\mathbf{s}}{\mathrm{d}p}\right)^T & \mathbf{0} \end{bmatrix}
$$

$$
\begin{aligned}
\frac{\mathrm{d}\mathbf{s}}{\mathrm{d}\mathbf{e}} &= 4\Big[\phi_1'\mathbf{A}_1' + \phi_2'\mathbf{A}_2' + \phi_{11}'\mathbf{n}_1'(\mathbf{n}_1')^T + \phi_{12}'\mathbf{n}_1'(\mathbf{n}_2')^T \\
&\quad + \phi_{21}'\mathbf{n}_2'(\mathbf{n}_1')^T + \phi_{22}'\mathbf{n}_2'(\mathbf{n}_2')^T\Big] \\
\frac{\mathrm{d}\mathbf{s}}{\mathrm{d}p} &= -\frac{1}{J}\mathbf{n}_3
\end{aligned} \tag{13.91}
$$

Note the negative sign in the lower left entry of **D***.

EXAMPLE 13.9

In undeformed coordinates and Kronecker product (*VEC*) notation, the second Piola–Kirchhoff stress for an incompressible hyperelastic materials may be written as

$$
\mathbf{s} = \left(\frac{\partial w'}{\partial \mathbf{e}}\right)^T = 2\varphi_1'\mathbf{n}_1' + 2\varphi_2'\mathbf{n}_2' - p\frac{1}{J}\mathbf{n}_3
$$

Find the corresponding expression in deformed coordinates. Derive ψ_1' and ψ_2', in which direct transformation furnishes

$$
\mathbf{t} = 2\psi_1'\mathbf{m}_1' + 2\psi_2'\mathbf{m}_2' - p\mathbf{i}
$$

Solution

The Cauchy stress is related to the second Piola–Kirchhoff stress by the relation $\mathbf{T} = \frac{1}{J}\mathbf{F}\mathbf{S}\mathbf{F}^T$. Upon invoking Kronecker product relations

$$
\mathbf{t} = VEC(\mathbf{T}) = \tfrac{1}{J}\mathbf{F}\otimes\mathbf{F}VEC(\mathbf{S}) = \tfrac{1}{J}\mathbf{F}\otimes\mathbf{F}\mathbf{s}
$$

Applying the transformation to the stress–strain relation results in

$$\mathbf{t} = \frac{1}{J}\mathbf{F} \otimes \mathbf{F}(2\varphi_1'\mathbf{n}_1' + 2\varphi_2'\mathbf{n}_2' - pI_3\mathbf{n}_3) = \frac{2\varphi_1'}{J}\mathbf{F} \otimes \mathbf{F}\mathbf{n}_1' + \frac{2\varphi_2'}{J}\mathbf{F} \otimes \mathbf{F}\mathbf{n}_2' - p\frac{1}{J^2}\mathbf{F} \otimes \mathbf{F}\mathbf{n}_3$$

Note that $\frac{1}{J^2}\mathbf{F} \otimes \mathbf{F}\mathbf{n}_3 = \mathbf{i}$, since $IVEC\left(\frac{1}{J^2}\mathbf{F} \otimes \mathbf{F}\mathbf{n}_3\right) = \frac{I_3}{J^2}\mathbf{F}\mathbf{C}^{-1}\mathbf{F}^T = \mathbf{I}$.

Finally, let $\psi_1' = \frac{\varphi_1'}{J}$, $\psi_2' = \frac{\varphi_2'}{J}$, $\mathbf{m}_1' = \mathbf{F} \otimes \mathbf{F}\mathbf{n}_1'$, $\mathbf{m}_2' = \mathbf{F} \otimes \mathbf{F}\mathbf{n}_2'$.

EXAMPLE 13.10

Uniaxial tension

Consider the Neo-Hookean elastomer satisfying

$\mathbf{w} = \alpha[I_1 - 1]$ subject to $I_3 = 1$

We seek the relation between s_1 and E_1. The solution will be obtained twice, once by enforcing the incompressibility constraint a priori and the second by enforcing the constraint a posteriori.

A priori: We assume for the sake of brevity that $E_2 = E_3$. Now $I_3 = 1$ implies that $c_2 = 1/\sqrt{c_1}$. The strain energy function now is $w = \alpha\left[c_1 + \frac{2}{\sqrt{c_1}} - 3\right]$. The stress S_1 is now found as

$$s_1 = 2\frac{dw}{dc_1} = 2\alpha\left[1 - \frac{1}{c_1^{3/2}}\right]$$

A posteriori: Now use the augmented function

$$w' = \alpha[II_1 - 1] - \tfrac{p}{2}[I_3 - 1]$$

and

$$s_1 = 2\frac{dw'}{dc_1} = 2\alpha - p/c_1$$

$$0 = s_2 = 2\frac{dw'}{dc_2} = 2\alpha - p/c_2$$

$$0 = s_3 = 2\frac{dw'}{dc_3} = 2\alpha - p/c_3$$

$$0 = \frac{dw'}{dp} = 0 \rightarrow I_3 = 1$$

It follows that $c_2 = c_3 = 1/\sqrt{c_1}$ and $p/c_2 = 2\alpha$. We conclude that

$$\begin{aligned} s_1 &= 2\alpha - p/c_1 \\ &= 2\alpha - \frac{c_2}{c_1}p/c_2 \\ &= 2\alpha\left[1 - \frac{c_2}{c_1}\right] \\ &= 2\alpha\left[1 - \frac{1}{c_1^{3/2}}\right] \end{aligned}$$

which agrees with the relation obtained by the a priori argument.

A posteriori with deviatoric invariants: Finally consider the augmented function with deviatoric invariants:

$$w' = \alpha[I_1/I_3^{1/3} - 3] - \tfrac{p}{2}[I_3 - 1] \tag{13.92}$$

$$s_1 = 2\frac{dw'}{dc_1} = 2\alpha\left[\frac{1}{I_3^{1/3}} - \frac{1}{3}\frac{I_1}{I_3^{4/3}}\frac{I_3}{c_1}\right] - p\frac{I_3}{c_1} \tag{13.93}$$

$$0 = s_2 = 2\frac{dw'}{dc_2} = 2\alpha\left[\frac{1}{I_3^{1/3}} - \frac{1}{3}\frac{I_1}{I_3^{4/3}}\frac{I_3}{c_2}\right] - p\frac{I_3}{c_2} \tag{13.94}$$

$$0 = s_3 = 2\frac{dw'}{dc_3} = 2\alpha\left[\frac{1}{I_3^{1/3}} - \frac{1}{3}\frac{I_1}{I_3^{4/3}}\frac{I_3}{c_3}\right] - p\frac{I_3}{c_3} \tag{13.95}$$

Note that Equations 13.94 and 13.95 imply that $c_2 = c_3$. Further,

$$\frac{dw'}{dp} = 0 \rightarrow I_3 = 1$$

demonstrating that incompressibility is satisfied. Now $I_1' = I_1 = c_1 + 2c_2$, and using Equation 13.94 gives

$$\frac{2\alpha}{3}\left[\frac{c_2 - c_1}{c_2}\right] = \frac{p}{c_2}$$

Substitution into Equation 13.92 results in

$$s_1 = 2\alpha\left[1 - \frac{1}{c_1^{3/2}}\right]$$

as in the a priori case and in the first a posteriori case. The Lagrange multiplier p is again revealed as the hydrostatic pressure referred to current coordinates.

13.5.3.2 Near-Incompressibility

As will be seen in Chapter 15, in the augmented strain energy function

$$w'' = w(I_1', I_2') - p[J - 1] - \frac{p^2}{2\kappa} \tag{13.96}$$

the pressure p serves to enforce the generalized constraint

$$p = -\kappa[J - 1] \tag{13.97}$$

Here κ is the bulk modulus $(-\partial p/\partial J)_{I_1', I_2'}$, and it is assumed to be very large compared to, for example, the small strain shear modulus. The tangent modulus tensor is now

$$TEN22(\mathbf{D}^*) = \begin{bmatrix} \frac{\mathrm{ds}}{\mathrm{de}} & \frac{\mathrm{ds}}{\mathrm{d}p} \\ -\left(\frac{\mathrm{ds}}{\mathrm{d}p}\right)^T & \frac{1}{\kappa} \end{bmatrix} \tag{13.98}$$

13.5.4 Nonlinear Materials at Large Deformation

Suppose that the constitutive relations are measured at constant temperature in the current configuration and are found to obey the form

$$\overset{\circ}{\mathbf{T}} = \mathbf{D}\mathbf{D} \tag{13.99}$$

in which the fourth-order tangent modulus tensor $\mathbf{D}$ may in general be a function of stress, strain, temperature, and internal state variables to be discussed in subsequent chapters. Recall that $\overset{\circ}{\mathbf{T}}$ is the Truesdell stress flux. The form in Equation 13.99 is sensible since $\overset{\circ}{\mathbf{T}}$ and $\mathbf{D}$ are both objective, and it encompasses the rate-constitutive models typical of hypoelasticity and plasticity. Conversion to undeformed coordinates is realized by

$$\begin{aligned} \dot{\mathbf{S}} &= J\mathbf{F}^{-1}\mathbf{D}\mathbf{D}\mathbf{F}^{-T} \\ &= J\mathbf{F}^{-1}\mathbf{D}\mathbf{F}^{-T}\dot{\mathbf{E}}\mathbf{F}^{-1}\mathbf{F}^{-T} \end{aligned} \tag{13.100}$$

With $\mathbf{s} = VEC(\mathbf{S})$ and $\mathbf{e} = VEC(\mathbf{E})$,

$$\begin{aligned} \dot{\mathbf{s}} &= J\mathbf{I} \otimes \mathbf{F}^{-1} VEC(\mathbf{D}\mathbf{F}^{-T}\dot{\mathbf{E}}\mathbf{F}^{-1}\mathbf{F}^{-T}) \\ &= J\mathbf{I} \otimes \mathbf{F}^{-1}(\mathbf{F}^{-1} \otimes \mathbf{I}) TEN22(\mathbf{D}) VEC(\mathbf{F}^{-T}\dot{\mathbf{E}}\mathbf{F}^{-1}) \\ &= J\mathbf{F}^{-1} \otimes \mathbf{F}^{-1} TEN22(\mathbf{D})(\mathbf{F}^{-T} \otimes \mathbf{I})(\mathbf{I} \otimes \mathbf{F}^{-T}) VEC(\dot{\mathbf{E}}) \\ &= J\mathbf{F}^{-1} \otimes \mathbf{F}^{-1} TEN22(\mathbf{D})\mathbf{F}^{-T} \otimes \mathbf{F}^{-T}\dot{\mathbf{e}} \\ &= TEN22(\mathbf{D}_0)\dot{\mathbf{e}} \end{aligned} \tag{13.101}$$

in which

$$\mathbf{D}_0 = ITEN22(J\mathbf{F}^{-1} \otimes \mathbf{F}^{-1} TEN22(\mathbf{D})\mathbf{F}^{-T} \otimes \mathbf{F}^{-T}) \tag{13.102}$$

Since

$$\dot{\mathbf{s}} = TEN22(\mathbf{D}_0)\dot{\mathbf{e}} \tag{13.103}$$

we may regard $\mathbf{D}_0$ as the *tangent modulus tensor referred to undeformed coordinates*.

The use of the Jaumann stress flux, or any flux other than the Truesdell flux, will not lead to a counterpart of $\mathbf{D}_0$ which is proportional *only to* a tangent modulus tensor measured in the current configuration. Recall the relation derived in Example 13.5:

$$JF^{-1}\overset{\circ}{T}F^{-T} = \dot{S} - JS\,tr\left(\dot{E}(2E+I)^{-1}\right) \\ - \{(2E+I)^{-1}\dot{E}S + S\dot{E}(2E+I)^{-1}\} \tag{13.104}$$

If we assume that $\overset{\circ}{T} = \hat{\mathbf{D}}D$, some effort suffices to obtain a relation of the form

$$\dot{s} = TEN22(\mathbf{D}_{01} + \mathbf{D}_{02})\dot{e} \tag{13.105}$$

in which

$$\mathbf{D}_{01} = ITEN22\left(JF^{-1} \otimes F^{-1}TEN22(\hat{\mathbf{D}})F^{-T} \otimes F^{-T}\right)$$
$$\mathbf{D}_{02} = ITEN22\left(J\left[sVEC^T\left((2E+I)^{-1}\right)\right] + [S \otimes (2E+I)^{-1} + (2E+I)^{-1} \otimes S]\right)$$

Observe that $\mathbf{D}_{01}$ is proportional to $\hat{\mathbf{D}}$, *but that* $\mathbf{D}_{02}$ *is not.*

14 Introduction to Nonlinear FEA

14.1 INTRODUCTION

Chapters 2 through 12 addressed linear solid mechanics and heat transfer, and corresponding finite element methods for linear problems. Applications that linear methods serve to analyze include structures under mild loads, disks and rotors spinning at modest angular velocities, and heated plates. However, a large number of problems of interest are nonlinear. For one example, plasticity is a nonlinear materials theory suited for metals in metal forming, vehicle crash, and ballistics applications. In problems with high levels of heat input, mechanical properties such as the elastic modulus, and thermal properties such as the coefficient of specific heat, may be strongly temperature dependent. Rubber seals and gaskets commonly experience strains exceeding 50%. Soft biological tissues typically are modeled as rubber-like. Many problems involve variable contact, for example, meshing gear teeth. Heat conducted across electrical contacts may be strongly dependent on normal pressures. Fortunately, much of the linear finite element method can be extended to nonlinear problems, as explained in this chapter.

In the following sections, attention is confined to isothermal problems. The extension to thermomechanical problems will be presented in the subsequent chapter. In particular, we will see that the finite element equations in the nonlinear case are formally similar to the linear finite element equations *if the displacements and forces (and/or temperatures and heat fluxes) are replaced by incremental counterparts.* The *tangent stiffness matrix* is now a function of the nodal displacements (and temperatures). In addition, it will be seen to possess the extremely important property of serving as the Jacobian matrix in Newton iteration, which is an extremely attractive (arguably optimal) method for solving nonlinear algebraic equations.

14.2 TYPES OF NONLINEARITY

There are three major types of nonlinearity in thermomechanical boundary value problems: (i) *material nonlinearity*, (ii) *geometric nonlinearity*, and (iii) *boundary condition nonlinearity*. Of course these effects can and do occur in together. Nonlinearities may also be present if the formulation is referred to deformed coordinates, possibly introducing stress fluxes and convected coordinates.

Material nonlinearity may occur through nonlinear dependence of the stress on the strain and/or temperature, including strain and temperature dependence of the tangent modulus tensor. Nonlinear material behavior may also ensue from history dependence, for example, a nonlinear dependence on the total "plastic work."

Geometric nonlinearity occurs because of large deformation, especially in problems referring to undeformed coordinates. Rubber components typically exhibit large deformation and require nonlinear kinematic descriptions. In this situation, a choice needs to be made of a strain measure and of the stress *conjugate* to it.

Boundary condition nonlinearity occurs because of nonlinear mechanical and inertial supports on the boundary. An example of a nonlinear support is a rubber pad under a machine, to absorb vibrations. It may also occur if the contact area between two bodies is an unknown to be determined as part of the solution.

For finite element methods for nonlinear problems, the loads are often viewed as applied in *increments*. Then *incremental variational principles* together with interpolation models for *incremental displacements and incremental temperatures* (the primary variables) furnish algebraic (static) or ordinary algebraic-differential equations (dynamic) in terms of vector-valued incremental displacements and/or incremental temperatures. For mechanical systems a typical equation is

$$\mathbf{K}(\boldsymbol{\gamma})\Delta\boldsymbol{\gamma} + \mathbf{M}(\boldsymbol{\gamma})\Delta\ddot{\boldsymbol{\gamma}} = \Delta\mathbf{f} \tag{14.1}$$

in which $\Delta\boldsymbol{\gamma}$ is the *incremental displacement vector* (to be defined shortly), $\Delta\mathbf{f}$ is the *incremental force vector*, $\mathbf{K}(\boldsymbol{\gamma})$ is the *tangent stiffness matrix*, and $\mathbf{M}(\boldsymbol{\gamma})$ is the (*tangent*) *mass matrix*. It will be seen that this type of equation is a realization of Newton iteration method for nonlinear equations.

14.3 NEWTON ITERATION

By virtue of its property of quadratic convergence, Newton iteration is arguably an optimal method for solving nonlinear algebraic equations. It is introduced in the current chapter, and its application to finite element analysis is scrutinized. In recent years very effective methods for solving nonlinear finite element problems, known as arc length methods, have been introduced. In Chapter 18 we present an arc length method which in fact is a particular realization of Newton iteration. In it, *arc length constraints* are accommodated by expanding the solution space to a dimension greater than in the conventional finite element method (e.g., greater than the number of incremental displacement degrees of freedom).

Letting f and x denote scalars, consider the nonlinear algebraic equation $f(x;\lambda)=0$ in which λ is a parameter we will call the load intensity. Such equations are often solved numerically by a two-stage process. The first is load incrementation: the load intensity λ is increased progressively using small increments. The second is iteration: at each increment the unknown x is computed using an iteration procedure. Suppose that at the nth increment of λ an accurate solution has been achieved as x_n. We now further suppose for simplicity that x_n is "close" to the actual solution x_{n+1} at the $(n+1)$st increment of λ. Using $x_{n+1}^{(0)} = x_n$ as the starting iterate, Newton iteration provides subsequent iterates according to the scheme

$$x^{(j+1)} = x^{(j)} - \left[\frac{\mathrm{d}f}{\mathrm{d}x}\right]^{-1}_{|x^{(j)}} f(x^{(j)}) \tag{14.2}$$

Let $\Delta_{n+1,j}$ denote $x_{n+1}^{(j+1)} - x_{n+1}^{(j)}$, which is the difference between two iterates. Then, to first order in the Taylor series

$$\begin{aligned}
\Delta_{n+1,j} - \Delta_{n+1,j-1} &= -\left[\left[\frac{\mathrm{d}f}{\mathrm{d}x}\right]^{-1}_{|x^{(j)}} f(x^{(j)}) - \left[\frac{\mathrm{d}f}{\mathrm{d}x}\right]^{-1}_{|x^{(j-1)}} f(x^{(j-1)})\right] \\
&\approx -\left[\frac{\mathrm{d}f}{\mathrm{d}x}\right]^{-1}_{|x^{(j)}} \left[f(x^{(j)}) - f(x^{(j-1)})\right] \\
&\approx -\left[\frac{\mathrm{d}f}{\mathrm{d}x}\right]^{-1}_{|x^{(j)}} \left[\frac{\mathrm{d}f}{\mathrm{d}x}\right]_{|x^{(j)}} \Delta_{n+1,j-1} + 0^2 \\
&\approx -\Delta_{n+1,j-1} + 0^2
\end{aligned} \tag{14.3}$$

in which 0^2 refers to second-order terms in increments. It follows that $\Delta_{n+1,j} \approx 0^2$. For this reason Newton iteration is observed to converge *quadratically* (presumably to the correct solution if the initial iterate is "sufficiently close"). After the iteration scheme has converged to the solution, the load intensity is incremented again.

EXAMPLE 14.1

Consider $f(x) = (x-1)^2$. Newton iteration is realized as the iteration scheme

$$x^{(j+1)} = x^{(j)} - \tfrac{1}{2}(x^{(j)} - 1)$$

The solution is unity, twice. If the initial iterate $x^{(0)}$ is taken to be 1/2, the iterates become 1/2, 3/4, 7/8, 15/16, ..., converging to unity. If $x^{(0)} = 2$, the iterates are 3/2, 5/4, 9/8, 17/16, likewise converging to unity. In both cases the error is halved in each iteration.

Returning to the main development, the nonlinear finite element method under static conditions will frequently be seen to pose nonlinear algebraic equations of the form

$$\delta\mathbf{u}^T[\boldsymbol{\varphi}(\mathbf{u}) - \lambda\mathbf{v}] = 0 \tag{14.4}$$

in which $\mathbf{u}$ and $\boldsymbol{\varphi}$ are $n \times 1$ vectors, $\mathbf{v}$ is a known constant $n \times 1$ unit vector, and λ represents "load intensity." The Newton iteration scheme provides the $(j+1)$st iterate for $\mathbf{u}_{n+1}$ as

$$\Delta_{n+1,j} = -\left[\left[\frac{\partial\boldsymbol{\varphi}}{\partial\mathbf{u}}\right]_{|\mathbf{u}_{n+1}^{(j)}}\right]^{-1}\left[\boldsymbol{\varphi}(\mathbf{u}_{n+1}^{(j)}) - \lambda_{j+1}\mathbf{v}\right], \quad \mathbf{u}_{n+1}^{(j+1)} - \mathbf{u}_{n+1}^{(j)} = \Delta_{n+1,j} \tag{14.5}$$

in which, for example, the initial iterate is $\mathbf{u}_n^{(0)}$. The use of an explicit matrix inverse is avoided by solving the linear system

$$\left[\frac{\partial \boldsymbol{\varphi}}{\partial \mathbf{u}}\right]\bigg|_{\mathbf{u}_{n+1}^{(j)}} \Delta_{n+1,j} = \boldsymbol{\varphi}(\mathbf{u}_{n+1}^{(j)}) - \mathbf{v}_{j+1}, \quad \mathbf{u}_{n+1}^{(j+1)} = \mathbf{u}_{n+1}^{(j)} + \Delta_{n+1,j} \tag{14.6}$$

Note that $\left[\frac{\partial \boldsymbol{\varphi}}{\partial \mathbf{u}}\right]\bigg|_{\mathbf{u}_{n+1}^{(j)}}$ in Newton iteration is a *Jacobian matrix.*

14.4 COMBINED INCREMENTAL AND ITERATIVE METHODS: A SIMPLE EXAMPLE

As illustrated in Figure 14.1, consider a one-dimensional rod of nonlinear material under small deformation, in which the elastic modulus is a linear function of strain: $\mathsf{E} = \mathsf{E}_0(1 + \alpha\varepsilon)$, $\varepsilon = \mathrm{E}_{11}$. The cross-sectional area A_0 and the length L_0 are constants, on the understanding that the deformation is expressed in terms of the undeformed configuration. Suppose that under static loading the equilibrium equation is

$$\frac{\mathsf{E}_0 A_0}{L_0}\left(1 + \alpha\frac{\gamma}{L_0}\right)\gamma = P \tag{14.7}$$

The load is applied in increments $\Delta_j P = P_{j+1} - P_j$ and the load after the $(n-1)$st increment is applied is denoted as P_n. Suppose next that the solution γ_n has been computed accurately at P_n. We now consider the actions necessary to determine the solution γ_{n+1} at load P_{n+1}. This is done by first computing the value of $\Delta_n\gamma = \gamma_{n+1} - \gamma_n$. Subtracting Equation 14.2 at the nth increment from the same equation but at the nth increment, the *incremental equilibrium equation* is now

$$\frac{\mathsf{E}_0 A_0}{L_0}\left(1 + \alpha\frac{2\gamma_n + \Delta_n\gamma}{L}\right)\Delta_n\gamma = \Delta_n P \tag{14.8}$$

Equation 14.8 is quadratic in the increment $\Delta_n\gamma$-in fact geometric nonlinearity generally leads to a quadratic function of increments. The error of neglecting the quadratic term may be small if the load increment is sufficiently small. However, we will retain the nonlinear term for the sake of accuracy and of illustrating the use of iterative procedures. In particular, the foregoing equation may be written in the form

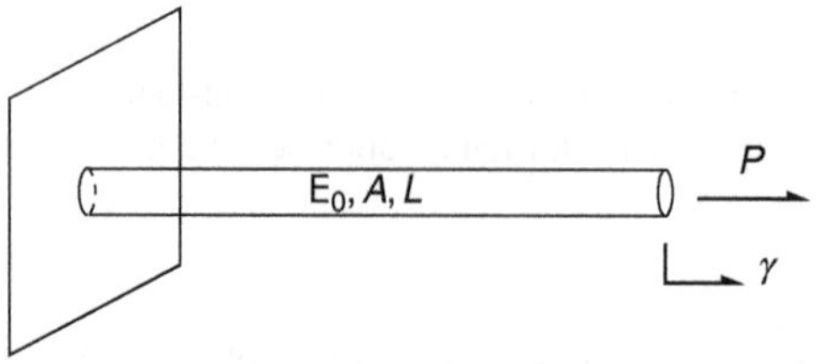

FIGURE 14.1 Stretching of a rod of nonlinear material.

$$g(x) = \beta x + \zeta x^2 - \eta = 0 \tag{14.9}$$

in which $x = \Delta_n \gamma$, $\eta = \frac{L_0 \Delta_n P}{\mathsf{E}_0 A_0}$, $\beta = 1 + 2\alpha\gamma_n / L_0$, and $\zeta = \frac{\alpha}{L_0}$. Newton iteration furnishes the *quadratically convergent* iteration scheme for the νth iterate, namely:

$$x^{\nu+1} = x^{\nu} - \frac{[\beta x^{\nu} + \zeta x^{\nu 2} - \eta]}{\beta + 2\zeta x^{\nu}}, \quad x^0 = \eta/\beta \tag{14.10}$$

As an example we use the values $\mathsf{E} = 10^7$ psi, $A = 1$ in^2, $L = 10$ in, $\alpha = 2$, $\gamma_n = 1$, and $\Delta_n P = 3$. These values imply significant nonlinearity and 10% strain. A simple program written in double precision produces the following values, which demonstrate a convergence in two iterations.

Iterate	Value
0	0
1	0.21428571e-5
2	0.21428565e-5
3	0.21428565e-5
4	0.21428565e-5

EXAMPLE 14.2

The efficiency of this scheme is addressed as follows. Consider $\zeta = 1$, $\beta = 12$, and $\eta = 1$. The correct solution is $x = 0.0828$. Starting with the initial value $x = 0$, the first two iterates are, approximately, $\frac{1}{12} = 0.833$, and $\frac{1}{12}(1 - \frac{1}{144}) = 0.0828$.

14.5 FINITE STRETCHING OF A RUBBER ROD UNDER GRAVITY

14.5.1 Model Problem

Figure 14.2 shows a rubber rod element under gravity—it is assumed to attain finite strain and to experience uniaxial tension. The figure exhibits the undeformed configuration, with an element occupying the interval (X_e, X_{e+1}), with X denoting the

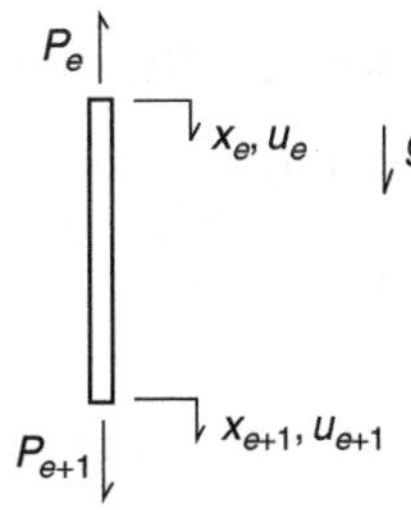

FIGURE 14.2 Rubber rod element under gravity.

downward direction. The element's length is l_e, its cross-sectional area is A_e, and its mass density is ρ. It is composed of rubber and is stretched axially by the loads P_e and P_{e+1}. Prior to stretching, a given material particle is located at X. After deformation it is located at $x(X)$, and the displacement $u(X)$ is given by $u(X) = x(X) - X$.

14.5.2 Nonlinear Strain–Displacement Relations

The element in Figure 14.2 is assumed to be short enough that a satisfactory approximation for the displacement $u(X)$ is provided by the linear interpolation model

$$\begin{aligned} u(X) &= u_e + (X - X_e)(u_{e+1} - u_e)/l_e \\ &= \mathbf{N}^T\mathbf{a}_e, \quad \mathbf{N}^T = \frac{1}{l_e}\{x_e - x \quad x - x_e\} \end{aligned} \tag{14.11}$$

in which $\mathbf{a}_e^T = \{u_e \quad u_{e+1}\}$. Now $u(X_e) = u_e$ and $u(X_{e+1}) = u_{e+1}$ are viewed as the unknowns to be determined using the finite element method. The Lagrangian strain $\mathrm{E} = \mathrm{E}_{xx}$ is approximated in the element as

$$\begin{aligned} \mathrm{E} &= \frac{\partial u}{\partial x} + \frac{1}{2}\left(\frac{\partial u}{\partial x}\right)^2 \\ &= \frac{u_{e+1} - u_e}{l_e} + \frac{1}{2}\left(\frac{u_{e+1} - u_e}{l_e}\right)^2 \\ &= \frac{1}{l_e}\{-1 \quad 1\}\mathbf{a}_e + \frac{1}{2}\mathbf{a}_e^T\frac{1}{l_e^2}\begin{bmatrix} 1 & -1 \\ -1 & 1 \end{bmatrix}\mathbf{a}_e \end{aligned} \tag{14.12}$$

In alternative notation Equation 14.12 is written as

$$\mathrm{dE}_{xx} = \mathbf{B}^T\mathrm{d}\mathbf{a}_e, \quad \mathbf{B} = \mathbf{B}_L + \mathbf{B}_{NL}\mathbf{a}_e \tag{14.13}$$

$$\mathbf{B}_L = \frac{1}{l_e}\begin{Bmatrix} -1 \\ 1 \end{Bmatrix}, \quad \mathbf{B}_{NL} = \frac{1}{l_e^2}\begin{bmatrix} 1 & -1 \\ -1 & 1 \end{bmatrix}$$

The vector $\mathbf{B}_L$ and the matrix $\mathbf{B}_{NL}$ are called the *linear and nonlinear strain–displacement matrices.*

14.5.3 Stress and Tangent Modulus Relations

The Neo–Hookean strain energy density function w was previously encountered in Chapter 13. It accommodates incompressibility and is stated in terms of the eigenvalues c_1, c_2, and c_3 of $\mathbf{C} = 2\mathbf{E} + \mathbf{I}$ as follows:

$$w = \frac{D}{6}(c_1 + c_2 + c_3 - 3), \quad \text{subject to } c_1c_2c_3 - 1 = 0 \tag{14.14}$$

in which D is the (small strain) elastic modulus. We saw in Chapter 13 that $c_2 = c_3$.

We first enforce the incompressibility constraint a priori by using the substitution

$$c_2 = c_3 = \frac{1}{\sqrt{c_1}} \tag{14.15}$$

After elementary manipulations and using $c_1 = 2\mathrm{E}_{xx} + 1$, the strain energy function emerges as

$$w = \frac{D}{6}\left[2\mathrm{E}_{xx} + 1 + 2/(2\mathrm{E}_{xx} + 1)^{-1/2}\right] \tag{14.16}$$

The (second Piola–Kirchhoff) stress S_{xx}, defined in general in Chapter 13, is now obtained as

$$\begin{aligned} \mathrm{S}_{xx} &= \frac{\partial w}{\partial \mathrm{E}_{xx}} \\ &= \frac{D}{3}\left[1 - (2\mathrm{E}_{xx} + 1)^{-3/2}\right] \end{aligned} \tag{14.17}$$

For small strains $(2\mathrm{E}_{xx} + 1)^{-3/2} \approx 1 - 3\mathrm{E}_{xx}$ in which case $\mathrm{S}_{xx} \approx D\mathrm{E}_{xx}$.

The tangent modulus D_T is also required:

$$\begin{aligned} D_T &= \frac{\partial \mathrm{S}_{xx}}{\partial \mathrm{E}_{xx}} \\ &= D(2\mathrm{E}_{xx} + 1)^{-5/2} \end{aligned} \tag{14.18}$$

If the strain E_{xx} is small compared to unity, D_T reduces to D.

We next satisfy the incompressibility constraint a posteriori, which we will see is how the constraint is generally satisfied in finite element analysis. An augmented strain energy function w^* is introduced by

$$w^* = \frac{D}{6}[c_1 + c_2 + c_3 - 3] - \frac{p}{2}(c_1c_2c_3 - 1) \tag{14.19}$$

in which the Lagrange multiplier p can be shown to be the (true) hydrostatic pressure, as was shown in a similar situation in Chapter 13. The augmented energy is stationary with respect to p as well as c_1, c_2, and c_3, from which it follows that $c_1c_2c_3 - 1 = 0$ (incompressibility).

The second Piola–Kirchhoff stresses satisfy

$$\begin{aligned} \mathrm{S}_{xx} &= \frac{\partial w^*}{\partial \mathrm{E}_{xx}} = 2\frac{\partial w^*}{\partial c_1} = \frac{D}{3} - \frac{p}{2}\frac{c_1c_2c_3}{c_1} \\ \mathrm{S}_{yy} &= \frac{\partial w^*}{\partial \mathrm{E}_{yy}} = 2\frac{\partial w^*}{\partial c_2} = \frac{D}{3} - \frac{p}{2}\frac{c_1c_2c_3}{c_2} = 0 \\ \mathrm{S}_{zz} &= \frac{\partial w^*}{\partial \mathrm{E}_{zz}} = 2\frac{\partial w^*}{\partial c_3} = \frac{D}{3} - \frac{p}{2}\frac{c_1c_2c_3}{c_3} = 0 \end{aligned} \tag{14.20}$$

The second and third equations in Equation 14.20 are immediately seen to imply $c_2 = c_3$, as previously stated. Enforcement of the stationarity condition ($\delta w^* = 0$) for $p(c_1c_2c_3 - 1 = 0)$ with $S_{yy} = S_{zz} = 0 = 0$ now furnishes that $p = \frac{2}{3}D/\sqrt{c_1}$. It follows that $S_{xx} = D\left[1 - c_1^{-3/2}\right]/3$, in agreement with Equation 14.7.

14.5.4 Incremental Equilibrium Relation

The Principle of Virtual Work states the condition for static equilibrium of the rod as

$$\boldsymbol{\phi}(\mathbf{a}_e; \mathbf{P}^e) = \int_{X_e}^{X_{e+1}} \mathbf{B}S_{xx}A_e\,dX - \mathbf{P}^e - \left[\rho g \int_{X_e}^{X_{e+1}} \mathbf{N}^T A_e\,dX\right]\mathbf{a}_e = \mathbf{0}, \quad \mathbf{P}^e = \begin{pmatrix} P_e^e \\ P_{e+1}^e \end{pmatrix} \tag{14.21}$$

in which the third left-hand term represents the weight of the element, while P_e represents the forces from the adjacent elements. In an incremental formulation, we replace the loads and displacements by their differential forms. In particular

$$\begin{aligned} \mathbf{0} = d\boldsymbol{\phi} &= \int_{X_e}^{X_{e+1}} \mathbf{B}\,dS_{xx}\,A_e\,dX + \int_{X_e}^{X_{e+1}} d\mathbf{B}\,SA_e\,dX - d\mathbf{P}^e \\ &= \mathbf{K}_e\,d\mathbf{a}_e - d\mathbf{P}^e \\ &= \left\{\mathbf{K}_{1e} + \mathbf{K}_{2e} + \mathbf{K}_{3e} + \mathbf{K}_{4e}\right\}d\mathbf{a}_e - d\mathbf{P}^e \end{aligned} \tag{14.22}$$

in which

$$\begin{aligned} \mathbf{K}_{1e} &= A_e \int_{X_e}^{X_{e+1}} \mathbf{B}_L D_T \mathbf{B}_L^T\,dX \\ &= \frac{\overline{D}_T A_e}{l_e}\begin{bmatrix} 1 & -1 \\ -1 & 1 \end{bmatrix} \end{aligned} \tag{14.23a}$$

$$\begin{aligned} \mathbf{K}_{2e} &= A_e \int_{X_e}^{X_{e+1}} D_T(\mathbf{B}_L\mathbf{a}_e^T\mathbf{B}_{NL} + \mathbf{B}_{NL}\mathbf{a}_e\mathbf{B}_L^T)\,dX \\ &= \frac{\overline{D}_T A_e}{l_e} 2\frac{[u_{e+1} - u_e]}{l_e}\begin{bmatrix} 1 & -1 \\ -1 & 1 \end{bmatrix} \end{aligned} \tag{14.23b}$$

$$\begin{aligned} \mathbf{K}_{3e} &= A_e \int_{X_e}^{X_{e+1}} \mathbf{B}_{NL}\mathbf{a}_e\mathbf{a}_e^T\mathbf{B}_{NL}^T\,dX \\ &= \frac{\overline{D}_T A_e}{l_e}\left(\frac{u_{e+1} - u_e}{l_e}\right)^2\begin{bmatrix} 1 & -1 \\ -1 & 1 \end{bmatrix} \end{aligned} \tag{14.23c}$$

$$\mathbf{K}_{4e} = A_e \int_{X_e}^{X_{e+1}} \mathbf{B}_L S \, \mathrm{d}X$$
$$= \frac{\overline{\mathrm{S}}_{xx} A_e}{l_e} \begin{bmatrix} 1 & -1 \\ -1 & 1 \end{bmatrix} \tag{14.23d}$$

and

$$\overline{D}_T = \frac{1}{l_e} \int_{X_e}^{X_{e+1}} D_T \, \mathrm{d}X, \quad \overline{\mathrm{S}}_{xx} = \frac{1}{l_e} \int_{X_e}^{X_{e+1}} \mathrm{S}_{xx} \, \mathrm{d}X \tag{14.23e}$$

Combining Equations 14.23a through 14.23e produces a simple relation for the *tangent stiffness matrix*:

$$\mathbf{K}_e = \kappa \begin{bmatrix} 1 & -1 \\ -1 & 1 \end{bmatrix}, \quad \kappa = \frac{\overline{D}_T A_e}{l_e} \left\{ 1 + 2\frac{u_{e+1} - u_e}{l_e} + \left(\frac{u_{e+1} - u_e}{l_e} \right)^2 \right\} + \frac{\overline{\mathrm{S}}_{xx} A_e}{l_e} \tag{14.24}$$

Suppose $(u_{e+1} - u_e)/l_e$ is small compared to unity, with the consequence that E_{xx} is also small compared to unity. It follows in this case that $\overline{D}_T \approx D$, $\overline{\mathrm{S}}_{xx} \approx \mathrm{S}_{xx}$. In consequence, $\mathbf{K}_e$ reduces to the stiffness matrix for a rod element of linearly elastic material experiencing small strain:

$$\mathbf{K}_e = \frac{\mathsf{E}A_e}{l_e} \begin{bmatrix} 1 & -1 \\ -1 & 1 \end{bmatrix} \tag{14.25}$$

Returning to the nonlinear problem, several special cases are now used to illuminate additional aspects of finite element modeling.

14.5.5 Single Element Built-in at One End

Figure 14.3 depicts a single element model of a rod which is built in at $X=0$: $X_e = X_0 = 0$. At the opposite end, $X_{e+1} = X_1 = 1$, it is submitted to the load P. The displacement at $X=0$ is subject to the constraint $u(0) = u_0 = 0$, so that Equation 14.24 becomes

$$\kappa(u_1) \begin{bmatrix} 1 & -1 \\ -1 & 1 \end{bmatrix} \begin{pmatrix} 0 \\ \mathrm{d}u_1 \end{pmatrix} = \begin{pmatrix} -\mathrm{d}P_r \\ \mathrm{d}P \end{pmatrix} \tag{14.26}$$

in which $\mathrm{d}P_r$ is an incremental reaction force which is considered unknown (of course from equilibrium $\mathrm{d}P_r = \mathrm{d}P$). Enforcing the constraint at the top of the rod causes Equation 14.26 to "condense" to

$$k \, \mathrm{d}u_1 = \mathrm{d}P \tag{14.27}$$

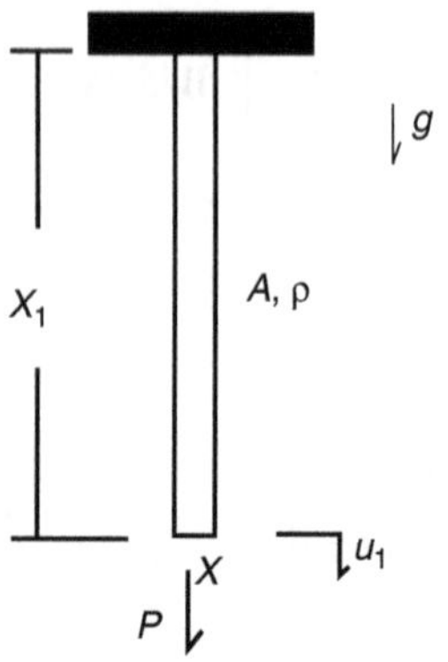

FIGURE 14.3 Rubber rod element under gravity: built in at top.

Also, the current shape function degenerates to the expression $\mathbf{N} \rightarrow N = X/X_1$.

The (Lagrangian) strain is given by

$$\mathrm{E}(u_1) = \frac{u_1}{X_1} + \frac{1}{2}\left(\frac{u_1}{X_1}\right)^2 \tag{14.28}$$

and the strain–displacement matrix reduces to

$$\mathbf{B} \rightarrow B = \frac{1}{X_1} + \frac{u_1}{X_1^2} \tag{14.29}$$

The (second Piola–Kirchhoff) stress S_{xx} is now obtained as a function of u_1:

$$\mathrm{S}_{xx}(u_1) = \frac{1}{3}D\left[1 - \left[2\left\{\frac{u_1}{X_1} + \frac{1}{2}\left(\frac{u_1}{X_1}\right)^2\right\} - 1\right]^{-\frac{3}{2}}\right] \tag{14.30}$$

14.5.6 On Numerical Solution by Newton Iteration

We see to solve Equation 14.21 numerically by Newton iteration, sometimes called "load balancing" in this context, as explained below. The Principle of Virtual Work implies that

$$\begin{aligned}\phi(u_1) &= \int_0^{X_1} \mathbf{B}\mathrm{S}_{xx}A\,\mathrm{d}X - P - \left[\rho g \int_0^{X_1} \mathbf{N}^T A\,\mathrm{d}X\right]\mathbf{a}_e \\ &= \left[\frac{1}{l_e} + \frac{u_1}{l_e^2}\right]AX_1\left(\frac{1}{3}D\left[1 - \left[2\left\{\frac{u_1}{X_1} + \frac{1}{2}\left(\frac{u_1}{X_1}\right)^2\right\} - 1\right]^{-\frac{3}{2}}\right]\right) - P - \rho g A\frac{X_1}{2}u_1 \\ &= 0\end{aligned} \tag{14.31}$$

in which the residual function $\phi(u_1)$ has been introduced. Consider an iteration process in which the jth iterate u_1^j has been determined. Newton iteration determines the next iterate u_1^{j+1} using

$$\begin{aligned} \kappa(u_1^j)\Delta_j u_1 &= -\varphi(u_1^j) \\ u_1^{j+1} &= u_1^j + \Delta_j u_1 \end{aligned} \tag{14.32}$$

in which $\kappa(u_1^j) = \partial\phi(u_1^j)/\partial u_1^j$ has already been given in Equation 14.24 and is now seen to be the *Jacobian matrix* of Newton iteration. Convergence to the correct value u_1 is usually rapid provided that the initial iterate u_1^0 is sufficiently close to u_1.

A satisfactory starting iterate for the current load step can be obtained by extrapolating from the previous solution values. As illustration, suppose the solution, say, is known at the load $P_j = j\Delta P$, in which ΔP is a load increment. The load is now incremented to produce $P_{j+1} = (j+1)\Delta P$, and the a starting iterate is needed to converge to the solution $u_{1,j+1}$ at this load using Newton iteration. It is frequently satisfactory to use $u_{1,j+1}^0 = u_{1,j}$. Another possibility is to use $u_{1,j+1}^0 = 2u_{1,j} - u_{1,j-1}$, or more generally "line search" using two or more previous solution values.

14.5.7 Assembled Stiffness Matrix for a Two-Element Model of the Rubber Rod under Gravity

A two-element model of the rubber rod under gravity is shown in Figure 14.4. *Assemblage* procedures are now illustrated for combining the element equilibrium relations to obtain the global equilibrium relation holding for an assemblage of elements. We will see that they are simple extensions of assembly procedures in linear finite element analysis. For elements "e" and "$e+1$" the element equilibrium relations are expanded as

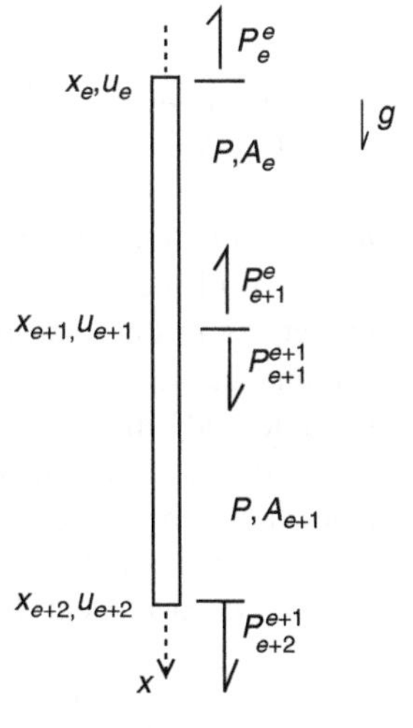

FIGURE 14.4 Two rubber rod elements under gravity.

$$k_{11}^e\, du_e + k_{12}^e\, du_{e+1} = dP_e^e \tag{14.33a}$$

$$k_{21}^e\, du_e + k_{22}^e\, du_{e+1} = dP_{e+1}^e \tag{14.33b}$$

$$k_{11}^{e+1}\, du_{e+1} + k_{12}^{e+1}\, du_{e+2} = dP_{e+1}^{e+1} \tag{14.33c}$$

$$k_{21}^{e+1}\, du_{e+1} + k_{22}^{e+1}\, du_{e+2} = dP_{e+2}^{e+1} \tag{14.33d}$$

The superscript indicates the element index, and the subscript indicates the node index. If no external force is applied at x_{e+1}, the interelement incremental force balance is expressed as

$$dP_{e+1}^e + dP_{e+1}^{e+1} = 0 \tag{14.34}$$

Adding Equations 14.33b and 14.33c furnishes

$$k_{21}^e\, du_e + \left(k_{22}^e + k_{11}^{e+1}\right) du_{e+1} + k_{12}^{e+1}\, du_{e+2} = 0 \tag{14.35}$$

Equations 14.33a, 14.33d, and 14.35 are expressed in matrix form as

$$\begin{bmatrix} k_{11}^e & k_{12}^e & 0 \\ k_{21}^e & k_{22}^e + k_{11}^{e+1} & k_{12}^{e+1} \\ 0 & k_{21}^{e+1} & k_{22}^{e+1} \end{bmatrix} \begin{pmatrix} du_e \\ du_{e+1} \\ du_{e+2} \end{pmatrix} = \begin{pmatrix} dP_e^e \\ 0 \\ dP_{e+2}^{e+1} \end{pmatrix} \tag{14.36}$$

Equation 14.36 illustrates that the (incremental) global stiffness matrix is formed by "overlaying" $\mathbf{K}_e$ and $\mathbf{K}_{e+1}$, with the entries added in the intersection (the 2–2 element).

Two identical elements under gravity, under equal end loads: If $\mathbf{K}_e = \mathbf{K}_{e+1}$ and $dP_e^e = -dP_{e+1}^{e+1} = -dP$, overlaying the element matrices leads to the global (two element) relation

$$\kappa \begin{bmatrix} 1 & -1 & 0 \\ -1 & 2 & -1 \\ 0 & -1 & 1 \end{bmatrix} \begin{pmatrix} du_1 \\ du_2 \\ du_3 \end{pmatrix} = \begin{pmatrix} -dP \\ 0 \\ dP \end{pmatrix} \tag{14.37}$$

Note that Equation 14.37 has *no solution* since the global stiffness matrix has no inverse: the second row is the negative of the sum of the first and last rows. This suggests that, owing to numerical errors in the load increments, the condition for static equilibrium is not satisfied numerically, and therefore the body is predicted to accelerate indefinitely (undergoes rigid body motion). However, we also know that this configuration, under equal and opposite incremental loads, is symmetric, implying a constraint $du_2 = 0$. This constraint permits "condensation," that is, reducing Equation 14.37 to a system with two unknowns by eliminating rows and columns associated with the middle incremental displacement. In particular,

$$\kappa \begin{bmatrix} 1 & -1 \\ -1 & 1 \end{bmatrix} \begin{pmatrix} du_1 \\ du_3 \end{pmatrix} = \begin{pmatrix} -dP \\ dP \end{pmatrix} \tag{14.38}$$

The condensed matrix is now proportional to the identity matrix, and the system has a solution. More generally, stiffness matrices may easily be singular or nearly singular (with a large condition number) unless constraints are used or introduced to suppress "rigid body modes."

14.6 NEWTON ITERATION NEAR A CRITICAL POINT

Now the Jacobian matrix is likely to be ill-conditioned whenever the tangent modulus tensor is nearly singular, which is said to occur at *critical points*. Unfortunately, since elastomers or metals experiencing plasticity are typically very compliant at large deformation, the analyst attempting to perform computations into large strain ranges may well encounter a critical point. Buckling also represents an example of a critical point. Since the tangent stiffness matrix is the Jacobian matrix for Newton iteration, convergence problems arise when the tangent stiffness becomes singular or ill-conditioned, and special methods are invoked to continue computations near and past critical points.

Several ways to continue computation in the vicinity of critical points are now listed.

1. Increase stiffness, such as by introducing additional constraints if available
2. Reduce load step sizes, and reform the stiffness matrix after each iteration
3. Switch to displacement control rather than load control
4. Using an arc length method, of which a particular method is described in Chapter 17

We now illustrate such a convergence problem using an equation of the form $\psi(x)=0$. As previously stated, Newton iteration seeks a solution through an iterative process given by

$$x_{j+1} = x_j - \left[\frac{\mathrm{d}\psi(x_j)}{\mathrm{d}x}\right]^{-1} \psi(x_j) \tag{14.39}$$

Clearly, recalling Equation14.3 using sequential iterates, with $\Delta_{n+1,j} = x_{n+1}^{(j+1)} - x_{n+1}^{(j)}$,

$$\Delta_{n+1,j} - \Delta_{n+1,j-1} \approx -\left[\frac{\mathrm{d}\psi}{\mathrm{d}x}\right]^{-1}_{|x^{(j)}} \left[\frac{\mathrm{d}\psi}{\mathrm{d}x}\right]_{|x^{(j)}} \Delta_{n+1,j-1} + 0^2 \tag{14.40}$$

Strictly speaking the argument of $\left[\frac{\mathrm{d}\psi}{\mathrm{d}x}\right]$ is not $x^{(j)}$. Rather, thanks to the Mean Value Theorem, the argument is an unknown value x^* between x_n and the converged value of x_{n+1}. Clearly, $\left[\frac{\mathrm{d}\psi}{\mathrm{d}x}\right]_{x^*}$ does not cancel the ill-conditioned matrix $\left[\frac{\mathrm{d}\psi}{\mathrm{d}x}\right]^{-1}_{|x^{(j)}}$, in which case the iterates will grow rapidly.

EXAMPLE 14.3

The performance of the Newton iteration procedure near a critical point is now illustrated using a simple example showing a slope (analogous to the tangent modulus tensor) which is asymptotically approaching zero. Consider

$$\phi(x) = \frac{2}{\pi}\tan^{-1}(x) - y = 0 \tag{14.41}$$

Equation 14.41 is depicted in Figure 14.5. The goal is to find the solution of x as y is incremented in the range (0,1). Clearly x approaches infinity as y approaches unity, so that the goal is to generate the $x(y)$ relationship accurately as close as possible to $y = 1$. The curve will appear as in Figure 14.5. When y is plotted against x, the curve asymptotically approaches unity, with the slope (stiffness) approaching zero. As y is incremented by small amounts just below unity, the differences in x due to the increment are large, so that the solution value at a nth load step, if used as the initial iterate, is not close to the solution at the $(n+1)$st load step.

Suppose that y is incremented such that the nth value of y is $y_n = n\Delta y$. To obtain the solution of the $(n+1)$th step, the Newton iteration procedure generates the $(\nu+1)$st iterate from the νth iterate as follows:

$$\begin{aligned} x_{n+1}^{(\nu+1)} &= x_{n+1}^{(\nu)} - \left[\frac{\mathrm{d}\phi\left(x_{n+1}^{(\nu)}\right)}{\mathrm{d}x}\right]^{-1} \phi\left(x_{n+1}^{(\nu)}\right) \\ &= x_{n+1}^{(\nu)} - \left(1 + \left(x_{n+1}^{(\nu)}\right)^2\right)\tan^{-1}\left(x_{n+1}^{(\nu)}\right) - \frac{\pi}{2} y_{n+1} \end{aligned} \tag{14.42}$$

in which $x_{n+1}^{(\nu)}$ is the νth iterate for the solution x_{n+1}. Of course a starting iterate x_{n+1}^{0} is needed. An attractive candidate is x_n. However, this may not be good enough when convergence difficulties appear.

Newton iteration for Equation 14.42 was implemented for this example in a simple double precision program. Numerical results are shown in Table 14.1. The increment Δy was reduced significantly as the asymptote was approached. Despite this reduction the iteration count increased noticeably near $y = 0.9999$, illustrating the onset of convergence difficulties.

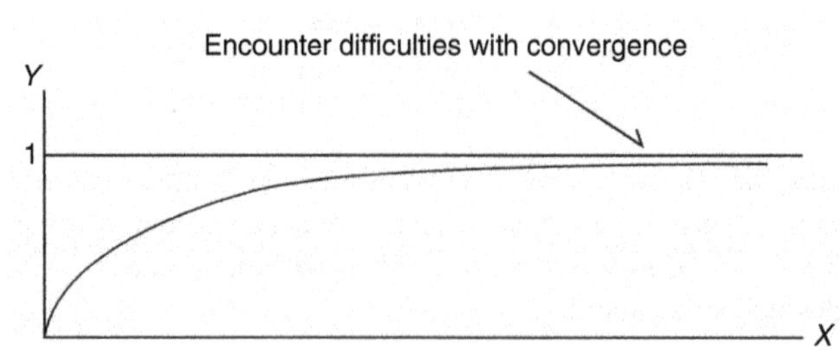

FIGURE 14.5 Illustration of the inverse tangent function.

TABLE 14.1
Convergence of Newton Iteration for $y = 2\tan^{-1}(x)/\pi$

Iteration Count	x	y	Δy
2	12.7062083	0.95	0.0001
3	63.6568343	0.99	0.0001
3	70.7309329	0.991	0.0001
3	90.9422071	0.993	0.0001
3	127.321711	0.995	0.0001
4	212.206062	0.997	0.0001
4	636.628644	0.999	0.0001
4	909.475631	0.9993	0.0001
5	1591.60798	0.9996	0.0001
6	6367.13818	0.9999	0.0001

14.7 INTRODUCTION TO THE ARC LENGTH METHOD

In, say, uniaxial tension, the Principle of Virtual Work leads to an equation of the form $\phi(u) = f$, $\frac{d\phi(u)}{du} = k(u)$ (incremental stiffness matrix). The arc length method to be discussed here is helpful if $k(u)$ is near zero (i.e., near a critical point).

In nonlinear problems we increment the loads using a load intensity parameter λ,

$$f = \lambda f_0, \quad 0 \leq \lambda \leq 1 \tag{14.43}$$

in which f_0 is the final load to be attained. The equilibrium equation thereby becomes

$$\phi(u) = \lambda f_0 \tag{14.44}$$

Suppose the solution has been attained at $\lambda_n f_0$, denoted by u_n, and the solution is sought for the subsequent load increment, satisfying

$$\phi(u_{n+1}) = \lambda_{n+1} f_0 \tag{14.45}$$

Introducing the incremental relations $\phi(u_{n+1}) - \phi(u_n) \approx \frac{1}{2}\left(\frac{d\phi}{du}(u_{n+1}) + \frac{d\phi}{du}(u_n)\right)\Delta_n u = \frac{1}{2}(k(u_{n+1}) + k(u_n))\Delta_n u$, an approximate solution may be obtained from the equation

$$\frac{1}{2}(k(u_{n+1}) + k(u_n))\Delta_n u = (\Delta_n \lambda) f_0 \tag{14.46}$$

in which

$$\Delta_n \lambda = \lambda_{n+1} - \lambda_n, \quad \Delta_n u = u_{n+1} - u_n$$

An equation in this form is said to represent *load control* since the solution is obtained at prescribed load levels.

If we make the additional approximation that

$$\frac{1}{2}(k(u_{n+1}) + k(u_n)) \approx k(u_n) \tag{14.47}$$

the equation is "explicit" and its solution requires no iteration. Otherwise, the method is implicit and iteration is required.

In problems in buckling, plasticity and hyperelasticity, it is quite possible for $\frac{1}{2}(k(u_{n+1}) + k(u_n))$ to be singular or nearly singular in which case there are severe difficulties achieving convergence and accuracy.

An alternative is "displacement control," in which increments of **u** are specified, and the conjugate force increments become reaction forces to be computed as part of the solution. We next discuss a combination of load control and displacement control, known as *arc length control.*

The arc length method is helpful in situations in which load control fails owing to a singular tangent stiffness matrix. It augments the Principle of Virtual Work with an additional equation imposing a constraint on a function of the magnitudes of the load and displacement increments. In the uniaxial tension case being considered here, an example of the resulting equations is

$$\begin{aligned} \hat{\phi}(u,\lambda) &= \phi(u) - \lambda f_0 = 0 \quad \text{equilibrium} \\ \psi(u,\lambda) &= \alpha(u - u_n) + \beta(\lambda - \lambda_n) - \Delta S = 0 \quad \text{arc length constraint} \end{aligned} \tag{14.48}$$

Here ΔS is a small positive number representing the "length" of the increment in the u–λ plane, and the coefficients α and β will be chosen subsequently to promote convergence.

Newton iteration is now applied to this pair of equations

$$\begin{aligned} \begin{Bmatrix} u_{n+1}^{(v+1)} \\ \lambda_{n+1}^{(v+1)} \end{Bmatrix} &= \begin{Bmatrix} u_{n+1}^{(v)} \\ \lambda_{n+1}^{(v)} \end{Bmatrix} - \begin{bmatrix} \dfrac{\partial \hat{\phi}\left(u_{n+1}^{(v)}, \lambda_{n+1}^{(v)}\right)}{\partial u} & \dfrac{\partial \hat{\phi}\left(u_{n+1}^{(v)}, \lambda_{n+1}^{(v)}\right)}{\partial \lambda} \\ \dfrac{\partial \psi\left(u_{n+1}^{(v)}, \lambda_{n+1}^{(v)}\right)}{\partial u} & \dfrac{\partial \psi\left(u_{n+1}^{(v)}, \lambda_{n+1}^{(v)}\right)}{\partial \lambda} \end{bmatrix}^{-1} \begin{Bmatrix} \hat{\phi}\left(u_{n+1}^{(v)}, \lambda_{n+1}^{(v)}\right) \\ \psi\left(u_{n+1}^{(v)}, \lambda_{n+1}^{(v)}\right) \end{Bmatrix} \\ &= \begin{Bmatrix} u_{n+1}^{(v)} \\ \lambda_{n+1}^{(v)} \end{Bmatrix} - \begin{bmatrix} k\left(u_{n+1}^{(v)}\right) & -f_0 \\ \alpha & \beta \end{bmatrix}^{-1} \begin{Bmatrix} \hat{\phi}\left(u_{n+1}^{(v)}, \lambda_{n+1}^{(v)}\right) \\ \psi\left(u_{n+1}^{(v)}, \lambda_{n+1}^{(v)}\right) \end{Bmatrix} \end{aligned} \tag{14.49}$$

and, consequently, Newton iteration with the arc length constraint is given by the iteration scheme

$$\begin{bmatrix} k\left(u_{n+1}^{(v)}\right) & -f_0 \\ \alpha & \beta \end{bmatrix} \begin{Bmatrix} u_{n+1}^{(v+1)} - u_{n+1}^{(v)} \\ \lambda_{n+1}^{(v+1)} - \lambda_{n+1}^{(v)} \end{Bmatrix} = -\begin{Bmatrix} \hat{\phi}\left(u_{n+1}^{(v)}, \lambda_{n+1}^{(v)}\right) \\ \psi\left(u_{n+1}^{(v)}, \lambda_{n+1}^{(v)}\right) \end{Bmatrix} \tag{14.50}$$

Suppose $k\left(u_{n+1}^{(v)}\right) = 0$, and choose $\alpha = f_0$. The eigenvalues $\mu_{1,2}$ of the matrix **A** are now given by the equation

$$\mu_j^2 - \beta\mu_j + f_0^2 = 0 \tag{14.51}$$

with the two roots

$$\mu_{1,2} = \frac{1}{2}\left(\beta \pm \sqrt{\beta^2 - 4f_0^2}\right) \tag{14.52}$$

The choice of β which maximizes the smaller eigenvalue is $\beta = 2f_0$. This choice renders the two eigenvalues equal and positive with the value $\mu_1 = \mu_1 = f_0$, and thereby circumvents the singularity arising if $k\left(u_{n+1}^{(v)}\right) = 0$.

15 Incremental Principle of Virtual Work

15.1 INCREMENTAL KINEMATICS

The Principle of Virtual Work is extended to nonlinear solid mechanics by restating the kinematic and equilibrium relations in incremental form and by applying variational principles using the *displacement increments* as the primary variables. Issues such as thermal effects and incompressibility are addressed in Chapters 16 and 17 dealing with thermohyperelasticity and thermoinelasticity.

Recall that the displacement vector $\mathbf{u}(\mathbf{X})$ has been assumed to admit a satisfactory approximation at the element level in the form $\mathbf{u}(\mathbf{X}) = \boldsymbol{\varphi}^T(\mathbf{X})\ \boldsymbol{\Phi}\boldsymbol{\gamma}(t)$. Also recall that the deformation gradient tensor is given by $\mathbf{F} = \frac{\partial \mathbf{u}}{\partial \mathbf{X}}$. Suppose that the body under study is subjected to a load vector $\mathbf{P}$, which is applied incrementally via load increments $\Delta_j\mathbf{P} = \mathbf{P}_{j+1} - \mathbf{P}_j$. The load after the $(n-1)$th load step is denoted as $\mathbf{P}_n$. The solution $\mathbf{P}_n$ is considered to have been determined computationally, and the solution of the current displacement increments is sought. Let $\Delta_n\mathbf{u} = \mathbf{u}_{n+1} - \mathbf{u}_n$, implying the *incremental interpolation model*

$$\Delta_n\mathbf{u} = \boldsymbol{\varphi}^T(\mathbf{X})\boldsymbol{\Phi}\Delta_n\boldsymbol{\gamma} \tag{15.1}$$

By suitably arranging the derivatives of $\Delta_n\mathbf{u}$ with respect to $\mathbf{X}$, a matrix $\mathsf{M}(\mathbf{X})$ may easily be determined for which $VEC(\Delta_n\mathbf{F}) = \mathsf{M}(\mathbf{X})\Delta_n\boldsymbol{\gamma}$, to be demonstrated in Example 15.1.

We next consider the Lagrangian strain tensor $\mathbf{E}(\mathbf{X}) = \frac{1}{2}(\mathbf{F}^T\mathbf{F} - \mathbf{I})$. Using Kronecker product algebra introduced in Chapter 3, we readily find that to first order in increments

$$\begin{aligned}
\Delta_n\mathbf{e} &= VEC(\Delta_n\mathbf{E}) \\
&= VEC\left(\tfrac{1}{2}\left[\mathbf{F}^T\Delta_n\mathbf{F} + \Delta_n\mathbf{F}^T\mathbf{F}\right]\right) \\
&= \tfrac{1}{2}\left[\mathbf{I}\otimes\mathbf{F}^T + \mathbf{F}\otimes\mathbf{I}\mathbf{U}\right]VEC(\Delta_n\mathbf{F}) \\
&= \mathbf{G}^T\Delta_n\boldsymbol{\gamma}, \quad \mathbf{G}^T = \tfrac{1}{2}\left[\mathbf{I}\otimes\mathbf{F}^T + \mathbf{F}\otimes\mathbf{I}\mathbf{U}\right]\mathsf{M}(\mathbf{X})\Delta_n\boldsymbol{\gamma}
\end{aligned} \tag{15.2}$$

The form in Equation 15.2 shows the convenience of Kronecker product notation. Namely it enables moving the incremental displacement vector to the end of the expression, with the consequence that it can be placed outside the domain integrals we will encounter subsequently.

Alternatively, for the current configuration an alternative strain measure is the Eulerian strain $\exists = \frac{1}{2}\left(\mathbf{I} - \mathbf{F}^{-T}\mathbf{F}^{-1}\right)$, which refers to deformed coordinates. Note that, since $\Delta_n(\mathbf{F}\mathbf{F}^{-1}) = \mathbf{0}$, $\Delta_n\mathbf{F}^{-1} = -\mathbf{F}^{-1}\Delta_n\mathbf{F}\mathbf{F}^{-1}$. Similarly, $\Delta_n\mathbf{F}^{-T} = -\mathbf{F}^{-T}\Delta_n\mathbf{F}^T\mathbf{F}^{-T}$. Simple manipulation furnishes the incremental strain–displacement relation for $\exists$ as

$$VEC(\Delta_n\exists) = \tfrac{1}{2}\left[\mathbf{F}^{-T}\mathbf{F}^{-1} \otimes \mathbf{F}^{-T}\mathbf{U} + \mathbf{F}^{-1} \otimes \mathbf{F}^{-T}\mathbf{F}^{-1}\right]\mathbf{M}\Delta_n\gamma \tag{15.3}$$

There also are geometric changes for which an incremental representation is useful. For example, since the Jacobian $J = det(\mathbf{F})$ satisfies $\mathrm{d}J = J\,tr(\mathbf{F}^{-1}\mathrm{d}\mathbf{F})$, we obtain the approximate formula

$$\begin{aligned}\Delta_n J &= J\,tr(\mathbf{F}^{-1}\Delta_n\mathbf{F}) \\ &= J VEC^T(\mathbf{F}^{-T})VEC(\Delta_n\mathbf{F}) \\ &= VEC^T(\mathbf{F}^{-T})J\mathbf{M}\Delta_n\boldsymbol{\gamma}\end{aligned} \tag{15.4}$$

Also of interest are the incremental counterparts for the directed area of surface element, the surface normal vector, and the surface area of an element. Chapter 13 reported the relations

$$\frac{\mathrm{d}[\mathbf{n}\,\mathrm{d}S]}{\mathrm{d}t} = \left[tr(\mathbf{D})\mathbf{I} - \mathbf{L}^T\right]\mathbf{n}\,\mathrm{d}S \tag{15.5a}$$

$$\frac{\mathrm{d}\mathbf{n}}{\mathrm{d}t} = \left[\left(\mathbf{n}^T\mathbf{D}\mathbf{n}\right)\mathbf{I} - \mathbf{L}^T\right]\mathbf{n} \tag{15.5b}$$

$$\frac{\mathrm{d}}{\mathrm{d}t}\mathrm{d}S = \left[tr(\mathbf{D}) - \mathbf{n}^T\mathbf{D}\mathbf{n}\right]\mathrm{d}S \tag{15.5c}$$

and we directly obtain the incremental counterparts as

$$\Delta_n[\mathbf{n}\,\mathrm{d}S] = \mathrm{d}S\left[\mathbf{n}VEC^T\left(\mathbf{F}^{-1}\right) - \mathbf{n}^T \otimes \mathbf{F}^{-T}\mathbf{U}\right]\mathbf{M}\Delta_n\boldsymbol{\gamma} \tag{15.6a}$$

$$\Delta_n\mathbf{n} = \left[\mathbf{n}\left[\left(\mathbf{n}^T\mathbf{F}^{-T}\right) \otimes \mathbf{n}^T\right] - \mathbf{n}^T \otimes \mathbf{F}^{-T}\mathbf{U}\right]\mathbf{M}\Delta_n\boldsymbol{\gamma} \tag{15.6b}$$

$$\Delta_n\,\mathrm{d}S = \mathrm{d}S\left[\mathbf{n}VEC^T\left(\mathbf{F}^{-T}\right) - \mathbf{n}\left(\mathbf{n}^T\mathbf{F}^{-T}\right) \otimes \mathbf{n}^T\right]\mathbf{M}\Delta_n\boldsymbol{\gamma} \tag{15.6c}$$

EXAMPLE 15.1

The derivation of Equation 15.6b from Equation 15.5b is now presented. First note that $\Delta_n\mathbf{n} \approx \frac{\mathrm{d}\mathbf{n}}{\mathrm{d}t}\Delta_n t$. Next

$$\begin{aligned}(\mathbf{n}^T\mathbf{D}\mathbf{n})\mathbf{n}\Delta_n t &\approx \mathbf{n}\left(\mathbf{n}^T\Delta_n\mathbf{F}\mathbf{F}^{-1}\mathbf{n}\right)\\ &= \mathbf{n}\left(\mathbf{n}^T\otimes\mathbf{n}^T\right)VEC\left(\Delta_n\mathbf{F}\mathbf{F}^{-1}\right)\\ &= \mathbf{n}\left(\mathbf{n}^T\otimes\mathbf{n}^T\right)\mathbf{F}^{-T}\otimes\mathbf{I}VEC(\Delta_n\mathbf{F})\\ &= \mathbf{n}\left(\mathbf{n}^T\mathbf{F}^{-T}\right)\otimes\mathbf{n}^T\mathbf{M}\Delta_n\boldsymbol{\gamma}\end{aligned}$$

Finally,

$$\begin{aligned}-\mathbf{L}^T\mathbf{n}\Delta_n t &= -\mathbf{F}^{-T}\Delta\mathbf{F}^T\mathbf{n}\\ &= -VEC\left(\mathbf{F}^{-T}\Delta_n\mathbf{F}^T\mathbf{n}\right)\\ &= -\mathbf{n}^T\otimes\mathbf{F}^{-T}\mathbf{U}\mathbf{M}\Delta_n\boldsymbol{\gamma}\end{aligned}$$

Several classical texts (e.g., Zienkiewicz and Taylor, 1991) on FEA have decomposed the strain–displacement relations into a linear portion and a residual nonlinear portion. In the current notation this is written as $\mathbf{G}=\mathbf{B}_L+\mathbf{B}_{NL}$. If we write $\mathbf{F}=\mathbf{I}+\mathbf{F}_u$ in which $\mathbf{F}_u=\frac{\partial\mathbf{u}}{\partial\mathbf{X}}$, it is immediate from Equation 15.3 that

$$\begin{aligned}\delta\mathbf{e} &= \left[\mathbf{B}_L^T+\mathbf{B}_{NL}^T(\boldsymbol{\gamma})\right]\delta\boldsymbol{\gamma}\\ \mathbf{B}_L^T &= \tfrac{1}{2}(\mathbf{I}\otimes\mathbf{I}+\mathbf{I}\otimes\mathbf{I}\mathbf{U})\mathbf{M}(\mathrm{X})\\ \mathbf{B}_{NL}^T &= \tfrac{1}{2}(\mathbf{I}\otimes\mathbf{F}_u^T+\mathbf{F}_u\otimes\mathbf{I}\mathbf{U})\mathbf{M}(\mathrm{X})\end{aligned} \tag{15.7}$$

EXAMPLE 15.2

Assuming linear interpolation models for u, v in a plane triangular membrane element with vertices (0,0), (1,0), (0,1), obtain the matrices **M**, **G**, $\mathbf{B}_L$, and $\mathbf{B}_{NL}$.

Solution

The interpolation model is

$$\begin{Bmatrix}u\\ v\end{Bmatrix}=\begin{bmatrix}1 & x & y & 0 & 0 & 0\\ 0 & 0 & 0 & 1 & x & y\end{bmatrix}\begin{bmatrix}\boldsymbol{\Phi} & \mathbf{0}\\ \mathbf{0} & \boldsymbol{\Phi}\end{bmatrix}\begin{Bmatrix}u_1\\ u_2\\ u_3\\ v_1\\ v_2\\ v_3\end{Bmatrix}$$

$$\boldsymbol{\Phi}=\begin{bmatrix}1 & 0 & 0\\ 1 & 1 & 0\\ 1 & 0 & 1\end{bmatrix}^{-1}=\begin{bmatrix}1 & 0 & 0\\ -1 & 1 & 0\\ -1 & 0 & 1\end{bmatrix}$$

We first formulate the deformation gradient tensor **F**.

Let $\boldsymbol{\gamma}_1 = \begin{Bmatrix} u_1 \\ u_2 \\ u_3 \end{Bmatrix}$ and $\boldsymbol{\gamma}_2 = \begin{Bmatrix} v_1 \\ v_2 \\ v_3 \end{Bmatrix}$. Now $\mathbf{F} = \mathbf{I} + \frac{\partial \mathbf{u}}{\partial \mathbf{x}}$, and $\frac{\partial \mathbf{u}}{\partial \mathbf{x}} = \begin{bmatrix} \frac{\partial u}{\partial x} & \frac{\partial u}{\partial y} \\ \frac{\partial v}{\partial x} & \frac{\partial v}{\partial y} \end{bmatrix}$. After some effort we may write $\frac{\partial \mathbf{u}}{\partial \mathbf{x}} = \mathbf{I} + \mathfrak{B}\boldsymbol{\Upsilon}\boldsymbol{\Gamma}$ and $\Delta_n \mathbf{F} = \mathfrak{B}\boldsymbol{\Upsilon}\Delta_n\boldsymbol{\Gamma}$ in which

$$\mathfrak{B} = \begin{bmatrix} \boldsymbol{\beta}_{11}^T & \boldsymbol{\beta}_{12}^T & \mathbf{0}^T & \mathbf{0}^T \\ \mathbf{0}^T & \mathbf{0}^T & \boldsymbol{\beta}_{21}^T & \boldsymbol{\beta}_{22}^T \end{bmatrix}, \quad \boldsymbol{\Upsilon} = \begin{bmatrix} \boldsymbol{\Phi} & \mathbf{0} & \mathbf{0} & \mathbf{0} \\ \mathbf{0} & \boldsymbol{\Phi} & \mathbf{0} & \mathbf{0} \\ \mathbf{0} & \mathbf{0} & \boldsymbol{\Phi} & \mathbf{0} \\ \mathbf{0} & \mathbf{0} & \mathbf{0} & \boldsymbol{\Phi} \end{bmatrix}, \quad \boldsymbol{\Gamma} = \begin{bmatrix} \boldsymbol{\gamma}_1 & \mathbf{0} \\ \mathbf{0} & \boldsymbol{\gamma}_1 \\ \boldsymbol{\gamma}_2 & \mathbf{0} \\ \mathbf{0} & \boldsymbol{\gamma}_2 \end{bmatrix}$$

with

$$\boldsymbol{\beta}_{11}^T = \{0\ 1\ 0\}, \quad \boldsymbol{\beta}_{12}^T = \{0\ 0\ 1\}, \quad \boldsymbol{\beta}_{21}^T = \{0\ 1\ 0\}, \quad \boldsymbol{\beta}_{22}^T = \{0\ 0\ 1\}$$

Now

$$VEC(\Delta_n\boldsymbol{\Gamma}) = (\mathbf{I} \otimes \mathfrak{B}\boldsymbol{\Upsilon})\begin{Bmatrix} \Delta_n\boldsymbol{\gamma}_1 \\ \Delta_n\boldsymbol{\gamma}_2 \end{Bmatrix}$$

$$VEC(\Delta_n\mathbf{F}) = (\mathbf{I} \otimes \mathfrak{B}\boldsymbol{\Upsilon})VEC(\Delta_n\boldsymbol{\Gamma}) = (\mathbf{I} \otimes \mathfrak{B}\boldsymbol{\Upsilon})\mathfrak{I}\Delta_n\boldsymbol{\gamma}$$

in which

$$\mathfrak{I} = \begin{bmatrix} \mathbf{I} & \mathbf{0} \\ \mathbf{0} & \mathbf{0} \\ \mathbf{0} & \mathbf{I} \\ \mathbf{0} & \mathbf{0} \\ \mathbf{0} & \mathbf{0} \\ \mathbf{I} & \mathbf{0} \\ \mathbf{0} & \mathbf{0} \\ \mathbf{0} & \mathbf{I} \end{bmatrix}$$

We conclude that $\mathbf{M} = (\mathbf{I} \otimes \mathfrak{B}\boldsymbol{\Upsilon})\mathfrak{I}$.

Next the strain–displacement matrix $\mathbf{G}$ is given by

$$\mathbf{G}^T = \tfrac{1}{2}\left[\mathbf{I} \otimes (\mathbf{I} + \mathfrak{B}\boldsymbol{\Upsilon}\boldsymbol{\Gamma})^T + [(\mathbf{I} + \mathfrak{B}\boldsymbol{\Upsilon}\boldsymbol{\Gamma}) \otimes \mathbf{I}]\mathbf{U}\right](\mathbf{I} \otimes \mathfrak{B}\boldsymbol{\Upsilon})\mathfrak{I}$$

Finally, the linear and nonlinear portions of the strain–displacement matrices are obtained as

$$\mathbf{B}_L^T = \tfrac{1}{2}[\mathbf{I} \otimes \mathbf{I} + (\mathbf{I} \otimes \mathbf{I})\mathbf{U}](\mathbf{I} \otimes \mathfrak{B}\boldsymbol{\Upsilon})\mathfrak{I}$$

$$\mathbf{B}_{NL}^T = \tfrac{1}{2}\left[\mathbf{I} \otimes (\mathfrak{B}\boldsymbol{\Upsilon}\boldsymbol{\Gamma})^T + [(\mathfrak{B}\boldsymbol{\Upsilon}\boldsymbol{\Gamma}) \otimes \mathbf{I}]\mathbf{U}\right](\mathbf{I} \otimes \mathfrak{B}\boldsymbol{\Upsilon})\mathfrak{I}$$

EXAMPLE 15.3

Repeat Example 15.2 with linear interpolation models for u, v, and w in a tetrahedral element with vertices (0,0,0), (1,0,0), (0,1,0), (0,0,1).

Solution

The interpolation model is

$$u(x, y, z) = \{1 \quad x \quad y \quad z\}\mathbf{\Phi}\begin{Bmatrix} u_1 \\ u_2 \\ u_3 \\ u_4 \end{Bmatrix}, \quad v(x, y, z) = \{1 \quad x \quad y \quad z\}\mathbf{\Phi}\begin{Bmatrix} v_1 \\ v_2 \\ v_3 \\ v_4 \end{Bmatrix}$$

$$w(x, y, z) = \{1 \quad x \quad y \quad z\}\mathbf{\Phi}\begin{Bmatrix} w_1 \\ w_2 \\ w_3 \\ w_4 \end{Bmatrix}$$

$$\mathbf{\Phi} = \begin{bmatrix} 1 & 0 & 0 & 0 \\ 1 & 1 & 0 & 0 \\ 1 & 0 & 1 & 0 \\ 1 & 0 & 0 & 1 \end{bmatrix}^{-1} = \begin{bmatrix} 1 & 0 & 0 & 0 \\ -1 & 1 & 0 & 0 \\ -1 & 0 & 1 & 0 \\ -1 & 0 & 0 & 1 \end{bmatrix}$$

Following Example 15.2, we again formulate the deformation gradient tensor.

$$\mathbf{F} = \mathbf{I} + \frac{\partial \mathbf{u}}{\partial \mathbf{x}}, \quad \Delta_n\mathbf{F} = \frac{\partial \Delta_n \mathbf{u}}{\partial \mathbf{x}}, \quad \frac{\partial \mathbf{u}}{\partial \mathbf{x}} = \begin{bmatrix} \frac{\partial u}{\partial x} & \frac{\partial u}{\partial y} & \frac{\partial u}{\partial z} \\ \frac{\partial v}{\partial x} & \frac{\partial v}{\partial y} & \frac{\partial v}{\partial z} \\ \frac{\partial w}{\partial x} & \frac{\partial w}{\partial y} & \frac{\partial w}{\partial z} \end{bmatrix} = \mathfrak{B}\mathbf{\Upsilon}\mathbf{\Gamma}$$

$$\mathbf{\Gamma} = \begin{bmatrix} \boldsymbol{\gamma}_1 & 0 & \mathbf{0} \\ \mathbf{0} & \boldsymbol{\gamma}_1 & \mathbf{0} \\ \mathbf{0} & \mathbf{0} & \boldsymbol{\gamma}_1 \\ \boldsymbol{\gamma}_2 & \mathbf{0} & \mathbf{0} \\ \mathbf{0} & \boldsymbol{\gamma}_2 & \mathbf{0} \\ \mathbf{0} & \mathbf{0} & \boldsymbol{\gamma}_2 \\ \boldsymbol{\gamma}_3 & \mathbf{0} & \mathbf{0} \\ \mathbf{0} & \boldsymbol{\gamma}_3 & \mathbf{0} \\ \mathbf{0} & \mathbf{0} & \boldsymbol{\gamma}_3 \end{bmatrix}$$

Now

$$\mathfrak{B} = \begin{bmatrix} \boldsymbol{\beta}_{11}^T & \boldsymbol{\beta}_{12}^T & \boldsymbol{\beta}_{13}^T & \mathbf{0}^T & \mathbf{0}^T & \mathbf{0}^T & \mathbf{0}^T & \mathbf{0}^T & \mathbf{0}^T \\ \mathbf{0}^T & \mathbf{0}^T & \mathbf{0}^T & \boldsymbol{\beta}_{21}^T & \boldsymbol{\beta}_{22}^T & \boldsymbol{\beta}_{23}^T & \mathbf{0}^T & \mathbf{0}^T & \mathbf{0}^T \\ \mathbf{0}^T & \mathbf{0}^T & \mathbf{0}^T & \mathbf{0}^T & \mathbf{0}^T & \mathbf{0}^T & \boldsymbol{\beta}_{21}^T & \boldsymbol{\beta}_{32}^T & \boldsymbol{\beta}_{33}^T \end{bmatrix}$$

$$\mathbf{\Upsilon} = \begin{bmatrix} \mathbf{\Phi} & \mathbf{0} & \mathbf{0} & \mathbf{0} & \mathbf{0} & \mathbf{0} & \mathbf{0} & \mathbf{0} & \mathbf{0} \\ \mathbf{0} & \mathbf{\Phi} & \mathbf{0} & \mathbf{0} & \mathbf{0} & \mathbf{0} & \mathbf{0} & \mathbf{0} & \mathbf{0} \\ \mathbf{0} & \mathbf{0} & \mathbf{\Phi} & \mathbf{0} & \mathbf{0} & \mathbf{0} & \mathbf{0} & \mathbf{0} & \mathbf{0} \\ \mathbf{0} & \mathbf{0} & \mathbf{0} & \mathbf{\Phi} & \mathbf{0} & \mathbf{0} & \mathbf{0} & \mathbf{0} & \mathbf{0} \\ \mathbf{0} & \mathbf{0} & \mathbf{0} & \mathbf{0} & \mathbf{\Phi} & \mathbf{0} & \mathbf{0} & \mathbf{0} & \mathbf{0} \\ \mathbf{0} & \mathbf{0} & \mathbf{0} & \mathbf{0} & \mathbf{0} & \mathbf{\Phi} & \mathbf{0} & \mathbf{0} & \mathbf{0} \\ \mathbf{0} & \mathbf{0} & \mathbf{0} & \mathbf{0} & \mathbf{0} & \mathbf{0} & \mathbf{\Phi} & \mathbf{0} & \mathbf{0} \\ \mathbf{0} & \mathbf{0} & \mathbf{0} & \mathbf{0} & \mathbf{0} & \mathbf{0} & \mathbf{0} & \mathbf{\Phi} & \mathbf{0} \\ \mathbf{0} & \mathbf{0} & \mathbf{0} & \mathbf{0} & \mathbf{0} & \mathbf{0} & \mathbf{0} & \mathbf{0} & \mathbf{\Phi} \end{bmatrix}$$

$$\boldsymbol{\beta}_{11}^T = \boldsymbol{\beta}_{21}^T = \boldsymbol{\beta}_{31}^T = \{0 \quad 1 \quad 0 \quad 0\}, \quad \boldsymbol{\beta}_{12}^T = \boldsymbol{\beta}_{22}^T = \boldsymbol{\beta}_{32}^T = \{0 \quad 0 \quad 1 \quad 0\}$$

$$\boldsymbol{\beta}_{13}^T = \boldsymbol{\beta}_{23}^T = \boldsymbol{\beta}_{33}^T = \{0 \quad 0 \quad 1 \quad 0\}$$

Finally, the incremental displacements are

$$\Delta_n \boldsymbol{\Gamma} = \mathfrak{I} \begin{Bmatrix} \Delta_n \gamma_1 \\ \Delta_n \gamma_2 \\ \Delta_n \gamma_3 \end{Bmatrix} \text{ in which } \mathfrak{I} \text{ is given below.}$$

As in Example 15.2,

$$\begin{aligned}
\mathbf{M} &= (\mathbf{I} \otimes \mathfrak{B}\mathbf{Y})\mathfrak{I} \\
\mathbf{G}^T &= \tfrac{1}{2}\left[\mathbf{I} \otimes (\mathbf{I} + \mathfrak{B}\mathbf{Y}\boldsymbol{\Gamma})^T + \left[(\mathbf{I} + \mathfrak{B}\mathbf{Y}\boldsymbol{\Gamma}) \otimes \mathbf{I}\right]\mathbf{U}\right](\mathbf{I} \otimes \mathfrak{B}\mathbf{Y})\mathfrak{I} \\
\mathbf{B}_L^T &= \tfrac{1}{2}[\mathbf{I} \otimes \mathbf{I} + (\mathbf{I} \otimes \mathbf{I})\mathbf{U}](\mathbf{I} \otimes \mathfrak{B}\mathbf{Y})\mathfrak{I} \\
\mathbf{B}_{NL}^T &= \tfrac{1}{2}\left[\mathbf{I} \otimes (\mathfrak{B}\mathbf{Y}\boldsymbol{\Gamma})^T + [(\mathfrak{B}\mathbf{Y}\boldsymbol{\Gamma}) \otimes \mathbf{I}]\mathbf{U}\right](\mathbf{I} \otimes \mathfrak{B}\mathbf{Y})\mathfrak{I}
\end{aligned}$$

and the matrix $\mathfrak{I}$ is now obtained as

$$\mathfrak{I} = \begin{bmatrix} \mathbf{I} & \mathbf{0} & \mathbf{0} \\ \mathbf{0} & \mathbf{0} & \mathbf{0} \\ \mathbf{0} & \mathbf{0} & \mathbf{0} \\ \mathbf{0} & \mathbf{I} & \mathbf{0} \\ \mathbf{0} & \mathbf{0} & \mathbf{0} \\ \mathbf{0} & \mathbf{0} & \mathbf{0} \\ \mathbf{0} & \mathbf{0} & \mathbf{I} \\ \mathbf{0} & \mathbf{0} & \mathbf{0} \\ \mathbf{0} & \mathbf{0} & \mathbf{0} \\ \mathbf{0} & \mathbf{0} & \mathbf{0} \\ \mathbf{I} & \mathbf{0} & \mathbf{0} \\ \mathbf{0} & \mathbf{0} & \mathbf{0} \\ \mathbf{0} & \mathbf{0} & \mathbf{0} \\ \mathbf{0} & \mathbf{I} & \mathbf{0} \\ \mathbf{0} & \mathbf{0} & \mathbf{0} \\ \mathbf{0} & \mathbf{0} & \mathbf{0} \\ \mathbf{0} & \mathbf{0} & \mathbf{I} \\ \mathbf{0} & \mathbf{0} & \mathbf{0} \\ \mathbf{0} & \mathbf{0} & \mathbf{0} \\ \mathbf{0} & \mathbf{0} & \mathbf{0} \\ \mathbf{I} & \mathbf{0} & \mathbf{0} \\ \mathbf{0} & \mathbf{0} & \mathbf{0} \\ \mathbf{0} & \mathbf{0} & \mathbf{0} \\ \mathbf{0} & \mathbf{I} & \mathbf{0} \\ \mathbf{0} & \mathbf{0} & \mathbf{0} \\ \mathbf{0} & \mathbf{0} & \mathbf{0} \\ \mathbf{0} & \mathbf{0} & \mathbf{I} \end{bmatrix}$$

15.2 STRESS INCREMENTS

For the purposes of deriving an incremental variational principle we shall see that the incremental 1st Piola–Kirchhoff stress $\Delta_n\bar{\mathbf{S}}$ is the starting point. However, to formulate mechanical properties, the objective increment of the Cauchy stress, based on the Truesdell stress flux, $\mathring{\Delta}_n\mathbf{T}$, is the starting point. Furthermore, in the resulting variational statement, which we called the *Incremental Principle of Virtual Work*, we shall find that quantity which appears is the increment of the 2nd Piola–Kirchhoff stress, $\Delta_n\mathbf{S}$.

From Chapter 5, $\bar{\mathbf{S}} = \mathbf{S}\mathbf{F}^T$, from which to first order

$$\Delta_n\bar{\mathbf{S}} = \Delta_n\mathbf{S}\mathbf{F}^T + \mathbf{S}\Delta_n\mathbf{F}^T \tag{15.8}$$

For the Cauchy stress, the increment must take account of the rotation of the underlying coordinate system and thereby be objective. We recall the objective Truesdell stress flux $\partial\mathring{\mathbf{T}}/\partial t$ introduced in Chapter 13:

$$\partial\mathring{\mathbf{T}}/\partial t = \partial\mathbf{T}/\partial t + \mathbf{T}\,tr(\mathrm{D}) - \mathbf{L}\mathbf{T} - \mathbf{T}\mathbf{L}^T \tag{15.9}$$

Among the possible stress fluxes, it is unique in being proportional to the rate of the 2nd Piola–Kirchhoff stress, namely

$$\partial\mathbf{S}/\partial t = J\mathbf{F}^{-1}\left(\partial\mathring{\mathbf{T}}/\partial t\right)\mathbf{F}^{-T} \tag{15.10}$$

An objective Truesdell stress increment $\mathring{\Delta}_n\mathbf{T} \approx dt(\partial\mathring{\mathbf{T}}/\partial t)$ is readily obtained as

$$VEC\left(\mathring{\Delta}_n\mathbf{T}\right) = \frac{1}{J}\mathbf{F}^T \otimes \mathbf{F}\,VEC(\Delta_n\mathbf{S}) \tag{15.11}$$

Further, once $VEC\left(\mathring{\Delta}_n\mathbf{T}\right)$ has been determined, the (nonobjective) increment $\Delta_n\mathbf{T}$ of the Cauchy stress may be computed using

$$\mathring{\Delta}_n\mathbf{T} = \Delta_n\mathbf{T} + \mathbf{T}\,tr\left((\Delta_n\mathbf{F})\mathbf{F}^{-1}\right) - \Delta_n\mathbf{F}\mathbf{F}^{-1}\mathbf{T} - \mathbf{T}\mathbf{F}^{-T}\Delta_n\mathbf{F}^T \tag{15.12}$$

from which

$$VEC(\Delta_n\mathbf{T}) = VEC\left(\mathring{\Delta}_n\mathbf{T}\right) - \left[\mathbf{T}VEC^T\left(\mathbf{F}^{-T}\right) - \left(\mathbf{T}\mathbf{F}^{-T}\right)\otimes\mathbf{I} - \mathbf{I}\otimes\left(\mathbf{T}\mathbf{F}^{-T}\right)\right]\mathbf{M}\Delta_n\gamma \tag{15.13}$$

15.3 INCREMENTAL EQUATION OF BALANCE OF LINEAR MOMENTUM

We now formulate the incremental equilibrium equation of nonlinear solid mechanics (assuming that the body has a fixed point). In the deformed (Eulerian) configuration, equilibrium at t_n requires

$$\int \mathbf{T}^T \mathbf{n}\, \mathrm{d}S = \int \rho \ddot{\mathbf{u}}\, \mathrm{d}V \tag{15.14}$$

Referred to the undeformed (Lagrangian) configuration, this equation becomes

$$\int \bar{\mathbf{S}}^T \mathbf{n}_0\, \mathrm{d}S_0 = \int \rho_0 \ddot{\mathbf{u}}\, \mathrm{d}V_0 \tag{15.15}$$

S_0 denotes the surface (boundary) in the undeformed configuration, and $\mathbf{n}_0$ is the surface normal vector in the undeformed configuration. Suppose the solution for $\bar{\mathbf{S}}$ is known as $\bar{\mathbf{S}}_n$ at time t_n and is sought at t_{n+1}. As usual, we introduce the increment $\Delta_n \bar{\mathbf{S}}$ to denote $\bar{\mathbf{S}}_{n+1} - \bar{\mathbf{S}}_n$. Now subtracting the equilibrium equation at time t_n from that at t_{n+1} furnishes the *incremental equilibrium equation*

$$\int \Delta_n \bar{\mathbf{S}}^T \mathbf{n}_0\, \mathrm{d}S_0 = \int \rho_0 \Delta_n \ddot{\mathbf{u}}\, \mathrm{d}V_0 \tag{15.16}$$

Application of the divergence theorem furnishes the differential equation

$$\left(\nabla^T \Delta_n \bar{\mathbf{S}}^T\right)^T = \rho_0 \Delta_n \ddot{\mathbf{u}} \tag{15.17}$$

which is the *local form of the incremental equilibrium equation.*

15.4 INCREMENTAL PRINCIPLE OF VIRTUAL WORK

To derive a variational principle for the current formulation, the quantity to be varied is the *incremental displacement vector* which accordingly is the *primary variable.* Following Chapter 4,

(i) Equation 15.17 is multiplied by $(\delta \Delta_n \mathbf{u})^T$
(ii) Integration is performed over the domain
(iii) The divergence theorem is invoked once
(iv) Terms appearing on the boundary are identified as primary and secondary variables
(v) Boundary conditions and constraints are applied

The reasoning process is very similar to that in the derivation of the Principle of Virtual work in finite deformation in which $\mathbf{u}$ is the unknown, and furnishes

$$\int tr(\delta \Delta_n \mathbf{E}^T \Delta_n \mathbf{S})\, \mathrm{d}V_0 + \int \delta \Delta_n \mathbf{F} \mathbf{S} \Delta_n \mathbf{F}^T\, \mathrm{d}V_0 + \int \delta \Delta_n \mathbf{u}^T \rho_0 \Delta_n \ddot{\mathbf{u}}\, \mathrm{d}V_0 = \int \delta \Delta_n \mathbf{u}^T \Delta_n \mathbf{t}_0\, \mathrm{d}S_0 \tag{15.18}$$

in which $\mathbf{t}_0$ is the traction experienced by dS_0. The increment $\Delta_n\mathbf{t}_0$ is momentarily assumed to be specified on the undeformed boundary. However, in a subsequent section it will be derived in the case in which the traction increment is specified in the current configuration.

The fourth term describes the virtual external work of the traction increments. The first term may be said to describe the virtual internal work of the stress increments, referred to the undeformed configuration. The third term describes the virtual internal work of the inertial force increments. The second term has no counterpart in the previously formulated Principle of Virtual Work in Chapter 13 and arises because of nonlinear geometric effects in the incremental formulation. We simply call it the geometric stiffness integral.

Owing to the importance of the Incremental Principle of Virtual Work, it is derived in detail below. It is convenient to perform the derivation using tensor-indicial notation. The incremental equilibrium equation referred to the undeformed configuration is restated as

$$\begin{aligned}\int \delta\Delta_n u_i \frac{\partial}{\partial X_j}(\Delta_n \bar{S}_{ij})\,dV_0 &= \int \frac{\partial}{\partial X_j}[\delta\Delta_n u_i(\Delta_n\bar{S}_{ij})]\,dV_0 - \int \frac{\partial}{\partial X_j}[\delta\Delta_n u_i]\Delta_n\bar{S}_{ij}\,dV_0 \\ &= \int \delta\Delta_n u_i \rho_0 \Delta_n \ddot{u}_i\,dV_0 \end{aligned} \tag{15.19}$$

The divergence theorem is invoked to convert the first right-hand term to

$$\begin{aligned}\int \frac{\partial}{\partial X_j}[\delta\Delta_n u_i(\Delta_n\bar{S}_{ij})]\,dV_0 &= \int \delta\Delta_n u_i(n_j\Delta_n\bar{S}_{ij})\,dS_0 \\ &= \int \delta\Delta_n u_i \Delta_n t_{0j}\,dS_0 \end{aligned} \tag{15.20}$$

which is recognized as the fourth term in Equation 15.18.

To first order in increments and now using tensor notation, the second right-hand term is written as

$$\begin{aligned}\int \frac{\partial}{\partial X_j}[\delta\Delta_n u_i]\Delta_n\bar{S}_{ij}\,dV_0 &= \int tr(\delta\Delta_n\mathbf{F}\Delta_n\bar{\mathbf{S}})\,dV_0 \\ &= \int tr\left(\delta\Delta_n\mathbf{F}\left[\Delta_n\mathbf{S}\mathbf{F}^T + \mathbf{S}\Delta_n\mathbf{F}^T\right]\right)dV_0 \\ &= \int tr\left(\mathbf{F}^T\delta\Delta_n\mathbf{F}\Delta_n\mathbf{S}\right)dV_0 + \int tr\left(\delta\Delta_n\mathbf{F}\mathbf{S}\Delta_n\mathbf{F}^T\right)dV_0 \end{aligned} \tag{15.21}$$

The second term is recognized as the second term in Equation 15.18. Recalling that the variational operator is applied to the primary variable which is an incremental displacement vector, the first term now becomes

$$\begin{aligned}\int tr\left(\mathbf{F}^T\delta\Delta_n\mathbf{F}\Delta_n\mathbf{S}\right)dV_0 &= \int tr\left(\tfrac{1}{2}\left[\mathbf{F}^T\delta\Delta_n\mathbf{F} + \delta\Delta_n\mathbf{F}^T\mathbf{F}\right]\Delta_n\mathbf{S}\right)dV_0 \\ &= \int tr\left(\delta\Delta_n\mathbf{E}^T\Delta_n\mathbf{S}\right)dV_0 \end{aligned} \tag{15.22}$$

which is recognized as the first term in Equation 15.18. Finally, the third term in Equation 15.18 is recognized as the incremental virtual work of the incremental inertial forces.

15.5 INCREMENTAL FINITE ELEMENT EQUATION

For present purposes let us suppose constitutive relations in the form

$$\Delta_n \mathbf{S} = \mathbf{D}_0(\mathbf{X}, \boldsymbol{\gamma}_n)\Delta_n \mathbf{E} + \boldsymbol{\Omega}\Delta_n t \tag{15.23}$$

in which $\mathbf{D}_0(\mathbf{X}, \boldsymbol{\gamma})$ is the *fourth-order tangent modulus tensor*. Equation 15.23 is capable of describing combined rate-independent and linearly rate-dependent response, for example, combined plasticity and viscoplasticity.

To take advantage of Kronecker product notation, the equation is rewritten as

$$\Delta_n \mathbf{s} = \boldsymbol{\chi}_0(\mathbf{X}, \boldsymbol{\gamma}_n)\Delta_n \mathbf{e} + \boldsymbol{\omega}\Delta_n t \tag{15.24}$$

$$\mathbf{s} = VEC(\mathbf{S}), \quad \mathbf{e} = VEC(\mathbf{E}), \quad \boldsymbol{\chi}_0 = TEN22(\mathbf{D}_0), \quad \boldsymbol{\omega} = VEC(\boldsymbol{\Omega})$$

Using the interpolation model (Equation 15.1), Equation 15.18 becomes

$$\delta\Delta_n \boldsymbol{\gamma}^T[(\mathbf{K}_T + \mathbf{K}_G)\Delta_n \boldsymbol{\gamma} + \mathbf{M}\Delta_n \ddot{\boldsymbol{\gamma}} - [\Delta_n \mathbf{f} - \mathbf{f}_{\omega}\Delta_n t]] = 0 \tag{15.25}$$

$$\mathbf{K}_T = \int \mathbf{M}^T \mathbf{G}\boldsymbol{\chi}_0 \mathbf{G}^T \mathbf{M}\, dV_0, \quad \mathbf{K}_G = \int \mathbf{M}^T \mathbf{S} \otimes \mathbf{M}\, dV_0$$

$$\mathbf{M} = \int \rho_0 \boldsymbol{\Phi}^T \boldsymbol{\varphi}\boldsymbol{\varphi}^T \boldsymbol{\Phi}\, dV, \quad \Delta_n \mathbf{f} = \int \boldsymbol{\Phi}^T \boldsymbol{\varphi}\Delta \mathbf{t}_0\, dS_0$$

$$\mathbf{f}_{\omega} = \int \mathbf{M}^T \mathbf{G}\boldsymbol{\omega}\, dV_0$$

$\mathbf{K}_T$ is now called the tangent modulus matrix, $\mathbf{K}_G$ the geometric stiffness matrix, $\mathbf{M}$ the (incremental) mass matrix, $\Delta_n \mathbf{f}$ is called the incremental (surface) force vector, and we call $\Delta_n \mathbf{f}_{\omega}$ the incremental viscous force.

15.6 CONTRIBUTIONS FROM NONLINEAR BOUNDARY CONDITIONS

Recalling Chapter 13, let I_i denote the principal invariants of $\mathbf{C}$, and let $\mathbf{i} = VEC(\mathbf{I})$, $c_2 = VEC(\mathbf{C}^2)$, $\mathbf{n}_i^T = \partial I_i/\partial \mathbf{c}$, and $\mathbf{A}_i = \partial \mathbf{n}_i/\partial \mathbf{c}$. Recall from Chapter 13 that

$$\begin{aligned} &\mathbf{n}_1 = \mathbf{i}, \quad \mathbf{n}_2 = I_1\mathbf{i} - \mathbf{c}, \quad \mathbf{n}_3 = I_2\mathbf{i} - I_1\mathbf{c} + \mathbf{c}_2 = I_3 VEC(\mathbf{C}^{-1}), \quad I_9 = \mathbf{I} \otimes \mathbf{I} \\ &\mathbf{A}_1 = \mathbf{0}, \quad \mathbf{A}_2 = \mathbf{i}\mathbf{i}^T - \mathbf{I}_9, \quad \mathbf{A}_3 = \mathbf{I} \otimes \mathbf{C} + \mathbf{C} \otimes \mathbf{I} - (\mathbf{i}\mathbf{c}^T + \mathbf{c}\mathbf{i}^T) + I_1(\mathbf{i}\mathbf{i}^T - \mathbf{I}_9) \end{aligned} \tag{15.26}$$

Equation 15.18 can be applied if increments of tractions are prescribed on the undeformed surface S_0. We now consider the more complex situation in which Δ_n is referred to the deformed surface S, on which they are prescribed functions of $\mathbf{u}$. This situation may occur under, for example, "dead loading." Using the relations of Chapter 13 conversion to undeformed coordinates is obtained using the relations

$$\mathbf{t}\,\mathrm{d}S = \mathbf{t_0}\,\mathrm{d}S_0, \quad \mathrm{d}S = \mu\,\mathrm{d}S_0, \quad \mu = J\sqrt{\mathbf{n}_0^T\mathbf{C}^{-1}\mathbf{n}_0} = \sqrt{\mathbf{n}_0^T \otimes \mathbf{n}_0^T\mathbf{n}_3} \tag{15.27}$$

and from Nicholson and Lin (1997b)

$$\Delta_n\mu \approx \mathrm{d}\mu = \mathbf{m}^T\,\mathrm{d}\mathbf{c} \approx \mathbf{m}^T\Delta_n\mathbf{c}, \quad \mathbf{m}^T = \mathbf{n}_0^T \otimes \mathbf{n}_0^T\mathbf{A}_3/2\mu \tag{15.28}$$

Suppose that $\Delta_n\mathbf{t}$ is expressed on S as follows:

$$\Delta_n\mathbf{t} = \Delta_n\underline{\mathbf{t}} - \mathbf{A}_M^T\Delta_n\mathbf{u} \tag{15.29}$$

Here $\Delta_n\underline{\mathbf{t}}$ is prescribed, while $\mathbf{A}_M$ is a known function of $\mathbf{u}$. Also S_0 is the undeformed counterpart of S. Owing to the presence of $\mathbf{A}_M$ the expression in Equation 15.29 is capable of modeling boundary conditions such as support by a nonlinear elastic foundation.

From the fact that $\mathbf{t}\,\mathrm{d}S = \mathbf{t}_0\,\mathrm{d}S_0$, we conclude that $\mathbf{t} = \mathbf{t}_0/\mu$. It follows that

$$\begin{aligned}\Delta_n\mathbf{t}_0 &= \mu\Delta_n\mathbf{t} + \mathbf{t}\mathbf{m}^T\Delta_n\mathbf{c}\\ &= \mu(\Delta_{\underline{n}}\mathbf{t} - \mathbf{A}_M^T\Delta_n\mathbf{u}) + \mathbf{t}\mathbf{m}^T\Delta_n\mathbf{c}\end{aligned} \tag{15.30}$$

From the Incremental Principle of Virtual Work, the right-hand term may be written as

$$\int \delta\Delta\mathbf{u}^T\Delta\mathbf{t}_0\,\mathrm{d}S_0 = \int \delta\Delta\mathbf{u}^T[\mu(\Delta\underline{\mathbf{t}} - \mathbf{A}_M^T\Delta\mathbf{u}) - \mathbf{t}\mathbf{m}^T\Delta\mathbf{c}]\,\mathrm{d}S_0 \tag{15.31}$$

Now recalling the interpolation models for the increments we obtain an incremental force vector plus two boundary contributions to the stiffness terms. In particular

$$\int \delta\Delta_n\mathbf{u}^T\Delta_n\mathbf{t}_0\,\mathrm{d}S_0 = \delta\Delta_n\boldsymbol{\gamma}^T\Delta_n\mathbf{f} - \delta\Delta_n\boldsymbol{\gamma}^T[\mathbf{K}_{BF} + \mathbf{K}_{BN}]\Delta_n\boldsymbol{\gamma} \tag{15.32}$$

$$\Delta_n\mathbf{f} = \int \mu\Delta_n\underline{\mathbf{t}}\,\mathrm{d}S_0, \quad \mathbf{K}_{BF} = \int \boldsymbol{\Phi}^T\boldsymbol{\varphi}\mu\mathbf{A}_M^T\boldsymbol{\varphi}^T\boldsymbol{\Phi}\,\mathrm{d}S_0, \quad \mathbf{K}_{BN} = 2\int \boldsymbol{\Phi}^T\boldsymbol{\varphi}(\mathbf{t}\mathbf{m}^T)\mathbf{G}^T\mathbf{M}\,\mathrm{d}S_0$$

The first boundary contribution is from the nonlinear elastic foundation coupling the traction and displacement increments on the boundary. The second arises from geometric nonlinearity in which the traction increment is prescribed on the *current configuration*.

15.7 EFFECT OF VARIABLE CONTACT

In many, if not most, "real world" problems, loads are transmitted to the member of interest via contact with other members, for example in gear teeth. The extent of the contact zone is now an unknown to be determined as part of the solution process. Solution of contact problems, discussed in an introductory way in Chapter 12, is a difficult problem that has absorbed the attention of many investigators. Some algorithms are suited primarily for linear kinematics. Here a development is given of an example which explicitly addresses the effect of large deformation.

Figure 15.1 shows a contactor moving toward a foundation, assumed to be rigid and fixed. We seek to follow the development of the contact area and the tractions arising throughout it. Recalling Chapter 12, corresponding to a point **x** on the contactor surface there is a target point **y**(**x**) on the foundation to which the normal **n**(**x**) at **x** points. As the contactor starts to deform **n**(**x**) rotates and points toward a new value **y**(**x**). As the point **x** approaches contact, the point **y**(**x**) approaches the foundation point which will come into contact with the contactor point at x.

We define a gap function $g(\mathbf{u},\mathbf{x})$ using $\mathbf{y}(\mathbf{x}) = \mathbf{x} + g\mathbf{n}$. Let **m** be the surface normal vector to the target (foundation) at **y**(**x**). Also, let S_c be the *candidate contact surface* on the contactor, whose undeformed counterpart is S_{0c}. There likewise is a *candidate contact surface* S_f on the foundation.

We limit attention to bonded contact, in which particles coming into contact with each other remain in contact and do not slide away from each other. Algorithms for sliding contact with and without friction are available. For simplicity, for the moment we also assume that shear tractions, in the osculating plane of point of interest, are

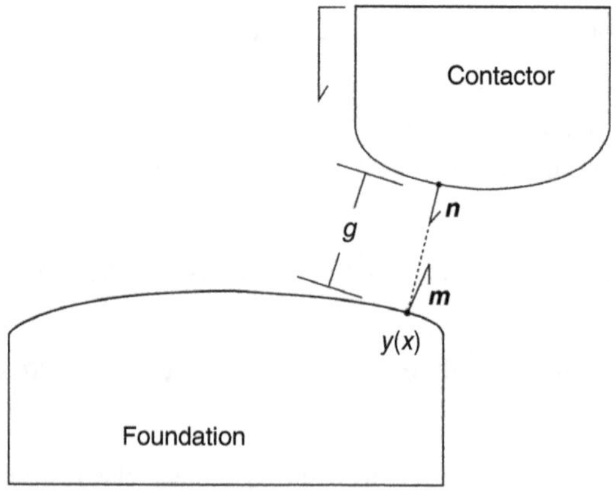

FIGURE 15.1 Contact scenario.

negligible. Suppose now that the interface can be represented by an elastic foundation satisfying the incremental relation

$$\Delta_n t_n = -k(g)\Delta_n u_n \tag{15.33}$$

Here $t_n = \mathbf{n}^T\mathbf{t}$ and $u_n = \mathbf{n}^T\mathbf{u}$ are the normal components of the traction and displacement vectors at $\mathbf{x}$. Since the only traction being considered is the normal traction (to the contactor surface), the transverse components of $\Delta\mathbf{u}$ are not needed (do not result from work). Also, $k(g)$ is a nonlinear stiffness function given in terms of the gap by, for example,

$$k(g) = \frac{k_H}{\pi}\left[\frac{\pi}{2} - \arctan(\alpha_k g - \varepsilon_r)\right] + k_L, \quad k_H/k_L \gg 1 \tag{15.34}$$

As in Chapter 12, when g is positive, the gap is open and k approaches k_L which should be chosen as a small number, theoretically zero. As g becomes negative, the gap is closing and k thereafter rapidly approaches k_H which should be chosen as a large number (theoretically infinite to prevent penetration of the rigid body). A function similar to this was also discussed in Chapter 14, and its characteristics were illustrated graphically.

Under the assumption that only the normal traction on the contactor surface in important, it likewise follows that we may use the relation $\mathbf{t} = t_n\mathbf{n}$, with the consequence that

$$\Delta_n\mathbf{t} = \Delta_n t_n\mathbf{n} + t_n\Delta_n\mathbf{n} \tag{15.35}$$

The contact model contributes the matrix $\mathbf{K}_c$ to the total stiffness matrix as follows (Nicholson and Lin, 1997b):

$$\begin{aligned}\int \delta\Delta\mathbf{u}^T\Delta\mathbf{t}\mu\,\mathrm{d}S_0 &= \int \delta\Delta u_n^T\Delta t_n\mu\,\mathrm{d}S_0 \\ &= -\delta\Delta\boldsymbol{\gamma}^T\mathbf{K}_c\Delta\boldsymbol{\gamma}\end{aligned} \tag{15.36}$$

$$\begin{aligned}\mathbf{K}_c = -2\int \boldsymbol{\Phi}^T\boldsymbol{\varphi}\mathbf{n}t_n\mathbf{m}^T\boldsymbol{\beta}^T\boldsymbol{\Phi}\,\mathrm{d}S_{c0} + \int k_c(g)\boldsymbol{\Phi}^T\boldsymbol{\varphi}\mathbf{n}\mathbf{n}^T\boldsymbol{\varphi}^T\boldsymbol{\Phi}\mu\,\mathrm{d}S_{c0} \\ + \int \boldsymbol{\Phi}^T\boldsymbol{\varphi}t_n\mu\mathbf{h}^T\,\mathrm{d}S_{c0}\end{aligned}$$

in which $\mathbf{h}^T$ is presented below. To update the gap, use may be made of the following relations reported in Nicholson and Lin (1997b). The differential vector $\mathrm{d}\mathbf{y}$ is tangent to the foundation surface, and hence $\mathbf{m}^T\mathrm{d}\mathbf{y} = 0$. It follows that

$$\begin{aligned}0 &\approx \mathbf{m}^T\Delta_n\mathbf{u} + g\mathbf{m}^T\Delta_n\mathbf{n} + \mathbf{m}^T\mathbf{n}\Delta_n g \\ \Delta g &= -\frac{\mathbf{m}^T\Delta\mathbf{u} + g\mathbf{m}^T\Delta\mathbf{n}}{\mathbf{m}^T\mathbf{n}}\end{aligned} \tag{15.37}$$

Using Equations 15.7 and 15.37 we may derive with some effort that

$$\Delta g = \mathbf{\Gamma}^T \Delta \boldsymbol{\gamma}, \quad \mathbf{\Gamma}^T = -\frac{\mathbf{m}^T \mathbf{\Phi}^T \boldsymbol{\varphi} + g\mathbf{h}^T}{\mathbf{m}^T \mathbf{n}}$$
$$\mathbf{h}^T = \mathbf{m}^T \left[\mathbf{n}\left((\mathbf{n}^T \mathbf{F}^{-1}) \otimes \mathbf{n}^T\right) - \mathbf{n}^T \otimes \mathbf{F}^{-T}\right]\mathbf{M}^T \tag{15.38}$$

15.8 INTERPRETATION AS NEWTON ITERATION

The (nonincremental) Principle of Virtual Work is restated in Lagrangian coordinates as

$$\int tr(\delta \mathbf{E}\mathbf{S})\, \mathrm{d}V_0 + \int \delta \mathbf{u}^T \rho \ddot{\mathbf{u}}\, \mathrm{d}V_0 = \int \delta \mathbf{u}^T \mathbf{t}_0\, \mathrm{d}S_0 \tag{15.39}$$

We assume for convenience that $\mathbf{t}_0$ is prescribed on S_0. The interpolation model for $\mathbf{e}$ was shown in Equation 15.8 to have the form $\delta \mathbf{e} = \left[\mathbf{B}_L^T + \mathbf{B}_{NL}^T(\boldsymbol{\gamma})\right]\delta\boldsymbol{\gamma}$. Upon cancellation of the variation $\delta\boldsymbol{\gamma}^T$, an algebraic equation in $\boldsymbol{\gamma}$ is obtained as

$$\mathbf{0} = \mathbf{\Psi}(\boldsymbol{\gamma}, \mathbf{f}) = \int [\mathbf{B}_L + \mathbf{B}_{NL}(\boldsymbol{\gamma})]\mathbf{s}\, \mathrm{d}V_0 + \int \mathbf{\Phi}^T \boldsymbol{\varphi}\rho_0 \ddot{\mathbf{u}}\, \mathrm{d}V_0, \quad \mathbf{f} = \int \mathbf{\Phi}^T \boldsymbol{\varphi}\mathbf{t}_0\, \mathrm{d}S_0, \quad \mathbf{s} = VEC(\mathbf{S}) \tag{15.40}$$

At the nth load step Newton iteration is expressed as

$$\boldsymbol{\gamma}_{n+1}^{(\nu+1)} = \boldsymbol{\gamma}_{n+1}^{(\nu)} - \mathbf{J}^{-1}\mathbf{\Phi}\left(\boldsymbol{\gamma}_{n+1}^{(\nu)}, \mathbf{f}_{n+1}\right), \quad \mathbf{J} = \frac{\partial \mathbf{\Phi}}{\partial \boldsymbol{\gamma}}\left(\boldsymbol{\gamma}_{n+1}^{(\nu)}, \mathbf{f}_{n+1}\right) \tag{15.41}$$

or alternatively as a linear system

$$\mathbf{J}\left(\boldsymbol{\gamma}_{n+1}^{(\nu+1)} - \boldsymbol{\gamma}_{n+1}^{(\nu)}\right) = \mathbf{\Phi}\left(\boldsymbol{\gamma}_{n+1}^{(\nu)}, \mathbf{f}_{n+1}\right)$$
$$\boldsymbol{\gamma}_{n+1}^{(\nu+1)} = \boldsymbol{\gamma}_{n+1}^{(\nu)} + \left[\boldsymbol{\gamma}_{n+1}^{(\nu+1)} - \boldsymbol{\gamma}_{n+1}^{(\nu)}\right] \tag{15.42}$$

If the load increments are small enough, the starting iterate may be estimated as the solution from the nth load step. Also a stopping (convergence) criterion is needed to determine when the effort to generate additional iterates is not rewarded by increased accuracy.

Careful examination of the relations from this and the incremental formulations uncovers that

$$\mathbf{J} = \mathbf{K}_T + \mathbf{K}_G \text{ (static)}$$
$$= \mathbf{K}_G + \mathbf{K}_T + 2\mathbf{M}/h^2 \text{ (dynamic)} \tag{15.43}$$

presuming, for example, that the trapezoidal rule is used in the dynamic case to model the time derivatives. Clearly *the incremental stiffness matrix is the same as the Jacobian matrix in Newton iteration*. Equation 15.43 is, of course, a very satisfying result—it reveals that the Jacobian matrix of Newton Iteration may be calculated by conventional finite element procedures at the element level followed by conventional assembly procedures. If the incremental equation is only solved once at each load increment, the ensuing solution may be viewed as the first iterate in a Newton iteration scheme. The one-time incremental solution can potentially be improved by additional iterations following Equation 15.44, but at the cost of additional effort. However, it is also possible that the use of multiple iterations will enable larger load steps, thereby compensating for the computational effort.

15.9 BUCKLING

Finite element equations based on classical linear buckling equations for beams and plates were treated in Chapter 11. In the classical equations, what strictly are geometrically nonlinear terms appear through what may be considered a linear correction term, furnishing linear equations. Here, in the absence of inertia and nonlinearity in the boundary conditions, we briefly present a more general viewpoint based on the incremental equilibrium equation

$$(\mathbf{K}_T + \mathbf{K}_G)\Delta_n\boldsymbol{\gamma} = \Delta_n\mathbf{f} \tag{15.44}$$

This solution will predict a very large incremental displacement if the stiffness matrix $\mathbf{K}_T+\mathbf{K}_G$ is ill-conditioned or outright singular. Of course, in elastic media, $\mathbf{K}_T$ is positive definite. However, in the presence of in-plane compression, we will see that $\mathbf{K}_G$ may have a negative eigenvalue whose magnitude is comparable to the smallest positive eigenvalue of $\mathbf{K}_T$. To see this recall that

$$\mathbf{K}_T = \int \mathbf{M}^T\mathbf{G}\boldsymbol{\chi}\mathbf{G}^T\mathbf{M}\,\mathrm{d}V_0, \quad \mathbf{K}_G = \int \mathbf{M}^T\mathbf{S}\otimes\mathbf{I}\mathbf{M}\,\mathrm{d}V_0 \tag{15.45}$$

We suppose that the element in question is thin in a local z (out-of-plane direction), corresponding to plane stress. Now in plate and shell theory, it is a common practice to add a transverse shear stress on the element boundaries to allow the element to support transverse loads. We assume that the transverse shear stresses only appear in the incremental force term and the tangent stiffness term, and that the geometric stiffness term strictly satisfies the plane stress assumption. It follows that, if the three-direction is out of the plane, the geometric stiffness term $\mathbf{K}_G$ contains the expression

$$\mathbf{S}\otimes\mathbf{I} = \begin{bmatrix} \mathrm{S}_{11}\mathrm{I} & \mathrm{S}_{12}\mathrm{I} & 0\mathrm{I} \\ \mathrm{S}_{12}\mathrm{I} & \mathrm{S}_{22}\mathrm{I} & 0\mathrm{I} \\ 0\mathrm{I} & 0\mathrm{I} & 0\mathrm{I} \end{bmatrix} \tag{15.46}$$

in which $\mathbf{S}$ is of course the 2nd Piola–Kirchhoff stress. In classical (linear) buckling theory loads which are applied proportionately induce proportionate in-plane stresses (i.e., the stress components have constant ratios). Accordingly, for a given load path, only one parameter, the length of the straight line the stress point traverses in the space of in-plane stresses, arises in the eigenvalue problem for the critical buckling load. In nonlinear problems, there is no assurance that the stress point follows a straight line even if the loads are proportionate. Instead, if λ denotes the distance along the line followed by the load point in proportional loading in load space, the stresses become numerical functions of λ.

As a simple alternative to the classical case addressing a complete member, we consider buckling of a single element, and suppose that the *stresses* appearing in Equation 15.46 are applied in a compressive sense along the faces of the element and in a proportional manner, i.e.,

$$\mathbf{S} \otimes \mathbf{I} \rightarrow \lambda \begin{bmatrix} (-\hat{\mathrm{S}}_{11})\mathrm{I} & (-\hat{\mathrm{S}}_{12})\mathrm{I} & 0\mathrm{I} \\ (-\hat{\mathrm{S}}_{12})\mathrm{I} & (-\hat{\mathrm{S}}_{22})\mathrm{I} & 0\mathrm{I} \\ 0\mathrm{I} & 0\mathrm{I} & 0\mathrm{I} \end{bmatrix} \tag{15.47}$$

in which the circumflex implies a reference value along the stress path at which $\lambda = 1$. The negative signs on the stresses are present since buckling is associated with compressive stresses, although the sign is not needed for the shear stress. At the element level, the equation now becomes

$$\left(\mathbf{K}_T - \lambda \hat{\mathbf{K}}_G\right)\Delta\boldsymbol{\gamma}_{n+1} = \Delta f_{n+1} \tag{15.48}$$

At a given load increment, the critical stress intensity for the current stress path, as a function of an two (spherical) angles determining the path in the stress space illustrated in Figure 11.11, is obtained by computing the λ value rendering $(\mathbf{K}_T - \lambda\hat{\mathbf{K}}_G)$ singular. To an extent, the integration in computing $\hat{\mathbf{K}}_G$ can be made independent of the stress path by adapting Equation 11.70 as follows:

$$\hat{\mathbf{K}}_G = IVEC\left(\left(\int \mathbf{M}^T \otimes \mathbf{M}^T \,\mathrm{d}V_0\right) VEC\left(\hat{\mathbf{S}} \otimes \mathbf{I}\right)\right) \tag{15.49}$$

16 Tangent Modulus Tensors for Thermomechanical Response of Elastomers

16.1 INTRODUCTION

Elastomeric materials embrace natural and synthetic rubber as well as biological tissues. Attention in this chapter is restricted to isotropic elastomers. Their characteristic is that deformation is recoverable even up to very large strains, and the stress is a nonlinear function of strain. Accordingly, they pose issues of material and geometric nonlinearity, potentially also of boundary condition nonlinearity. They pose two additional issues not addressed in Chapter 15. One is the presence of thermal fields coupled (weakly) to the mechanical field, and the second is the presence of a pressure field arising to enforce the constraint of incompressibility or near-incompressibility and serving as an additional primary variable.

Within an element the finite element method makes use of interpolation models for the displacement vector $\mathbf{u}(\mathbf{X},t)$ and temperature $T(\mathbf{X},t)$, and pressure $p = -\mathrm{tr}(\boldsymbol{\tau})/3$ in incompressible or near-incompressible materials:

$$\mathbf{u}(\mathbf{X},t) = N^T(\mathbf{X})\boldsymbol{\gamma}(t), \quad \mathrm{T}(\mathbf{X},t) - \mathrm{T}_0 = \nu^T(\mathbf{X})\boldsymbol{\theta}(t), \quad p = \boldsymbol{\xi}^T(\mathbf{X})\boldsymbol{\psi}(t) \tag{16.1}$$

in which T_0 is the temperature in the reference configuration, assumed constant.

Here $\mathbf{N}(\mathbf{X}) = \boldsymbol{\varphi}^T(\mathbf{X})\boldsymbol{\Phi}$, $\nu(\mathbf{X})$, and $\boldsymbol{\xi}(\mathbf{X})$ are shape functions and $\boldsymbol{\gamma}$, $\boldsymbol{\theta}$, and $\boldsymbol{\psi}$ are vectors of nodal values of displacement, temperature $(\mathrm{T} - \mathrm{T}_0)$, and pressure, respectively. Application of the strain displacement relations and their thermal analogs furnishes

$$\mathbf{f}_1 = VEC(\mathbf{F} - \mathbf{I}) = \mathbf{M}_1\boldsymbol{\gamma} = \mathbf{U}^T\mathbf{M}_2\boldsymbol{\gamma}, \quad \mathbf{f}_2 = VEC(\mathbf{F}^T - \mathbf{I}) = \mathbf{M}_2\boldsymbol{\gamma}, \quad \delta\mathbf{e} = \boldsymbol{\beta}^T\delta\boldsymbol{\gamma},$$
$$\boldsymbol{\beta} = \mathbf{M}_2\mathbf{G}^T, \quad \mathbf{G}^T = \tfrac{1}{2}(\mathbf{F}^T \otimes \mathbf{I} + \mathbf{I} \otimes \mathbf{F}^T\mathbf{U}), \quad \nabla_0\mathrm{T} = \boldsymbol{\beta}_T^T\boldsymbol{\theta} \tag{16.2}$$

in which $\mathbf{e} = VEC(\mathbf{E})$ is the Lagrangian strain vector. Also, $\nabla_0 = \mathbf{F}^T\nabla$ is the gradient operator referred to the deformed configuration. Of course, the matrix $\boldsymbol{\beta}$ and the vector $\boldsymbol{\beta}_T$ are typically expressed in terms of natural coordinates.

16.2 COMPRESSIBLE ELASTOMERS

The Helmholtz potential was introduced in Chapter 6 and shown to underlie the relations of classical coupled thermoelasticity. The thermohyperelastic properties of compressible elastomers may likewise be derived from the Helmholtz free energy density ϕ (per unit mass), which is a function of T and **E**. Under isothermal conditions it is conventional to introduce the strain energy density $w(\mathbf{E}) = \rho_0\phi$ (T,**E**) (T constant), in which ρ_0 is the density in the undeformed configuration. Typically, the elastomer is assumed to be isotropic, in which case ϕ can be expressed as a function of T, I_1, I_2, and I_3. Alternatively, it may be expressed as a function of T and the stretch ratios λ_1, λ_2, and λ_3, which are the eigenvalues of $\sqrt{\mathbf{C}}$, the square root of the right Cauchy–Green strain tensor.

With ϕ specified as a function of T, I_1, I_2, and I_3, the entropy density η per unit mass and the specific heat c_e at constant (Lagrangian) strain are quoted from Chapter 6 as

$$c_e = \mathrm{T}\left.\frac{\partial \eta}{\partial \mathrm{T}}\right|_{\mathbf{E}}, \qquad \eta = -\left.\frac{\partial \phi}{\partial \mathrm{T}}\right|_{\mathbf{E}} \tag{16.3}$$

The second Piola–Kirchhoff stress satisfies the relation (cf. Chapter 6) $\mathbf{s}^T = VEC^T(\mathbf{S}) = \rho_0 \left.\frac{\partial \phi}{\partial \mathbf{e}}\right|_T$ and is obtained as

$$\mathbf{s}^T = 2\sum_i \rho_0 \phi_i \mathbf{n}_i^T, \quad \phi_i = \frac{\partial \phi}{\partial I_i} \tag{16.4}$$

Also of importance is the (isothermal) tangent modulus matrix

$$\mathbf{D}_T = \left.\frac{\partial \mathbf{s}}{\partial \mathbf{e}}\right|_T = 4\sum_i \sum_j \rho_0 \phi_{ij} \mathbf{n}_i \mathbf{n}_j^T + 4\sum_i \rho_0 \phi_i \mathbf{A}_i, \quad \phi_{ij} = \frac{\partial^2 \phi}{\partial I_i \partial I_j} \tag{16.5}$$

An expression for $\mathbf{D}_T$ has also been derived by Nicholson and Lin (1997c) for compressible, incompressible, and near-incompressible elastomers described by strain energy functions (Helmholtz free energy functions) based on the use of stretch ratios rather than invariants.

16.3 INCOMPRESSIBLE AND NEAR-INCOMPRESSIBLE ELASTOMERS

When the temperature T is held constant, elastomers can often be considered to satisfy the *internal constraint* of incompressibility or near-incompressibility. To satisfy the constraint a posteriori, ϕ is augmented with terms involving a new parameter playing the role of a Lagrange multiplier. Typically, this new parameter may be interpreted as the pressure p, referred to the undeformed configuration. Consequently, the thermo-hyperelastic properties of incompressible and near-incompressible elastomers may be derived from the augmented Helmholtz free energy, which is a function of **E**, T, and p. The constraint introduces additional terms into the governing finite element equations and requires an interpolation model for the new primary variable p.

If the elastomer is incompressible at constant temperature, the augmented Helmholtz function ϕ may be written as

$$\phi = \phi_d(J_1, J_2, \mathrm{T}) - \lambda\xi(J, \mathrm{T})/\rho_0, \quad J_1 = I_1/I_3^{1/3}, \quad J_2 = I_2/I_3^{2/3} \tag{16.6}$$

where ξ is a material function satisfying the constraint $\xi(J,\mathrm{T}) = 0$ and $J = I_3^{1/2} = \det(\mathbf{F})$. It is easily shown that ϕ_d depends on the *deviatoric Lagrangian strain* $\mathbf{E}_d(=\frac{1}{2}(\mathbf{C}/I_3^{1/3} - \mathbf{I})$ owing to the introduction of the deviatoric invariants J_2 and J_3, previously encountered in Chapter 13, Equation 13.76. The "thermodynamic" pressure is given by

$$\lambda = p = -tr(\mathbf{T})/3 = \left.\frac{\partial\xi}{\partial J}\right|_{\mathrm{T}} \tag{16.7}$$

For an elastomer which is near-incompressible at constant temperature, ϕ may be written as

$$\rho_0\phi = \rho_0\phi_d(J_1, J_2, \mathrm{T}) - p\xi(J, \mathrm{T}) - p^2/2\kappa_0 \tag{16.8}$$

in which κ_0 is a constant. The near-incompressibility constraint is expressed by $\partial\phi/\partial p = 0$, which implies

$$p = -\kappa_0\xi(J, \mathrm{T}) \tag{16.9}$$

The bulk modulus κ defined by $\kappa = -\left.\frac{\partial p}{\partial J}\right|_{\mathrm{T}}$, and we conclude that

$$\kappa = \kappa_0\left.\frac{\partial\xi}{\partial J}\right|_{\mathrm{T}} \tag{16.10}$$

Chen et al. (1997) presented sufficient conditions under which near-incompressible models reduce to the incompressible case as $\kappa \to \infty$. Nicholson and Lin (1996) proposed the relations

$$\xi(J, \mathrm{T}) = f^3(\mathrm{T})J - 1, \quad \phi_d = \phi_1(J_1, J_2) + \phi_2(\mathrm{T}), \quad \phi_2(\mathrm{T}) = c_e\mathrm{T}(1 - \ln(\mathrm{T}/\mathrm{T}_0)) \tag{16.11}$$

with the consequence that

$$p = -\kappa_0(f^3(\mathrm{T})J - 1), \quad \kappa = f^3(\mathrm{T})\kappa_0 \tag{16.12}$$

Equation 16.12 provides a linear pressure–volume relation in which thermomechanical effects are confined to thermal expansion expressed using a constant volume coefficient α. It directly generalizes the pressure–volume relation of classical linear isotropic elasticity. If the near-incompressibility constraint is assumed to be satisfied a priori, the Helmholtz free energy is recovered as

$$\phi(I_1,I_2,I_3,\mathrm{T}) = \phi_d(J_1,J_2,\mathrm{T}) + \kappa_0(f^3(\mathrm{T}) - 1)^2/2\rho_0 \tag{16.13}$$

The last term in Equation 16.13 results from retaining the lowest nonvanishing term in a Taylor series representation of ϕ about $f^3(\mathrm{T})J{-}1$.

Assuming Equation 16.13, the entropy density now includes a term involving p:

$$\eta = -\frac{\partial\phi}{\partial\mathrm{T}} = -\frac{\partial\phi_d}{\partial\mathrm{T}} + \pi\alpha f^4(\mathrm{T})/\rho_0, \quad \pi = p/f^3(\mathrm{T}) \tag{16.14}$$

The stress and the (9 × 9) tangent modulus tensor are correspondingly modified from the compressible case to accommodate near-incompressibility

$$\begin{aligned}\mathbf{s}^T &= \rho_0\frac{\partial\phi_d}{\partial\mathbf{e}}\bigg|_{\mathrm{T},\pi} - \pi f^3(\mathrm{T})\mathbf{n}_3^T/J \\ \mathbf{D}_{TP} &= \frac{\partial\mathbf{s}}{\partial\mathbf{e}}\bigg|_{\mathrm{T},\pi} = \rho_0\left(\frac{\partial}{\partial\mathbf{e}}\right)^T\frac{\partial\phi_d}{\partial\mathbf{e}} - \pi f^3(\mathrm{T})\left[2\mathbf{A}_3/J - \mathbf{n}_3\mathbf{n}_3^T/J^3\right]\end{aligned} \tag{16.15}$$

16.3.1 Examples of Expressions for the Helmholtz Potential

There are two broad approaches to the formulation of Helmholtz potential:

1. To express ϕ as a function of I_1, I_2, and I_3, and T (and p)
2. To express ϕ as a function of the principal stretches λ_1, λ_2, and λ_3, and T (and p)

The latter approach is thought to possess the convenient feature of allowing direct use of test data, say from uniaxial tension. We distinguish several cases.

16.3.1.1 Invariant-Based Incompressible Models: Isothermal Problems

In the entitled case, the strain energy function depends only on I_1, I_2 and incompressibility is expressed by the constraint $I_3 = 1$, assumed to be satisfied a priori. In this category, the most widely used models include the Neo-Hookean material (a):

$$\phi = C_1(I_1 - 3), \quad I_3 = 1 \tag{16.16}$$

and the (two-term) Mooney–Rivlin material (b):

$$\phi = C_1(I_1 - 3) + C_2(I_2 - 3), \quad I_3 = 1 \tag{16.17}$$

in which C_1 and C_2 are material constants. Most finite element codes with hyperelastic elements support the Mooney–Rivlin model. In principle, Mooney–Rivlin coefficients C_1 and C_2 can be determined independently by "fitting" suitable load–deflection curves, for example, uniaxial tension. Values for several different rubber compounds are listed in Nicholson and Nelson (1990).

16.3.1.2 Invariant-Based Models for Compressible Elastomers under Isothermal Conditions

Two widely studied strain energy functions are due to Blatz and Ko (1962). Let G_0 be the shear modulus and ν_0 the Poisson's ratio, referred to the undeformed configuration. The two models are:

$$\rho_0\phi_1 = \frac{1}{2}G_0\left(I_1 + \frac{1-2\nu_0}{\nu_0}I_3 - I_3^{\frac{\nu_0}{1-2\nu_0}} - \frac{1+\nu_0}{\nu_0}\right)$$

$$\rho_0\phi_2 = \frac{1}{2}G_0\left(\frac{I_2}{I_3} + 2I_3 - 5\right) \tag{16.18}$$

Let w denote the Helmholtz free energy evaluated at a constant temperature, in which case it reduces to the strain energy. We note a general expression for w which is implemented in several commercial finite element codes (e.g., ANSYS, 2000):

$$w(J_1,J_2,J) = \sum_i \sum_j C_{ij}(J_1-3)^i\,(J_2-3)^j + \sum_k (J_r-1)^k/D_k,$$

$$J_r = J/(1+\mathrm{E}_{th}) \tag{16.19}$$

in which E_{th} is called the thermal expansion strain, while C_{ij} and D_k are material constants. Several codes provide software routines for estimating the model coefficients from user-supplied data.

Several authors have attempted to uncouple the response into isochoric (volume conserving) and volumetric parts even in the compressible range, giving rise to functions of the form $w = w_1(J_1,J_2) + w_2(J)$, and recall that J_1, J_2 are the deviatoric invariants. A number of proposed forms for ϕ_2 are discussed in Holzappel (1996).

16.3.1.3 Thermomechanical Behavior under Non-Isothermal Conditions

Finally we illustrate the accommodation of coupled thermomechanical effects. A more detailed presentation is given in Section 16.5. Simple extensions of, say, the Mooney–Rivlin material have been proposed by Dillon (1962), Nicholson and Nelson (1990), and Nicholson (1995) for compressible elastomers, and in Nicholson and Lin (1996) for incompressible and near-incompressible elastomers. From the latter reference, the model for near-incompressible elastomers is

$$\rho_0\phi = C_1(J_1-2) + C_2(J_2-3) + \rho_0 c_e \mathrm{T}(1-\ln(\mathrm{T}/\mathrm{T}_0)) - (f^3(\mathrm{T})J-1)\pi - \pi^2/2\kappa_0 \tag{16.20}$$

in which, as before, $\pi = p/f^3(\mathrm{T})$. The model assumes that the coefficient of specific heat at constant strain is a constant. As previously mentioned a model similar to

Nicholson and Lin (1996) has been proposed by Holzappel and Simo (1996) for compressible elastomers described using stretch ratios.

16.4 STRETCH-RATIO-BASED MODELS: ISOTHERMAL CONDITIONS

For compressible elastomers, Valanis and Landel (1967) proposed a strain energy function based on the decomposition

$$\phi(\lambda_1,\lambda_2,\lambda_3,\mathrm{T}) = \phi(\lambda_1,\mathrm{T}) + \phi(\lambda_2,\mathrm{T}) + \phi(\lambda_3,\mathrm{T}), \quad \mathrm{T}\ \text{fixed} \tag{16.21}$$

Ogden (1986) has proposed the form

$$\rho_0\phi(\lambda,\mathrm{T}) = \sum_1^N \mu_p(\lambda^{\alpha p} - 1), \quad \mathrm{T}\ \text{fixed} \tag{16.22}$$

In principle, in incompressible isotropic elastomers stretch-ratio-based models have the advantage of permitting direct use of "archival" data from single stress tests, for example uniaxial tension.

We now illustrate the application of Kronecker product algebra to thermohyperelastic materials under isothermal conditions. We then accommodate thermal effects. From Nicholson and Lin (1997c), we invoke the expression for the differential of a tensor-valued isotropic function of a tensor. Namely let $\mathbf{A}$ denote a nonsingular $n \times n$ tensor with distinct eigenvalues, and let $\mathbf{F}(\mathbf{A})$ be a tensor-valued isotropic function of $\mathbf{A}$, admitting representation as a convergent polynomial:

$$\mathbf{F}(\mathbf{A}) = \sum_0^{\propto} \phi_j \mathbf{A}^j \tag{16.23}$$

Here ϕ_j are constants. A compact expression for the differential $\mathrm{d}\mathbf{F}(\mathbf{A})$ is presented using Kronecker product notation.

The reader is referred to Nicholson and Lin (1997c) for the derivation of the following expression. With $\mathbf{f} = VEC(\mathbf{F})$ and $\mathbf{a} = VEC(\mathbf{A})$,

$$\begin{gathered}
\mathrm{d}\mathbf{f}(\mathbf{a}) = \tfrac{1}{2}\mathbf{F}'^T \oplus \mathbf{F}'\,\mathrm{d}\mathbf{a} + \mathbf{W}\,\mathrm{d}\boldsymbol{\omega} \\
\mathbf{F}'(\mathbf{A}) = \sum_0^{\propto} j\phi_j \mathbf{A}^{j-1}, \quad \frac{\mathrm{d}\mathbf{F}}{\mathrm{d}\mathbf{A}} = ITEN22\left(\frac{\mathrm{d}\mathbf{f}}{\mathrm{d}\mathbf{a}}\right) \\
\mathbf{W} = -(\mathbf{F} - \mathbf{A}\mathbf{F}'/2)^T \ominus (\mathbf{F} - \mathbf{A}\mathbf{F}'/2) + \tfrac{1}{2}(\mathbf{A}^{\mathrm{T}} \otimes \mathbf{F}' - \mathbf{F}'^T \otimes \mathbf{A})
\end{gathered} \tag{16.24}$$

Also, $\mathrm{d}\boldsymbol{\omega} = VEC(\mathrm{d}\boldsymbol{\Omega})$ in which $\mathrm{d}\boldsymbol{\Omega}$ is an antisymmetric tensor representing the rate of rotation of the principal directions. The critical step is to determine a matrix $\mathbf{J}$ such that $\mathbf{W}\,\mathrm{d}\omega = -\mathbf{J}\,\mathrm{d}\mathbf{a}$. It is shown in Nicholson and Lin (1997c) that $\mathbf{J} = -[\mathbf{A}^T \ominus \mathbf{A}]^{-\mathrm{I}}\mathbf{W}$, in which $[\mathbf{A}^T \ominus \mathbf{A}]^I$ is the Morse–Penrose inverse (Dahlquist and Bjork, 1974). Accordingly

$$\mathbf{df/da} = \mathbf{F}'^T \oplus \mathbf{F}'/2 - [\mathbf{A}^T \ominus \mathbf{A}]^I \mathbf{W} \tag{16.25}$$

We now apply the tensor derivative to elastomers modeled using stretch ratios, especially in the model due to Ogden (1972). In particular a strain energy function w was proposed which for compressible elastomers and for isothermal response is equivalent to the form

$$w = tr\left(\sum_i \xi_i \left[\mathbf{C}^{\zeta_i} - \mathbf{I}\right]\right) \tag{16.26}$$

in which ξ_i, ζ_i are material properties. The (9×9) tangent modulus tensor $\boldsymbol{\chi}_0$ appearing in Chapter 15 for the incremental form of the Principle of Virtual Work is obtained as

$$\boldsymbol{\chi}_0 = 4\sum_i \zeta_i \xi_i (\zeta_i - 1)\mathbf{C}^{\zeta_i - 2} \oplus \mathbf{C}^{\zeta_i - 2}/2 + 4\sum_i \zeta_i \xi_i [\mathbf{A}^T \ominus \mathbf{A}]^I \mathbf{W}_i \tag{16.27}$$

$$\mathbf{W}_i = -\frac{3-\zeta_i}{2}\mathbf{C}^{\zeta_i} \otimes \mathbf{C}^{\zeta_i} + \frac{\zeta_i - 1}{2}[\mathbf{C}^{\zeta_i - 2} \otimes \mathbf{C} - \mathbf{C} \otimes \mathbf{C}^{\zeta_i - 2}] \tag{16.28}$$

16.5 EXTENSION TO THERMOHYPERELASTIC MATERIALS

A development of the thermohyperelastic model in Equation 16.20 is now given, following Nicholson and Lin (1996). The body initially experiences temperature T_0 uniformly. It is assumed that temperature effects occur primarily as thermal expansion, that volume changes are small, and that volume changes depend linearly on temperature. Thus materials of present interest may be described as mechanically nonlinear but thermally linear.

Owing to the role of thermal expansion, it is desirable to uncouple dilatational and deviatoric effects as much as possible. To this end we invoke the deviatoric Cauchy–Green strain $\hat{C} = C/I_3^{1/3}$ in which I_3 is the third principal invariant of $\mathbf{C}$. Upon modifying w and expanding it in $J - 1$, $\left(J = I_3^{1/2}\right)$ and retaining lowest order terms, we obtain

$$w = tr\left(\sum_i \xi_i \left[\hat{C}^{\zeta_i} - I\right]\right) + \frac{1}{2}\kappa(J-1)^2 \tag{16.29}$$

in which κ is the bulk modulus. The expression for $\boldsymbol{\chi}_0$ in Equation 16.27 is affected by these modifications.

To accommodate thermal effects it is necessary to recognize that w is simply the Helmholtz free energy density $\rho_0\phi$ under isothermal conditions, in which ρ_0 is the mass density in the undeformed configuration. It is assumed that $\phi = 0$ in the undeformed configuration. As for invariant-based models, we may obtain a function

ϕ with three terms: a purely mechanical term ϕ_M, a purely thermal term ϕ_T, and a mixed term ϕ_{TM}. Now with entropy denoted by η, ϕ satisfies the relations

$$\mathbf{s}^T = \rho_0 \frac{\cdot\partial\varphi}{\partial\mathbf{e}|_{\mathrm{T}}}, \quad \eta = -\frac{\partial\varphi}{\partial\mathrm{T}|_e} \tag{16.30}$$

As previously stated, the specific heat at constant strain, $c_e = \mathrm{T}\partial\eta/\partial\mathrm{T}|_e$, is assumed to be constant, from which we obtain

$$\phi_T = c_e\mathrm{T}[1 - \ln(\mathrm{T}/\mathrm{T}_0)] \tag{16.31}$$

On the assumption that thermal effects in shear (i.e., deviatoric effects) can be neglected relative to thermal effects in dilatation, the purely mechanical effect is equated with the deviatoric term in Equation 16.29.

$$\phi_M = tr\left(\sum_i \xi_i\left[\hat{\mathbf{C}}^{\zeta_i} - \mathbf{I}\right]\right) \tag{16.32}$$

Of greatest present interest in the current context is ϕ_{TM}. The development of Nicholson and Lin (1996) furnishes

$$\phi_{TM} = \frac{\kappa[\beta^3(\mathrm{T})J - 1]^2}{2\rho}, \quad \beta(\mathrm{T}) = (1 + \alpha(\mathrm{T}/\mathrm{T}_0)/3)^{-1} \tag{16.33}$$

The tangent modulus tensor $\chi_0 = \partial\mathbf{s}/\partial\mathbf{e}$ now has two parts: $\chi_M + \chi_{TM}$, in which χ_M is recognized as χ_0, derived in Equation 16.27. Omitting the details, Kronecker product algebra serves to derive the following expression for the thermomechanical position of the tangent modulus tensor.

$$\chi_{TM} = \frac{\kappa}{J\rho}\beta^3\left[\frac{\beta^3}{J^2}\mathbf{n}_3\mathbf{n}_3^T + (\beta^3 J - 1)\left[\mathbf{A}_3 - \frac{\mathbf{n}_3\mathbf{n}_3^T}{J^2}\right]\right] \tag{16.34}$$

The foregoing discussion of thermohyperelastic models has been limited to compressible elastomers. However, many elastomers used in applications such as seals are incompressible or near-incompressible. For such applications, as we have seen that an additional field variable is introduced, namely the hydrostatic pressure (referred to deformed coordinates). It serves as a Lagrange multiplier enforcing the incompressibility and near-incompressibility constraints. Following the approach for invariant-based models, Equation 16.34 may be extended to incorporate the constraints of incompressibility and near-incompressibility.

The tangent modulus tensor presented here purely addresses the differential of stress with respect to strain. However, if coupled heat transfer (conduction and radiation) is considered, a more general expression for the tangent modulus tensor is required, expressing increments of stress and entropy in terms of increments of strain and temperature. A development accommodating heat transfer for invariant based elastomers is given in Nicholson and Lin (1997a).

EXAMPLE 16.1

Derive explicit forms of the stress and tangent modulus tensors using the Helmoltz potential for a near-incompressible thermohyperelastic material.

SOLUTION

Enforcing the near-incompressibility constraint a priori, the Helmholtz potential function of interest is

$$\phi(\mathrm{T},J_1,J_2,\pi)=\frac{1}{\rho_0}[C_1(J_1-2)+C_2(J_2-3)]+c_e\mathrm{T}[1-\ln(\mathrm{T}/\mathrm{T}_0)]+\frac{1}{\rho_0}\frac{\kappa_0(f^3(\mathrm{T})J-1)^2}{2}$$

Now the stress is to be obtained using the relations

$$\mathbf{s}^T=\rho_0\frac{\partial\phi}{\partial\mathbf{e}}\bigg|_{\mathrm{T},\pi}=C_1\frac{\mathrm{d}J_1}{\mathrm{d}\mathbf{e}}+C_2\frac{\mathrm{d}J_2}{\mathrm{d}\mathbf{e}}-\pi f^3(\mathrm{T})\frac{\mathrm{d}J}{\mathrm{d}\mathbf{e}},\quad \pi=-\kappa_0(f^3(\mathrm{T})J-1)$$

In terms of $\mathbf{m}_1$ and $\mathbf{m}_2$ presented subsequently in Equation 16.63

$$\frac{\mathrm{d}J_1}{\mathrm{d}\mathbf{e}}=\mathbf{m}_1^T,\quad \frac{\mathrm{d}J_2}{\mathrm{d}\mathbf{e}}=\mathbf{m}_2^T$$

Also

$$\frac{\mathrm{d}J}{\mathrm{d}\mathbf{e}}=2\frac{\mathrm{d}}{\mathrm{d}\mathbf{c}}\left(\sqrt{I_3}\right)=\frac{1}{\sqrt{I_3}}\frac{\mathrm{d}I_3}{\mathrm{d}\mathbf{c}}=\frac{\mathbf{n}_3^T}{J}$$

Consequently,

$$\mathbf{s}^T=C_1\mathbf{m}_1^T+C_2\mathbf{m}_2^T-\kappa\pi f^3(\mathrm{T})\frac{\mathbf{n}_3^T}{J}$$

The tangent modulus tensor may now be stated as

$$\mathbf{D}_T=\frac{\partial\mathbf{s}}{\partial\mathbf{e}}\bigg|_{\mathrm{T},\pi}=C_1\frac{\partial\mathbf{m}_1}{\partial\mathbf{e}}+C_2\frac{\partial\mathbf{m}_2}{\partial\mathbf{e}}-\pi f^3(\mathrm{T})\left[2\frac{\mathbf{A}_3}{J}-\frac{\mathbf{n}_3\mathbf{n}_3^T}{J^3}\right]$$

Since $\mathbf{m}_1=2\left(\mathbf{i}-\frac{1}{3}\frac{I_1}{I_3}\mathbf{n}_3\right)/I_3^{1/3}$ and $\mathbf{m}_2=2\left(\mathbf{n}_2-\frac{2}{3}\frac{I_2}{I_3}\mathbf{n}_3\right)/I_3^{2/3}$ (Equation 16.63) some manipulation serves to derive that

$$\frac{\mathrm{d}\mathbf{m}_1}{\mathrm{d}\mathbf{e}}=2\frac{\mathrm{d}\mathbf{m}_1}{\mathrm{d}\mathbf{c}}=-\frac{4}{3I_3}\left[-\mathbf{m}_1\mathbf{n}_3^T+\frac{1}{I_3{}^{1/3}}\left(\mathbf{n}_3\mathbf{i}^T-\frac{I_1}{I_3}\mathbf{n}_3\mathbf{n}_3^T+I_1\mathbf{A}_3\right)\right]$$

$$\frac{\mathrm{d}\mathbf{m}_2}{\mathrm{d}\mathbf{e}}=-\frac{8}{3}\frac{\mathbf{m}_2\mathbf{n}_3^T}{I_3}+\frac{4}{I_3^{2/3}}\left(\mathbf{A}_2+\frac{2}{3}\frac{I_2}{I_3{}^2}\mathbf{n}_3\mathbf{n}_3^T-\frac{2}{3}\frac{\mathbf{n}_3\mathbf{n}_2^T}{I_3}-\frac{2}{3}\frac{I_2}{I_3}\mathbf{A}_3\right)$$

EXAMPLE 16.2

Recover c_e upon differentiating $\phi_{rt} = c_e\mathrm{T}[1 - \ln(\mathrm{T}/\mathrm{T}_0)]$ twice with respect to T.

SOLUTION

Differentiation once gives

$$\begin{aligned}\frac{\partial \phi_{rt}}{\partial \mathrm{T}} &= c_e[1 - \ln(\mathrm{T}/\mathrm{T}_0)] + c_e\mathrm{T}\left[-\frac{1}{\mathrm{T}/\mathrm{T}_0}\frac{1}{\mathrm{T}_0}\right] \\ &= -c_e \ln(\mathrm{T}/\mathrm{T}_0)\end{aligned}$$

Differentiating a second time gives

$$\frac{\partial^2 \phi_{rt}}{\partial \mathrm{T}^2} = -c_e\frac{1}{\mathrm{T}/\mathrm{T}_0}\frac{1}{\mathrm{T}_0} = -\frac{c_e}{\mathrm{T}}$$

Hence

$$c_e = -\mathrm{T}\frac{\partial^2 \phi_{rt}}{\partial \mathrm{T}^2}$$

But we know that $c_e = \mathrm{T}\frac{\partial \eta}{\partial T}\Big|_{\mathbf{E}}$, in which $\eta = -\frac{\partial \phi_{rt}}{\partial \mathrm{T}}$. It follows that

$$c_e = -\mathrm{T}\frac{\partial^2 \phi_{rt}}{\partial \mathrm{T}^2}$$

verifying that c_e is recovered from $c_e\mathrm{T}[1 - \ln(\mathrm{T}/\mathrm{T}_0)]$.

16.6 THERMOMECHANICS OF DAMPED ELASTOMERS

Thermoviscohyperelasticity is a topic central to important applications such as rubber mounts used in hot engines for vibration isolation. The current section describes a simple thermoviscohyperelastic constitutive model thought to be suitable for near-incompressible elastomers exhibiting modest levels of viscous damping following a Voigt-type of model. Two potential functions are used to provide a systematic treatment of reversible and irreversible effects. One is the familiar Helmholtz free energy in terms of the strain and the temperature; it describes reversible, thermohyperelastic effects. The second potential function, based on the model of Ziegler and Wehrli (1987), incorporates elements for modeling viscous dissipation and arises directly from the entropy production inequality. It provides a consistent thermodynamic framework for describing damping in terms of a *viscosity tensor* depending on strain and temperature.

The formulation leads to a simple energy balance equation, which is used to derive a rate variational principle. Together with the Principle of Virtual Work, variational equations governing coupled thermal and mechanical effects are presented. Finite element equations are derived from the thermal equilibrium equation and from the Principle of Virtual Work. Several quantities such as internal energy density χ have reversible and irreversible portions, indicated by the subscripts r and i: $\chi = \chi_r + \chi_i$. The thermodynamic formulation in the succeeding paragraphs is referred to undeformed coordinates.

There are several types of viscoelastic behavior in elastomers, especially if they contain fillers such as carbon black. For example, under load elastomers experience stress softening and compression set, which are long-term viscoelastic phenomena. Of interest here is the type of damping which is usually assumed in vibration isolation, in which the stresses have an elastic and a viscous portion reminiscent of the classical Voigt model, and the viscous portion is proportional to strain rates. The time constants are small. It is viewed as arising in small motions superimposed on the large strains which already reflect long-term viscoelastic effects.

16.6.1 Balance of Energy

The conventional equation for the balance of energy is expressed as

$$\begin{aligned} \rho_0 \dot{\chi} &= \mathbf{s}^T \dot{\mathbf{e}} - \nabla_0^T \mathbf{q}_0 + \rho_0 h \\ &= \mathbf{s}_r^T \dot{\mathbf{e}} + \mathbf{s}_i^T \dot{\mathbf{e}} - \nabla_0^T \mathbf{q}_0 + \rho_0 h \end{aligned} \tag{16.35}$$

in which $\mathbf{s} = VEC(\mathbf{S})$ and $\mathbf{e} = VEC(\mathbf{E})$. Here χ is the internal energy per unit mass, $\mathbf{q}_0$ is the heat flux vector referred to undeformed coordinates, ∇_0 is the divergence operator referred to undeformed coordinates, and h is the heat input per unit mass, for simplicity assumed independent of temperature. The state variables are recognized to be $\mathbf{e}$ and T. The Helmholtz free energy ϕ_r per unit mass (which is regarded as reversible) and the entropy η per unit mass are introduced using

$$\phi_r = \chi - \mathrm{T}\eta \tag{16.36}$$

Upon obvious rearrangement,

$$\nabla_0^T \mathbf{q}_0 - \rho_0 h = \mathbf{s}_r^T \dot{e} + \mathbf{s}_i^T \dot{e} - \rho_0 \mathrm{T}\dot{\eta} - \rho_0 \eta \dot{\mathrm{T}} - \rho_0 \dot{\phi}_r \tag{16.37}$$

16.6.2 Entropy Production Inequality

The entropy production inequality is stated as

$$\begin{aligned} \rho_0 \mathrm{T}\dot{\eta} &\geq -\nabla_0^T q_0 + \rho_0 h + \mathbf{q}_0^T \nabla \mathrm{T}/\mathrm{T} \\ &\geq \rho_0 \dot{\phi}_r - \mathbf{s}_r^T \dot{\mathrm{e}} - \mathbf{s}_i^T \dot{\mathrm{e}} + \rho_0 \mathrm{T}\dot{\eta} + \rho_0 \eta \dot{\mathrm{T}} + \mathbf{q}_0^T \nabla \mathrm{T}/\mathrm{T} \end{aligned} \tag{16.38}$$

As previously indicated, the Helmholtz potential is assumed to represent only *reversible* thermohyperelastic effects. However, we decompose η into reversible and irreversible portions: $\eta = \eta_r + \eta_i$. Also, ϕ_r, η_r, and η_i are assumed to be differentiable functions of $\mathbf{E}$ and T. We further suppose that $\eta_i = \eta_{i1} + \eta_{i2}$ in which

$$\rho_0 \mathrm{T}\dot{\eta}_{i2} = \left[-\nabla_0^T \mathbf{q}_0 + \rho_0 h\right]_i \tag{16.39}$$

This may be interpreted as saying that part of the viscous dissipation is to "absorbed" as heat. We finally suppose that reversible effects are absorbed as a reversible portion of the heat input, as follows:

$$\rho_0 \mathrm{T}\dot{\eta}_r = \left[-\nabla_0^T \mathbf{q}_0 + \rho_0 h\right]_r \tag{16.40}$$

In addition, from conventional arguments using Maxwell relations,

$$\rho\partial\phi_r/\partial\mathbf{e} = \mathbf{s}_r^T, \quad \partial\phi_r/\partial\mathrm{T} = -\eta_r \tag{16.41}$$

The consequence of the foregoing assumptions now emerges as the inequality

$$\mathbf{s}_i^T\dot{\mathbf{e}} - \mathbf{q}_0^T\nabla_0\mathrm{T}/\mathrm{T} \geq \rho_0\eta_{i1}\dot{\mathrm{T}} \tag{16.42}$$

Inequality (Equation 16.42) is satisfied if $\rho_0\eta_{i1}\dot{\mathrm{T}} \leq 0$ and also if

$$\mathbf{s}_i^T\dot{\mathbf{e}} \geq 0 \tag{16.43a}$$

$$-q_0^T\nabla\mathrm{T}/\mathrm{T} \geq 0 \tag{16.43b}$$

Inequality (Equation 16.43b) is conventionally assumed to express the fact that heat flows *irreversibly* from hot to cold zones. Inequality (Equation 16.43a) states that "viscous work" is *dissipative*. The statement $\rho_0\eta_{i1}\,\dot{\mathrm{T}} \leq 0$ is difficult to justify except by appealing to its consequence in the form of the physically appealing inequalities in Equation 16.43. Of course it may be true that $\eta_{i2} = 0$, in which case the irreversible entropy is related to the irreversible heat input in precisely the same way as the reversible entropy is related to the reversible heat input.

16.6.3 Dissipation Potential

Following Ziegler and Wehrli (1987), the specific dissipation rate potential $\boldsymbol{\Psi}(\mathbf{q}_0,\dot{\mathbf{e}},\mathbf{e},\mathrm{T}) = \rho_0\eta_i\dot{\mathrm{T}}$ is introduced and assumed to serve as a *rate potential* for the irreversible stress and temperature gradient as follows:

$$\mathbf{s}_i^T = \rho_0\Lambda_i\partial\boldsymbol{\Psi}/\partial\dot{\mathbf{e}} \tag{16.44a}$$

$$-\nabla_0^T\mathrm{T}/\mathrm{T} = \Lambda_t\rho_0\partial\boldsymbol{\Psi}/\partial\mathbf{q}_0 \tag{16.44b}$$

The function $\boldsymbol{\Psi}$ is selected such that Λ_i and Λ_t are positive scalars, in which case inequalities (Equation 16.44a and 16.44b) require that

$$(\partial\boldsymbol{\Psi}/\partial\dot{\mathbf{e}})\dot{\mathbf{e}} \geq 0 \tag{16.45a}$$

$$(\partial\boldsymbol{\Psi}/\partial\mathbf{q}_0)\mathbf{q}_0 \geq 0 \tag{16.45b}$$

This may be interpreted as indicating the convexity of a dissipation surface in ($\dot{\mathbf{e}}$,q_0) space. Finally, to state the constitutive relations for a thermoviscohyperelastic material it is sufficient to specify ϕ_r and $\boldsymbol{\Psi}$.

A simple example is now presented to illustrate how inequality (Equation 16.45) provides a "framework" for describing dissipative effects. On the expectation that properties governing heat transfer are not affected by strain, we introduce the decomposition

$$\boldsymbol{\Psi} = \boldsymbol{\Psi}_i + \boldsymbol{\Psi}_t, \quad \rho_0 \boldsymbol{\Psi}_t = \tfrac{1}{2}\Lambda_t \mathbf{q}_0^T \mathbf{q}_0 \tag{16.46}$$

Here, $\boldsymbol{\Psi}_t$ represents thermal effects. Equation 16.44b implies that

$$-\nabla_0 \mathrm{T}/\mathrm{T} = \Lambda_t \mathbf{q}_0 \tag{16.47}$$

This is essentially the conventional Fourier law of heat conduction, with $1/\Lambda_t$ recognized as the thermal conductivity.

As an elementary example of viscous dissipation, suppose that

$$\boldsymbol{\Psi}_i = \mu(\mathrm{T}, J_1, J_2)\dot{\mathbf{e}}^T \dot{\mathbf{e}}/2, \quad \Delta_i = 1 \tag{16.48}$$

in which $\mu(\mathrm{T},J_1,J_2)$ is the viscosity. Application of Equation 16.44a now gives

$$\mathbf{s}_i = \mu(\mathrm{T},J_1,J_2)\dot{\mathbf{e}} \tag{16.49}$$

and inequality (Equation 16.45a) requires that the viscosity function μ be positive.

16.6.4 Thermal Field Equation for Damped Elastomers

The energy balance equations of thermohyperelasticity (i.e., the reversible response) are now reappearing in terms of a balance law among reversible portions of the stress, entropy, and internal energy. Equation 16.41 implies that

$$\rho_0 \dot{\phi}_r = \mathbf{s}_r^T \dot{\mathbf{e}} - \rho_0 \eta_r \dot{\mathrm{T}} \tag{16.50}$$

The ensuing Maxwell reciprocity relation is

$$\partial \mathbf{s}_r^T / \partial \mathrm{T} = -\rho_0 \partial \eta_r / \partial \mathbf{e} \tag{16.51}$$

Now familiar operations furnish the reversible part of the equation of thermal equilibrium (balance of energy).

$$\left[-\nabla_0^T \mathbf{q}_0 + \rho_0 h\right]_r = -\mathrm{T}\left(\partial \mathbf{s}_r^T / \partial \mathrm{T}\right)\dot{\mathbf{e}} + \rho_0 c_e \dot{\mathrm{T}}, \quad c_e = \mathrm{T}\partial \eta_r / \partial \mathrm{T} \tag{16.52}$$

For the irreversible part, we recall the relation

$$-\left(\nabla_0^T \mathbf{q}_0 - \rho_0 h\right)_i = -\mathbf{s}_i^T \dot{\mathbf{e}} + \rho_0 c_i \dot{\mathrm{T}} \tag{16.53}$$

and

$$\rho_0 \mathrm{T}\dot{\eta}_{i2} = \rho_0 \mathrm{T}\frac{\partial \eta_{i2}}{\partial \mathbf{e}}\dot{\mathbf{e}} + \rho_0 \mathrm{T}\frac{\partial \eta_{i2}}{\partial \mathrm{T}}\dot{\mathrm{T}} \tag{16.54}$$

and

$$-\mathbf{s}_i^T = \rho_0 \mathrm{T}\frac{\partial \eta_{i2}}{\partial \mathbf{e}}, \quad c_i = \mathrm{T}\frac{\partial \eta_{i2}}{\partial \mathrm{T}} \tag{16.55}$$

Upon adding the reversible and irreversible portions of the heat input we obtain the thermal field equation

$$-\nabla_0^T \mathbf{q}_0 + \rho_0 h = -\mathrm{T}\partial \mathbf{s}_r^T/\partial \mathrm{T}\dot{\mathbf{e}} - \mathbf{s}_i^T \dot{\mathbf{e}} + \rho_0(c_e + c_i)\dot{\mathrm{T}} \tag{16.56}$$

It is easily seen that Equation 16.62 directly reduces to a well-known expression in classical linear thermoelasticity when irreversible terms are suppressed. In addition, under adiabatic conditions in which $-\nabla_0^T \mathbf{q}_0 + \rho_0 h = 0$, most of the "viscous" work $\mathbf{s}_i^T \dot{\mathbf{e}}$ is absorbed as a temperature increase controlled by $\rho_0(c_e + c_i)$, while a smaller portion is absorbed into the elastic strain energy field. Finally, note that the specific heat at constant strain possesses a reversible and an irreversible portion.

16.7 CONSTITUTIVE MODEL IN THERMOVISCOHYPERELASTICITY

If the current formulation is followed, it is sufficient to introduce the *Helmholtz free energy density* and the *dissipation potential* to characterize thermal, mechanical, and viscous behavior

16.7.1 Helmholtz Free Energy Density

In the moderately damped thermohyperelastic material the reversible stress is assumed to satisfy a *thermohyperelastic* constitutive relation suitable for near-incompressible elastomers. Following the earlier development for an undamped elastomer, the Helmholtz free energy is introduced using

$$\phi_r = \phi_{rm}(J_1,J_2,I_3) + \phi_{rt}(\mathrm{T}) + \phi_{rtm}(\mathrm{T},I_3) + \phi_{ro} \tag{16.57}$$

Here ϕ_{rm} represents the purely mechanical response and can be identified as the conventional isothermal strain energy density function associated, for example, with the Mooney–Rivlin model. The formulation can easily be adapted to stretch ratio-based models such as the Ogden model (Ogden, 1986). The function $\phi_{rt}(\mathrm{T})$ represents the purely thermal portion of the Helmholtz free energy density. Finally, $\phi_{rtm}(\mathrm{T},I_3)$ represents thermomechanical effects, again based on the assumption that the primary coupling is through volumetric expansion. The quantity ϕ_{ro} represents the Helmholtz free energy in the reference state, and for simplicity it is assumed to vanish. The forms of ϕ_{rt} and ϕ_{rtm} developed in the previous sections of this chapter are recalled.

$$\phi_{rt}(\mathrm{T}) = c_e\mathrm{T}[1 - \ln(\mathrm{T}/\mathrm{T}_0)] \tag{16.58}$$

$$\phi_{rtm}(\mathrm{T},I_3) = \frac{\kappa}{2\rho_0}[f^3(\mathrm{T})J - 1]^2 \tag{16.59}$$

$$J = I_3^{1/2} = \det(\mathbf{F}), \quad f(\mathrm{T}) = \left[1 + \frac{\alpha}{3}(\mathrm{T} - \mathrm{T}_0)\right]^{-1}$$

and α is of course the volumetric coefficient of thermal expansion. For the sake of illustration, for $\phi_{rm}(J_1,J_2,I_3)$ we display the classical two-term Mooney–Rivlin model (Gent, 1992):

$$\phi_{rm}(J_1,J_2,I_3) = C_1(J_1 - 1) + C_2(J_2 - 1), \qquad I_3 = 1 \tag{16.60}$$

The reversible stress is now stated as

$$\mathbf{s}_r = 2\phi_j\mathbf{n}_j, \quad \phi_j = \partial\phi_r/\partial I_j \tag{16.61}$$

$$\mathbf{n}_1 = \mathbf{i}, \quad \mathbf{i} = VEC(\mathbf{I}), \quad \mathbf{n}_1 = I_1\mathbf{i} - \mathbf{c}, \quad \mathbf{c} = VEC(\mathbf{C}), \quad \mathbf{n}_3 = I_3 VEC(\mathbf{C}^{-1})$$

16.7.2 Dissipation Potential

Fourier's law of heat conduction is recalled as from

$$\rho\boldsymbol{\Psi}_t = \frac{1}{2k_t}[\mathbf{q}_0^T\mathbf{q}_0] \tag{16.62}$$

The viscous stress $\mathbf{s}_i$ depends on the shear part of the strain rate as well as the temperature. However, since the elastomers of interest are nearly incompressible, to good approximation $\mathbf{s}_i$ can be taken as a function of the (total) Lagrangian strain rate.

The current framework admits several possible expressions for $\boldsymbol{\Psi}_i$, of which an example was already given in Section 16.6.3. Here, taking a more general viewpoint, we seek expressions of the form $\boldsymbol{\Psi}_i = \frac{1}{2}\dot{\mathbf{e}}^T\mathbf{D}_v(\mathbf{e},\mathrm{T})\dot{\mathbf{e}}$ in which $\mathbf{D}_v(\mathbf{e},\mathrm{T})$ is called the viscosity tensor; it is symmetric and positive definite to satisfy Equation 16.45b. (Of course, the correct expression is determined by experiments.) The simple example in Section 16.6.3 corresponds to $\mathbf{D}_v(\mathbf{e},\mathrm{T}) = \mu(\mathbf{e},\mathrm{T})\mathbf{I}$. As a second example, to ensure isotropy suppose that $\boldsymbol{\Psi}_i$ depends on $\dot{J}_1$ and $\dot{J}_2$ through a relation of the form $\boldsymbol{\Psi}_i(\dot{J}_1,\dot{J}_2,\mathbf{e},\mathrm{T})$, and note that

$$\dot{J}_1 = \mathbf{m}_1^T\dot{\mathbf{e}}, \quad \mathbf{m}_1^T = 2\left[\mathbf{i}^T - \frac{1}{3}\frac{I_1}{I_3}\mathbf{n}_3^T\right]\Big/I_3^{1/3}, \quad \dot{J}_2 = \mathbf{m}_2^T\dot{\mathbf{e}}, \quad \mathbf{m}_2^T = 2\left[\mathbf{n}_2^T - \frac{2}{3}\frac{I_2}{I_3}\mathbf{n}_3^T\right]\Big/I_3^{2/3} \tag{16.63}$$

For an expression reminiscent of the two-term Mooney–Rivlin strain energy function let us consider the specific form:

$$\boldsymbol{\Psi}_i = \mu_1(\mathrm{T})\left[C_{1v}\dot{J}_1^2/2 + C_{2v}\dot{J}_2^2\right] \tag{16.64}$$

in which C_{1v} and C_{2v} are constant positive material coefficients. We now obtain the viscosity tensor

$$\mathbf{D}_v = \mu(\mathrm{T})\left[C_{1v}\mathbf{m}_1\mathbf{m}_1^T + C_{2v}\mathbf{m}_2\mathbf{m}_2^T\right] \tag{16.65}$$

Unfortunately, this tensor is problematic since it is only positive semi-definite. To see this consider whether there exists a nonvanishing vector $\mathbf{b}$ for which $(C_{1v}\mathbf{m}_1\mathbf{m}_1^T + C_{2v}\mathbf{m}_2\mathbf{m}_2^T)\mathbf{b} = 0$.

But this is certainly true since $\mathbf{b}$ need only lie in a subspace exterior to the two-dimensional subspace spanned by $\mathbf{m}_1$ and $\mathbf{m}_2$.

As a third example, suppose that the dissipation potential is expressed in terms of the deformation rate tensor $\mathbf{D}$ as $\boldsymbol{\Psi}_i = \mu(\mathrm{T})tr(\mathbf{D}^2)/2$, which has the advantage that the deformation rate tensor $\mathbf{D}$ is in the observed (current) configuration in which measurements are performed. But now, with $\mathbf{d} = VEC(\mathbf{D})$,

$$\begin{aligned}
\mathbf{d} &= \mathbf{F}^{-T} \otimes \mathbf{F}^{-T}\dot{\mathbf{e}} \\
\mathbf{D}_v &= \mu(T)\mathbf{F}^{-1} \otimes \mathbf{F}^{-1}\mathbf{F}^{-T} \otimes \mathbf{F}^{-T} \\
&= \mu(T)(\mathbf{F}^{-1}\mathbf{F}^{-T}) \otimes (\mathbf{F}^{-1}\mathbf{F}^{-T}) \\
&= \mu(T)(2\mathbf{E} + \mathbf{I})^{-1} \otimes (2\mathbf{E} + \mathbf{I})^{-1}
\end{aligned} \tag{16.66}$$

which is always positive definite if $\mu(\mathrm{T}) > 0$.

16.8 VARIATIONAL PRINCIPLES AND FINITE ELEMENT EQUATIONS FOR THERMOVISCOHYPERELASTIC MATERIALS

16.8.1 Mechanical Equilibrium

In this section we present one of the several possible formulations for the finite element equations of a thermoviscohyperelastic medium, neglecting inertia. Application of variational methods to the mechanical field equation (Balance of Linear Momentum) furnishes the Principle of Virtual Work in the form

$$\int tr(\delta\mathbf{E}\mathbf{S}_i)\,\mathrm{d}V_0 = \int \delta\mathbf{u}^T\mathbf{t}_0\,\mathrm{d}S_0 - \int tr(\delta\mathbf{E}\mathbf{S}_r)\,\mathrm{d}V_0 \tag{16.67}$$

in which $\mathbf{t}_0$ as usual denotes the traction vector on the undeformed surface S_0. As illustrated in the examples of the foregoing section, we expect that the dissipation potential has the form $\boldsymbol{\Psi}_i = \frac{1}{2}\dot{\mathbf{e}}^T\mathbf{D}_v(\mathbf{e},T)\dot{\mathbf{e}}$, from which

$$\mathbf{s}_i = \mathbf{D}_v\dot{\mathbf{e}} \tag{16.68}$$

and $\mathbf{D}_v$ is of course the viscosity tensor and is symmetric. Furthermore, it is positive definite since $\mathbf{s}_i^T\dot{\mathbf{e}} \geq 0$ for all $\dot{\mathbf{e}}$. Equation 16.67 is thus rewritten as

$$\int \delta\mathbf{e}^T\mathbf{D}_v\dot{\mathbf{e}}\,\mathrm{d}V_0 = \int \delta\mathbf{u}^T\mathbf{t}_0\,\mathrm{d}S_0 - \int \delta\mathbf{e}^T\mathbf{s}_r\,\mathrm{d}V_0 \tag{16.69}$$

With interpolation models of the form $\dot{\mathbf{e}}(\mathbf{X},t) = \boldsymbol{\beta}^T(\mathbf{X})\boldsymbol{\Phi}\dot{\boldsymbol{\gamma}}(t)$, $\mathbf{u}(\mathbf{X},t) = \boldsymbol{\varphi}^T(\mathbf{X})\boldsymbol{\Phi}\boldsymbol{\gamma}(t)$, and $\mathrm{T}(\mathbf{X},t) - \mathrm{T}_0 = \boldsymbol{\nu}^T(\mathbf{X})\boldsymbol{\Psi}\boldsymbol{\theta}(t)$, at the element level the finite element equation for the mechanical field now becomes

$$\mathbf{K}_v(\boldsymbol{\gamma},\mathrm{T})\dot{\boldsymbol{\gamma}} = \mathbf{f} - \mathbf{f}_r(\boldsymbol{\gamma}),\ \mathbf{K}_v = \int \boldsymbol{\Phi}^T\boldsymbol{\beta}\mathbf{D}_v(\mathbf{X},\boldsymbol{\gamma},\boldsymbol{\theta})\boldsymbol{\beta}^T\boldsymbol{\Phi}\,dV_0,\ \mathbf{f}_r(\boldsymbol{\gamma},t) = \int \boldsymbol{\Phi}^T\boldsymbol{\beta}\mathbf{s}_r(\mathbf{X},t,\boldsymbol{\gamma})\,dV_0 \tag{16.70}$$

The tangent viscous matrix $\mathbf{K}_v$ now plays the role of the tangent modulus matrix in hyperelasticity. Hyperelastic effects now appear in the force term $\mathbf{f}_r(\boldsymbol{\gamma},t)$.

16.8.2 Thermal Equilibrium Equation

The equation for thermal equilibrium is rewritten as

$$\rho_0(\dot{\eta}_r + \dot{\eta}_{i2}) = \left[-\nabla_0^T\mathbf{q}_0 + \rho_0 h\right]/\mathrm{T} \tag{16.71}$$

With some effort using Equation 16.56, a rate (incremental) variational principle may be obtained in the form

$$\begin{aligned}&\int \delta\mathrm{T}\left[-\mathrm{T}\partial\mathbf{s}_r^T/\partial\mathrm{T}\dot{\mathbf{e}} - \mathbf{s}_i^T\dot{\mathbf{e}} + \rho_0(c_e + c_i)\dot{\mathrm{T}}\right]/\mathrm{T}\,dV_0\\ &= \int \delta\mathrm{T}\left[\left[-\nabla_0^T\mathbf{q}_0 + \rho_0 h\right]/\mathrm{T}\right]dV_0\end{aligned} \tag{16.72}$$

Upon approximating T in the denominator by T_0, letting k denote the thermal conductivity, and invoking the isotropic Fourier law of heat conduction, we may obtain the thermal equilibrium equation in the form

$$\begin{aligned}&\int \delta\mathrm{T}\left[-\partial\mathbf{s}_r^T/\partial\mathrm{T}\dot{\mathbf{e}} - \mathbf{s}_i^T\dot{\mathbf{e}} + \rho_0(c_e + c_i)\dot{\mathrm{T}}\right]dV_0 + \int k(\nabla_0\delta\mathrm{T})^T\nabla_0\mathrm{T}\,dV_0\\ &= \int \delta\mathrm{T}\rho_0 h\,dV_0 - \int \delta\mathrm{T}\mathbf{n}^T\mathbf{q}_0\,dS_0\end{aligned} \tag{16.73}$$

Using the usual interpolation models for displacement and temperature, Equation 16.73 reduces directly to finite element equation for the thermal field. The equation assumes the form

$$\begin{aligned}&\mathbf{M}_T\dot{\boldsymbol{\theta}} + \boldsymbol{\Sigma}_{TM}(\boldsymbol{\gamma},\boldsymbol{\theta})\dot{\boldsymbol{\gamma}} + \mathbf{K}_T\boldsymbol{\theta} = \mathbf{f}_h + \mathbf{f}_q\\ &\boldsymbol{\Sigma}_{TM}(\boldsymbol{\gamma},\boldsymbol{\theta}) = \int \boldsymbol{\Psi}^T\boldsymbol{\nu}\left[-\partial\mathbf{s}_r^T/\partial\mathrm{T} - \mathbf{s}_i^T\right]\boldsymbol{\beta}^T\boldsymbol{\Phi}\,dV_0\end{aligned} \tag{16.74}$$

17 Tangent Modulus Tensors for Inelastic and Thermoinelastic Materials

17.1 PLASTICITY

The theories of plasticity and thermoplasticity model material behavior in important applications such as metal forming, ballistics, and welding. The main goals of the current section are to present an example of constitutive models in plasticity, viscoplasticity, thermoplasticity, and damage mechanics, to derive the corresponding tangent modulus tensors, and to formulate variational and finite element statements, all while accommodating the challenging problems of finite strain and kinematic hardening. But we first start with a presentation of a constitutive model in small strain isothermal plasticity.

17.2 TANGENT MODULUS TENSOR IN SMALL STRAIN ISOTHERMAL PLASTICITY

The basic assumptions are given below.

1. **Kinematic Decomposition**

   The strain rate decomposes into an elastic (recoverable, reversible) portion and an inelastic (permanent, irreversible) portion as follows:

$$\dot{\mathbf{E}} = \dot{\mathbf{E}}_e + \dot{\mathbf{E}}_i \tag{17.1}$$

   in which $\dot{\mathbf{E}}_e$ satisfies small strain isotropic elastic relations (inverse of the Lamé form)

$$\dot{\mathbf{E}}_e = \frac{1}{2\mu}\left(\dot{\mathbf{S}} - \dot{\mathbf{S}}^* - \frac{\lambda}{2\mu + 3\lambda} tr(\dot{\mathbf{S}})\mathbf{I}\right) \tag{17.2a}$$

in which $\mathbf{S}_d^*$ is a traceless reference stress (backstress) to be introduced shortly. The dilatational and deviatoric portions of this relation satisfy

$$\dot{\mathbf{E}}_{ed} = \frac{1}{2\mu}\left(\dot{\mathbf{S}}_d - \dot{\mathbf{S}}^*\right), \quad tr(\dot{\mathbf{E}}_e) = \frac{1}{2\mu + 3\lambda} tr(\dot{\mathbf{S}})\mathbf{I} \tag{17.2b}$$

In *VEC* notation, Equation 17.2 is restated as

$$\dot{\mathbf{e}}_e = \mathbf{C}_e(\dot{\mathbf{s}} - \dot{\mathbf{s}}^*), \quad \mathbf{C}_e = \frac{1}{2\mu}\left(\mathbf{I}_9 - \frac{\lambda}{2\mu + 3\lambda}\mathbf{i}\mathbf{i}^T\right) \tag{17.3a}$$

and $\mathbf{C}_e$ is the elastic compliance tensor (inverse of the elastic modulus tensor). For the deviatoric and dilatational parts the corresponding relations are

$$\dot{\mathbf{e}}_{ed} = \mathbf{C}'_e(\dot{\mathbf{s}}_d - \dot{\mathbf{s}}^*), \quad \mathbf{C}'_e = \frac{\mathbf{I}_9}{2\mu}$$

$$\mathbf{i}^T\dot{\mathbf{e}}_e = \frac{1}{2\mu + 3\lambda}\mathbf{i}^T\dot{\mathbf{s}} \tag{17.3b}$$

Typically, plastic strain is viewed as permanent strain. As illustrated in Figure 17.1, in a uniaxial tensile specimen the stress S_{11} may be increased to the point A, and then unloaded along the path AB. The slope of the unloading portion is E, the same as that of the initial elastic portion. When the stress becomes zero, there still is a residual strain E_i, which may be identified as the inelastic strain. However, if the stress had instead been increased to point C, it would encounter *reversed loading* at point D, which reflects the fact that the elastic region need not include the zero-stress value.

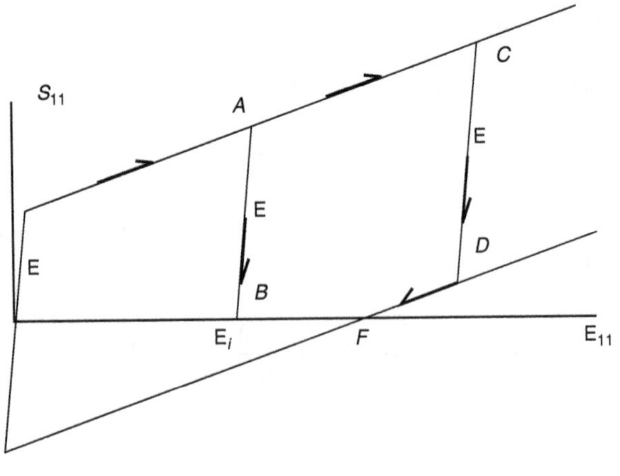

FIGURE 17.1 Illustration of inelastic strain.

2. **Inelastic Incompressibility**

The inelastic strain does not contribute to the volume strain.

$$tr(\dot{\mathbf{E}}_i) = \mathbf{i}^T \dot{\mathbf{e}}_i = 0 \tag{17.4}$$

It follows that inelastic strain only contributes to the shear (deviatoric) portion of the strain, and hence is said to occur in shear.

3. **Independence of the Hydrostatic Stress**

In plasticity inelastic strain is assumed not to be affected by the hydrostatic (isotropic) portion of the stress, $tr(\mathbf{S})$, and to depend only on the deviatoric (shear) portion of the stress,

$$\mathbf{S}_d = \mathbf{S} - \tfrac{1}{3} tr(\mathbf{S})\mathbf{I} \quad \text{or} \quad \mathbf{s}_d = \left(\mathbf{I}_9 - \tfrac{1}{3}\mathbf{i}\mathbf{i}^T\right)\mathbf{s}$$

4. **Yield and Loading Conditions**

There exists a function $\Psi(\mathbf{s}_d - \mathbf{s}_d^*, \mathbf{e}_i, \mathbf{k}) = 0$, which represents a closed convex surface in stress space, called the *yield surface*. Here $\mathbf{s}^*$ is a point interior to the yield surface, sometimes called the *backstress*. It changes if inelastic strain changes, referred to as *inelastic flow*, and the elastic strain vanishes when $\mathbf{s} = \mathbf{s}^*$. If flow is occurring the current stress point must be located on this surface. In Section 17.5, we will consider viscoplasticity, in which the notion of the yield surface is modified such that the current stress point is now exterior to the yield surface during flow. The current stress point may then be viewed as being on a loading surface while the yield surface is now a reference surface.

In plasticity, there are three possibilities referred to as the *loading conditions*.

$$\text{(a)} \quad \Psi(\mathbf{s}_d - \mathbf{s}_d^*, \mathbf{e}_i, \mathbf{k}) < 0$$

The stress point is interior to the yield surface and only elastic strain is changing. The stress changes are proportional to strain changes through elastic relations.

$$\text{(b)} \quad \Psi(\mathbf{s}_d - \mathbf{s}_d^*, \mathbf{e}_i, \mathbf{k}) = 0 \quad \text{and} \quad \frac{\partial \Psi}{\partial \mathbf{s}_d} \dot{\mathbf{s}}_d = 0$$

The stress point is located on the yield surface, but is moving tangentially to the surface. In this event inelastic strain is unchanging, and the stress changes are related to the strain changes by elastic relations. This situation is referred to as *neutral loading*.

$$\text{(c)} \quad \Psi(\mathbf{s}_d - \mathbf{s}_d^*, \mathbf{e}_i, \mathbf{k}) = 0 \quad \text{and} \quad \frac{\partial \Psi}{\partial \mathbf{s}_d} \dot{\mathbf{s}}_d > 0$$

The stress point is located on the yield surface and is moving toward the exterior of the surface. Inelastic flow occurs in this case, which is referred to as *loading*.

5. **Hardening**

The vector $\mathbf{k}$ represents the "hardening" effect of inelastic deformation history, reflecting irreversibility of plastic deformation. Visually hardening occurs as shape, size, and location changes of the yield surface. The hardening vector $\mathbf{k}$ is assumed to satisfy an evolution relation of the form

$$\dot{\mathbf{k}} = \mathbf{H}_1(\mathbf{s}_d - \mathbf{s}_d^*, \mathbf{e}_i, \mathbf{k})\dot{\mathbf{e}}_i \tag{17.5}$$

in which $\mathbf{H}_1$ is an experimentally determined matrix.

Work hardening is a commonly made assumption. Here it is expressed by the relation

$$\dot{\mathbf{k}} = k_1(\mathbf{s} - \mathbf{s}^*)\dot{\mathbf{e}}_i \tag{17.6a}$$

in which k_1 is a material constant.

An example of a work hardening model for evolution of the backstress is given by

$$\dot{\mathbf{s}}^* = k_2(\mathbf{s} - \mathbf{s}^*)\dot{\mathbf{e}}_i \tag{17.6b}$$

and k_2 is a material constant. A relation of this form will play a role in our later treatment of kinematic hardening.

6. **Associated Flow Rule**

The inelastic strain rate vector in plasticity is assumed to be normal to the yield surface at the current stress point.

$$\dot{\mathbf{e}}_i = \Lambda\left(\frac{\partial\Psi}{\partial\mathbf{s}}\right)^T \tag{17.7}$$

Note that $\dot{\mathbf{s}}^T\dot{\mathbf{e}}_i = \Lambda\dot{\mathbf{s}}^T\left(\frac{\partial\Psi}{\partial\mathbf{s}}\right)^T$. If $\Lambda > 0$, the associated flow rule then implies that $\dot{\mathbf{s}}^T\dot{\mathbf{e}}_i > 0$ during plastic deformation, which is known as Drucker's criterion for stability in the small (cf. Rowe et al., 1991). It also follows from the assumption that the yield surface is convex that $(\mathbf{s} - \mathbf{s}^*)\dot{\mathbf{e}}_i \geq 0$.

7. **Consistency Condition**

The consistency condition states that the stress point remains on the yield surface *during loading*, in which event the yield function satisfies the relation

$$\dot{\Psi} = \frac{\partial\Psi}{\partial(\mathbf{s}_d - \mathbf{s}_d^*)}(\dot{\mathbf{s}}_d - \dot{\mathbf{s}}_d^*) + \left(\frac{\partial\Psi}{\partial\mathbf{e}_i} + \frac{\partial\Psi}{\partial\mathbf{k}}\mathbf{H}_1\right)\dot{\mathbf{e}}_i = 0 \tag{17.8}$$

Thanks to hardening, the yield surface deforms or moves such that the stress point remains on it if plastic flow is occurring, even as the stress point is moving toward the exterior of the yield surface.

8. **Constitutive Relation for the Inelastic Strain Rate**
Equations 17.7 and 17.8 imply that

$$\frac{\partial\Psi}{\partial(\mathbf{s}_d - \mathbf{s}_d^*)}(\mathbf{s}_d - \mathbf{s}_d^*)^\bullet + \Lambda\left(\frac{\partial\Psi}{\partial\mathbf{e}_i} + \frac{\partial\Psi}{\partial\mathbf{k}}\mathbf{H}_1\right)\left(\frac{\partial\Psi}{\partial\mathbf{s}_d}\right)^T = 0 \tag{17.9}$$

The parameter Λ is immediately seen to be

$$\Lambda = \frac{\dfrac{\partial\Psi}{\partial(\mathbf{s}_d - \mathbf{s}_d^*)}(\dot{\mathbf{s}}_d - \dot{\mathbf{s}}_d^*)}{-\left(\dfrac{\partial\Psi}{\partial\mathbf{e}_i} + \dfrac{\partial\Psi}{\partial\mathbf{k}}\mathbf{H}_1\right)\left(\dfrac{\partial\Psi}{\partial(\mathbf{s}_d - \mathbf{s}_d^*)}\right)^T} \tag{17.10}$$

The inelastic strain rate is now seen to be given by

$$\dot{\mathbf{e}}_i = \mathbf{C}_i(\dot{\mathbf{s}}_d - \dot{\mathbf{s}}_d^*)$$

in which

$$\mathbf{C}_i = \frac{\left(\dfrac{\partial\Psi}{\partial(\mathbf{s}_d - \mathbf{s}_d^*)}\right)^T \dfrac{\partial\Psi}{\partial(\mathbf{s}_d - \mathbf{s}_d^*)}\dot{\mathbf{s}}_d}{-\left(\dfrac{\partial\Psi}{\partial\mathbf{e}_i} + \dfrac{\partial\Psi}{\partial\mathbf{k}}\mathbf{H}_1\right)\left(\dfrac{\partial\Psi}{\partial(\mathbf{s}_d - \mathbf{s}_d^*)}\right)^T} \tag{17.11}$$

The requirement that $\Lambda > 0$ implies that $-\left(\frac{\partial\Psi}{\partial\mathbf{e}_i} + \frac{\partial\Psi}{\partial\mathbf{k}}\mathbf{H}_i\right)\left(\frac{\partial\Psi}{\partial(\mathbf{s}_d - \mathbf{s}_d^*)}\right)^T > 0$, in which event $\mathbf{C}_i$ becomes *positive semidefinite*. In terms of the full (as opposed to deviatoric) stress tensor $\mathbf{S} = IVEC(\mathbf{s})$, the relation governing the inelastic strain rate is now

$$\dot{\mathbf{e}}_i = \mathbf{C}_i\left(\mathbf{I}_9 - \tfrac{1}{3}\mathbf{i}\mathbf{i}^T\right)(\dot{\mathbf{s}} - \dot{\mathbf{s}}^*) \tag{17.12}$$

9. **Tangent Modulus Tensor**
We first suppose that the evolution of the backstress follows a hardening model of the form

$$\dot{\mathbf{s}}_d^* = \begin{cases} \mathbf{H}_2\dot{\mathbf{e}}_i & \text{loading} \\ \mathbf{0} & \text{otherwise} \end{cases} \tag{17.13}$$

But now $\dot{\mathbf{e}}_i = \mathbf{C}_i(\dot{\mathbf{s}}_d - \dot{\mathbf{s}}_d^*) = \mathbf{C}_i\dot{\mathbf{s}}_d - \mathbf{C}_i\mathbf{H}_2\dot{\mathbf{e}}_i$, with the consequence that

$$\dot{\mathbf{e}}_i = (\mathbf{I} + \mathbf{C}_i\mathbf{H}_2)^{-1}\mathbf{C}_i\dot{\mathbf{s}}_d \tag{17.14}$$

assuming that $\mathbf{I}+\mathbf{C}_i\mathbf{H}_2$ is nonsingular, as seems very reasonable owing to the typically small magnitude of $\mathbf{C}_i$. It follows that $\dot{\mathbf{s}}^* = \mathbf{H}_2\dot{\mathbf{e}}_i = \mathbf{H}_2\mathbf{C}_i(\dot{\mathbf{s}}_d - \dot{\mathbf{s}}^*)$, and hence

$$\dot{\mathbf{s}}^* = (\mathbf{I}+\mathbf{H}_2\mathbf{C}_i)^{-1}\mathbf{H}_2\mathbf{C}_i\dot{\mathbf{s}}_d \tag{17.15}$$

provided that plastic flow is occurring. We now add the deviatoric elastic and inelastic strain rates to obtain the total deviatoric strain rate.

$$\begin{aligned}\dot{\mathbf{e}}_d = \dot{\mathbf{e}}_{ed} + \dot{\mathbf{e}}_i &= (\mathbf{C}'_e + \mathbf{C}_i)(\dot{\mathbf{s}}_d - \dot{\mathbf{s}}^*)\\ &= \left(\frac{\mathbf{I}_9}{2\mu} + \mathbf{C}_i\right)(\mathbf{I}+\mathbf{H}_2\mathbf{C}_i)^{-1}\dot{\mathbf{s}}_d\end{aligned} \tag{17.16}$$

It follows that $\dot{\mathbf{s}}_d = (\mathbf{I}+\mathbf{H}_2\mathbf{C}_i)\left(\frac{\mathbf{I}_9}{2\mu} + \mathbf{C}_i\right)^{-1}\dot{\mathbf{e}}_d$ during plastic flow.

The dilatational portion of the strain rate is elastic and satisfies Equation 17.3b: $\mathbf{i}^T\dot{\mathbf{s}} = (2\mu + 3\lambda)\mathbf{i}^T\dot{\mathbf{e}}$. The relation between the total stress rate and the total strain rate may be derived to obtain the elastic–plastic tangent modulus tensor $\mathbf{D}_{\text{ep}}$:

$$\dot{\mathbf{s}} = \dot{\mathbf{s}}_d + \mathbf{i}\mathbf{i}^T\dot{\mathbf{s}} = \mathbf{D}_{\text{ep}}\dot{\mathbf{e}}$$

$$\mathbf{D}_{\text{ep}} = (\mathbf{I}+\mathbf{H}_2\mathbf{C}_i)\left(\frac{\mathbf{I}_9}{2\mu} + \mathbf{C}_i\right)^{-1}\left(\mathbf{I}_9 - \frac{\mathbf{i}\mathbf{i}^T}{3}\right) + (2\mu + 3\lambda)\mathbf{i}\mathbf{i}^T \tag{17.17}$$

Recall that the Incremental Principle of Virtual Work requires the tensor relating the stress increment to the strain increment. We may now say that

$$\Delta_n\mathbf{s} \approx \mathbf{D}_{\text{ep}}\Delta_n\mathbf{e} \tag{17.18}$$

Equation 17.18 indicates the rate independent (inviscid) nature of plasticity since the strain increment is proportional to the stress increment *no matter how rapidly or slowly the stress is applied.*

EXAMPLE 17.1

Von Mises yield surface with kinematic and isotropic hardening

The entitled yield function is given by

$$\Psi_i = \sqrt{(\mathbf{s}_d - k_1\mathbf{e}_i)^T(\mathbf{s}_d - k_1\mathbf{e}_i)} - \left(k_0 + k_2\int \mathbf{s}_d^T\dot{\mathbf{e}}_i\,dt\right) = 0$$

Observe that work hardening is present since $\mathbf{k} = k_2\int \mathbf{s}_d^T\dot{\mathbf{e}}_i\,dt$. Also $\frac{\partial\Psi_i}{\partial\mathbf{k}} \to k_2$ and $\mathbf{H}_1 \to k_2\mathbf{s}_d^T$. In addition, $\mathbf{s}^* = k_1\mathbf{e}_i$, so that $\mathbf{H}_2 \to k_1$. After straighforward manipulation,

$$\frac{\partial \Psi_i}{\partial \mathbf{s}_d} = \mathbf{n}^T, \quad \mathbf{n}^T = \frac{(\mathbf{s}_d - k_1 \mathbf{e}_i)^T}{\sqrt{(\mathbf{s}_d - k_1 \mathbf{e}_i)^T (\mathbf{s}_d - k_1 \mathbf{e}_i)}}$$

$$\frac{\partial \Psi_i}{\partial \mathbf{e}_i} + \frac{\partial \Psi_i}{\partial \mathbf{k}} \mathbf{H}_1 = -k_1 \mathbf{n}^T + k_2 \mathbf{s}_d^T$$

from which

$$\mathbf{C}_i = \frac{\mathbf{n}\mathbf{n}^T}{k_1 + k_2 \mathbf{s}_d^T \mathbf{n}}$$

This matrix is positive semidefinite assuming that $k_1 + k_2 \mathbf{s}_d^T \mathbf{n} > 0$. Substitution into Equation 17.17 immediately yields $\mathbf{D}_{\mathrm{ep}}$.

EXAMPLE 17.2

Yield surface with strain hardening: small deformation and uniaxial loading

In isothermal plasticity, assuming the following yield function, find the stress–strain curve under uniaxial loading.

$$\Psi_i = \sqrt{(\mathbf{s}_d - k_1 \mathbf{e}_i)^T (\mathbf{s}_d - k_1 \mathbf{e}_i)} - \left(k_0 + k_2 \int_0^t \sqrt{\dot{\mathbf{e}}_i^T \dot{\mathbf{e}}_i}\, \mathrm{d}t \right)$$

We again assume that plastic strain is incompressible: $tr(\mathbf{E}_i) = 0$. This yield surface exhibits strain hardening in that the radius of the yield surface depends on the arc length traversed in inelastic strain space.

Solution

In the case of uniaxial stress and isotropy, $\mathrm{S}_{yy} = \mathrm{S}_{zz} = 0$, with the consequence that

$$\mathrm{S}_{dxx} = \mathrm{S}_{xx} - \tfrac{1}{3}(\mathrm{S}_{xx} + 0 + 0) = \tfrac{2}{3}\,\mathrm{S}_{xx}$$

$$\mathrm{S}_{dyy} = \mathrm{S}_{dzz} = -\tfrac{1}{3}\,\mathrm{S}_{xx}$$

For plastic incompressibility, $tr(\mathbf{E}_i) = 0$. Hence, $\mathrm{E}^p_{xx} + \mathrm{E}^p_{yy} + \mathrm{E}^p_{zz} = 0$. And so for uniaxial loading $\mathrm{E}^p_{yy} = \mathrm{E}^p_{zz} = -\tfrac{1}{2}\mathrm{E}^p_{xx}$. The consistency condition $\Psi_i = 0$ now implies that

$$\sqrt{\left(\tfrac{2}{3}\,\sigma_{xx} - k_1 e^p_{xx}\right)^2 + \left(-\tfrac{1}{3}\,\sigma_{xx} + \tfrac{1}{2}\,k_1 e^p_{xx}\right)^2} = k_0 + k_2 \int_0^t \sqrt{\dot{e}^{p2}_{xx} + \left(\tfrac{1}{2}\,\dot{e}^p_{xx}\right)^2 + \left(\tfrac{1}{2}\,\dot{e}^p_{xx}\right)^2}\, \mathrm{d}t$$

and hence

$$\sqrt{\tfrac{2}{3}}\left(\sigma_{xx} - \tfrac{3}{2}\,k_1 e^p_{xx}\right) = k_0 + k_2 \sqrt{\tfrac{3}{2}}\, e^p_{xx} \quad \text{so that} \quad \sigma_{xx} = \sqrt{\tfrac{3}{2}}\, k_0 + \tfrac{3}{2}(k_1 + k_2) e^p_{xx}$$

The ensuing uniaxial stress–strain curve is depicted in Figure 17.2.

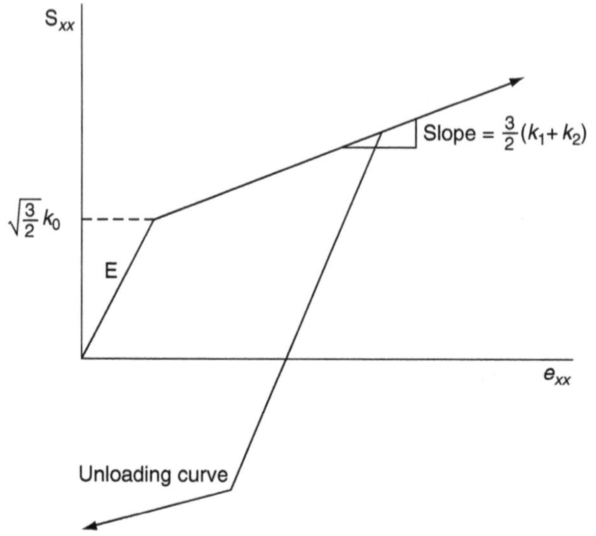

FIGURE 17.2 Stress–strain path with strain hardening.

17.3 PLASTICITY UNDER FINITE STRAIN

17.3.1 Kinematics

The deformation rate tensor admits an additive decomposition into elastic and inelastic portions.

$$\mathbf{D} = \mathbf{D}_e + \mathbf{D}_i \tag{17.19}$$

The Lagrangian strain tensor $\mathbf{E}$ satisfies the relation $\dot{\mathbf{E}} = \mathbf{F}^T\mathbf{D}\mathbf{F}$, from which we formally introduce the elastic and inelastic strains (for large deformation) as

$$\mathbf{E}_e = \int \mathbf{F}^T\mathbf{D}_e\mathbf{F}\,\mathrm{d}t, \quad \mathbf{E}_i = \int \mathbf{F}^T\mathbf{D}_i\mathbf{F}\,\mathrm{d}t \tag{17.20}$$

Of course there are alternatives to this type of decomposition, for example, the logarithmic plastic strain (Xiao et al., 1997).

17.3.2 Plasticity

We present an example of a constitutive equation for plasticity at large deformation to illustrate how the tangent modulus tensor is formulated. For simplicity we ignore the difference between the stress and strain and their deviatoric counterparts (i.e., ignore elastic strain), and we also assume that the backstress remains at the origin in stress space (no kinematic hardening). With $\boldsymbol{\chi}_e$ the tangent modulus tensor relating $\dot{\mathbf{s}}$ to $\dot{\mathbf{e}}$ under elastic conditions (retaining the assumption of rate independence), the constitutive equation of interest is obtained from the previous section using

$$\dot{\mathbf{e}}_i = \mathbf{C}_i\dot{\mathbf{s}}, \quad \dot{\mathbf{e}}_e = \mathbf{C}_e\dot{\mathbf{s}}, \quad \mathbf{C}_e = \chi_e^{-1}, \quad \dot{\mathbf{k}} = \mathbf{H}(\mathbf{s},\mathbf{e}_i,\mathbf{k})\dot{\mathbf{e}}_i$$

$$\mathbf{C}_i(\mathbf{s},\mathbf{e}_i,\mathbf{k}) = \left(\frac{\partial \Psi_i}{\partial \mathbf{s}}\right)^T \frac{\partial \Psi_i}{\partial \mathbf{s}} \bigg/ h, \quad h = -\left[\left(\frac{\partial \Psi_i}{\partial \mathbf{e}_i} + \frac{\partial \Psi_i}{\partial \mathbf{k}}\mathbf{H}\right)\left(\frac{\partial \Psi_i}{\partial \mathbf{s}}\right)^T\right] \tag{17.21}$$

As before Ψ_i is called the yield function, but now it is a function of the inelastic portion of the Lagrangian strain, as well as of the history of inelastic deformation represented by, say, work hardening.

Combining the elastic and inelastic portions furnishes the tangent modulus tensor as

$$\chi = \left[\chi_e^{-1} + \mathbf{C}_i\right]^{-1} = [\mathrm{I} + \chi_e\mathbf{C}_i]^{-1}\chi_e \tag{17.22}$$

Suppose that in uniaxial tension the elastic portion of the tangent modulus is $\chi_e \to \mathrm{E}_e$, and that the inelastic portion relating the stress increment and the inelastic strain increments is $\mathbf{C}_i^{-1} \to \mathrm{E}_i$. Typically $\mathrm{E}_i \ll \mathrm{E}_e$. The total uniaxial tangent modulus is then $\frac{\mathrm{E}_i}{(1+\mathrm{E}_i/\mathrm{E}_e)}$.

For the sake of visualization we illustrate several possible behaviors of the yield surface. It is distorted by the history of plastic strain through hardening. In Figure 17.3, the conventional model of isotropic hardening is illustrated in which the yield surface expands as a result of plastic deformation. The principal values of the second Piola–Kirchhoff stress are shown on the axes, and the yield surface is shown in the S_I–S_{II} plane. This model is unrealistic in predicting a growing elastic region—reversed plastic loading is typically encountered at much higher stresses than isotropic hardening predicts. An alternative is kinematic hardening (Figure 17.4), in which the yield surface moves with the stress point. After a few percent of plastic strain have developed, the yield surface may cease to encircle the origin. A reference point interior to the yield surface, previously encountered as the backstress, is assumed to serve as the point at which the elastic strain vanishes.

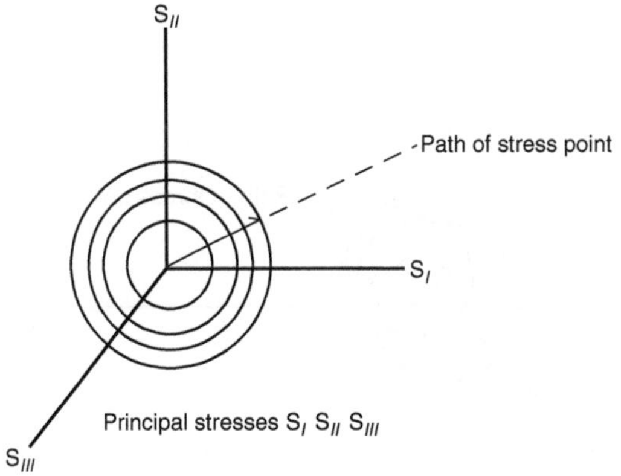

FIGURE 17.3 Illustration of yield surface expansion under isotropic hardening.

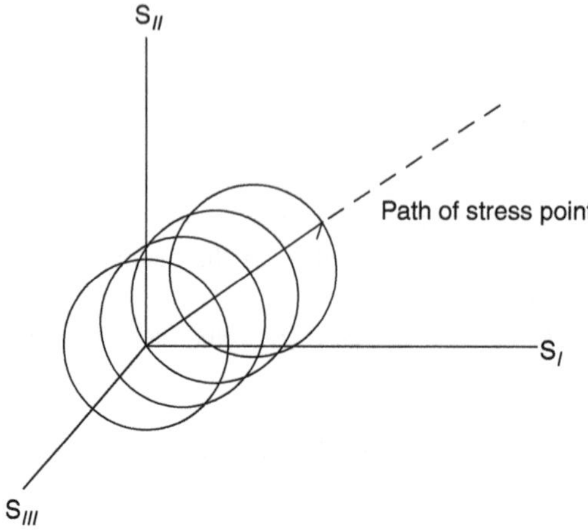

FIGURE 17.4 Illustration of yield surface motion under kinematic hardening.

Combined isotropic and kinematic hardening is shown in Figure 17.5. However, this figure shows that the yield surface contracts, which is consistent with actual observations (e.g., Ellyin, 1997). The rate of movement must in some sense exceed the rate of contraction for the material to remain stable, with a positive definite tangent modulus tensor.

17.4 THERMOPLASTICITY

If plastic work occurs over a sufficiently short time period there is insufficient time for heat to escape, with the consequence that some or all of the plastic work is converted into heat and gives rise to an increased temperature. Of course, it is also

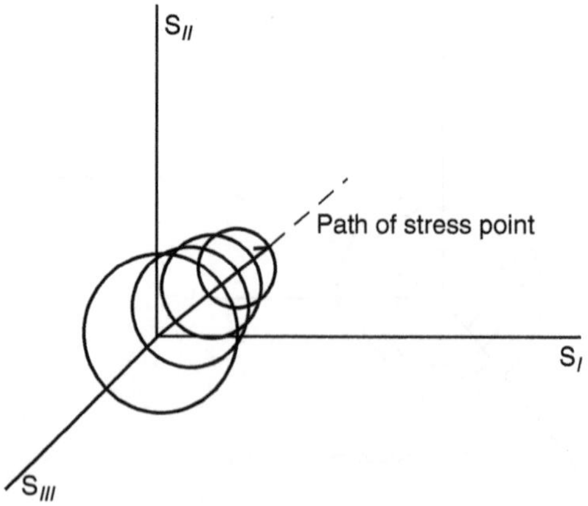

FIGURE 17.5 Illustration of combined kinematic and isotropic hardening.

possible that plastic work occurs in the presence of externally introduced heat. Both effects represent instances of *thermoplasticity*.

As in Chapter 16, following Ziegler and Wehrli (1987) two potential functions are introduced to provide a systematic way to give separate descriptions to reversible and dissipative (irreversible) effects. The first is interpreted as the Helmholtz free energy density and the second is a dissipation potential. To accommodate kinematic hardening we also assume an extension of the Green and Naghdi (GN) (1965) formulation, in which the Helmholtz free energy decomposes into reversible and irreversible parts, with the irreversible part depending on the plastic strain. Here it also depends on the temperature and a workless *internal state variable*.

Application of thermodynamic concepts to inelastic deformation is much more challenging than for damped elastomers treated in Chapter 16. The reason is that inelasticity, for example, via the yield surface, is usually described in stress space while the Helmholtz free energy uses the strain as a state variable.

17.4.1 Balance of Energy

The conventional equation for energy balance is augmented using a vector-valued workless internal variable $\boldsymbol{\alpha}_0$ regarded as representing "microstructural rearrangements."

$$\rho_0\dot{\chi}_0 = \mathbf{s}^T\dot{\mathbf{e}}_r + \mathbf{s}^T\dot{\mathbf{e}}_i - \nabla_0^T\mathbf{q}_0 + \rho_0 h + \boldsymbol{\beta}_0^T\dot{\boldsymbol{\alpha}}_0 \tag{17.23}$$

where χ_0 is the internal energy per unit mass in the undeformed configuration, $\mathbf{s} = VEC(\mathbf{S})$, $\mathbf{e} = VEC(\mathbf{E})$, and $\boldsymbol{\beta}_0$ is the "flux" per unit mass associated with $\boldsymbol{\alpha}_0$. However, note that $\boldsymbol{\beta}_0 = \mathbf{0}$ for $\boldsymbol{\alpha}_0$ to be workless, and hence its reversible and irreversible portions are related by $\boldsymbol{\beta}_{0i} = -\boldsymbol{\beta}_{0r}$. Also $\mathbf{q}_0$ is the heat flux vector referred to undeformed coordinates and h is the heat input per unit mass, for simplicity assumed to be independent of temperature. For use in the Helmholtz free energy, the state variables are recognized to be $\mathbf{E}_r$, $\mathbf{E}_i$, T, and $\boldsymbol{\alpha}_0$.

The next few paragraphs will go over some of the same ground as for damped elastomers in Chapter 16, except for two major points. In Chapter 16, the stress was assumed to decompose into reversible and irreversible portions, in the spirit of elementary Voigt models of viscoelasticity. In the current context, the strain shows a corresponding decomposition, in the spirit of the classical Maxwell models of viscoelasticity. In addition, introducing a workless internal variable $\boldsymbol{\alpha}_0$ will be seen to give the model the flexibility to accommodate phenomena such as kinematic hardening.

The Helmholtz free energy ϕ_0 per unit mass and the entropy η per unit mass are again introduced using

$$\phi_0 = \chi_0 - \mathrm{T}\eta_0 \tag{17.24}$$

The balance of energy, which is Equation 17.23 governing the thermal field, is now rewritten as

$$\nabla_0^T\mathbf{q}_0 - \rho_0 h = \mathbf{s}^T\dot{\mathbf{e}}_r + \mathbf{s}^T\dot{\mathbf{e}}_i - \rho_0\mathrm{T}\dot{\eta}_0 - \rho_0\eta_0\dot{\mathrm{T}} - \rho_0\dot{\phi}_0 + \boldsymbol{\beta}_0^T\dot{\boldsymbol{\alpha}}_0 \tag{17.25}$$

17.4.2 Entropy Production Inequality

Entropy production now is governed by the inequality

$$\begin{aligned}\rho_0 \mathrm{T}\dot{\eta}_0 &\geq -\nabla_0^T \mathbf{q}_0 + \rho_0 h + \mathbf{q}_0^T \nabla \mathrm{T}/\mathrm{T} \\ &\geq \rho_0 \dot{\phi}_0 - \mathbf{s}^T \dot{\mathbf{e}}_r - \mathbf{s}^T \dot{\mathbf{e}}_i + \rho_0 \mathrm{T}\dot{\eta}_0 + \rho_0 \eta_0 \dot{\mathrm{T}} + \mathbf{q}_0^T \nabla \mathrm{T}/\mathrm{T} - \boldsymbol{\beta}_0^T \dot{\boldsymbol{\alpha}}_0 \end{aligned} \tag{17.26}$$

Now viewing ϕ_{0r} as a differentiable function of $\mathbf{e}_r$, T, and $\boldsymbol{\alpha}_0$, we conclude that

$$\partial\phi_{0r}/\partial\mathbf{e}_r = \mathbf{s}^T, \quad \partial\phi_{0r}/\partial\mathrm{T} = -\rho_0 \eta_{r0}, \quad \partial\phi_{0r}/\partial\boldsymbol{\alpha}_0 = \boldsymbol{\beta}_{0r}^T \tag{17.27}$$

Extending the GN formulation, we introduce the stress $\mathbf{s}^*$ using $\mathbf{s}^{*T} = \rho_0 \partial\phi_i/\partial\mathbf{e}_i$ and assume that $\eta_0 = -\partial\phi_0/\partial\mathrm{T}$ and $\rho_0 \partial\phi_{0i}/\partial\boldsymbol{\alpha}_0 = \boldsymbol{\beta}_{0i}^T$:

$$\begin{aligned}\mathbf{s}^{*T} = \rho_0 \partial\phi_i/\partial\mathbf{e}_i, \quad \eta_{0i} = -\partial\phi_i/\partial\mathrm{T}, \quad \eta_{0r} = -\partial\phi_r/\partial\mathrm{T} \\ \rho_0 \partial\phi_{0i}/\partial\boldsymbol{\alpha}_0 = \boldsymbol{\beta}_{0i}^T\end{aligned} \tag{17.28}$$

The entropy production inequality (Equation 17.26) now reduces to

$$(\mathbf{s}^T - \mathbf{s}^{*T})\dot{\mathbf{e}}_i - \mathbf{q}_0^T \nabla_0 \mathrm{T}/\mathrm{T} \geq 0 \tag{17.29}$$

Inequality (Equation 17.29) is satisfied if

$$(\mathbf{s}^T - \mathbf{s}^{*T})\dot{\mathbf{e}}_i \geq 0 \tag{17.30a}$$

$$-\mathbf{q}_0^T \nabla_0 \mathrm{T}/\mathrm{T} \geq 0 \tag{17.30b}$$

Inequality Equation 17.30 states that the inelastic work done by the reduced stress $\mathbf{s}^T - \mathbf{s}^{*T}$ is positive and that heat flows from hot to cold. The first inequality exhibits the quantity $\mathbf{s}^* = VEC(\mathbf{S}^*)$ with dimensions of stress. In the subsequent sections $\mathbf{s}^*$ will be viewed as the previously mentioned backstress: it is interior to a yield surface and can be used to characterize the motion of the yield surface in stress space. Clearly, the present formulation gives a thermodynamic interpretation to the back stress. In classical kinematic hardening in which the hyperspherical yield surface does not change size or shape but just moves, the reference stress is simply the geometric center. If kinematic hardening occurs, as stated before, the yield surface need not include the origin even with small amounts of plastic deformation. Thus, there is no reason in general to regard $\dot{\mathbf{e}}_r$ as vanishing at the origin. Instead $\dot{\mathbf{e}}_r = \mathbf{0}$ is now assumed to hold when $\mathbf{s} = \mathbf{s}^*$.

17.4.3 Dissipation Potential

Following the approach developed in Chapter 16 we introduce a specific irreversible potential Ψ for which

$$\dot{\mathbf{e}}_i^T = \rho_0 \Lambda_i \partial\Psi/\partial\boldsymbol{\ni}, \quad -\nabla_0^T \mathrm{T}/\mathrm{T} = \Lambda_t \rho_0 \partial\Psi/\partial\mathbf{q}_0, \quad \boldsymbol{\ni} = \mathbf{s} - \mathbf{s}^* \tag{17.31a}$$

from which, with $\Lambda_i > 0$ and $\Lambda_t > 0$, Equation 17.30 becomes

$$\rho_0 \Lambda_i (\partial \Psi / \partial \boldsymbol{\backepsilon}) \boldsymbol{\backepsilon} + \rho_0 \Lambda_t (\partial \Psi / \partial \mathbf{q}_0) \mathbf{q}_0 > 0 \tag{17.31b}$$

Partly on the expectation that properties governing heat transfer are not affected by strain, we introduce the following decomposition into inelastic and thermal portions:

$$\Psi = \Psi_i + \Psi_t, \quad \rho_0 \Psi_t = \frac{\Lambda_t}{2} q_0^T q_0 \tag{17.32}$$

and Ψ_i will be seen in subsequent sections to represent mechanical effects. The thermal constitutive relation derived from Ψ_t implies Fourier's law:

$$-\nabla_0 \mathrm{T}/\mathrm{T} = \Lambda_t q_0 \tag{17.33}$$

17.4.4 Thermoinelastic Tangent Modulus Tensor

The elastic strain rate is assumed to correspond to small elastic deformation superimposed on the finite inelastic deformation, and to satisfy a linear *thermohypoelastic* constitutive relation

$$\dot{\mathbf{e}}_r = \mathbf{C}_r (\mathbf{s} - \mathbf{s}^*)^{\boldsymbol{\cdot}} + \mathbf{a}_r \dot{\mathrm{T}} \tag{17.34}$$

Of course $\mathbf{C}_r$ is a 9×9 second-order elastic compliance tensor, and $\mathbf{a}_r$ is the 9×1 thermoelastic expansion vector, with both presumed to be constant and known from measurements. Analogously, for rate-independent thermoplasticity we seek tensors $\mathbf{C}_i$ and $\mathbf{a}_i$, depending on $\boldsymbol{\backepsilon}$, $\mathbf{e}_i$, and T, such that

$$\dot{\mathbf{e}}_i = \mathbf{C}_i (\mathbf{s} - \mathbf{s}^*)^{\boldsymbol{\cdot}} + \mathbf{a}_i \dot{\mathrm{T}} \tag{17.35a}$$

$$\dot{\mathbf{e}} = [\mathbf{C}_r + \mathbf{C}_i](\mathbf{s} - \mathbf{s}^*)^{\boldsymbol{\cdot}} + (\mathbf{a}_r + \mathbf{a}_i)\dot{\mathrm{T}} \tag{17.35b}$$

During thermoplastic deformation the stress and temperature satisfy a thermoplastic *yield condition* of the form

$$\Pi_i(\boldsymbol{\backepsilon}, \mathbf{e}_i, \mathbf{k}, \mathrm{T}, \eta_{0i2}) = 0 \tag{17.36}$$

and Π_i is called the *yield function*. Here the vector $\mathbf{k}$ is introduced to represent the effect of the history of inelastic strain $\mathbf{e}_i$, for example, through work hardening. To embrace dependence on the temperature, it is now assumed to be given by a relation of the form,

$$\dot{\mathbf{k}} = \mathbf{K}(\mathbf{e}_i, \mathbf{k}, \mathrm{T}) \dot{\mathbf{e}}_i \tag{17.37}$$

The "consistency condition" requires that $\dot{\Pi}_i = 0$ during thermoplastic flow, and accordingly

$$\frac{d\Pi_i}{d\boldsymbol{\backepsilon}} \dot{\boldsymbol{\backepsilon}} + \frac{d\Pi_i}{d\mathbf{e}_i} \dot{\mathbf{e}}_i + \frac{d\Pi_i}{d\mathbf{k}} \dot{\mathbf{k}} + \frac{d\Pi_i}{d\mathrm{T}} \dot{\mathrm{T}} + \frac{d\Pi_i}{d\eta_{0i2}} \dot{\eta}_{0i2} = 0 \tag{17.38}$$

We introduce a thermoplastic extension of the conventional associated flow rule, whereby the inelastic strain rate vector is normal to the yield surface at the current stress point. Here we add an analogous assumption regarding the entropy.

$$\dot{\mathbf{e}}_i = \Lambda_i \left(\frac{\mathrm{d}\Pi_i}{\mathrm{d}\backepsilon} \right)^T \tag{17.39a}$$

$$\dot{\eta}_{0i2} = \Lambda_i \frac{\mathrm{d}\Pi_i}{\mathrm{dT}} \tag{17.39b}$$

Equation 17.39 suggests that *the yield function may be identified as the dissipation potential*: $\Pi_i = \rho_0 \Psi_i$. Upon making this identification, standard manipulation furnishes

$$\begin{aligned}
&\dot{\mathbf{e}}_i = \mathbf{C}_i \dot{\backepsilon} + \mathbf{a}_i \dot{\mathrm{T}}, \qquad \dot{\eta}_{0i2} = \mathbf{b}_i^T \dot{\backepsilon} + c_i \dot{\mathrm{T}} \\
&\mathbf{C}_i = \left(\frac{\partial \Psi_i}{\partial \backepsilon} \right)^T \frac{\partial \Psi_i}{\partial \backepsilon} \Big/ H, \quad \mathbf{a}_i = \mathbf{b}_i = \left(\frac{\partial \Psi_i}{\partial \backepsilon} \right)^T \frac{\partial \Psi_i}{\partial \mathrm{T}} \Big/ H \\
&c_i = \left(\frac{\partial \Psi_i}{\partial \mathrm{T}} \right)^2 \Big/ H, \qquad H = -\left[\left(\frac{\partial \Psi_i}{\partial \mathbf{e}_i} + \frac{\partial \Psi_i}{\partial \mathbf{K}} \mathbf{K} \right) \left(\frac{\partial \Psi_i}{\partial \backepsilon} \right)^T + \frac{\partial \Psi_i}{\partial \eta_{0i2}} \frac{\partial \Psi_i}{\partial \mathrm{T}} \right]
\end{aligned} \tag{17.40}$$

and H must be positive for Λ_i to be positive. Note that the dependence of the yield function on temperature accounts for c_i in the current formulation. The dissipation inequalities (Equation 17.30) are now satisfied if $H > 0$.

Next, recall that $\mathbf{s}^*$ depends on $\mathbf{e}_i$, T, and $\boldsymbol{\alpha}_0$ since $\mathbf{s}^{*T} = \rho_0 \partial \phi_{0i} / \partial \mathbf{e}_i$. For simplicity we now neglect dependence on $\boldsymbol{\alpha}_0$ and assume that a relation of the following form can be measured for $\mathbf{s}^*$:

$$\dot{\mathbf{s}}^* = \boldsymbol{\Gamma} \dot{\mathbf{e}}_i + \boldsymbol{\vartheta} \dot{\mathrm{T}}, \quad \boldsymbol{\Gamma} = \frac{\partial}{\partial \mathbf{e}_i} \left(\frac{\partial}{\partial \mathbf{e}_i} \right)^T \Psi_i, \quad \boldsymbol{\vartheta}^T = \partial^2 \Psi_i / \partial \mathbf{e}_i \partial \mathrm{T} \tag{17.41}$$

The *thermoinelastic tangent compliance tensor* and *thermomechanical vector* are obtained after simple additional manipulation as

$$\dot{\mathbf{e}} = \mathbf{C} \dot{\mathbf{s}} + \mathbf{a} \dot{\mathrm{T}} \tag{17.42a}$$

$$\begin{aligned}
\mathbf{C} &= (\mathbf{C}_r + \mathbf{C}_i) \left[\mathbf{I} - (\mathbf{I} + \boldsymbol{\Gamma} \mathbf{C}_i)^{-1} \boldsymbol{\Gamma} \mathbf{C}_{ii} \right] \\
\mathbf{a} &= \left[\mathbf{a}_r + \mathbf{a}_i + (\mathbf{C}_r + \mathbf{C}_i) \boldsymbol{\Gamma} \mathbf{a}_i - (\mathbf{C}_r + \mathbf{C}_i)(\mathbf{I} + \boldsymbol{\Gamma} \mathbf{C}_i)^{-1} \boldsymbol{\vartheta} \right]
\end{aligned} \tag{17.42b}$$

Of course the tangent modulus tensor is the inverse of $\mathbf{C}$.

The foregoing formulation may be extended to enforce plastic incompressibility.

EXAMPLE 17.3

Thermoplastic Helmholtz free energy and dissipation function

We now provide a simple example using the Helmholtz free energy density function and the dissipation potential function to derive constitutive relations. The expression assumed below involves a Von Mises yield function, linear kinematic hardening, linear work hardening, and linear thermal softening.

1. *Helmholtz free energy density*

$$\phi_0 = \phi_{0r} + \phi_{0i}, \quad \rho_0\phi_{0i} = k_3\mathbf{e}_i^T\mathbf{e}_i$$
$$\rho_0\phi_{0r} = \mathbf{e}_r^T\mathbf{C}_r^{-1}\mathbf{e}_r/2 - \mathbf{a}^T\mathbf{C}_r^{-1}(\mathrm{T}-\mathrm{T}_0)\mathbf{e}_r + \rho_0 c_r'\mathrm{T}(1-\ln(\mathrm{T}/\mathrm{T}_0))$$

in which c_r' is a known constant. Applying the previous relations furnishes

$$\boldsymbol{\backepsilon} = \rho_0(\partial\phi_{0r}/\partial\mathbf{e}_r)^T = \mathbf{C}_r^{-1}[\mathbf{e}_r - \mathbf{a}_r(\mathrm{T}-\mathrm{T}_0)]$$

and

$$c_r = -\mathrm{T}\frac{\partial^2\phi_{0r}}{\partial\mathrm{T}^2} = c_r'$$

Of course the last two relations between the reduced stress, the elastic strain, and the temperature are the same as in linear thermoelasticity except for the presence of the backstress.

2. *Dissipation potential*

The dissipation potential is again assumed to have a decomposition into mechanical and thermal portions, and the following specific forms are introduced:

$$\Psi = \Psi_i + \Psi_t, \quad \rho_0\Psi_t = \frac{\Lambda_t}{2}\mathbf{q}_0^T\mathbf{q}_0$$

$$\Psi_i = \sqrt{\boldsymbol{\backepsilon}^T\boldsymbol{\backepsilon}} - [k_0 + k_1k - k_2(\mathrm{T}-\mathrm{T}_0)] = 0, \quad \dot{k} = \boldsymbol{\backepsilon}^T\dot{\mathbf{e}}_i$$

Straightforward manipulations serve to derive

$$H = k_1\sqrt{\boldsymbol{\backepsilon}^T\boldsymbol{\backepsilon}} = k_1[K_0 + k_1k - k_2(\mathrm{T}-\mathrm{T}_0)]$$

$$\mathbf{C}_i = \frac{\boldsymbol{\backepsilon}\boldsymbol{\backepsilon}^T}{\boldsymbol{\backepsilon}^T\boldsymbol{\backepsilon}}\Big/H, \quad c_i = k_2^2\Big/H, \quad \mathbf{a}_i = \mathbf{b}_i = \frac{k_2\boldsymbol{\backepsilon}}{\sqrt{\boldsymbol{\backepsilon}^T\boldsymbol{\backepsilon}}}\Big/H$$

Consider a two-stage thermomechanical loading illustrated schematically in Figure 17.6. Let S_I, S_{II}, and S_{III} denote the principal values of the second Piola–Kirchhoff stress, and suppose that $S_{III} = 0$. In the first stage, with the temperature held fixed at T_0 the stresses are applied proportionally well into the plastic range. The center of the yield surface moves along a line in the (S_I, S_{II}) plane, and the yield surface expands as it moves. In the second stage, suppose that the stresses S_1 and S_2 are fixed but that the temperature increases to T_1 and then to T_2, T_3, and T_4. The plastic strain must increase and hence the center of the yield surface moves. In addition, in the assumed yield function strain hardening tends to cause the yield surface to expand isotropically around $\mathbf{s}^*$ while the

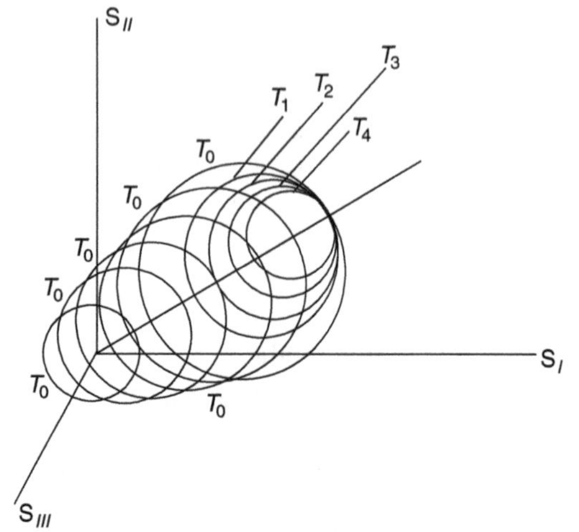

FIGURE 17.6 Effect of load and temperature on yield surface.

increased temperature tends to make it contract. However, in this case thermal softening must dominate strain hardening and contraction must occur since the center of the yield surface must move further along the path shown even as the yield surface continues to "kiss" the fixed stresses S_I and S_{II}.

Unfortunately, accurate finite element computations in plasticity and thermoplasticity often require close attention to the location of the front of the yielded zone. This front will usually occur interior to elements, essentially reducing the continuity order of the fields (discontinuity in strain gradients). In addition in computations the stress point will initially deviate from the yield surface. Special procedures such as *Return Mapping* (cf. Belytschko et al., 2000) have been developed in some codes to coerce the stress point onto the yield surface.

The shrinkage of the yield surface with temperature may provide the explanation of the phenomenon of adiabatic shear banding, which is commonly encountered in some materials during impact or metal forming. In rapid processes such as high-speed metalworking, plastic work is mostly converted into heat thereby causing high temperatures—there is not enough time for the heat to flow away from a spot experiencing high plastic deformation. However, under some conditions the process is unstable even as the stress level is maintained. In particular, as the material gets hotter the rate of plastic work accelerates, thanks to the softening evident in Figure 17.6. The instability is manifested in small periodically spaced bands in the center of which the material has melted and resolidified, usually in a much more brittle form than before. The *adiabatic shear bands* thereby formed can nucleate brittle fracture.

17.5 TANGENT MODULUS TENSOR IN VISCOPLASTICITY

17.5.1 Mechanical Field

The thermodynamic discussion of Section 17.4 applies to thermoinelastic deformation, for which the first example given concerned quasi-static plasticity and thermoplasticity. However, it is equally applicable when rate sensitivity is present, in which case viscoplasticity and thermoviscoplasticity are attractive models. An example of a constitutive model, for example, following Perzyna (1971), is given in undeformed coordinates as

$$\mu_v \dot{\mathbf{e}}_i = \left\langle 1 - \frac{k}{\Psi_i} \right\rangle \frac{\left(\dfrac{\partial \Psi_\mathrm{i}}{\partial \boldsymbol{\ni}}\right)^T}{\sqrt{\dfrac{\partial \Psi_i}{\partial \boldsymbol{\ni}}\left(\dfrac{\partial \Psi_i}{\partial \boldsymbol{\ni}}\right)^T}}$$

$$\left\langle 1 - \frac{k}{\Psi_i} \right\rangle = \begin{cases} 1 - \dfrac{k}{\Psi_i} & \text{if } 1 - \dfrac{k}{\Psi_i} \geq 0 \\ 0 & \text{otherwise} \end{cases} \tag{17.43}$$

and the dissipation function $\Psi_i(\boldsymbol{\ni}, \mathbf{e}_i, \mathbf{k}, \mathrm{T}, \eta_{0i})$ now is also a *loading surface* function (to be explained shortly); μ_v is called the *viscosity*. The inelastic strain rate vanishes if $1 - \frac{k}{\Psi_i} < 0$, which is interpreted to mean that the stress point (i.e., $\boldsymbol{\ni}$) is interior to the reference surface determined by points $\hat{\mathbf{s}}$ satisfying $\Psi_i(\hat{\mathbf{s}} - \mathbf{s}^*, \mathbf{e}_i, \mathbf{k}, \mathrm{T}, \eta_{0i}) = 0$. Inelastic flow is occurring if $1 - \frac{k}{\Psi_i} > 0$, in which case the stress point $\boldsymbol{\ni}$ is exterior to the reference surface. The equation of the loading surface is $\Psi_i(\boldsymbol{\ni}, \mathbf{e}_i, \mathbf{k}, \mathrm{T}, \eta_{0i}) = k \Big/ \left(1 - \mu_v \sqrt{\dot{\mathbf{e}}_i^T \dot{\mathbf{e}}_i}\right)$, and clearly the model must be restricted to strain rates satisfying $\mu_v \sqrt{\dot{\mathbf{e}}_i^T \dot{\mathbf{e}}_i} < 1$.

The elastic response is still considered linear and in the form

$$\dot{\mathbf{e}}_r = \boldsymbol{\chi}_r^{-1} \dot{\boldsymbol{\ni}} + \alpha_r \dot{\mathrm{T}} \tag{17.44}$$

Also from thermoplasticity we retain the relations

$$\dot{\mathbf{s}}^* = \boldsymbol{\Gamma} \dot{\mathbf{e}}_i + \boldsymbol{\vartheta} \dot{\mathrm{T}}, \quad \boldsymbol{\Gamma} = \frac{\partial}{\partial \mathbf{e}_i} \left(\frac{\partial}{\partial \mathbf{e}_i} \right)^T \Psi_i, \quad \boldsymbol{\vartheta}^T = \partial^2 \Psi_i / \partial \mathbf{e}_i \partial \mathrm{T} \tag{17.45}$$

Corresponding to $\boldsymbol{\ni}$ there is a new reference stress $\mathbf{s}'$ and a corresponding reference vector $\boldsymbol{\ni}' = \mathbf{s}' - \mathbf{s}^*$ defined as follows. The vector $\boldsymbol{\ni}'$ lies on the previously mentioned quasi-static reference yield surface such the vectors $\boldsymbol{\ni}$ and $\boldsymbol{\ni}'$ have the same origin and direction. The latter terminates on the reference surface while the former terminates outside the reference surface (at the current value of $\boldsymbol{\ni}$) if inelastic flow is occurring. Alternatively stated, $\boldsymbol{\ni}'$ is located at the intersection of the reference yield surface and the line correcting $\mathbf{s}^*$ to $\mathbf{s}$. This situation is illustrated in Figure 17.7.

Figure 17.7 suggests that viscoplasticity and thermoviscoplasticity can be formulated to accommodate phenomena such as kinematic hardening and thermal shrinkage of the reference yield surface.

As in plasticity we assume that the backstress has an evolution law of the form $\dot{\mathbf{s}}^* = \mathbf{H}_b \dot{\mathbf{e}}_i$, from which we find

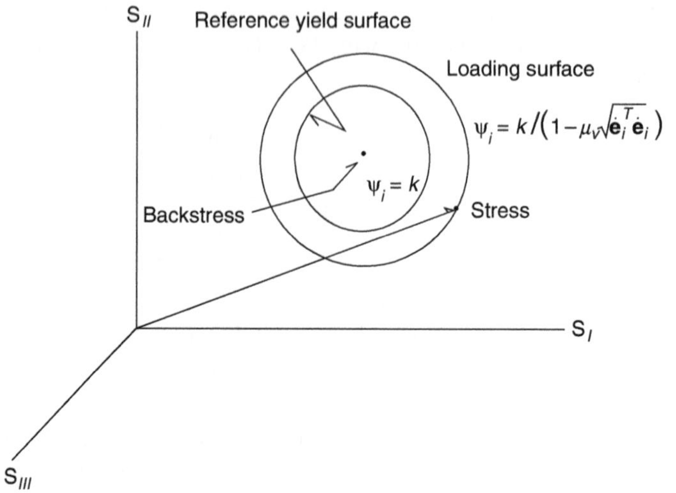

FIGURE 17.7 Illustration of loading surface and reference surface in viscoplasticity.

$$\dot{\mathbf{s}}^* = \mu_v\left\langle 1 - \frac{k}{\Psi}\right\rangle \mathbf{H}_b \frac{(\partial\Psi/\partial \ni)^T}{\sqrt{(\partial\Psi/\partial \ni)(\partial\Psi/\partial \ni)^T}} \tag{17.46}$$

The tangent modulus tensor now reduces to the constant elastic tensor $\boldsymbol{\chi}_r$, and viscoplastic effects appear in a force denoted $\mathbf{f}_v$, which depends on the current values of the state variables but not the current values of their rates. From Equation 17.44 $\dot{\mathbf{s}} - \dot{\mathbf{s}}^* = \boldsymbol{\chi}_r\dot{\mathbf{e}}_r - \boldsymbol{\chi}_r\boldsymbol{\alpha}_r\dot{\mathrm{T}}$ and so

$$\begin{aligned}\dot{\mathbf{s}} &= \mathbf{H}\dot{\mathbf{e}}_i + \boldsymbol{\chi}_r(\dot{\mathbf{e}} - \dot{\mathbf{e}}_i) - \boldsymbol{\chi}_r\boldsymbol{\alpha}_r\dot{\mathrm{T}} \\ &= \boldsymbol{\chi}_r\dot{\mathbf{e}} - \boldsymbol{\chi}_r\boldsymbol{\alpha}_r\dot{\mathrm{T}} + \mathbf{f}_v, \quad \mathbf{f}_v = \mu_v(\mathbf{H}_b - \boldsymbol{\chi}_r)\left(1 - \frac{k}{\Psi_i}\right)\frac{(\partial\Psi/\partial \ni)^T}{\sqrt{(\partial\Psi/\partial \ni)(\partial\Psi/\partial \ni)^T}}\end{aligned} \tag{17.47}$$

The Incremental Principle of Virtual Work and the corresponding finite element equation are now stated to first order in increments as

$$\begin{aligned}&\int \delta\Delta_n\mathbf{e}^T\boldsymbol{\chi}_r\Delta_n\mathbf{e}\,\mathrm{d}V_0 + \int \delta\Delta_n\mathbf{e}^T\boldsymbol{\chi}_r\boldsymbol{\alpha}_r\Delta_n\mathrm{T}\,\mathrm{d}V_0 + \int \delta\Delta_n\mathbf{u}^T\rho_0\Delta_n\ddot{\mathbf{u}}\,\mathrm{d}V_0 \\ &\quad = \int \delta\Delta_n\mathbf{u}^T\Delta_n\mathbf{t}\,\mathrm{d}S_0 - \int \delta\Delta_n\mathbf{e}^T\mathbf{f}_v\,\mathrm{d}V_0\end{aligned} \tag{17.48}$$

and

$$\mathbf{K}\Delta_n\boldsymbol{\gamma} + \boldsymbol{\Sigma}_{TM}\Delta_n\boldsymbol{\theta} + \mathbf{M}\Delta_n\ddot{\boldsymbol{\gamma}} = \mathbf{F} - \mathbf{F}_v$$

$$\boldsymbol{\Sigma}_{TM} = \int (\boldsymbol{\Phi}^T\boldsymbol{\beta})(\boldsymbol{\chi}_r\boldsymbol{\alpha}_r)\boldsymbol{\nu}\boldsymbol{\Psi}\,\mathrm{d}V_0, \quad \mathbf{F}_v = \int \boldsymbol{\Phi}^T\boldsymbol{\beta}\mathbf{f}_v\,\mathrm{d}V_0$$

in which $\boldsymbol{\beta}$ appears in the incremental strain–displacement relation.

EXAMPLE 17.4

Find the stress–strain curve under uniaxial tension if a constant strain rate is imposed in the linear, small strain, isothermal viscoplasticity model in which the yield surface is given by

$$\Psi_i = \sqrt{(\mathbf{s}_d - k_1\mathbf{e}_i)^T(\mathbf{s}_d - k_1\mathbf{e}_i)} - \left(k_0 + k_2\int_0^t \sqrt{\dot{\mathbf{e}}_i^T\dot{\mathbf{e}}_i}\ \mathrm{d}t\right)$$

Of course this surface reflects strain hardening. Referring to the previous exercise in plasticity Ψ_i may be written as

$$\Psi_i = \sqrt{\tfrac{2}{3}}\mathrm{S}_{xx} - \sqrt{\tfrac{3}{2}}(k_1 + k_2)\ \mathrm{E}_{xx}^{vp} - k_0$$

To formulate the finite element relations, use is made of the fact that

$$\frac{\partial\Psi_i}{\partial \mathrm{S}_{xx}}\Bigg/\sqrt{\frac{\partial\Psi_i}{\partial \mathrm{S}_{xx}}\left(\frac{\partial\Psi_i}{\partial \mathrm{S}_{xx}}\right)^T} = 1$$

The constitutive model for viscoplasticity reduces to the uniaxial relations

$$\mu_v\dot{\mathrm{E}}_{xx}^{vp} = \langle\Psi_i - k\rangle = \sqrt{\tfrac{2}{3}}\left[\mathrm{S}_{xx} - \tfrac{3}{2}(k_1 + k_2)\mathrm{E}_{xx}^{vp} - \sqrt{\tfrac{3}{2}}(k_0 + k)\right]$$

Of course the strain rate may be decomposed into elastic and viscoelastic parts as $\dot{\mathbf{e}} = \dot{\mathbf{e}}^e + \dot{\mathbf{e}}^{vp}$. Also $\dot{\mathrm{E}}_{xx}^{e} = \dot{\mathrm{S}}_{xx}/\mathbf{E}$. Accordingly

$$\dot{\mathrm{E}}_{xx} = \frac{\dot{\mathrm{S}}_{xx}}{\mathbf{E}} + \frac{1}{\mu_v}\sqrt{\tfrac{2}{3}}\left[\mathrm{S}_{xx} - \tfrac{3}{2}(k_1 + k_2)\mathrm{E}_{xx}^{vp} - \sqrt{\tfrac{3}{2}}(k_0 + k)\right]$$

Now $\dfrac{\mathrm{d}}{\mathrm{d}t} = \dfrac{\mathrm{dE}_{xx}}{\mathrm{d}t}\dfrac{\mathrm{d}}{\mathrm{dE}_{xx}}$, and so

$$\dot{\mathrm{E}}_{xx} = \frac{\dot{\mathrm{E}}_{xx}}{\mathbf{E}}\frac{\mathrm{dS}_{xx}}{\mathrm{dE}_{xx}} + \frac{1}{\mu_v}\sqrt{\tfrac{2}{3}}\left[\mathrm{S}_{xx} - \tfrac{3}{2}(k_1 + k_2)\mathrm{E}_{xx}^{vp} - \sqrt{\tfrac{3}{2}}(k_0 + k)\right]$$

from which emerges a simple differential equation for the uniaxial stress–strain relation at the constant strain rate $\dot{\mathrm{E}}_{xx}$:

$$\begin{aligned}\frac{\mathrm{dS}_{xx}}{\mathrm{dE}_{xx}} + \sqrt{\frac{2}{3}}\frac{\mathbf{E}}{\mu_v\dot{\mathrm{E}}_{xx}}\mathrm{S}_{xx} &= \sqrt{\frac{3}{2}}\frac{(k_1 + k_2)\mathbf{E}}{\mu_v\dot{\mathrm{E}}_{xx}}\mathrm{E}_{xx}^{vp} + \mathbf{E}\left[1 + \frac{(k_0 + k)}{\mu_v\dot{\mathrm{E}}_{xx}}\right]\\ &= \sqrt{\frac{3}{2}}\frac{(k_1 + k_2)\mathbf{E}}{\mu_v\dot{\mathrm{E}}_{xx}}\left(\mathrm{E}_{xx} - \frac{\mathrm{S}_{xx}}{\mathbf{E}}\right) + \mathbf{E}\left[1 + \frac{(k_0 + k)}{\mu_v\dot{\mathrm{E}}_{xx}}\right]\end{aligned}$$

so that

$$\frac{\mathrm{dS}_{xx}}{\mathrm{dE}_{xx}} + \left(\sqrt{\frac{2}{3}}\frac{\mathbf{E}}{\mu_v\dot{\mathrm{E}}_{xx}} + \sqrt{\frac{3}{2}}\frac{(k_1 + k_2)}{\mu_v\dot{\mathrm{E}}_{xx}}\right)\mathrm{S}_{xx} = \sqrt{\frac{3}{2}}\frac{(k_1 + k_2)\mathbf{E}}{\mu_v\dot{\mathrm{E}}_{xx}}\mathrm{E}_{xx} + \mathbf{E}\left[1 + \frac{(k_0 + k)}{\mu_v\dot{\mathrm{E}}_{xx}}\right]$$

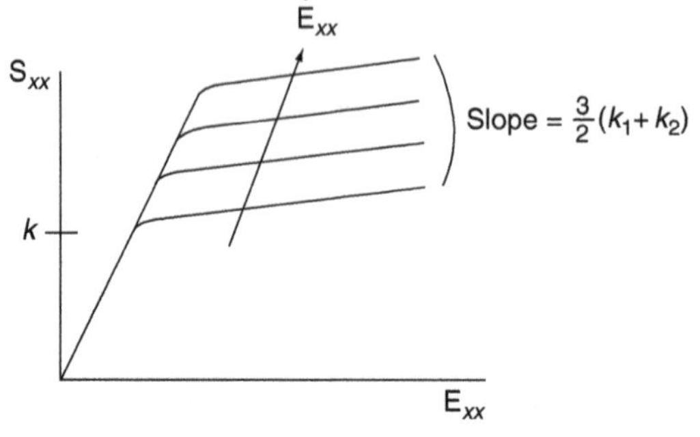

FIGURE 17.8 Stress vs. strain at constant strain rate in a viscoplastic material.

The transient part of the solution is given by

$$S_{xx} = A\exp\left(-\left(\sqrt{\frac{2}{3}}\frac{\mathbf{E}}{\mu_v \dot{E}_{xx}} + \sqrt{\frac{3}{2}}\frac{(k_1 + k_2)}{\mu_v \dot{E}_{xx}}\right)E_{xx}\right)$$

in which A is determined by imposing initial values on the combined transient and steady state solutions. The steady state part of the solution is given by

$$S_{xx} = \frac{\frac{3}{2}(k_1 + k_2)}{\left(1 + \frac{2}{3}\frac{(k_1 + k_2)}{\mathbf{E}}\right)}E_{xx} + \sqrt{\frac{3}{2}}\frac{[\mu_v \dot{E}_{xx} + (k_0 + k)]}{\left(1 + \frac{3}{2}\frac{(k_1 + k_2)}{\mathbf{E}}\right)}$$

The slope of the asymptote is approximately $\frac{3}{2}(k_1 + k_2)$ and its intercept is approximately $\sqrt{\frac{3}{2}}[(k_0 + k) + \mu_v \dot{E}_{xx}]$.

The ensuing uniaxial stress–strain curve is illustrated in Figure 17.8.

17.5.2 Thermoinelasticity: Thermal Field

Of course a variational equation and a finite element equation are needed for the thermal field Equation 17.25, and are formulated as was done in Chapter 16. To illustrate the process of formulating the finite element equation for the thermal field, we make the simplifying assumption that the irreversible entropy only depends on temperature.

The equation for the reversible portion of the energy balance is

$$\left(-\nabla_0^T \mathbf{q}_0 + \rho_0 h\right)_r = \rho_0 c_e \dot{T} - T\frac{\partial(\mathbf{s} - \mathbf{s}^*)^T}{\partial T}\dot{\mathbf{e}}_r \tag{17.49}$$

which of course reduces to the thermal field equation of classical thermoelasticity when the backstress is removed and the reversible strain is equated with the total strain.

The equation for the irreversible portion of the energy balance is now

$$\left(-\nabla_0^T \mathbf{q}_0 + \rho_0 h\right)_i = \rho_0 c_i \dot{\mathrm{T}} - (\mathbf{s} - \mathbf{s}^*)\dot{\mathbf{e}}_i \tag{17.50}$$

Adding the reversible and irreversible portions now gives

$$-\nabla_0^T \mathbf{q}_0 + \rho_0 h = \rho_0 (c_e + c_i)\dot{\mathrm{T}} - (\mathbf{s} - \mathbf{s}^*)\dot{\mathbf{e}}_i - \mathrm{T}\frac{\partial(\mathbf{s} - \mathbf{s}^*)}{\partial \mathrm{T}} \tag{17.51}$$

In adiabatic situations such as short-term response after impact, the left-hand side of Equation 17.51 vanishes. The right side then suggests that the inelastic work $(\mathbf{s} - \mathbf{s}^*)\dot{\mathbf{e}}_i$ is converted into temperature increase in accordance with $\rho_0(c_e + c_i)\dot{\mathrm{T}}$, except for a small portion (proportional to the thermal expansion coefficient) represented by $\mathrm{T}\frac{\partial(\mathbf{s}-\mathbf{s}^*)}{\partial \mathrm{T}}$.

We now wish to formulate the incremental finite element equation for the thermal field. We use the constitutive model introduced in Section 17.5.1 for thermoplasticity, and the counterparts for thermoviscoplasticity can be recovered by obvious substitutions of the constitutive relations for the inelastic strain rate. For simplicity we assume that there is no internal generation of heat, i.e., $\dot{h} = 0$. We also assume Fourier's law in the form $\mathbf{q}_0 = -\kappa_0 \nabla \mathrm{T}$, and that $(c_e + c_i)$ is constant. The governing equation now becomes

$$-\kappa_0 \nabla^2 \mathrm{T} + \rho_0 (c_e + c_i)\dot{\mathrm{T}} + \lambda\alpha(\mathbf{i}^T \dot{\mathbf{e}}) - (\mathbf{s} - \mathbf{s}^*)^T(\mathbf{C}\dot{\mathbf{s}} + \mathbf{a}\dot{\mathrm{T}}) = 0 \tag{17.52}$$

Applying Fourier's law now gives

$$-\kappa_0 \nabla^2 \mathrm{T} + \rho_0 (c_e + c_i)\dot{\mathrm{T}} + \lambda\alpha(\mathbf{i}^T \dot{\mathbf{e}}) - (\mathbf{s} - \mathbf{s}^*)^T \mathbf{C}\boldsymbol{\chi}\dot{\mathbf{e}} - (\mathbf{s} - \mathbf{s}^*)^T \boldsymbol{\alpha}_i \dot{\mathrm{T}} = 0 \tag{17.53}$$

Equations 17.52 and 17.53 provide a way of avoiding using the inelastic strain increment explicitly in the variational principle.

The finite element equation is now sought. As in Chapter 16 we use

$$\kappa_0 \nabla^2 \mathrm{T} \approx \kappa_0 \nabla^2 \Delta_n \mathrm{T} + \kappa_0 \nabla^2 \mathrm{T}_n \tag{17.54}$$

In terms of increments the thermal field equation is now

$$\begin{aligned} &- h\kappa_0 \nabla^2 \Delta_n \mathrm{T} + \left[\rho_0 (c_e + c_i) - (\mathbf{s} - \mathbf{s}^*)^T \boldsymbol{\alpha}_i\right]\Delta_n \mathrm{T} \\ &+ \left[\lambda\alpha\mathbf{i}^T - (\mathbf{s} - \mathbf{s}^*)^T \mathbf{C}\boldsymbol{\chi}\right]\Delta_n \mathbf{e} = h\kappa_0 \nabla^2 \mathrm{T}_n \end{aligned} \tag{17.55}$$

Standard finite element procedures now provide the element-level finite element equation for the thermal field.

$$[h\mathbf{K}_\theta + (\mathbf{M}_{\theta 1} + \mathbf{M}_{\theta 2})]\Delta_n \boldsymbol{\theta} + \boldsymbol{\Sigma}_2^T \Delta_n \boldsymbol{\gamma} = \Delta_n \mathbf{f}_{T1} - \Delta_n \mathbf{f}_{T2}$$

$$\mathbf{K}_\theta = \int \mathbf{\Psi}^T \mathbf{\beta}_T \kappa_0 \mathbf{\beta}_T^T \mathbf{\Psi}\,\mathrm{d}V_0, \qquad \mathbf{M}_{\theta 1} = \int \mathbf{\Psi}^T \nu[\rho_0(c_e + c_i)]\nu^T \mathbf{\Psi}\,\mathrm{d}V_0,$$

$$\mathbf{M}_{\theta 2} = \int \mathbf{\Psi}^T \nu\left[(\mathbf{s}-\mathbf{s}^*)^T \boldsymbol{\alpha}_i\right]\nu^T \mathbf{\Psi}\,\mathrm{d}V_0, \quad \mathbf{\Sigma}_2^T = \int \mathbf{\Psi}^T \nu\left[\lambda\alpha\mathbf{i}^T - (\mathbf{s}-\mathbf{s}^*)^T \mathbf{C}\boldsymbol{\chi}\right]\boldsymbol{\beta}\boldsymbol{\Phi}\,\mathrm{d}V_0,$$

$$\Delta_n \mathbf{f}_{T1} = -\int \mathbf{\Psi}^T \nu\left(\mathbf{n}_0^T \mathbf{q}_0\right)\mathrm{d}S_0, \qquad \Delta_n \mathbf{f}_{T2} = -h\int \mathbf{\Psi}^T \nu \kappa_0 \nabla^2 T_n\,\mathrm{d}S_0 \tag{17.56}$$

Finally, we recapitulate the incremental finite element equations of the mechanical and thermal fields in the form

$$\begin{aligned} &\mathbf{M}\Delta_n\ddot{\boldsymbol{\gamma}} + (\mathbf{K}_T + \mathbf{K}_G)\Delta_n\boldsymbol{\gamma} - \mathbf{\Sigma}_1\Delta_n\boldsymbol{\theta} = \Delta_n\mathbf{f}_m &&: \text{mechanical field} \\ &[h\mathbf{K}_\theta + (\mathbf{M}_{\theta 1} + \mathbf{M}_{\theta 2})]\Delta_n\boldsymbol{\theta} + \mathbf{\Sigma}_2^T\Delta_n\boldsymbol{\gamma} = \Delta_n\mathbf{f}_{T1} - \Delta_n\mathbf{f}_{T2} &&: \text{thermal field} \end{aligned} \tag{17.57}$$

17.6 CONTINUUM DAMAGE MECHANICS

Ductile fracture occurs by processes which are associated with the notion of damage. A damage parameter is also introduced to explain tertiary creep, in which the strain grows rapidly at a fixed stress and temperature. An internal damage variable is introduced which accumulates with inelastic deformation. It also is manifested in reductions in properties such as the experimental values of the elastic modulus and yield stress. When the damage parameter in a given element reaches a known or assumed critical value, the element is considered to have failed. The element may then be removed from the mesh (the element is considered to be no longer supporting the load). If so, the displacement and temperature fields are recalculated to accommodate the element deletion.

There are two different "schools" of thought on the suitable notion of a damage parameter. One, associated with Gurson (1977), Tvergaard (1981), and Thomason (1990), considers damage to involve a specific mechanism occurring in a three-stage process: nucleation of voids, their subsequent growth, and finally their coalescence to form a macroscopic defect. The coalescence event can be used as a criterion for element failure. The parameter used to measure damage is the void volume fraction f. Models and criteria for the three processes have been formulated. For both nucleation and growth, evolution of f is governed by a constitutive equation of the form

$$\mathbf{f} = \Xi(\mathbf{f}, \mathbf{e}_i, \mathrm{T}) \tag{17.58}$$

for which several specific forms have been proposed. To this point, a nominal stress is used in the sense that the reduced ability of material to support stress is not accommodated.

The second school of thought is more phenomenological in nature and is not dependent on a specific micromechanical mechanism. It uses the parameter D, which is interpreted as the fraction of damaged area A_d to total area A_0 that the stress (traction) acts on. Consider a uniaxial tensile specimen which has experienced damage and is now exhibiting elastic behavior. Suppose that damaged area A_d can no longer support a load. For a given load P, the *true* stress at a point in the undamaged zone is $S = \frac{P}{A_0 - A_D} = \frac{1}{1-D}\frac{P}{A_0} = \frac{1}{1-D}S'$. Here S' is a nominal stress, *but is also the measured stress*. If E is the elastic modulus measured in an undamaged specimen, the modulus measured in the current specimen will be $\mathsf{E}' = \mathsf{E}/(1 - \mathrm{D})$, demonstrating that damage is manifest in small changes in properties, in particular $\mathrm{D} = 1 - \mathsf{E}/\mathsf{E}'$.

As an illustration of damage, suppose specimens are loaded into the plastic range, unloaded, and then loaded again. Without the notion of damage the stress–strain curve should return to its original path. However, owing to damage there are slight changes in the elastic slope, in the yield stress, and in the slope after yield (exaggerated in Figure 17.9).

From the standpoint of thermodynamics, damage is an internal variable representing irreversible effects and as such is dissipative. In reality the amount of mechanical or thermal energy absorbed by damage is probably small, so that its role in the energy balance equation is often neglected. Even so, for the sake of a consistent framework for treating dissipation associated with damage, a dissipation potential Ψ_d may be introduced for damage, as has been done, for example, by Bonora (1997). As an example, the contribution to the irreversible entropy production may be assumed in the rate form $\mathfrak{D}\dot{\mathrm{D}} \geq 0$, in which $\mathfrak{D}$ is the "force" associated with "flux" D. Positive dissipation is assured if relations are used such as

$$\mathfrak{D} = \frac{\partial \Psi_d}{\partial \dot{\mathrm{D}}}, \quad \Psi_d = \frac{1}{2}\Lambda_d(\mathbf{e}_i, \mathrm{T}, \mathbf{k})\dot{\mathrm{D}}^2, \quad \Lambda_d(\mathbf{e}_i, \mathrm{T}, \mathbf{k}) > 0 \tag{17.59}$$

An example of a potentially satisfactory function tying damage to inelastic work is

$$\Lambda_d(\mathbf{e}_i, \mathrm{T}, \mathbf{k}) = \Lambda_{d0} \int (\mathbf{s} - \mathbf{s}^*)\dot{\mathbf{e}}_i \, \mathrm{d}t, \quad \Lambda_{d0} \text{ a positive constant} \tag{17.60}$$

Specific examples of constitutive relations for damage are given, for example, in Bonora (1997).

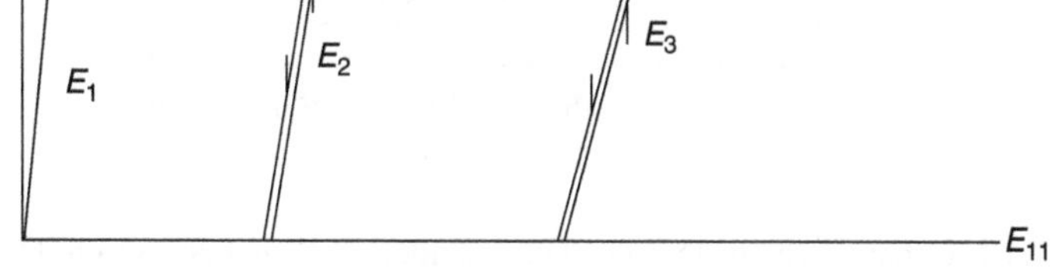

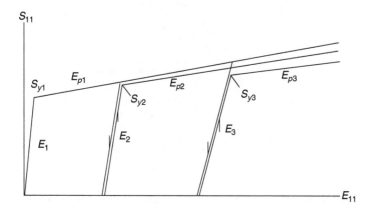

FIGURE 17.9 Illustration of effect of damage on elastic–plastic properties.

At the current values of the damage parameter, the finite element equations are solved for the nodal displacements, from which may be computed the inelastic strains and the inelastic work done in the current load or time increment. This information may then be used to update the damage parameter values at each element using the damage evolution equation (Equation 17.59). Upon doing so, the damage parameter values are compared to critical values. As stated previously, if the critical damage parameter value is attained, the element is deleted. In many cases the string of deleted elements may be viewed as a crack (Al-Grafi, 2003).

The finite element code LS_DYNA version 9.5 (2000) incorporates a material model which includes viscoplasticity and damage mechanics. It can easily be upgraded to include thermal effects in which all viscoplastic work is turned into heat. Such a model has been shown to reproduce the location and path of a crack in a dynamically loaded welded structure (Moraes and Nicholson, 2002).

EXAMPLE 17.5

Tertiary Creep of IN 617 at high temperature

An example of how a damage variable is useful in modeling material behavior is provided by tertiary creep of IN 617 (Gordon and Nicholson, 2006). Creep may be regarded as a type of viscoplasticity except that the reference yield surface is sometimes neglected. The steel alloy IN 617 is used in turbomachinery and experiences rapid growth of strain

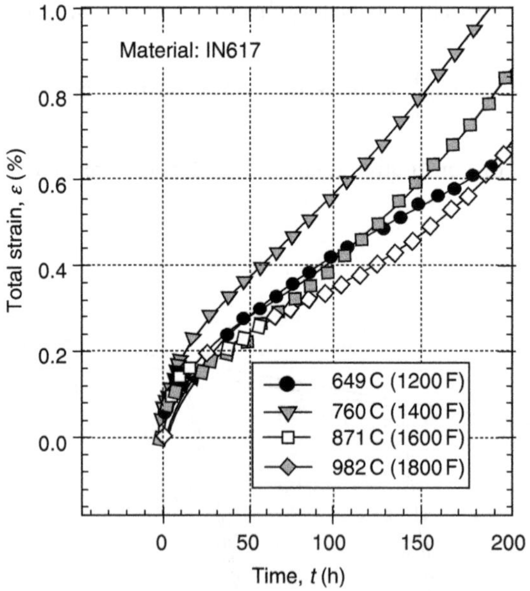

FIGURE 17.10 Tertiary creep of IN 617. (From Gordon, A.P. and Nicholson, D.W., Finite Element Analysis of IN 617 Tertiary Creep, Report, University of Central Florida, Orlando, FL, 2006.)

under fixed stress and at high temperature. The tertiary creep behavior is shown experimentally in Figure 17.10, in which lower stresses were used at the higher temperatures.

A Norton-type isotropic constitutive model has been formulated in which the Von Mises creep strain rate $\dot{\varepsilon}_{cr}$ is modeled as

$$\dot{\varepsilon}_{cr} = B(\exp -\exists/RT)\left(\frac{\sigma}{1 - \mathrm{D}}\right)^n \tag{17.61}$$

Here σ is the Von Mises stress and $\exists$ is the activation energy, while the constants B and n are selected for a best fit up to 5% strain. The damage evolution equation, based on the classical treatment of Rabotnov (1969), is given by

$$\dot{D} = \frac{M\sigma^{\chi}}{(1 - D)^{\phi}} \tag{17.62}$$

in which M, χ, and ϕ are likewise constants to be chosen to match experiments up to 5% strain.

Equations 17.61 and 17.62 were implemented by Gordon and Nicholson (2006) in ANSYS in a user-material model and used to find the best-fit values of B, M, n, χ, and ϕ. An example of the comparison between the model and the experimental data, up to 5% strain, is shown in Figure 17.11.

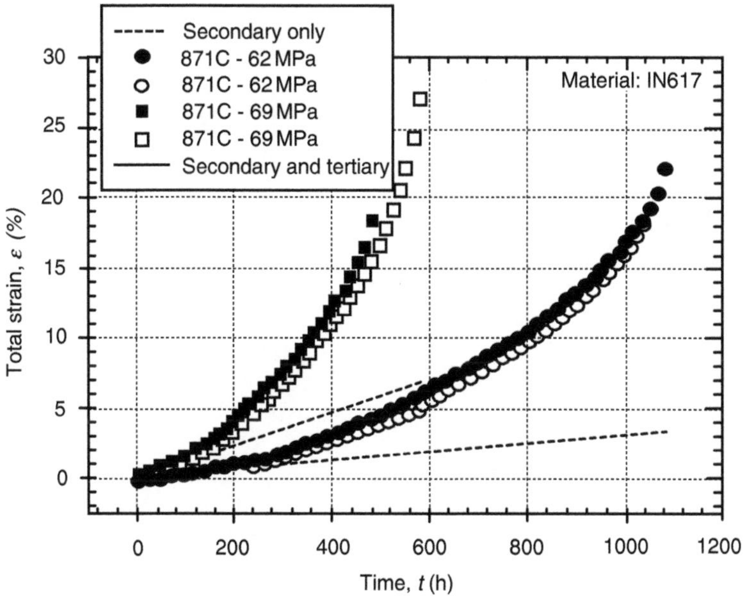

FIGURE 17.11 Comparison of computations and experimental values up to 5% strain. (From Gordon, A.P. and Nicholson, D.W., Finite Element Analysis of IN 617 Tertiary Creep, Report, University of Central Florida, Orlando, FL, 2006.)

18 Selected Advanced Numerical Methods in FEA

In nonlinear finite element analysis, a solution is typically sought using Newton iteration, in classical form or augmented as an arc length method to bypass critical points in the load–deflection behavior. Another important topic is the treatment of incompressibility in nonlinear problems. Here three additional treatments of numerical methods are presented.

18.1 ITERATIVE TRIANGULARIZATION OF PERTURBED MATRICES

18.1.1 INTRODUCTION

The finite element method applied to nonlinear problems typically gives rise to a large linear system of the form $\mathbf{K}_0(\boldsymbol{\gamma}_0)\Delta\boldsymbol{\gamma}_0 = \Delta\mathbf{f}_0$ in which the stiffness matrix $\mathbf{K}_0$ is positive definite and symmetric and banded. Also, $\Delta\boldsymbol{\gamma}_0$ is the incremental nodal displacement vector and $\Delta\mathbf{f}_0$ is the incremental nodal force vector. The stiffness matrix depends on nodal displacements and is updated during the incremental solution process, leading to a perturbed matrix $\mathbf{K} = \mathbf{K}_0 + \Delta\mathbf{K}$, in which $\Delta\mathbf{K}$ is assumed to be very small in norm (e.g., magnitude of largest eigenvalue) compared to $\mathbf{K}_0$. Given the fact that triangular factors $\mathbf{L}_0$ and $\mathbf{L}_0^T$ have been obtained for $\mathbf{K}_0$, it is attractive to formulate and employ an iteration procedure for the perturbed matrix $\mathbf{K}$ using $\mathbf{L}_0$ as the initial iterate. The procedure should not involve solving intermediate linear systems except by forward and backward substitution using already known triangular factors. A procedure is presented in the section below following Nicholson (2005a) and shown, in three simple examples, to produce accurate estimates within a few iterations. In the scalar case, the iterates examined "track" the Taylor series exactly. The author is unaware of any previously established and widely implemented iterative procedure for matrix triangularization.

It should be noted that updating triangular factors is relevant to important finite element applications other than the nonlinear solid mechanics. For example, in fracture mechanics suppose that the crack front advances. There is a local change in the mesh in the vicinity of the crack tip, corresponding to a perturbation of the stiffness matrix. As a second example, if the finite element method is used in an optimal design study, the search procedure sets the design parameters and an FEA is performed. Then the search procedure moves the search point locally in space of design parameters, thereby perturbing the structural model and the stiffness matrix.

18.1.2 Incremental Finite Element Equation

To set the problem under study in the appropriate context, we consider dynamic response of a nonlinear solid. Application of the Incremental Principle of Virtual Work and introduction of suitable interpolation models, following Chapter 15, furnish incremental finite element relations:

$$\mathbf{M}\Delta\ddot{\boldsymbol{\gamma}} + \mathbf{K}(\boldsymbol{\gamma})\Delta\boldsymbol{\gamma} = \Delta\mathbf{f} \tag{18.1}$$

$\mathbf{M}$ mass matrix, $n \times n$ and positive definite, assumed constant
$\mathbf{K}(\boldsymbol{\gamma})$ incremental stiffness matrix, $n \times n$ and positive definite
$\Delta\boldsymbol{\gamma}$ incremental nodal displacement vector
$\Delta\mathbf{f}$ incremental consistent nodal force vector

We assume that Equation 18.1 is integrated using a one-step procedure based on the trapezoidal rule (Newmark's method). Let h denote the time step and let $\gamma_n = \gamma(t_n)$. At time $t_{n+1} = (n+1)h$, Equation 18.1 becomes, following (Zienkiewicz and Taylor, 1989),

$$\left[\mathbf{M} + \tfrac{h^2}{4}\mathbf{K}(\boldsymbol{\gamma}_n)\right]\Delta_{n+1}\boldsymbol{\gamma} = \Delta_{n+1}\mathbf{g} \tag{18.2}$$

in which

$$\Delta_{n+1}\mathbf{g} = \tfrac{h^2}{4}(\Delta_{n+1}\mathbf{f} + \Delta_n\mathbf{f} - \mathbf{K}\Delta_n\boldsymbol{\gamma}) + \mathbf{M}(\Delta_n\boldsymbol{\gamma} + h\Delta_n\mathbf{q}), \quad \Delta_n\mathbf{q} = \Delta_n\dot{\boldsymbol{\gamma}}$$
$$\Delta_{n+1}\boldsymbol{\gamma} = \boldsymbol{\gamma}_{n+1} - \boldsymbol{\gamma}_n$$

We say that the stiffness matrix at the $(n+1)$st load or time step is perturbed relative to the stiffness matrix at the nth load step. The solution of perturbed linear systems has been the subject of many investigations. Schemes based on explicit matrix inversion include the Sherman–Morrison–Woodbury formulae (cf. Golub and Van Loan, 1986). An alternate method is to carry bothersome terms to the right-hand side and then to iterate. For example, the perturbed linear system may be approximated to first order in increments as

$$\mathbf{K}_0\Delta\boldsymbol{\gamma} = \Delta\mathbf{f} - \Delta\mathbf{K}\boldsymbol{\gamma}_0 \tag{18.3}$$

and an iterative solution procedure, *assuming convergence*, may then be employed as

$$\mathbf{K}_0\Delta\boldsymbol{\gamma}^{(j+1)} = \Delta\mathbf{f} - \Delta\mathbf{K}\left(\boldsymbol{\gamma}^{(j)}\right)\boldsymbol{\gamma}_0, \quad \boldsymbol{\gamma}^{(j+1)} = \boldsymbol{\gamma}_0 + \Delta\boldsymbol{\gamma}^{(j+1)} \tag{18.4}$$

Unfortunately, in a typical nonlinear problem, especially in systems with decreasing stiffness, it is necessary to update the triangular factors after several increments. If plasticity, hyperelasticity, or buckling is of concern, it is often wise to update the stiffness matrix after each increment.

18.1.3 Iterative Triangularization Procedure

A square matrix is said to be *lower triangular* if all super-diagonal entries vanish. Similarly a square matrix is said to be upper triangular if all subdiagonal entries vanish. Consider a nonsingular real matrix $\mathbf{A}$. It may be decomposed as

$$\mathbf{A} = \mathbf{A}_l + \text{diag}(\mathbf{A}) + \mathbf{A}_u \tag{18.5}$$

in which diag($\mathbf{A}$) consists of the diagonal entries of $\mathbf{A}$, with zeroes elsewhere, $\mathbf{A}_l$ coincides with $\mathbf{A}$ below the diagonal with all other entries set to zero, and $\mathbf{A}_u$ coincides with $\mathbf{A}$ above the diagonal with all other entries set to zero.

Now limiting attention to symmetric matrices, for later use we introduce the matrix functions

$$\text{lower}(\mathbf{A}) = \mathbf{A}_l + \tfrac{1}{2}\text{diag}(\mathbf{A}), \quad \text{upper}(\mathbf{A}) = \mathbf{A}_u + \tfrac{1}{2}\text{diag}(\mathbf{A}) \tag{18.6}$$

The reader may readily verify that

1. The product of two lower (upper) triangular matrices is also lower (upper) triangular
2. The inverse of a nonsingular lower (upper) triangular matrix is also lower (upper) triangular

To formulate the iteration procedure, let $\mathbf{K}_0$ denote a symmetric positive definite matrix for which the unique triangular factors $\mathbf{L}_0$ and $\mathbf{L}_0^T$ have already been computed. If $\mathbf{K}_0$ is banded, the maximum width of its rows (the bandwidth) equals $2b - 1$, in which b is the bandwidth of $\mathbf{L}_0$. The factors of the perturbed matrix $\mathbf{K}$ may be written as

$$[\mathbf{K}_0 + \Delta\mathbf{K}] = [\mathbf{L}_0 + \Delta\mathbf{L}]\left[\mathbf{L}_0^T + \Delta\mathbf{L}^T\right] \tag{18.7}$$

We may rewrite Equation 18.7 as

$$\left[\mathbf{I} + \mathbf{L}_0^{-1}\Delta\mathbf{L}\right]\left[\mathrm{I} + \Delta\mathbf{L}^T\mathbf{L}_0^{-T}\right] = \mathbf{L}_0^{-1}[\mathbf{K}_0 + \Delta\mathbf{K}]\mathbf{L}_0^{-T} \tag{18.8}$$

from which

$$\mathbf{L}_0^{-1}\Delta\mathbf{L} + \Delta\mathbf{L}^T\mathbf{L}_0^{-T} = \mathbf{L}_0^{-1}\Delta\mathbf{K}\mathbf{L}_0^{-T} - \mathbf{L}_0^{-1}\Delta\mathbf{L}\Delta\mathbf{L}^T\mathbf{L}_0^{-T} \tag{18.9}$$

Note that $\mathbf{L}_0^{-1}\Delta\mathbf{L}$ is *lower triangular*. It follows that

$$\Delta\mathbf{L} = \mathbf{L}_0\text{lower}\left(\mathbf{L}_0^{-1}\Delta\mathbf{K}\mathbf{L}_0^{-T} - \mathbf{L}_0^{-1}\Delta\mathbf{L}\Delta\mathbf{L}^T\mathbf{L}_0^{-T}\right) \tag{18.10}$$

The factor of 1/2 in the definition of lower(*) and upper(*) functions is motivated by the fact that the diagonal entries of $\mathbf{L}_0^{-1}\Delta\mathbf{L}$ and $\Delta\mathbf{L}^T\mathbf{L}_0^{-T}$ are the same.

An iteration procedure based on Equation 18.10 is proposed as

$$\begin{aligned}\Delta\mathbf{L}^{(j+1)} &= \mathbf{L}_0\text{lower}\left(\mathbf{L}_0^{-1}\Delta\mathbf{K}\mathbf{L}_0^{-T} - \mathbf{L}_0^{-1}\Delta\mathbf{L}^{(j)}\Delta\mathbf{L}^{(j)T}\mathbf{L}_0^{-T}\right)\\ \Delta\mathbf{L}^{(1)} &= \mathbf{L}_0\text{lower}\left(\mathbf{L}_0^{-1}\Delta\mathbf{K}\mathbf{L}_0^{-T}\right)\end{aligned} \tag{18.11}$$

Note that the computations in Equation 18.11 may be performed using only forward substitution involving $\mathbf{L}_0$. As demonstration, we introduce the matrix $\mathbf{C}$ using $\mathbf{L}_0\mathbf{C} = \Delta\mathbf{K} - \Delta\mathbf{L}\Delta\mathbf{L}^T$. ($\mathbf{C}$ is not the right Cauchy–Green strain tensor.) Clearly $\mathbf{C}$ may be computed using forward substitution. Now introduce $\mathbf{D}$ using $\mathbf{D}\mathbf{L}_0^T = \mathbf{C}$, so that $\mathbf{L}_0\mathbf{D}^T = \mathbf{C}^T$. Of course $\mathbf{D}^T$ and hence also $\mathbf{D}$ are computed using forward substitution. The last step is to compute $\Delta\mathbf{L} = \mathbf{L}_0\text{lower}(\mathbf{D})$.

We now introduce an approximate convergence argument. For an approximate convergence criterion, we study the similar relation

$$\Delta\mathbf{A} = \mathbf{A}^{-1}\left[\Delta\mathbf{K} - (\Delta\mathbf{A})_\infty^T(\Delta\mathbf{A})\right] \tag{18.12}$$

in which $(\Delta\mathbf{A})_\infty$ is the solution (converged iterate) for $\Delta\mathbf{A}$. Consider the iteration procedure

$$\Delta\mathbf{A}^{(j+1)} = \mathbf{A}^{-1}\left[\Delta\mathbf{K} - (\Delta\mathbf{A})_\infty^T\Delta\mathbf{A}^j\right] \tag{18.13}$$

which is very similar to Equation 18.11. Subtraction of two successive iterates and application of matrix norm inequalities furnish

$$\Delta\mathbf{A}^{(j+2)} - \Delta\mathbf{A}^{(j+1)} = -\mathbf{A}^{-1}\Delta\mathbf{A}_\infty\left[\Delta\mathbf{A}^{(j+1)} - \Delta\mathbf{A}^{(j)}\right] \tag{18.14}$$

An example of a matrix norm is the Euclidean norm $\text{norm}(\mathbf{A}) = tr^{1/2}(\mathbf{A}^T\mathbf{A})$. Application of matrix norm properties furnishes

$$\text{norm}\left(\Delta\mathbf{A}^{(j+2)} - \Delta\mathbf{A}^{(j+1)}\right) \le \text{norm}\left(\mathbf{A}^{-1}\Delta\mathbf{A}_\infty\right)\text{norm}\left(\Delta\mathbf{A}^{(j+1)} - \Delta\mathbf{A}^{(j)}\right) \tag{18.15}$$

Convergence is assured in this example if $\sigma(\mathbf{A}^{-1}\Delta\mathbf{A}_\infty) < 1$, in which σ denotes the spectral radius (e.g., Dahlquist and Bjork, 1974; magnitude of the largest eigenvalue) and serves as a greatest lower bound on matrix norms (Varga, 1962). But then, recalling that $\mathbf{A}$ is positive definite, we recognize that

$$\sigma\left(\mathbf{A}^{-1}\Delta\mathbf{A}_\infty\right) \le \sigma\left(\mathbf{A}^{-1}\right)\sigma(\Delta\mathbf{A}_\infty) = \frac{\sigma(\Delta\mathbf{A}_\infty)}{\lambda_{\min}(\mathbf{A})} \tag{18.16}$$

A (conservative) convergence criterion is now revealed as

$$\max_j |\lambda_j(\Delta\mathbf{A}_\infty)| < \min_k |\lambda_k(\mathbf{A})| \tag{18.17}$$

in which $\lambda_j(\Delta\mathbf{A}_\infty)$ denotes the jth eigenvalue of an $n \times n$ matrix $\Delta\mathbf{A}_\infty$. Clearly, convergence is expected if the perturbation matrix $\Delta\mathbf{A}_\infty$ has a sufficiently small norm. Applied to the current problem, we likewise expect convergence to occur if $\max_j |\lambda_j(\Delta\mathbf{L}_\infty)| < \min_k |\lambda_k(\mathbf{L}_0)|$.

The convergence criterion in Equation 18.17 appears to be discouraging if $\mathbf{K}_0$ is ill-conditioned since $\min_k |\lambda_k(\mathbf{L}_0)|$ is then very small. Ill-conditioned stiffness matrices are a common occurrence in the nonlinear finite element method, for example, when the structural material exhibits plasticity or hyperelasticity, or when buckling occurs. However, in such situations the equilibrium equation can usually be adjoined to an "arc length" constraint of which an example is presented in subsequent sections. Doing so results in an augmented linear system in which the matrix is no longer ill-conditioned (Kleiber, 1989).

EXAMPLE 18.1

Demonstrate convergence in the scalar equation

$$2L\Delta L + \Delta L^2 = \Delta K \tag{18.18}$$

Solution

A Taylor series representation for the solution ΔL exists in the form

$$\Delta L = a_0\Delta K + a_1\Delta K^2 + a_2\Delta K^3 + a_3\Delta K^4 + \cdots \tag{18.19}$$

Upon substituting (18.18) into (18.19) and making suitable identifications, the first few terms of the Taylor series are obtained as

$$\Delta L = \frac{1}{2L}\Delta K - \frac{1}{8L^3}\Delta K^2 + \frac{1}{16L^5}\Delta K^3 - \frac{5}{128L^7}\Delta K^3 + O\left(\frac{\Delta K^3}{L^9}\right)$$

The scalar version of the iteration procedure (Equation 18.11) is

$$2L\Delta L^{(j+1)} = \Delta K - \left(\Delta L^{(j)}\right)^2, \quad \Delta L^{(0)} = 0$$

Omitting the manipulations, the first three iterates are obtained as

$$\begin{aligned}
\Delta L^{(1)} &= \frac{1}{2L}\Delta K \\
\Delta L^{(2)} &= \underbrace{\frac{1}{2L}\Delta K - \frac{1}{8L^3}\Delta K^2}_{\text{Taylor series}} \\
\Delta L^{(3)} &= \underbrace{\frac{1}{2L}\Delta K - \frac{1}{8L^3}\Delta K^2 + \frac{1}{16L^5}\Delta K^3 - \frac{5}{128L^7}\Delta K^4}_{\text{Taylor series}}
\end{aligned}$$

and are seen to "track" the Taylor series exactly.

EXAMPLE 18.2

Generate the first two iterates for the matrices

$$\mathbf{K}_0 = \begin{bmatrix} a & 0 & 0 \\ b & c & 0 \\ d & e & f \end{bmatrix} \begin{bmatrix} a & b & d \\ 0 & c & e \\ 0 & 0 & f \end{bmatrix} = \begin{bmatrix} a^2 & ab & ad \\ ab & b^2+c^2 & bd+ce \\ ad & bd+ce & d^2+e^2+f^2 \end{bmatrix}$$

and

$$\mathbf{K}_0 + \Delta\mathbf{K} = \begin{bmatrix} a & 0 & 0 \\ b & c & 0 \\ d & e+g & f \end{bmatrix} \begin{bmatrix} a & b & d \\ 0 & c & e+g \\ 0 & 0 & f \end{bmatrix}$$

$$= \begin{bmatrix} a^2 & ab & ad \\ ab & b^2+c^2 & bd+c(e+g) \\ ad & bd+c(e+g) & d^2+(e+g)^2+f^2 \end{bmatrix}$$

$$\Delta\mathbf{K} = \begin{bmatrix} 0 & 0 & 0 \\ 0 & 0 & cg \\ 0 & cg & g(2e+g) \end{bmatrix}$$

SOLUTION

The initial triangular factor satisfies.

$$\mathbf{L}_0 = \begin{bmatrix} a & 0 & 0 \\ b & c & 0 \\ d & e & f \end{bmatrix}, \quad \mathbf{L}_0^{-1} = \frac{1}{acf} \begin{bmatrix} cf & 0 & 0 \\ -bf & af & 0 \\ be-cd & -ae & ac \end{bmatrix}$$

The exact (converged) triangular factor is

$$\mathbf{L}_0 + \Delta\mathbf{L}_\infty = \begin{bmatrix} a & 0 & 0 \\ b & c & 0 \\ d & e+g & f \end{bmatrix}$$

from which

$$\Delta\mathbf{L}_\infty = \begin{bmatrix} 0 & 0 & 0 \\ 0 & 0 & 0 \\ 0 & g & 0 \end{bmatrix}$$

reflecting the error in $\mathbf{L}_0$ as the initial approximation to $\mathbf{L}_0 + \Delta\mathbf{L}_\infty$.

Upon performing the iteration procedure we find

$$\mathbf{L}_0^{-1}\Delta\mathbf{K}\mathbf{L}_0^{-T} = \frac{1}{(acf)^2}\begin{bmatrix} cf & 0 & 0 \\ -bf & af & 0 \\ be-cd & -ae & ac \end{bmatrix}\begin{bmatrix} 0 & 0 & 0 \\ 0 & 0 & cg \\ 0 & cg & g(2e+g) \end{bmatrix}\begin{bmatrix} cf & -bf & be-cd \\ 0 & af & -ae \\ 0 & 0 & ac \end{bmatrix}$$

$$= \frac{1}{(acf)^2}\begin{bmatrix} 0 & 0 & 0 \\ 0 & 0 & a^2c^2fg \\ 0 & a^2c^2fg & a^2c^2g^2 \end{bmatrix}$$

Accordingly,

$$\text{lower}(\mathbf{L}_0^{-1}\Delta\mathbf{K}\mathbf{L}_0^{-T}) = \begin{bmatrix} 0 & 0 & 0 \\ 0 & 0 & 0 \\ 0 & \frac{g}{f} & \frac{1}{2}\frac{g^2}{f^2} \end{bmatrix}$$

and we obtain the first iterate as

$$\Delta\mathbf{L}^{(1)} = \mathbf{L}_0\text{lower}(\mathbf{L}_0^{-1}\Delta\mathbf{K}\mathbf{L}_0^{-T}) = \begin{bmatrix} 0 & 0 & 0 \\ 0 & 0 & 0 \\ 0 & g & \frac{1}{2}\frac{g^2}{f} \end{bmatrix}$$

In terms of the Euclidean matrix norm, a measure of relative error is introduced as

$$\text{error}(\Delta\mathbf{L}^{(1)}) = \frac{\text{norm}(\Delta\mathbf{L}_\infty - \Delta\mathbf{L}^{(1)})}{\text{norm}(\Delta\mathbf{L}_\infty)} = \frac{1}{2}\left|\frac{g}{f}\right|$$

Suppose $\frac{g}{f} = 0.1$. The relative error for the initial iterate is then $\text{error}(\Delta\mathbf{L}^{(1)}) = 5\%$. *Otherwise stated, in this example, the error in the triangular factors was reduced 95% in one iterate.*

We seek the second iterate and examine the additional error reduction. Now

$$\mathbf{L}_0^{-1}\Delta\mathbf{L}^{(1)}\Delta\mathbf{L}^{(1)\mathrm{T}}\mathbf{L}_0^{-T} = \frac{1}{(acf)^2}\begin{bmatrix} cf & 0 & 0 \\ -bf & af & 0 \\ be-cd & -ae & ac \end{bmatrix}\begin{bmatrix} 0 & 0 & 0 \\ 0 & 0 & 0 \\ 0 & g & \frac{1}{2}\frac{g^2}{f} \end{bmatrix}\begin{bmatrix} 0 & 0 & 0 \\ 0 & 0 & g \\ 0 & 0 & \frac{1}{2}\frac{g^2}{f} \end{bmatrix}$$

$$\times \begin{bmatrix} cf & -bf & be-cd \\ 0 & af & -ae \\ 0 & 0 & ac \end{bmatrix} = \begin{bmatrix} 0 & 0 & 0 \\ 0 & 0 & 0 \\ 0 & 0 & \frac{g^2}{f^2}(1+\frac{1}{4}\frac{g^2}{f^2}) \end{bmatrix}$$

Continuing,

$$\mathbf{L}_0^{-1}\Delta\mathbf{K}\mathbf{L}_0^{-T} - \mathbf{L}_0^{-1}\Delta\mathbf{L}^{(1)}\Delta\mathbf{L}^{(1)T}\mathbf{L}_0^{-T} = \begin{bmatrix} 0 & 0 & 0 \\ 0 & 0 & \frac{g}{f} \\ 0 & \frac{g}{f} & -\frac{1}{4}\frac{g^4}{f^4} \end{bmatrix}$$

$$\text{lower}(\mathbf{L}^{-1}\Delta\mathbf{K}\mathbf{L}^{-T} - \mathbf{L}^{-1}\Delta\mathbf{L}^{(1)}\Delta\mathbf{L}^{(1)T}\mathbf{L}^{-T}) = \begin{bmatrix} 0 & 0 & 0 \\ 0 & 0 & 0 \\ 0 & \frac{g}{f} & -\frac{1}{8}\frac{g^4}{f^4} \end{bmatrix}$$

The second iterate is thereby given by

$$\begin{aligned} \Delta\mathbf{L}^{(2)} &= \mathbf{L}_0\text{lower}(\mathbf{L}_0^{-1}\Delta\mathbf{K}\mathbf{L}_0^{-T} - \mathbf{L}_0^{-1}\Delta\mathbf{L}^{(1)}\Delta\mathbf{L}^{(1)T}\mathbf{L}_0^{-T}) \\ &= \begin{bmatrix} a & 0 & 0 \\ b & c & 0 \\ d & e & f \end{bmatrix}\begin{bmatrix} 0 & 0 & 0 \\ 0 & 0 & 0 \\ 0 & \frac{g}{f} & -\frac{1}{8}\frac{g^4}{f^4} \end{bmatrix} \\ &= \begin{bmatrix} 0 & 0 & 0 \\ 0 & 0 & 0 \\ 0 & g & -\frac{1}{8}\frac{g^4}{f^3} \end{bmatrix} \end{aligned}$$

The relative error for the second iterate now is

$$\text{error}\left(\Delta\mathbf{L}^{(2)}\right) = \frac{\text{norm}\left(\Delta\mathbf{L}_\infty - \Delta\mathbf{L}^{(2)}\right)}{\text{norm}(\Delta\mathbf{L}_\infty)} = \frac{\frac{1}{8}\frac{g^4}{f^3}}{g} = 0.0125\%$$

Evidently, in the first iterate the error is reduced to $1/20$th of the initial value, and in the second iterate to $1/8000$th of the initial value, representing a 400-fold improvement from the first to the second iterate. The convergence rate is much faster than linear convergence characteristic of "fixed point iteration."

EXAMPLE 18.3

Illustrate the effect of the perturbation on the rate of convergence.

SOLUTION

Suppose

$$\mathbf{K} = \begin{bmatrix} 1 & \frac{1}{2} \\ \frac{1}{2} & \frac{10+n}{30} \end{bmatrix}, \quad \mathbf{K}_0 = \begin{bmatrix} 1 & \frac{1}{2} \\ \frac{1}{2} & \frac{1}{3} \end{bmatrix}, \quad \Delta\mathbf{K} = \begin{bmatrix} 0 & 0 \\ 0 & \frac{n}{30} \end{bmatrix}$$

in which $n > -5/2$. The exact solutions are

$$\mathbf{L} = \begin{bmatrix} 1 & 0 \\ \frac{1}{2} & \sqrt{\frac{10+n}{30} - \frac{1}{4}} \end{bmatrix}, \quad \Delta\mathbf{L}_\infty = \begin{bmatrix} 0 & 0 \\ 0 & \sqrt{\frac{10+n}{30} - \frac{1}{4}} - \frac{1}{\sqrt{12}} \end{bmatrix}$$

Clearly, the method fails if $n \leq -5/2$, in which case $\mathbf{K}$ is no longer positive definite. To ensure real numbers attention is restricted to the range $-5/2 \leq n \leq 5/2$.
Following the procedures in Example 18.2, the first two iterates are obtained as

TABLE 18.1
Error Reduction with First and Second Iterations

n	Error($\Delta L^{(1)}$)	1 − Error($\Delta L^{(1)}$)	Error($\Delta L^{(2)}$)	1 − Error($\Delta L^{(2)}$)
0.5	0.053	0.947	0.0042	0.9958
−0.5	0.057	0.943	0.0094	0.9916
+2.0	0.176	0.824	0.059	0.941
−2.0	0.277	0.75	0.132	0.916

$$\Delta \mathbf{L}_1 = \begin{bmatrix} 0 & 0 \\ 0 & \frac{n}{10\sqrt{3}} \end{bmatrix}, \quad \Delta \mathbf{L}_2 = \begin{bmatrix} 0 & 0 \\ 0 & \frac{n}{10\sqrt{3}} - \frac{n^2}{100\sqrt{3}} \end{bmatrix}$$

The previously defined relative error measures are now given by

$$\text{error}\left(\Delta \mathbf{L}^{(1)}\right) = \frac{\left|\frac{n}{5} - \left(\sqrt{1+\frac{2n}{5}} - 1\right)\right|}{\left|\sqrt{1+\frac{2n}{5}} - 1\right|}, \quad \text{error}(\Delta \mathbf{L}^{(2)}) = \frac{\left|\frac{n}{5} - \frac{n^2}{50} - \left(\sqrt{1+\frac{2n}{5}} - 1\right)\right|}{\left|\sqrt{1+\frac{2n}{5}} - 1\right|}$$

The effect of the perturbation is thus reduced as illustrated in Table 18.1.

For the relatively small perturbation represented by $n = \pm 0.5$, the first iteration removes approximately 95% of the initial error. The second iteration removes over 99% of the error, for at least a fivefold improvement over the first iteration. For the relatively large perturbation represented by $n = \pm 2.0$, the first iterate removes more than 70% of the initial error. Over 90% is removed by the second iteration, for at least a twofold improvement over the first iteration.

18.2 STIFF ARC LENGTH CONSTRAINT IN NONLINEAR FEA

18.2.1 Introduction

Arc length constraints enable iterative solution procedures in nonlinear FEA to converge even at critical points, such as occur in buckling problems, and enable computation to continue beyond the critical points, for example, in postbuckling. They were briefly introduced in a simple example in Chapter 14. The arc length constraint replaces the conventional $m \times m$ stiffness matrix with an augmented $(m+1) \times (m+1)$ stiffness matrix. Its use is referred to as *arc length control*, in contrast to *load control* which furnishes the conventional stiffness matrix. It also contrasts with displacement control in which displacements are introduced in increments and the corresponding forces are computed as reactions. In the current chapter, an example of an arc length constraint is introduced following Nicholson (2005b). It identifies arc length parameters maximizing the stiffness (absolute value of the determinant) of the augmented matrix. The parameters, viewed as a vector, must be perpendicular to the rows of the stiffness matrix, likewise considered vectors. The augmented stiffness matrix is nonsymmetric and lacks the small bandwidth of the conventional stiffness matrix. However, using a block triangularization, it is demonstrated that solution may be attained by standard finite element operations, namely triangularization of a banded nonsingular portion of the stiffness matrix followed by forward and backward substitutions involving banded lower and upper triangular matrices. The proposed constraint is expected to permit convergence under longer arc lengths than currently implemented methods. A simple example is presented to illustrate application of the constraint.

In conventional load-control-based finite element modeling of nonlinear problems in solid mechanics, a combined incremental and iterative solution procedure is typically followed. The iteration procedure is a realization of Newton iteration, in which the stiffness matrix (combined geometric and tangent) serves as the Jacobian matrix. Frequently, due to nonlinear geometry under compression or to softening material behavior such as plasticity or hyperelasticity, the stiffness matrix appearing in the incremental equations exhibits a "critical point" at (near) which it is singular (ill-conditioned). Arc length constraints, implementing arc length control, have been introduced to permit calculation at critical points. A recent review of arc length methods has been authored by Memon and Su (2004). Three different implementations have been compared in Ragon et al. (2001). The initial method is attributed to Riks (1979) and Wempner (1971), and modified by Crisfield (1981), Ramm (1981), and Fafard and Massicotte (1993). Arc length constraints are widely implemented in finite element codes, e.g., Moharir (1998).

Arc length methods adjoin a constraint to the conventional incremental finite element equation arising under load control to furnish an augmented stiffness matrix. Doing so enables iterations to converge even at critical points. The goal of the current investigation is to identify parameters of the arc length method ensuring that the augmented stiffness matrix is not only nonsingular but *optimally stiffened*. It will be seen that the best choice is to select a vector arising in the arc length constraint, appearing in the bottom row of the augmented stiffness matrix, such that it is orthogonal to the rows (considered as vectors) of the conventional stiffness matrix.

One issue raised by previous investigators (e.g., Crisfield, 1981) has been that the augmented stiffness matrix is nonsymmetric and unbanded. However, by introducing a block triangularization, the solution procedure will be shown to reduce to conventional finite element operations, namely triangularization of a banded symmetric nonsingular matrix, followed by forward and backward substitution involving banded lower and upper triangular matrices.

The notion of a critical point is illustrated in Figure 18.1. Consider a body submitted to loads following a path in load space to a final load $\mathbf{f}_0$, the magnitude of which is the maximum attained on the path. Along the path the magnitude of the load may be written as $\lambda|\mathbf{f}_0|$, in which the load intensity λ satisfies $0 \leq \lambda \leq 1$. The magnitude of the global displacement vector is denoted as $|\boldsymbol{\gamma}|$. Figure 18.1 illustrates a critical point followed by a zone of decreasing load intensity and negative stiffness, such as occurs in postbuckling. In the current investigation, critical points may be maxima, minima, or saddle points.

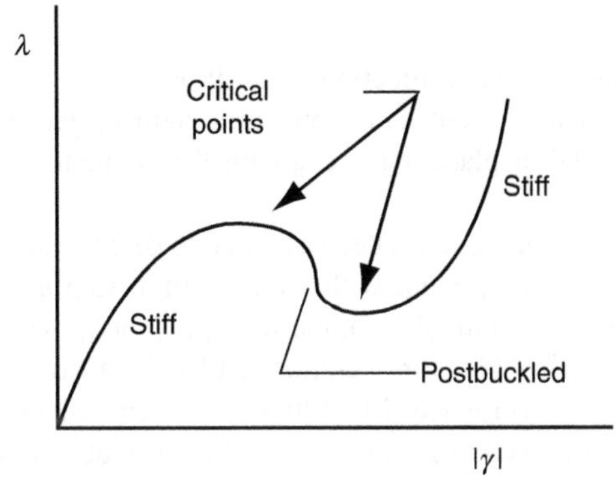

FIGURE 18.1 Load–deflection characteristic exhibiting critical points and buckling.

18.2.2 Newton Iteration for Nonlinear Finite Element Equations

For the sake of explaining how, in the instance being presented, the arc length method is a special case of Newton iteration, we briefly recapitulate Newton iteration in nonlinear FEA, expanding on the presentation in Chapter 14. Consider a solid body referred to the undeformed configuration, with volume V and boundary S. It experiences large deformation and nonlinear material behavior under boundary tractions $\mathbf{t}$ prescribed on S. Balance of linear momentum (load balancing) is fulfilled by the Principle of Virtual Work (Chapter 15) as

$$\boldsymbol{\psi}(\boldsymbol{\gamma}(t)) = \mathbf{0}, \quad \delta\boldsymbol{\gamma}^T\boldsymbol{\psi}(\boldsymbol{\gamma}) = \int \text{trace}(\delta\boldsymbol{\varepsilon}\boldsymbol{\sigma})\,\mathrm{d}V + \int \delta\mathbf{u}^T\rho\ddot{\mathbf{u}}\,\mathrm{dV} - \int \delta\mathbf{u}^T\mathbf{t}\,\mathrm{d}S \quad (18.20)$$

$\boldsymbol{\psi}$ $m \times 1$ vector representing unbalanced loads when nonvanishing

$\boldsymbol{\gamma}(t)$ $m \times 1$ time dependent vector of nodal displacements

$\mathbf{u}(\mathbf{X},t)$ 3×1 displacement vector

$\mathbf{X}$ 3×1 position vector in undeformed configuration

$\boldsymbol{\varepsilon}$ 3×3 Lagrangian strain tensor

$\boldsymbol{\sigma}$ 3×3 second Piola–Kirchhoff stress tensor

ρ mass density

and δ denotes the variational operator. The goal is to compute the displacement vector $\mathbf{u}(\mathbf{X},t)$.

The body is assumed to be "discretized" using a mesh of small elements connected at nodes. In FEA, it is assumed that the displacement vector in the eth element may be approximated to satisfactory accuracy using an interpolation model of the form

$$\mathbf{u}(\mathbf{X},t) \approx \boldsymbol{\varphi}_e^T(\mathbf{X})\boldsymbol{\Phi}_e\boldsymbol{\gamma}_e(t) \quad (18.21)$$

in which

$\boldsymbol{\varphi}_e(\mathbf{X}) = m_e \times 1$ vector-valued function of position
$\boldsymbol{\Phi}_e \quad = m_e \times m_e$ matrix of constants reflecting element geometry
$\boldsymbol{\gamma}_e(t) \ = m_e \times 1$ nodal displacement vector for the element

Suppose that the solution process has been invoked to determine $\boldsymbol{\gamma}_n$, the *global* nodal displacement vector at time $t_n = nh$, with "small" time step h. The task is now to formulate a scheme for iteratively computing $\boldsymbol{\gamma}_{n+1}$ at $t_{n+1} = (n+1)h$. For this purpose, we first describe Newton iteration under load control, which is well-known to experience convergence difficulties near "critical points" at which the dynamic stiffness matrix becomes singular. We then introduce a recently introduced arc length method Nicholson (2005b) with what will be called a *stiff constraint.* It will be seen to enable circumventing the critical point.

18.2.3 Newton Iteration without Arc Length Constraint

Let $\Delta_n \boldsymbol{\gamma} = \boldsymbol{\gamma}_{n+1} - \boldsymbol{\gamma}_n$ denote the incremental nodal displacement vector, with similar definitions for stress, strain, displacement, and traction. From Chapter 15, we know that at dynamic equilibrium

$$\begin{aligned}
\Delta_n \boldsymbol{\psi} &= [\mathbf{K}_G + \mathbf{K}_T]\Delta_n\boldsymbol{\gamma} + \mathbf{M}\Delta_n\ddot{\boldsymbol{\gamma}} - \Delta_n\mathbf{f}_m = \mathbf{0} \\
\mathbf{K}_T(\boldsymbol{\gamma}_{n+1}) &= \int \mathbf{M}^T\mathbf{G}\boldsymbol{\chi}\mathbf{G}^T\mathbf{M}\,\mathrm{d}V_0, \quad \mathbf{K}_G(\boldsymbol{\gamma}_{n+1}) = \int \mathbf{M}^T\boldsymbol{\sigma}\otimes\mathbf{I}\mathbf{M}\,\mathrm{d}V_0 \\
\mathbf{M} &= \int \rho_0\boldsymbol{\Phi}^T\boldsymbol{\varphi}\boldsymbol{\varphi}^T\boldsymbol{\Phi}\,\mathrm{d}V_0, \qquad \Delta_n\mathbf{f}_m = \int \rho_0\boldsymbol{\Phi}^T\boldsymbol{\varphi}\Delta\mathbf{t}_0\,\mathrm{d}S_0 \\
VEC(\Delta_n\mathbf{F}) &= \mathbf{M}(\mathbf{X})\Delta_n\boldsymbol{\gamma}, \qquad \mathbf{G}^T = \tfrac{1}{2}[\mathrm{I}\otimes\mathbf{F}^\mathbf{T} + \mathbf{F}\otimes\mathbf{I}]\mathbf{M}(\mathbf{X})
\end{aligned} \tag{18.22}$$

$\mathbf{F}$	3×3 deformation gradient tensor
$\mathbf{K}_T$	$m \times m$ tangent stiffness matrix
$\mathbf{K}_G$	$m \times m$ geometric stiffness matrix
$\boldsymbol{\chi}$	9×9 tangent modulus tensor
$\mathbf{U}$	9×9 permutation tensor
VEC	VEC operator
$\otimes$	Kronecker product symbol

The Newmark method serves to express $\Delta_n\ddot{\boldsymbol{\gamma}}$ in terms of $\Delta_n\boldsymbol{\gamma}$ and thereby furnishes

$$\begin{aligned}
\Delta_n\boldsymbol{\psi} &= \left[\mathbf{K}_G + \mathbf{K}_T + \tfrac{4}{h^2}\mathbf{M}\right]\Delta_n\boldsymbol{\gamma} - \Delta_n\mathbf{f}_m - \Delta_n\mathbf{r}_m = 0 \\
\Delta_n\mathbf{r}_m &= \Delta_{n-1}\mathbf{f}_m + \tfrac{2}{h}\mathbf{M}\Delta_{n-1}\dot{\boldsymbol{\gamma}} - \left[\mathbf{K}_G + \mathbf{K}_T - \tfrac{4}{h^2}\mathbf{M}\right]\Delta_{n-1}\boldsymbol{\gamma}
\end{aligned} \tag{18.23}$$

Of course the residual vector $\Delta_n\mathbf{r}_m$ is known from the previous (nth) time step.

The Newton iteration scheme may now be stated as

$$\begin{aligned}&\mathbf{K}_D\left[\boldsymbol{\gamma}_{n+1}^{(v+1)}-\boldsymbol{\gamma}_{n+1}^{(v)}\right]=-\boldsymbol{\psi}\left(\boldsymbol{\gamma}_{n+1}^{(v)}\right),\quad \boldsymbol{\gamma}_{n+1}^{(0)}=\boldsymbol{\gamma}_n\\&\mathbf{K}_D=\mathbf{K}_G+\mathbf{K}_T+\tfrac{4}{h^2}\mathbf{M}\end{aligned} \tag{18.24}$$

Note that $\mathbf{K}_D$ is real and symmetric. Under load control, Equation 18.24 is the equation to solved.

The matrix $\mathbf{K}_D(\boldsymbol{\gamma}_n)$ in many problems exhibits critical points, at (near) which it becomes singular (ill-conditioned). This difficulty may, for example, be produced (a) by geometric nonlinearity under compression (e.g., buckling), associated with the geometric stiffness matrix $\mathbf{K}_G(\boldsymbol{\gamma}_n)$ or (b) by material softness, such as plasticity or hyperelasticity, associated with the tangent stiffness matrix $\mathbf{K}_T(\boldsymbol{\gamma})$. Beyond the critical point the stiffness matrix may be indefinite, as in postbuckling (cf. Figure 18.1). Clearly, computation is impossible (or very difficult) in such cases if direct solution of Equation 18.5 is attempted.

18.2.4 Arc Length Method

However, as stated previously arc length methods have been used, by now for several decades, to circumvent the numerical difficulties posed by critical points. They involve two major features:

(a) Writing $\Delta_n\mathbf{f}_m=(\Delta_n\lambda)\mathbf{f}_m$, in which $\mathbf{f}_m$ is the final load vector to be attained, and $\Delta_n\lambda$ is the incremental load intensity
(b) Adjoining an *arc length constraint* to Equation 18.24 in the form

$$\zeta(\boldsymbol{\gamma}_{n+1},\lambda_{n+1})=\mathbf{g}_m^T\Delta_n\boldsymbol{\gamma}+g\Delta_n\lambda-\Delta S=0 \tag{18.25}$$

g_m an $m\times 1$ vector to be determined
g a scalar to be determined
ΔS arc length, a positive scalar parameter

We refer to $\begin{Bmatrix}\mathbf{g}_m\\ g\end{Bmatrix}$ as the *arc length vector.*

It appears that the most commonly implemented methods (e.g., Riks, 1979) involve the conventional choices

$$\begin{aligned}\mathbf{g}_m&=\tfrac{1}{2}\Delta_n\boldsymbol{\gamma}\\ g&=\tfrac{1}{2}\Delta_n\lambda\end{aligned} \tag{18.26}$$

and the arc length constraint is expressed using ΔS^2: $\frac{1}{2}\Delta_n\boldsymbol{\gamma}^T\Delta_n\boldsymbol{\gamma}+\frac{1}{2}(\Delta_n\lambda)^2-\Delta S^2=0$. With this choice, the arc length constraint can be visualized in terms of an $m+1$ dimensional *hypersphere* surrounding the solution points at the nth load step.

A search procedure is then followed on the hypersphere until the equilibrium equation is satisfied (load is balanced).

Returning to the general case represented by Equation 18.25, the $m \times m$ incremental finite element equation is now supplanted by an $(m+1) \times (m+1)$ augmented system of equations. To apply Newton iteration to the augmented system, note that

$$\begin{aligned} d\boldsymbol{\psi}(\boldsymbol{\gamma}_{n+1}, \lambda_{n+1}) &= \mathbf{K}_D \, d\boldsymbol{\gamma}_{n+1} - \mathbf{f}_0 \, d\lambda_{n+1} \\ d\zeta(\boldsymbol{\gamma}_{n+1}, \lambda_{n+1}) &= \mathbf{g}_m^T \, d\boldsymbol{\gamma}_{n+1} + g \, d\lambda_{n+1} \end{aligned} \tag{18.27}$$

Newton iteration applied to Equation 18.27 now gives rise to the $m \times 1$ linear system

$$\mathbf{K}^* \begin{Bmatrix} \boldsymbol{\gamma}_{n+1}^{(\upsilon+1)} - \boldsymbol{\gamma}_{n+1}^{(\upsilon)} \\ \lambda_{n+1}^{(\upsilon+1)} - \lambda_{n+1}^{(\upsilon)} \end{Bmatrix} = \begin{Bmatrix} -\boldsymbol{\psi}\left(\boldsymbol{\gamma}_{n+1}^{(\upsilon)}, \lambda_{n+1}^{(\upsilon)}\right) \\ -\zeta\left(\boldsymbol{\gamma}_{n+1}^{(\upsilon)}, \lambda_{n+1}^{(\upsilon)}\right) \end{Bmatrix}, \quad \mathbf{K}^* = \begin{bmatrix} \mathbf{K}_D & -\mathbf{f}_m \\ \mathbf{g}^T & g \end{bmatrix}, \quad \lambda_{n+1}^{(0)} = \lambda_n \tag{18.28}$$

Choices $\mathbf{g}_m$ and g which *ensure* that $\mathbf{K}^*$ is nonsingular, and even more which maximize the magnitude of the determinant of $\mathbf{K}^*$, will be said to render the arc length constraint *stiff*. Our primary task now is to consider choices for $\mathbf{g}_m$ and g to maximize stiffness (defined below as the magnitude of the determinant). Doing so is expected to permit convergence using larger values of the arc length parameter than in the current method. To find the optimal parameters we restrict attention to the case in which $\mathbf{K}_D$ has rank $m-1$ (rank deficiency of unity). The case of rank lower than $m-1$ will be addressed in a subsequent investigation.

18.2.5 Stiff Arc Length Constraint

18.2.5.1 Stiffness of K*

We assume that $\mathbf{K}_D$ is of rank $m-1$ (rank deficiency $= 1$), and rewrite it as

$$\mathbf{K}_D = \begin{bmatrix} \mathbf{K}_D^{(m-1)} & \boldsymbol{\kappa}_{m-1} \\ \boldsymbol{\kappa}_{m-1}^T & \mathbf{k}_m \end{bmatrix} \tag{18.29}$$

Owing to unit rank deficiency, any row of $\mathbf{K}_D$ can be expressed as a linear combination of the other rows. Furthermore the rows and columns can be ordered such that the upper $(m-1) \times (m-1)$ block is nonsingular. To illustrate this fact consider the reordering $\begin{bmatrix} 1 & 0 & 3 \\ 0 & 0 & 0 \\ 3 & 0 & 6 \end{bmatrix} \rightarrow \begin{bmatrix} \begin{bmatrix} 1 & 3 \\ 3 & 6 \end{bmatrix} & \begin{matrix} 0 \\ 0 \end{matrix} \\ \begin{matrix} 0 & 0 \end{matrix} & 0 \end{bmatrix}$. The 3×3 matrix is of rank 2. The upper left-hand 2×2 block of the reordered matrix is nonsingular. (It appears that reordering will not be necessary if none of the rows of $\mathbf{K}_D$ is null; for example, no reordering is needed in $\begin{bmatrix} 1 & -1 & 0 \\ -1 & 2 & -1 \\ 0 & -1 & 1 \end{bmatrix}$.)

We assume that the above-mentioned ordering either is not needed or has been performed. Accordingly, $\mathbf{K}_D^{(m-1)}$ is nonsingular and $\boldsymbol{\kappa}_{m-1}^T$ is a linear combination of the $\mu - 1$ rows of $\mathbf{K}_D^{(m-1)}$: there exists an $(m-1) \times 1$ vector $\boldsymbol{\alpha}$ such that

$$\begin{aligned}
\boldsymbol{\kappa}_{m-1} &= \mathbf{K}_D^{(m-1)}\boldsymbol{\alpha} \\
k_m &= \boldsymbol{\kappa}_{m-1}^T\boldsymbol{\alpha} \\
&= \boldsymbol{\kappa}_{m-1}^T\left[\mathbf{K}_D^{(m-1)}\right]^{-1}\boldsymbol{\kappa}_{m-1} \\
&= \boldsymbol{\alpha}^T\mathbf{K}_D^{(m-1)}\boldsymbol{\alpha}
\end{aligned} \tag{18.30}$$

If $\mathbf{K}_D$ is positive semidefinite, $\mathbf{K}_D^{(m-1)}$ must be positive definite and k_m must be a positive number.

Now consider the augmented stiffness matrix incorporating the arc length constraint.

$$\mathbf{K}^* = \begin{bmatrix} \mathbf{K}_D^{(m-1)} & \boldsymbol{\kappa}_{m-1} & -\mathbf{f}_{m-1} \\ \boldsymbol{\kappa}_{m-1}^T & k_m & -f_m \\ \mathbf{g}_{m-1}^T & g_m & g \end{bmatrix}, \quad \mathbf{g}_m = \begin{Bmatrix} \mathbf{g}_{m-1} \\ g_m \end{Bmatrix}, \quad \mathbf{f}_m = \begin{Bmatrix} \mathbf{f}_{m-1} \\ f_m \end{Bmatrix} \tag{18.31}$$

The matrix is singular if the third column is a linear combination of the first two (block) columns. If so there exist a vector $\boldsymbol{\mu}$ and a scalar ν for which

$$-\mathbf{f}_{m-1} = \mathbf{K}_D^{(m-1)}\boldsymbol{\mu} + \nu\boldsymbol{\kappa}_{m-1} \tag{18.32a}$$

$$-f_m = \boldsymbol{\kappa}_{m-1}^T\boldsymbol{\mu} + \nu k_{mm} \tag{18.32b}$$

$$= \boldsymbol{\alpha}^T\mathbf{K}_D^{(m-1)}\boldsymbol{\mu} + \nu\boldsymbol{\alpha}^T\boldsymbol{\kappa}_{m-1} \tag{18.32c}$$

$$= -\boldsymbol{\alpha}^T\mathbf{f}_{m-1} \tag{18.32d}$$

and

$$g = \mathbf{g}_{m-1}^T\boldsymbol{\mu} + \nu g_m \tag{18.32e}$$

In the particular situation in which $f_m = \boldsymbol{\kappa}_{m-1}^T\mathbf{K}_D^{(m-1)^{-1}}\mathbf{f}_{m-1}$, the matrix is singular regardless of the choice of $\mathbf{g}_{m-1}^T$, g_m, and g, since the second row (of blocks) is a linear combination of the first row. Since the finite element code contains the matrix $\mathbf{K}_D^{(m-1)}$ and a solver, this difficulty can readily be detected. Suppose, for example, that $\mathbf{K}_D$ is positive semidefinite, in which case $\mathbf{K}_D^{(m-1)}$ is positive definite. The linear system $\mathbf{K}_D^{(m-1)}\boldsymbol{\eta} = \mathbf{f}_{m-1}$ may be solved numerically using triangularization followed by forward and backward substitution. Next $\boldsymbol{\kappa}_{m-1}^T\boldsymbol{\eta}$ is compared to f_m. If they are equal, a different path in load space should be followed to attain the final load.

Alternatively, in the much more likely situation in which $f_m \neq \boldsymbol{\kappa}_{m-1}^T \mathbf{K}_D^{(m-1)^{-1}}\mathbf{f}_{m-1}$, the matrix is nonsingular provided that $\mathbf{g}_{m-1}^T$, g_m, and g are chosen such that the bottom row is *not* a linear combination of the upper two (block) rows: i.e., such that there do *not* exist a vector $\hat{\boldsymbol{\mu}}$ and a scalar $\hat{\nu}$ for which

$$\begin{aligned}
\mathbf{g}_{m-1}^T &= \hat{\boldsymbol{\mu}}^T\mathbf{K}_D^{(m-1)} + \hat{\nu}\boldsymbol{\kappa}_{m-1}^T \\
g_m &= \hat{\boldsymbol{\mu}}^T\boldsymbol{\kappa}_{m-1} + \hat{\nu}k_m \\
&= \hat{\boldsymbol{\mu}}^T\mathbf{K}_D^{(m-1)}\boldsymbol{\alpha} + \hat{\nu}\boldsymbol{\kappa}_{m-1}^T\boldsymbol{\alpha} \\
&= \mathbf{g}_{m-1}^T\boldsymbol{\alpha} \\
g &= -\hat{\boldsymbol{\mu}}^T\mathbf{f}_{m-1} - \hat{\nu}f_m
\end{aligned} \tag{18.33}$$

We assume that the choice of $\mathbf{g}_{m-1}^T$, g_m, and g does not satisfy Equation 18.33, and proceed to consider the choice which renders $\mathbf{K}^*$ stiff.

18.2.5.2 Arc Length Vector Which Maximizes Stiffness: Examples

The *stiffness* $\chi(\mathbf{A})$ of a square matrix $\mathbf{A}$ is defined as the magnitude of its determinant: $\chi(\mathbf{A}) = |\det(\mathbf{A})|$. The matrix is said to be stiff if $\chi(\mathbf{A}) > 0$ (i.e., is nonsingular). Clearly, if stiffness is near zero, the matrix is nearly singular. If the arc length parameters are chosen to maximize the stiffness, the matrix $\mathbf{A}$ is "farther away" from being singular than for other choices.

The first m rows of $\mathbf{K}^*$, considered as row vectors, span an m dimensional space, whose (possibly complex) base vectors we denote by $\mathbf{e}_j$, $j = 1, 2, \ldots, m$. It will be seen that g does not affect stiffness, and that the magnitudes of $\mathbf{g}_m = \begin{Bmatrix} \mathbf{g}_{m-1} \\ g_m \end{Bmatrix}$ and g can both be set to unity since it is later necessary to manipulate the arc length ΔS to attain convergence. (It can easily be easily seen in Equation 18.26 that a similar magnitude choice is implicit in the conventional arc length formulation.)

The direction of the lower row, viewed as a (row) vector, is of greatest interest. As will be shown in the following, it should be chosen to coincide with the null eigenvector of $\mathbf{K}_D$, and thereby to be orthogonal to the m rows of $\mathbf{K}_D$.

EXAMPLE 18.4

As a simple example, consider the matrix

$$\mathbf{A} = \begin{bmatrix} 1 & 0 & 0 \\ 0 & 1 & 0 \\ \dfrac{g_1}{\sqrt{g_1^2+g_2^2+g_3^2}} & \dfrac{g_2}{\sqrt{g_1^2+g_2^2+g_3^2}} & \dfrac{g_3}{\sqrt{g_1^2+g_2^2+g_3^2}} \end{bmatrix}$$

The determinant of $\mathbf{A}$ is $\dfrac{g_3}{\sqrt{g_1^2+g_2^2+g_3^2}}$. The stiffness attains its maximum value if $g_1 = 0$, $g_2 = 0$, $g_3 = 1$. Considering the rows as row vectors, with this choice the third row is orthogonal to the first two rows.

EXAMPLE 18.5

As another simple example, consider the matrix

$$\mathbf{A} = \begin{bmatrix} a_{11} & a_{12} & 0 \\ a_{12} & a_{22} & 0 \\ \dfrac{g_1}{\sqrt{g_1^2+g_2^2+g_3^2}} & \dfrac{g_2}{\sqrt{g_1^2+g_2^2+g_3^2}} & \dfrac{g_3}{\sqrt{g_1^2+g_2^2+g_3^2}} \end{bmatrix}$$

$$= \begin{bmatrix} \hat{\mathbf{A}} & \mathbf{0} \\ \mathbf{g}^T & \dfrac{g_3}{\sqrt{g_1^2+g_2^2+g_3^2}} \end{bmatrix}, \quad \hat{\mathbf{A}} = \begin{bmatrix} a_{11} & a_{12} \\ a_{12} & a_{22} \end{bmatrix}, \quad \mathbf{g} = \begin{Bmatrix} \dfrac{g_1}{\sqrt{g_1^2+g_2^2+g_3^2}} \\ \dfrac{g_2}{\sqrt{g_1^2+g_2^2+g_3^2}} \end{Bmatrix}$$

Since $\hat{\mathbf{A}}$ is symmetric, there exists an orthogonal matrix $\mathbf{Q}$ for which

$$\mathbf{Q}\hat{\mathbf{A}}\mathbf{Q}^T = \mathbf{\Lambda} = \begin{bmatrix} \lambda_1(\hat{\mathbf{A}}) & 0 \\ 0 & \lambda_2(\hat{\mathbf{A}}) \end{bmatrix}, \quad \lambda_{1,2} = \frac{a_{11}+a_{22}}{2} \pm \sqrt{\left(\frac{a_{11}-a_{22}}{2}\right)^2 + a_{12}^2}$$

We now subject $\mathbf{A}$ to a similarity transformation as follows:

$$\mathbf{\Gamma} = \begin{bmatrix} \mathbf{Q} & \mathbf{0} \\ \mathbf{0}^T & 1 \end{bmatrix} \begin{bmatrix} \hat{\mathbf{A}} & \mathbf{0} \\ \mathbf{g}^T & \dfrac{g_3}{\sqrt{g_1^2+g_2^2+g_3^2}} \end{bmatrix} \begin{bmatrix} \mathbf{Q}^T & \mathbf{0} \\ \mathbf{0}^T & 1 \end{bmatrix} = \begin{bmatrix} \mathbf{\Lambda} & \mathbf{0} \\ \mathbf{g}^T\mathbf{Q}^T & \dfrac{g_3}{\sqrt{g_1^2+g_2^2+g_3^2}} \end{bmatrix}$$

and $\mathbf{\Gamma}$ possesses the same eigenvalues as $\mathbf{A}$, namely λ_1, λ_2, and $\dfrac{g_3}{\sqrt{g_1^2+g_2^2+g_3^2}}$.

Clearly, the choice which maximizes the stiffness is, again, $g_1 = 0$, $g_2 = 0$, $g_3 = 1$. With this choice, considering the rows of $\mathbf{A}$ to be (row) vectors, the bottom row of $\mathbf{A}$ is again orthogonal to the first two rows.

EXAMPLE 18.6

Finally, consider the matrix

$$\mathbf{A} = \begin{bmatrix} \cos\theta & \sin\theta \\ \cos\psi & \sin\psi \end{bmatrix}$$

The determinant of $\mathbf{A}$ is given by

$$\det(\mathbf{A}) = \cos\theta\sin\psi - \sin\theta\cos\psi = \sin(\psi - \theta)$$

and the maximum stiffness clearly is attained when $\psi - \theta = \pm\pi/2$. With $+\pi/2$, $\mathbf{A} = \begin{bmatrix} \cos\theta & \sin\theta \\ \sin\theta & -\cos\theta \end{bmatrix}$, and note that the bottom row is orthogonal to the top row. With $-\pi/2$, we obtain $\mathbf{A} = \begin{bmatrix} \cos\theta & \sin\theta \\ -\sin\theta & \cos\theta \end{bmatrix}$. The second row is again orthogonal to the first row, *but points in the opposite direction from the positive choice.*

18.2.5.3 Arc Length Vector Which Maximizes Stiffness: General Argument

A proof is now given generalizing the observations from examples (18.4 through 18.6). The symmetric matrix $\mathbf{K}_D$ is again assumed to be $m \times m$ and to have rank $m-1$. Given an $m \times 1$ vector $\mathbf{f}_m$, as stated previously our task is to determine the $m \times 1$ vector $\left\{\begin{smallmatrix} \mathbf{g}_m \\ g_{m+1} \end{smallmatrix}\right\}$, with $\mathbf{g}_m$ and g_{m+1} both *of unit magnitude*, to maximize $\chi(\mathbf{K}^*)$.

Since $\mathbf{K}_D$ is symmetric, there exist m orthonormal eigenvectors $\boldsymbol{\xi}_\varphi$, one of which corresponds to the null eigenvalue. We set the order such that $\boldsymbol{\xi}_m$ corresponds to the null eigenvalue and form the proper orthogonal matrix $\mathbf{Q} = [\boldsymbol{\xi}_1\ \boldsymbol{\xi}_2 \cdots \boldsymbol{\xi}_m]$. The determinant of $\mathbf{K}^*$ is unaffected by the similarity transformation, giving rise to $\mathbf{K}^\#$ as follows:

$$\mathbf{K}^\# = \begin{bmatrix} \mathbf{Q} & \mathbf{0} \\ \mathbf{0}^T & 1 \end{bmatrix} \begin{bmatrix} \mathbf{K}_D & -\mathbf{f} \\ \mathbf{g}_m^T & g \end{bmatrix} \begin{bmatrix} \mathbf{Q}^T & \mathbf{0} \\ \mathbf{0}^T & 1 \end{bmatrix}$$

$$= \begin{bmatrix} \boldsymbol{\Lambda}_m & -\hat{\mathbf{f}} \\ \hat{\mathbf{g}}_m^T & \hat{g} \end{bmatrix}, \quad \boldsymbol{\Lambda}_m = \mathbf{Q}\mathbf{K}_D\mathbf{Q}^T, \ \hat{\mathbf{f}} = \mathbf{Q}\mathbf{f}, \ \hat{\mathbf{g}}_m = \mathbf{Q}\mathbf{g}_m, \ \hat{g} = g \tag{18.34}$$

Since the eigenvectors serve to diagonalize $\mathbf{K}_D$, $\mathbf{K}^\#$ now becomes

$$\mathbf{K}^\# = \begin{bmatrix} \lambda_1(\mathbf{K}_D) & 0 & 0 & \cdot & \cdot & \cdot & \cdot & \cdot & 0 & -\hat{f}_1 \\ 0 & \lambda_2(\mathbf{K}_D) & 0 & \cdot & \cdot & \cdot & \cdot & \cdot & 0 & -\hat{f}_2 \\ 0 & 0 & \lambda_3(\mathbf{K}_D) & \cdot & \cdot & \cdot & \cdot & \cdot & 0 & -\hat{f}_3 \\ \cdot & \cdot & \cdot & \cdot & \cdot & \cdot & \cdot & \cdot & \cdot & \cdot \\ \cdot & \cdot & \cdot & \cdot & \cdot & \cdot & \cdot & \cdot & \cdot & \cdot \\ \cdot & \cdot & \cdot & \cdot & \cdot & \cdot & \cdot & \cdot & \cdot & \cdot \\ \cdot & \cdot & \cdot & \cdot & \cdot & \cdot & \cdot & \cdot & \cdot & \cdot \\ \cdot & \cdot & \cdot & \cdot & \cdot & \cdot & \cdot & \lambda_{m-1}(\mathbf{K}_D) & \cdot & -\hat{f}_{m-1} \\ 0 & 0 & 0 & \cdot & \cdot & \cdot & \cdot & \cdot & 0 & -\hat{f}_m \\ \hat{g}_1 & \hat{g}_2 & \hat{g}_3 & \cdot & \cdot & \cdot & \cdot & \hat{g}_{m-1} & \hat{g}_m & \hat{g} \end{bmatrix} \tag{18.35}$$

The determinant of $\mathbf{K}^\#$ is given by

$$\det(\mathbf{K}^\#) = \det(\mathbf{K}^*) = -\hat{g}_m \hat{f}_m \prod_{j=1}^{m-1} \lambda_j(\mathbf{K}_D) \tag{18.36}$$

Note that it is independent of $g = \hat{g}$, which has been previously set to unit magnitude.

Recalling that $\{\mathbf{g}_{m-1}^T \quad g_m\}$ has unit magnitude, the stiffness $\chi(\mathbf{K}^*)$ is maximized by the choice

$$\hat{g}_m = 1 \quad \text{and} \quad \hat{g}_j = 0 \quad \text{if } j = 1, 2, 3, \ldots, m-2, m-1 \tag{18.37}$$

$\mathbf{K}^{\#}$ is now given by

$$\mathbf{K}^{\#} = \begin{bmatrix} \lambda_1(\mathbf{K}_m) & 0 & 0 & \cdot\;\cdot\;\cdot\;\cdot & \cdot & 0 & -\hat{f}_1 \\ 0 & \lambda_2(\mathbf{K}_m) & 0 & \cdot\;\cdot\;\cdot\;\cdot & \cdot & 0 & -\hat{f}_2 \\ 0 & 0 & \lambda_3(\mathbf{K}_m) & \cdot\;\cdot\;\cdot\;\cdot & \cdot & 0 & -\hat{f}_3 \\ \cdot & \cdot & \cdot & \cdot\;\cdot\;\cdot\;\cdot & \cdot & \cdot & \cdot \\ \cdot & \cdot & \cdot & \cdot\;\cdot\;\cdot\;\cdot & \cdot & \cdot & \cdot \\ \cdot & \cdot & \cdot & \cdot\;\cdot\;\cdot\;\cdot & \cdot & \cdot & \cdot \\ \cdot & \cdot & \cdot & \cdot\;\cdot\;\cdot\;\cdot & \cdot & \cdot & \cdot \\ \cdot & \cdot & \cdot & \cdot\;\cdot\;\cdot\;\cdot & \lambda_{m-1}(\mathbf{K}_m) & \cdot & -\hat{f}_{m-1} \\ 0 & 0 & 0 & \cdot\;\cdot\;\cdot\;\cdot & \cdot & 0 & -\hat{f}_m \\ 0 & 0 & 0 & \cdot\;\cdot\;\cdot\;\cdot & 0 & 1 & 1 \end{bmatrix} \tag{18.38}$$

If the rows of $\mathbf{K}^*$ are viewed as vectors, it is clear that $\hat{\mathbf{g}}_m$ is orthogonal to the first $\mu - 1$ rows of $\mathbf{K}_D$. In fact, it is orthogonal to all m rows since the mth row is a linear combination of the previous rows owing to the unit rank deficiency of $\mathbf{K}_D$. Hence $\mathbf{Q}\mathbf{K}_D\mathbf{Q}^T\hat{\mathbf{g}}_m = 0$.

But note that

$$\begin{aligned} \mathbf{0} &= \mathbf{Q}\mathbf{K}_D\mathbf{Q}^T\hat{\mathbf{g}}_m \\ &= \mathbf{Q}\mathbf{K}_D\mathbf{Q}^T\mathbf{Q}\mathbf{g}_m \\ &= \mathbf{Q}\mathbf{K}_D\mathbf{g}_m \end{aligned} \tag{18.39}$$

from which we conclude that $\mathbf{K}_D\mathbf{g}_m = \mathbf{0}$, that $\mathbf{g}_m$ is orthogonal to the rows of $\mathbf{K}_D$, and finally that that $\mathbf{g}_m$ *is collinear with the null eigenvector of* $\mathbf{K}_D$.

18.2.5.4 Numerical Determination of the Optimal Arc Length Vector

The process of identifying a unit vector orthogonal to $\mu - 1$ vectors is well known as Gram–Schmidt orthogonalization and is briefly summarized in the current context. Let $\mathbf{a}_1^T$ denote the first row of $\mathbf{K}^*$, and let $\mathbf{a}_1' = \mathbf{a}_1/\sqrt{\mathbf{a}_1^T\mathbf{a}_1}$. A sequence of $\mu - 1$ orthonormal eigenvectors $\mathbf{a}_j'$ is then generated sequentially according to

$$\begin{aligned} \mathbf{a}_j'' &= \mathbf{a}_j - \sum_{i=1}^{j-1}\left(\mathbf{a}_j^T\mathbf{a}_i'\right)\mathbf{a}_i' \\ \mathbf{a}_i' &= \mathbf{a}_i''/\sqrt{\mathbf{a}_i''^T\mathbf{a}_i''} \end{aligned} \tag{18.40}$$

in which $\mathbf{a}_j^T$ denotes the jth row of $\mathbf{K}^*$. Clearly,

$$\mathbf{a}_k'^T\mathbf{a}_j'' = \mathbf{a}_k'^T\mathbf{a}_j - \sum_{i=1}^{j-1}\left(\mathbf{a}_j^T\mathbf{a}_i'\right)\delta_{ik} = 0, \quad \delta_{ik} = \begin{cases} 1, & k = i \\ 0, & k \neq i \end{cases} \tag{18.41}$$

The arc length vector is now found as

$$\mathbf{g}'_m = \mathbf{b}_m - \sum_{i=1}^{m-1} (\mathbf{b}_m^T \mathbf{a}'_i)\mathbf{a}'_i$$

and

$$\mathbf{g}_m = \pm \mathbf{g}'_m \Big/ \sqrt{\mathbf{g}'^T_m \mathbf{g}'_m} \tag{18.42}$$

in which $\mathbf{b}_m$ is a trial vector, for example, $\mathbf{f}_m$.

Note that the arc length vector $\mathbf{g}_m$ can point in a positive or negative sense along the underlying base vector, and g may also be positive or negative. To determine the sense of the $\mathbf{g}_m$ and g, we recall Figure 18.1. For consistency with the positive arc length ΔS, the sum of $\mathbf{g}_m^T \Delta_n \boldsymbol{\gamma}$ and $g\Delta_n\lambda$ must each be nonnegative. If these two terms were of opposite sign, an unstable numerical process would ensue. For example, if $g\Delta_n\lambda$ is negative and grows in magnitude, the arc length constraint requires $\mathbf{g}_m^T \Delta_n \boldsymbol{\gamma}$ to be positive and to likewise grow in magnitude. There is no impediment to this process continuing until both terms attain very large magnitudes, despite the small value of ΔS. Consequently the arc length relation (Equation 18.24) serves as a constraint only if both terms are nonnegative. In the stiff segments of the load–deflection curve, in which the load intensity and the magnitude of the displacement vector are both increasing, this requires that $g \geq 0$ and that $-\frac{\pi}{2} \leq \cos^{-1}\left(\frac{\mathbf{g}_m^T \Delta_n \boldsymbol{\gamma}}{\sqrt{(\Delta_n \boldsymbol{\gamma})^T \Delta_n \boldsymbol{\gamma}}}\right) \leq \frac{\pi}{2}$. In the postbuckled segments, in which the load intensity decreases but the magnitude of the displacements increases, it follows that $g < 0$ and $-\frac{\pi}{2} \leq \cos^{-1}\left(\frac{\mathbf{g}_m^T \Delta_n \boldsymbol{\gamma}}{\sqrt{(\Delta_n \boldsymbol{\gamma})^T \Delta_n \boldsymbol{\gamma}}}\right) \leq \frac{\pi}{2}$. If the $(n+1)$st load step occurs at a critical point, then $\Delta_n\lambda = 0$ and $\mathbf{g}_m^T \Delta_n \boldsymbol{\gamma} = \Delta S$.

Near a critical point, an appealing way to predict the sense of $\mathbf{g}_m$ and g *before computation* at the current step is select them to satisfy

$$g\Delta_{n-1}\lambda > 0, \quad -\frac{\pi}{2} \leq \cos^{-1}\left(\frac{\mathbf{g}_m^T \Delta_{n-1} \boldsymbol{\gamma}}{\sqrt{(\Delta_{n-1} \boldsymbol{\gamma})^T \Delta_{n-1} \boldsymbol{\gamma}}}\right) \leq \frac{\pi}{2} \tag{18.43}$$

thereby making use of the solution at the previous time step. Doing so should be safe away from critical points. Across the critical point, the sign of $\Delta_{n+1}\lambda$ may or may not change from that $\Delta_{n-1}\lambda$ of depending on whether it is an extremum or a saddle point. In this case it appears wise to compute one iterate with $g > 0$ and one with $g < 0$, and to continue with the value for which $g\Delta_{n+1}\lambda > 0$.

18.2.6 Solution Procedure

At first glance, the matrix $\mathbf{K}^*$ appears problematic in that it is asymmetric and unbanded, as is true in the conventional arc length method. However asymmetry and unbandedness are consequences of the $(m+1)$st row and column. The upper left $m \times m$ block is symmetric and can be stored accordingly. As will be seen, after a block triangularization is invoked the solution procedure can be reduced to conventional finite element operations including triangularization of the nonsingular symmetric matrix $\mathbf{K}_D^{(m-1)}$, followed by forward and backward substitutions using banded lower and upper triangular matrices, respectively. The procedure is established below.

18.2.6.1 Block Triangularization

The reader may verify the block triangularization

$$\begin{bmatrix} \mathbf{K}_D^{(m-1)} & \boldsymbol{\kappa}_m & -\mathbf{f}_{m-1} \\ \boldsymbol{\kappa}_m^T & k_m & -f_m \\ \mathbf{g}_{m-1}^T & g_m & g \end{bmatrix} = \begin{bmatrix} \mathbf{L}_{m-1} & 0 \\ \begin{Bmatrix} \boldsymbol{\kappa}_m^T \\ \mathbf{g}_{m-1}^T \end{Bmatrix} \mathbf{U}_{m-1}^{-1} & \begin{bmatrix} 1 & 0 \\ 0 & 1 \end{bmatrix} \end{bmatrix} \begin{bmatrix} \mathbf{U}_{m-1} & \mathbf{L}_{m-1}^{-1}\{\boldsymbol{\kappa}_m \;\; -\mathbf{f}_{m-1}\} \\ \begin{Bmatrix} \mathbf{0}^T \\ \mathbf{0}^T \end{Bmatrix} & \begin{bmatrix} k_m & -f_m \\ g_m & g \end{bmatrix} - \boldsymbol{\Gamma} \end{bmatrix} \tag{18.44}$$

in which $\boldsymbol{\Gamma} = \begin{Bmatrix} \boldsymbol{\kappa}_m^T \\ \mathbf{g}_{m-1}^T \end{Bmatrix} \mathbf{U}_{m-1}^{-1}\mathbf{L}_{m-1}^{-1}\{\boldsymbol{\kappa}_m \;\; -\;\mathbf{f}_{m-1}\}$. Also conventional **LU** triangularization is invoked to furnish $\mathbf{K}_D^{(m-1)} = \mathbf{L}_{m-1}\mathbf{U}_{m-1}$, in which $\mathbf{L}_{m-1}$ is lower triangular and $\mathbf{U}_{m-1}$ is upper triangular.

If $\mathbf{K}_D^{(m-1)}$ is symmetric and positive definite (i.e., $\mathbf{K}_D$ is positive semidefinite with rank $m-1$), then $\mathbf{U}_{m-1} = \mathbf{L}_{m-1}^T$. If we introduce $\{\mathbf{w}_{11} \;\; \mathbf{w}_{12}\}$ obtained from $\mathbf{U}_{m-1}^T\{\mathbf{w}_{11} \;\; \mathbf{w}_{12}\} = \{\boldsymbol{\kappa}_m \;\; -\mathbf{f}_{m-1}\}$, forward substitution and transposition furnishes $\begin{Bmatrix} \mathbf{w}_{11}^T \\ \mathbf{w}_{12}^T \end{Bmatrix} = \begin{Bmatrix} \boldsymbol{\kappa}_m^T \\ \mathbf{g}_{m-1}^T \end{Bmatrix} \mathbf{U}_{m-1}^{-1}$. Also, writing $\mathbf{L}_{m-1}\{\mathbf{w}_{21} \;\; \mathbf{w}_{22}\} = \{\boldsymbol{\kappa}_m \;\; -\mathbf{f}_{m-1}\}$ and using forward substitution furnishes the vector $\{\mathbf{w}_{21} \;\; \mathbf{w}_{22}\} = \mathbf{L}_{m-1}^{-1}\{\boldsymbol{\kappa}_m \;\; -\mathbf{f}_{m-1}\}$. Note that $\boldsymbol{\Gamma} = \begin{Bmatrix} \mathbf{w}_{11}^T \\ \mathbf{w}_{12}^T \end{Bmatrix}\{\mathbf{w}_{21} \;\; \mathbf{w}_{22}\}$ is a 2×2 matrix.

18.2.6.2 Solution of the Outer Problem

We first describe the solution process for the outer problem expressed as

$$\begin{bmatrix} \mathbf{L}_{m-1} & \{0 \;\; 0\} \\ \begin{Bmatrix} \mathbf{w}_{11}^T \\ \mathbf{w}_{12}^T \end{Bmatrix} & \begin{bmatrix} 1 & 0 \\ 0 & 1 \end{bmatrix} \end{bmatrix} \begin{Bmatrix} z_{m-1} \\ \begin{Bmatrix} z_m \\ z_{m+1} \end{Bmatrix} \end{Bmatrix} = -\begin{Bmatrix} \boldsymbol{\psi}_{m-1} \\ \begin{Bmatrix} \psi_m \\ \zeta \end{Bmatrix} \end{Bmatrix} \tag{18.45}$$

Forward substitution is used to solve $\mathbf{L}_{m-1}z_{m-1} = -\boldsymbol{\psi}_{m-1}$ for z_{m-1}, from which

$$\begin{Bmatrix} z_m \\ z_{m+1} \end{Bmatrix} = -\begin{Bmatrix} \psi_m \\ \zeta \end{Bmatrix} - \begin{Bmatrix} \mathbf{w}_{11}^T \\ \mathbf{w}_{12}^T \end{Bmatrix} z_{m-1} \tag{18.46}$$

18.2.6.3 Solution of the Inner Problem

The inner problem is now expressed as

$$\begin{bmatrix} \mathbf{U}_{m-1} & \{\mathbf{w}_{21} \quad \mathbf{w}_{22}\} \\ \begin{Bmatrix} \mathbf{0}^T \\ \mathbf{0}^T \end{Bmatrix} & \begin{bmatrix} k_m & -f_m \\ g_m & 1 \end{bmatrix} \end{bmatrix} \begin{Bmatrix} \Delta_{n+1}\Upsilon_{m-1} \\ \begin{Bmatrix} \Delta_{n+1}\Upsilon_m \\ \Delta_{n+1}\zeta \end{Bmatrix} \end{Bmatrix} = \begin{Bmatrix} z_{m-1} \\ \begin{Bmatrix} z_m \\ z_{m+1} \end{Bmatrix} \end{Bmatrix} \tag{18.47}$$

Next

$$\begin{Bmatrix} \Delta_{n+1}\Upsilon_m \\ \Delta_{n+1}\zeta \end{Bmatrix} = \frac{\begin{bmatrix} 1 & f_m \\ -g_m & k_m \end{bmatrix}}{k_m + f_m g_m} \begin{Bmatrix} z_m \\ z_{m+1} \end{Bmatrix} \tag{18.48}$$

Finally backward substitution serves to solve for $\Delta_{n+1}\Upsilon_{m-1}$ using

$$\mathbf{U}_{m-1}\Delta_{n+1}\Upsilon_{m-1} = z_{m-1} - \{\mathbf{w}_{21} \quad \mathbf{w}_{22}\} \frac{\begin{bmatrix} 1 & f_m \\ -g_m & k_m \end{bmatrix}}{k_m + f_m g_m} \begin{Bmatrix} z_m \\ z_{m+1} \end{Bmatrix} \tag{18.49}$$

EXAMPLE 18.7

Illustrate the orthogonalization and solution procedure using

$$\mathbf{K}_D = \begin{bmatrix} 1 & -1 & 0 \\ -1 & 2 & -1 \\ 0 & -1 & 1 \end{bmatrix}, \quad \mathbf{f} = \begin{Bmatrix} 1 \\ 2 \\ 3 \end{Bmatrix}$$

$$\mathbf{K}^* = \begin{bmatrix} 1 & -1 & 0 & -1 \\ -1 & 2 & -1 & -2 \\ 0 & -1 & 1 & -3 \\ g_1 & g_2 & g_3 & g_4 \end{bmatrix}, \quad \begin{Bmatrix} \Delta_n\psi_1 \\ \Delta_n\psi_2 \\ \Delta_n\psi_3 \\ \Delta_n\varsigma \end{Bmatrix} = 10^{-3}\begin{Bmatrix} 4 \\ 3 \\ 2 \\ 1 \end{Bmatrix}$$

in which $|g_4| = 1$ and $\sqrt{g_1^2 + g_2^2 + g_3^2} = 1$.

Solution

Elementary algebra serves to establish that $\det(\mathbf{K}^*) = 6(g_1+g_2+g_3)$, independently of g_4.

It remains to determine g_1, g_2, and g_3. We first *analytically* determine the values of g_1, g_2, and g_3, which maximize the determinant subject to the magnitude restriction. This is equivalent to maximizing the augmented function

$$\Pi = 6(g_1 + g_2 + g_3) + \exists(g_1^2 + g_2^2 + g_3^2 - 1)$$

in which $\exists$ is a Lagrange multiplier. Elementary manipulation furnishes both a maximum and a minimum:

$$\frac{\partial \Pi}{\partial \Lambda} = g_1^2 + g_2^2 + g_3^2 - 1 = 0$$

$$\frac{\partial \Pi}{\partial g_1} = 6 + 2\exists g_1 = 0$$

$$\frac{\partial \Pi}{\partial g_2} = 6 + 2\exists g_2 = 0$$

$$\frac{\partial \Pi}{\partial g_3} = 6 + 2\exists g_3 = 0$$

Clearly, the extrema satisfy the magnitude condition, and $g_1 = g_2 = g_3 = \pm 1/\sqrt{3}$. The maximum value of the determinant is $6\sqrt{3}$, the minimum value is $-6\sqrt{3}$, and the stiffness is $6\sqrt{3}$.

To determine g_1, g_2, and g_3 using the orthogonalization procedure, we set $g_4 = 1$ and seek $\begin{Bmatrix} g_1 \\ g_2 \\ g_3 \end{Bmatrix}$ to lie exterior to the subspace spanned occupied by the vectors $\mathbf{a}_1 = \begin{Bmatrix} 1 \\ -1 \\ 0 \end{Bmatrix}$ and $\mathbf{a}_2 = \begin{Bmatrix} 0 \\ -1 \\ 1 \end{Bmatrix}$. First,

$$a_1' = \frac{1}{\sqrt{2}} \begin{Bmatrix} 1 \\ -1 \\ 0 \end{Bmatrix}$$

and

$$a_2'' = \begin{Bmatrix} 0 \\ -1 \\ 1 \end{Bmatrix} - \frac{1}{2}\left[\{1 \quad -1 \quad 0\} \begin{Bmatrix} 0 \\ -1 \\ 1 \end{Bmatrix} \right] \begin{Bmatrix} 1 \\ -1 \\ 0 \end{Bmatrix} = \frac{1}{2} \begin{Bmatrix} -1 \\ -1 \\ 2 \end{Bmatrix}$$

Consequently,

$$a_2' = \frac{1}{\sqrt{6}} \begin{Bmatrix} -1 \\ -1 \\ 2 \end{Bmatrix}$$

Next, using $\mathbf{f}$ as a trial vector,

$$\begin{Bmatrix} \hat{g}_1 \\ \hat{g}_2 \\ \hat{g}_3 \end{Bmatrix} = \begin{Bmatrix} 1 \\ 2 \\ 3 \end{Bmatrix} - \frac{1}{2}\left[\{1 \quad -1 \quad 0\} \begin{Bmatrix} 1 \\ 2 \\ 3 \end{Bmatrix} \right] \begin{Bmatrix} 1 \\ -1 \\ 0 \end{Bmatrix} - \frac{1}{6}\left[\{-1 \quad -1 \quad 2\} \begin{Bmatrix} 1 \\ 2 \\ 3 \end{Bmatrix} \right] \begin{Bmatrix} -1 \\ -1 \\ 2 \end{Bmatrix}$$

$$= 2 \begin{Bmatrix} 1 \\ 1 \\ 1 \end{Bmatrix}$$

and, setting the magnitude equal to unity, the desired vector is found as

$$\mathbf{g}_m = \begin{Bmatrix} g_1 \\ g_2 \\ g_3 \end{Bmatrix} = \pm \frac{1}{\sqrt{3}} \begin{Bmatrix} 1 \\ 1 \\ 1 \end{Bmatrix}$$

in agreement with the analytical maximum. Taking the positive sense, the augmented stiffness matrix is now

$$\mathbf{K}^* = \begin{bmatrix} 1 & -1 & 0 & -1 \\ -1 & 2 & -1 & -2 \\ 0 & -1 & 1 & -3 \\ \frac{1}{\sqrt{3}} & \frac{1}{\sqrt{3}} & \frac{1}{\sqrt{3}} & 1 \end{bmatrix}$$

To illustrate the solution procedure, a glance at Equation 18.43 reveals the identifications

$$\mathbf{K}_D^{(m-1)} = \begin{bmatrix} 1 & -1 \\ -1 & 2 \end{bmatrix}, \quad \mathbf{L}_{m-1} = \begin{bmatrix} 1 & 0 \\ -1 & 1 \end{bmatrix}, \quad \mathbf{U}_{m-1} = \begin{bmatrix} 1 & -1 \\ 0 & 1 \end{bmatrix}$$

$$\boldsymbol{\kappa}_m = \begin{Bmatrix} 0 \\ -1 \end{Bmatrix}, \qquad \mathbf{g}_{m-1} = \frac{1}{\sqrt{3}} \begin{Bmatrix} 1 \\ 1 \end{Bmatrix}, \quad \mathbf{f}_{m-1} = \begin{Bmatrix} 1 \\ 2 \end{Bmatrix}$$

Elementary manipulation gives

$$\mathbf{U}_{m-1}^{-1} = \begin{bmatrix} 1 & 1 \\ 0 & 1 \end{bmatrix}, \quad \mathbf{L}_{m=1}^{-1} = \begin{bmatrix} 1 & 0 \\ 1 & 1 \end{bmatrix}, \quad \begin{Bmatrix} \boldsymbol{\kappa}_m^T \\ \mathbf{g}_{m-1}^T \end{Bmatrix} \mathbf{U}_{m-1}^{-1} = \begin{bmatrix} 0 & -1 \\ 1/\sqrt{3} & 2/\sqrt{3} \end{bmatrix}$$

$$\mathbf{L}_{m-1}^{-1}\{\boldsymbol{\kappa}_m \ - \mathbf{f}_{m-1}\} = \begin{bmatrix} 0 & -1 \\ -1 & -3 \end{bmatrix}$$

and

$$\boldsymbol{\Gamma} = \begin{Bmatrix} \boldsymbol{\kappa}_m^T \\ \mathbf{g}_{m-1}^T \end{Bmatrix} \mathbf{U}_{m-1}^{-1} \mathbf{L}_{m-1}^{-1} \{\boldsymbol{\kappa}_m \ - \mathbf{f}_{m-1}\} = \begin{bmatrix} 1 & 3 \\ -2/\sqrt{3} & -7/\sqrt{3} \end{bmatrix}$$

The block triangularization now results in

$$\mathbf{K}^* = \begin{bmatrix} \begin{bmatrix} 1 & 0 \\ -1 & 1 \end{bmatrix} & \begin{bmatrix} 0 & 0 \\ 0 & 0 \end{bmatrix} \\ \begin{bmatrix} 0 & -1 \\ \frac{1}{\sqrt{3}} & \frac{2}{\sqrt{3}} \end{bmatrix} & \begin{bmatrix} 1 & 0 \\ 0 & 1 \end{bmatrix} \end{bmatrix} \begin{bmatrix} \begin{bmatrix} 1 & -1 \\ 0 & 1 \end{bmatrix} & \begin{bmatrix} 0 & -1 \\ -1 & -3 \end{bmatrix} \\ \begin{bmatrix} 0 & 0 \\ 0 & 0 \end{bmatrix} & \begin{bmatrix} 0 & -6 \\ \sqrt{3} & 1 + \frac{8}{\sqrt{3}} \end{bmatrix} \end{bmatrix}$$

The outer and inner problems are expressed as

$$\begin{bmatrix} \begin{bmatrix} 1 & 0 \\ -1 & 1 \end{bmatrix} & \begin{bmatrix} 0 & 0 \\ 0 & 0 \end{bmatrix} \\ \begin{bmatrix} 0 & -1 \\ \frac{1}{\sqrt{3}} & \frac{2}{\sqrt{3}} \end{bmatrix} & \begin{bmatrix} 1 & 0 \\ 0 & 1 \end{bmatrix} \end{bmatrix} \begin{Bmatrix} \Delta_n z_1 \\ \Delta_n z_2 \\ \Delta_n z_3 \\ \Delta_n z_1 \end{Bmatrix} = -10^{-3} \begin{Bmatrix} 4 \\ 3 \\ 2 \\ 1 \end{Bmatrix}$$

$$\begin{bmatrix} \begin{bmatrix} 1 & -1 \\ 0 & 1 \end{bmatrix} & \begin{bmatrix} 0 & -1 \\ -1 & -3 \end{bmatrix} \\ \begin{bmatrix} 0 & 0 \\ 0 & 0 \end{bmatrix} & \begin{bmatrix} 0 & -6 \\ \sqrt{3} & 1+\frac{7}{\sqrt{3}} \end{bmatrix} \end{bmatrix} \begin{Bmatrix} \Delta_n \gamma_1 \\ \Delta_n \gamma_2 \\ \Delta_n \gamma_3 \\ \Delta_n \varsigma \end{Bmatrix} = \begin{Bmatrix} \Delta_n z_1 \\ \Delta_n z_2 \\ \Delta_n z_3 \\ \Delta_n z_1 \end{Bmatrix}$$

Upon applying forward substitution in the outer problem followed by backward substitution in the inner problem, the solution is determined to be

$$\begin{Bmatrix} \Delta_n z_1 \\ \Delta_n z_2 \\ \Delta_n z_3 \\ \Delta_n z_4 \end{Bmatrix} = 10^{-3} \begin{Bmatrix} -4 \\ -7 \\ -9 \\ -1+18\sqrt{3} \end{Bmatrix}, \quad \begin{Bmatrix} \Delta_n \gamma_1 \\ \Delta_n \gamma_2 \\ \Delta_n \gamma_3 \\ \Delta_n \varsigma \end{Bmatrix} = 10^{-3} \begin{Bmatrix} \frac{1}{2}(1+5/\sqrt{3}) \\ 5\left(-1+\frac{1}{2\sqrt{3}}\right) \\ \frac{5}{2\sqrt{3}}\left(-1+\sqrt{3}\right) \\ 3/2 \end{Bmatrix}$$

In the current example, it is evident that the stiffness matrix $\mathbf{K}_D$ arising under load control is singular. But the augmented matrix arising under stiff arc length control is well behaved and a solution is readily attained by a procedure combining triangularization, forward substitution, and backward substitution.

18.3 NON-ITERATIVE SOLUTION OF FINITE ELEMENT EQUATIONS IN INCOMPRESSIBLE SOLIDS

18.3.1 INTRODUCTION

Finite element equations for incompressible and near-incompressible media give rise to a matrix with a diagonal block of zeroes or very small numbers. The matrices are not amenable to conventional techniques involving pivoting on diagonal entries. Uzawa methods (Arrow et al., 1959) have been applied to the associated linear systems. They are iterative and converge when the matrix is nonsingular. In the current study an alternate form of the matrix is used which is amenable to solution *without iteration*. It likewise is applicable whenever the matrix is nonsingular. The solution process consists of a block **LU** factorization, followed by Cholesky decomposition (triangularization) of a positive definite diagonal block together with several forward and backward substitution operations. Two illustrative examples are developed.

In compressible solids, the finite element stiffness matrix typically is positive definite. The governing equation in finite element form can frequently be solved by triangularization, consisting of Cholesky decomposition followed by forward and backward substitution. However, suppose the material is incompressible or near-incompressible. The strains are now subject to an internal constraint (are not independent), and serve to determine stresses only to within an indeterminate pressure. The pressure field serves as an unknown Lagrange multiplier in an auxiliary finite element equation to satisfy the incompressibility constraint.

The structure of the finite element equations (equilibrium plus constraint) at first glance poses a computational problem, since a block of the stiffness matrix is null. Traditional triangularization and solution methods based on pivoting are not directly applicable to such a matrix. Much of the literature on this problem is based on the Uzawa method, which attains the solution through an iteration scheme. (cf. In the current investigation, a block triangularization solution is formulated in which the blocks are obtained by Cholesky decomposition, as well as forward and backward substitutions. This scheme obviates the need for iteration while using real variables.)

18.3.2 Finite Element Equation for an Incompressible Medium

To set the problem under study in a context, we consider dynamic response of a near-incompressible nonlinear solid. The special cases of static response, incompressible media, and linear behavior can be retrieved from this case. Application of the Incremental Principle of Virtual Work and introduction of suitable interpolation models (Chapter 15) furnish the finite element relations

$$\begin{aligned} \mathbf{M}\Delta\ddot{\boldsymbol{\gamma}} + \mathbf{K}(\boldsymbol{\gamma})\Delta\boldsymbol{\gamma} - \boldsymbol{\Omega}\Delta\boldsymbol{\pi} &= \Delta\mathbf{f} \\ \boldsymbol{\Omega}^T\Delta\boldsymbol{\gamma} + \frac{\Delta\boldsymbol{\pi}}{\kappa} &= 0 \end{aligned} \qquad (18.50)$$

$\mathbf{M}$ mass matrix, $n \times n$ and positive definite
$\mathbf{K}$ incremental stiffness matrix, $n \times n$ and positive definite
$\boldsymbol{\Omega}$ pressure–displacement matrix, $n \times p$ of rank p, $p < n$
$\Delta\boldsymbol{\gamma}$ incremental nodal displacement vector
$\Delta\boldsymbol{\pi}$ incremental nodal pressure vector
$\Delta\mathbf{f}$ incremental consistent nodal force vector

In particular, as usual $\Delta\boldsymbol{\gamma}$ is the difference between the nodal displacement vectors at two load or time steps. We assume that the mass and stiffness matrices are positive definite and $n \times n$. The pressure–displacement matrix $\boldsymbol{\Omega}$ is $n \times p$ and is restricted to have rank p, $p < n$.

The first equation is a realization of the balance of linear momentum, while the second represents the a posteriori enforcement of the near-incompressibility constraint. If $\kappa \to \infty$, the incompressible case is recovered. If $\mathbf{M} = 0$, the static case is recovered. Finally, to specialize to the linear case, the incremental symbol Δ may be removed.

We assume that Equation 18.50 is integrated using a one-step procedure based on the trapezoidal rule (Newmark's method). As before let h denote the time step and let $\boldsymbol{\gamma}_n = \boldsymbol{\gamma}(t_n)$. At time $t_{n+1} = (n+1)h$ Equation 18.50 becomes, following Zienkiewicz and Taylor (1989),

$$\mathbf{A}\begin{Bmatrix} \Delta_{n+1}\boldsymbol{\gamma} \\ \Delta_{n+1}\boldsymbol{\pi} \end{Bmatrix} = \begin{Bmatrix} \Delta_{n+1}\mathbf{g} \\ 0 \end{Bmatrix}, \quad \mathbf{A} = \begin{bmatrix} \mathbf{M} + \frac{h^2}{4}\mathbf{K}(\boldsymbol{\gamma}_n) & -\frac{h^2}{2}\boldsymbol{\Omega} \\ -\frac{h^2}{2}\boldsymbol{\Omega}^T & -\frac{h^2}{2}\frac{\mathbf{I}_p}{\kappa} \end{bmatrix} \tag{18.51}$$

in which

$\mathbf{I}_p$ is the $p \times p$ identity matrix

$$\Delta_{n+1}\mathbf{g} = \frac{h^2}{4}(\Delta_{n+1}\mathbf{f} + \Delta_n\mathbf{f} - \mathbf{K}\Delta_n\boldsymbol{\gamma}) + \mathbf{M}(\Delta_n\boldsymbol{\gamma} + h\Delta_n\mathbf{q}), \Delta_n\mathbf{q} = \Delta_n\dot{\boldsymbol{\gamma}}$$

$$\Delta_{n+1}\boldsymbol{\gamma} = \boldsymbol{\gamma}_{n+1} - \boldsymbol{\gamma}_n.$$

Note that $\mathbf{A}$ is symmetric, with the advantage of saving computer storage.

A comment is in order on the restriction that the $n \times p$ matrix $\boldsymbol{\Omega}$ have rank p. We consider whether the matrix $\begin{bmatrix} \mathbf{K} & -\boldsymbol{\Omega} \\ -\boldsymbol{\Omega}^T & \mathbf{0} \end{bmatrix}$ is singular, if $\mathbf{K}$ is $n \times n$ and positive definite while $\boldsymbol{\Omega}$ is $n \times p$ *but* of rank $\pi - 1$ (*or less*). The matrix is singular if, and only if, $\begin{bmatrix} \mathbf{I} & -\mathbf{K}^{-1/2}\boldsymbol{\Omega} \\ -\boldsymbol{\Omega}^T\mathbf{K}^{-1/2} & \mathbf{0} \end{bmatrix}$ is singular. But this new matrix is singular if, and only if, there exists a nontrivial vector whose product with this matrix vanishes. This is simply the condition that there exist a nonzero $p \times 1$ vector $\boldsymbol{x}$ for which $\boldsymbol{\Omega}^T\mathbf{K}^{-1}\boldsymbol{\Omega}\boldsymbol{x} = 0$. Such a vector exists since $\boldsymbol{\Omega}^T\mathbf{K}^{-1}\boldsymbol{\Omega}$ at most has rank $\pi - 1$. It follows that $\begin{bmatrix} \mathbf{K} & -\boldsymbol{\Omega} \\ -\boldsymbol{\Omega}^T & 0 \end{bmatrix}$ is singular if $\boldsymbol{\Omega}$ has rank less than p.

Next consider the matrix $\begin{bmatrix} \mathbf{K} & -\boldsymbol{\Omega} \\ -\boldsymbol{\Omega}^T & -\mathbf{I}_p/\kappa \end{bmatrix}$ in which the bulk modulus is a very large positive number (for near-incompressibility). Again $\boldsymbol{\Omega}$ has rank $\pi - 1$. The matrix cannot be singular since only the zero vector satisfies $[\boldsymbol{\Omega}^T\mathbf{K}^{-1}\boldsymbol{\Omega} + \mathbf{I}_p/\kappa]\boldsymbol{x} = 0$. The smallest eigenvalue of this latter matrix is $1/\kappa$, which is a very small number. But the condition number of this matrix is $\kappa \max_{j=1,p} [\lambda_j(\boldsymbol{\Omega}\mathbf{K}^{-1}\boldsymbol{\Omega}^T)] + 1$, and will typically be a very large number for near-incompressible materials. Consequently, if $\boldsymbol{\Omega}$ has rank $p - 1$ (*or less*), the matrix is expected to be ill-conditioned and convergence will be very difficult to achieve.

18.3.3 Uzawa's Method

To explain Uzawa's method we follow the development in Zienkiewicz and Taylor (1989) for the case in which

$$\mathbf{C}\begin{Bmatrix} \boldsymbol{\gamma} \\ \boldsymbol{\pi} \end{Bmatrix} = \begin{Bmatrix} \mathbf{f} \\ 0 \end{Bmatrix}, \quad \mathbf{C} = \begin{bmatrix} \mathbf{K} & -\boldsymbol{\Omega} \\ -\boldsymbol{\Omega}^T & 0 \end{bmatrix} \tag{18.52}$$

There is an extensive and continuing literature on the Uzawa method (Hu and Zou, 2001).

The term $\frac{\boldsymbol{\pi}}{\rho}$ is subtracted from both sides to furnish

$$\begin{bmatrix} \mathbf{K} & -\boldsymbol{\Omega} \\ -\boldsymbol{\Omega}^T & -\frac{\mathbf{I}_p}{\rho} \end{bmatrix}\begin{Bmatrix} \boldsymbol{\gamma} \\ \boldsymbol{\pi} \end{Bmatrix} = \begin{Bmatrix} \mathbf{f} \\ -\frac{\boldsymbol{\pi}}{\rho} \end{Bmatrix} \tag{18.53}$$

The Uzawa method is realized as the iteration scheme

$$\mathbf{B}\begin{Bmatrix}\boldsymbol{\gamma}\\ \boldsymbol{\pi}\end{Bmatrix}^{j+1}=\begin{Bmatrix}\mathbf{f}\\ -\dfrac{\boldsymbol{\pi}^j}{\rho}\end{Bmatrix},\quad \mathbf{B}=\begin{bmatrix}\mathbf{K} & -\boldsymbol{\Omega}\\ -\boldsymbol{\Omega}^T & -\dfrac{\mathbf{I}_p}{\rho}\end{bmatrix} \tag{18.54}$$

in which $\rho > 0$ is an acceleration parameter. Successive iterates of $\boldsymbol{\pi}$ satisfy

$$\left[\mathbf{I}_p+\rho\boldsymbol{\Omega}^T\mathbf{K}^{-1}\boldsymbol{\Omega}\right]\boldsymbol{\pi}^{j+1}=\boldsymbol{\pi}^j-\rho\boldsymbol{\Omega}^T\mathbf{K}^{-1}\mathbf{f} \tag{18.55}$$

in which the superscript is a counter for the iterate.

This sequence converges if the maximum eigenvalue of $[\mathbf{I}_p+\rho\boldsymbol{\Omega}^T\mathbf{K}^{-1}\boldsymbol{\Omega}]^{-1}$ is less than unity. But note that $\boldsymbol{\Omega}^T\mathbf{K}^{-1}\boldsymbol{\Omega}$ is positive definite and of rank p. Hence the eigenvalues of $\left[\mathbf{I}_p+\rho\boldsymbol{\Omega}^T\mathbf{K}^{-1}\boldsymbol{\Omega}\right]$ all exceed unity. It follows that the eigenvalues of $\left[\mathbf{I}_p+\rho\boldsymbol{\Omega}^T\mathbf{K}^{-1}\boldsymbol{\Omega}\right]^{-1}$ are less than unity, implying convergence. This scheme represents "fixed point iteration," for which the convergence rate is *linear*.

We note that $\mathbf{B}$ can be triangularized using complex variables as follows:

$$\mathbf{B}=\begin{bmatrix}\mathbf{L} & \mathbf{0}\\ -\boldsymbol{\Omega}^T\mathbf{L}^{-T} & i\mathbf{I}_p\end{bmatrix}\begin{bmatrix}\mathbf{L}^T & -\mathbf{L}^{-1}\boldsymbol{\Omega}\\ \mathbf{0} & i\left[\frac{\mathbf{I}_p}{\rho}+\mathbf{W}^T\mathbf{W}\right]\end{bmatrix} \tag{18.56}$$

$$\mathbf{L}\mathbf{L}^T=\mathbf{K},\quad i=\sqrt{-1},$$

and $\mathbf{W}$ is obtained by solving the linear system $\mathbf{LW}=\boldsymbol{\Omega}$ using forward substitution.

The triangularization is only performed once in linear problems. Forward and backward substitution then are repeated at each iteration to attain the solution. In particular, the decomposition is used

$$\begin{bmatrix}\mathbf{L} & \mathbf{0}\\ -\boldsymbol{\Omega}^T\mathbf{L}^{-T} & i\mathbf{I}_p\end{bmatrix}\begin{Bmatrix}z_1^{j+1}\\ z_2^{j+1}\end{Bmatrix}=\begin{Bmatrix}f\\ -\frac{\boldsymbol{\pi}^j}{\rho}\end{Bmatrix},\quad \begin{bmatrix}\mathbf{L}^T & -\mathbf{L}^{-1}\boldsymbol{\Omega}\\ \mathbf{0} & i\left[\frac{\mathbf{I}_p}{\rho}+\mathbf{W}^T\mathbf{W}\right]\end{bmatrix}\begin{Bmatrix}\boldsymbol{\gamma}^{j+1}\\ \boldsymbol{\pi}^{j+1}\end{Bmatrix}=\begin{Bmatrix}z_1^{j+1}\\ z_2^{j+1}\end{Bmatrix} \tag{18.57a}$$

followed by readily performed operations on block submatrices:

$$\mathbf{L}z_1^{j+1}=\mathbf{f}\quad \text{forward substitution} \tag{18.57b}$$

$$z_2^{j+1}=i\frac{\boldsymbol{\pi}^j}{\rho}-i\boldsymbol{\Omega}^T\mathbf{L}^{-T}z_1^{j+1}\quad \text{backward substitution} \tag{18.57c}$$

$$i\left[\frac{I_p}{\rho}+\mathbf{W}^T\mathbf{W}\right]\boldsymbol{\pi}^{j+1}=z_2^{j+1}\quad p\times p \text{ triangularization} \tag{18.57d}$$

$$\mathbf{L}^T\boldsymbol{\gamma}^{j+1}=z_1^{j+1}+\mathbf{L}^{-1}\boldsymbol{\Omega}\boldsymbol{\pi}^{j+1}\quad \text{backward substitution} \tag{18.57e}$$

Note that $\left[\frac{\mathbf{I}_p}{\rho}+\mathbf{W}^T\mathbf{W}\right]$ in Equation 18.57d is positive definite, so that Cholesky decomposition using real mathematics applicable.

We note that it is also problematic for the Uzawa method if the rank of $\boldsymbol{\Omega}$ equals $p-1$ (or less). Recall from Equation 18.54 the convergence criterion that the eigenvalues of $[\mathbf{I}_p+\rho\boldsymbol{\Omega}^T\mathbf{K}^{-1}\boldsymbol{\Omega}]^{-1}$ must be less than unity. However, if rank $\boldsymbol{\Omega}=p-1$, $\boldsymbol{\Omega}^T\mathbf{K}^{-1}\boldsymbol{\Omega}$ is singular (rank less than p) and $[\mathbf{I}_p+\rho\boldsymbol{\Omega}^T\mathbf{K}^{-1}\boldsymbol{\Omega}]^{-1}$ has an eigenvalue equal to unity. Consequently, convergence will not occur. (The existence of a difficulty is not surprising since the matrix in Equation 18.52 is singular if **rank** $(\boldsymbol{\Omega}) < \boldsymbol{p}$.)

18.3.4 Modification to Avoid Iteration

A simple modification to the Uzawa method is introduced which furnishes solution without iteration while using real mathematics. Returning to the incremental finite element formulation, Equation 18.50 may be rewritten in the equivalent form

$$\mathbf{C}\begin{Bmatrix}\Delta_{n+1}\boldsymbol{\gamma}\\ \Delta_{n+1}\boldsymbol{\pi}\end{Bmatrix}=\begin{Bmatrix}\Delta_{n+1}\mathbf{g}\\ \mathbf{0}\end{Bmatrix},\quad \mathbf{C}=\begin{bmatrix}\mathbf{M}+\frac{h^2}{4}\mathbf{K}(\boldsymbol{\gamma}_n) & -\frac{h^2}{2}\boldsymbol{\Omega}\\ \frac{h^2}{2}\boldsymbol{\Omega}^T & \frac{h^2}{2}\frac{\mathbf{I}_p}{\kappa}\end{bmatrix} \tag{18.58}$$

Note that the sign of the lower row in the matrix has been changed. This forfeits the symmetry of the matrix (which has no real significance for computer storage), but will prove to permit a non-iterative solution based on triangularization of positive definite submatrices (and real number operations).

Observe that block triangular factorization in Equation 18.58 furnishes

$$\begin{bmatrix}\mathbf{M}+\frac{h^2}{4}\mathbf{K}(\boldsymbol{\gamma}_n) & -\frac{h^2}{2}\boldsymbol{\Omega}\\ \frac{h^2}{2}\boldsymbol{\Omega}^T & \frac{h^2}{2}\frac{\mathbf{I}_p}{\kappa}\end{bmatrix}=\begin{bmatrix}\mathbf{L}_s & \mathbf{0}\\ \frac{h^2}{2}\boldsymbol{\Omega}^T\mathbf{L}_s^{-T} & \mathbf{I}_p\end{bmatrix}\begin{bmatrix}\mathbf{L}_s^T & -\mathbf{L}_s^{-1}\frac{h^2}{2}\boldsymbol{\Omega}\\ \mathbf{0}^T & \frac{h^4}{4}\mathbf{W}^T\mathbf{W}+\frac{h^2}{2}\frac{\mathbf{I}_p}{\kappa}\end{bmatrix} \tag{18.59}$$

$$\mathbf{L}_s\mathbf{L}_s^T=\mathbf{M}+\frac{h^2}{4}\mathbf{K}(\boldsymbol{\gamma}_n)$$

The solution is attained by the following decomposition and readily performed operations of block submatrices.

$$\begin{gathered}\begin{bmatrix}\mathbf{L}_s & \mathbf{0}\\ \frac{h^2}{2}\boldsymbol{\Omega}^T\mathbf{L}_s^{-T} & \mathbf{I}_p\end{bmatrix}\begin{Bmatrix}\Delta_{n+1}\mathbf{z}_1\\ \Delta_{n+1}\mathbf{z}_2\end{Bmatrix}=\begin{Bmatrix}\Delta_{n+1}\mathbf{g}\\ \mathbf{0}\end{Bmatrix}\\ \begin{bmatrix}\mathbf{L}_s^T & -\mathbf{L}_s^{-1}\frac{h^2}{2}\boldsymbol{\Omega}\\ \mathbf{0}^T & \frac{h^4}{4}\mathbf{W}^T\mathbf{W}+\frac{h^2}{2}\frac{\mathbf{I}_p}{\kappa}\end{bmatrix}\begin{Bmatrix}\Delta_{n+1}\boldsymbol{\gamma}\\ \Delta_{n+1}\boldsymbol{\pi}\end{Bmatrix}=\begin{Bmatrix}\Delta_{n+1}\mathbf{z}_1\\ \Delta_{n+1}\mathbf{z}_2\end{Bmatrix}\end{gathered} \tag{18.60}$$

$$\begin{aligned}\mathbf{L}_s\Delta_{n+1}\mathbf{z}_1&=\Delta_{n+1}\mathbf{g} && \text{forward substitution}\\ \mathbf{w}&=\mathbf{L}_s^T\Delta_{n+1}\mathbf{z}_1 && \text{backward substitution}\\ \Delta_{n+1}\mathbf{z}_2&=-\frac{h^2}{2}\boldsymbol{\Omega}^T\mathbf{w} && \\ \mathbf{L}_s\mathbf{W}&=\boldsymbol{\Omega} && \text{forward substitution}\end{aligned}$$

$$\left[\frac{h^4}{4}\mathbf{W}^T\mathbf{W} + \frac{h^2}{2}\mathbf{I}_p/\kappa\right]\Delta_{n+1}\boldsymbol{\pi} = \Delta_{n+1}\mathbf{z}_2 \quad \text{triangularization}$$

$$\mathbf{L}_s^T\Delta_{n+1}\boldsymbol{\gamma} = \Delta_{n+1}\mathbf{z}_1 + \frac{h^2}{2}\mathbf{W}\Delta_{n+1}\boldsymbol{\pi} \quad \text{backward substitution}$$

This solution procedure is enabled by the fact that $\left[\frac{h^4}{4}\mathbf{W}^T\mathbf{W} + \mathbf{I}_p/\kappa\right]$ is positive definite and hence can be triangularized.

It is worth noting that a non-iterative solution can be achieved using the symmetric form in the Uzawa method (Equation 18.54), following the procedure presented in Equation 18.57 involving complex numbers.

EXAMPLE 18.8

Demonstrate that the method works in the following example.

Consider a single element model for an incompressible isotropic elastic rod, shown in Figure 18.2. The rod is of length L with a square Y by Y cross section, and $A = Y^2$.

Shear modulus is μ. Interpolation models are assumed in the form

$$u(x) = xu(L)/L, \quad v(y) = w(y) = yv(Y)/Y, \quad p = p_0(\text{constant})$$

Omitting the details, the Principle of Virtual Work together with the variational form of the incompressibility constraint may be readily shown to furnish

$$\begin{bmatrix} \frac{2\mu A}{L} & 0 & -A \\ 0 & \frac{4\mu AL}{Y^2} & -2A\frac{L}{Y} \\ A & 2\frac{AL}{Y} & 0 \end{bmatrix} \begin{Bmatrix} u(L) \\ v(Y) \\ p_0 \end{Bmatrix} = \begin{Bmatrix} f \\ 0 \\ 0 \end{Bmatrix}$$

This equation was encountered and solved in Chapter 11. Simple manipulation following the foregoing procedures furnishes $f = \frac{3\mu A}{L}$. This is the exact answer for incompressible isotropic linear elasticity, since the Young's modulus E in this case satisfies $\mu = \frac{E}{2(1+\nu)}$, and the Poisson's ratio ν equals 1/2.

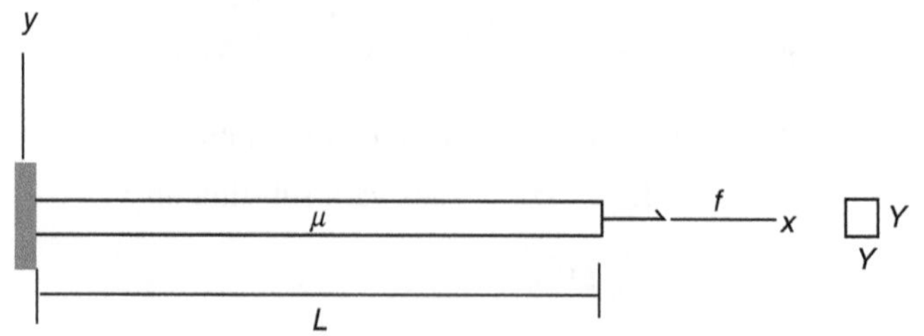

FIGURE 18.2 Rod of incompressible elastic material.

References

MONOGRAPHS AND TEXTS

Arrow, K.J., Hurwitz, L., and Uzawa, H., 1958, *Studies in Nonlinear Programming*, Stanford University Press, Palo Alto, CA.

Bathe, K.-J., 1996, *Finite Element Procedures*, Prentice-Hall, Englewood Cliffs, NJ.

Belytschko, T., Liu, W.K., and Moran, B., 2000, *Nonlinear Finite Elements for Continua and Structures*, Wiley, New York.

Bonet, J. and Wood, R., 1997, *Nonlinear Continuum Mechanics for Finite Element Analysis*, Cambridge University Press, Cambridge.

Brush, D. and Almroth, B., 1975, *Buckling of Bars. Plates and Shells*, McGraw-Hill, New York.

Callen, H.B., 1985, *Thermodynamics and an Introduction to Thermostatistics*, 2nd Ed, Wiley, New York.

Chandrasekharaiah, D. and Debnath, L., 1994, *Continuum Mechanics*, Academic Press, San Diego, CA.

Chung, T.J., 1988, *Continuum Mechanics*, Prentice-Hall, Englewood Cliffs, NJ.

Crisfield, M.A., 1991, *Nonlinear Finite Element Analysis of Solids and Structures*, Wiley, New York.

Dahlquist, G. and Bjork, A., 1974, *Numerical Methods*, Prentice-Hall, Englewood Cliffs, NJ.

Ellyin, F., 1997, *Fatigue Damage, Crack Growth and Life Prediction*, Chapman & Hall, London.

Eringen, A.C., 1962, *Nonlinear Theory of Continuous Media*, McGraw-Hill, New York.

Ewing, G.M., 1985, *Calculus of Variations with Applications*, Ellis Horwood, Chichester, UK.

Gear, C.W 1971, *Numerical Initial Value Problems in Ordinary Differential Equations*, Prentice-Hall, Englewood Cliffs, NJ.

Gent, A.N., ed, 1992, *Engineering with Rubber: How to Design Rubber Components*, Hanser, New York.

Golub, G. and Van Loan, C., 1996, *Matrix Computations*, Johns Hopkins University Press, Baltimore, MD.

Graham, A., 1981, *Kronecker Products and Matrix Calculus with Applications*, Ellis Horwood, Chichester.

Hildebrand, F.B., 1976, *Advanced Calculus with Applications*, 2nd Ed, Prentice-Hall, Englewood Cliffs, NJ.

Hughes, T., 2000, *The Finite Element Method: Linear Static and Dynamic Analysis*, Dover, New York.

Kleiber, M., 1989, *Incremental Finite Element Modeling in Non-Linear Solid Mechanics*, Ellis Horwood, Chichester.

Oden, J.T., 1972, *Finite Elements of Nonlinear Continua*, McGraw-Hill, New York.

Rabotnov, Y.N., 1969, *Creep Problems in Structural Mechanics*, North Holland, Amsterdam.

Reddy, J.N., 2004, *An Introduction to Nonlinear Finite Element Analysis*, Oxford University Press, Oxford.

Rowe, G.W., Hartley, P., Pillinger, I., and Sturgess, C.E.N., 1991, *Finite Element Plasticity and Metalforming Analysis*, Cambridge University Press, Cambridge.

Schey, H.M., 1973, *DIV, GRAD, CURL and All That*, Norton, New York.

Thomason, P.F. 1990, *Ductile Fracture of Metals*, Pergamon Press, Oxford.
Varga, R.S., 1962, *Matrix Iteration Analysis*, Prentice-Hall, Englewood Cliffs, NJ.
Wang, C.-T., 1953, *Applied Elasticity*, Maple Press, York, PA.
Zienkiewicz, O. and Taylor, R.L., 1989, *The Finite Element Method*, Vol I, 4th Ed, McGraw-Hill, New York.
Zienkiewicz, O. and Taylor, R.L., 1991 *The Finite Element Method*, Vol II, 4th Ed, McGraw-Hill, New York.

ARTICLES AND OTHER SOURCES

Al-Grafi, M., 2003, *Unified Damage Softening Model for Ductile Fracture, Doctoral Dissertation*, University of Central Florida, Orlando, FL.
ANSYS User Manual Ver 6.0, 2000, Swanson Analysis Systems.
Arrow, K.J., Hurwitz, L., and Uzawa, H., 1958, *Studies in Nonlinear Programming*, Stanford University Press, Palo Alto, CA.
Bettayeb, F., Haciane, S., and Aoudia, S., 2004, Nonlinear filtering of ultrasonic signals using time-scale debauches decomposition, available at http://www.ndt.net/article/wcndt2004/html/htmltxt/627_bettayeb.htm.
Blatz, P.J. and Ko, W.L., 1962, Application of finite elastic theory to the deformation of rubbery materials, *Trans Soc Rheol*, Vol 6, 223.
Bonora, N., 1997, A nonlinear CDM model for ductile failure, *Eng Fract Mech*, Vol 58, No 1–2, 11–28.
Chen, J., Wan, W., Wu, C.T., and Duan, W., 1997, On the perturbed Lagrangian formulation for nearly incompressible and incompressible hyperelasticity, *Comp Methods Appl Mech Eng*, Vol 142, 335.
Crisfield, M.A., 1981, A fast incremental/iterative solution procedure that handles 'snap-through,' *Comp Struct*, Vol 13, 55–62.
Dennis, B.H., Dulikravitch, G.S., and Yoshimura, S., 2004, A finite element formulation for the determination of unknown boundary conditions for three dimensional steady thermoelastic problems, *J Heat Transfer-Trans ASME*, Vol 126, 110–118.
Dillon, O.W., 1962, A nonlinear thermoelasticity theory, *J Mech Phys Solids*, Vol 10, 123.
Fafard, M. and Massicotte, B., 1993, Geometrical interpretation of the arc length method, *Comp Struct*, Vol 46, No 4, 603–616.
Gordon, A.P. and Nicholson, D.W., 2006, Finite Element Analysis of IN 617 Tertiary Creep, Report, University of Central Florida, Orlando, FL.
Green, A. and Naghdi, P., 1965, A general theory of an elastic–plastic continuum, *Arch Rat Mech Anal*, Vol 18, 19.
Gurson, A.L., 1977, Continuum theory of ductile rupture by void nucleation and growth. Part 1. Yield criteria and flow rules for porous ductile media, *J Eng Mater Technol*, Vol 99, 2.
Holzappel, G., 1996, On large strain viscoelasticity: continuum applications and finite element applications to elastomeric structures, *Int J Numer Meth Eng*, Vol 39, 3903.
Holzappel, G. and Simo, J., 1996, Entropy elasticity of isotropic rubber-like solids at finite strain, *Comp Methods Appl Mech Eng*, Vol 132, 17.
Hu, Q. and Zou, J., 2001, An iterative method with variable relaxation parameters for saddle point problems, *SIAM J Matrix Anal Appl*, Vol 23, No 2, 317–338.
Kaplan, I., 2002, "Spectral analysis and filtering with the wavelet transform," available at http://www.bearcave.com/misl/misl_tech/wavelets/freq/index.html.
Knott, R., 2005. A general pattern for arctan(1/n) = arctan(1/(n + 1)) + arctan(1/Z), available at http://www.mcs.surrey.ac.uk/Personal/R.Knott/Fibonacci/arctanproof1.html.

LS-DYNA Ver 95, 2000, Livermore Software Technology Corporation, Livermore CA, www.lstc.com.

Memon, B.A., and Su, X., 2004, Arc-length technique for nonlinear finite element analysis, *J Zhejiang U SCIENCE*, Vol 5, No 5, 618–628.

Moharir, M.M. 1998, Theoretical validation and design application of MSC/NASTRAN snap-through buckling capability, available at http://www.mscsoftware/support/library, /conf/amuc98/p01598.pdf.

Moraes, R. and Nicholson, D.W. 2002, Local damage criterion for ductile fracture with application to welds under dynamic loads, *Advances in Fracture and Damage Mechanics II*, eds M. Guagliano and M.H. Aliabadi, Hoggar Press, Geneva, 277.

Nicholson, D.W., 1995, Tangent modulus matrix for finite element analysis of hyperelastic materials, *Acta Mechanica*, Vol 112, 187.

Nicholson, D.W., 2005a, Iterative triangularization of updated finite element stiffness matrices, *Acta Mechanica*, Vol 174, 241–249.

Nicholson, D.W., 2005b, Stiff arc length method for nonlinear FEA, *Acta Mechanica*, Vol 175, 123–137.

Nicholson, D.W. and Lin, B., 1996, Theory of thermohyperelasticity for near-incompressible elastomers, *Acta Mechanica*, Vol 116, 15.

Nicholson, D.W. and Lin, B., 1997a, Finite element method for thermomechanical response of near-incompressible elastomers, *Acta Mechanica*, Vol 124, 181.

Nicholson, D.W. and Lin, B., 1997b, Incremental finite element equations for thermomechanical of elastomers: Effect of boundary conditions including contact, *Acta Mechanica*, Vol 128, No 1–2, 81.

Nicholson, D.W. and Lin, B., 1997c, On the tangent modulus tensor in hyperelasticity, *Acta Mechanica*, Vol 131, 121–132.

Nicholson, D.W. and Lin, B., 2005, On a fourth order FEA multistep time integration method for lightly damped media, *Acta Mechanica*, Vol 183, 23–40.

Nicholson, D.W. and Lin, B., 2006, Accelerated eigenstructure computation in FEA, *Developments in Theoretical and Applied Mechanics*, Proceedings of SECTAM XXIII, Mayaguez, PR.

Nicholson, D.W. and Nelson, N., 1990, Finite element analysis in design with rubber, Rubber Reviews, *Rubber Chem Technol*, Vol 63, 638.

Ogden, R.W., 1986, Recent advances in the phenomenological theory of rubber elasticity, *Rubber Chem Technol*, Vol 59, 361.

Perzyna, P., 1971, Thermodynamic theory of viscoplasticity, in *Advances in Applied Mechanics*, Vol 11, Academic Press, New York.

Ramm, E., 1981, Strategies for tracing nonlinear response near critical points, nonlinear finite element analysis in structural mechanics, Proceedings of the Europe–US Workshop, Springer, Berlin, 63–89.

Ragon, S.A., Guerdal, Z., and Watson, L.T., 2001, A comparison of three algorithms for tracing non-linear equilibrium paths of structural systems, available at http://eprints.cs.vt.edu:8000/archive/00000529/01/nflowIJ5501.pdf.

Riks, E., 1979, An incremental approach to the solution of snapping and buckling problems, *Int J Solids Struct*, Vol 15, 524–551.

Truesdell, C. and Noll, W., 1965, The non-linear field theories of mechanics, in *Encyclopedia of Physics*, Vol III/3, ed. S. Flugge, Springer, Berlin.

Tvergaard, V., 1981, Influence of voids on shear band instabilities under plane strain conditions, *Int J Fract*, Vol 17, 389–407.

Valanis, K. and Landel, R.F., 1967, The strain energy function of a hyperelastic material in terms of the extension ratios, *J Appl Phys*, Vol 38, 2997.

Wempner, G.A., 1971, Discrete approximation related to nonlinear theory of solids, *Int J Solids Struct*, Vol 17, 1581–1599.

Xiao, H., Bruhns, O.T., and Meyers, A. 1997, Logarithmic strain, logarithmic spin and logarithmic rate, *Acta Mechanica*, Vol 124, No 104, 89.

Ziegler, H. and Wehrli, C., 1987, The Derivation of Constitutive Relations from the Free Energy and the Dissipation Function, *Adv Appl Mech*, Vol 25, 187.

Index

A

B

C

D

E

F

G

U

V

W

Y

Milton Keynes UK
Ingram Content Group UK Ltd.
UKHW021904071024
449327UK00021B/1617

9 780367 387433